T0122624

# Lecture Notes in Electrical Engineering

**Volume 28**

## Lecture Notes in Electrical Engineering

Proceedings of the European Computing Conference: Volume 2
Mastorakis, Nikos, Mladenov, Valeri (Eds.)
978-0-387-84818-1, Vol. 28

Proceedings of the European Computing Conference: Volume 1
Mastorakis, Nikos, Mladenov, Valeri (Eds.)
978-0-387-84813-6, Vol. 27

Electronics System Design Techniques for Safety Critical Applications
Sterpone, Luca
978-1-4020-8978-7, Vol. 26

Data Mining and Applications in Genomics
Ao, Sio-Iong
978-1-4020-8974-9, Vol. 25

Informatics in Control, Automation and Robotics: Selected Papers from the International Conference on Informatics in Control, Automation and Robótics 2007
Filipe, J.B.; Ferrier, Jean-Louis; Andrade-Cetto, Juan (Eds.)
978-3-540-85639-9, Vol. 24

Digital Terrestrial Broadcasting Networks
Beutler, Roland
ISBN 978-0-387-09634-6, Vol. 23

Logic Synthesis for Compositional Microprogram Control Units
Barkalov, Alexander, Titarenko, Larysa
ISBN: 978-3-540-69283-6, Vol. 22

Sensors: Advancements in Modeling, Design Issues, Fabrication and Practical Applications
Mukhopadhyay, Subhas Chandra; Huang, Yueh-Min (Eds.)
ISBN: 978-3-540-69030-6, Vol. 21

Smart Sensors and Sensing Technology
Mukhopadhyay, Subhas Chandra; Sen Gupta, Gourab (Eds.)
ISBN: 978-3-540-79589-6 , Vol. 20

Basic Principles of Fresnel Antenna Arrays
Minin, Igor V., Minin, Oleg V.
ISBN: 978-3-540-79558-2, Vol. 19

*(Continued after index)*

Nikos Mastorakis · Valeri Mladenov ·
Vassiliki T. Kontargyri
Editors

# Proceedings of the European Computing Conference

Volume 2

*Editors*
Nikos Mastorakis
Hellenic Naval Academy, Military Institutes of University Education
Leoforos Chatzikyriakou
185 39 Piraeus, Greece
mastor@hna.gr

Valeri Mladenov
Department of Theoretical Electrical Engineering
Technical University of Sofia
Kl. Ohridski St. 8, 1000 Sofia, Bulgaria
valerim@tu-sofia.bg

Vassiliki T. Kontargyri
National Technical University of Athens
Roumelis 6, 152 35 Vrilissia
Athens, Greece
vkont@central.ntua.gr

ISSN 1876-1100
e-ISSN 1876-1119
ISBN 978-1-4899-7924-7
e-ISBN 978-0-387-85437-3
DOI 10.1007/978-0-387-85437-3


Softcover re-print of the Hardcover 1st edition 2009


Printed on acid-free paper

springer.com

# Contents

# Contributors

**Ali Abbasi** Department of Civil Engineering, K.N. Toosi University of Technology, 19697 Tehran, Iran, AliAbbasi.civileng@gmail.com

**A. Abdollahzadeh** Department of Computer Engineering, Amir Kabir University, Tehran, Iran

**M.T. Aboul-Ela** Civil Engineering Department, Minia University, Egypt

**Ulrich Albrecht** Department of Mathematics, Auburn University, Auburn, AL 36849, USA, albreuf@mail.auburn.edu

**José Antonio Álvarez** Departamento de Arquitectura de Computadores y Electrónica, Universidad de Almería, Almeria, Spain, jaberme@ace.ual.es

**David Anderson** School of Computing, University of Portsmouth, Portsmouth PO1 2EG, UK, david.anderson@port.ac.uk

**Hamed Arabi** Science and Research, Islamic Azad University, Tehran, Iran

**Osamu Arai** Faculty of Software and Information Science, Iwate Prefectural University, Morioka, Iwate, Japan, arai@fujita.soft.iwate-pu.ac.jp

**Francisco Araque** Dpto. De Lenguajes y Sistemas Informáticos, University of Granada, 18071, Granada (Andalucía), Spain, faraque@ugr.es

**G. Atsalakis** Department of Production Engineering and Management, Technical University of Crete, Chania, Greece, atsalak@otenet.gr

**A.R. Awad** Department of Environmental Engineering, Tishreen University, Lattakia, Syria

**Ali-Reza Bagheri** Department of Civil Engineering, K.N. Toosi University of Technology, 1346 Valiasr St. 19697 Tehran, Iran

**Daniel Baldovin** Institute of Solid Mechanics, Romanian Academy, Bucharest 010141, Romania

**Stefan Balint**

**Youssef Balouki** Department of Mathematics and Computer Science, Mohammed V University, Ibn Bettouta, Rabat Morocco

**Lucian Bărbulescu** Faculty of Automation, Computer Science and Electronics, University of Craiova, 200646 Craiova, Romania, luci@cs.ucv.ro

**Cristea Boboila** Faculty of Mathematics and Computer Science, University of Craiova, Craiova, Romania, boboila@central.ucv.ro

**Simona Boboila** Computer Science Department, Stony Brook University, Stony Brook, NY, USA, mboboila@cs.sunsysb.edu

**Ágnes Bogárdi-Mészöly** Department of Automation and Applied Informatics, Budapest University of Technology and Economics, Goldmann György tér 3. IV. em., 1111 Budapest, Hungary

**Mohamed Bouhdadi** Department of Mathematics and Computer Science, Mohammed V University, Ibn Bettouta, Rabat Morocco

**Liliana Braescu** Department of Computer Science, West University of Timisoara, Blv. V. Parvan 4, Timisoara 300223, Romania

**Esra Celik** Department of Computer Engineering, Yildiz Technical University, Besiktas/Istanbul, Turkey, esra@ce.yildiz.edu.tr

**Paulo R. M. Cereda** Computer Science Department, Federal University of São Carlos, Rod. Washington Luis, km 235 - 13565-905, São Carlos – SP – Brasil, paulo_cereda@dc.ufscar.br

**Ming-Jun Chan** National Kaohsiung Normal University, 116 Heping First Road, Kaohsiung, Taiwan (R.O.C), taiwanka@yahoo.com.tw

**Hassan Charaf** Department of Automation and Applied Informatics, Budapest University of Technology and Economics, Goldmann György tér 3. IV. em., 1111 Budapest, Hungary

**Veturia Chiroiu** Continuum Mechanics Department, Institute of Solid Mechanics, Bucharest 010141, Romania

**Chih-Hung Chung** National Kaohsiung Normal University, Kaohsiung City, Taiwan (R.O.C), bbc160@yahoo.com.tw

**C. Ciufudean** Faculty of Electrical Engineering and Computer Science, University “Stefan cel Mare”, Suceava, Romania, calin@eed.usv.ro

**Aldo Cumani** Istituto Nazionale di Ricerca Metrologica, Str. delle Cacce, 91, I-10135 Torino, Italy, cumani@inrim.it

**Adrian Curaj** Institute of Solid Mechanics, Romanian Academy, Bucharest, Romania

**Ciprian Cuzmin** Department of Manufacturing Engineering, "Dunărea de Jos" University of Galaţi, Galaţi, 800201, Romania, ciprian.cuzmin@ugal.ro

**Claudius Dan** Faculty of Electronics, Telecommunication and Information Technologies, Politehnica University of Bucharest, blvd. Iuliu Maniu, Bucharest 77209, Romania, claus@mESsnet.pub.ro

**Cecilia Delgado** Dpto. De Lenguajes y Sistemas Informáticos, University of Granada, 18071, Granada (Andalucía), Spain, cdelgado@ugr.es

**Carlo dell'Aquila** Dipartimento di Informatica, Università di Bari, Via E. Orabona 4, I-70125 Bari, Italy, dellaquila@di.uniba.it

**Elena Doicaru** Faculty of Automation, Computer Science and Electronics, University of Craiova, Craiova 200646, Romania, dmilena@electronics.ucv.ro

**Matevž Dolenc** Faculty of Civil and Geodetic Engineering, University of Ljubljana, Jamova 2, 1000 Ljubljana, Slovenia, mdolenc@itc.fgg.uni-lj.si

**Ştefania Donescu** Mathematics Department, Technical University of Civil Engineering, Bd. Lacul Tei nr.122-124, Bucharest 020396, Romania

**Dan Dumitriu** Continuum Mechanics Department, Institute of Solid Mechanics, Bucharest 010141, Romania

**Sherif El-Kassas** Department of Computer Science, The American University in Cairo, Cairo, Egypt

**Alexandru Epureanu** Department of Manufacturing Engineering, "Dunărea de Jos" University of Galaţi, Galaţi, 800201, Romania

**Rong-Jyue Fang** Department of Information Management, Southern Taiwan University of Technology, Kaohsiung City 806, Taiwan (R.O.C.), rxf26@mail.stut.edu.tw

**Edite M. G. P. Fernandes** Minho University, Campus de Gualtar, 4710-057 Braga, Portugal, emgpf@dps.uminho.pt

**Jose Jesús Fernández** Departamento de Arquitectura de Computadores y Electrónica, Universidad de Almería, Almeria, Spain, jose@ace.ual.es

**C. Filote** Faculty of Electrical Engineering and Computer Science, University "Stefan cel Mare", Suceava, Romania, calin@eed.usv.ro

**Hamido Fujita** Faculty of Software and Information Science, Iwate Prefectural University, Morioka, Iwate, Japan, issam@soft.iwate-pu.ac.jp

**L. D. Genete** Business Information System Department, "Alexandru Ioan Cuza" University of Iasi, 22 Carol I Blvd., Ro-700505, Romania

**Thomas F. George** Office of the Chancellor and Center for Nanoscience, Departments of Chemistry & Biochemistry and Physics and Astronomy, University of Missouri-St. Louis, St. Louis, MO 63121, USA

**Majid Ghandchi** Department Electrical Engineering and Computer, Ahar Branch, Islamic Azad University, Ahar, Iran

**Rahim Ghayour** Department of Electrical Engineering, School of Engineering, Shiraz University, Shiraz, Iran, rghayour@shirazu.ac.ir

**M. Goodarzi**

**Reginaldo A. Gotardo** Computer Science Department, Federal University of São Carlos, Rod. Washington Luis, km 235 - 13565-905, São Carlos – SP – Brasil, reginaldo_gotardo@dc.ufscar.br

**Robson E. Grande** Computer Science Department, Federal University of São Carlos, Rod. Washington Luis, km 235 - 13565-905, São Carlos – SP – Brasil, robson_grande@dc.ufscar.br

**Antonio Guiducci** Istituto Nazionale di Ricerca Metrologica, Str. delle Cacce, 91, I-10135 Torino, Italy, guiducci@inrim.it

**Jianda Han** State Key Laboratory of Robotics, Shenyang Institute of Automation, Academia Sinica. 110016 Shenyang, China

**Takeshi Hashimoto** Department of Automation and Applied Informatics, Budapest University of Technology and Economics, Goldmann György tér 3. IV. em., 1111 Budapest, Hungary

**Frank S. Haug** Opus College of Business, University of St. Thomas, St. Paul, MN 55105, USA

**Shivanand B. Hiremath** National Institute of Industrial Engineering, Mumbai, India

**Stefan Holban** Departament of Computer Science, University "Politehnica" of Timisoara, 300006 Timisoara Pta Victoriei No. 2, Timis, Romania, stefan@aspc.cs.upt.ro

**Chong-Cheng Hong** National Kaohsiung Normal University, Kaohsiung City, Taiwan (R.O.C), t771580@yahoo.com.tw

**Jason Howarth** School of Computing and Mathematics, Charles Sturt University, Boorooma St. Wagga Wagga, NSW, 2678 Australia

**Fun-Chi Hsu** National Kaohsiung Normal University, Kaohsiung City, Taiwan (R.O.C.), kyohikaru@msn.com

**Chia-Jui Hsu** National Kaohsiung Normal University, Kaohsiung City, Taiwan (R.O.C.), chiajuimr@hotmail.com

**Chia-Jung Hsu** National Kaohsiung Normal University, Kaohsiung, Taiwan (R.O.C), z3charles@msn.com

**Chia-Hsun Wu** National Kaohsiung Normal University, Kaohsiung City, Taiwan (R.O.C.), der54367@msn.com

**Jhen-Wei Huang** Department of Computer and Information Science, National Taichung University, Taiwan (R.O.C)

**Ion N. Ion** Institute of Solid Mechanics, Romanian Academy, Bucharest, Romania

**Zhe Jiang** State Key Laboratory of Robotics, Shenyang Institute of Automation, Academia Sinica, 110016 Shenyang, China

**Mingyan Jiang** Centre Tecnològic de Telecomunicacions de Catalunya (CTTC), Av. Del, Canal Olímpic s/n, 08860 Castelldefels (Barcelona), Spain; School of Information Science and Engineering, Shandong University, Jinan, 250100 China, jiangmingyan@sdu.edu.cn, mingyan.jiang@cttc.es

**E.S. Karapidakis** Technological Educational Institute of Crete, 3 Romanou str, 73133, Chania, Greece

**Alireza Kargar** Department of Electrical Engineering, School of Engineering, Shiraz University, Shiraz, Iran

**M. Elif Karsligil** Department of Computer Engineering, Yildiz Technical University, 34349 Besiktas/Istanbul, Turkey, elif@ce.yildiz.edu.tr

**V.N. Kasyanov** A.P. Ershov Institute of Informatics Systems/Novosibirsk State University, Novosibirsk, 630090 Russia

**Hiroshi Kawaguchi** Graduate School, Kobe University, Kobe, Hyogo, 657-8507 Japan, kawapy@cs28.cs.kobe-u.ac.jp

**Kentaro Kawakami** Graduate School, Kobe University, Kobe, Hyogo, 657-8507 Japan, kawakami@cs28.cs.kobe-u.ac.jp

**Omid Khayat** Department of Biomedical Engineering, Amirkabir University, Iran

**Bruno Y. L. Kimura** Computer Science Department, Federal University of São Carlos, Rod. Washington Luis, km 235 - 13565-905, São Carlos – SP – Brasil, bruno_kimura@dc.ufscar.br

**Robert Klinc** Faculty of Civil and Geodetic Engineering, University of Ljubljana, Jamova 2, 1000 Ljubljana, Slovenia, rklinc@itc.fgg.uni-lj.si

**Edita Kolářová** Brno University of Technology, Technická 8, 616 00 Brno, Czech Republic

**Zeyneb Kurt** Department of Computer Engineering, Yildiz Technical University, 34349 Besiktas/Istanbul, Turkey, zeyneb@ce.yildiz.edu.tr

**Miguel Angel Lagunas** Centre Tecnològic de Telecomunicacions de Catalunya (CTTC), Av. Del, Canal Olímpic s/n, 08860 Castelldefels (Barcelona), Spain, m.a.lagunas@cttc.es

**Kuan-Chou Lai** Department of Computer and Information Science, National Taichung University, Kaohsiung City, Taiwan (R.O.C)

**N. Lassouaoui** University Paris X, Nanterre, Pôle Scientifique et Technique de Ville d'Avray, Groupe Electromagnétisme Appliqué, 50 rue de Sèvre 92410, Ville d'Avray, France

**Ljubomir Lazić** School of Electrical Engineering, Vojvode Stepe 283, Belgrade, Serbia, llazic@vets.edu.yu

**Chung-Ping Lee** Department of Industrial Technology Education, National Kaohsiung Normal University, Kaohsiung City, Taiwan (R.O.C.), cl87369@yahoo.com.tw

**Chungping Lee** Department of Industrial Technology Education, National Kaohsiung Normal University, Kaohsiung City, Taiwan (R.O.C.), cl87369@yahoo.com.tw

**Ezio Lefons** Dipartimento di Informatica, Università di Bari, Via E. Orabona 4, I-70125 Bari, Italy, lefons, tangorra@di.uniba.it

**Tihamér Levendovszky** Department of Electrical and Electronics Engineering, Shizuoka University, 5-1, 3-chome Johoku, 432-8561 Hamamatsu, Japan

**Che-Chern Lin** Department of Industrial Technology Education, National Kaohsiung Normal University, Kaohsiung City 806, Taiwan (R.O.C.), cclin@njknucc.nknu.edu.tw, cclin@nknucc.nknu.edu.tw

**James N.K. Liu** Department of Computing, Hong Kong Polytechnic University, Hung Hom, Kowloon, Hong Kong, csnkliu@comp.polyu.edu.hk

**Howard Lo** Department of Industrial Technology Education, National Kaohsiung Normal University, Kaohsiung City 806, Taiwan (R.O.C.), howardlab@yahoo.com

**Constantin Lupsoiu** Faculty of Mathematics and Computer Science, University of Craiova, Romania, clupsoiu@central.ucv.ro

**Kavi Mahesh** Department of Computer Science and Engineering, PES Institute of Technology, Bangalore 560085, India, drkavimahesh@gmail.com

**Abdalla Mahmoud** Department of Computer Science, The American University in Cairo, Cairo, Egypt

**Sunil Mallya** Department of Computer Science and Engineering, PES Institute of Technology, Bangalore 560085, India, mallya16@gmail.com

**Ghuan-Hsien Mao** Shu-Te University, Yen-Chau, Kaohsiung County, Taiwan (R.O.C.), js.mao@msa.hinet.net

**Vasile Marinescu** Department of Manufacturing Engineering, "Dunărea de Jos" University of Galaţi, Galaţi, 800201, Romania

**Olga Martin** University "Politehnica" of Bucharest, 313, Splaiul Independentei, Romania, omartin_ro@yahoo.com

**Nikos E Mastorakis** Military Insitutes of University Education (ASEI) Hellenic Naval Academy, Terma Chatzikyriakou, 18539, Piraues, Greece, mastor@wseas.org, mastor@softlab.ece.ntua.gr, mastor@ieee.org

**Farzad Meysami-Azad** Department of Civil Engineering, K.N. Toosi University of Technology, 19697 Tehran, Iran

**Marcel Migdalovici** Institute of Solid Mechanics, Romanian Academy, Bucharest, Romania

**Sabina-Cristiana Mihalache** Business Information Systems Department, Alexandru Ioan Cuza University of Iaşi, Carol I Blvd, no. 700505, Iaşi, Romania, sabina.mihalache@gmail.com

**Ana Maria Mitu** Institute of Solid Mechanics, Romanian Academy, Bucharest 010141, Romania

**Radu Munteanu** Tehnical University of Cluj Napoca, Politehnica University of Bucharest, Romania

**Ligia Munteanu** Continuum Mechanics Department, Institute of Solid Mechanics, Romanian Academy, Bucharest 010141, Romania

**YiYong Nie** State Key Laboratory of Robotics, Shenyang Institute of Automation, Academia Sinica, 110016 Shenyang, China

**Nitin** Member IEEE and ACM, Department of CSE and IT, Jaypee University of Information Technology, Waknaghat, Solan–173215, Himachal Pradesh, India, delnitin@gmail.com

**Nicolae Oancea** Department of Manufacturing Engineering, "Dunărea de Jos" University of Galaţi, Galaţi, 800201, Romania

**H. Hafdallah Ouslimani** University Paris X, Nanterre, Pôle Scientifique et Technique de Ville d'Avray, Groupe Electromagnétisme Appliqué, 50 rue de Sèvre 92410, Ville d'Avray, France

**Zeljko Panian** Faculty of Economics and Business, University of Zagreb, Croatia

**Serena Pastore** INAF – Astronomical Observatory of Padova, Vicolo Osservatorio 5 – 35122 – Padova, Italy

**Vasile Daniel Păvăloaia** Business Information Systems Department, Alexandru Ioan Cuza University of Iasi, Bdul Carol I nr.22, Iasi 700505, Romania, danpav@uaic.ro

**I. Von Poser** Ingenierurtechnik, Merck KGaA., Darmstadt, Germany

**A. Priou** University Paris X, Nanterre, Pôle Scientifique et Technique de Ville d'Avray, Groupe Electromagnétisme Appliqué, 50 rue de Sèvre 92410, Ville d'Avray, France

**B. Punantapong** Biomaterials and Nanotechnology Research Unit, Department of Industrial Physics and Medical Instrumentation, King Mongkut's Institute of Technology North Bangkok, Bangsue, Bangkok 10800, Thailand

**S. Albert Rabara** Department of Computer Science, St.Joseph's College (Autonomous), Tiruchirappalli – 620 002, Tamilnadu, India

**Saeed K. Rahimi** Opus College of Business, University of St. Thomas, St. Paul, MN 55105, USA

**Youssef Rezvan** Department Electrical Engineering and Computer, Ahar Branch, Islamic Azad University, Ahar, Iran, y-rezvan@iau-ahar.ac.ir

**Farshad Rezvan** Department of Mathematical Sciences, University of Tabriz, Tabriz, Iran, frezvan@tabrizu.ac.ir

**Ricardo A. Rios** Computer Science Department, Federal University of São Carlos, Rod. Washington Luis, km 235 - 13565-905, São Carlos – SP – Brasil, ricardo_rios@dc.ufscar.br

**Javier Roca** Departamento de Arquitectura de Computadores y Electrónica, Universidad de Almería, Almeria, Spain, jroca@ace.ual.es

**Ana Maria A. C. Rocha** Minho University, Campus de Gualtar, 4710-057 Braga, Portugal, arocha@dps.uminho.pt

**Saeed-Reza Sabbagh-Yazdi** Department of Civil Engineering, K. N. Toosi University of Technology, 19697 Tehran, Iran, SYazdi@kntu.ac.ir

**Sameer. S. Saigaonkar** Diploma in Industrial Engineering, National Institute of Industrial Engineering, Mumbai, India

**Yoshinori Sakata** Graduate School, Kobe University, Kobe, Hyogo, 657-8507 Japan, yoshi@cs28.cs.kobe-u.ac.jp

**Alberto Salguero** Dpto. De Lenguajes y Sistemas Informáticos, University of Granada, 18071, Granada (Andalucía), Spain, agsh@ugr.es

**Ahmed Sameh** Department of Computer Science, The American University in Cairo, Cairo, Egypt, sameh@aucegypt.edu

**B. Satco** Faculty of Electrical Engineering and Computer Science, University “Stefan cel Mare”, Suceava, Romania, calin@eed.usv.ro

**Cornel Secara** Institute of Solid Mechanics, Romanian Academy, Bucharest 010141, Romania

**Vivek Kumar Sehgal** Member IEEE and ACM, Department of ECE, Jaypee University of Information Technology, Waknaghat, Solan–173215, Himachal Pradesh, India, vivekseh@gmail.com

**M. Sepehri** Department of Computer Engineering and IT, Islamic Azad University, Fars Sciences and Research Branch, Fars, Iran

**R. Serban** Instituite dÌngenierie Informatique de Limoges et de Rodez, Limoges, France, rares.serban@3il.fr

**Hamid Reza Shahdoosti** Department of Biomedical Engineering, Amirkabir University, Iran

**Simon C.K. Shiu** Department of Computing, Hong Kong Polytechnic University, Hung Hom, Kowloon, Hong Kong, csckshiu@comp.polyu.edu.hk

**Alex Soiguine** SOiGUINE LLC, 1333 N. Camino Alto, Vallejo, CA 94589, USA

**A.P. Stasenko** A.P. Ershov Institute of Informatics Systems/Novosibirsk State University, Novosibirsk, 630090, Russia

**Ion Stefanescu** Department of Electronics, Communications and Computers, University of Pitesti, 110040 Pitesti Str. Targul din Vale, nr.1, Arges, Romania, jean@upit.ro, ionjean@yahoo.com

**Edward Stow** School of Computing and Mathematics, Charles Sturt University, Boorooma St. Wagga Wagga, NSW, 2678 Australia

**Filippo Tangorra** Dipartimento di Informatica, Università di Bari, Via E. Orabona 4, I-70125 Bari, Italy, tangorra@di.uniba.it

**Z. Cihan Taysi** Department of Computer Engineering, Yildiz Technical University, Besiktas/Istanbul, Turkey, cihan@ce.yildiz.edu.tr

**Virgil Teodor** Department of Manufacturing Engineering, “Dunărea de Jos” University of Galaţi, Galaţi, 800201 Romania, virgil.teodor@ugal.ro Marian Cucu

**Csaba Török** Technical University Košice, Letná 9, 042 00 Košice, Slovakia

**Al. Ţugui** Business Information System Department, “Alexandru Ioan Cuza” University of Iasi, 22 Carol I Blvd., Ro-700505, Romania

**H. Irem Turkmen** Department of Computer Engineering, Yildiz Technical University, 34349 Besiktas/Istanbul, Turkey, irem@ce.yildiz.edu.tr

**Constantin Udrişte** Department of Mathematics, University Politehnica of Bucharest, Splaiul Independentei 313, 060042 Bucharest, Romania, udriste@mathem.pub.ro

**Valuchandhar Vadivelu** Department of Computer Science, St.Joseph's College (Autonomous), Tiruchirappalli – 620 002, Tamilnadu, India

**Ioannis O. Vardiambasis** Microwave Communications and Electromagnetic Applications Laboratory, Division of Telecommunications, Department of Electronics, Technological Educational Institute (T.E.I.) of Crete – Chania Branch, Romanou 3, Chalepa, 73133 Chania Crete, Greece, ivardia@chania.teicrete.gr

**Luige Vladareanu** Institute of Solid Mechanics, Romanian Academy, Bucharest, Romania, luigiv@arexim.ro

**Mirela-Catrinel Voicu** Faculty of Economic Sciences, West University of Timişoara, Romania, mirela.voicu@fse.uvt.ro

**Yu-Chen Weng** Department of Industrial Technology Education, National Kaohsiung Normal University, Kaohsiung City 806 Taiwan, jane2080@gmail.com

**Chao-Chin Wu** Department of Computer Science and Information Engineering, National Changhua University of Education, Taiwan, R.O.C.

**Chun-Hung Wu** National Kaohsiung Normal University, Kaohsiung City, Taiwan (R.O.C.), hans_wu@hotmail.com

**Ming-Chih Wu** Graduate Institute of Educational Technology, National Pingtung University of Education, Taiwan (R.O.C.), michywu@yahoo.com.tw

**Hsieh-Hua Yang** Department of Health Care Administration, Chang Jung Christian University, Kway Jen, Tainan, Taiwan (R.O.C), yangsnow@mail.cjcu.edu.tw

**Hung-Jen Yang** Department of Industrial Technology Education, National Kaohsiung Normal University, Kaohsiung City, Taiwan (R.O.C), hjyang@nknucc.nknu.edu.tw

**A. Gokhan Yavuz** Department of Computer Engineering, Yildiz Technical University, Besiktas/Istanbul, Turkey gokhan@ce.yildiz.edu.tr

**Masahiko Yoshimoto** Graduate School, Kobe University, Kobe, Hyogo, 657-8507 Japan, yosimoto@cs28.cs.kobe-u.ac.jp

**Jane You** Department of Computing, Hong Kong Polytechnic University, Hung Hom, Kowloon, Hong Kong, csyjia@comp.polyu.edu.hk

**Jui-Chen Yu** National Science and Technology Museum, #720 Ju-Ru 1st Rd., Kaohsiung, Taiwan, R.O.C., raisin@mail.nstm.gov.tw

**Dongfeng Yuan** School of Information Science and Engineering, Shandong University, Jinan, 250100 China, dfyuan@sdu.edu.cn

**Adrian Zafiu** Department of Electronics, Communications and Computers, University of Pitesti, 110040 Pitesti Str. Targul din Vale, nr.1, Arges, Romania, adrian_zafiu@yahoo.com, adrian.zafiu@upit.ro

**YiWen Zhao** State Key Laboratory of Robotics, Shenyang Institute of Automation, Academia Sinica, 110016 Shenyang, China

**Sérgio D. Zorzo** Computer Science Department, Federal University of São Carlos, Rod. Washington Luis, km 235 – 13565-905, São Carlos – SP – Brasil, zorzo@dc.ufscar.br

# Part I
# Computer Applications in Power Systems

In order to meet the increasing demand for reliable design and expansion of power networks, the use of computers plays a leading role in the power industry's future R&D programs. Digital computers are already widely used in the analysis and design of large power networks. Hybrid computers offer a convenient alternative, especially where fast and accurate simulation of complex nonlinearities is required. Computer models can be used to simulate the changing states of electrical power systems. Such simulations enable the power engineer to study performance and predict disturbances. Computer models include mathematical background, algorithms and the basic tools needed to study complex power systems, their interaction and likely response to different types of network pathologies or disturbances. In this part we present some computer modeling techniques that constitute the framework of modern power system analysis. The papers include wind energy production forecasting, operation management on autonomous power system and a low-power portable H.264/AVC decoder using elastic pipeline.

# Chapter 1
# Wind Energy Production Forecasting

**G. Atsalakis**

**Abstract** This paper presents a forecasting system of short-term wind energy production. The system is a feedforward neural network that uses genetic algorithm as a learning method of forecasting the wind energy production of the following day. The data concerns four different wind power plants. The neural network forecasts are compared with the predictions made by a neural network that uses Quasi-Newton learning algorithm and are evaluated based on the four common statistical measures. Both models perform satisfactorily as far as the prediction of the following day's wind energy production is concerned.

## 1.1 Introduction

Wind energy is the energy source that contributes mostly to the renewable energy mix in Europe. The rapid expansion of the usage of wind energy within most countries requires the development of functional and accurate tools for wind energy production. The importance of wind power forecasting up to 48 hours ahead has been recognized by the wind plant operators, utilities, transmission system operators and other end-users. The short-range prediction (10–48 hours ahead) is required, if wind energy should become a substitute for other energy sources, for example coal or gas. The accuracy requirements are not as high as for shorter ranges and depend on the electricity trading mechanisms. An important problem arises by the intermittency of the wind for the deployment of wind energy into the electricity networks.

To ensure grid security a Transmission System Operator needs today for each kilowatt of wind energy either an equal amount of spinning reserve or a forecasting system that can predict the amount of energy that will be produced from wind over a period of 1 to 48 hours. In the range from 5 to 15 m/s a wind

G. Atsalakis (✉)
Department of Production Engineering and Management, Technical University of Crete, Chania, Greece
e-mail: atsalak@otenet.gr

N. Mastorakis et al. (eds.), *Proceedings of the European Computing Conference*, Lecture Notes in Electrical Engineering 28,
DOI 10.1007/978-0-387-85437-3_1, 

turbine's production increases with a power of three. For this reason, a Transmission System Operator requires accuracy for wind speed forecasts of 1 m/s in this wind speed range. The major alternatives to wind energy forecasting are backup storage capacities and a strong interconnectivity between the electrical grids. Both alternatives are associated with technical and economical restrictions, which can be barriers to the installation of larger amounts of wind energy. Forecasting can therefore be considered as a more efficient way of increasing wind energy penetration.

## 1.2 Related Work

Many studies focus on predicting the wind speed rather than the power output. Traditionally, short-range forecasts have utilized on-site observations with various persistence based on statistical forecast models, including autoregressive time series techniques (Brown [7], Poggi [30], Weisser [35]). Other traditional methods are used by Youcef Ettoumi [36]. Giebel [12] surveys the literature on short-range wind speed and wind power forecasting. Similarly Sfetsos [34] compares various techniques applied in wind energy forecasting. Atsalakis [2–5] presents neuro-fuzzy models (ANFIS) to energy forecasting. Also fuzzy theory has been applied by Chedid [8], Sanders [33] and Michalik [27].

Artificial neural networks are being applied in the energy forecasting issue by Datta [10], Khotanzad [19], Papalexopolous [29], Kretzschmar [21], Czernichow [9], Javeed Nizami [17], Rummeihart [32], Abdel-Aal [1], Mohandes [28], Kalogirou [18], Mandal [24] and Metaxiotis [26].

## 1.3 Theoretical Approach

### *1.3.1 Neural Networks*

Artificial neural networks reproduce the behavior of the brain and nervous systems in a simplified computational form. They are constituted of highly interconnected simple elements, called artificial neurons, which receive information, elaborate them through mathematical functions and pass them to other artificial neurons. Particularly, in multilayer perceptron feedforward networks the artificial neurons are organized in layers: an input layer, one or more hidden layers and an output layer [31, 14].

Neural networks may be considered as a "multivariate nonlinear nonparametric inference technique that is data driven and model free" [6]. In fact they are information-processing models designed to capture the underlying relationships among various data sets autonomously. They were first developed in the 1950s, and indeed the analogies to human intelligence are significant. Neural network systems are constructed along the same organizational principles as in the human brain, performing mathematical mapping from one domain or layer to another by using algorithmic techniques.

Neural networks will not only discover the relationships among the data sets and use these to predict the future values, but will use the actual results as they become known to adjust the formulas to predict more accurately in the years to come. Essentially, they learn from their mistakes. Over time, as they continue to accumulate data and feedback, the neural networks will retest and hone their formulas to achieve ever more accurate predictions.

There are several advantages in using Neural Networks. The most vital issue which needs to be mentioned is the ability to learn from data and thus the potential to generalize, i.e. produce an acceptable output for previously unseen input data (important in prediction tasks). This even holds (to a certain extent) when input series contain low-quality or missing data. Another valuable quality is the non-linear nature of a neural network. Potentially a vast amount of problems may be solved. Furthermore no expert system is needed which makes the network extremely flexible to the changes in the environment. Only one has to retrain the system. Regarding downsides, the black-box-property first springs to mind. When relating one single outcome of the network to a specific internal decision turns out to be extremely difficult. Noisy data also reinforces the negative implications of establishing incorrect causalities, overtraining (or over-fitting), which will damage generalization. Finally, a certain degree of knowledge in the subject is required as it is not trivial to assess the relevance of the chosen input series. In this study, one hidden layer is considered, since it is believed that this type of network can approximate any function [16, 22].

### 1.3.2 Quasi-Newton Method for Training the Neural Network

The non-linear optimization algorithms are based on minimizing the sum of squared errors in respect with the parameter set $\Omega$ (see Eq. 1.1):

$$\psi = \sum_{t=1}^{T-1} \{Q_t - \hat{\psi}_t\}^2 \tag{1.1}$$

where $T$ is the number of observations in the simulation run, $\hat{\psi}_t$ is the approximated value of $Q_t$, conditional on information at time $t$.

To minimize a non-linear function, the parameter vector $\Omega$ must be randomly initialized, and then an iterating process is used on the coefficient set $\Omega$ until $\psi$ is minimized. Such a process is based on the Quasi-Newton algorithm. Starting with the initial set of sum of squared errors $\psi(\Omega_0)$, based on the initial coefficient vector $\Omega_0$ and a second order Taylor expansion is used to find $\psi(\Omega_1)$ (see Eq. 1.2):

$$\psi(\Omega_1) = \psi(\Omega_0) + \nabla_0(\Omega_1 - \Omega_0) + \frac{1}{2}(\Omega_1 - \Omega_0)' H_0(\Omega_1 - \Omega_0) \tag{1.2}$$

where $\nabla_0$ is the gradient of the error function with respect to the parameter set $\Omega_0$ and $H_0$ is the Hessian of the error function. To find the direction of the parameter set from iteration 0 to iteration 1, the error function $\psi(\Omega_1)$ in respect with $(\Omega_1 - \Omega_0)$ must be minimized (see Eq. 1.3):

$$(\Omega_1 - \Omega_0) = -H_0^{-1}\nabla_0 \tag{1.3}$$

The iteration is continued until the error function is minimized according to the tolerance value or when the maximum number of iterations is reached. The main disadvantage of this method is that one may not get results of global solution, but local or saddle point solution.

### *1.3.3 Genetic Algorithm for Training the Neural Network*

The inclusion of Genetic Algorithm (GA) search techniques was undertaken for two reasons. The first relates to the potential that GAs offer in terms of adaptiveness. The flexibility, robustness, and simplicity which GAs offer renders them very attractive in that respect. The second reason stems from the difficulty in optimizing neural network applications. Genetic Algorithms, by operating on entire populations of candid solutions on a parallel way, are much less likely to get stuck at a local optimum. Since GAs search the fitness landscape stochastically, they are more likely than traditional techniques to estimate the true global optimum. In this paper, the optimization task is defined as an improvement of forecasting the ability of the network. Genetic Algorithms are a class of probabilistic search techniques based on biological evolution [11, 13, 15]. The algorithms mimic biological evolution by applying a computationally simulated version of "survival of the fittest" [20].

A Genetic Algorithm starts with a population of random coefficient vectors. Then, two pairs of coefficients are selected, with a replacement. The best fitness of the coefficient vectors is given by the minimization of the sum of the squared errors. The winner of each pair has the best fitness. These two winning vectors are retained for "breeding" purposes. The next step is the crossover, where the two parents "breed" two children. The following step is the mutation of the children. With some small probability, which decreases over time, each element or coefficient of the two children's vectors is subject to a mutation. Following the mutation operation, the four members of the "family" engage in a fitness tournament. The children are evaluated by the same fitness criteria which are used to evaluate the parents. The two vectors with the best fitness, whether parents or children, survive and pass to the next generation, whilst the two with the worst fitness value are extinguished. The above process is repeated, with parents returning to the population pool for possible selection again, until the next generation is populated [25].

## 1.4 Computational Results

### 1.4.1 Parameter Selection

The quantitative parameters used in this implementation were the minimum, the average, and the maximum speed of the wind, the minimum, the average, the maximum speed of helix, the minimum, the average and the maximum attributed force and the daily hours of operation.

They were then modified and acted upon scaled input values as received from the input layer. The scaled parameters can be considered as a fluctuation rate. Table 1.1 presents the parameters that the model uses as inputs.

The processing layer (hidden) consists of three neurons; each one of which carries a different computation. The formulas for each neuron are composed from the modified parameters and are then interpreted through an activation function.

### 1.4.2 Network Parameter Issues

After comprehensive testing over both static and dynamic topologies, it was decided to go for a three-layered approach, as this was the most efficient computationally. The first layer was designated as the input layer where ten input neurons would take ten scaled input values of the series in the historic data.

The second layer is the hidden layer, which does all the computations, and it is comprised by three neurons. The three neurons in the hidden layer process the input in a parallel fashion to improve the predictions by using the sigmoid transfer function.

The third layer (the output layer) consists of one neuron with combinations of different weights on the inputs received from the hidden layer. The weighted mean of these inputs is the next predicted value and it is scaled back to give the exact value of the predictions.

**Table 1.1** List of inputs

| Input index | Symbol | Description of inputs |
|---|---|---|
| 1 | M_S_MIN | Minimum wind speed |
| 2 | M_S_AVG | Mean wind speed |
| 3 | M_S_MAX | Maximum wind speed |
| 4 | RPM_MIN | Minimum helix speed |
| 5 | RPM_AVG | Mean helix speed |
| 6 | RPM_MAX | Maximum helix speed |
| 7 | KW_MIN | Minimum attributed force |
| 8 | KW_AVG | Mean attributed force |
| 9 | KW_MAX | Maximum attributed force |
| 10 | H | Operation hours |

The sigmoid transfer function is ideally used in this study rather than the gaussian. The sigmoid function acts as an output gate that can be opened when it takes the value one or closed when it takes the value zero. Since the function is continuous, it is also possible for the gate to be partially opened (i.e. somewhere between zero and one). Models incorporating sigmoid transfer functions often help generalized learning characteristics and yield models with improved accuracy; however the use of sigmoid transfer functions can also lead to longer training times. The gaussian transfer function significantly alters the learning dynamics of a neural network model. While the sigmoid function acts as a gate (opened, closed or somewhere in-between) for a neuron's output response, the gaussian function acts like a probabilistic output controller. Like the sigmoid function, the output response is normalized between zero and one, but the gaussian transfer function is more likely to produce the "in-between state". It would be far less likely, for example, for the neuron's output gate to be fully opened (i.e. an output of one). Given a set of inputs to a neuron, the output will normally be some type of partial response. That means that the output gate will open partially. Gaussian based networks tend to learn quicker than sigmoid counterparts, but can be prone to memorization.

### *1.4.3 Data Series Partitioning*

The data concerns four wind power plants that are established in a Greek island. The data sample of each plant contains 365 daily time series observations. One typical method for training a network is to firstly separate the data series into two disjoint sets: the training set and the test set. On this aspect 80% of the data is used for training and the remaining 20% for the evaluation of the models.

The network is trained directly on the training set and its ability to forecast is measured on the test set. A network's generalization ability indirectly measures how well the network can deal with unforeseen inputs, in other words, inputs on which it was not trained in. A network that produces high forecasting errors on unforeseen inputs, but low errors on training inputs, is said to have an over fit of the training data. Overfiting occurs when the network is blindly trained to a minimum in the total squared errors and is based on the training set. A network that has over fit the training data is said to have poor generalization ability.

The data is transformed before is fed to the neural network for training and testing. It is necessary to have all training targets normalized between 0 and 1, as the output neuron is restricted to a signal of values of only 0 or 1. As a result all inputs have also been normalized to the 0 to 1 range. It is accepted that by conducting this normalizing process the characteristics of the training process are enhanced [28]. The data is scaled in order to facilitate the nonlinear estimation process and to avoid the underflow or overflow problems. Helge Petersohn's scaling function is used to scale the data from zero to one, denoted [0, 1],

[25]. To compare the models estimations with the actual values the data is transformed again in no scaling status.

### *1.4.4 Learning and Training*

In this model, wind energy production depends on a set of current variables that are described above in the network's parameters. Two types of neural models were created. One type has been trained with Genetic Algorithms. The Genetic Algorithms do not involve taking gradients or second derivatives and is a global and evolutionary search process. The population size is set to 50 and the number of generation is set to 20.

The other type was developed in order to compare the first type, and uses the Quasi-Newton optimization method for training. Using this training method there is a strong danger of getting stuck in a local rather than a global minimum. Furthermore the Hessian matrix may fail to invert, or become "near-singular" leading to imprecise or even absurd results, for the coefficient vector of the neural network. The number of iterations is set to 20,000.

Learning is accomplished by using the two above mentioned supervised methods. In supervised learning, the network is provided with the actual outputs which in turn are compared with the produced output values, so that the network can note the difference and begin to adjust itself or learn from certain actions. The weights of the arcs between the nodes are adjusted based on the above learning algorithms and aim to decrease the difference between produced and actual values. The network then receives the new input and repeats the procedure, seeks further reduction. After training, weights associated with the best performance on the training set were selected and were applied to the evaluation of the models.

### *1.4.5 Out of Sample Evaluation*

This section focuses on the out-of-sample forecasting ability of the two competing models. Each of the forecasting models described in the last section are estimated by in-sample data. Once the training process of the models is completed, they are ready to predict the series, which they are trained upon. The prediction process is very similar to the training process, except from the evaluation and the parameter adjustment steps. The prediction horizon here is one step ahead of forecasts. Since the prediction is one step ahead it is implied that the models will predict the next day's energy production. The out-of-sample forecasting is referred to a sample of 73 daily observations, which is the testing sample of the models. On the basis of the estimated coefficients of the models, one-step-ahead forecasts were generated from both models.

At this stage, the relative performance of the models is measured by four frequently used measures. To name them: root mean squared error (RMSE),

**Table 1.2** NN forecasts using Genetic Algorithm as learning method

| | Plant 1 | Plant 2 | Plant 3 | Plant 4 | Average |
|---|---|---|---|---|---|
| RMSE | 2.090 | 2.587 | 2.745 | 2.963 | 2.596 |
| MAE | 1.686 | 1.925 | 2.259 | 2.211 | 2.020 |
| MAPE | 15.772 | 18.285 | 24.143 | 22.239 | 20.105 |
| MSE | 5.671 | 6.692 | 13.99 | 8.787 | 5.287 |

mean absolute error (MAE), mean absolute percentage error (MAPE) and mean squared error (MSE) [23]. Table 1.2 reports the out-of-sample forecasting errors of the four plants that are calculated by the neural network that has been trained with the Genetic Algorithms.

Table 1.3 reports the out-of-sample forecasting errors of the same four plants that are calculated by the neural network, which in turn have been trained with the Quasi-Newton method. A plant by plant comparison of the two different types of neural networks when the above learning approaches are used yields slightly different results.

When the neural network model is using the Genetic Algorithms approach it has lower RMSE, in the three of the four plants, than those of the neural network model when applying the Quasi-Newton approach. Also it has lower MAE in two of the four plants, lower MAPE in three of the four plants and lower MSE in two out of four plants. Slight but clear improvement on the forecasts, in terms of the four plants' mean error, produces the neural network that uses the Genetic Algorithm approach. It gives mean values lower than in all of the above statistical performance measures (RMSE, MAE, MAPE and MSE). This implies that the forecasting values given by a neural network that uses the Genetic Algorithms approach are in general more accurate than the forecasting that is based on the Quasi-Newton approach.

Figure 1.1 illustrates the actual values and the out-of-sample wind power production forecasting of the neural network that is trained with Genetic Algorithm for the first plant. In many cases the two lines coexist.

Figure 1.2 illustrates the out-of-sample forecasts of the neural network, which is trained by the Quasi-Newton algorithm, as well as the actual values of the first plant. As one can identify from the graph, the models produced a high level of accuracy in the prediction of 24 hours average in wind power production.

**Table 1.3** NN forecasts using Quasi-Newton as learning method

| | Plant 1 | Plant 2 | Plant 3 | Plant 4 | Average |
|---|---|---|---|---|---|
| RMSE | 2.631 | 2.397 | 2.911 | 3.504 | 2.860 |
| MAE | 2.085 | 1.727 | 2.253 | 2.750 | 2.203 |
| MAPE | 17.716 | 16.182 | 20.418 | 32.011 | 21.581 |
| MSE | 6.926 | 5.747 | 8.477 | 12.284 | 8.358 |

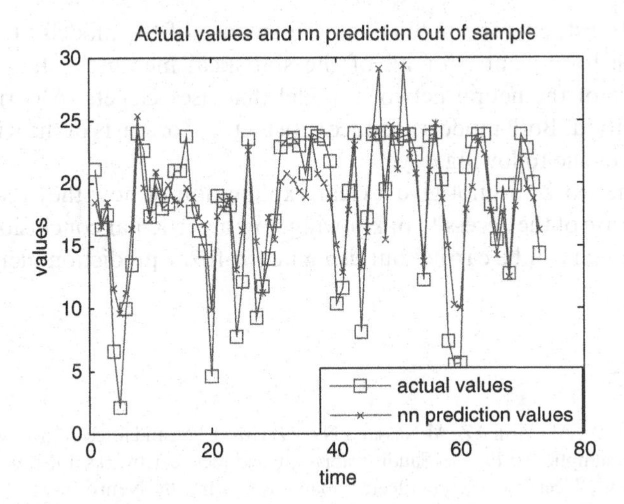


Fig. 1.1 Genetic Algorithm approach

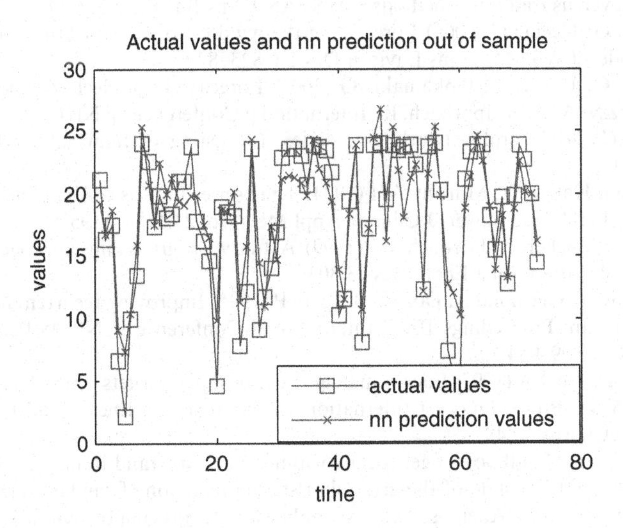


Fig. 1.2 Quasi-Newton approach

## 1.5 Conclusion and Further Research

To sum up, this paper studied the predictive performance of neural networks that applied Genetic Algorithms for training, using data from four real-word wind power plants. The results were compared with a neural network that uses the Quasi-Newton method for training. The data contains 365 daily observations. It was scaled and 80% of the samples were used for training, and the remaining 20% were used for testing the models. Four statistical measures were calculated to estimate the predictive performance of the models. The average error of the four plants, for all of the statistical measures, shows the best performance of the neural network model that uses Genetic Algorithms as a training method. Both models produce satisfactory forecasts of the wind energy production of the following day.

Neural networks do not give a clear explanation of how they reach a conclusion. In case of the necessity of knowing how a particular conclusion is made, further research can be carried out using neuro-fuzzy prediction methods.

## References

1. Abdel-Aal RE, Al-Garni AZ, Al-Nassar YN (1997) Modeling and forecasting monthly electric energy consumption in Eastern Saudi Arabia using adductive networks. Energy 22:911–921
2. Atsalakis G, Ucenic C (2006) Electric load forecasting by Neuro-fuzzy approach. In: WSEAS International Conference on Energy and Environmental Systems, Evia, Greece
3. Atsalakis G, Ucenic C (2006) Forecasting the electricity demand using a Neuro-fuzzy approach versus traditions methods. J WSEAS Trans Bus Econ 3:9–17
4. Atsalakis G, Ucenic C (2006) Forecasting the wind energy production using a Neuro-fuzzy model. J WSEAS Trans Environ Dev 2:6 823–829
5. Atsalakis G, Ucenic C, Plokamakis G (2005) Forecasting of electricity demand using Neuro-fuzzy (ANFIS) approach. In: International Conference on NHIBE, Corfu, Greece
6. Azoff E (1994) Neural network time series forecasting of financial markets. Wiley, New York
7. Brown BG, Katz RW, Murphy AH (1984) Time series models to simulate and forecast wind speed and wind power. J Climate Appl Meteorol 23:1184–1195
8. Chedid R, Mezher T, Jarrouche C (1999) A fuzzy programming approach to energy resource allocation. Int J Energ Res 23:303–317
9. Czernichow T, Germond A, Dorizzi B, Caire P (1995) Improving recurrent network load forecasting. In: Proceedings IEEE International Conference ICNN'95 Perth (Western Australia), 2:899–904
10. Datta D, Tassou SA (1997) Energy management in supermarkets through electrical load prediction. In: Proceedings 1st International Conference on Energy and Environment, Limassol (Cyprus), 2:493–587
11. Davis L (1991) Handbook of genetic algorithms. Van Nostrand Reinhold, New York
12. Giebel G (2003) The state-of-the-art in short-term prediction of wind power, Deliverable Report D1.1, Project Anemos, Available online at anemos.cma.fr/ modules.php?name= Downloads& dop = viewdownload&cid = 3
13. Goldberg DE (1989) Genetic algorithms in search optimization and machine learning. Addison-Wesley, Reading MA
14. Hagan MT, Demuth HB, Beale M (1996) Neural network design. PWS Publishing Company, Boston
15. Holland JH (1975) Adaptation in natural and artificial systems. University of Michigan Press, Ann Arbor
16. Hornik K, Stinchombe M, White H (1989) Multilayer feedforward networks are universal approximators. Neural Networks 2:359–366
17. Javeed Nizami S, Al-Garni AG (1995) Forecasting electric energy consumption using neural networks. Energ Policy 23:1097–1104
18. Kalogirou SA (2000) Applications of artificial neural networks for energy systems. Appl Energ 67:17–35

19. Khotanzad A, Hwang RC, Maratukulam D (1993) Hourly load forecasting by Neural networks. Proceedings of the IEEE PES Winter Meeting, Ohio
20. Kingdon J (1997) Intelligent systems and financial forecasting. Springer-Verlag, Berlin
21. Kretzschmar R, Eckert P, Cattani D, Eggimann F (2004) Neural network classifiers for local wind prediction. J Appl Meteorol 43:727–738
22. Maier HR, Dandy GC (2000) Neural networks for prediction and forecasting of water re- resources variables: a review of modeling issue and applications. Environ Model Source Softw 15:101–124
23. Makridakis S, Weelwright SE, McGee VE (1983) Forecasting: methods and applications. Wiley, New York
24. Mandal JK, Sinha AK, Parthasarathy G (1995) Application of recurrent neural network for short term load forecasting in electric power system. In: Proceedings IEEE International Conference ICNN'95, Perth (Western Australia), 5:2694–2698
25. McNelis DP (2005) Neural networks in finance: gaining predictive edge in the market. Elsevier Academic Press, San Diego, CA
26. Metaxiotis K, Kagiannas A, Ashounis D, Psarras J (2003) Artificial intelligence in short term electric load forecasting: a state-of-the-art survey for the researcher. Energ Convers Manage 44:1525–1534
27. Michalik G, Khan ME, Bonwick WJ, Mielczarski W (1997) Structural modeling of energy demand in the residential sector: the use of linguistic variables to include uncertainty of customer's behavior. Energy 22:949–958
28. Mohandes AM, Rehman S, Halawani TO (1998) A neural network approach for wind speed prediction. Renew Energ 13:345–354
29. Papalexopoulos AD, Shangyou H, Peng TM (1994) An implementation of neural network based load forecasting models for the EMS. IEEE Trans Power Syst 9(4)
30. Poggi P, Muselli M, Notton G, Cristofari C, Louche A (2003) Forecasting and simulating wind speed in Corsica by using an autoregressive model. Energ Convers Manage 44:3177–3196
31. Rosenblatt F (1958) The perceptron: a probabilistic model for information storage and organization in the brain. Psychol Rev 65:386–408
32. Rummeihart DE, Hinton BE, Williams RJ (1986) Learning internal representations by error propagation. In: Rumelhart DE, McClelland JL (eds) Parallel distributed processing: explorations in the microstructures of cognition.MIT Press, Cambridge, MA
33. Sanders I, Batty WJ, Hagino K (1993) Supply of and demand for a resource: fuzzy logistical optimization technique. Appl Energ 46:285–302
34. Sfetsos A (2000) A comparison of various forecasting techniques applied to mean hourly wind speed time series. Renew Energ 21:23–35
35. Weisser D (2003) A wind energy analysis of Grenada: an estimation using the 'Weibull' density function. Renew Energ 28:1803–1812
36. Youcef Ettoumi F, Sauvageot H, Adane AEH (2003) Statistical bivariate modeling of wind using first-order Markov chain and Weibull distribution. Renew Energ 28:1787–1802

# Chapter 2
# Operation Management on Autonomous Power System

**E.S. Karapidakis**

**Abstract** The running developments in the field of power systems have, as a result, the maximization of complexity and their marginal operation with regard to their dynamic security. This becomes more perceptible in autonomous power systems. Consequently, in the modern energy environment, the use of enhanced operation management and monitoring programs is judged necessary. The results that are arrived at in the present chapter were developed via a concretization of algorithms, which were incorporated in an implemented operation and planning management system. The presented algorithms have as object to improve the level of combination between unit commitment, economic dispatch and dynamic security assessment (DSA).

## 2.1 Introduction

Energy Management System (EMS) is one of the main control center's programs. The high processing power, which is provided by modern computers, creates the requirement for significant fast and reliable energy management programs in power system's control centers [13]. An EMS also includes dynamic security assessment algorithms for real-time performance [4], which considered as critical support factor for rational planning and power system operation. Specifically, any EMS is separated into two basic categories, depending on their implementation time:

- The real-time planning and operation programs, where the load and corresponding wind power production forecasting (when such possibility exists), supply the necessary data entry for economic dispatch (ED) and unit commitment (UC) algorithm integration. Then, DSA algorithms provide the safety level of a given profile under pre-selected contingencies, in order for a final and optimal operation decision to be retrieved.

E.S. Karapidakis (✉)
Technological Educational Institute of Crete, 3 Romanou str, 73133, Chania, Greece

N. Mastorakis et al. (eds.), *Proceedings of the European Computing Conference*, Lecture Notes in Electrical Engineering 28,
DOI 10.1007/978-0-387-85437-3_2, 

- The short-term planning and system operation programs, where, off-line, the system administrator analyzes and estimates the next most optimal possible operation system profiles via different scripts.

The results that are arrived at in the present chapter were developed via the concretization of algorithms, which were incorporated in an implemented operation and planning management system. The presented algorithms have as object to improve the level of combination between unit commitment, economic dispatch and dynamic security assessment.

## 2.2 Algorithm Implementation

The main modules of the implemented program are related either to real-time operation management, or to short-term power system planning, as is shown in Fig. 2.1. Specifically, the following sub-algorithms were integrated and examined.

### *2.2.1 Load and wind power forecasting*

The mathematic methods for load demand forecasting, and in certain cases, for wind power generation forecasting, are divided into three basic categories:

- Time series methods, where load demand curves follow given seasonal, monthly, weekly and daily periodicity, [2].
- Regression based methods, where load demand forecasting is calculated by the linear combination of previous hours, [9].

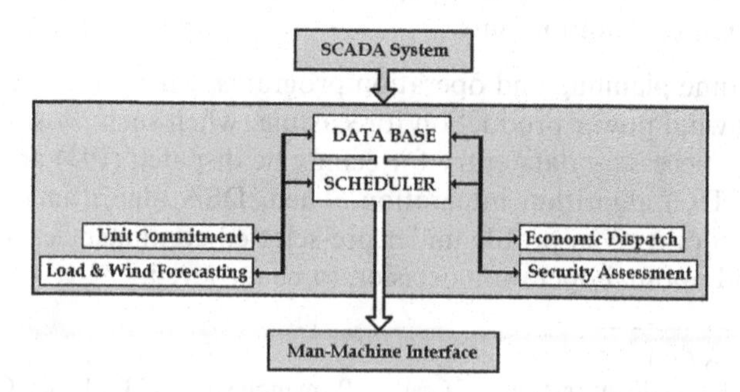


**Fig. 2.1** EMS functions & procedures

- Computational intelligence based methods, through supervised learning, where large number of historical data are used for artificial intelligence architectures training [19, 18, 12, 15].

### 2.2.2 *Unit Commitment*

Electrical power generation Units Commitment (UC) problem is a complex optimization problem with many restrictions [10, 1, 16]. Specifically, the mathematic expression is given as:

$$\min \sum_{t=0}^{T-1} \sum_{i=1}^{N} u_i^t C_i(g_i^t) + (1 - u_i^{t-1}) u_i^t s_i \tag{2.1}$$

with the following restrictions, as load demand on the examined time horizon (2.2), the required spinning reserve (2.3) and the technical minimum and maximum limits of units generation (2.4).

$$\sum_{i=1}^{N} g_i^t = D^t \tag{2.2}$$

$$\sum_{i=1}^{N} r(g_i^t) \geq R^t \tag{2.3}$$

$$g_i^{\min} \leq g_i^t \leq g_i^{\max} \tag{2.4}$$

The previous equations represent the problem of economically optimal power units generation profile in power systems. In Table 2.1 below, all the parameters are individually described.

**Table 2.1** Parameter description

| Symbol | Description |
|---|---|
| N | Total number of power units |
| T | Time |
| $D^t$ | Load demand |
| $R^t$ | Required spinning reserve |
| $g^t$ | Power generation |
| $g^{min}$ | Minimum power generation |
| $g^{max}$ | Maximum power generation |
| $u^t$ | Decision of unit i operation |
| $C_i(g_i)$ | Fuel cost function of unit i |
| $r(g_i)$ | Spinning reserve function of unit i |
| $s_i$ | Start-up cost of unit i |
| $r_{max}$ | Maximum spinning reserve |

### 2.2.3 Economic Dispatch

Economic analysis of each of the conventional generation units is based on the following general economic parameters:

- Hourly Fuel Consumption (Kcal/h): the hourly fuel consumption is given as a function of generating power P of the corresponding unit.
- Hourly Fuel Cost (Euro/h): the hourly fuel cost results from the hourly fuel consumption and the corresponding cost, according to the relation:

$$F = \frac{\text{kcal}}{h} \cdot \frac{\text{Euro}}{\text{kcal}} = \frac{\text{Euro}}{h} \tag{2.5}$$

- Differential Hourly Fuel Consumption d(Kcal/h)/d(MW)): the differential hourly fuel consumption represents the difference of hourly consumption fuel (d(Kcal/h)), which results from produced power change (d(MW)), as it appears in:

$$\text{d(kcal/h)/d(MW)} \tag{2.6}$$

- Differential Production Cost (Euro/MWh): the differential production cost results from the differential hourly consumption and the corresponding cost, according to the relation:

$$\frac{\text{d(kcal/h)}}{\text{d(MW)}} \cdot \frac{\text{Euro} \cdot 10^6}{10^6 \cdot \text{kcal}} = \frac{\text{Euro}}{\text{MWh}} \tag{2.7}$$

An energy system, which allocates a significant number of power generation units, with different technical and economical characteristics, the total production cost varies sensibly, according to production combinations of the available generators. Accordingly, economic dispatch of power generation units is a classic optimization problem with restrictions [8, 3]. The mathematic expression of the previous optimization problem is:

$$\min\left(\sum_{t=1}^{N} C_i(P_i)\right) \tag{2.8}$$

with the following restrictions, as load demand on the examined time horizon (2.9), the required spinning reserve (2.10) and the technical minimum and maximum limits of units generation (2.11).

$$\sum_{i=1}^{N} P_i = D + L \tag{2.9}$$

**Table 2.2** Parameter description

| Symbol | Description |
|---|---|
| N | Total number of power units |
| D | Load demand |
| R | Required spinning reserve |
| $P_i$ | Power generation of unit i |
| $P^{min}$ | Minimum power generation |
| $P^{max}$ | Maximum power generation |
| $C(P_i)$ | Fuel cost function of unit i |
| $R_i$ | Spinning reserve of unit i |

$$\sum_{i=1}^{N} P_i \geq P \tag{2.10}$$

$$P_i^{min} \leq P_i \leq P_i^{max} \tag{2.11}$$

The previous equations (2.8), (2.9), (2.10) and (2.11) represent the optimal economic dispatch problem of power generation for the running load demand. In Table 2.2, all the corresponding parameters are individually described.

The fuel consumption curve of each unit is given in the majority of cases from a second degree polynomial function:

$$C(P_i) = a \cdot P^2 + b \cdot P + c \tag{2.12}$$

Previous relation is used in the proposed algorithm, while it is considered as analytically significant and representative of the real production cost of the corresponding units.

### 2.2.4 Dynamic Security Assessment

A reliable dynamic secure operation monitoring requires the analysis of tens or even hundreds of possible operating profiles (steady and transient), in desirable and practically feasible computational time, [17].

The mathematic methods for dynamic security assessment of a power system under pre-selected disturbances (as short-circuits and unit's trip) are divided into two categories:

- Conventional Mathematic Methods, where the results are given via analytic recurrent mathematic functions or corresponding mathematic models, which describe in detail the examined system. However, the analytic mathematic solution of this complex problem requires the availability of a prohibitorily long period of time, taking into account the modern EMS real-time operation [6].

- Expert systems and computational intelligence based methods, where the previous experience of operators, combined with a considerably large number of historical data, is used for AI architectures training [20, 11], in order for satisfactorily fast and precise DSA results to be achieved.

## 2.3 Operation Management

A short-term generation planning and power system operation program was developed, taking into account the technical restrictions and the economic characteristics of each generation unit. Initially, the program determines an available unit commitment for a given load demand curve, while the selected generators' set points are recalculated via individual economic dispatch algorithms. Then, the dynamic safety level of the proposed operation profile is estimated under various pre-selected transient disturbances, in order for an acceptance or a redefinition of new security margins to be retrieved. More precisely, the main program processes are:

1. Twenty-four hours' Economic Unit Commitment.
2. Dynamic Security Assessment of the previous Unit Commitment.
3. New Unit Commitment with new Dynamic Security Criteria.
4. Final Dynamic Security Level Verification.

The software package (algorithms and interface) has been implemented with Visual Basic (version 6) in multi-document interface, while UC and DSA modules have been developed independently, as dynamic link libraries (dlls), in order to provide an easy replacement. The database is a relational database (RDBMS) via dynamic access object (DAO 3.6).

### *2.3.1 Data Entry*

In Fig. 2.2 below, a data entry sample of load demand curve and wind power generation is depicted.

### *2.3.2 Unit Commitment*

Under the selected load demand curve and the corresponding wind power production, the UC algorithm provides the most optimal calculated system generation profile.

The technical characteristics (min/max), cost curves, start-up costs and fuel consumptions, as well as the response type of each conventional generation unit, contribute to corresponding unit set points determination.

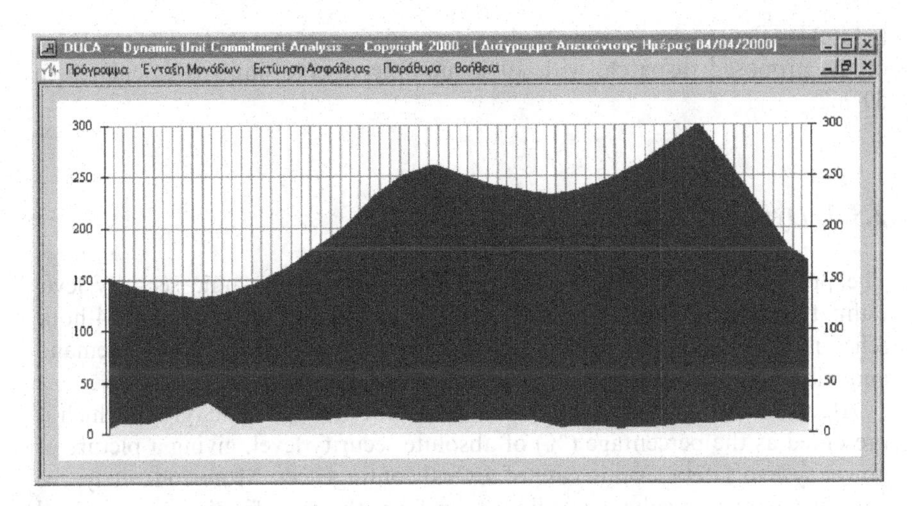


**Fig. 2.2** Load and wind power data entry

One more algorithm parameter that influences considerably the generation dispatch profile is the desirable spinning reserve. In Fig. 2.3, UC results for the examined time horizon (24 hours), is presented, taking into account all the previous data.

According to the previous Fig. 2.3, load demand is mainly covered by the "base" power units, as six steam units (green) and the combined cycle unit (orange). Continuously, high loads are covered by the four "fast" diesel units (mauve), while the main daily peaks are served by the available gas units (red).

Part of total load, as was already reported, covers the selected wind power production (yellow) for the examined time period, while the spinning reserve

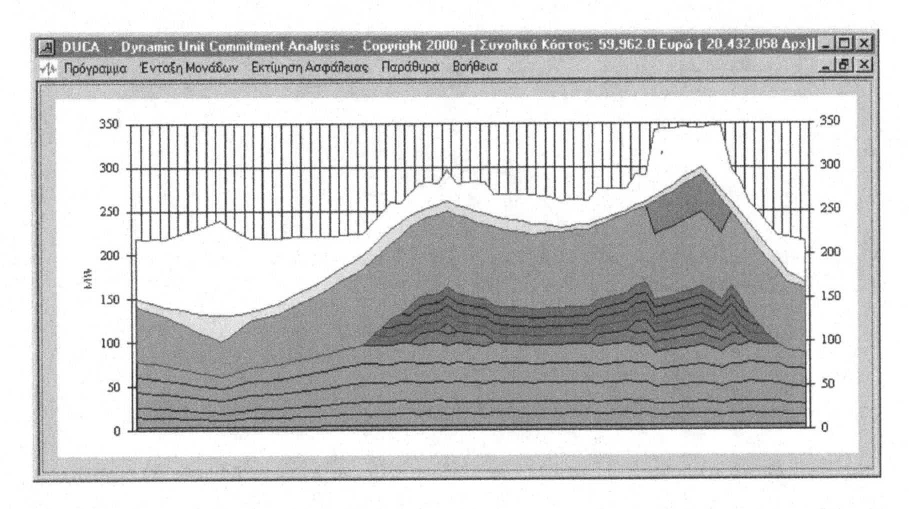


**Fig. 2.3** Unit commitment monitoring

(white) is calculated according to the proposed operation profile. The total cost of the proposed dispatch, with 10% spinning reserve limit, is calculated in 59,962.

### 2.3.3 *Dynamic Security Assessment*

The previous UC results provide the data entry of dynamic security level estimation algorithm under pre-selected disturbances of the examined 24-hour schedule. In Fig. 2.4, the level of dynamic safety, in parallel with load demand and wind power production, is presented.

Algorithms results are combined in one total indicator line (grey), which is presented as the percentage (%) of absolute security level, giving a picture of power system safety level in case of several contingences. More concretely, the current indicator results as medium term from the classification of examined functional points (zero unsafe and one safe) via five decision trees of pre-selected common disturbances.

### 2.3.4 *Preventive Security Assessment*

Unit commitment module is executed once again, receiving as additional entry the dynamic security restrictions, which were exported by the DSA algorithm [14]. In the following Fig. 2.5, final preventive security assessment results are plotted, where all the corresponding diagrams of the examined power system for short-term planning and operation are in one screen.

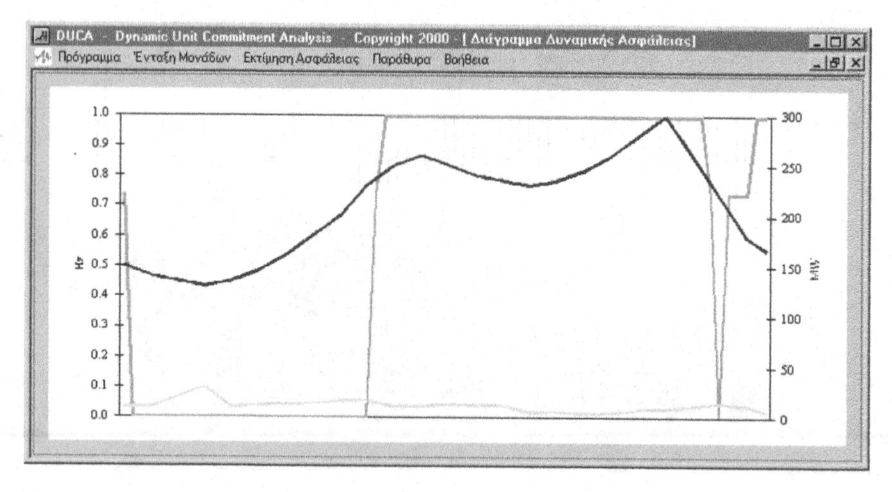


**Fig. 2.4** Dynamic security level monitoring

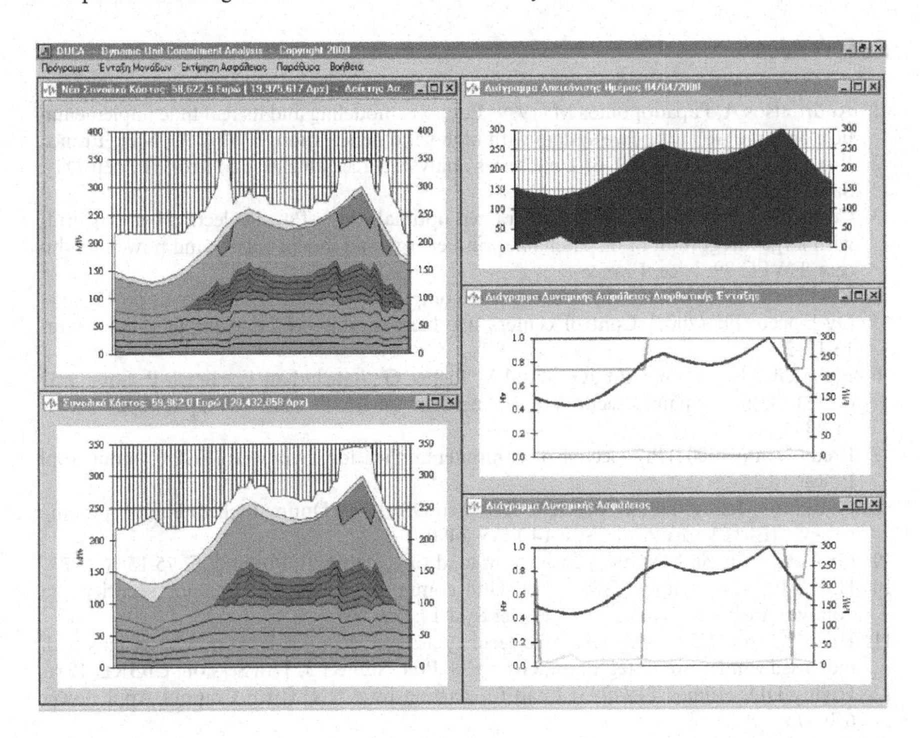


**Fig. 2.5** Final results screen

The improvement of dynamic security level is confirmed by safety index. As is shown in Fig. 2.5, the difference between the previous security level without restrictions and the final proposed profile easily becomes obvious.

## 2.4 Conclusion

From the start of electrical energy systems, human operators (system dispatcher) had the whole responsibility of power generation and distribution control. During the 1980 s, Nathan Cohn (IEEE Computer Application in Power, January 1988) described a precocious energy management application of power generation and distribution control. Since then, the programs that are used in power systems control centers present significant improvement, based on the modern computer's increasing processing power [5, 7].

The present chapter presents a real-time energy management program for short-term operation planning, which was initially applied in Crete's autonomous power system. The efficiency of the program's calculations are judged to be sufficiently satisfactory, taking into account the evaluations results in real operation conditions.

## References

1. Androutsos A, Papadopoulos M (1999) Logistics modeling and its real time implementation on large isolated powe system with high wind penetration. EWEC 99, Nice, France
2. Box GE, Jenkins GM (1976) Time series analysis forecasting and control. Holden-Day, San Francisco.
3. Contaxis G, Vlachos A (1999) Constrained optimal power flow in electrical energy grids with large integration of dispatchable wind energy and independent wind power producers. EWEC 99, Nice, France
4. Dy-Liacco TE (1988) System security: the computer's role. IEEE Spectrum, pp 45–50
5. Dy-Liacco TE (2002) Control centers are here to stay. IEEE Comput Appl Power 15:18–23
6. Ejebe GC, Jing C, Waight JG, Vittal V, Pieper G, Jamshidian F, Hirsch P, Sobajic D (1998) Online dynamic security assessment in an EMS. IEEE Comput Appl Power 11:43–47
7. Fred I Denny (2002) Prospective on computer applications in power. IEEE Comput Appl Power 15:24–29
8. Gooi HB, Mendes DP, W.Bell KR, Kirschen DS (1999) Optimal scheduling of spinning reserve. IEEE Trans Power Syst 14:1485–1490
9. Gross G, Galiana FD (1987) Short term load forecasting. In Proc IEEE, 75:1558–1573
10. Haili Ma, Shahidehpour SM (1999) Unit commitment with transmission security and voltage constraints. IEEE Trans Power Syst 14:757–764
11. Hatziargyriou ND (1998) Dynamic security assessment of isolated power systems with increased wind power integration. Group 38, Pref. Subject 3, 37th Session, CIGRE, Paris
12. Highly DD, Hilmes TJ (1993) Load forecasting by ANN. IEEE Comput Appl Power 6:10–15
13. Jamnicsky L (1996) EMS network security applications of the future. IEEE Comput Appl Power 9(2):42–46
14. Karapidakis ES, Hatziargyriou ND (2002) On-line preventive dynamic security of isolated power systems using decision trees. IEEE Trans Power Syst 17:297–304
15. Kariniotakis G, Matos M, Miranda V (1999) Assessment of benefits from advanced load and wind forecasting in autonomous power systems. EWEC 99, Nice, France
16. Kazarlis K, Bakirtzis A, Petridis V (1996) A genetic algorithm solution to the unit commitment problem. IEEE Trans Power Syst 11:83–90
17. Kumar ABR, Brandwajn V, Ipakchi A, Rambabu Adapa (1998) Integrated framework for dynamic security analysis. IEEE Trans Power Syst 13(3):816–821
18. Lee KY, Cha YT, Park JH (1992) Short term load forecasting using an artificial neural network. IEEE Trans PAS 7:124–131
19. Park DC, El Sharkawi MA, Marks RJ, Atlasm LE, Damborg J (1991) Electric load forecasting using an artificial neural network. IEEE Trans PAS 6:442–448
20. Pecas Lopes JA (1998) Application of neural network based stability assessment tools to an operational environment of large autonomous power systems. Group 38, Pref. Subject 1, 37th Session, CIGRE, Paris

# Chapter 3
# A Low-Power Portable H.264/AVC Decoder Using Elastic Pipeline

**Yoshinori Sakata, Kentaro Kawakami, Hiroshi Kawaguchi, and Masahiko Yoshimoto**

**Abstract** We propose an elastic pipeline architecture that can apply dynamic voltage scaling (DVS) to a dedicated hardware, and implement the elastic pipeline to a portable H.264/AVC decoder LSI with embedded frame buffer SRAM. A supply voltage and operating frequency are decreased by a feedback-type voltage/frequency control algorithm. In a portable H.264/AVC decoder, embedded SARM can be utilized as frame buffer since the frame buffer is not so large that an external DRAM is required. In the proposed pipeline architecture, the power in the embedded SRAM and even in a local bus connecting with the frame buffer SRAM can be controlled by dynamic voltage scaling (DVS). We carried out simulation in the 320 × 180 pixels baseline profile and 320 × 240 pixels mail profile. The total power reduction in 320 × 180 pixels and 320 × 240 pixels are 30 and 31%, respectively.

## 3.1 Introduction

Dynamic voltage scaling (DVS) is a low-power technique that controls an operating frequency and a supply voltage on an LSI, according to an application workload. DVS is well utilized on general-purpose processors to achieve both high peak performance and low average power [1, 2].

Figure 3.1 shows a relationship between an operating frequency and a power, with DVS and without DVS (= clock gating). In DVS, if a workload does not need a high operating frequency, we can choose a combination of a lower operating frequency and a lower supply voltage. Due to this power optimization, the power becomes lower than the conventional scheme without DVS, when a workload is low. If the maximum performance is instantaneously needed, the highest supply voltage and highest operating frequency are utilized so that DVS can accommodate the peak performance.

Y. Sakata (✉)
Graduate School, Kobe University, Kobe, Hyogo, 657-8507, Japan
e-mail: yoshi@cs28.cs.kobe-u.ac.jp

N. Mastorakis et al. (eds.), *Proceedings of the European Computing Conference*, Lecture Notes in Electrical Engineering 28,
DOI 10.1007/978-0-387-85437-3_3, 

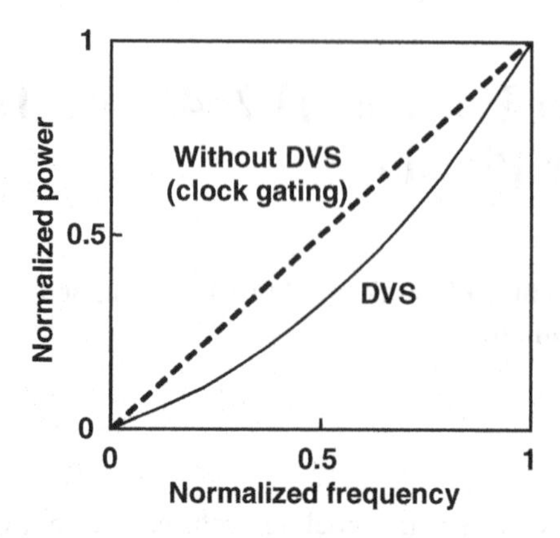


**Fig. 3.1** Relationships between power and frequency in DVS and clock gating

In a case of an application to hardwired logic circuits for real-time processing, there are a few problems; a dedicated hardware is often built with pipeline architecture for high performance. Considering the likely worst-case workload, the starting time of a pipeline process is segmented into the worst-case execution cycles (WCEC). Thus, the required operating frequency is uniquely fixed, and there is no room to apply DVS in the hardwired logic circuits.

To realize DVS in hardwired logic circuits, we propose an elastic pipeline architecture. Depending on the characteristics of the input data, this architecture can conserve process cycles in the pipeline operation. The slack time is exploited for DVS, which achieves lower power in hardwired logic circuits.

The rest of this paper is organized as follows. Section 3.2 mentions the conventional pipeline architecture. Section 3.3 describes the proposed elastic pipeline architecture for DVS in hardwired logic circuits. In Section 3.4 we exhibit the experimental results of the proposed architecture. Section 3.5 summarizes this paper.

## 3.2 Conventional Pipeline Architecture

Figure 3.2 illustrates a timing diagram of the conventional pipeline architecture. The WCEC is the maximum number of execution cycles required for one pipeline process. A gray area in the figure shows the number of processing cycles that a pipeline stage processes a datum. A hatching area means common idle cycles in a pipeline process after all pipeline stages were completed. Considering the worst-case workload, a starting time of each pipeline process is fixed to the WCEC in the conventional pipeline architecture. Hence, all the pipeline stages have to idle until the next starting time even if all the pipeline stages finish earlier than the WCEC.

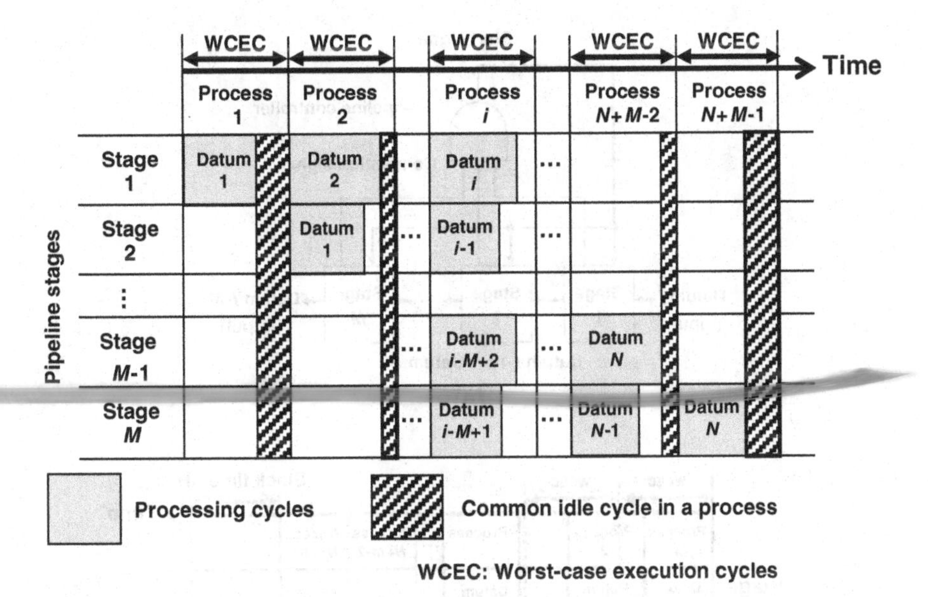


**Fig. 3.2** Timing diagram of the conventional pipeline

## 3.3 Elastic Pipeline Architecture

### *3.3.1 Concept*

We propose the elastic pipeline architecture as the solution to the issue of the conventional pipeline architecture [3]. Figure 3.3(a) and (b) shows a concept and a timing diagram of the proposed elastic pipeline architecture. After each stage in the elastic pipeline is completed, it sends a completion signal to the pipeline controller. As soon as the pipeline controller collects all the completion signals from all the pipeline stages, each pipeline stage proceeds to the next pipeline process with the start signal.

In the proposed architecture, the common idle cycles are built up in pipeline processes, and become a lump of time. As illustrated in Fig. 3.3(b), a pipeline process in the elastic pipeline requires less time than the conventional pipeline since the common idle cycles are put off. Thereby, the elastic pipeline architecture produces the slack time, $\Delta H$, compared to the conventional pipeline architecture.

### *3.3.2 Feedback-Type Voltage/Frequency Control Algorithm*

For DVS in the proposed elastic pipeline, a supply voltage and an operating frequency are changed by a feedback-type voltage/frequency control algorithm

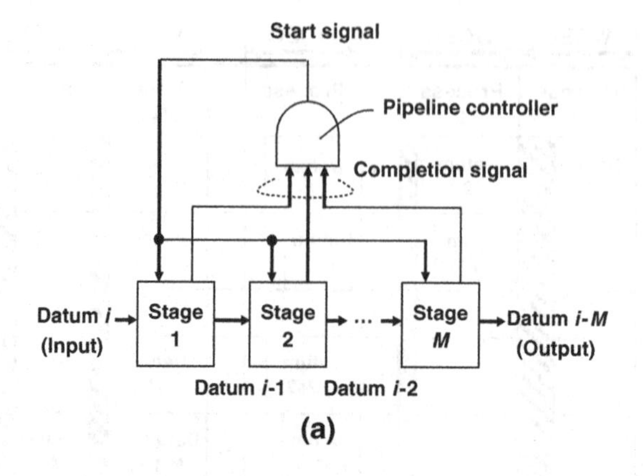


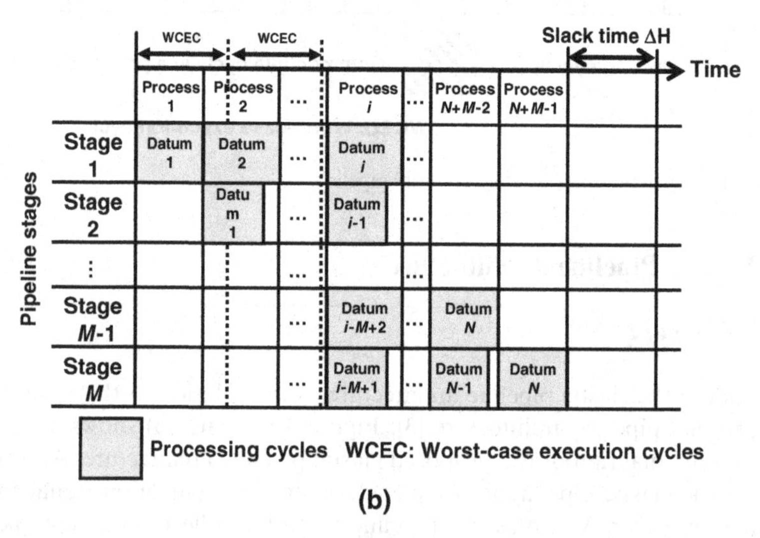


**Fig. 3.3** Proposed elastic pipeline architecture: (a) concept and (b) timing diagram

as illustrated in Fig. 3.4 [4]. In an H.264 codec, data are processed in every single macro block (MB: 16 × 16 pixels). In this algorithm, a frame is divided into some slots; a set of MBs are assigned to a slot. The first and second slots are always processed with the maximum frequency ($=f$ in Fig. 3.4). However, these slots are potentially completed earlier since the elastic pipeline reduces the number of processing cycles. Now, pay attention to the third slots, where the slack time, ΔH, is left. Even considering a voltage/frequency transition time, $T_{td}$, the third slot has twice as long a time as $T_{slot}$ (a processing time for a slot), which allows the third slot to be processed at a half of $f$. Note that a real-time operation is guaranteed in this feedback algorithm. As described, we prepare the two kinds of operating frequencies, $f$ and $f/2$ in this study [3, 4].

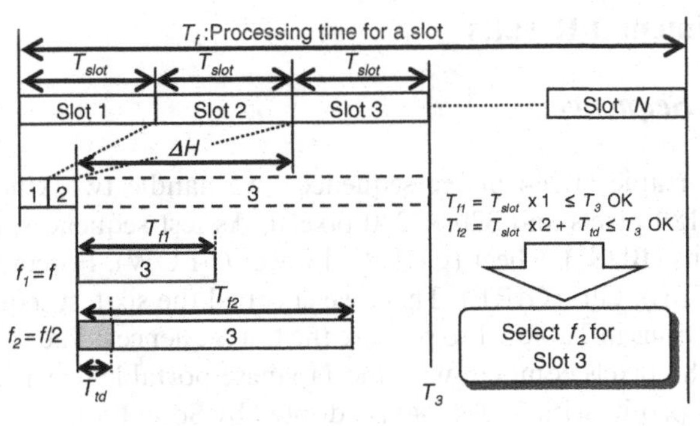


**Fig. 3.4** Feedback-type voltage/frequency control algorithm

### *3.3.3 Architecture*

To estimate the effect of the power reduction in the proposed pipeline architecture, we designed a H.264 decoder architecture as shown in Fig. 3.5. SRAM utilized for reference images is embedded on a chip. The external DRAM is used as bit stream buffer, if the resolution is very large as HDTV. In this case, a supply voltage and operating frequency should be fixed since it is preferable that the DRAM interface operates at a fixed supply voltage and operating frequency for compatibility with other hardware cores. But in this study, we handle small resolution which is used for the portable product. So SRAM which utilized for bit stream buffer is embedded. We can adapt DVS for not only decoder core but SRAM and local bus connecting with the frame buffer SRAM.

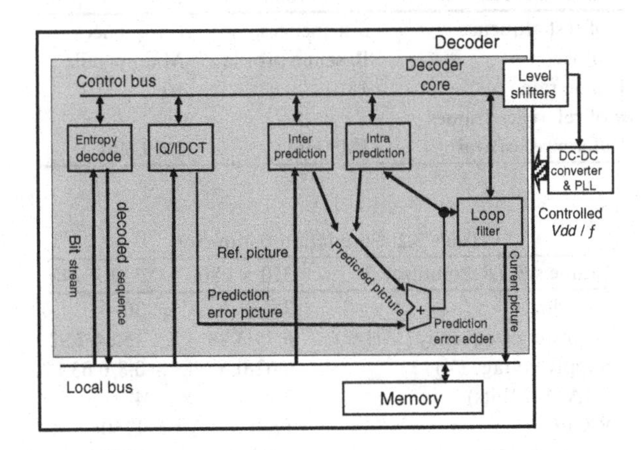


**Fig. 3.5** H.264 decoder block diagram

## 3.4 Experimental Results

### 3.4.1 Test Sequence

Assuming portable H.264 image sequences, we handle two kinds of resolution (320 × 180 pixels and 320 × 240 pixels). As test sequences, we chose six sequences: Bus (BUS1), Cheer (CHER), Flower (FLOW), Foreman (FORE2), Football (FTBL), Girl (GIRL). Then, we encoded the six test sequences under the configurations in Table 3.1 to prepare the test sequences: The baseline profile with 320 × 180 pixels complies with the Japanese portable television standard, and the main profile with 320 × 240 is adopted by Sony PSP [5].

Table 3.2 illustrates the simulation parameters. The respective supply voltage and operating frequency are prepared for the two kinds of resolutions. The capacities of the embedded SRAMs are 2 M bits and 4 M bits, respectively. The SRAM capacity of the 320 × 240 pixel main profile is twice as large since the number of reference frames is two. However, note that the WCECs and the numbers of logic gates are equal between the two kinds of resolutions. In other words, the operating frequencies are different, but the sizes of the decoder cores are the same.

### 3.4.2 The Optimum Number of Slots Per Frame

Since the elastic pipeline architecture reduces the processing cycle, we can apply DVS to the decoder LSI. In this subsection, the optimum number of slots is discussed.

Figure 3.6 illustrates the simulation result of the relationship between the power of the decoder core and the number of slots, using the BUS1 sequence.

**Table 3.1** Encoding configuration

| Frame size (Resolution) | 320 × 180 pixels | 320 × 240 pixels |
|---|---|---|
| # of test sequence | 6 sequences | 6 sequences |
| Profile | Baseline profile | Main profile |
| Frame rate | 15 | 30 |
| # of reference frames | 1 | 2 |
| Reference software | JM9.6[6] | JM9.6[6] |

**Table 3.2** Simulation parameter

| Frame size (Resolution) | 320 × 180 | 320 × 240 |
|---|---|---|
| # of slots | 75 | 30 |
| Operating frequency (MHz) | 6.75/3.38 | 15.84/7.92 |
| Supply voltage (V) | 0.62/0.6 | 0.8/0.63 |
| SRAM (Mbits) | 2 | 4 |
| WCEC | 1760 | 1760 |
| # of logic gates | 600329 | 600329 |

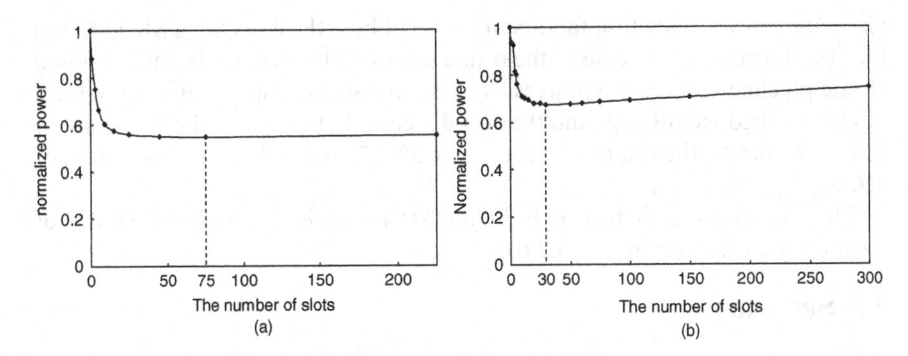


Fig. 3.6 The number of slots in a frame vs. power using the test sequence "Bus": (a) 320 × 180 pixels and (b) 320 × 240 pixels

The power reduction factor depends on the number of slots. In this simulation, the transition time is assumed to be 50 μs [1]. The baseline profile with 320 × 180 pixels has the power minimum when the number of slices is 75. On the other hand, the optimum number of slots is 30 in the main profile with 320 × 240 pixels.

If the number of slots is smaller than the optimum point, the power reduction drastically becomes worse. For instance, if there are merely several slots, there are few chances to make the operating frequency and the supply voltage lower, which causes high power consumption. Alternatively, if there are many slots, there are many chances to change the operating frequency and supply voltage. The voltage/frequency transition time, however, becomes longer. The power reduction gradually becomes worse with the increase of the slice number.

### 3.4.3 Power Saving

As well as the decoder core, we can apply DVS to the embedded frame buffer SRAM and the local bus connecting to the SRAM. Figure 3.7 (a) shows the

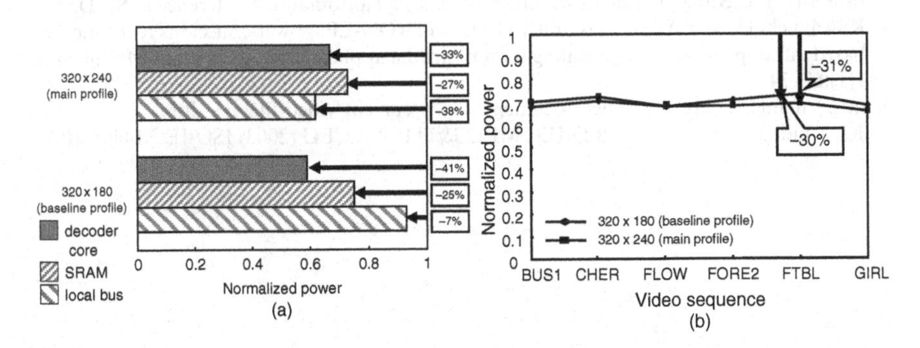


Fig. 3.7 Power reduction ratio: (a) decode core, embedded SRAM, local bus, (b) overall decoder

respective power reduction factors in the local bus, the embedded SRAM, and the decoder core. In the case of the frame size of 320 × 180 pixels, the proposed elastic pipeline reduces the powers by 7, 25, and 41%, respectively, in the local bus, the embedded SRAM, and the decoder core. In the case of the frame size of 320 × 240 pixels, the respective factors are 38, 27, and 33% in the resolution of 320 × 180 pixels.

The overall power reduction is 30 and 31% on average, in the two kinds of resolutions, as shown in Fig. 3.7 (b).

## 3.5 Summary

We proposed the elastic pipeline architecture that can apply DVS to a hard-wired circuit. We implemented a H.264 decoder LSI, and controlled the embedded frame buffer SRAM and the local bus connecting to the embedded SRAM with DVS, as well as the decoder core.

We verified that the proposed elastic pipeline reduces the power on the H.264 decoder LSI by 7% in the local bus, by 25% in the frame buffer SRAM, and by 41% in a decoder core in a 320 × 180 pixel baseline profile. In a case that 320 × 240 pixel main profile, the power is reduced by 38, 27, and 33% in the local bus, the frame buffer SRAM, and the decoder core, respectively. The total power reductions in the baseline profile and the mail profile pixels are 30 and 31%, respectively

## References


1. Nowka KJ, Carpenter GD, MacDonald EW, Ngo HC, Brock BC, Ishii KI, Nguyen TY, Burns JL (2002) A 32-bit PowerPC system-on-a-chip with support for dynamic voltage scaling and dynamic frequency scaling. IEEE J Solid-State Circuits 37(11):1441–1447
2. Kawakami K, Kanamori M, Morita Y, Takemura J, Miyama M, Yoshimoto M (2005) Power-minimum frequency/voltage cooperative management method for VLSI processor in leakage-dominant technology era. IEICE Trans Fundamentals E88-A(12):3290–3297
3. Kawakami K, Kuroda M, Kawaguchi H, Yoshimoto M (2007) Power and memory bandwidth reduction of an H.264/AVC HDTV decoder LSI with elastic pipeline architecture. In: Proceeding of Asia and South Pacific design automation conference (ASP-DAC)
4. Kawaguchi H, Shin Y, and Sakurai T (2005) μITRON-LP: power-conscious real-time OS based on cooperative voltage scaling for multimedia applications. IEEE Trans Multimedia 7(1):67–74
5. http://manuals.playstation.net/document/jp/psp/current/video/filetypes.html
6. Joint Video Team (JVT) of ISO/IEC MPEG&ITU-T VCEG (2003) ISO/IEC 14496-10.

# Part II
# Computational Techniques and Algorithms

In this part we consider some computational techniques, tools and algorithms for solving different problems. We present theoretical and philosophical discussions, pseudo-codes for algorithms, and computing paradigms that illustrate how computational techniques can be used to solve complex problems, simulate and explain natural phenomena, and possibly allow the development of new computing technologies. The topics include computational methods for detecting types of nonlinear dynamics, the worm cutter tool profiling based on discreet surfaces representation, suggesting correct words algorithms developing in FarsiTeX, analytical and numerical computation in mechanical modeling, real genetic algorithm for traveling salesman problem optimization, the constrictive test effort cost model, a semantics of core computational model for ODP specifications, computer simulation via direct modeling, image segmentation with improved artificial fish swarm algorithm, propagation of elastic-plastic waves in bars, Shahyat algorithm as a clustering method, a modified electromagnetism-like algorithm based on a pattern search method, latency hiding and adaptability for parallel iterative algorithms, clustering the linearly inseparable clusters and a top-down minimization algorithm for non-deterministic systems. The results can serve an interdisciplinary community which needs to use mathematical techniques in the development of computational analysis and engineering methodology, in order to better solve real life problems such as described above. It also aims to bridge the subject areas of mathematics, classical computer science, engineering, and future generation computational analysis.

# Chapter 4
# Computational Methods for Detecting Types of Nonlinear Dynamics

**Mirela-Catrinel Voicu**

**Abstract** In this paper we present implementation methods used to detect the dynamics displayed by nonlinear determinist discrete systems. Our study focuses on attractors. The most complex type of behavior for a nonlinear system is that of chaotic dynamics. It is known that the Lyapunov exponents are a tool used to establish the type of behavior in nonlinear dynamics. For this reason, in this paper we also present an implementation method used in order to obtain the Lyapunov exponents. In order to exemplify our study, we will present some numerical results for a certain system of the second order. Our implementation methods can be used for any nonlinear determinist discrete system and for any order of the system.

## 4.1 Introduction

In the study of nonlinear dynamics, situations occur where analytical tools alone are not sufficient. However, in many cases we can use numerical simulation to detect the type of dynamics. All numerical algorithms are based on the mathematical theory. Here are some important aspects: the computer performance, the software used for implementation and the implementation methods. All these must imply a very short time to detect the results. If the results are obtained quickly, many simulations can be made, thus leading to a good conclusion to the study.

The aim of this paper is to emphasize the importance of implementation methods of computational study. As an example, we present some results obtained for a nonlinear determinist discrete dynamical system. In this paper we focus on chaotic behavior.

---

M.-C. Voicu (✉)
Faculty of Economic Sciences, West University of Timi oara, Romania
e-mail: mirela.voicu@fse.uvt.ro

N. Mastorakis et al. (eds.), *Proceedings of the European Computing Conference*, Lecture Notes in Electrical Engineering 28,
DOI 10.1007/978-0-387-85437-3_4, 

## 4.2 Mathematical Framework Presentation

### 4.2.1 Model Presentation

We consider the following model:

$$s_t = \left(\frac{(c+\alpha)b}{1+\beta(e^{s_{t-1}}-1)^2} + (1-\alpha)b\right)s_{t-1} + \left(\frac{-cb}{1+\beta(e^{s_{t-1}}-1)^2}\right)s_{t-2} \tag{4.1}$$

where $\alpha > 0, \beta > 0, b \in (0,1), c > 1, t \geq 0, s_t \in R$ and $e^{s_t}$ is the exchange rate at the moment $t$.

The model presented in equation (4.1) is a model proposed in [1] and describes the exchange rate evolution.

The vectorial form of equation (4.1) is given by the following expression:

$$(s_t, s_{t+1}) = F(s_{t-1}, s_t) \tag{4.2}$$

where $F: R^2 \to R^2$, with

$$F(x,y) = (F_1(x,y), F_2(x,y)), \qquad F_1(x,y) = y$$

and

$$F_2(x,y) = \left(\frac{(c+\alpha)b}{1+\beta(e^y-1)^2} + (1-\alpha)b\right)y + \left(\frac{-cb}{1+\beta(e^y-1)^2}\right)x.$$

In our numerical study we will use both forms. For certain values of parameters ($\alpha$, $\beta$, $b$ and $c$) and the initial condition, we have obtained some qualitative results (see [6]), but in many situations the analytical tools alone are not sufficient. For this reason, in order to detect different system behavior types, we have made numerical simulations. Fixing the values of parameters ($\alpha$, $\beta$, $b$ and $c$) and the initial condition, using software applications, we can generate an unlimited number of iterations of the system and a graphical representation for their values. When the number of iterations increases, generally, we can observe the trajectory tendency.

During this study stage, we want to make some observations:

- The value ranges of parameters and of the initial condition cover sets with an infinite number of values, and a numerical study can be made only for a limited number of such values. The selection (for simulation) of certain parameters and initial condition values is influenced by different reasons (e.g., the value signification for the phenomenon which is modelled by the system).
- Using numerical simulations, we can easily detect only attractors (e.g., in the case of a periodical repulsive cycle, we must have the initial conditions of the cycle, and this is practically impossible).

We can classify the different attractors: attracting fixed point, attracting periodical-$p$ cycle, quasiperiodic attractor and strange attractor.

The sensitive dependence on initial condition is one of the most essential aspects for identifying the chaos. We recall that the sensitive dependence on initial conditions means that two trajectories starting very close together will rapidly diverge from each other. The strange attractor is associated with a chaotic state of time evolution and is characterized by the sensitive dependence on initial conditions.

A measure of the average rate of exponential divergence exhibited by a chaotic system is given by the Lyapunov exponents of the system; the positivity of such exponents can suggest the presence of chaos.

The Lyapunov exponents $_{1,2}$ are given by

$$\{e^{\lambda_1}, e^{\lambda_2}\} = \lim_{n \to \infty} \left\{ eigenvalues\ of \left( \prod_{t=0}^{n-1} J(F(s_t, s_{t+1})) \right)^{\frac{1}{n}} \right\} \tag{4.3}$$

where $J(F(s_t, s_{t+1}))$ represents the Jacobian matrix of the function $F$. For a period-$p$ point the Lyapunov exponents are given by

$$\{e^{\lambda_1}, e^{\lambda_2}\} = \left\{ eigenvalues\ of \left( \prod_{t=0}^{p-1} J(F(s_t, s_{t+1})) \right)^{\frac{1}{p}} \right\} \tag{4.4}$$

We recall now that for an attracting period-$p$ cycle the Lyapunov exponents are negative; in case of a bifurcation point, at least one Lyapunov exponent is zero; for a limit cycle one Lyapunov exponent is zero and the others are negative and for a chaotic behavior the highest Lyapunov exponent is positive while the sum of the all Lyapunov exponents is negative.

When system (4.2) displays a chaotic behavior, in order to compute the Lyapunov exponents, we use the method proposed in [2], based on the Householder QR factorization.

### 4.2.2 *The Householder QR Factorization for a Matrix $A \in R^2 x R^2$*

We recall the Householder QR factorization (see [7]). For $A = \begin{pmatrix} a_{11} & a_{12} \\ a_{21} & a_{22} \end{pmatrix}$ we choose the vector $x = \begin{pmatrix} a_{11} \\ a_{21} \end{pmatrix}$. The Householder vector according to $x$ is $v = x - \|x\|_2 \begin{pmatrix} 1 \\ 0 \end{pmatrix}$, where $\|x\| = \sqrt{a_{11}^2 + a_{21}^2}$. We obtain $v = \begin{pmatrix} v_1 \\ v_2 \end{pmatrix} = \begin{pmatrix} a_{11} - \sqrt{a_{11}^2 + a_{21}^2} \\ a_{21} \end{pmatrix}$. The Householder matrix according to the vector $v$ is $H = I - \frac{2}{v^T v} v v^T$, where

$$v^T v = \left(a_{11} - \sqrt{a_{11}^2 + a_{21}^2} \quad a_{21}\right)\begin{pmatrix} a_{11} - \sqrt{a_{11}^2 + a_{21}^2} \\ a_{21} \end{pmatrix} = \left(a_{11} - \sqrt{a_{11}^2 + a_{21}^2}\right)^2 + a_{21}^2$$

and

$$vv^T = \begin{pmatrix} a_{11} - \sqrt{a_{11}^2 + a_{21}^2} \\ a_{21} \end{pmatrix} . . \left(a_{11} - \sqrt{a_{11}^2 + a_{21}^2} \quad a_{21}\right)$$
$$= \begin{pmatrix} \left(a_{11} - \sqrt{a_{11}^2 + a_{21}^2}\right)^2 & a_{21}\left(a_{11} - \sqrt{a_{11}^2 + a_{21}^2}\right) \\ a_{21}\left(a_{11} - \sqrt{a_{11}^2 + a_{21}^2}\right) & a_{21}^2 \end{pmatrix}.$$

We obtain $H = \begin{pmatrix} H_{11} & H_{12} \\ H_{21} & H_{22} \end{pmatrix}$, where

$$H_{11} = 1 - \frac{2}{\left(a_{11} - \sqrt{a_{11}^2 + a_{21}^2}\right)^2 + a_{21}^2}\left(a_{11} - \sqrt{a_{11}^2 + a_{21}^2}\right)^2$$

$$H_{12} = -\frac{2}{\left(a_{11} - \sqrt{a_{11}^2 + a_{21}^2}\right)^2 + a_{21}^2} a_{21}\left(a_{11} - \sqrt{a_{11}^2 + a_{21}^2}\right)$$

$$H_{21} = -\frac{2}{\left(a_{11} - \sqrt{a_{11}^2 + a_{21}^2}\right)^2 + a_{21}^2} a_{21}\left(a_{11} - \sqrt{a_{11}^2 + a_{21}^2}\right)$$

$$H_{22} = 1 - \frac{2}{\left(a_{11} - \sqrt{a_{11}^2 + a_{21}^2}\right)^2 + a_{21}^2} a_{21}^2.$$

The matrix $H$ has the property $H = H^T = H^{-1}$ because $HH = \left(I - \frac{2}{v^T v} vv^T\right)^2 = I - \frac{4}{v^T v} vv^T + \frac{4}{v^T v} vv^T = I$ implies $H = H^{-1}$. The property $H^T = H^{-1}$ means that the matrix $H$ is an orthogonal matrix.

The matrix $HA$ is upper triungular because

$$HA_{21} = -\frac{2}{\left(a_{11} - \sqrt{a_{11}^2 + a_{21}^2}\right)^2 + a_{21}^2} a_{21}\left(a_{11} - \sqrt{a_{11}^2 + a_{21}^2}\right) a_{11}$$
$$+ \left(1 - \frac{2}{\left(a_1 1 - \sqrt{a_{11}^2 + a_{21}^2}\right)^2} a_{21}^2\right) a_{21} - 2a_{21}a_{11}^2 + 2a_{21}a_{11}\sqrt{a_{11}^2 + a_{21}^2}$$
$$+ \left(a_{11} - \sqrt{a_{11}^2 + a_{21}^2}\right)^2 a_{21} + a_{21}^3 - 2a_{21}^3 = 0.$$

If we choose $Q = H$ and $R = HA$ we obtain $A = HHA = QR$, this means that the matrix $A$ is decomposed into an orthogonal matrix $Q$ and an upper triangular matrix $R$.

If we denote $J_i = J(F(s_{i-1}, s_i))$, $A_1 = J_1, A_i = J_i Q_{i-1}$ for $i = \overline{2,n}$, then $\prod_{i=1}^{n} J_i = Q_n R_n \ldots R_1$. It is shown, see [2], that the diagonal elements $\lambda_{jj}^n$, j = 1,2 of the upper triungular matrix product $R_n \ldots R_1$ obtained from this algorithm verify the relation: $\lim_{n\to\infty} \ln\left|\lambda_{jj}^n\right| = \lambda_j$.

If $r_{jj}^i$ are the diagonal elements of the matrix $R_i$, then $\lambda_j = \lim_{n\to\infty} \frac{1}{n} \sum_{i=1}^{n} \ln\left|r_{jj}^i\right|$.

## 4.3 Implementation Methods for Numerical Simulations

### *4.3.1 Implementation Method Used to Detect the Behavior Type of System (4.1)*

In our computational study we are concentrated on some objectives:

1. to determine the behavior type for certain cases;
2. to obtain images for attractors;
3. to calculate the Lyapunov exponents;
4. to obtain animation.

The time necessary to obtain the results is very important for our objectives. In order to detect the behavior type of the equation (4.1), we will generate a number of iterations.

In the case in which the trajectory moves in time towards an attractor (fixed point, periodical cycle, limit cycle or chaotic attractor), the iterations number, which is necessary to observe the behavior type, is different from one case to another. However, in the case of a fixed point or a periodical cycle (with a small period) this number is small, but for other situations, this number can be considerably higher (some being hundreds of thousands or even greater).

In such a study, it is important to have simultaneously an image for the generated iterations, because in the moment in which we observe a tendency we can stop generating other iterations.

Moreover, in the case in which the trajectory moves in time towards an attractor, generally, the first iterations are not necessary in the graphical representation of the attractor. For example, in Fig. 4.1 we have the image for the first 4,000 points of a trajectory (in $R^2$). In Fig. 4.2, we have the image for the points corresponding to the iterations (200,000–204,000).

From the observations presented before, we propose the following way of implementation:

- We generate a fixed number of iterations (for example, 20,000) and their graphical representation.

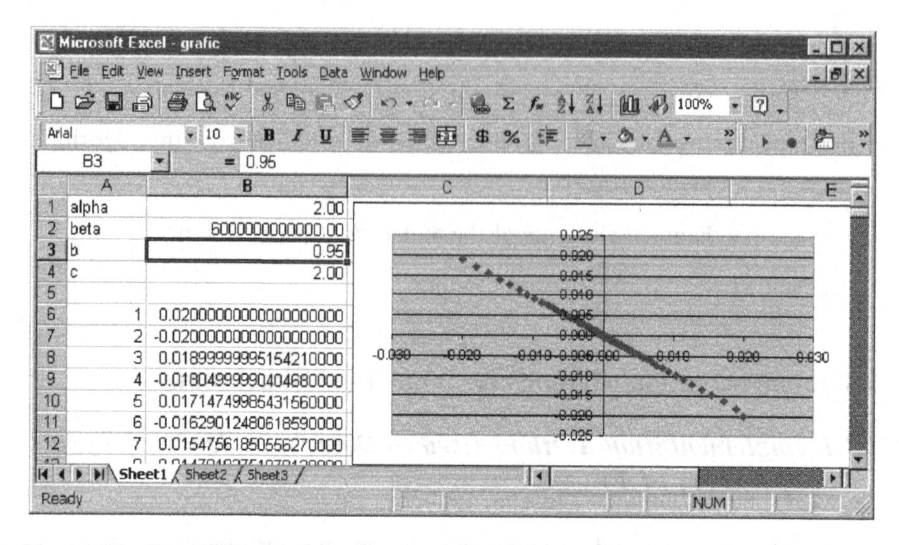


Fig. 4.1 The first 4000 points from the trajectory

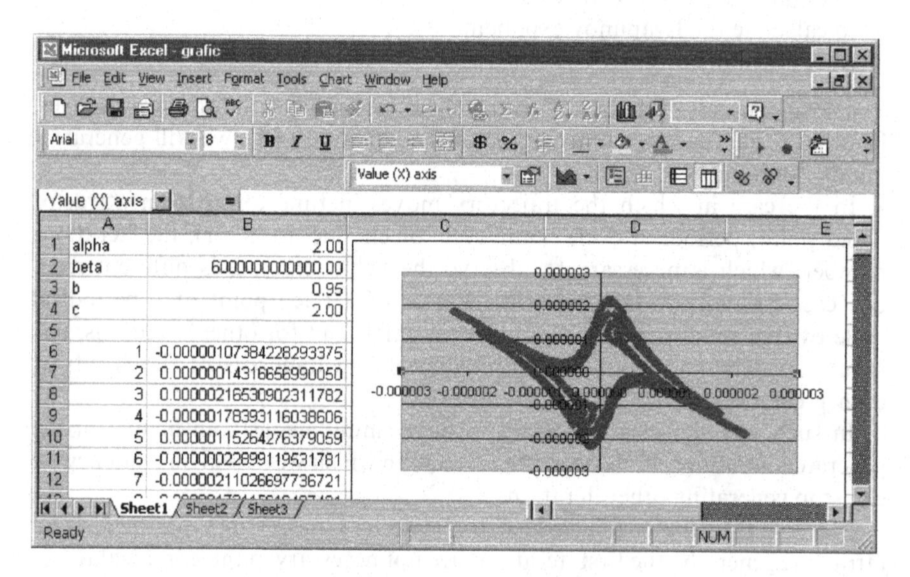


Fig. 4.2 Some points (from a trajectory) which provide the image of an attractor

- If the image suggests a certain behavior type, we stop generating other iterations.
- If this is not the case, we will generate the following group of iterations (here we generate groups of 20,000 iterations – in fact, we will generate some

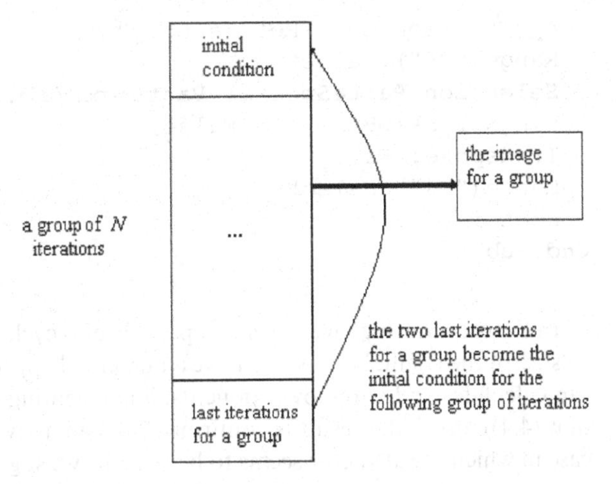


**Fig. 4.3** Generating group of iterations

groups with the same number of iterations, which makes the implementation easier). In this moment, we will also destroy the last group.

For the last generated group of iterations, we will repeat the same algorithms.

This implementation manner enables the successive observation of an unlimited number of points of the trajectory (see Fig. 4.3).

In order to provide such an implementation, we propose to work (for example) in *Excel* using macros.

In the cells *B1: B4* we introduce the values of the parameters and in the cells *B6:B7* we introduce the initial condition (see Fig. 4.1).

In the cell *B8* we use the following *Excel* formula:

```
=(($B$4+$B$1)*$B$3/(1+$B$2*(EXP(B7)-1)^2)+(1-$B$1)*$B$3)*B7-
 $B$4*$B$3/(1+$B$2*(EXP(B7)-1)^2)*B6
```

and we copy this formula in the domain *B9:B20005*.

In order to obtain an image for the graphical representation of these points of the trajectory, we make a chart in the space $(s_t, s_{t+1})$ (see Fig. 4.2). Here, we use 20,000 cells, in which the initial condition is changed with the two last iterations from the group, and we will repeat this change until we can observe a tendency in dynamics. In order to obtain the results quickly, we also use macros (which automatically execute these changes of values).

```
Sub initial_condition_changing()
  For i=1 To 10
  Application.Goto Reference:="R20005C1
  "Range("B20004:B20005").Select
  Selection.Copy
```

```
    Application.Goto Reference:="R1C1"
    Range("B6").Select
    Selection.PasteSpecial Paste:=xlValues, Operation:=
    xlNone, SkipBlanks:=_False,
    Transpose:=False
    Range("E4").Select
    Next i
End Sub
```

In the case in which we observe a periodical-$p$ cycle, theoretically we cannot be sure of the result, because we have a numerical approximation. However, we can calculate the Lyapunov exponents, implementing the definition (see equation (4.4)) and if the result is confirmed, we can arrive at a conclusion. In the case in which the attractor seems to be chaotic, we suggest to verify first if there is a sensitive dependence on the initial conditions (see Figs. 4.4 and 4.5), and if so, to calculate the Lyapunov exponents using the method proposed in [2].

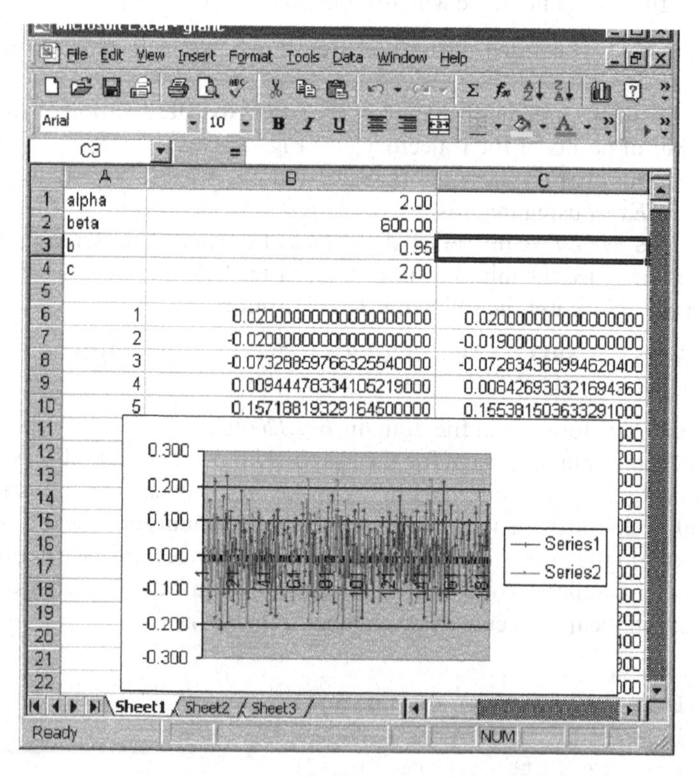


Fig. 4.4 A case in which we have sensitive dependence on the initial conditions

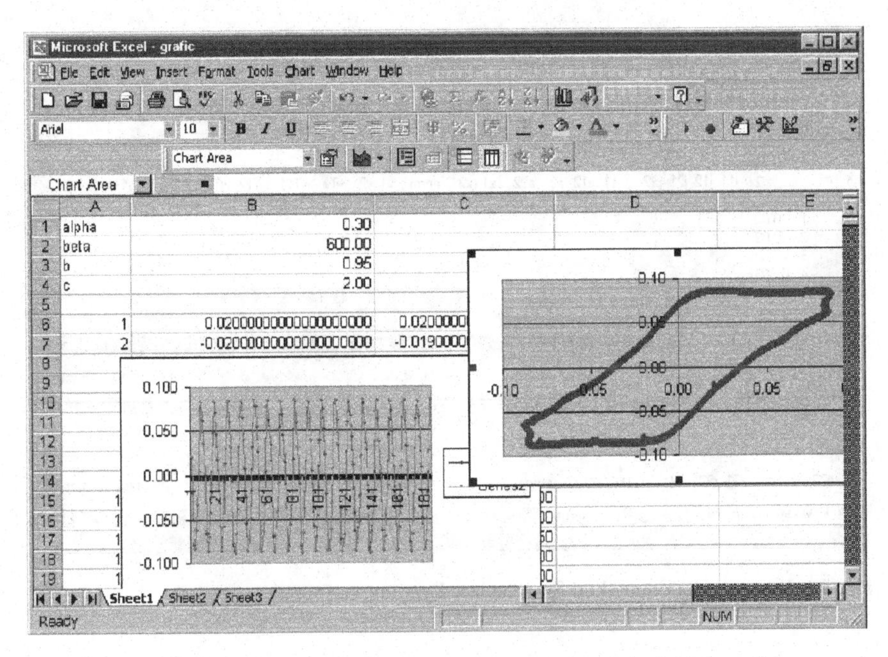


Fig. 4.5 A case in which we do not have sensitive dependence on the initial conditions

### 4.3.2 Implementation Method Used to Calculate the Lyapunov Exponents, Using Householder QR Factorization

In order to obtain the Lyapunov exponents in the case in which we have a chaotic attractor (the trajectory image suggests a chaotic attractor) or a limit cycle, we cannot use their definitions and for this reason we suggest for implementation the method proposed in [2], which uses the Householder QR factorization. In this case we must calculate approximations for Lyapunov exponents until $t$ has a high value (generally, we can conclude only after 200,000–300,000 iterations).

We propose to calculate the Lyapunov exponents approximations for groups of iterations (for example, 6000), and in the same way like in Section 4.3.1, we save only the values necessary to continue the algorithm for the following group of iterations. All the values, which are not necessary, will be deleted. In this way, we can quickly obtain a good approximation of the Lyapunov exponents. In the case in which we do not use groups of iterations (independently of the used software application) and we calculate the approximations for a greater number of iterations, the time necessary to obtain the results can be a serious problem.

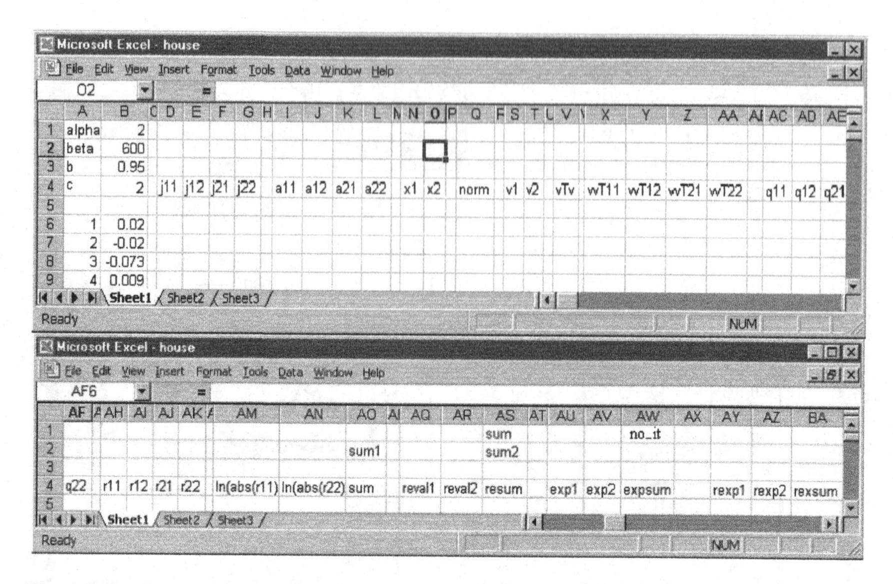


**Fig. 4.6** Implementation method for calculate Lyapunov exponents

In order to obtain quickly the results, our implementation method proposes that we follow some steps.

Here we will use the algorithm presented in the Section 4.2.2 and the dates presented in Fig. 4.6.

Now we present the steps of our implementation:

1. In columns *D:G* we calculate the elements of matrix *J*.
2. In columns *I:AO* we calculate: the elements of matrix *A;* the vector $x$ and its norm; the vector $v$; the value $v^T v$; the matrices $vv^T$, $Q$ and $R$; natural logarithm of the absolute value of diagonal elements of the matrix $R$ and their sum.
3. We make all these calculations for groups of $N$ iterations (we have choose $N$ to be 6000). At this step, for the two last iterations, we save all the values from the domain from columns *B:AO* in another domain, like example in columns *CX:EL*.
4. In the domain from columns *AM:AO* we have formulas. We save all these results as values in the domain from columns *AQ:AS*.
5. We delete all values calculated in the *step 2*.
6. We delete all values calculated in the *step 1*.
7. In the cells *AT1*(*AU1*, respectively *AV1*) we save the sum for the values from columns *AQ*(*AR*, respectively *AS*), corresponding to the current and previous groups of iterations.
8. In the cells *AX1* we save the value of iterations number for which we have calculated all values in the domain from the columns *B:AS*.

9. The last iterations from a group (see the *step 3*) become the initial condition for the following group of iterations.

   We will repeat all the *steps 1–9* for some groups of iterations (e.g., 20–30 groups).
10. Now, we will calculate *N*(6000) approximations for Lyapunov exponents corresponding to the last group of *N* iterations. In the domain from the columns *AU:AW* we have formulas and for a high speed of calculation we make copies of values in the domain from the columns *AY:BA*. In the moment in which we have the copies in the domain from the columns *AY:BA*, we will delete the formulas from the domain from columns *AU:AW*.

If the approximations for Lyapunov exponents corresponding for the last group of *N* iterations do not lead to a good conclusion on Lyapunov exponents values, we will repeat the *step 9–10* for another number of iterations groups and for the last group we will calculate the new approximations for Lyapunov exponents.

In all our steps, we have used *VBA* (*Visual Basic for Applications*) macros in *Excel*. The use of macros increases considerably the speed with which we can obtain results. This aspect is very important in order to arrive to a better conclusion regarding our numerical results.

## 4.4 Numerical Simulations

We have made many numerical simulations and we have found many situations in which the system displays a chaotic behavior.

For the particular case where $\alpha = 2$, $b = 0.95$, $c = 2$ and the initial condition $(s_0, s_1) = (0.02, -0.02)$, we investigate the ranges of parameter $\beta$ for which system (4.2) presents a chaotic or a non chaotic behavior.

We observe different intervals of values for $\beta$ for which system (4.2), in general, displays a chaotic behavior. These intervals are [46, $\infty$), [16,28], [6.2,12], [4,4.6], [2.9,3.8], [1.7,1.8]. Every two intervals presented here are separated by an interval of values of $\beta$ which characterizes a sequence of period-doubling bifurcations for system (4.2). We present in Fig. 4.7 the strange attractors (in space $(s_t, s_{t+1})$) which characterize different types of intervals of values for $\beta$ for which the system displays a chaotic behavior.

In Table 4.1 we give the values of Lyapunov exponents in the case of the strange attractors present in Fig. 4.7.

The images from Fig. 4.7 seem to represent the same attractor which increases and is deformed, but probably that is not so clear. The transformation image cannot be obtained statically. In order to obtain an animation we can use different applications (e.g., in *Flash*). Our animation is presented on the page: http://www.catrinelvoicu.home.ro/chaos.html.

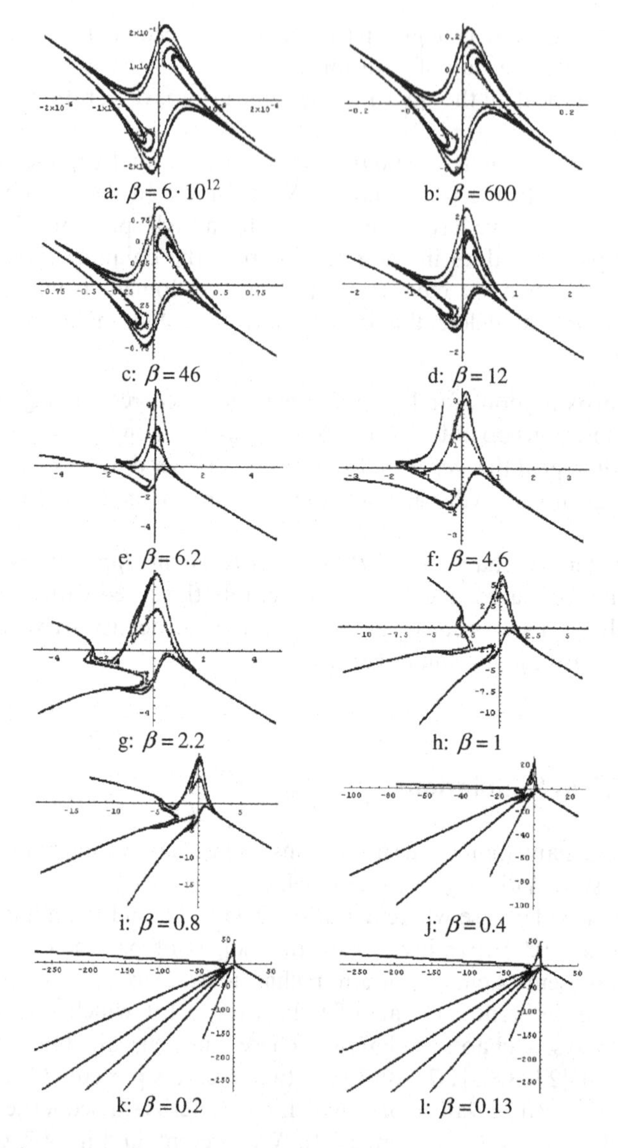


**Fig. 4.7** Chaotic attractors in the case c=2, $s_0 = 0.02, s_1 = -0.02, b = 0.95, \lambda = 2$

In Table 4.2, we present an example of a sequence of period-doubling bifurcations, displayed by system (4.2). We use the same initial conditions (0.02, –0.02). Here $(s_0, s_1)$ means a periodical point from the periodical cycle towards which tends the trajectory starting from point (0.02, –0.02).

**Table 4.1** Lyapunov exponents in the case $s_0 = 0.02, = -0.02, b = 0.95, \lambda = 2$

| $\beta$ | $\lambda_1$ | $\lambda_2$ |
|---|---|---|
| $6 \cdot 10^{12}$ | 0.3728 | –1.2735 |
| 600 | 0.4029 | –1.2158 |
| 46 | 0.2685 | –1.1734 |
| 12 | 0.3299 | –1.341 |
| 6.2 | 0.2636 | –1.9012 |
| 4.6 | 0.2244 | –1.4617 |
| 2.2 | 0.2033 | –1.5614 |
| 1 | 0.09489 | –1.424 |
| 0.8 | 0.143 | –1.5599 |
| 0.4 | 0.1054 | –3.3219 |
| 0.2 | 0.0787 | –2.5733 |
| 0.13 | 0.0132 | –4.1387 |

**Table 4.2** A sequence of period-doubling bifurcations (the first period is 4)

| | $s_0$ | $s_1$ | Period | $\lambda_1$ | $\lambda_2$ |
|---|---|---|---|---|---|
| 40.5528 | 0.22816 | –0.10586 | 1024 | –0.0012 | –0.6448 |
| 40.5513 | 0.49251 | 0.23373 | 512 | –0.0063 | –0.6387 |
| 40.52 | –0.49253 | 0.2336 | 128 | –0.0177 | –0.6272 |
| 40.51 | –0.48868 | 0.228887 | 128 | –0.0486 | –0.5963 |
| 40.38 | –0.49377 | 0.24238 | 64 | –0.0568 | –0.5882 |
| 40.3 | –0.49496 | 0.23532 | 32 | –0.0242 | –0.6208 |
| 39 | –0.50179 | 0.23606 | 16 | –0.0062 | –0.6398 |
| 36 | –0.52696 | 0.24804 | 8 | 0.0920 | –0.5538 |
| 32 | –0.5617 | 0.26258 | 4 | 0.2447 | –0.4125 |
| 28.3853 | –0.58727 | 0.26821 | 4 | 0.0002 | –0.6963 |

## 4.5 Conclusion

We have presented our implementation methods for detecting the dynamics of a nonlinear system and for the calculation of the Lyapunov exponents. These methods present the way in which we can obtain a very large number of observations with the computer, in a very short time, which, at the same time, is conclusive for the obtained results.

**Acknowledgments** Result paper within the research project CNCSIS AT code 67/2007.

## References

1. De Grauwe P, Dewachter H, Embrechts M (1993) Exchange rate theory: chaotic models of foreign exchange markets. Blackwell Publishers
2. Eckmann J-P, Ruelle D (1985) Ergodic theory of chaos and strange attractors. Rev Modern Phys 57(3), Part I

3 Juncu VD, Rafiei-Naeini M, Dudekb P (2006) Integrated circuit implementation of a compact discrete-time chaos generator analog integrated circuits and signal processing. Springer Netherlands, ISSN:0925-1030, 46(3):275–280
4. Ulrike F (2006) Generalized models as a tool to study the stability of nonlinear dynamical systems – Nonlinear dynamics, chaos, and applications. In: 6th Crimean School and Workshops
5. Voicu MC (2004) On the computational tools role in the nonlinear dynamics study. In: Proceedings of ICNPAA 5th, Timişoara, România, Cambridge Scientific Publishers Ltd, UK, pp 665–672
6. Voicu MC (2004) On the periodical solutions of a nonlinear dynamical system. In: Proceedings of ICNPAA 5th, Timişoara, România, Cambridge Scientific Publishers Ltd, UK, pp 657–664
7. http://www.cs.ut.ee/~toomas_1/linalg/lin2/node8.html

# Chapter 5
# The Worm Cutter Tool Profiling Based on Discreet Surfaces Representation

**Virgil Teodor, Marian Cucu, and Nicolae Oancea**

**Abstract** The worm cutter profiling for cylindrical or helical surfaces curl rolling generation represents a problem with the point contact between two surfaces reciprocally enwrapping. In this paper there is proposed an algorithm based on the tangents method and a software, developed in *java* programming language, based on the discreet representation of the surface to be generated. Applications of this algorithm for simple elementary profiles which may be assembled in compound profiles are presented.

## 5.1 Introduction

The problem of the worm cutter profiling is a problem of surface enwrapping on second degree – the contact between the surface to be generated, associated with a circular centrode, and the primary peripheral surface of the worm cutter, is a point.

Due to this, the profiling of these tools presumes the use of the intermediary surface method, in two successive steps [2–4]:

- determining of the rack-gear surface, as a cylindrical surface, reciprocally enveloping with the rack-gear flank, in conditions of linear contact between the two surfaces;
- the rack-gear flank being known, there is determined the primary peripheral surface of the worm cutter as reciprocally enveloping with the rack-gear flank, in the conditions of linear contact between these two surfaces.

So, the study of the reciprocally enwrapping surfaces is reduced to a succession of enwrapping problems with linear contact developed, based on Gohman's theorem [2, 3].

For both of these steps, the tangents method [1, 5] may be used for the representation of the profiles in discreet form.

V. Teodor (✉)
Department of Manufacturing Engineering, "Dun rea de Jos" University of Gala i,
Gala i, 111 Domneasc Street, 800201, Romania
e-mail: virgil.teodor@ugal.ro

N. Mastorakis et al. (eds.), *Proceedings of the European Computing Conference*, Lecture Notes in Electrical Engineering 28,
DOI 10.1007/978-0-387-85437-3_5, 

## 5.2 The Rack-Gear Tool Profiling

There are determined the relative movements between spaces associated with the blank centrode and the rack-tool centrode, Fig. 5.1

$$\xi = \omega_3^T(\varphi_1) \cdot X - a \tag{5.1}$$

and the reverse

$$X = \omega_3(\varphi_1)[\xi + a], \tag{5.2}$$

where

$$a = \| -R_{rp} \quad -\lambda \quad 0 \|^T. \tag{5.3}$$

A certain profile, from the curl of the profile to be generated, associated to a circular centrode having the radius Rrp, may be discreetly expressed through the matrix (5.4), referred to the XY system, associated with the whirl of profiles to be generated.

The number of coordinates to define the matrix (5.4) can be correlated to the precision of representing the $\Sigma$ profile:

$$\Sigma = \left\| \begin{matrix} X_A & X_2 & \cdots & X_i & \cdots & X_B \\ Y_A & Y_2 & \cdots & Y_i & \cdots & Y_B \end{matrix} \right\|^T. \tag{5.4}$$

The model used to define the $\Sigma$ matrix current point coordinates is:

$$\begin{cases} X = X_A + v \cdot \cos \beta_i; \\ Y = Y_A + v \cdot \sin \beta_i, \end{cases} \tag{5.5}$$

where $v$ is an incremental variable.

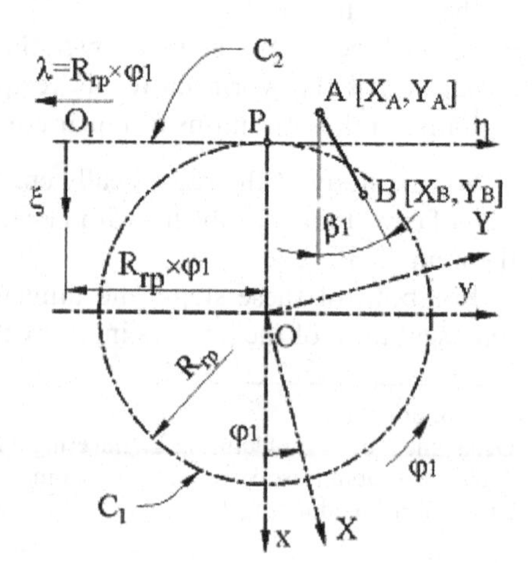


**Fig. 5.1** The profile of the curl to be generated

The $\Sigma$ family of profiles, described into the $\xi\eta$ reference system, associated with the worked piece, has the form:

$$(\Sigma)_{\varphi_1} \left| \begin{array}{l} \xi = X_i \cos\varphi_1 - Y_i \sin\varphi_1 + R_{rp}; \\ \eta = X_i \sin\varphi_1 + Y_i \cos\varphi_1 + R_{rp}\varphi_1. \end{array} \right. \quad (5.6)$$

The enveloping condition, determined to express the generated profile in the form (5.4), by using the tangents method, is

$$\left|\sin\beta_i\left[Y_i - R_{rp}\sin\varphi_1\right] - \cos\beta_i\left[-X_i - R_{rp}\cos\varphi_1\right]\right| \leq \varepsilon, \quad (5.7)$$

where $\varepsilon$ is small enough.

The ensemble formed by the family of profiles of discrete representation (5.6) and the specific enveloping condition (5.7) represents, in a discrete form, the profile of the rack-tool enwrapped to $\Sigma$ profile in form:

$$I = \left\| \begin{array}{cccc} \xi_1 & \xi_2 & \cdots & \xi_n \\ \eta_1 & \eta_2 & \cdots & \eta_n \end{array} \right\|^T. \quad (5.8)$$

Based on the algorithm briefly suggested above, an application written in java language was imagined, in order to allow us to find the rack-tool profile by starting from a numerical expression of the profile to be generated.

## 5.3 Reference Systems and Generation Movements

There are defined the references systems (see Fig. 5.2):

- $xyz$ is the global reference system, with the $z$ axis superposed on the rotation axis of the orderly surface curl to be generated;

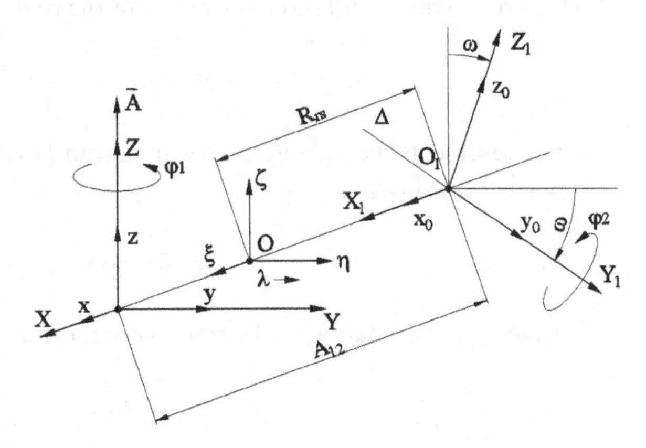


Fig. 5.2 References systems

- $x_0y_0z_0$ – global reference system, with $y_0$ axis superposed on the $\Delta$ axis, the worm cutter axis;
- $XYZ$ – relative reference system, joined with the orderly surfaces to be generated;
- $X_1Y_1Z_1$ – relative reference system, joined with the primary peripheral surface of the worm cutter;
- $\xi\eta\zeta$– relative reference system, joined with the intermediary surface (the generating rack cutter);
- $A_{12}$ – the distance between axis $A$ (the axis of the orderly surface curl to be generated) and axis $\Delta$ (the axis of the worm cutter).

There are defined the correlations between the moving parameters:

- $\varphi_1$ is the angular parameter of the rotation around the $Z$ axis;
- $\varphi_2$ – the angular parameter of the rotation around the $Y_1$ axis;
- $\lambda$ – the translation parameter along the $\eta$ axis,

$$\lambda = R_{rp} \cdot \varphi_1; \tag{5.9}$$

- $R_{rp}$ – the radius of the centrode associated with the curl to be generated;

$$\lambda = p \cdot \varphi_2 \cdot \cos\omega, \tag{5.10}$$

$$\sin\omega = \frac{R_{rp}}{R_{rs}} \cdot \frac{1}{z}, \tag{5.11}$$

where $z$ is the number of circular pitch of curl to be generated;

- $p$ – the helical parameter of the primary peripheral surface of the worm cutter.

There are known the movements:

$$x = \omega_3^T(\varphi_1) \cdot X, \tag{5.12}$$

which represent the rotation movement of the centrode with the $R_{rp}$ radius;

$$x = \xi + a, \tag{5.13}$$

which represents the $\xi\eta\zeta$ reference system translation, joined with the intermediary surface;

$$x_0 = \omega_2^T(\varphi_2) \cdot X_1 \tag{5.14}$$

representing the rotation of the worm cutter around its own axis;

$$x_0 = \beta(x - B); \tag{5.15}$$

$$B=\begin{Vmatrix}-A_{12}\\0\\0\end{Vmatrix};\beta=\begin{Vmatrix}1&0&0\\0&\cos\omega&-\sin\omega\\0&\sin\omega&\cos\omega\end{Vmatrix},\tag{5.16}$$

representing the transformation matrix between the global references systems.

From (5.12), (5.13), (5.14) and (5.15) the movement assembly is determined by means of the relative movement between relative reference systems $\xi\eta\zeta$ and $X_IY_IZ_I$:

$$X_1=\omega_2(\varphi_2)\beta[\xi+a-B]\tag{5.17}$$

and

$$\xi=\beta^T\omega_2^T(\varphi_2)X_1+B-a.\tag{5.18}$$

## 5.4 The Primary Peripheral Surface of the Worm Cutter

There are defined the principles of the intermediary surface profile determination based on the tangents method.

The intermediary surface may be represented by the matrix:

$$I=\left\{\begin{Vmatrix}\xi_i\\\eta_i\\t\cdot j\end{Vmatrix},i=1,2,\ldots,n\right\},\ j=(-m,\ldots,-1,0,1,\ldots m),\tag{5.19}$$

with the $\xi_i$, $\eta_i$ coordinates of the determinate intermediary surface profile, in correlation with the previous methodology; $t$ is the linear increment measured along the cylindrical surface generatrix.

### *5.4.1 The Intermediary Surface Family in the Reference System of the Worm Cutter Primary Peripheral Surface*

Based on the equation (5.17) and on the discreet known model of the cylindrical surface, there is determined the cylindrical surface family, in the worm cutter reference system.

After development, the equations are brought to form:

$$(I)_{\varphi_2}\left|\begin{array}{l}X_1=[\xi_i+R_{rs}]\cdot\cos\varphi_2-[\eta_i-\lambda]\cdot\sin\omega\cdot\sin\varphi_2-t\cdot j\cdot\cos\omega\cdot\sin\varphi_2;\\Y_1=[\eta_i-\lambda]\cdot\cos\omega-t\cdot j\cdot\sin\omega;\\Z_1=[\xi_i+R_{rs}]\cdot\sin\varphi_2-[\eta_i-\lambda]\cdot\sin\omega\cdot\cos\varphi_2-t\cdot j\cdot\cos\omega\cdot\cos\varphi_2;\end{array}\right.\tag{5.20}$$

with $(i = 1, 2, \ldots, n); (j = -m, (-m+1), \ldots, -1, 0, 1, 2, \ldots m)$.

The representation of the intermediary surface family in form (5.20) is a points array if a discreet representation of the$\varphi_2$ parameter is accepted. Obviously the link (5.10) between $\lambda$ and $\varphi_2$ is accepted.

The $(I)_{\varphi_2}$ surface family enveloping is the primary peripheral surface of the worm cutter.

### 5.4.2 The Enwrapping Condition

There is defined the enwrapping condition in the GOHMAN form,

$$\vec{N}_I \cdot \vec{R}_{\varphi_2} = 0 \tag{5.21}$$

where, from (5.18):

$$\vec{R}_{\varphi_2} = \frac{d\xi}{d\varphi_2} = \beta^T \cdot \dot{\omega}_2^T(\varphi_2) \cdot X_1 - \dot{a}_\lambda \cdot \frac{d\lambda}{d\varphi_2}. \tag{5.22}$$

Finally, the following matrix results:

$$R_{\varphi_2} = \left\| \begin{matrix} [\eta_i - \lambda] \cdot \sin\omega + t \cdot j \cdot \cos\omega \\ -[\xi_i + R_{rs}] \cdot \sin\omega + p \cdot \cos\omega \\ -[\xi_i + R_{rs}] \cdot \sin\omega \end{matrix} \right\| \tag{5.23}$$

$$(i = 1, 2, \ldots, n); (j = -m, \ldots, -1, 0, 1, 2, \ldots m).$$

The normal at intermediary surface is

$$\vec{N}_I = \vec{l}_{i,i+1} \times \vec{k}. \tag{5.24}$$

The assembly of the equations (5.20), (5.21), (5.23) and (5.24) represents the primary peripheral surface in worm cutter form.

### 5.4.3 The Characteristic Curve of the Conjugated Surfaces

The contact curve between the two reciprocally enwrapping surfaces, the $I$ intermediary surface and the S primary peripheral surface of the worm cutter, is defined by the assembly of the equations (5.20), (5.21), (5.23) and (5.24), for $\varphi_2 = 0$.

So, the characteristic curve becomes:

$$C_{I,S} \left| \begin{array}{l} X_1 = [\xi_i + R_{rs}]; \\ Y_1 = \eta_i \cdot \cos\omega - t \cdot j \cdot \sin\omega; \\ Z_1 = \eta_i \cdot \cos\omega - t \cdot j \cdot \sin\omega, \end{array} \right. \tag{5.25}$$

with $(i = 1, 2, \ldots, n); (j = -m, \ldots, -1, 0, 1, 2, \ldots m)$, at which there is added the enwrapping condition (5.21).

The characteristic curve results as the matrix

$$C_{I,S} = \| X_{1_{i,j}} \quad Y_{1_{i,j}} \quad Z_{1_{i,j}} \|^T$$
$$(i = 1, 2, \ldots, n); (j = -m, \ldots, -1, 0, 1, 2, \ldots m), \tag{5.26}$$

for points belonging to $I$ which satisfy the (5.21) enwrapping condition.

### 5.4.4 Primary Peripheral Surface of the Worm Cutter

If the characteristic curve (5.26) has a helical movement around the $Y_I$ axis, with p the helical parameter, a cylindrical helical surface with constant pitch is generated, a surface which represents the primary peripheral surface of the worm cutter,

$$S \left| \begin{array}{l} X_1 = X_{1_{i,j}} \cdot \cos v + Z_{1_{i,j}} \cdot \sin v; \\ Y_1 = Y_{1_{i,j}} + p \cdot v; \\ Z_1 = -X_{1_{i,j}} \cdot \sin v + Z_{1_{i,j}} \cdot \cos v, \end{array} \right. \tag{5.27}$$

with $v$ as the variable angular parameter.

The axial section of the primary peripheral surface of the worm cutter tool is obtained from equation (5.27) and the condition

$$Z_1 = 0. \tag{5.28}$$

The condition to determine the axial section is

$$-X_{1_{i,j}} \cdot \sin v + Z_{1_{i,j}} \cdot \cos v = 0. \tag{5.29}$$

In this way, the assembly of the equations (5.28) and (5.29) represents an in-plane curve – the axial profile of worm cutter, for all of the points which belong to the primary peripheral surface of the worm cutter.

## 5.5 The Software for the Rolling Generating Tool Profiling (RGTP)

The presented algorithm [4, 5] allows the synthesis of the dedicated software, in two successive steps, which allow:

- a. the rack-gear tool profiling, respectively the $I$ axial section primary peripheral surface profiling reciprocally enwrapping with an ordered surface curl associated with a circular centrode;
- b. the axial section profiling of the primary peripheral surface of the worm cutter, the $S$ cylindrical helical surface with constant pitch, based on the

discreet expression of the *I* rack-gear flank [2, 6]. There were elaborated the following two flowcharts:

- the flowchart for the rack-gear profiling reciprocally enwrapping with a $\Sigma$ profile, represented in a discreet form based on the tangents method, see Figs. 5.3 and 5.4 software which give the rack-gear crossing section coordinates and graphical form;

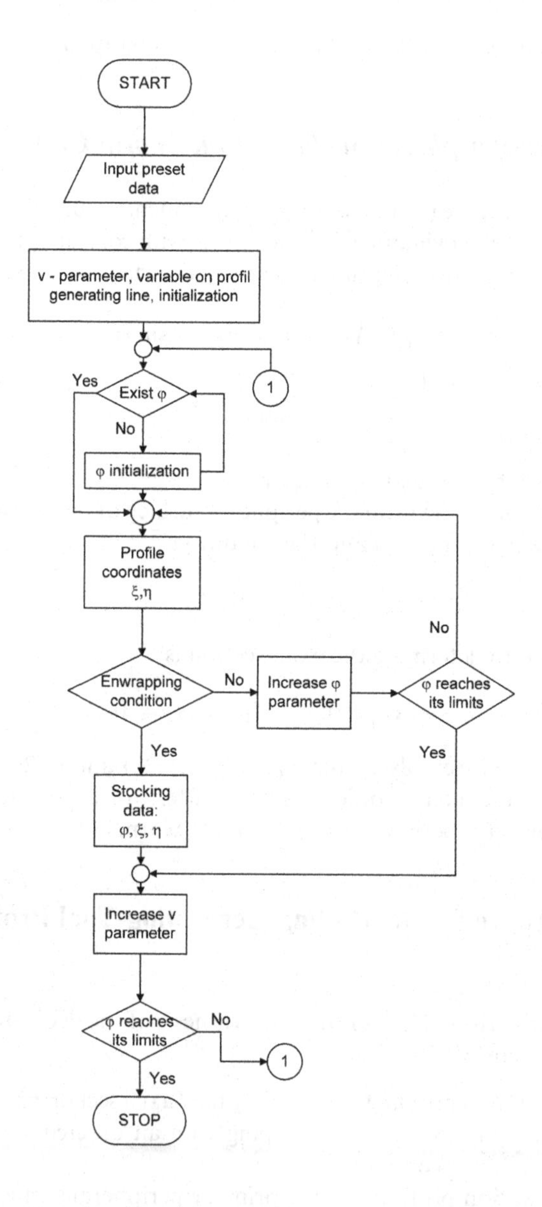


**Fig. 5.3** Flowchart for the rack-cutter profiling

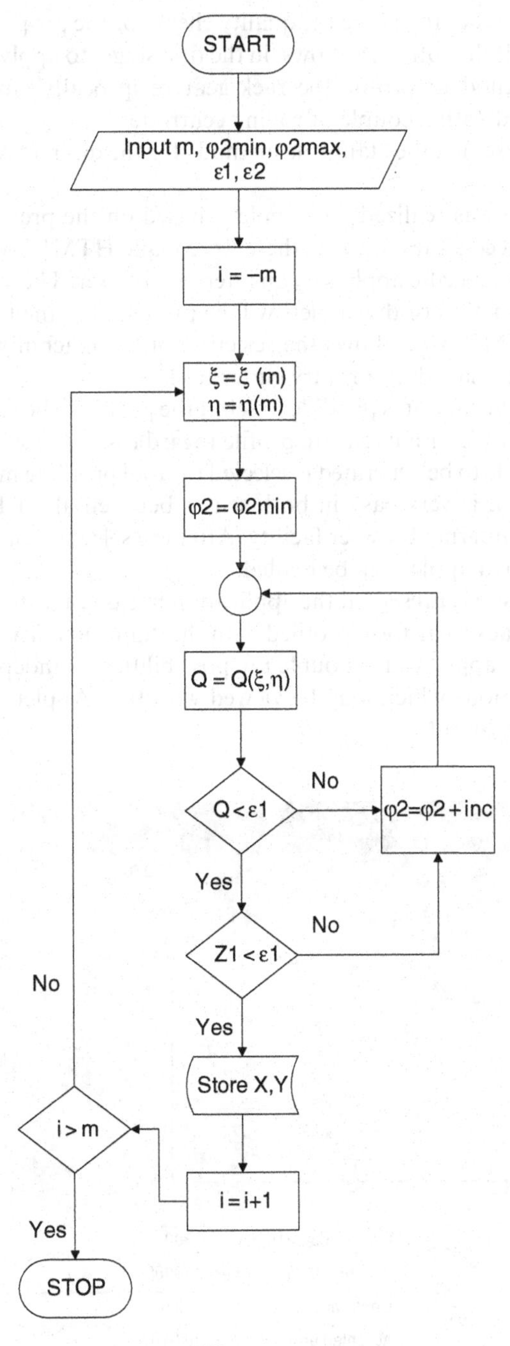


**Fig. 5.4** Flowchart for worm cutter profiling

Moreover, in order to ensure the quality check for the proposed method (the tangents method) the software allows, in the first stage, to apply another known enwrapping method to profile the rack-gear reciprocally enveloping with a profile associated with a couple of rolling centrodes.

There was presented the "tangents method" for the discreet representation of surfaces.

For this there was realized java applets, based on the previously presented flowcharts. To access these applets there were made HTML pages which allow the viewing of the specific applets by an internet browser. The structure of these pages and the function of the applets will be presented in the following.

First, the HTML page allows the selection of the machining mode, with a rack-gear, a gear-shaped or a rotary cutter tool.

By accessing the links it is possible to select the profile to be realized, selecting a straight line profile or a circle arc profile regardless of the selected tool.

After the profile to be generated is selected the tool profiling method is chosen.

The navigation is very easy in both senses between all of RGTP software pages, using the internet browser facility. After the selection of the method, the corresponding java applet will be loaded.

In Fig. 5.5, there is presented the application in case of a straight line profile generated by a rack-gear tool, profiled with the "tangents method".

The presented applets have double use possibilities, as independent applications, as applications which may be viewed with the "Appletviewer" software produced by Sun brand.

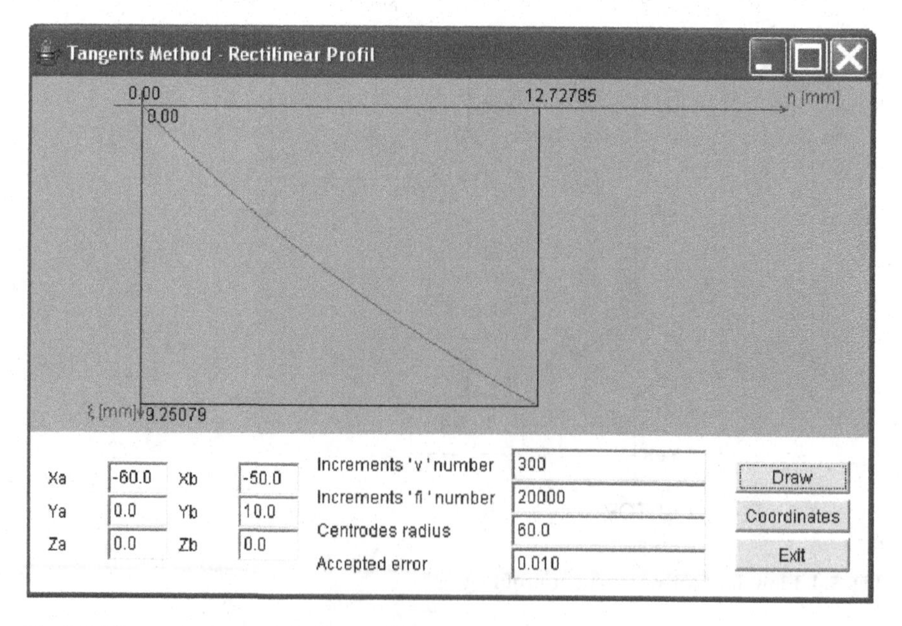


**Fig. 5.5** The applet for straight line profile

As it is possible to view in Fig. 5.5, the use of the applets is very simple and user friendly. The user may modify any of the input data by editing these inside edit box. After clicking on the "Draw" button, the application is launched with new input data and after finishing the calculus a tool graphical representation is made.

By selecting the option "Coordinates", a table is obtained with all the tool profile calculated coordinates, as it is presented in Fig. 5.6.

Because the number of coordinates is very large the viewing window has a scrolling bar. These coordinates may be saved for subsequent using.

The accepted error refers to the enwrapping condition error which is a criterion for tool profile determination.

The graphical representation of the calculated profiles is based on original algorithms. The application determines the maximum scale for the modeled profiles, without distortion on any axis, based on the determination of the display window dimensions. It is considered the obligatory representation of the reference system origin. After the scale determination, it is drawn the profile as a polyline which will include all of the modeled points and the reference system axes are drawn.

We have to notice that this drawing mode is in the present trend to make web-based interfaces software, very easy to use.

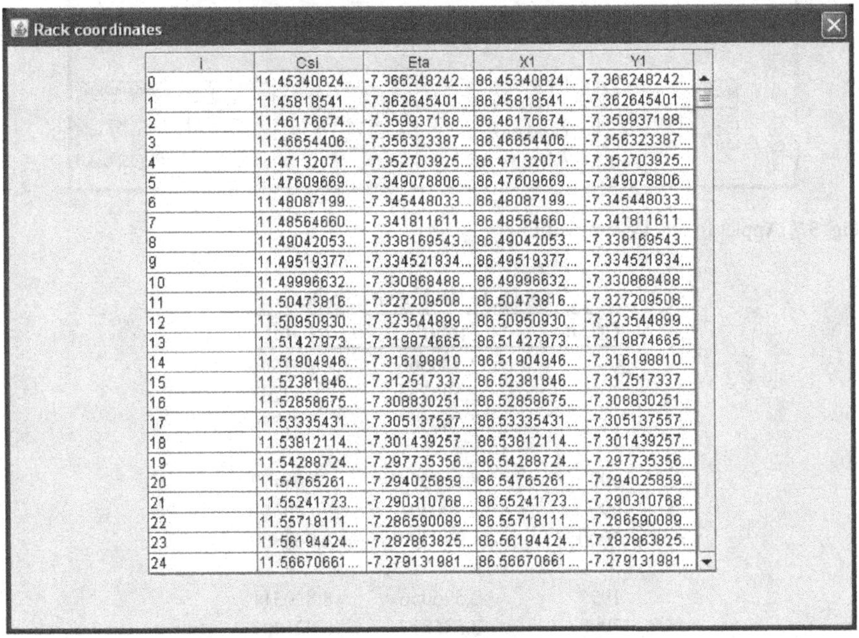

Rack coordinates

| i | Csi | Eta | X1 | Y1 |
|---|---|---|---|---|
| 0 | 11.45340824... | -7.366248242... | 86.45340824... | -7.366248242... |
| 1 | 11.45818541... | -7.362645401... | 86.45818541... | -7.362645401... |
| 2 | 11.46176674... | -7.359937188... | 86.46176674... | -7.359937188... |
| 3 | 11.46654406... | -7.356323387... | 86.46654406... | -7.356323387... |
| 4 | 11.47132071... | -7.352703925... | 86.47132071... | -7.352703925... |
| 5 | 11.47609669... | -7.349078806... | 86.47609669... | -7.349078806... |
| 6 | 11.48087199... | -7.345448033... | 86.48087199... | -7.345448033... |
| 7 | 11.48564660... | -7.341811611... | 86.48564660... | -7.341811611... |
| 8 | 11.49042053... | -7.338169543... | 86.49042053... | -7.338169543... |
| 9 | 11.49519377... | -7.334521834... | 86.49519377... | -7.334521834... |
| 10 | 11.49996632... | -7.330868488... | 86.49996632... | -7.330868488... |
| 11 | 11.50473816... | -7.327209508... | 86.50473816... | -7.327209508... |
| 12 | 11.50950930... | -7.323544899... | 86.50950930... | -7.323544899... |
| 13 | 11.51427973... | -7.319874665... | 86.51427973... | -7.319874665... |
| 14 | 11.51904946... | -7.316198810... | 86.51904946... | -7.316198810... |
| 15 | 11.52381846... | -7.312517337... | 86.52381846... | -7.312517337... |
| 16 | 11.52858675... | -7.308830251... | 86.52858675... | -7.308830251... |
| 17 | 11.53335431... | -7.305137557... | 86.53335431... | -7.305137557... |
| 18 | 11.53812114... | -7.301439257... | 86.53812114... | -7.301439257... |
| 19 | 11.54288724... | -7.297735356... | 86.54288724... | -7.297735356... |
| 20 | 11.54765261... | -7.294025859... | 86.54765261... | -7.294025859... |
| 21 | 11.55241723... | -7.290310768... | 86.55241723... | -7.290310768... |
| 22 | 11.55718111... | -7.286590089... | 86.55718111... | -7.286590089... |
| 23 | 11.56194424... | -7.282863825... | 86.56194424... | -7.282863825... |
| 24 | 11.56670661... | -7.279131981... | 86.56670661... | -7.279131981... |

**Fig. 5.6** Coordinate viewing

## 5.6 Applications

Next, it is presented the numerical and graphical results for the worm cutter profiling which is reciprocally enveloping with a circle arc profile.

In Fig. 5.7 and Table 5.1, there are shown the form and coordinates of the crossing profile of the rack-gear reciprocally enveloping with the shaft:

$$X_0 = -62 \text{ mm}; Y_0 = 0.0 \text{ mm}; Z_0 = 0.0 \text{ mm};$$

$$r = 10 \text{ mm}; v_{\min} = -60°; v_{\max} = 60°;$$

$$z = 16 \text{ curls}; \sin\omega = \frac{R_{rp}}{z(A_{12} - R_{rp})},$$

see Figs. 5.1 and 5.2.

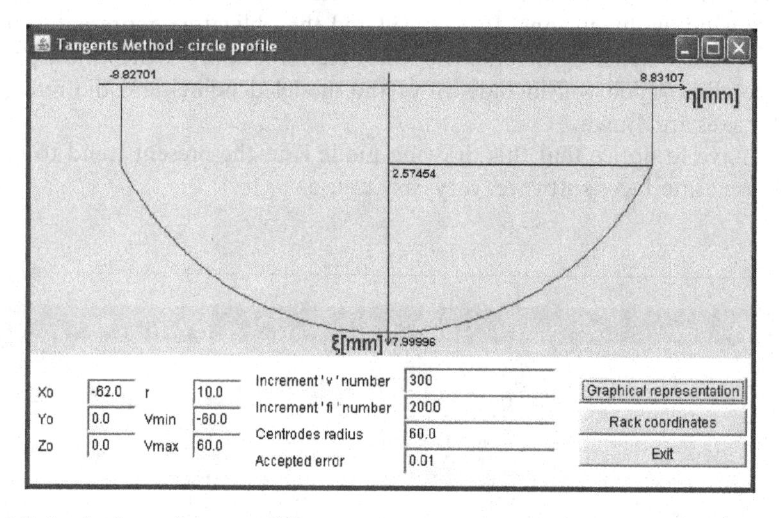


**Fig. 5.7** Applet for rack-gear profiling

**Table 5.1** Rack-gear profile

| Crt. no. | $\xi$ [mm] | $\eta$ [mm] |
|---|---|---|
| 0 | 2.580566 | –8.82806 |
| 1 | 2.590754 | –8.82297 |
| 2 | 2.600937 | –8.81786 |
| ⋮ | ⋮ | ⋮ |
| 998 | 7.99999 | –0.01477 |
| 999 | 8.000001 | –0.00367 |
| 1000 | 7.999999 | 0.006809 |
| ⋮ | ⋮ | ⋮ |
| 1997 | 2.596056 | 8.820311 |
| 1998 | 2.585877 | 8.825407 |
| 1999 | 2.575693 | 8.830489 |

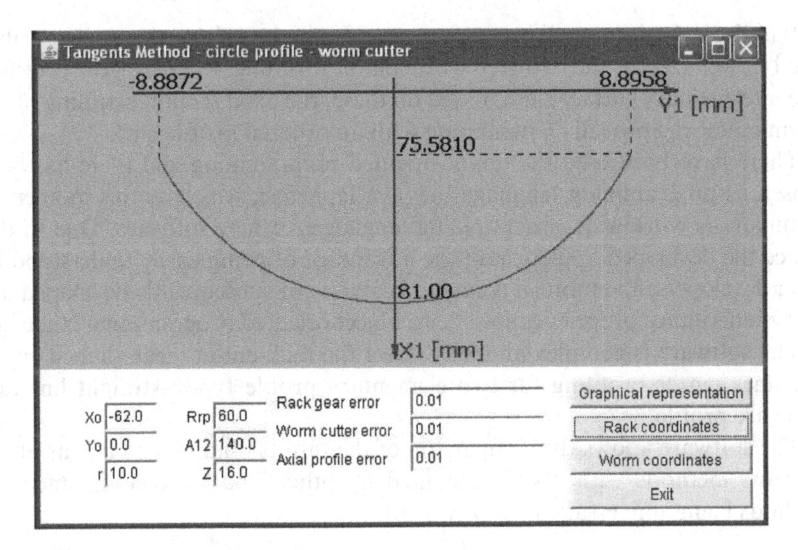


**Fig. 5.8** Applet for worm cutter profiling

**Table 5.2** Worm cutter profile

| Crt. no. | $X_1$ [mm] | $Y_1$ [mm] |
|---|---|---|
| 0 | 75.58057 | –8.82806 |
| 1 | 75.59075 | –8.82297 |
| 2 | 75.60094 | –8.81786 |
| ⋮ | ⋮ | ⋮ |
| 998 | 80.99999 | –0.01477 |
| 999 | 81 | –0.00367 |
| 1000 | 81 | 0.006809 |
| ⋮ | ⋮ | ⋮ |
| 1997 | 75.59606 | 8.820311 |
| 1998 | 75.58588 | 8.825407 |
| 1999 | 75.57569 | 8.830489 |

In Fig. 5.8 and Table 5.2, there is shown the axial profile of the worm cutter reciprocally enveloping with the square crossing shaft.

Both profiles are obtained based on original software as application for the presented algorithm.

## 5.7 Conclusions

The proposed methodology for rack-gear profiling reciprocally enwrapping with an orderly surface curl is based on a particular way to represent the surface – the tangents method.

Based on the intermediary surface specific methodology there were elaborated flowcharts for the two-step worm cutter profiling: the rack-gear profiling (the intermediary surface) and, based on these, the axial section profiling of the worm cutter reciprocally enwrapping with an ordered profile curl.

There have been used the object oriented programming and there has been chosen as programming language the java language, which at this moment is dominant as worldwide programming language for new software. Due to this choice the developed applets have the advantage of being easily understood by other developers, and more, it is easy to be reused or subsequently developed due to the inheritance proprieties specific to object oriented programming language.

The software is complex and may select the rack-cutter, gear-shaped cutter or rotary cutter profiling for two elementary profile types: straight line and circle arc profile.

The software allows the comparison of the results obtained by means of the tangents methods with results obtained by others' basic or complementary methods being universally acknowledged.

## References

1. Frumu anu G, Cucu M, Oancea N (2006) Tangents method to profile rack-tool to generate by wrapping. The Annals of "Dun rea de Jos" University of Gala i, Fasc V, Technologies in Machine Building, year XXIV (XXIX), ISSN 1221–4566, pp 70–75
2. Litvin FL (1994) Gear geometry and applied theory. Prentice Hall
3. Oancea N (2004) Generarea suprafe elor prin înf urare. Vol I: Teoreme fundamentale. "Dun rea de Jos" University Foundation Publishing House, ISBN 973-627-106-4, ISBN 973-627-107-2, Gala i
4. Oancea N (2004) Generarea suprafe elor prin înf urare. Vol II: Teoreme complementare. "Dunărea de Jos" University Foundation Publishing House, ISBN 973-627-106-4, ISBN 973-627-170-6, Gala i
5. Teodor V, Oancea N, Dima M (2006) Profilarea sculelor prin metode analitice. "Dun rea de Jos" University Foundation Publishing House, ISBN (10)973-627-333-4, ISBN (13)978-973-627-333-9, Gala i

# Chapter 6
# Suggesting Correct Words Algorithms Developing in FarsiTeX

**Youssef Rezvan, Majid Ghandchi, and Farshad Rezvan**

**Abstract** In this paper, methods and algorithms of suggesting correct words in a spell-checker for Farsi language have been developed, and complexity of them has been calculated. Some software covering these algorithms have been implemented and results of spell-checking and receiving suggestions has been discussed. We have concluded the paper by introducing some other techniques.

## 6.1 Introduction

Although it has long been the case that spell-checking for English texts is a common and easy practice, for Farsi Language yet, to our knowledge, there does not exist such a suitable system.

Recently some approaches to the design of a Farsi Spell-checker for FarsiTeX has been proposed [7, 6]. FarsiTeX (see Fig. 6.1) is an open-source typesetting program [1] based on LaTeX [5] version 2.09. LaTeX, a manuscript preparing program, has been developed over TeX, [2–4].

Here in this paper we have developed some algorithms for suggesting correct words in a Farsi spell-checker. This is mainly in Section 6.2. Increasing the speed of running and controlling memory usages are important issues, which have been covered in subsequent sections. We have implemented some spell-checking software covering our proposed algorithms to test our discussions. In the conclusion we have also discussed the ways to upgrade this Farsi Spell-checker version 0.1, as well as other techniques.

Y. Rezvan (✉)
Islamic Azad University, Ahar Branch, Dept. Electrical Eng. and Computer Ahar, Iran
e-mail: y-rezvan@iau-ahar.ac.ir

N. Mastorakis et al. (eds.), *Proceedings of the European Computing Conference*, Lecture Notes in Electrical Engineering 28,
DOI 10.1007/978-0-387-85437-3_6, 

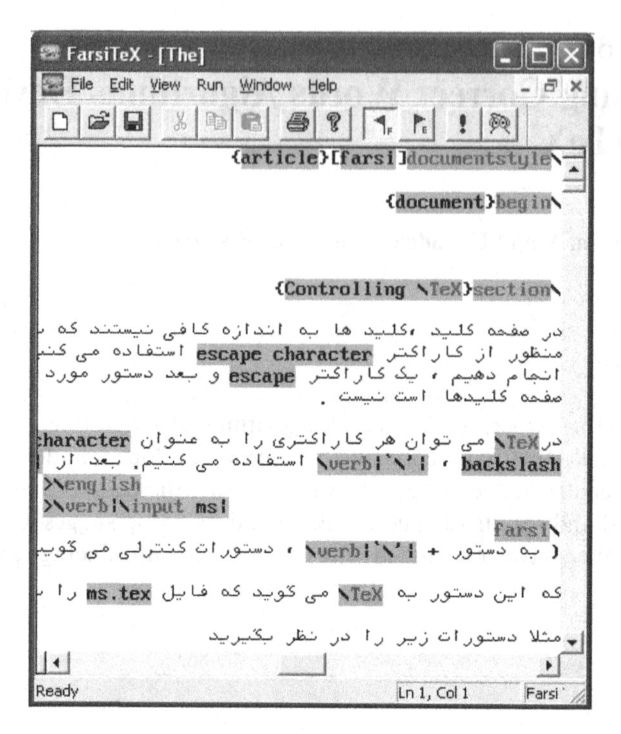


**Fig. 6.1** In this version of spell-checker we deal just with Farsi words, and ignore any English letters

When a text is fed to a spell-checker, by parsing the text, all of the words in it will be checked in a database of *correct words*. We have such a database of words which are known to be spelled correctly. This database can be increased. We have designed a special facility to do this, which is a program named *dict*. When a text is verified to have the correct spellings of all its words, it would be suitable to be fed to a *dict* program, which adds all its words to the database. This *dict* program has the intelligence of recognizing where a word is already in the database and avoiding repetitions.

To this end, we need a structure to our database. We call this database a dictionary file. It is obvious that to have a powerful spell-checker we need a large dictionary file.

When the spell-check program is initiated to check a text, it first loads a dictionary file. If there does not exist any dictionary file, it is better to run first the *dict* program to setup a dictionary file. To this end, we need some correctly spelled texts to cover almost all of the needed words. So to work with this program, at first we need a human spell-checking to create a dictionary file. This procedure is normal for any spell-checking program for any language.

## 6.2 Algorithms of Suggesting Correct Words

The spell-checker reads words of the input text one-by-one. This is a procedure performed by a parser and has its own algorithms and problems. On every word, the spell-checker stops to look it up inside the dictionary. If it is in the dictionary then it is correct, and the spell-checker moves on to the next word.

If it is a word that isn't inside the dictionary, then it is may be an incorrect word.

### 6.2.1 *On an Incorrect Word*

It will be suitable if we reach the correct word just by minimal changes being made to the incorrect word. This is a common facility among spell-checkers. To be a good spell-checker, it should have some auto-corrects or some suggestions as to the correct words which the user can then select from. Methods and algorithms to do this for Farsi language have some similarity to English spell-checkers, but there are also some differences.

Here we list common algorithms. They all will add words to the suggestion list, if they find a different but similar correct word.

- First we interchange the place of any two adjacent letters. In many cases this procedure finds a correct word because it is a common mistake of typists to press a button before a preceded letter.As an example, for a three-letter incorrect word, this method produces two other three-letter words. If these words are in the dictionary file, they will be placed in the suggestion list.
- Another routine suitable for incorrect words that have been created by missing a letter when typing, is the insertion of the 32 letters of the Farsi alphabet one-by-one in the places between each pairs of the letters of the incorrect word and also at the beginning and end of the word. For example, if a three-letter word is incorrect there are four places to add a letter to this word to make it a four letter word, so from this three-letter word the spell-checker produces four times 32, equalling 128 new four-letter words. All these new words are to be checked to be in the dictionary file, and if it is so, will be added to the suggestion list.
- The other routine is the dropping routine. This routine drops the letters of the incorrect word, one at a time, and then checks the new words. For a three-letter word, this algorithm would find three new two-letter words.
- The forth routine is the replacing one. This is suitable for the incorrect word which has one letter misspelled. By this algorithm, the program drops each letter and replaces it by all 32 letters of the Farsi alphabet, one by one. In this way, for an incorrect three-letter word, we come up with three times 32, equalling 96 new three-letter words.

We describe some different problems facing Farsi spell-checkers, especially in our Spell-checker.

## 6.3 Increasing Speed

How can we increase the speed at which the program runs? The speed drawbacks return to several points, including:

- loading the dictionary and,
- searching for every suggested word and other searches.

In spell-checker V0.1, Linear Search is a simple search algorithm that searches the whole dictionary every time when they were called. To enhance the process of searching we want to use an efficient binary search which is specialized to our application. For example, using size definite categorizes. For this we must save our words in objects that can have their own functions such as compare function and to perform searches on the list of these objects.

We should implement the new version of the spell-checker in a platform which supports more object orienting methods and has powerful functions in working with strings and lists.

## 6.4 Memory Usage Problems

The other problem is the memory usage, and now we present some ways to solve these issues.

### *6.4.1 Using Static Functions*

At first the words will be saved in objects, their functions being static ones. Because we have too many objects of this type (actually every Farsi word in the dictionary will be an object in the memory), and if each object has a non-static function, it itself will take place in the memory. So by making them static we create one instance of this type of functions and use this instance with all objects. This is illustrated by Figs. 6.2 and6.3.

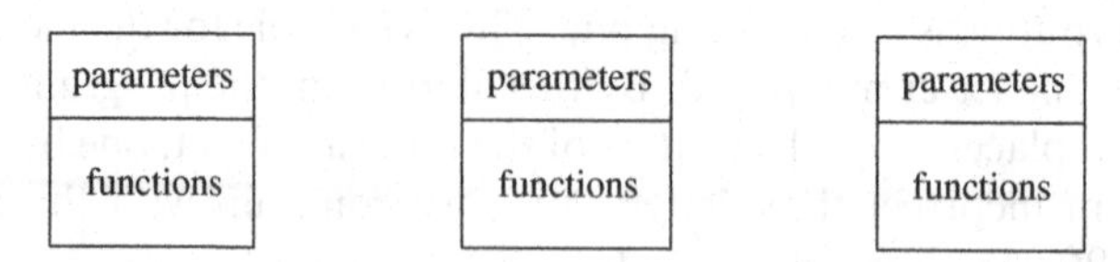


**Fig. 6.2** Objects with non-static functions

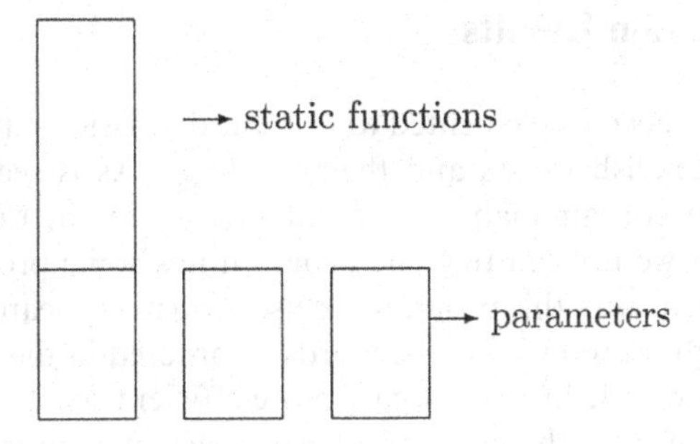


**Fig. 6.3** Objects with static functions

### *6.4.2 Using Multi Files*

The other way is to save and load these words in more files. Our suggestion for the method of the splitting is according to what we call weight, and it is related to the applications of the word. More applied words have more weights. At the time when the dictionary is being created, each word is to be counted and if it is used in more places we increase its weight. So we need fair texts to create dictionaries. It means that in our texts, to create dictionaries, the number of occurrences of a word should be in balance with the normal usage of it in overall texts.

In this way, we load at first those words with more weight in the memory, search for them and then the other file with lower weight. This novel process cause even faster searches.

Now our dictionary is in several files. At first we load the more weighted file. With this file, we look for the incorrect words in the text. We save these words in a list that we call *mistake list*.

The second time we load the other file of a dictionary that has a lower weight. With this file we search just inside the mistake list. If they were founded in this latter dictionary file, we omit them from the mistake list. In this way the mistake list gets shrunk. If now the mistake list were empty, the process has been finished. If not, we should load another file of the dictionary. This process can be continued to find the incorrect words of the input text.

### *6.4.3 The Roots of the Farsi Words Instead of the Whole Words*

The other way to decrease the memory usage is to save only the roots of the words. This is mentioned in an earlier paper [8]. It means we should have functions to omit the prefix and postfix of a word while creating the dictionary, and we should save just the root. Then while finding the mistakes we should have functions to add these postfixes and prefixes. Absolutely it causes the program to run slowly, but does save a lot of memory.

## 6.5 Implementation Results

Some software has been implemented to test our discussions. In the version 0.1 we ignored any English letters and their spellings. As is seen in Fig. 6.1, a FarsiTeX text may contain many commands written by the English alphabet.

During the test we noticed that the program has some problems: The first problem is the speed, as in this program we used too many searches so using not so suitable algorithms and saving the words unordered in the file resulted in a slow program. So we need to investigate more efficient search algorithms.

The second problem is the using of a lot of space in memory:

1. In the way to completing the dictionary, the mass of words that are to be loaded in the memory is growing. This may not be a problem in modern systems with a large amount of memory, however in systems with memories lower than 64 MB it causes the systems to work more slowly and forces the words to load in the slow virtual memory or the swap memory. We should highlight that this problem has occurred because of the saving of the all branching of a word in the dictionary file, without any attempt first to find the roots of the words and save them instead.
2. The other reason is due to the loading of the whole dictionary at once.

These problems were discussed in the previous Sections 6.3 and 6.4 on increasing speed and memory usage problems. We will present features of other versions of the spell-checker in another paper.

## 6.6 Other Techniques and Conclusion

One of the problems, which is related to the memory in this version 0.1 program, is the data structure of the words that are to be loaded into the memory.

In this program, the data structure that we have used is the C++ vector container that needs a connected memory. To have a better performance, we should get rid of this restriction. We can use better data structures in this program while we will write it with c# platform. These data structures are such as maps, lists and so on. Of course for these new data structures we should write their search algorithms.

We now describe some implementing techniques that we may use to upgrade the program.

For saving the objects in files we only save the parameters such as string itself. For implementing the binary search algorithm we need to save the words in an ordered way. These causes *make dictionary* application to be more complicated. For search we need a compare function. We put these functions inside objects and we can use more specialized functions for them and even enter the change functions into them too.

The other problem is in the suggestions. These suggestions are not in a good order. By using a multi-files system for weighting the dictionary, some of these problems will be solved, as it forces words more commonly used to be suggested first. For enhancing it at last we need to sort suggestions by their weight using fast sort methods.

The other problem of this program is in not havinga good interaction with the end-user. For solving this problem we propose to use an editor interface and for this we need to have an open-source editor, and having a hand over its codes. This new editor may show mistakes in other colors than the text color or underline them and may suggest words by tool tips. We hope to use OpenOffice.Org as our editor.

We developed some algorithms to give hints for the misspelled Farsi words in the texts. We also developed a novel spell-checker version 0.1. This can be an acceptable software, but it should be completed to be a user-friendly one.

As we saw, there are many novel approaches to follow and upgrade the *spell-checker* program. These methods are to be tested, measured and to be evaluated and weighted. The benefits of having a good interactive program are seemed to be greater than having speedy programs but we hope gradually that we find courage to follow the way.

**Acknowledgements** We say our best thanks to Dr Jahani, for his encouragement and for reading the manuscript. The research was supported in part by the Islamic Azad University, Ahar Branch; and in the case of the third author it was also supported by the Department of Mathematical Sciences of university of Tabriz.

## Appendix

Some view of our program illustrated by figures.

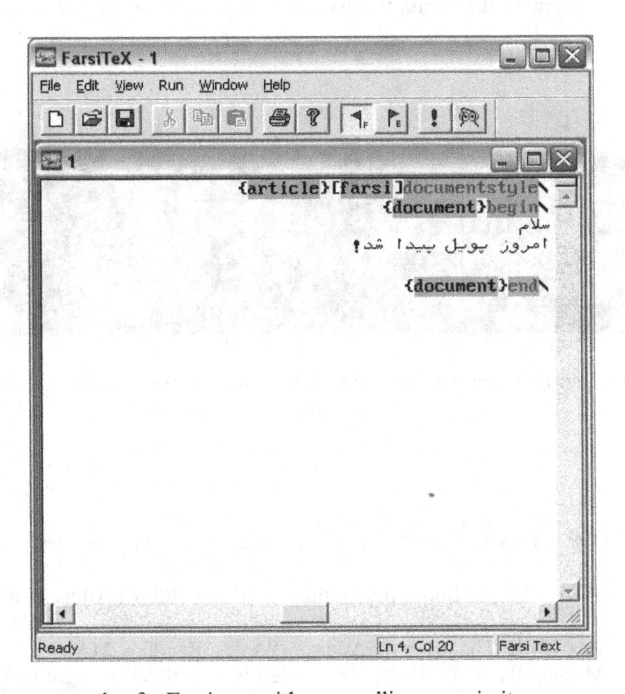


**Fig. 6.4** This an example of a Farsi text with one spelling error in it

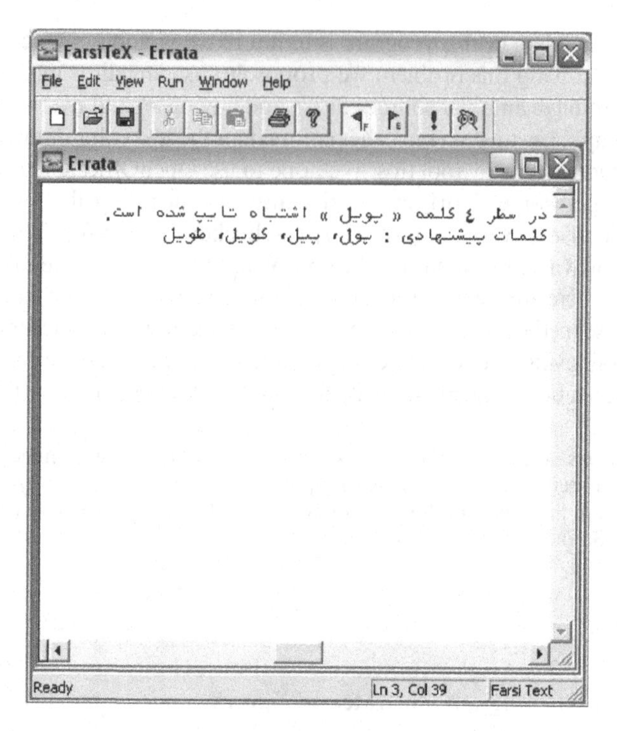


**Fig. 6.5** This is content of the Errata file, written by our Spell-checker V0.1

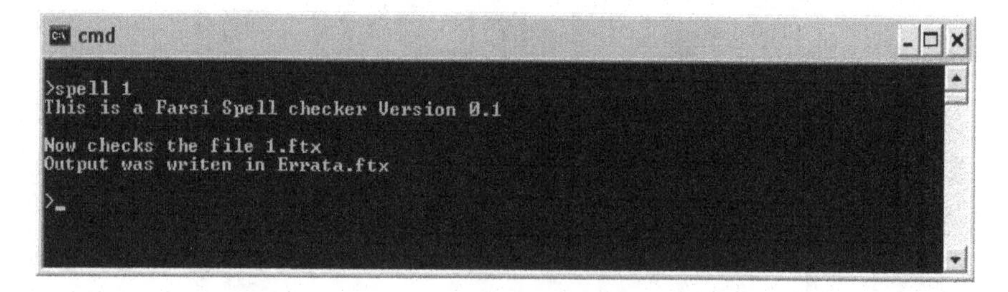


**Fig. 6.6** We call our Spell-checker in a command prompt as in this figure

## References

1. Ghodsi M (1997) FarsiTeX guide (in Farsi Language). Computer department, Sharif University
2. Knuth DE (1986) The TeXBook. Illustrations by Duane Bibby. Addison-Wesley, Reading, MA
3. Knuth DE (1986) TeX: The program.Addison-Wesley, Reading, MA
4. Knuth DE, Mackay P (1986) Mixing right-to-left texts with left-to-right texts. TUGboat 8(1)

5. Kopka H, Daly P (1993) A guide to LaTeX, document preparation for beginners and advanced users. Addison-Wesley
6. Rezvan Y, Ivaz K, Rezvan F (2006) Towards spell-checking in FarsiTeX. In: Proc 5th WSEAS Int. Conf. on Data Networks, Communications & Computers, Bucharest, Romania
7. Rezvan Y, Rezvan F, Ghandchi M (2007) FarsiTeX and a novel approach to its spell-checker and the dictionary file of it. WSEAS Transactions on Computers 6(2)
8. Rezvan Y, Rezvan F, Ghandchi M (to appear) Evaluation and development of FarsiTeX novel spell-checker algorithmsSpell-checkSpell-check

# Chapter 7
# On the Analytical and Numerical Computation in Mechanical Modeling

**Marcel Migdalovici**

**Abstract** In the paper are studied the possibilities of replacing the manual analytical calculation, which intervenes in the mechanical modeling, by an isomorphic numerical calculation which can be performed on digital computers. An algorithm is described for performing the greatest common divisor of two polynomials with several variables that may be used to determine the analytical inverse matrix for a matrix of such polynomials that are used in a mathematical modeling of mechanical phenomena. A new definition of the Euclidean ring is proposed. The question of how much can lead the formal (analytical) calculation in the modeling up to replacing with a numerical method of solution calculation is emphasized.

## 7.1 Introduction About Coding

We start the coding of the summing and the product operations in the set of polynomial of several variables with real or integer coefficients with the coding of the algebraic operations for two monomials as follows [11]:

$$\begin{aligned} &C_{a_1\dots a_m} X_1^{a_1} \dots X_m{}^{a_m} + C_{b_1\dots b_m} X_1^{b_1} \dots X_m{}^{b_m} \rightarrow \\ &\rightarrow \{(C_{a_1\dots a_m}, a_1, \dots, a_m), (C_{b_1\dots b_m}, b_1, \dots, b_m)\} \end{aligned} \tag{7.1}$$

$$\begin{aligned} &C_{a_1\dots a_m} X_1{}^{a_1} \dots X_m{}^{a_m} * C_{b_1\dots b_m} X_1{}^{b_1} \dots X_m{}^{b_m} \rightarrow \\ &\rightarrow (C_{a_1\dots a_m} * C_{b_1\dots b_m}, a_1 + b_1, \dots, a_m + b_m) \end{aligned} \tag{7.2}$$

$$C * C_{a_1\dots a_m} X_1{}^{a_1} \dots X_m{}^{a_m} \rightarrow (C * C_{a_1\dots a_m}, a_1, \dots, a_m) \tag{7.3}$$

For coding the summing and the product operations of two polynomials we codify the summing and the product polynomials for these polynomials.

M. Migdalovici (✉)
Romanian Academy, Institute of Solid Mechanics 15, C-tin Mille street, Bucharest, Romania

N. Mastorakis et al. (eds.), *Proceedings of the European Computing Conference*, Lecture Notes in Electrical Engineering 28,
DOI 10.1007/978-0-387-85437-3_7, © Springer Science+Business Media, LLC 2009

In this manner we see that it is possible to replace the manual analytical calculation by an isomorphic numerical calculation which can be performed on digital computers.

## 7.2 Notions on the Set of Polynomials

A unitary and commutative ring $K$ without divisors of zero is named an integral domain (i.d.).

Let $K$ a factorial ring (f.r.), therefore an integral domain with the property that every non-zero and non-invertible element of $K$ is a product of prime elements of $K$ [4, 6, 10].

If $a, b \in K$ we say that $a$ divide $b$ if $b = ac$ with $c \in K$. Will be denoted $a/b$.

A non-zero and non-invertible element $p \in K$ is named "prime" if for any $a, b \in K$ with $p/ab$ it follows $p/a$ or $p/b$.

An element $c \in K$ (if exist) is named a greatest common divisor (g.c.d.) of $a$ and $b$, denoted $c = (a, b)$, if $c/a, c/b$ and if $d/a$, $d/b$ then $d/c$.

The elements $a, b \in K$ such that $(a, b) = 1$ are named relatively prime.

Two elements $d1, d2 \in K$ such that exists $u \in K$ invertible, with $d1 = u\ d2$ are named adjoins in divisibility.

The ring of polynomials of one variable with coefficients in $K$ is denoted by $K[X]$ and the ring of polynomials of several variables $X_1, \ldots, X_n$ with coefficients in $K$ is denoted by $K[X_1, \ldots, X_n]$.

Let $K$ i.d. and $f \in K[X]$ of the form

$$f = a_o + a_1 X + \ldots + a_n X^n \tag{7.4}$$

The (g.c.d.) for the coefficients of the polynomial $f$ is denoted by $c(f) \in K$.

If $f \in K[X]$ is of the form (7.4) and $a \in K$ with $a/f$ then $a/a_i$, where $a_i \in K, i = 1, \ldots, n$.

If $g \in K[X]$ and $c(g) = 1$ we say that $g$ is primal polynomial.

The ring of polynomials in indeterminate $X_m$, $m \leq n$, over ring $K_0$ is denoted $K_0[X_m]$, where $K_0 = K[X_1, \ldots, X_{m-1}, X_{m+1}, \ldots, X_n]$.

A polynomial $g \in K_0(X_m)$ is of the form

$$g = b_0 + b_1 X_m + \ldots + b_n X_m^n \tag{7.5}$$

The coefficients $b_0, b_1, \ldots, b_n$ of polynomial $g$ are polynomials of the ring $K[X_1, \ldots, X_{m-1}, X_{m+1}, \ldots, X_n]$ of polynomials.

If $K$ i.d. then $K[X]$ i.d. and if $K$ f.r. then $K[X]$ f.r.

## 7.3 A Theorem of Division

Let $K$ factorial ring, $0 < m \leq n$, with $m, n \in N$ and $K[X_1, \ldots, X_n]$ the factorial ring of polynomials with several variables and with coefficients in $K$.

We formulate bellow the following [9]:

**Theorem 1** *If polynomials* $p_1, p_2 \in K[X_1, \ldots, X_n], p_1 \neq 0, p_2 \neq 0$, *for fixed* $m$, *there exists polynomials* $q_1, q_2, r \in K[X_1, \ldots, X_n]$, *unique without an adjoins in divisibility, such that*

$$p_1 q_1 = p_2 q_2 + r \tag{7.6}$$

*The polynomial* $r$ *have the alternative* $r = 0$ *or* $\deg r < \deg p_2$, *with degree referred to variable*$X_m$, *and the polynomials* $q_1, q_2, r$ *are relatively prime*, $q_1 \neq 0$.

*Let now* $p_1, p_2 \in K(X_1, \ldots, X_n), p_1 \neq 0, p_2 \neq 0$.

*The set of divisors with zero remainder for both polynomials* $p_1$ *and* $p_2$ *is denoted by* $D(p_1, p_2)$ *and briefly the set of divisors for polynomials* $p_1$ *and* $p_2$ *is named.*

*There is the following property* [9]:

**Theorem 2** *In the conditions of first theorem, for fixed* $m$, *the equality* $D(p_1, p_2) = D(p_2, r)$ *is true, where the remainder of the division of the polynomials* $p_1$ *and* $p_2$ *is* $r$.

*This theorem permits to give Euclid's type algorithm for performing the greatest common divisor of two polynomials of several variables with coefficients in factorial ring.*

## 7.4 A New Definition of Euclidean Ring

N. Jacobson, in his treatise [6], gives the following definition of Euclidean ring:

*A domain of integrity* $D$ *is called Euclidean if there exists a map* $\delta : D \to N$, *of* $D$ *into the set* $N$ *of non-negative integers, such that if* $a, b \neq 0 \in D$ *then there exist* $q, r \in D$ *such that* $a = b\ q + r$ *where* $\delta(r) < \delta(b)$.

We propose the following definition:

*A factorial ring* $D$ *is called Euclidean if there exists a map* $\delta : D \to N$, *of* $D$ *into the set* $N$ *of non-negative integers, such that if* $a, b \neq 0 \in D$ *then there exist* $q_1, q_2, r \in D$ *such that* $a\ q_1 = b\ q_2 + r$ *where* $\delta(r) < \delta(b)$ *and* $q_1, q_2, r$ *are relatively prime.*

## 7.5 The Model of Thin Plates and Coding

The thin plate is supposed homogeneous, isotropic and elastic linear. The thickness of the plate is supposedly constant and also the hypothesis Love-Kirchoff of the "non-deformed normal element" is considered [3]. We analyze the thin plate in hypothesis that the middle surface is of revolution.

The revolution thin plate is described by the vector of position for the point $P(\theta, z)$ situated on the middle surface:

$$\vec{R}\,(\theta, z) = r(z)\cos(\theta)\vec{i} + r(z)\sin(\theta)\vec{j} + z\vec{k} \tag{7.7}$$

The vector of displacements with his applied point on the surface is considered of the form:

$$\vec{U} = u\vec{t_z} + v\vec{t_\theta} - w\vec{n} \tag{7.8}$$

In (7.8), $\vec{t_z}$ and $\vec{t_\theta}$ are the unitary vectors attached to coordinate curves concerning the point $P(\theta, z)$.

The first fundamental form of the middle surface is of the form: $ds^2 = r(z)^2 d\theta^2 + (1 + r_z{}^2)dz^2$ with $r_z = d\,r(z)/dz$ and the coefficients of the fundamental form $A = r(z), B = (1 + r_z{}^2)^{1/2}$.

The coefficients of the second fundamental form of the surface are: $D = -r/B, D' = 0, D'' = r_{z2}/B$, where $r_{z2} = d^2r(z)/dz^2$.

The building of the codified thin plate model on computer is based on coding of the differentiation operation.

For deduction of the codified differentiation of the vector $\vec{U}$, firstly the codified formulas of differentiation for the unitary vectors $\vec{t_z}, \vec{t_\theta}, \vec{n}$ are deduced.

The vector equilibrium equations of the Goldenveizer model [3], attached to any middle surface point of the plate and to appropriate three dimensional local system of axis, are:

$$-\frac{\partial(r\,\vec{F}^{(z)})}{\partial z} - \frac{\partial(B\,\vec{F}^{(\theta)})}{\partial\theta} + rB\vec{P} = 0. \tag{7.9}$$

$$\begin{aligned} &-\frac{\partial(r\,\vec{C}^{(z)})}{\partial\,z} - \frac{\partial(B\,\vec{C}^{(\theta)})}{\partial\,\theta} - rB\vec{F}^{\theta}X\vec{t_z} - \\ &- rB\vec{F}^{z}X\vec{t_\theta}rB\vec{C} = 0 \end{aligned} \tag{7.10}$$

In (7.9) and (7.10), $\vec{P}$ is the external force which action on the plate, $\vec{C}$ is the external resultant moment and

$$\vec{F}^{(z)} = N^{z\theta}\vec{t_z} - N^{z}\vec{t_\theta} + Q^{z}\vec{n}, \vec{C}^{z} = M^{z}\vec{t_z} - M^{z\theta}\vec{t_\theta}, \tag{7.11}$$

$$\vec{F}^{(\theta)} = -N^{\theta}\vec{t_z} - N^{\theta z}\vec{t_\theta} + Q^{\theta}\vec{n}, \vec{C}^{\theta} = -M^{\theta z}\vec{t_z} - M^{\theta}\vec{t_\theta} \tag{7.12}$$

The coding of the operations for the deduction of the model of revolution thin plates, in displacements, is developed in the following fixed order of successive variable which intervenes in the model [11]:

$$\begin{gathered} E,\ h,\ \tfrac{1}{1+\nu},\ \tfrac{1}{1-\nu^2},\ \tfrac{1}{r},\ r,\ \ldots,\ r_{z5},\ \tfrac{1}{B},\ B,\ u,\ u_\theta,\ u_z, \\ u_{\theta\theta},\ u_{\theta z},\ u_{zz},\ u_{\theta\theta\theta},\ \ldots,\ u_{zzzz},\ v,\ \ldots,\ v_{zzzz},\ w,\ \ldots,\ w_{zzzz} \end{gathered} \tag{7.13}$$

In (7.13) the modulus of elasticity is $E$, the shell half-thickness is $h$, Poisson's coefficient is $\nu$ and the components of displacement of a point, situated on the middle surface of shell, are $u$, $v$, $w$.

Vector and scalar codified operations of addition, multiplication and differentiation subroutines have been performed.

All the results needed for the mathematical model, expressed in displacements, are deduced on the computer by decoding the analytical expressions and are used as input data for the program of static or dynamic calculus of thin plates [8].

The numerical method used for calculus of the revolution thin plates takes into account the boundary conditions at the extremities $z = z_1$ and $z = z_2$ as well as the development in series of vector displacement components concerning coordinate variables [5, 8].

In this mathematical model the research is compound of two stages, the analytical stage where are defined the equations of the model using the parameters of the physical model and the numerical stage where the values of the parameters are defined and a numerical method of performing the solution is applied.

Many authors (Knuth 1981, Buchberger et al. 1982, Davenport et al. 1987, etc) have analyzed the symbolic computation, but this research is not yet exhausted [1, 2, 7].

## 7.6 Conclusion

The possibilities of replacing the manual analytical calculation by an isomorphic numerical calculation are investigated.

A question about how much can lead the analytical calculation in the modeling up to replacing with a numerical method of the solutions calculation appeal. The large analytical phase developed in the modeling permit us to increase the precision of the final calculus of solution. Any steps are analyzed here. New strategies in this research are needed, assisted by the computer.

## References

1. Buchberger B, Collins GE, Loos R (1982) Symbolic & algebraic computation. Computing supplementum 4. Springer–Verlag, Wien-New York
2. Davenport J, Siret Y, Tournier E (1987) Formal calculus (in French). Masson, Paris, New York, Barcelone, Milan, Mexico, Sao Paulo
3. Goldenveizer AL (1953) Theory of elastic shells (in Russian). Gostehizdat, Moscow
4. Ion D Ion, Nita C, Nastasescu C (1984) Complements of algèbra (in Romanian). Scientific and Encyclopaedic Publishing House, Bucharest
5. Iovanescu VR (1980) Stress in shells with middle surface of mixed type (in Romanian). Ph.D. thesis, Bucharest University
6. Jacobson N (1973) Basic algebra vol I. Freeman, San Francisco

7. Knuth DE (1981) The art of computer programming vol II seminumerical algorithms. Addison–Wesley
8. Migdalovici M (1985) Automation of mechanical structures calculus with application to nuclear power plant (in Romanian). Ph.D. thesis, Bucharest University
9. Migdalovici M (2004) A theorem of division with a remainder in a set of polynomials with several variables. Creative Math 13:5–10
10. Nastasescu C, Nita C, Vraciu C (1986) Basic algebra vol I (in Romanian). Publishing House of the Romanian Academy, Bucharest
11. Visarion V, Migdalovici M (1979) On the transposition of the analytical calculation on computers. Rev Roum Sci Techn Mec Appl 24:847–853

# Chapter 8
# Real Genetic Algorithm for Traveling Salesman Problem Optimization

**A.R. Awad, I. Von Poser, and M.T. Aboul-Ela**

**Abstract** This work aims at solving the Traveling Salesman Problem (TSP) through developing an advanced intelligent technique based on real Genetic Algorithm (GA). The used GA comprises real-value coding with specific behavior taking each code as it is (whether binary, integer, or real), rank selection, and efficient uniform genetic operators. The results indicated, in comparison with the other applied optimization methods (linear, dynamic, Monte Carlo and heuristic search methods), that the real GA produces significantly the lowest distance (least cost tour) solution. It is concluded that the real GA approach is robust and it represents an efficient search method and is easily applied to nonlinear and complex problems of the TSP in the field of solid waste routing system in the large cities.

## 8.1 Introduction

The problem of solid waste and its management is complex in small towns and critical in large metropolitan areas [1]. The routing is one of the main components of solid waste management in the cities where the collection takes 85% of the solid waste system cost and only 15% for disposal [2]. The traveling salesman problem (TSP) is one of the most widely studied and most often cited problems in operations research. For over fifty years the study of the TSP has led to improving solution methods for a wide range of practical problems [3]. Those studies are based mainly on several modeling approaches like linear, dynamic programming techniques and heuristic techniques [4–6]. Genetic algorithms have also been used successfully to solve this NP-problem. However, in many of such works, a little effort was made to handle the nonlinear optimization problem of routing solid waste collection.

In general, the TSP has attracted a great deal of attention because it is simple to state but difficult to solve [7]. The exhaustive algorithms for solving the TSP

A.R. Awad (✉)
Department of Environmental Engineering, Tishreen University, Lattakia, Syria

N. Mastorakis et al. (eds.), *Proceedings of the European Computing Conference*, Lecture Notes in Electrical Engineering 28,
DOI 10.1007/978-0-387-85437-3_8, 

or node route problem are rarely very good in any sense [8]. They perform well for $n \leq 6$ and very badly for $n \geq 15$; it is time consuming and needs huge storage and memory time. It has been reported by Coney (1988) that a 21-location tour would require 77,100 years of computer time on a million operation per second (PC) [9]. Mathematical programming approaches have had rather limited success with this problem. According to Thieraut and Klekamp (1975) for a 20-node problem, integer linear programming requires 8,000 variables and 440 constraints while dynamic programming is limited to 13-node problems [10].

The aim of this paper is to find the optimal routing for solid waste collecting in cities, taking Irbid City in Jordan as an example problem, through developing an advanced intelligent technique based on real Genetic Algorithm (GA). In addition, a comparison has been made with other optimization techniques such as linear, dynamic, Monte Carlo simulation, and heuristic algorithms.

## 8.2 Encoding Schemes of GA

There are many variations of GAs but the following general description encompasses most of the important features. The analogy with nature requires creation within a computer of a set of solutions called a population. Each individual in a population is represented by a set of parameter values that completely propose a solution. These are encoded into chromosomes, which are originally sets of character strings, analogous to the chromosome found in DNA.

The GA search, sometimes with modification, has proved to perform efficiently in a large number of applications. This efficiency lies in the robustness of the search method that underlies the GA approach and in the flexibility of the formulation itself. In contrast to traditional optimization methods that track only a single pathway to the optimal solution, Genetic Algorithms cover a whole population of possible solutions. The selection of an appropriate chromosome representation of candidate solutions to the problem at hand is the foundation for the application of Genetic Algorithms to a specific problem. With a sufficiently large population of chromosomes, adequate representation will be achieved. In real-value coding, variables that can take on continuous values are represented as a real (i.e. continuous) variable in the string. In this case, the string consists of a series of real values.

Real-value chromosomes have been used successfully in multireservoir system operation by various authors [11, 12], and in pipe networks optimization by the authors [13]. In this paper a new GA based methodology for optimal solving of the TSP has been applied. This methodology takes by real representation with any code (binary, integer, or real) without any need to transfer from one to another, i.e. our GA, using chromosome as unit, works equally well with integer and non-integer decision variables. This is a specific behavior which distinguishes our real-coding GA from other (binary, Gray, or integer) coding GAs. Thus the encoding scheme used herein proved to have the property of matching the effective problem search space towards the GA search space. But here in our

**Table 8.1** Coding of the network

| A | 1 |
|---|---|
| B | 2 |
| C | 3 |
| D | 4 |
| E | 5 |
| F | 6 |
| G | 7 |
| H | 8 |
| I | 9 |
| J | 10 |
| K | 11 |
| L | 12 |
| M | 13 |
| N | 14 |
| O | 15 |
| P | 16 |

←Node No. 16 is added to indicate the ending point.

example the nodes are taken as integer quantities working with one chromosome which comprises all nodes encoded as A, B, C and so on (Table 8.1). Nevertheless, this choice does not limit our GA.

## 8.3 Example: A Real Problem of Solid Waste Routing

The use of GA to solve a traveling salesman problem will be illustrated by using real representation to solve the solid waste network studied initially in Reference [14].

Irbid is divided into six regions which are considered as separate solid waste generation areas. Each of these regions has its own department which regulates the solid waste services in the region. There is no specific routing basis for the vehicles being left to the driver's choice. Occasionally, one pick-up point may be missed. In regions which have two collection vehicles, they may meet at the same pick-up point several times. Once the solid waste is loaded into the vehicles, it is carried out of Irbid to the disposal site located far away.

The suggested procedure for solving the vehicle routing problem in the selected Region 2 of Irbid begins with a particular node closest to the garage or the previous region and ends with the nodes closest to the disposal site. This reduces the number of permutations considerably. Further detailed procedures can be found in Reference [15]. There are 31 pick-up points (nodes) in this Region 2 with about 50 containers distributed on it. Two major streets pass through the area of this region and divide the total area into three sectors, each having its own nodes and its own network, and each network has its distance matrix (Fig. 8.1). Network I has 9 nodes (pick-up points), network II has 7 and finally network III has 15 nodes. The network III was used as a case study. A set of 15 nodes with 1 or 2 waste containers at each node were in service by vehicles.

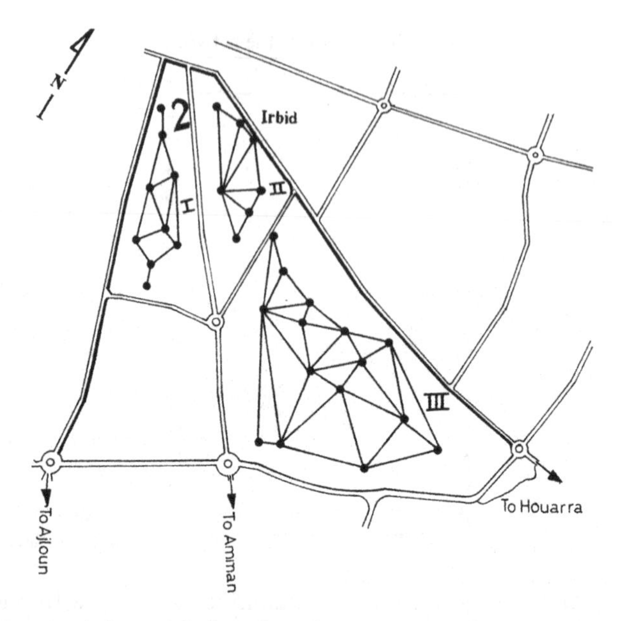


Fig. 8.1 The constructed networks in region two

The collecting vehicles are equipped with compactors and have to collect the contents of about 30 full containers. The distances between nodes in Network III are shown in Fig. 8.2. The routing problem in this part of Irbid is a node routing or traveling salesman problem (TSP).

Our problem has been the construction of a tour through n points with the minimum distance, where the vehicle does not have to return to the starting

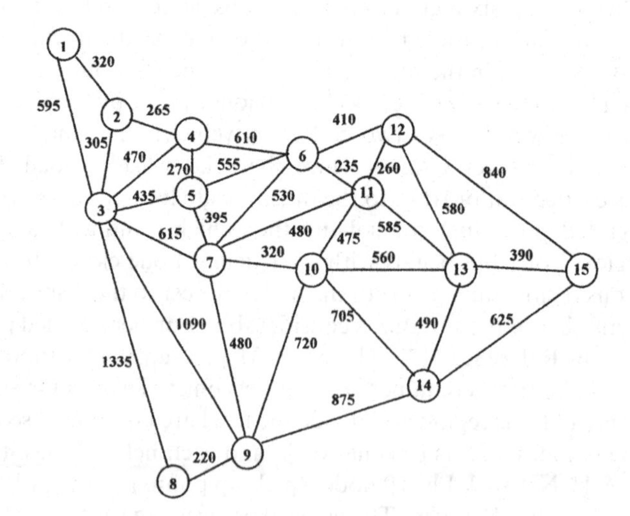


Fig. 8.2 Network (III) of region two

point, i.e. the number of possible tours is (n-2)! The application based on the fact that the shortest solid waste collection tour should begin with the first node No.1 (closest to the municipality garage), pass by all nodes once and only once, and end with the node No. 15 (closest to the disposal site out of the city).

## 8.4 GA Optimization

The following procedures are required for the formulation of GA:

### 8.4.1 Coding

The Genetic Algorithm requires that the decision variables describing trial solutions to the solid waste routing problem be represented by a unique coded string of nodes (pick-up points). This coded string is similar to the structure of a chromosome of genetic code. As for the example, there are 15 decision variables to be made about the network. Each of these decision variables can take one node (pick-up point). A real string made of 15 substrings is used for representing the problem into a suitable form for use within a GA (Table 8.1). This string (chromosome) of 15 genes represents a route design for the network consisting of 15 nodes.

It is worthy to mention that the node (No. 16 or P) has not been coded but was used (as pseudo node) to indicate the ending point of the studied network.

### 8.4.2 Fitness

The fitness of a coded string representing a solution for the traveling salesman problem of solid waste routing is determined by the shortest cycle/cost provided that the collection vehicle passes through all nodes just once for each.

The evaluation or objective function used is rather simple and is a determinant of the distance of a routing solution by summing the lengths of the nodes distances making up the network (Table 8.2). The value of 10,000 m is a hypothetical value which points to the node itself, but actually such a case never occurs. In the used coding scheme also an infeasible solution can appear. In the distance matrix every infeasible route selection such as "C to J" will be penalized with a high distance of 5,000 m. An example for fitness evaluation is shown in Fig. 8.3.

Infeasible solutions, which have failed to meet the aforementioned node requirements, are not removed from the population. Instead, they are allowed to join the population and help guide the search, but for a certain price. Although this technique does not exclude infeasible solutions from the

**Table 8.2** Distance matrix among the nodes

| | stri 0 | inte 1 | inte 2 | inte 3 | inte 4 | inte 5 | inte 6 | inte 7 | inte 8 | inte 9 | inte 10 | inte 11 | inte 12 | inte 13 | inte 14 | inte 15 | integer 16 |
|---|---|---|---|---|---|---|---|---|---|---|---|---|---|---|---|---|---|
| stri | | A | B | C | D | E | F | G | H | I | J | K | L | M | N | O | |
| 1 | A | 000 | 320 | 595 | 5000 | 5000 | 5000 | 5000 | 5000 | 5000 | 5000 | 5000 | 5000 | 5000 | 5000 | 5000 | |
| 2 | B | 320 | 10000 | 305 | 265 | 5000 | 5000 | 5000 | 5000 | 5000 | 5000 | 5000 | 5000 | 5000 | 5000 | 5000 | |
| 3 | C | 595 | 305 | 10000 | 470 | 435 | 5000 | 615 | 1335 | 1090 | 5000 | 5000 | 5000 | 5000 | 5000 | 5000 | |
| 4 | D | 5000 | 265 | 470 | 10000 | 270 | 610 | 5000 | 5000 | 5000 | 5000 | 5000 | 5000 | 5000 | 5000 | 5000 | |
| 5 | E | 5000 | 5000 | 435 | 270 | 10000 | 555 | 395 | 5000 | 5000 | 5000 | 5000 | 5000 | 5000 | 5000 | 5000 | |
| 6 | F | 5000 | 5000 | 5000 | 610 | 555 | 10000 | 530 | 5000 | 5000 | 5000 | 235 | 410 | 5000 | 5000 | 5000 | |
| 7 | G | 5000 | 5000 | 615 | 5000 | 395 | 530 | 10000 | 5000 | 480 | 320 | 480 | 5000 | 5000 | 5000 | 5000 | |
| 8 | H | 5000 | 5000 | 1335 | 5000 | 5000 | 5000 | 5000 | 10000 | 220 | 5000 | 5000 | 5000 | 5000 | 5000 | 5000 | |
| 9 | I | 5000 | 5000 | 1090 | 5000 | 5000 | 5000 | 480 | 220 | 10000 | 720 | 5000 | 5000 | 5000 | 875 | 5000 | |
| 10 | J | 5000 | 5000 | 5000 | 5000 | 5000 | 5000 | 320 | 5000 | 720 | 10000 | 475 | 5000 | 560 | 705 | 5000 | |
| 11 | K | 5000 | 5000 | 5000 | 5000 | 5000 | 235 | 480 | 5000 | 5000 | 475 | 10000 | 260 | 585 | 5000 | 5000 | |
| 12 | L | 5000 | 5000 | 5000 | 5000 | 5000 | 410 | 5000 | 5000 | 5000 | 5000 | 260 | 10000 | 580 | 5000 | 5000 | |
| 13 | M | 5000 | 5000 | 5000 | 5000 | 5000 | 5000 | 5000 | 5000 | 5000 | 560 | 585 | 580 | 10000 | 490 | 5000 | |
| 14 | N | 5000 | 5000 | 5000 | 5000 | 5000 | 5000 | 5000 | 5000 | 875 | 705 | 5000 | 5000 | 490 | 10000 | 5000 | |
| 15 | O | 5000 | 5000 | 5000 | 5000 | 5000 | 5000 | 5000 | 5000 | 5000 | 5000 | 5000 | 840 | 390 | 625 | 10000 | |
| 16 | P | 5000 | 5000 | 5000 | 5000 | 5000 | 5000 | 5000 | 5000 | 5000 | 5000 | 5000 | 5000 | 5000 | 5000 | 10 | |

population, it reduces the probability of infeasible solutions remaining in future generations and permits convergence from all regions of the parameter space.

### 8.4.3 *Selection, Crossover and Mutation*

Detailed explanation and analysis of the selection, crossover and mutation operators for reproduction process using real representation with examples, features and processes can be seen in Reference [14]. The foregoing procedures of the GA formulations are shown in Fig. 8.4. The used parameters for

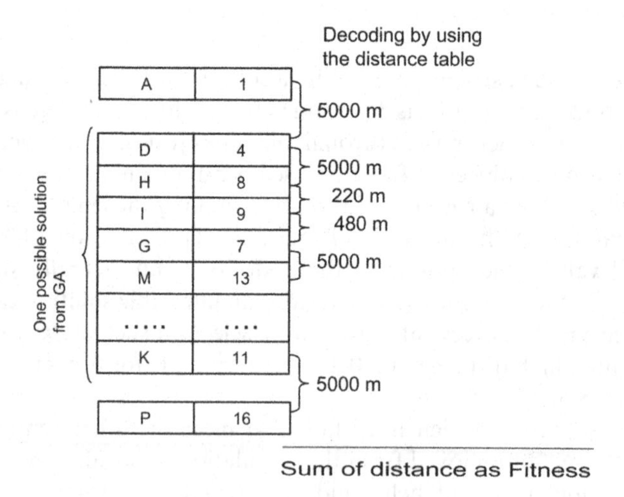


**Fig. 8.3** Example for fitness evaluation

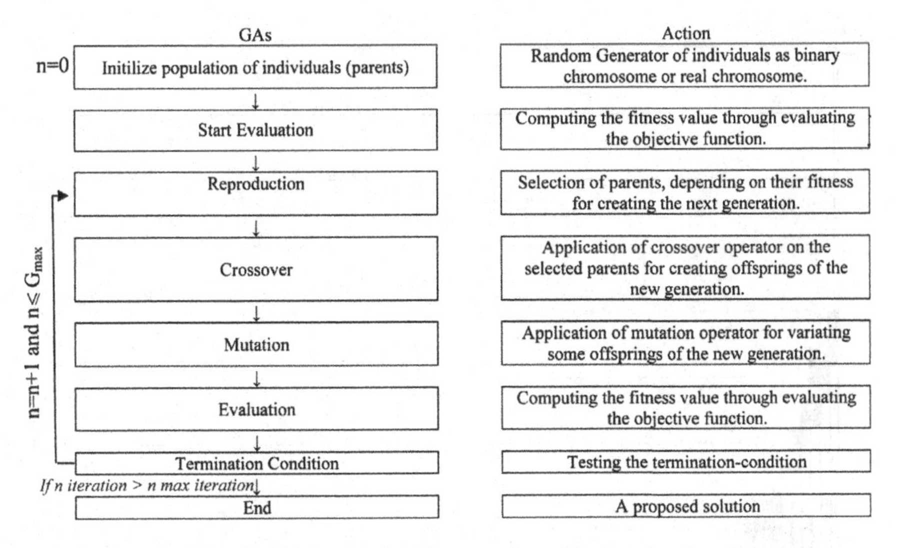


**Fig. 8.4** Flowchart for the GA implementation together with the related actions

implementing our GA technique in the real example are the population size of 40 with probabilities of 0.8 for crossover and 0.4 for mutation. The optimization was carried out over 80 generations.

Furthermore, it is worthy to mention here, that an important aspect of crossover operator in application to a multivariate problem in real-representation comes from the fact that the gene comprises a single allele and it is itself the parameter value; while in the binary or Gray coding the crossover occurs only at the boundaries of the gene which consists of alleles or bits, thus exposing the gene to split. Therefore, what distinguishes our GA work is that the crossover comes only in-between the genes, thus avoiding destroying any of them, so the coded information is not destroyed; contrary to what happens with most GAs based on binary coding and others.

## 8.5 Results

The present work, revealed by using linear and dynamic programming techniques, has shown that the mathematical modeling has limited success with the TSP, with limited number of nodes and with the need to a large number of constraints; and that for a case like ours it takes much longer. These results conform with the other reported results in the literature that mathematical programming approaches have had rather limited success with TSP.

By using GA, a computer program of three operators has been developed. The GA deals with a real-value chromosome comprising the 15 genes representing the nodes (route) network to be optimized. The fitness of the chromosome is computed through the objective or evaluation function, which determines the

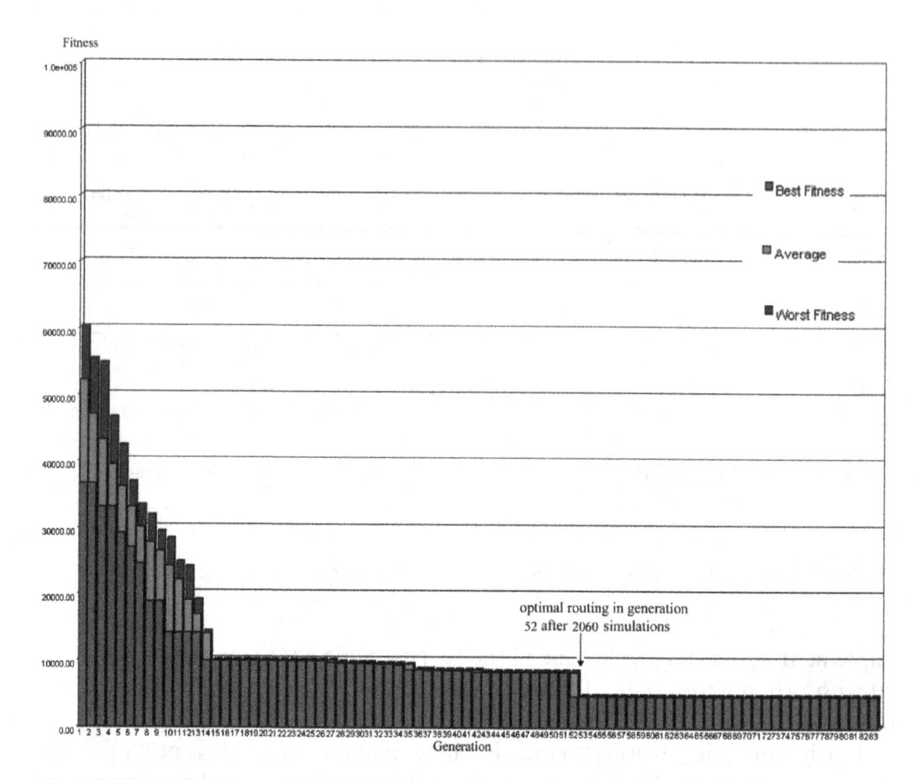


**Fig. 8.5** Worst, average and best of generation distances for solid waste routing in the studied network of TSP

cost (distance) solution by summing the lengths of the node distances making up the network.

The GA searches for the minimum length of the solid waste routing, thus the objective function is supposed to be minimized. The total search space is $6.2 \times 10^9$ possible network routes. The reached results are presented in Fig. 8.5, which shows a typical plot of the cost (distance) of the solution in each generation.

These results are presented as worst fitness, average fitness and best fitness to differentiate among them, and to show the optimality of the GA operators program. The GA found the best solution which led to the shortest – distance (lowest cost tour) of 6,585 m (6,595-10) with 2,060 simulations (evaluations) for collection vehicles routing in Irbid. Figure 8.6 shows the final route of this best solution. The computing time did not exceed 35 seconds CPU time on a PC (AMD 2.4 GHZ), while in the mathematical methods it takes much longer time and in the heuristic algorithms and their modification and Monte Carlo simulations it takes about 5 minutes.

The quality of the network routes solutions reached and the robustness of the method can be gauged by comparing with the other results from literature presented by the authors [15] who applied their own research work on the same

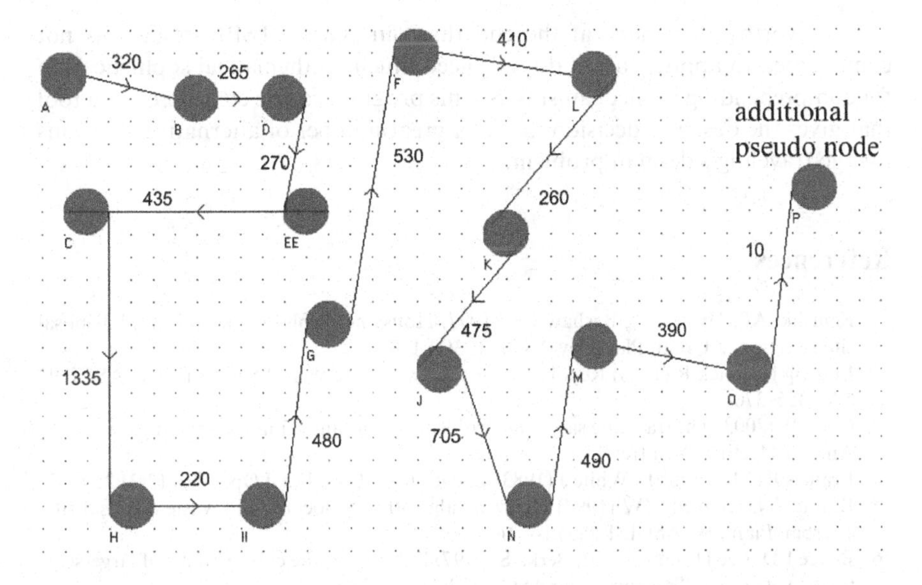


**Fig. 8.6** Shortest distance routing for solid waste collection in the studied network of TSP

example for modeling techniques of Monte Carlo Simulation and Heuristic Algorithms which resulted that the shortest tour was by Monte Carlo Simulation 6,715 m (one million random trials) and by modified heuristic algorithm was 7,945 m.

## 8.6 Conclusion

The results of our study, in comparison with the other applied optimization methods (linear, dynamic, Monte Carlo and heuristic search method), indicate that the real GA, through its specific behavior and through its efficient operators, presents significantly the lowest distance (cost tour) solution. Accordingly, it is concluded that the real GA approach is robust and it represents an efficient search method and is easily applied to nonlinear and complex problems of the TSP in the field of solid waste routing system in the large cities. Furthermore, it has certain advantages over other applied optimization techniques such as linear, dynamic, Monte Carlo simulation and heuristic search methods in both cost and computer time.

The addition of other real constraints such as pick-up points (nodes) or extension of the studied waste collecting network for bigger serviced cities and so on, as decision variables can also be incorporated into our real GA by an increase in the chromosome length by adding more genes for representing such additions in a route design for waste collecting problem in cities, i.e., more generations would be required to reach the optimum.

It is worthy to remark at the end that our (GA – TSP) program is not complicated to apply, since it doesn't need much mathematical sophistication for comprehending its mechanisms. So this program can be considered as a tool that gives the designer/decision maker a great number of alternative solutions for the traveling salesman problem.

## References

1. Koushki AP, Hulsey JL, Bashaw EK (1997) House hold solid waste traits and disposal site selection. J Urban Plan Dev ASCE 123(1):1–9
2. Ludwig H, Black R (1968) Report on the solid waste problem. J Sanit Eng Div ASCE 94 SA2:355–370
3. Cook W (2003) The traveling salesman problem and optimization on a grand scale. SIAM Annual Meeting, Montreal
4. Francis R, McGinnis L, White J (1983) Location analysis. Eur J Oper Res 12(3):220–252
5. Ramu VN, Kennedy JW (1994) Heuristic algorithm to locate solid waste disposal site. J Urban Plan Dev ASCE 120(1):14–21
6. Boyce ED, Lee H, Janson NB, Berka S (1997) Dynamic route choice model of large-scale traffic network. J Transport Eng ASCE 123(4):276–282
7. Rivett P, Ackoff RA (1963) Manager's guide to operational research.John Wiley and Sons Inc., New York
8. Goodman S, Hedetniemi S (1985) Introduction to design and analysis of algorithms. Mc Graw-Hill, New York
9. Coney W (1988) Traveling salesman problem solved with simulation techniques. Int J Syst Sci 19(10):2115–2122
10. Thieraut R, Klekamp R (1975) Decision making through operations research.John Wiley and Sons Inc., New York
11. Oliveira R, Loucks DP (1997) Operating rules for multireservoir systems. J Water Resour Res 33(4):839–852
12. Wardlaw R, Sharif M (1999) Evaluation of genetic algorithms for optimal reservoir system operation. J Water Resour Plan Manag ASCE 125(1):25–33
13. Awad A, Von Poser I (2005) Genetic algorithm optimization of water supply networks. Water Intelligence Online Journal (WIO), International Water Association (IWA) Publishing 8(4):1–25
14. Abu-Hassan R (1988) Routing of solid waste collection vehicles in Irbid, M.Sc. Thesis, Department of Civil Engineering, Jordan University of Science and Technology
15. Awad A, Aboul-Ela MT, Abu-Hassan R (2001) Development of a simplified procedure for routing solid waste collection. Int J Sci Technol (Scientia Iranica) 8(1):71–75

# Chapter 9
# The COTECOMO: COnstractive Test Effort COst MOdel

**Ljubomir Lazić and Nikos Mastorakis**

**Abstract** The primary purpose of Software Testing Process and Evaluation (STP&E) is to reduce risk. While there exists extensive literature on software cost estimation techniques, industry practice continues to rely upon standard regression-based algorithms. These software effort models are typically calibrated or tuned to local conditions using local data. This paper cautions that current approaches to model calibration often produce sub-optimal models because of the large variance problem which is inherent in cost data and by including far more effort multipliers than the data supports. Building optimal models requires that a wider range of models be considered while correctly calibrating these models requires rejection rules that prune variables and records and use multiple criteria for evaluating model performance. This article compares the approaches taken by three (COCOMO II, FP, UCP) widely used models for software cost and schedule estimation to develop COTECOMO (COnstractive Test Effort COst MOdel). It also documents what we call the large variance problem that is a leading cause of cost model brittleness or instability. This paper proposes Software/System Test Point (STP), a new metric for estimating overall software testing process. STP covers so-called black-box testing; an estimate for the test activities, which precede scenarios (threads) testing (white-box testing included), will already have been included in the estimate produced by function point analysis. Software test point is a useful metric for test managers interested in estimating software test effort, and the metric aids in the precise estimation of project effort and addresses the interests of metric group.

L. Lazić (✉)
School of Electrical Engineering, Vojvode Stepe 283, Belgrade, Serbia
e-mail: llazic@vets.edu.yu

N. Mastorakis et al. (eds.), *Proceedings of the European Computing Conference*, Lecture Notes in Electrical Engineering 28,
DOI 10.1007/978-0-387-85437-3_9, 

## 9.1 Introduction

Software testing has been considered so important that organizations can assign teams only for testing activities. Testing is an important activity to ensure software quality. Tests are usually run several times to certify that code maintenance did not accidentally insert defects into working parts of the software. In such situations, test teams shall be able to estimate the required effort to execute test cases in its schedules and to request more resources or negotiate deadlines when necessary. When regarding model-based testing approaches that we use in our documented Integrated and Optimized Software Testing methodology (IOSTP) [1–4], a high number of test cases can be automatically generated. As team resources are limited, it may be not practical to execute all generated test cases. Their complexity usually determines the effort required to execute them and it can be used for planning test resources and test suites. Several software development estimation models have been proposed over the years. However, these models do not estimate the effort for executing a given test suite, since their estimations are based on software development complexity instead of its test planning, test case design and test execution complexity. According to our reading of the literature (e.g. [5–14]), "best practices" in model-based effort estimation include:

- Local calibration (or LC); i.e. using local data to set two special tuning parameters;
- Stratification; i.e. given a database of past projects, and a current project to be estimated, restrict local calibration to just those records from similar projects.

These two approaches can easily be shown to fall short of what should be expected from a "best practice". If stratification improved performance, then subsets of the data should usually generate better models with lower error rates than models learned from all the data. Similarly, while LC sometimes produces the best models, often it does not. We compare standard LC with the multiple data mining algorithms in our COTECOMO (COnstractive Test Effort COst MOdel) which is an extension and adaption of T. Menzies' et al. [15] COSEEKMO effort estimation workbench applied to IOSTP [1]. COTECOMO is a general workbench containing multiple data mining tools, all of which are unleashed on a data set and a best method is selected on a dataset-by-dataset basis. While standard methods sometimes do well, data mining algorithms yielded models with overall smaller mean error and much smaller standard deviation. The most prominent results from this study are that comparing the performance of development efforts estimators to select a best estimator is problematic [5, 15]. The variance in the performance seen over these small data sets is very large, for example, the standard deviations are often much larger than the mean. Such large deviations complicate comparisons across different data sets. Hence, we must compare model performance using some heuristic rejection rules that compare more than just mean performance.

Note that this paper is about finding better practices for automatic model-based effort estimation. For a discussion on best practices in manual expert-based estimation, see [14]. There are few research articles which focus on the test effort estimation model [6, 9]. The proposed software test effort estimation metric, Software/System Test Point (STP), has been conceptualized based on the fact of COTECOMO effort estimation procedure and case study results that suggest that the accuracy of analysis increases when the system components are dealt with separately. Also, as testing plays a significant role in the software engineering community as a software reliability improvement activity and is being considered as a separate branch of software engineering, the need of estimating system test effort separately has been felt. Many formulae proposed for software size estimation [14] have pre-defined values. The values specified in the formulae vary from project to project and organization to organization. Hence, this paper proposes a framework which can be adopted by organizations to estimate software/system test effort. The proposed metric is a generalized metric for software/system testing which can be tailored to the operational or architectural processes of the organization or project. Menzies [15] argues that elaborate and expensive testing regimes will not yield much more than inexpensive manual or simple automatic testing schemes. Pol, Veenendaal [9] have developed and applied Test Points successfully to estimate test effort for black-box testing.

This paper proposes a framework for estimating the software/system test effort using a new metric called Software/System Test Points – STP. The organization deploying this metric initially has to conduct a survey to calculate the weights of the identified attributes, which will be used in estimating the software/system test points. Estimated STP should be compared against the actual STP to improve the estimation accuracy for the subsequent projects in adaptive manner. The metric proposed is suitable for the projects having a software/system testing team headed by a test manager or project manager. In scenarios, where the system testing team is an independent organization, it is assumed that the necessary information/data for estimating STP is provided to the test manager.

This paper starts with a brief introduction of the attributes identified for the proposed metric, which affect system test effort. The survey conducted for identifying the relationship between the 12 identified attributes on the system test effort has been explained. Simulation has been performed to calculate the correlation coefficients of the 12 identified attributes. The proposed methodology for estimating STP has been explained together with the COTECOMO methodology of evaluating STP by applying it for two projects. The two projects considered for estimating and evaluating STP have independent testing teams headed by a test manager. The two projects were operating at the same level of maturity.

This paper is organized as follows. Section 9.2 presents the Attributes Affecting Software/System Test Effort. Section 9.3 introduces the COTECOMO methodology context and new STP estimation method. The COTECOMO effort estimation procedure and case study are presented in Section 9.4. Finally, the paper is concluded in Section 9.5.

## 9.2 Attributes Affecting Software/System Test Effort

Test Engineering covers a large gamut of activities to ensure that the final product achieves some quality goal. These activities must be planned well in advance to ensure that these objectives are met. Plans are based on estimations. In the early years, the Waterfall model has been applied to software development. This model looks upon test engineering as merely a stage in the entire development lifecycle. When techniques evolved over the years for estimating development time and effort, the concept of estimating test-engineering time was overlooked completely. Test engineering is seldom planned for in most organizations and as a result, products enter the market insufficiently tested. Negative customer reactions and damage to the corporate image is the natural consequence. To avoid this, the correct development lifecycle must be chosen and planning should be done early on in the cycle as in our IOSTP [1–4]. Estimation is basically a four-step approach:

1. Estimate the size of the development product. This is either in LOC (Lines of Code) [10–14] or FP (Function Points) [8] or concept of using UCP (Use Case Points) [6, 7, 16] and our STP estimation model.
2. Estimate the effort in person-months or person-hours.
3. Estimate the schedule in calendar months.
4. Estimate the cost in currency.
5. Find best estimator from these test efforts estimation models by applying COTECOMO estimation rules.

### *9.2.1 Conventional Approach to Test Effort Estimation*

Test engineering managers use many different methods to estimate and schedule their test engineering efforts. Different organizations use different methods depending on the type of projects, the inherent risks in the project, the technologies involved etc. Most of the time, test effort estimations are clubbed with the development estimates and no separate figures are available. Here is a description of some conventional methods in use:

1. Ad-hoc method

   The test efforts are not based on any definitive timeframe. The efforts continue until some pre-decided timeline set by managerial or marketing personnel is reached. Alternatively, it is done until the budgeted finances run out.

   This is a practice prevalent in extremely immature organizations and has error margins of over 100% at times (see Fig. 9.1).
2. Percentage of development time

   The fundamental premise here is that test engineering efforts are dependent on the development time / effort. First, development effort is estimated

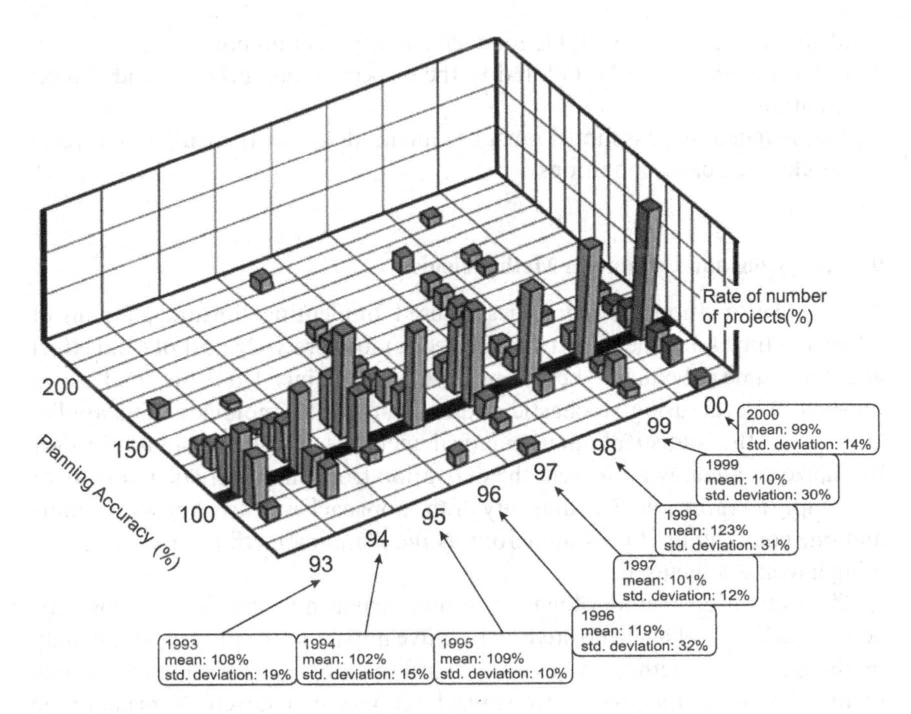


**Fig. 9.1** Trend of cost estimation accuracy [5]

using some techniques such as LOC or Function Points. The next step is using some heuristic to peg a value next to it. This varies widely and is usually based on previous experiences. This method is not defendable since it is not based on any scientific principles or techniques. Schedule overruns could range from 50 to 75% of estimated time. This method is also by far the most used.

3. From the Function Point estimates

   Capers Jones [5] estimates that the number of test cases can be determined by the function points estimate for the corresponding effort. The formula is Number of Test Cases = (Function Points) $^{1.2}$. The actual effort in person-hours is then calculated with a conversion factor obtained from previous project data. The disadvantage of using FP is that they require detailed requirements in advance. Another issue is that modern object-oriented systems are designed with Use Cases in mind and this technique is incompatible with them [6, 7, 16].

### *9.2.2 The Selection of the Sizing and Estimation Methodologies*

The selection of the Sizing and Estimation methodologies and the tailoring is a task of management knowing that:

- Ideal process models suitable for each environment do not exist
- Process models must be tailored to the respective organization and project situation
- The assigned process model must be enhanced and permanently adapted to the changed basic conditions

#### 9.2.2.1 Sizing and Estimation Methodologies

Sizing and estimation methodologies are used for identification and planning of effort and time schedule of software projects, more precisely, just of total effort and total time schedule. They were developed mainly from the analysis of finished projects, using stochastic and/or heuristic technologies. This applies at least to the statistical/ mathematical methods. Other methods however formalize an analogy approach, the derivation from characteristic numbers or use simply a regression. The diversity of the approaches nearly knows no limits and one could mine, it is as numerous as the planning methods and programming languages themselves.

The estimation methodologies determine environmental factors (cost drivers), which could have a positive or negative impact to the project. Depending on the estimation method there could be just one single factor e.g. productivity up to unlimited factors, which even must be defined by yourself. As result of the estimation you get quasi correction factors for the values of a software project. If these factors are applied to the base value from the sizing method, you get a good (real numbers) estimate of the total effort and the total time schedule.

Basic considerations to estimates and efforts influencing areas (see Fig. 9.2):

- Estimates are not absolute numbers and have always deviations
- The deviations can be calculated for each method, whereby each method possesses other deviations according to its model
- The deviation of an estimate can be very high depending upon implementation within or outside of its basic condition (specification, extensions, environmental factors/cost drivers etc.)

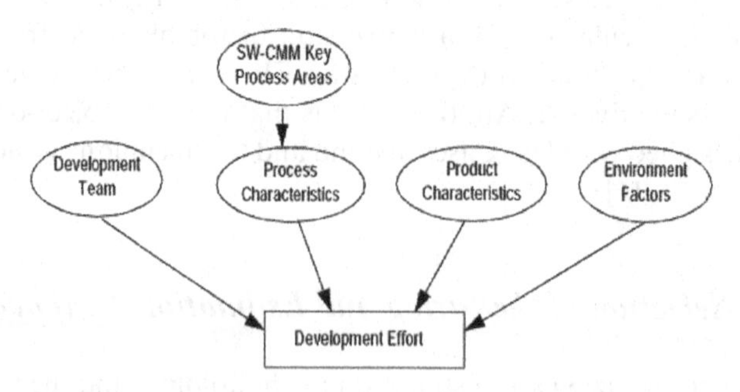


**Fig. 9.2** Efforts influencing areas

- The values for the deviation depend on the regarded method and the correct computation (average value simply relative error or squares of error etc.)
- A comparison require always resembles basic conditions

#### 9.2.2.2 Characteristics of the Cost Estimation

- The effort which have to be spend determines the costs of the project
- The determination of the environmental factors (cost drivers) is a major task of the cost analysis
- As basis of the computations and for comparison and/or determination of the environmental factors (cost drivers) different metrics/methods exists

#### 9.2.2.3 General Advantages of Estimation Methodologies:

- Substantial potential to:
  - Reduce the effort
  - Optimization of time schedule of software projects
  - Improvement of Quality even in field of frequently changed degree of characteristics and not standardized specifications
- Much more accurate than expert judgment (in relation of 25% : 75%)
- Deliver reproducible results
- Supply real numbers instead of random results
- Form the basis for a toolbox
- Support of numerous departments of an enterprise
- Show the real equivalent of a project (target costs)
- Create transparency
- Protect the trust ship between client and vendor by supplying a traceable and reproducible estimation
- The application of estimation methodologies is a critical success factor for software projects.

Many factors affect effort of testing, the accuracy of estimate increases when all the factors are considered.

Broadly, system test effort depends upon the process of testing, type of testing, and test tools used in the organization. The 12 attributes identified for the proposed metric that affect system test effort are as follows:

1. Observed requirement stability
2. Observed software stability
3. Test process tools used
4. Testing tools used
5. Type of testing (e.g., Black box, White Box, etc.) to be performed
6. Test artifacts required
7. System knowledge of the tester
8. Complexity of the system

9. Availability of reusable test cases and test artifacts
10. Environment (UNIX, Windows, Web based, Client-Server)
11. Application type (Real time systems, Database applications)
12. Required software reliability

The descriptions of the attributes identified are as follows:

### *9.2.3 The Software Test Effort Models – Critical Review*

Each of the different "schools of thought" has their own view of the prediction problem despite the interactions and subtle overlaps between process and product identified there. Furthermore each of these views model a *part*of the problem rather than the whole. Perhaps the most critical issue in any scientific endeavor is agreement on the constituent elements or variables of the problem under study. Models are developed to represent the salient features of the problem in a systemic fashion. This is as much the case in physical sciences as social sciences. Economists could not predict the behavior of an economy without an integrated, complex, macroeconomic model of all of the known, pertinent variables. Excluding key variables such as *savings rate* or *productivity* would make the whole exercise invalid. By taking the wider view we can construct a more accurate picture and explain supposedly puzzling and contradictory results. Our analysis of the studies shows how empirical results about test effort and defect density can make sense if we look for alternative explanations. Collecting data from case studies and subjecting it to isolated analysis is not enough because statistics on its own does not provide scientific explanations. We need compelling and sophisticated theories that have the power to explain the empirical observations. The isolated pursuit of these single issue perspectives on the quality prediction problem is, in the longer-term, fruitless. Part of the solution to many of the difficulties presented above is to develop prediction models that *unify* the key elements from the diverse software quality prediction models. We need models that predict software quality by taking into account information from the development process, problem complexity, defect detection process, and design complexity. We must understand the cause and effect relations between important variables in order to explain why certain design processes are more successful than others in terms of the products they produce. It seems that successful engineers already operate in a way that tacitly acknowledges these cause-effect relations. After all if they didn't how else could they control and deliver quality products? Project managers make decisions about software quality using best guesses; it seems to us that will always be the case and the best that researchers can do is:

1. Recognize this fact and
2. Improve the "guessing" process.

We, therefore, need to model the subjectivity and uncertainty that is pervasive in software development. Likewise, the challenge for researchers is in transforming this uncertain knowledge, which is already evident in elements of the various quality models already discussed, into a prediction model that other engineers can learn from and apply. Ultimately, this research is aiming to produce a method for the statistical process control (SPC) of software production implied by the SEI's Capability Maturity Model.

Effort estimation is a challenge in every software project. The estimates will impact costs and expectations on schedule, functionality and quality. While expert estimates are widely used, they are difficult to analyze and the estimation quality depends on the experience of experts from similar projects. Alternatively, ore formal estimation models can be used. Traditionally, software size estimated in the number of Source Lines of Code SLOC), Function Points (FP) and Object Points (OP) are used as input to these models, e.g. COCOMO and COCOMO II [12].

Because of difficulties in estimating SLOC, FP or OP, and because modern systems are often developed using the Unified Modeling Language (UML), UML-based software sizing approaches are proposed. Examples are effort estimation methods based on use cases [6, 16] and software size estimation in terms of FP from various UML diagrams [17, 16]. The *Use Case Points* (UCP) estimation method introduced in 1993 by Karner estimates effort in person-hours based on use cases that mainly specify functional requirements of a system [18]. Use cases are assumed to be developed from scratch, be sufficiently detailed and typically have less than 10–12 transactions. The method has earlier been used in several industrial software development projects (small projects compared to this study) and in student projects. There have been promising results and the method was more accurate than expert estimates in industrial trials [19, 20]. An early project estimate helps managers, developers, and testers plan for the resources a project requires. As the case studies indicate, the UCP method can produce an early estimate within 20% of the actual effort, and often, closer to the actual effort than experts and other estimation methodologies [21]. Moreover, in many traditional estimation methods, influential technical and environmental factors are not given enough consideration. The UCP method quantifies these subjective factors into equation variables that can be tweaked over time to produce more precise estimates. Finally, the UCP method is versatile and extensible to a variety of development and testing projects. It is easy to learn and quick to apply. The author encourages more projects to use the UCP method to help produce software on time and under budget.

#### 9.2.3.1 Industry Case Studies

From the time Karner produced his initial report in 1993, many case studies have been accomplished that validate the reasonableness of the UCP method. In the first case study in 2001, Suresh Nageswaran published the results of a UCP estimation effort for a product support Web site belonging to large North American software company [6]. Nageswaran, however, extended the UCP equation to include testing and project management coefficients to derive a more accurate estimate. While testing and project management might be considered non-functional requirements, nevertheless they can significantly increase the length of the project. Testing a Java 2 Enterprise Edition implementation, for example, may take longer than testing a Component Object Model component; it is not unusual to spend significant time coordinating, tracking, and reporting project status. *Nageswaran's extensions to the UCP equation produced an estimate of 367 man-days, a deviation of 6% of the actual effort of 390 man-days.*

In a recent research his author, Nageswaran said he had also applied the UCP method to performance testing, unit-level testing, and white box testing. In the second case study, research scientist Dr. Bente Anda [19] evaluated the UCP method in case studies from several companies and student projects from the Norwegian University of Science and Technology that varied across application domains, development tools, and team size. *For the above studies, the average UCP estimate is 19%; the average expert estimate is 20%*.Additionally, at the 2005 International Conference on Software Engineering, Anda, et al. [20] presented a paper that described the UCP estimate of an incremental, large-scale development project that was within 17% of the actual effort. In the third case study, Agilis Solutions and FPT Software partnered to produce an estimation method, based on the UCP method that produces very accurate estimates. In an article that was presented at the 2005 Object-Oriented, Programming, Systems, Languages, and Applications conference, Edward R. Carroll of Agilis Solutions stated:

After applying the process across hundreds of sizable (60 man-months average) software projects, we have demonstrated metrics that prove an estimating accuracy of less than 9% deviation from actual to estimated cost on 95% of our projects. Our process and this success factor are documented over a period of five years, and across more than 200 projects [21]. *To achieve greater accuracy, the Agilis Solutions/FPT Software estimation method includes a risk coefficient with the UCP equation.*

### 9.2.4 Adaptive Cyclic Pairwise Rejection Rules of Estimation Model Candidates Using performance measures (ACPREPM)

We consider COCOMO II, FP, UCP and STP estimation model candidates using standard performance measures as many authors do like Menzies [15] and Nagappan [22]. The rejection rules are the core of COTECOMO effort estimation workbench. In our own work, we revised the COSEEKMO effort estimation workbench for assessing different estimation models based upon COCOMO data sets using data mining techniques despite large deviations rules from [15, 22] until they satisfied certain sanity checks. Large deviations in effort estimation area require model performance comparison using some heuristic rejection rules that compare more than just mean performance data. The results for each treatment were compared using each treatment's of MMRE, SD, PRED(N) and correlation. MMRE comes from the magnitude of the relative error, or MRE, the absolute value of the relative error:

$$MRE = |predicted - actual| \tag{9.1}$$

The mean magnitude of the relative error, or MMRE, is the average percentage of the absolute values of the relative errors over an entire data set. Given T $\psi$ tests (estimators), the MMRE is:

$$MMRE = \frac{100}{T}\sum_{i}^{T}\frac{|predicted_i - actual_i|}{actual_i} \tag{9.2}$$

The standard deviation, or SD, is root mean square of predicted and actual effort deviation. Given a T tests (estimators), the SD is:

$$SD = \frac{100}{T}\sum_{i}^{T}(predicted_i - actual_i)^{1/2} \tag{9.3}$$

The PRED(N) reports the average percentage of estimates that were within N% of the actual effort candidates' estimated values. Given T test examples, then:

$$PRED(N) = \frac{100}{T}\sum_{i}^{T}\begin{cases} 1 \cdot if \cdot MRE_i \leq \frac{N}{100} \\ 0 \cdot otherwise \end{cases} \tag{9.4}$$

For example, a PRED(20) = 50% means that half the estimates are within 20% of the actual effort data.

Another performance measure of a model predicting numeric values is the correlation between predicted and actual values. Correlation ranges from + 1 to –1 and a correlation of + 1 means that there is a perfect positive linear relationship between variables.

We choose estimating model candidates, according to our reading of the literature (e.g. [5–22]), only those which satisfy starting criteria (SC) defined by:

$$\text{MMRE} \leq 10\% \text{ and SD} \leq 10\% \tag{9.5}$$

than we apply adaptive cyclic pairwise rejection of estimation model in our COTECOMO framework. This stratification of estimator candidates improve performance, then this estimator subsets of the estimators performance data [22, 15] should usually generate better models with lower error rates than models learned from all the collected estimators performance data.

#### 9.2.4.1 Adaptive Cyclic Pairwise Rejection of Estimation Model

The treatments were examined in pairs and if either seemed to perform worse, that one was rejected. This process repeated until no treatment could be shown to be worse than any other. The remaining treatments were called the survivors and were printed.

In addition to these sanity checks, the following model evaluation rules were used in this study:

- *Rule 1* is a statistical test condoned by standard statistical textbooks. If a two-tailed t-test reports that the mean of two treatments is statistically different, then we can check if one mean is less than the other.
- *Rule 2* (which checks for correlation) is a common check used at JPL: the best effort model tracks well between predicted and actual values. Without rule 2, many parts of our data produce multiple survivors.
- *Rule 3* is added in case there is more than one model with the same average performance and $R^2$. The model with the smallest MMRE is selected using equation (2).
- *Rule 4* was added because PRED(20) = 50% using equation (4) is our required performance measure for effort models.
- *Rule 5* This rule rejects treatments that have similar performance, but use more attributes.

Different rules are applied because they measure different aspects of model performance. Figure 9.3 shows adaptive cyclic pairwise rejection rules algorithm using standard statistical tests on MMRE, where an **error** is MMRE, **worse** function that apply **statistically Different** criteria comparing two MMREs ***x*** and ***y***, using, well known from statistics, a two-tailed t- test at the 95% confidence interval, i.e.

```
function worse(x,y)
  if    statisticallyDifferent(x,y)
  then
        if error(x) < error(y)                    then return y fi #rule1
        if error(y) < error(x)                    then return x fi #rule1
  else
        if correlation(x) < correlation(y)        then return x fi #rule2
        if correlation(y) < correlation(x)        then return y fi #rule2

        if sd(x)/mean(x)< sd(y)/mean(y)           then return y fi #rule3
        if sd(y)/mean(y)< sd(x)/mean(x)           then return x fi #rule3

        if pred(x)      < pred(y)                 then return x fi #rule4
        if pred(y)      < pred(x)                 then return y fi #rule4

        if |Subset(x)|  < |Subset(y)|             then return y fi #rule5
        if |Subset(y)|  < |Subset(x)|             then return x fi #rule5
  fi
  return 0 # if no reason to return true
```

**Fig. 9.3** Adaptive cyclic pairwise rejection rules algorithm

## 9.3 COTECOMO Methodology

We propose software test estimation "***The Integrated Methodology driven Estimation points out opportunities to optimize our business processes***" approach, over the our IOSTP [1–4], so called COnstructiveTestEffortCOstMOdel – COTECOMO methodology that provide a fully customized and tailored solution to perfectly match software testing needs. In some ways, End-to-End (E2E) Architecture Testing is essentially a "gray box" approach to testing – a combination of the strengths of white box and black box testing. In white box testing, a tester has access to, and knowledge of, the underlying system components. Although white box testing can provide very detailed and valuable results, it falls short in detecting many integration and system performance issues. In contrast, black box testing assumes little or no knowledge of the internal workings of the system, but instead focuses on the end-user experience – ensuring the user is getting the right results in a timely manner. Black box tests cannot typically pinpoint the cause of problems. Nor can they ensure that any particular piece of code has been executed, runs efficiently, and does not contain memory leaks or other similar problems. By merging white and black box testing techniques, End-To-End Architecture Testing eliminates the weaknesses inherent in each, while capitalizing on their respective advantages. This paper cautions that current approaches to model calibration often produce sub-optimal models because of the large variance problem inherent in cost data and by including far more effort multipliers than the data supports. Building optimal models requires that a wider range of models be considered while correctly calibrating these models requires rejection rules that prune variables and records and use multiple criteria for evaluating model performance. This article compares the approaches taken by three (COCOMO II, FP, UCP) widely used models for software cost and schedule estimation to develop COTECOMO (COnstractive Test Effort COst MOdel). It also documents what we call the large variance problem that is a leading cause of cost model brittleness or instability. We applied the E2E Test strategy in our Integrated and Optimized Software Testing framework (IOSTP). In determining the best source of data to support analyses, IOSTP with embedded RBOSTP considers credibility and cost of each test scenario i.e. concept. Resources for simulations and software test events are weighed against desired confidence levels and the limitations of both the resources and the analysis methods. The program manager works with the test engineers to use IOSTP with embedded RBOSTP [3] to develop a comprehensive evaluation strategy that uses data from the most cost-effective sources; this may be a combination of archived, simulation, and software test event data, each one contributing to addressing the issues for which it is best suited.

The central elements of IOSTP with embedded RBOSTP are: the acquisition of information that is credible; avoiding duplication throughout the life cycle; and the reuse of data, tools, and information. The system/software under test is described by objectives, parameters i.e. factors (business requirements – NR are indexed by $j$) in requirement specification matrix, where the major capabilities of subsystems being tested are documented and represent an independent i.e. input variable to optimization model. Information is sought under a number of test conditions or scenarios. Information may be gathered through feasible series of experiments (E): software test method, field test, through simulation, or through a combination, which represent test scenario indexed by$i$ i.e. sequence of test events. Objectives or parameters may vary in importance $\alpha_j$ or severity of defect impacts. Each M&S or test option may have $k$ models/tests called modes, at different level of credibility or probability to detect failure $\beta_{ijk}$ and provide a different level of computed test event information benefit $B_{ijkl}$ of experimental option for cell $(i,j)$, mode $k$, and indexed option $l$ for each feasible experiment depending on the nature of the method and structure of the test. Test event benefit $B_{ijkl}$ of feasible experiment can be simple ROI or design parameter solution or both etc. The cost $C_{ijkl}$, of each experimental option corresponding to $(i,j,k,l)$ combination must be estimated through standard cost analysis techniques and models. For every feasible experiment option, tester should estimate time duration $T_{jikl}$of experiment preparation end execution. The testers of each event, through historical experience and statistical calculations define the $E_{ijkl}$'s (binary variable 0 or 1) that identify options. The following objective function is structured to maximize benefits and investment in the most important test parameters and in the most credible options. The model maintains a budget, schedule and meets certain selection requirements and restrictions to provide feasible answers through maximization of benefit index – $B_{enefit}I_{ndex}$:

$$B_{enefit}I_{ndex} = \max_{i,j,k,l} \sum_{j} \sum_{i} \sum_{k} \sum_{l} \alpha_j \beta_{ijk} B_{ijkl} E_{ijkl} \tag{9.6}$$

Subject to:

$$\sum_{j} \sum_{i} \sum_{k} \sum_{l} C_{ijkl} E_{ijkl} \leq BUDGET \text{ (Budget constraint);}$$

$$\sum_{j} \sum_{i} \sum_{k} \sum_{l} T_{ijkl} E_{ijkl} \leq TIME\ SCHEDULE \text{ (Time-schedule constraint)}$$

$$\sum_{l} E_{ijkl} \leq 1 \text{ for all } i,j,k \text{ (at most one option selected per cell } i,j,k \text{ mode)}$$

$$\sum_{k} \sum_{l} E_{ijkl} \geq 1 \text{for all } i,j \text{ (at least one experiment option per cell } i,j\text{)}$$

### 9.3.1 *Test Effort Estimation Using System Test Points: A New Metric for Practitioners*

The ideal system test process considered for conceptualizing the software/ System Test Point (STP) metric which quantify effort for the various activities to be performed for software/system testing. The activities of system testers start from the requirements phase. It is a best practice to review the requirements document, which helps in addressing the non-testable requirements during the upstream software development. The preparation and review of test artifacts can be performed once the requirements are baselined. The preparation of test cases and its peer review can be performed during the design phase. Some of the test teams perform the review of design document as well. The test cases prepared will be executed during the integration and system testing phase. The phases, activities and artifacts, might vary from project to project and organization to organization. The process group or software manager needs to identify the relevant activities that will be performed in their project depending upon the project commitments. Many factors affect effort of testing, the accuracy of estimate increases when all the factors are considered. Our metrics group collects the data pertaining to projects such as system test effort, total effort of the project, size of the software and rating for 12 identified attributes, and calculates the correlation of the 12 identified attributes with respect to system test effort. The identified the 12 attributes (designated with TP) for the proposed metric that affect system test effort are as follows (also shown in Table 9.1):

**Table 9.1** STP applied for project#1 and project#2

| Test factor | Description | Weights | Rating Project #1 | Rating Project #2 | Project #1Wi*Ri | Project #2Wi*Ri |
|---|---|---|---|---|---|---|
| TP1 | Requirements stability | 6 | 8 | 2 | 49.32096 | 12.33024 |
| TP2 | Software stability | 10 | 6 | 1 | 61.72826 | 10.28804 |
| TP3 | Testing process tools | 12 | 3 | 8 | 35.60357 | 94.94285 |
| TP4 | Testing tools | 10 | 6 | 8 | 61.26588 | 81.68784 |
| TP5 | Type of testing | 7 | 8 | 4 | 55.79434 | 27.89717 |
| TP6 | Artifacts | 14 | 8 | 6 | 112.2052 | 84.15389 |
| TP7 | System knowledge of the tester | 9 | 4 | 4 | 35.60357 | 35.60357 |
| TP8 | Complexity | 8 | 8 | 9 | 63.19248 | 71.09154 |
| TP9 | Reusable test cases | 12 | 7 | 6 | 87.12085 | 74.67502 |
| TP10 | Environment | 10 | 4 | 2 | 38.06962 | 19.03481 |
| TP11 | Application | 12 | 4 | 4 | 48.85858 | 48.85858 |
| TP12 | Reliability | 13 | 4 | 4 | 52.8659 | 52.8659 |
| NR | | | | | 460 | 340 |
| ESTP | | | | | 21.74 | 14.05 |
| FP | | | | | 75 | 52 |
| Ratio | ESTP/FP= | | | | 0.289867 | 0.270192 |
| | | | | AverageRatio = | | 0.280 |

1. Observed requirement stability
2. Observed software stability
3. Test process tools used
4. Testing tools used
5. Type of testing (e.g., Black box, White Box, etc.) to be performed
6. Test artifacts required
7. System knowledge of the tester
8. Complexity of the system
9. Availability of reusable test cases and test artifacts
10. Environment (UNIX, Windows, Web based, Client-Server)
11. Application type (Real time systems, Database applications)
12. Required software reliability

In this paper, the top-down approach was followed for calculating the weights used in system test point estimation. A survey has been conducted for calculating the effect of 12 identified attributes on system testing effort. Eleven respondents were identified based upon their experience in software management, specifically in test management and testing. The experience of the respondents varied from 1 to 15 years. The mean and mode of the respondents experience was 5 years. Respondents have been provided with a questionnaire. The questionnaire used for the survey had the 12 attributes and brief description of the attributes to enhance understanding of the respondents. Weighting represents the influence of the attribute on system test effort. The weights have been calculated using the respondents rating. Data collected from respondents has been simulated using the RISK 4.0 software. Montecarlo simulation has been performed to calculate the correlation coefficients. Correlation coefficients of simulated data have been considered as weights of the attributes. The results represent that the identified attributes have a positive correlation on the system test effort with the magnitudes as shown in Table 9.1. Results show that the requirements stability and type of testing has a minimum effect when compared with the artifacts and reliability on system test effort. The other factors that have higher affect on system test effort include reusability of test cases, application type and usage of testing process tools. Weights in Table 9.1 might vary from organization to organization, due to the varied project settings. So, either the organization can have its own values or the values need to be calculated based on the data collected across the organizations.

Broadly, system test effort depends upon the process of testing, type of testing, and test tools used in the organization. The cognitive factors and skills are not considered for calculating STP. An STP estimation results for our case study of two projects #1 and 2 is shown in Table 9.1. The TP factors affecting the software/system testing have weights (designated with $\mathbf{W_i}$ in equation 9.7) that correspond to correlation coefficients and the manager has provided corresponding rating (designated with $\mathbf{R_i}$ in equation 9.7) as per the instructions for rating the attributes. The STP method has been applied to a Project #1 and Project #2 and it was observed that the values obtained from estimation were approximately the same as the actual results. Test managers provided ratings for Project #1 and Project #2 as shown in Table 9.1. Subsequently, the system test points and percentage of system test effort have been estimated. The total size of the project in function points is considered for calculating percentage of system test effort. Systems test points have been estimated using equation:

$$Estimated\ System\ Test\ Points\ (ESTP) = C_a dj\ {}^*NR^*(W_i^*\ R_i) \quad (9.7)$$

Where $\mathbf{C_{adj}} = 6.74 \cdot 10^{-5}$, is particular adjustment constant, which value is set in order to make **ESTP** and **FP** values to be comparable and expressed in percentage. Number of business requirements is denoted with **NR**. Percentage of System Test effort has been estimated using equation:

$$Percentage\ of\ Estimated\ System\ Test\ Effort\ (PESTE) = (ESTP/FP)^*\ 100 \quad (9.8)$$

The results of the estimation for Project#1 and Project#2 are also shown in Table 9.1. System test effort can be estimated based upon the percentage of system test effort and the project effort.

### 9.3.2 Software/System Test Point (STP) Evaluation

STP has been evaluated for two projects, Project #1 and Project #2. The actual system test points were calculated, based on the actual software/system test effort, but this data we will have when project ended. The actual system test effort is calculated using formulae:

*Actual Software/System Test Effort (in person-months) = Requirements review effort of test team +*
*Design review effort of test team + Test artifact development effort + Test artifact review effort + Manual testing +*
*Automated testing + Performance testing.*

The actual percentage of Software/System Test Effort is calculated using formula:

$$Percentage\ Actual\ Software/System\ Test\ Effort\ (PASTE) = (Actual\ system\ test\ effort\ /\ Total\ effort\ of\ the\ project)^*\ 100 \quad (9.9)$$

The actual system test points have been calculated using formula:

$$Actual\ Software/System\ Test\ Points\ (ASTP) = (Actual\ system\ test\ effort/Total\ effort\ of\ the\ project)^*\ Function\ points \quad (9.10)$$

The actual software/system test effort and software/system test points for Project #1 and Project #2 are listed in Table 9.2.

Table 9.2 Actual values of project #1 and project #2

| Actual test effort data | Project #1 | Project #2 |
|---|---|---|
| Actual software/system test efforts (person-months) | 7 | 24 |
| Percentage of Actual software/system test efforts (PASTE) | 22.58 | 32 |
| Actual software/system test points (ASTP) | 18.06 | 16.64 |

Table 9.3 Variation of actual and estimated values

| | PASTE | PESTE | $\delta$ | ASTP | ESTP | $\Delta$ |
|---|---|---|---|---|---|---|
| Project #1 | 22.58 | 27.17 | –4.58 | 18.06 | 21.738 | -3.67 |
| Project #2 | 32 | 27.02 | 4.98 | 16.64 | 14.05 | 2.59 |

The variation in the estimated and actual values was calculated. The percentage of effort spent for software/system testing is approximately the same as the estimated percentage of system test effort for the projects. The difference (PASTE-PESTE) between percentage actual system test effort and percentage estimated system test effort is represented by "$\delta$". The difference (ASTP-ESTP) between actual system test points and estimated system test points is represented by "$\Delta$". The "$\delta$" and "$\Delta$" values are calculated for projects #1 and #2, and the results are shown in Table 9.3.

The variation in the actual and estimated STP and percentage of test effort in the estimation accuracy i.e. "$\delta$" and "$\Delta$"is less than 5%, easy to learn and quick to apply. The managers may need to update the weights only when the actual system test points consistently vary from the estimated system test points. There, also, have been promising results and the method was more accurate than expert estimates in industrial trials [19, 20].

## 9.4 COTECOMO Effort Estimation Procedure and Case Study Results

Effort estimation is a challenge in every software project. The estimates will impact costs and expectations on schedule, functionality and quality. While expert estimates are widely used, they are difficult to analyze and the estimation quality depends on the experience of experts from similar projects. Alternatively, more formal estimation models can be used. Traditionally, software size estimated in the number of Source Lines of Code (SLOC), Function Points (FP) and Object Points (OP) are used as input to these models, e.g. COCOMO and COCOMO II. Because of difficulties in estimating SLOC, FP or OP, and because modern systems are often developed using the Unified Modeling Language (UML), UML-based software sizing approaches are proposed.

Examples are effort estimation methods based on Use Cases Points and software size estimation in terms of FP from various UML diagrams. All these estimation models are candidates in three steps of the COTECOMO effort estimation procedure, which we defined as follows:

**Step 1.** As a team (development and test), apply during planning phase:

1.1 The *Brooks rule* using current time expended on the project assuming we are half-way through the "design" phase i.e. 1/2 of total project effort in design phase ("My rule of thumb is 1/3 of the schedule for design, 1/6 for coding, 1/4 for component testing, and 1/4 for system testing.") and 1/2 for test phase.

or

1.2 Apply own company rule for planned activities during project realization, similar to nominal percentage given in Table 9.4 from [23].

**Table 9.4** Typical software development activities – six types of application [23]

| Activities Performed | Web | MIS | Outsource | Commercial | System | Military |
|---|---|---|---|---|---|---|
| 01 Requirements | 5.00% | 7.50% | 9.00% | 4.00% | 4.00% | 7.00% |
| 02 Prototyping | 10.00% | 2.00% | 2.50% | 1.00% | 2.00% | 2.00% |
| 03 Architecture | | 0.50% | 1.00% | 2.00% | 1.50% | 1.00% |
| 04 Project plans | | 1.00% | 1.50% | 1.00% | 2.00% | 1.00% |
| 05 Initial design | | 8.00% | 7.00% | 6.00% | 7.00% | 6.00% |
| 06 Detail design | | 7.00% | 8.00% | 5.00% | 6.00% | 7.00% |
| 07 Design reviews | | | 0.50% | 1.50% | 2.50% | 1.00% |
| 08 Coding | 30.00% | 20.00% | 16.00% | 23.00% | 20.00% | 16.00% |
| 09 Reuse acquisition | 5.00% | | 2.00% | 2.00% | 2.00% | 2.00% |
| 10 Package purchase | | 1.00% | 1.00% | | 1.00% | 1.00% |
| 11 Code inspections | | | | 1.50% | 1.50% | 1.00% |
| 12 Independent verification and validation | | | | | | 1.00% |
| 13 Configuration management | | 3.00% | 3.00% | 1.00% | 1.00% | 1.50% |
| 14 Formal integration | | 2.00% | 2.00% | 1.50% | 2.00% | 1.50% |
| 15 User documentation | 10.00% | 7.00% | 9.00% | 12.00% | 10.00% | 10.00% |
| 16 Unit testing | 30.00% | 4.00% | 3.50% | 2.50% | 5.00% | 3.00% |
| 17 Function testing | | 6.00% | 5.00% | 6.00% | 5.00% | 5.00% |
| 18 Integration testing | | 5.00% | 5.00% | 4.00% | 5.00% | 5.00% |
| 19 System testing | | 7.00% | 5.00% | 7.00% | 5.00% | 6.00% |
| 20 Field testing | | | | 6.00% | 1.50% | 3.00% |
| 21 Acceptance testing | | 5.00% | 3.00% | | 1.00% | 3.00% |
| 22 Independent testing | | | | | | 1.00% |
| 23 Quality assurance | | | 1.00% | 2.00% | 2.00% | 1.00% |
| 24 Installation/training | | 2.00% | 3.00% | | 1.00% | 1.00% |
| 25 Project management | 10.00% | 12.00% | 12.00% | 11.00% | 12.00% | 13.00% |
| Total | 100.00% | 100.00% | 100.00% | 100.00% | 100.00% | 100.00% |
| Activities | 7 | 18 | 21 | 20 | 23 | 25 |

**Step 2.** Apply Adaptive Cyclic Pairwise Rejection Rules of Estimation model candidates using performance measures (ACPREPM) to these estimation model candidates:

2.1 Compute SLOC by sizing components individually and Effort using COCOMO-II
   (a) As a developer individually compute SLOC and Effort
   (b) As a manager estimate SLOC and Effort
   (c) Find Low, High, and Average of (a) & (b) estimated effort

2.2 Apply OO Function Point analysis to estimate efforts and then calculate SLOC/FP ratio for the implementation language used:

   2.2.1 Compute Function Point estimations of SLOC as described in D. Garmus and D. Herron's book [8].
   2.2.2 Compute SLOC by module breakdown analysis as described in article [7] to:

      (a) Count Classes to Develop/Modify – Boundary, Entity, Control
      (b) Size each Class in (a) by counting methods to develop/modify
      (c) Size each method (using high, medium, low) and assign SLOC count to each size category.
      (d) Convert SLOC to hours, and hours to work weeks using your performance data.

   2.2.3 Find Low, High, and Average of 2.2.1 & 2.2.2 estimated effort

   - Apply the OO Function Point Analysis to Completed Code.

2.3 Apply UCP estimation method as described in articles [6, 16].
2.4 Apply STP estimation method as described in this article (Section 9.3).

**Step 3.** Compute average estimate of 2.1–2.4 data.
Finally, use the STP, FP and UCP methodology on your delivered implementation to arrive at actual STP, FP and UCP and then compute the SLOC/FP, ESTP/FP and FP/UCP ratio for the implementation language you used.

In our case study from Project#1 and Project#2, where Java implementation language is used, COTECOMO workbench estimated data are summarized in Table 9.5:

**Table 9.5** Estimated and actual values

| Estimated and Actual test effort data | Proj. #1 | Proj. #2 |
|---|---|---|
| ESTP | 21.74 | 14.05 |
| FP | 75 | 52 |
| UCP | 10 | 7 |
| Actual SLOC | 3750 | 2600 |

From Table 9.5 we computed:
1 FP = 50 Java SLOC
ESTP/FP = 0.28
FP/UCP = 7.514
These formulas are very useful to compare various estimation methods.

## 9.5 Conclusion and Future Work

An early project estimate helps managers, developers, and testers plan for the resources a project requires. As the case studies indicate, the our STP and UCP method can produce an early estimate within 20% of the actual effort, and often, closer to the actual effort than experts and other estimation methodologies [16, 23]. Building optimal models requires that a wider range of models be considered while correctly calibrating these models requires rejection rules that prune variables and records and use multiple criteria for evaluating model performance. This article compares the approaches taken by three (COCOMO II, FP, UCP) widely used models for software cost and schedule estimation to develop COTECOMO (COnstractive Test Effort COst MOdel). It also documents what we call the large variance problem that is a leading cause of cost model brittleness or instability.

In this paper we present a novel method for estimating OO software projects size exploiting UML diagrams and three steps COTECOMO procedure. The main features of our approach are:

1. the method can be totally automated;
2. it exploits all the principal UML diagrams in its estimation process;
3. its output is in terms of SLOC, UCP so can be compared with well known standard like FP.

In this very moment, we are setting an experiment to fix the formulas' constants which we introduced in the method.

## References

1. Lazić Lj, Velašević D, Mastorakis N (2003) A framework of integrated and optimized software testing process. In: WSEAS Conference, August 11–13, Crete, Greece, also in WSEAS TRANSACTIONS on COMPUTERS 2(1)
2. Lazić Lj, Velašević D (2004) Applying simulation and design of experiments to the embedded software testing process. Software Testing, Verification and Reliability, John Willey & Sons Ltd., 14(4):257–282
3. Lazić Lj, Mastorakis N (2005) RBOSTP: Risk-based optimization of software testing process-Part 2. WSEAS Transactions on Information Science and Applications, ISSN 1790-0832, 2(7):902–916

4. Lazić Lj, Mastorakis N (2007) A framework of software testing metrics – Part 1 and 2. In: WSEAS Engineering Education 2007 Multiconference, Agios Nikolaos, Crete Island, Greece, pp 23–28
5. Jones C (1998) Estimating software costs. McGraw-Hill
6. Nageswaran Suresh (2001) Test effort estimation using use case points. In: Quality Week 2001, San Francisco, Caifornia, USA, <www.Cognizant.com/ cogcommunity/presentations/Test_ Effort_Estimation.pdf>
7. Carbone M, Santucci G (2002) Fast&&Serious: a UML based metric for effort estimation. In: Proceedings of the 6th ECOOP Workshop on Quantitative Approaches in Object-Oriented Software Engineering (QAOOSE'02), Spain
8. Garmus D, Herron D (2001) Function point analysis. Addison-Wesley, ISBN 0-201-69944-3
9. Veenendaal EPWM van, Dekkers T (1999) Test point analysis: a method for test estimation. In: Kusters R, Cowderoy A, Heemstra F, Veenendaal E van (eds) Project control for software quality. Shaker Publishing BV, Maastricht, The Netherlands
10. Boehm B (1981) Software engineering economics. Prentice Hall
11. Boehm B (2000) Safe and simple software cost analysis. IEEE Software, September/ October 2000, pp 14–17, Available from http://www.computer.org /certification/beta/ Boehm Safe.pdf
12. Boehm B, Horowitz E, Madachy R, Reifer D, Clark BK, Steece B, Brown AW, Chulani S, and Abts C (2000) Software cost estimation with cocomo II. Prentice Hall
13. Hall M, Holmes G (2003) Benchmarking attribute selection techniques for discrete class data mining. IEEE Trans Knowl Data Eng 15(6):1437–1447
14. Jorgensen M (2004) A review of studies on expert estimation of software development effort. J Syst Soft 70(1–2):37–60
15. Menzies Tim, Chen Zhihao, Hihn Jairus, Lum Karen (2006) Selecting best practices for effort estimation. IEEE Trans Soft Eng 32(11)
16. Clemmons Roy K (2006) Project estimation with use case points. CrossTalk Feb .
17. Uemura T, Kusumoto S, Inoue K (1999) Function point measurement tool for UML design specification. In: Proceedings of the 6th Int'l IEEE Software Metrics Symposium, IEEE-CS Press, pp 62–69
18. Karner Gustav (1993) Resource estimation for objectory projects. Objective Systems SF AB
19. Anda Bente (2003) Improving estimation practices by applying use case models. www.cognizant.com/cogcommunity/presentations/Test_Effort_Esti-mation.pdf
20. Anda Bente et al. (2005) Effort estimation of use cases for incremental large-scale software development. In: 27th International Conference on Software Engineering, St Louis, MO, pp 303–311
21. Carroll Edward R (2005) Estimating software based on use case points. In: 2005 Object-Oriented, Programming, Systems, Languages, and Applications (OOPSLA) Conference, San Diego, CA
22. Nagappan N, Williams L, Vouk M, Osborne J (2005) Early estimation of software quality using in-process testing metrics: a controlled case study. In: Third Software Quality Workshop, co-located with the International Conference on Software Engineering (ICSE 2005), pp 46–52
23. Carper Jones (2004) Software project management practices: failure versus success. CrossTalk Oct.

# Chapter 10
# A Semantics of Core Computational Model for ODP Specifications

**Youssef Balouki and Mohamed Bouhdadi**

**Abstract** The Reference Model for Open Distributed Processing (RM-ODP) defines a framework within which support of distribution, interoperability and portability can be integrated. However other ODP standards have to be defined. We treat in this paper the need for formal notation for concepts in the computational language. Indeed, the ODP viewpoint languages are abstract in the sense that they define what concepts should be supported not how these concepts should be represented. Using the denotational semantics in the context of UML we define in this paper syntax and semantics for core ODP structural concepts defined in the RM-ODP foundations part and in the computational language namely object action, computational object, and computational interface. These specification concepts are suitable for describing and constraining ODP computational viewpoint specifications.

## 10.1 Introduction

The rapid growth of distributed processing has led to a need for coordinating framework for the standardization of Open Distributed Processing (ODP). The Reference Model for Open Distributed Processing (RM-ODP) [1–4] provides a framework within which support of distribution, networking and portability can be integrated. It consists of four parts. The foundations part [2] contains the definition of the concepts and analytical framework for normalized description of (arbitrary) distributed processing systems. These concepts are grouped in several categories. The architecture part [3] defines a framework comprising five viewpoints, five viewpoint languages, ODP functions and ODP transparencies.

However, RM-ODP [3] only provides a framework for the definition of new ODP standards. These standards include standards for ODP functions [6–7]; standards for modelling and specifying ODP systems; standards for testing

Y. Balouki (⊠)
Mohammed V University, Department of Mathematics and Computer Science,
B.P. 1014 Av. Ibn Bettouta, Rabat Morocco

N. Mastorakis et al. (eds.), *Proceedings of the European Computing Conference*, Lecture Notes in Electrical Engineering 28,
DOI 10.1007/978-0-387-85437-3_10, © Springer Science+Business Media, LLC 2009

ODP systems. Also RM-ODP recommends defining ODP types for ODP systems [8].

In this paper we treat the need for formal notation of ODP viewpoint languages. We use the UML/OCL denotational meta-modelling, semantics to define semantics for structural specification concepts in ODP computational language. The part of RM-ODP considered consists of modelling and specifying concepts defined in the RM-ODP foundations part and concepts in the computational language.

The paper is organized as follows. Section 10.2 describes related works. Section 10.3 we introduce the subset of concepts considered in this paper namely the object model and main structural concepts in the enterprise language. Section 10.4 defines the meta-model for the language of the model: object template, interface template, action template, type, and role. The meta-model syntax for considered concepts consists of class diagrams and OCL constraints. Section 10.5 describes the UML/OCL meta-model for instances of models Section 10.6 makes the connection between models and their instances using OCL. A conclusion end perspectives end the paper.

## 10.2 Related Work

The languages Z [9], SDL [10], LOTOS [11] and, Esterel [12] are used in RM-ODP architectural semantics part [4] for the specification of ODP concepts. However, no formal method is likely to be suitable for specifying every aspect of an ODP system. In fact, these methods have been developed for hardware design and protocol engineering. The inherent characteristics of ODP systems imply the need to integrate different specification languages and different verification methods.

Elsewhere, this challenge in formal methods world is the same in software methods. Indeed, there had been an amount of research for applying the Unified Modelling Languages UML [13] as a notation for the definition of the syntax of UML itself [14–16]. This is defined in terms of three views: the abstract syntax, well-formedness rules, and modeling elements semantics. The abstract syntax is expressed using a subset of UML static modelling notations that is class diagrams. The well-formedness rules are expressed in Object Constrains Language OCL [17]. OCL is used for expressing constraints on object structure which cannot be expressed by class diagrams alone. A part of UML meta-model itself has a precise semantics [18, 19] defined using denotational meta-modelling semantics approach. A denotational approach [20] is realized by a definition of the form of an instance of every language element and a set of rules which determine which instances are and are not denoted by a particular language element. The three main steps of the approach are: (1) define the meta-model for the language for models (syntax domain), (2) define the meta-model for the language of instances (semantics domain), and (3) define the mapping between these two languages.

Furthermore, for testing ODP systems [2–3], the current testing techniques [21, 22] are not widely accepted and especially for the enterprise viewpoint specifications. A new approach for testing, namely agile programming [23, 24] or test first approach [25] is being increasingly adopted. The principle is the integration of the system model and the testing model using UML meta-modelling approach [26, 27]. This approach is based on the executable UML [28]. In this context OCL is used to specify the invariants [19] and the properties to be tested [24]. The OCL invariants are defined using the UML denotational semantics and OCL itself.

In this context we used the meta-modelling syntax and semantics approach in the context of ODP languages. We defined syntax of a sub-language for the ODP QoS-aware enterprise viewpoint specifications [29]. We also defined a UML meta-model semantics for structural concepts in ODP enterprise [30].

## 10.3 The RM-ODP

We overview in this section the core structural concepts for ODP enterprise language. These concepts are sufficient to demonstrate the general principle of denotational semantics in the context of ODP viewpoint languages.

### *10.3.1 The RM-ODP Foundations Part*

The RM-ODP object model [3] corresponds closely to the use of the term data-model in the relational data model. To avoid misunderstandings, the RM-ODP defines each of the concepts commonly encountered in object oriented models. It underlines a basic object model which is unified in the sense that it has successfully to serve each of the five ODP viewpoints. It defines the basic concepts concerned with existence and activity: the expression of what exists, where it is and what it does. The core concepts defined in the object model are object and action.

An object is the unit of encapsulation: a model of an entity. It is characterized by its behavior and, dually, by its states.

Encapsulation means that changes in an object state can occur only as a result of internal actions or interactions.

An action is a concept for modeling something which happens. ODP actions may have duration and may overlap in time. All actions are associated with at least one object: internal actions are associated with a single object; interactions are actions associated with several objects.

Objects have an identity, which means that each object is distinct from any other object. Object identity implies that there exists a reliable way to refer to objects in a model. When the emphasis is placed on behavior, an object is informally said to perform functions and offer services; these functions are specified in terms of interfaces. It interacts with its environment at its interaction points which are its interfaces. An interface is a subset of the interactions in which an object can participate. An ODP object can have multiple interfaces. Like objects, interfaces can be instantiated.

The other concepts defined in the object model are derived from the concepts of object and action; those are class, template, type, subtype/supertype, subclass/superclass, composition, and behavioral compatibility.

Composition of objects is a combination of two or more objects yielding a new object.

An object is behaviorally compatible with a second object with respect to a set of criteria if the first object can replace the second object without the environment being able to notice the difference in the object behavior on the basis of the set of criteria.

A type (of an $<x>) is a predicate characterizing a collection of <x>s. Objects and interfaces can be typed with any predicate. The ODP notion of type is much more general than of most object models. Also ODP allows ODP to have several types, and to dynamically change types.

A class (of an <x>) defines the set of all <x>s satisfying a type. An object class, in the ODP meaning, represents the collection of objects that satisfy a given type. ODP makes the distinction template and class explicit. The class concept corresponds to the OMG extension concept; the extension of a type is the set of values that satisfy the type at a particular time. A subclass is a subset of a class. A subtype is therefore a predicate that defines a subclass. ODP subtype and subclass hierarchies are thus completely isomorphic.

A <x> template is the specification of the common features of a collection x in a sufficient detail that an x can be instantiated using it.

Types, classes, templates are needed for object, interface, and action.

## 10.3.2 The RM-ODP Computational Language

The definition of a language for each viewpoint describes the concepts and rules for specifying ODP systems from the corresponding viewpoint. The object concepts defined in each viewpoint language are specializations of those defined in the foundations part of RM-ODP.

A computational specification defines the functional decomposition of an ODP system into objects which interact at interfaces. The basic concepts of the computational language are the computational interface, the computational object, the interaction and the binding object. The binding object is a computational object which supports a binding between a set of other computational objects.

An interaction is a signal, an operation or a flow.

A signal is an atomic shared action resulting in one-way communication from an initiating object to a responding object.

An operation is an interaction between a client object and a server object.

A flow is an abstraction of a sequence of interactions resulting in a conveyance of information from a producer object to a consumer object.

Like interaction kinds, an interface is signal, operation or flow.

A signal interface is an interface in which all the interactions are signals.

In an operation interface all the interactions are operations.

All the interactions are flows in a flow interface.

A computational object template comprises a set of computational interface templates which the object can instantiate, a behavior specification and an environment contract specification. The behavior of the object and the environment contract are specified in terms of a set of properties (attributes).

A computational interface template is associated to each kind of interface. It comprises a signal, a stream, or an operation interface signature as appropriate, a behavior specification and an environment contract specification. Like for the object template, the behavior and the environment contract are defined as a set of properties.

## 10.4 The Syntax Domain

We define in this section the meta-models for the concepts presented in the previous section. Figure 10.1 defines the context free syntax for the core object concepts and Fig. 10.2 defines the context free syntax for concepts in the computational language.

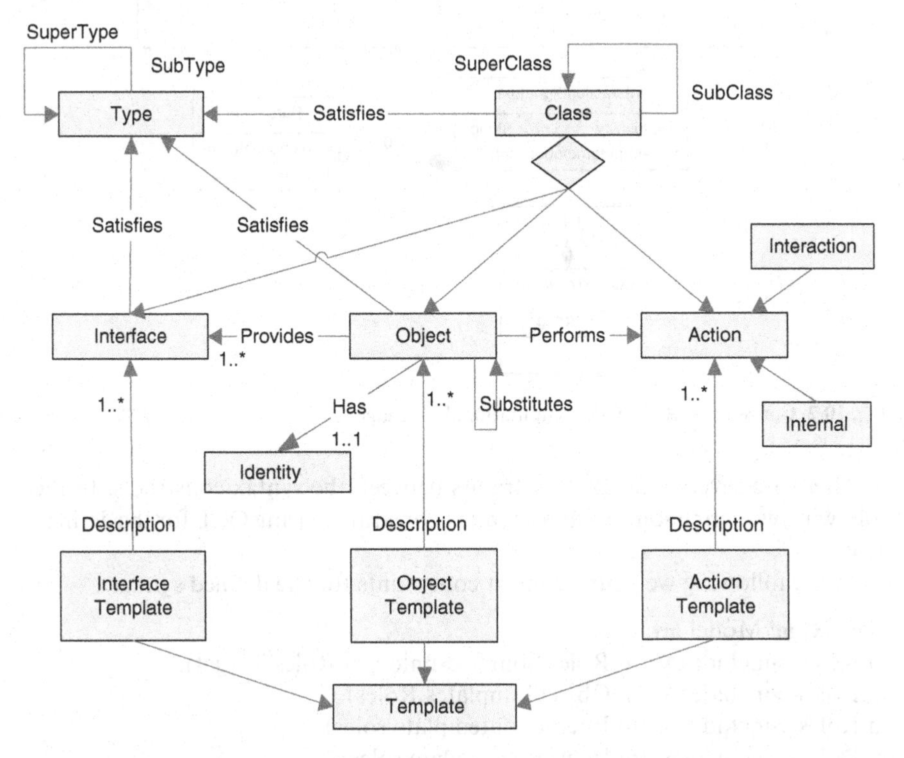


Fig. 10.1 RM-ODP foundation object model

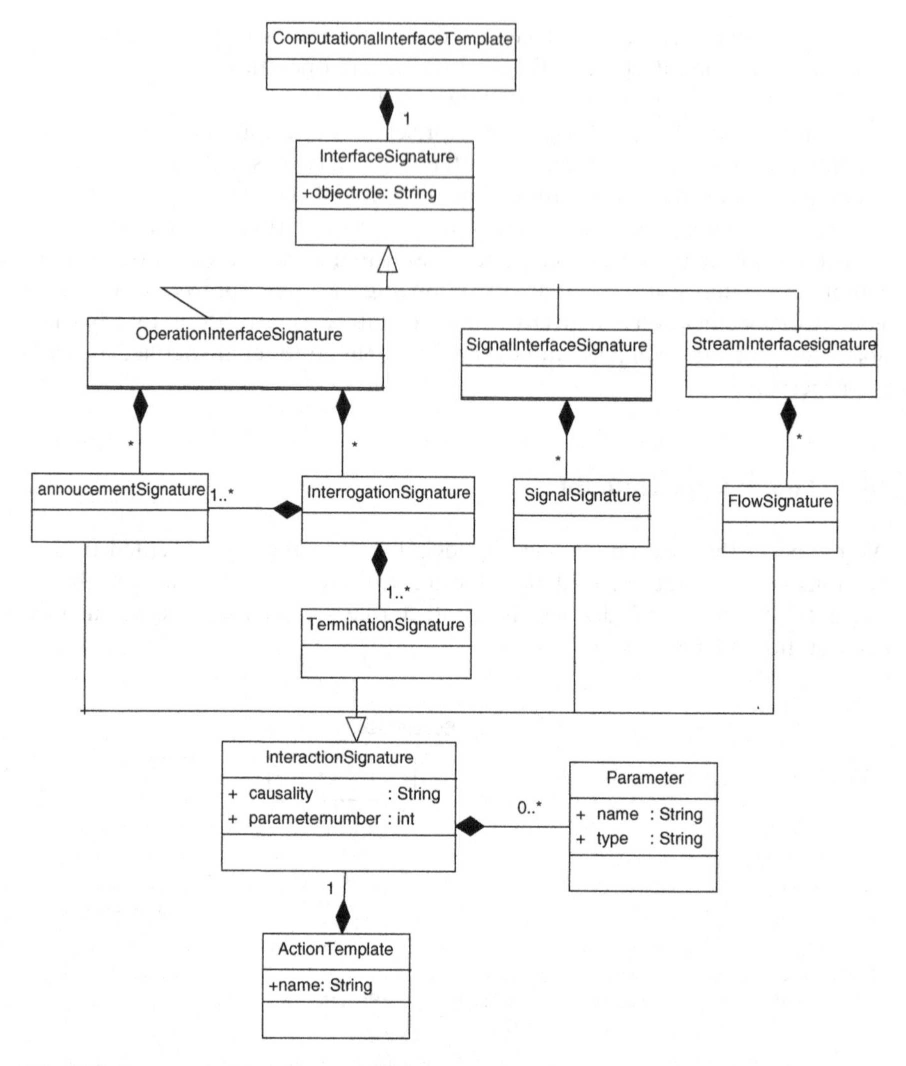


**Fig. 10.2** Concepts in RM-ODP computational language

There are several context constraints between the syntax constructs. In the following we define some of these context constraints using OCL for the defined syntax.

In the following we define context constraints for the defined syntax.

**Context m**: Model **inv**:
m.Roles->includesAll(m.Roles.Source->union(m.Roles.Target)
m.Roles->includesAll(m.ObjectTemplates.Roles)
m.Roles->includesAll(m.Interactiontemplate.roles)
m.Roles->includesAll(m.InterfaceTemplate.roles)

m.InteractionTemplates->includesAll(m.ObjectTemplates.Interactiontemplates)
m.InteractionTemplates.Types->includesAll(m.Types)
m.ObjectTemplates.InterfaceTemplates->includesAll(m.InterfaceTemplates)
m.ObjectTemplates.InterfaceTemplates->includesAll(m.InterfaceTemplates)
includesAll(m.ObjectTemplates.Interactiontemplates)

## 10.5 Semantics Domain

The semantics of a UML model is given by constraining the relationship between a model and possible instances of that model. That is, constraining the relationship between expressions of the UML abstract syntax for models and expressions of the UML abstract syntax for instances. The latter constitute the semantics domain, which is integrated with the syntax domain within the same meta-model (Figs. 10.1 and 10.2).This defines the UML context free abstract syntax. We give in the following a context constraint between the instances of semantics domain. These constrains are relatively simple but demonstrate the general principle.

We consider the concepts of subtype/supertype (RM-ODP 2-9.9) and subclass/superclass (RM-ODP 2-9.10) as relations between types and classes correspondingly.

**Context m** : model **inv**
m.types-> forall( t1: Type, t2: Type | t2.subtype -> includes(t1) implies t1.valid_for.satisfies_type = t2)
m.types-> forall( t1: Type, t2: Type | t1.supertype ->includes(t2) implies t1.valid_for.satisfies_type = t2)

## 10.6 Meaning Function

The semantics for the UML-based language defined focuses on the relationship between a system model and its possible instances (systems). Both of the domains are defined using UML/OCL. The association between the instances of the two domains (syntax and semantics) is defined using the same UML meta-models. The OCL constraints complete the meaning function the context of UML [20, 31]. We give in the following some constraints which are relatively simple, but they demonstrate the general principle. We use the semantics of invariants as defined in [19].

Clause 7.1.9 of part 3 “computational object template: An object template which comprises a set of Computational interface template which the object can instantiate, a behaviour specification and an environment contact specification.”

**Context m** : Model **inv** :
m.ComputationalObjectTemplate->
includesall(m.objectTemplate.Conputational.InterfaceTemplate)->
union (m.behaviorTemplate)

Clause 7.1.10 of part 3 "computational interface template: An interface template for either a signal interface, a stream interface or operation interface. A computational interface template comprises a signal, a stream or an operation signature as appropriate, a behaviour specification and an environment contract specification."

**Context m** : Model **inv** :

m.ComputationalInterfaceTemplate

->includesall(m.objectTemplate.Signal.InterfaceTemplate)

-> union (m.objectTemplate.StreamInterfaceTemplate)

-> union (m.objectTemplate.OperationInterfaceTemplate)

-> union (m.behaviorTemplate)

Each interface has a signature:

**Context** cit: ComputationelInterfaceTemplate **inv** :
cit.OperatioInterfaceTemplate -> inclaudesAll(cit.AnnoucementSignature)
->union(cit.IntegrationSignature)
and cit.SignalInterfaceTemplate -> inclaudesAll(cit.SignalSignature)
and cit.StreamInterfaceTemplate -> inclaudesAll(cit.StreamSignature)

Clause 7.1.11 of part 3 "signal interface signature: An interface signature for signal interface . A signal interface signature comprises a finite set of action template, one for each signal type in interface."

Each action template comprises the name for signal, the number, names and types of its parameters and indication of causality (initiating or responding, but not both) with respect to the objects which instantiates the template.

**Context**: Signal InterfaceTemplate **inv** :
self.signaltype -> forall(t | self.siganture->includes(set (t.actiontemplate)) ) and
Forall( at : actiontemplate | at.names ->includesall(self.signal.name)->union
(self.parmeter.name) and
at.types ->inclaudesall(self.parmeter.type)) and
(self.causality = #Intiating xor self.causality = #Responding)

Clause 7.1.12 of part 3 "operation interface signature: An interface signature for operation interface. A operation interface signature Comprises a set of announcement and interrogation signatures as appropriate, one for each operation type in the interface, together with an indication of causality (client or server, but not both) for the interface as whole, with respect to the object which instantiates the template. Each announcement signature is an action template containing the name of the invocation and the number, names and types of its parameters."

**Context** OperationInterfaceTemplate **inv** :

Sel.oclIsTypeOf (Signal InterfaceTemplate )

self.signature -> includes (set(annoucementsignature)) -> union

set (integrationsignature)
self.annoucementsignature -> includes(set (actiontemplate)) and
Forall( at : actiontemplate | at.names ->includesall(self.invocation.name)
->union(self.parmeter.name) and
at.types ->inclaudesall(self.parmeter.type))

Each interrogation signature comprises an action template with the following elements: (1) the name of the invocation; (2) the number, names and types of its parameters; and (3) a finite, non-empty set of action templates, one for each possible termination type of invocation each containing both name of the termination and the number, names and types of its parameters.

**Context** InterrogationInterface **inv**:
self.signature -> includes (set(actiontemplate)) and Forall( at : actiontemplate | at.names ->includesall(invocation.name) ->union(parmeter.name) and at.types ->inclaudesall(parmeter.type)) and forall (I:invoction | exists( at :actiontemplate |at.names ->includesall(I.termination.name)->union(termination.parmeter.name) and at.types ->inclaudesall(termination.parmeter.type))

**Context** OperationTntemplate **inv** :
Self.interogationTemplate -> includes(set(self.InvocationTemplate)->union (self.TerminationTemplate) and Forall( I:invocationTemplate |exists(T : terminationtemplate | self.I.possible_termination = T)

Clause 7.1.13 of part 3 "stream interface signature: An interface signature for stream interface. A stream interface comprises a set of action templates, one for each flow type in the stream interface. Each action template for a flow contains the name for flow, the information type of the flow, and indication of causality for flow (i.e. producer or consumer but not both) with respect to the object which instantiates the template."

**Context** Stream InterfaceTemplate **inv** :
Self.oclIsTypeOf (Signal InterfaceTemplate )
Self.streamtype -> forall(t | includes(Self.signature -> set (t.actiontemplate)) ) and
Forall( at : actiontemplate | at.names ->includesall(self.flow.name) and
at.types ->inclaudesall(self.inforamtion.type)) and (self.causality =#Producer xor self.causality =#Consumer)

## 10.7 Conclusion

We address in this paper the need of formal ODP viewpoint languages. Indeed, the ODP viewpoint languages define what concepts should be supported – not how these concepts should be represented. One approach to defining the formal semantics of a language is denotational: elaborating the value or instance denoted by an expression of the language in a particular context. Using the denotational meta-modeling semantics, we define in this paper the UML/OCL based syntax and semantics for a core Open Distributed Processing concepts defined in the foundations part and in the computational viewpoint language. These concepts are suitable for describing and constraining ODP computational viewpoint specifications. We are applying the same denotational semantics to define semantics for concepts characterizing dynamic behavior in the enterprise and computational languages.

## References

1. ISO/IEC (1994) Basic reference model of open distributed processing – Part 1: Overview and guide to use. ISO/IEC CD 10746-1
2. ISO/IEC (1994) RM-ODP-Part 2: Descriptive model. ISO/IEC DIS 10746-2
3. ISO/IEC (1994) RM-ODP-Part 3: Prescriptive model. ISO/IEC DIS 10746-3
4. ISO/IEC (1994) RM-ODP-Part 4: Architectural semantics. ISO/IEC DIS 10746-4
5. OMG (1991) The object management architecture. OMG
6. ISO/IEC (1999) ODP type repository function. ISO/IEC JTC1/SC7 N2057
7. ISO/IEC (1995) The ODP trading function. ISO/IEC JTC1/SC21
8. Bouhdadi M et al. (2000) An informational object model for ODP applications. Malay J Comput Sci 13(2):21–32
9. Spivey J.M (1992) The Z reference manual. Prentice Hall
10. IUT (1992) SDL: Specification and description language. IUT-T-Rec. Z.100
11. ISO/IUT (1998) LOTOS: A formal description technique based on the temporal ordering of observational behavior. ISO/IEC 8807
12. Bowman H et al. (1995) FDTs for ODP. Comput Stand Interf J Elsevier 17(5–6):457–479
13. Rumbaugh J et al. (1999). The unified modeling language. Addison Wesley
14. Rumpe B (1998) A note on semantics with an emphasis on UML. In: Second ECOOP Workshop on Precise Behavioral Semantics, Springer LNCS 1543:167–188.
15. Evans A et al. (1998) Making UML precise. Object Oriented Programming, Systems languages and Applications, (OOPSLA'98), Vancouver, Canada, ACM Press
16. Evans A et al. (1999) The UML as a formal modeling notation. LNCS 1618, Springer, pp 349–364
17. Warmer J, Kleppe A (1998) The object constraint language: precise modeling with UML. Addison Wesley
18. Kent S et al. (1999) A meta-model semantics for structural constraints in UML. In: Kilov H, Rumpe B, Simmonds I (eds) Behavioral specifications for businesses and systems, Kluwer, Chapter 9
19. Evans E et al. (1999) Meta-modeling semantics of UML. In: Kilov H, Rumpe B, Simmonds I (eds) Behavioral specifications for businesses and systems, Kluwer, Chapter 4.
20. Schmidt DA (1986) Denotational semantics: A methodology for language development. Allyn and Bacon, Massachusetts
21. Myers G (1979) The art of software testing. John Wiley & Sons
22. Binder R (1999) Testing object oriented systems. Models patterns and tools. Addison-Wesley
23. Cockburn A (2002) Agile software development. Addison-Wesley
24. Rumpe B (2004) Agile Modeling with UML. Springer LNCS 2941:297–309
25. Beck K (2001) Column on test-first approach. IEEE Software 18(5):87–89
26. Briand L (2001) A UML-based approach to system testing. Springer LNCS 2185:194–208

27. Rumpe B (2003) Model-based testing of object-oriented systems. Springer LNCS 2852:380–402
28. Rumpe B (2002) Executable modeling UML. A vision or a nightmare? In: Issues and Trends of Information technology management in Contemporary Associations, Seattle, Idea Group, London, pp 697–701
29. Bouhdadi M et al. (2002) An UML-based meta-language for the QoS-aware enterprise specification of open distributed systems. Collaborative Business Ecosystems and Virtual Enterprises, FIP Series, Springer 85:255–264
30. Bouhdadi M, Balouki Y, Chabbar E (2007) Meta-modelling syntax and semantics for structural concepts for open networked enterprises. LNCS 4707:45–54
31. France R, Kent S, Evans A, France R (1999) What does the term semantics mean in the context of UML. In: Moreira Ana MD, Demeyer S (eds) ECOOP Workshops 1999, ECOOP'99 Workshop Reader, Springer, LNCS 1743:34–36

# Chapter 11
# Computer Simulation Via Direct Modeling

**Alex Soiguine**

**Abstract** The article introduces a design of a novel computer simulation environment that is based on an original approach, a method of direct computer modeling. This "triple helix"-like method is comprised of basic physical laws of element interactions, data type class hierarchy of an object-oriented computer language, and an executing computer program that can potentially run on a network of parallel processors. The approach is universally applicable for modeling a wide spectrum of physical, and other, processes and phenomena. The method is directed towards consistent elimination of solving equations as a main object of computer numerical algorithms in the simulation problem. New computer simulation for complex adaptive systems of interactive parts that change in response to local inputs will be used in tight conjunction with the C++ object-oriented computer language.

## 11.1 Introduction

Computer modeling and simulation are the key elements in achieving progress in engineering and science. New simulation components require complex algorithms and must function efficiently on an evolving range of architectures designed for large-scale parallel computations.

The current work is about bringing something positive and practical into that troubled area. The proposed method is directed at consistent elimination of simulation methods that have evolved over the last 300 years from sequential pencil-and-paper procedures as a main object of computer numerical algorithms in physical processes simulation problems. The objective is to demonstrate the feasibility of the Direct Computer Modeling paradigm for developing modern computational, numerical techniques and sophisticated computer modeling methods with the simulation capability for a wide spectrum of physical and other processes. The Direct Computer Modeling (DiCOMO) paradigm has, in its

A. Soiguine (✉)
SOiGUINE LLC, 1333 N. Camino Alto, #227, Vallejo, CA 94589, USA

N. Mastorakis et al. (eds.), *Proceedings of the European Computing Conference*, Lecture Notes in Electrical Engineering 28,
DOI 10.1007/978-0-387-85437-3_11, 

very nature, potential intrinsic parallelism of the numerical algorithms, implying easy and direct application to high performance computing. The DiCOMO method is about new behavior-oriented computations supported by contemporary software and hardware resources. It is an elegantly simple mathematical approach to capturing physics, making computer algorithms adequate for evaluating real world phenomena. The DiCOMO approach to creating computer programs for simulating physical phenomena can be successfully applied to an entire spectrum of multi-physics systems. This work demonstrates the ability of the DiCOMO to explicitly define the necessary class type variables corresponding to truly involved physical entities and ways to establish their communication, taking protocols as class variables pointer structures, and embedding the overall physical processes into a complex adaptive system of interactive parts and elements, which behave and change their states in response to local inputs defined by their neighbors' interactions.

A. Soiguine formulated the original idea in "The Method of Direct Computer Modeling" [1]. It was successfully applied to some computer simulation problems in traditional single processor computational environment and conventional programming languages [2]. The DiCOMO approach to simulation of physical processes is an effective alternative to numerical methods based on centuries-old ideas. A physical process model in the DiCOMO context is "triple helix"-like, comprising of basic physical laws of element interactions, data type class hierarchy of an object-oriented computer language, and hardware, executing computer program that can run on a network of parallel processors.

## 11.2 Background

There is an acute need for developing an effective computer simulation, taking advantage of the parallel nature of the proposed numerical schemes for easy and direct application to high performance computing. To realize that the DiCOMO can successfully and very effectively work in creating necessary software environment, we are considering its potential implementations in some practically important examples.

### *11.2.1 The DiCOMO General Scheme as Applied to Continuous Media*

The PDEs remain the basic models of real physical phenomena in continuous media, while computers play the role of powerful resources that numerically solve equations.

Figure 11.1 shows the logic of the traditional approach. PDEs are derived through a discretezation procedure of real physical media, with a subsequent mathematical limit process, accompanied generally by many non-physical

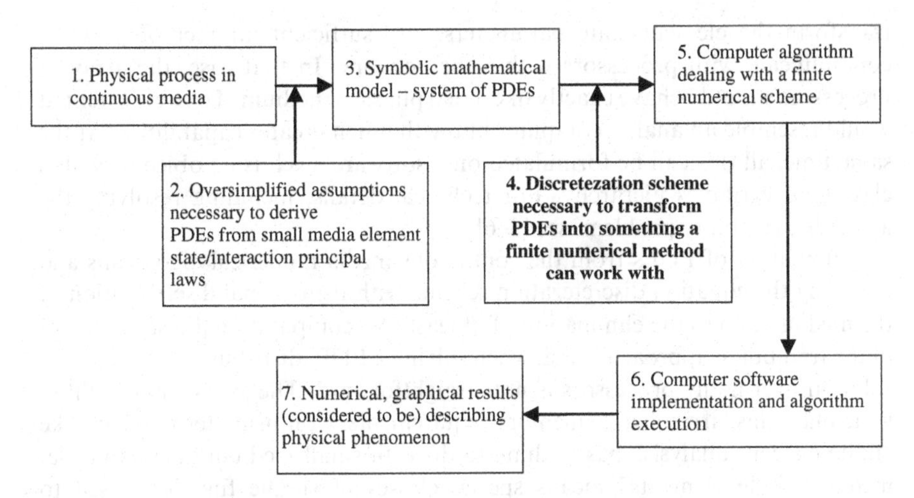


**Fig. 11.1** The logic of the traditional approach

assumptions. Any PDE is just an expression that symbolically encodes an unnecessarily oversimplified process that is meant to portray a real physical process. Would it not be more logical to run a discrete algorithm of interactions/state evolutions of small media elements? In other words, it is more physically consistent and computationally effective to exclude blocks 3 and 4 from the traditional scheme and remove the statement "oversimplified assumptions necessary to derive PDE from" from block 2 (Fig. 11.2).

An option exists to use a large array of microprocessors, corresponding to elements in a physical medium [1]. Each microprocessor would save all numerical parameters defining the state of an element, all functions necessary to

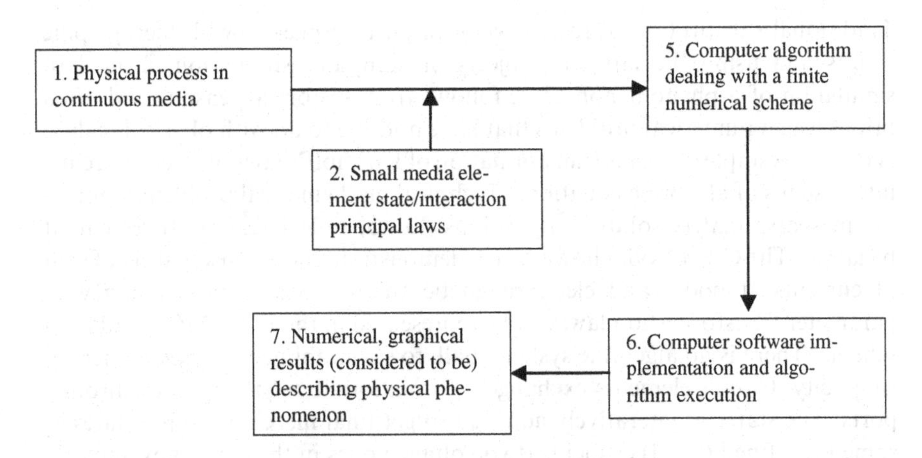


**Fig. 11.2** The logic of the modified traditional approach

transform the element state parameters, and sufficient number of ports to communicate with processors within the network. In that case, the array of processors should behave exactly like a real physical medium. To some extent, it would resemble an analog computer but with much greater capabilities. At the same time, all this can be formulated on a software level using object-oriented class type variables hierarchy. For technical details, including resolving the algorithms stability problem, see [3–6].

Elimination of PDEs from the continuous media numerical algorithms and replacing the equation discretezation scheme with the original discretezation of the medium allows the elimination of at least two computational disadvantages of the traditional approach. First, when solving a PDE, distribution of equations solution over many processors is not straightforward. The sparse matrix fill-in issue may cause irresolvable memory requirements. Confusing terminology like "finite element analysis" has nothing to do with small medium interacting elements. "Finite elements" means special classes of simple functions used to approximate unknown solution of a PDE. In that method, data inevitably must be shared among processors, causing the second problem: data traffic, resulting from data sharing, fully diminishes positive effects of parallelization.

One other approach should be mentioned in connection with the current proposal: the cellular automata approach [7]. Although some expectations have existed about the applicability of that method to continuous media simulation with multiprocessor parallelization [1], it did not realize to be a proper tool. DiCOMO is much more general and promising because it does not suffer from the cellular automata principal assumption: a "cell" should be a device with a finite number of states.

### *11.2.2 IC Simulation Algorithms*

Traditional substitution of real flows of physical processes with hieroglyphic scripts and using the latter as objects of computer simulation, instead of simulation of a phenomenon as is, follows from historical reasons and often brings tough numerical problems that have nothing to do with physical reality. A typical example of that is transformation of Kirhhoff's laws of electric circuits into systems of algebraic equations. Technical mathematical problems appearing in sparse matrix solution algorithms do not reflect real electrical circuit behavior. The C++ code shown below demonstrates how conservation of sum of currents at nodes and electromagnetic energy conservation, along with parameter transformation law, can be represented in the DiCOMO simulation scheme. There is no algebraic system at all, to say nothing of terrific numerical simplicity. Circuit elements exchange information about their states through ports. The states are iteratively adjusted to get final matching to the states at some predefined (fixed) elements (two voltage nodes in the current example).

The base class for an arbitrary electronic device is formally represented by a C++ class type variable:

```
class el_dev
{
public:
string el_deva_type; // like "RES" for resistors, "CAP"
                     // for capacitors, "TRA " for
                     // transistors, etc. (not as
                     // critical as for reading SPICE
                     // deck)
string el_deva_name; // mainly, el_deva_type plus index
                     // (number)
long instN;          // device number (index)
long nmbPorts;       // number of ports
    state * el_deva_par;
                     // name and value for such items as
                     // resistor resistance, capacitor
                     // capacitances, temperatures, etc.
    state * el_deva_stapar;
                     // name and value for physical
                     // entities defining current device
                     // state, say, drain current,
                     // voltage at gate (to some extend,
                     // duplicated by port_par values)
    port * el_deva_port;
                     // array of the device ports
    port * empty_port;
    el_dev(void);
    ~el_dev(void);
    bool equal(port p1, port p2);
                     // checks if ports are equal (just
                     // by name and owner's name
    port* get_port (char* name);
                     // returns pointer to a port by its
                     // name
int connect2(el_dev* neighbor, string mine, string his);
           //connects current port (name "mine")
           // to a "neighbor"'s device port of name
               "his"
  double get_port_curr(string mine);
                     // returns electric current at the
                     // current device's port
                     // named "mine"
```

```
double get_port_volt(string mine);
                    // returns voltage at that port
double get_2port_curr(string mine);
                    // returns the sum of all electric
                    // currents at port "mine"
                    // from all ports it is connected
                       to
virtual state* get_new_state () = 0;
}
```

**The *state* and *port* structures are defined as:**

```
typedef struct astate
                    // structure to keep all information
                    // about numerical parameter
{
string sta_name;    // name of parameter
double sta_value;   // its numerical value
bool set;           // has been set
bool fixed;         // must be kept as given
}state;
typedef struct aport
{
 string port_name;   // may be like "RIGHT" or "LEFT"
                        for
                    // resistor or capacitor, or
                       "SOURCE",
                    // "GATE", "DRAIN"
                    // for transistor, etc.
 string el_deva_own; // name of the device port belongs to,
                    // that's "el_deva_name" comprising
                    // device type and index (number)
 state* port_par;   // what physical entitis may be read
                    // and saved at the port: mainly,
                       at the
                   // port and exiting (entering)
                      current
aport* port_nei;   // array of ports the current port
                      talks to
long  nmbNei;     // number of other ports it is
                     connected to
bool connected;   // is connected?
} port;
```

Defined class type variables can fully describe arbitrary network device states, and interconnect topologies and communication protocols.

The class, representing a resistor, inherits el_dev and acquires extra functions. With the necessary implementations of the member functions the whole simulation program for the resistor chain connecting two given voltage sources works through the code containing just several lines and only executing simple arithmetical operations:

```
for (int i = 0;i <NUM_RES;i++)
{
        if  (resres[i]->  get_port("LEFT")->port_par[0].
            fixed)
               l=resres[i]-> get_port_volt("LEFT")
        if  (resres[i]->get_port  ("RIGHT")->port_par[0].
            fixed
               r=resres[i]->get_port_volt("RIGHT");
}
        double act;
        act = r-l;
do
{
        resid = 0.0;
        double re = 0.0;
        for (int i=0;i  <NUM_RES;i++)
        {
               re = resres[i]->get_new_state();
               if (re!=BAD_DOUBLE)
               resid+=re;
        }
               resid -=l;
               resid =act/resid;
               for (int i=0;i  <NUM_RES;i++)
               {
                     resres[i]-> scale_curr_val(resid);
               }
} while (fabs(1-resid)>1.0e-20);
```

Only two passes between given (defined) voltage nodes are necessary. Although the example is quite simple, it contains all features that should be used for an arbitrary device network. The important ideas of the DiCOMO implementation in the case of electrical circuits, and any boundary problem, are to define paths between predefined elements, propagate states between predefined sets of elements by state transformation laws and adjust elements state parameters at defined nodes (boolean parameter *fixed* in the *state* structure). In the case of ICs, the Kirhhoff's laws support stability. In the case of continuous media we need to use time step adjustments by applying energy conservation conditions. We see in this example that a system (network) evolves by exchanging information about elements states through their ports, rapidly achieving a stable overall network state.

All that remains to do in the case of a network of arbitrary electronic devices is specific implementation of functions `get_new_state()` for each type of a device. When an integration scheme is necessary inside that function, for example in the cases of active devices or transistors or transistors containing internal devices like nanoobjects, a specific relation between overall and device specific click timers will be used. That idea of factorizing a sequence of clicking timers is also very specific and powerful in the DiCOMO simulation scheme.

Comparing the speed of the above code with publicly available SPICE3 code in the case of 100 serial resistors in the loop of 1000 cycles showed that DiCOMO runs about 30 times faster.

The extension of the C++ language followed from the DiCOMO, when applied to integrated circuit simulation, does not have the restrictions of FPGA (or FPOA) languages, which resulted from their applicability to only specific and simple device types. On the opposite side of the spectrum of networks of arbitrary devices, the C++ language extension is significantly more flexible than SystemC®, which attempts to work in similar situations [8].

### *11.2.3 The Molecular Binding Simulation*

An intriguing application for the DiCOMO approach is the "molecular Lego" – simulation of combining molecules into complex structures. It is worth noticing that parallelization in DiCOMO is quite different from the parallel algorithms appearing in the PDE solutions [9].

The idea is to consider, with the help of DiCOMO approach, a simulation of capturing metal atoms by carbon nanotube structure. The resulting molecular composition may be used for hydrogen storage. The main requirement set by DOE is that the storage capacity must be above 6wt% of hydrogen. Preliminary first-principle calculations, based on quantum mechanical density functional theory [10], demonstrated that a high-transition metal, such as Ti, absorbed on a single wall carbon nanotube, can dissociate hydrogen molecules and strongly bond four $H_2$ molecules, creating a reversible system with up to 8wt% of hydrogen.

The principal difference between the DiCOMO approach for that kind of processes and previously considered examples is that connections between interacting elements change with time. A molecule, as a 3D element with directed vectors of interaction forces, will have in the C++ class type representation the `aport* port_nei` array from the structure port, varying in time, depending of the molecular orientation in space. Technically, it will be a more difficult programming task, although nothing unexpected for the approach applicability.

### *11.2.4 DiCOMO Approach Compatibility with the Sun's Fortress Object-Oriented Language for Parallelization*

As stated in "The Method of Direct Computer Modeling" [1], the ideal situation for the DiCOMO implementation would be a multiprocessor system with the number of CPUs equal to the number of interacting elements, representing the modeled physical media. Since multiprocessor systems are becoming more and more available, an urgent demand for an adequate programming language arises.

Sun Microsystems's new language, Fortress, is a part of an effort to come up with a design for the supercomputer of the future through the program on High Productivity Computing Systems. The language has been under development for more than three years. The most important thing in Fortress is that it is parallel. Most popular languages, such as Fortran and C++, have no direct concurrency, which must be grafted on top of them. With Fortress, program structure will reflect the science. The language machine assumptions include features critically important to the DiCOMO paradigm: not worrying about the number of processors, vast amount of parallelism (millions of simultaneous threads), fine-grained threading (any loop is parallel by default), shared global address space, transactional access to shared variables. Fortress will be offered to the community at large via an open source format. DiCOMO will profit immensely from the existence of Fortress.

## 11.3 Conclusions

The DiCOMO object-oriented based approach allows the development of algorithms that, being parallel in nature, are independent of a platform multiprocessor/ network designs and node-CPU architecture. The abstractions that represent physics of the simulation remain untouched. One can think of the approach as an inverse of C. Mead's "neuromorphic" network modeling approach [11]. Physical interactions between elements and element states evolution are reproduced on the level of object-oriented class type variable states/interconnections/ communication and communication protocol implementations.

The overall program is to design a computer simulation environment that would be universally applicable for modeling a wide spectrum of physical, and other, processes and phenomena based on an original approach, a method of direct computer modeling, tightly connected to object-oriented computer languages, physically natural model representations and potentially straightforward implementations in parallel processor environments. The specific objectives of the program are as follows:

1. Create templates of basic C++ class type variables representing general physical media entities following the DiCOMO philosophy.

2. Create derived class hierarchy representing different types of physical systems: general continuous media, electrical circuits, transistors bearing nanoscale elements (carbon nanotubes) and molecular dynamics with special attention to molecular binding problems.
3. Demonstrate feasibility of the approach to simulation of networks comprising traditional devices (passive, MOSFETs, etc.).
4. Demonstrate superiority of the approach in speed/number of operations compared with traditional matrix solution based simulation tools, such as SPICE.
5. Demonstrate feasibility of the approach to simulation of networks of novel types of transistors comprising nanoscale elements.
6. Check feasibility of the method in applying to the "molecular Lego" problems.
7. Formulate basic problems of the standardization of C++ language extension necessary for effective applicability of the DiCOMO paradigm.

## References


1. Alexandrov VV, Soiguine AM (1989) The method of direct computer modeling. Preprint #102, Institute of Informatics of the USSR Academy of Science (in Russian), http://www.soiguine.com/dicomo.pdf
2. Backus J (1978) Can programming be liberated from the von neumann style? A functional style and its algebra of programs. Comm ACM 21:613–641
3. Soiguine A (2004) The direct computer modeling approach to continuous media dynamics: C+ + implementation. In: Proceedings of the 1st International Conference "From Scientific Computing to Computational Engineering", Athens, http://www.soiguine.com/DirectComputerModelingApproach.pdf
4. Soiguine A (2005) The direct computer modeling paradigm applied to string simulation: An example of intellectual medium. In: 1st International Conference on Experiments/Process/System Modeling/Simulation/Optimization. Athens, Greece, http://www.soiguine.com/DirectComputerModelingApplied ToString.pdf
5. Soiguine A (2005) Continuous media simulation in C+ + class type variables environment. In: 3rd WSEAS International Conference on Heat Transfer, Thermal Engineering & Environment. Corfu, Greece, http://www.soiguine. com/Continuous_Media_Simulation.pdf
6. Soiguine A (2007) Direct computer modeling – New paradigm for continuous media simulation. Accepted for publication in Journal of Fluid Dynamics (Riadh, KSA), http://www.fluiddynamics.edu.sa
7. Wolfram S (2002) A new kind of science. Wolfram Media Inc.
8. IEEE Standard SystemC® Language Reference Manual (2006)
9. Plimpton S, Hendrickson B (1995)Parallel molecular dynamics algorithms for simulation of molecular systems. In: T. G. Mattson (ed) Parallel Computing in Computational Chemistry, pp 114–136
10. Le Bris C, Lions P-L (2005) From atoms to crystals: A mathematical journey. Bull Am Math Soc 42:291–363
11. Mead C (1989) Analog VLSI and neural systems. Addison-Wesley

# Chapter 12
# Image Segmentation with Improved Artificial Fish Swarm Algorithm

**Mingyan Jiang, Nikos E. Mastorakis, Dongfeng Yuan, and Miguel Angel Lagunas**

**Abstract** Some improved adaptive methods about step length are proposed in the Artificial Fish Swarm Algorithm (AFSA), which is a new heuristic intelligent optimization algorithm. The experimental results show that proposed methods have better performances such as good and fast global convergence, strong robustness, insensitivity to initial values, and simplicity of implementation. We apply the method in the image processing for the multi-threshold image segmentation compared with Genetic Algorithm (GA) and Particle Swarm Optimization (PSO). The properties are discussed and analysed at the end.

## 12.1 Introduction

In recent years, with the rise of artificial intelligence and artificial life, research into swarm intelligence has aroused great concern amongst numerous scholars, and some new-type bionic algorithms with swarm intelligence have become hot research topics, such as the Genetic Algorithm (GA) [1], the Particle Swarm Optimization (PSO) algorithm [2], the Ant Colony Optimization (ACO) algorithm [3], the Bees Algorithm (BA) [4], and the Artificial Fish Swarm Algorithm (AFSA) [5]. Their applications in many kinds of scientific research fields show their good properties and practical value. They have some common characteristics and also have their own unique characteristics. AFSA as a new optimization algorithm has become a very hot topic as it offers new ideas to solve the optimization problem in signal processing [6], complex function optimization [7], neural network classifiers [8], network combinatorial optimization [9], multi-user detection in communication [10, 11], sequence code estimation [12], and some applications [13]. In these applications, the algorithm reflects good

M. Jiang (✉)
Centre Tecnològic de Telecomunicacions de Catalunya (CTTC), Av. Del, Canal Olímpic s/n, 08860 Castelldefels (Barcelona), Spain School of Information Science and Engineering, Shandong University, Jinan, 250100 China
e-mail: jiangmingyan@sdu.edu.cn, mingyan.jiang@cttc.es

N. Mastorakis et al. (eds.), *Proceedings of the European Computing Conference*, Lecture Notes in Electrical Engineering 28,
DOI 10.1007/978-0-387-85437-3_12, 

performances and becomes a prospective method in solving optimization problems. Its basic idea is to imitate the fish behaviours such as praying, swarming, following with local search of fish individual for reaching the global optimum, it is random and parallel search algorithm, it has the good ability to overcome local extrema, obtain the global extrema and has fast convergence speed. In this paper, we propose some adaptive step methods and apply one in image processing.

## 12.2 The Artificial Fish Swarm Algorithm

### *12.2.1 The Artificial Fish*

The Artificial Fish (AF) realizes external perception by its vision. $X$ is the current state of an AF, *Visual* is the visual distance, and $X_v$ is the visual position at some moment. If the state at the visual position is better than the current state, it goes forward a step in this direction, and arrives at the $X_{nest}$ state; otherwise, it continues an inspecting tour in the vision. The greater number of inspecting tours the AF does, the more knowledge about the overall states of the vision the AF obtains. Certainly, it does not need to travel throughout complex or infinite states, which is helpful to find the global optimum by allowing certain local optimum with some uncertainty.

### *12.2.2 Five Basic Behaviors of AF*

Fish usually stay in the place with a lot of food, so we simulate the behaviors of fish based on this characteristic to find the global optimum, which is the basic idea of the AFSA. Five basic behaviors of AF are defined [5, 12] as follows:

(1). AF_Prey: This is a basic biological behavior that tends to the food; generally the fish perceives the concentration of food in water to determine the movement by vision or sense and then chooses the tendency. (2). AF_Swarm: The fish will assemble in groups naturally in the moving process, which is a kind of living habit to guarantee the existence of the colony and avoid dangers. (3). AF_Follow: In the moving process of the fish swarm, when a single fish or several ones find food, the neighborhood partners will trail and reach the food quickly. (4). AF_Move: Fish swim randomly in water; in fact, they are seeking food or companions in larger ranges. (5). AF_Leap: Fish stop somewhere in water, every AF's result will gradually be the same, the difference of objective values (food concentration, FC) becomes smaller within some iterations, it might fall into local extrema, change the parameters randomly to the still states for leaping out of the current state. The detail behaviors pseudo-code can be seen in [6].

### 12.2.3 The Improved Adaptive Step Length in AFSA

In the AFSA, there are many parameters that have impacts on the final optimization result. In this paper, we only consider the parameter *Step*. With the increase of the *Step*, the speed of convergence is accelerated. However, when the increase of the *Step* is out of a range, the speed of convergence is decelerated, and sometimes the emergence of vibration can influence the speed of convergence greatly. Using the adaptive step may prevent the emergence of vibration, increase the convergence speed and enhance the optimization precision.

In the behaviors of AF_Prey, AF_Swarm and AF_Follow, which use the-*Step* parameter in every iteration, the optimized variables (vector) have the various quantity of *Step**rand( ); *Step* is a fixed parameter, rand( ) is a uniformly distributed function. We give some adaptive *Step* methods as follows ($t$ means iteration time)

1. $Step_{t+1} = \alpha \bullet Step_t, \alpha = 0.9 \sim 0.99$ and $Step_{t=1} = Step$
2. Prior period of iteration, using *Step**rand( ); later period, using method (1).
3. $Step_{t+1} = \frac{\beta \bullet N-t}{N}$ $Step, \beta = (1.1 \sim 1.5)$, N is the all iteration times.

The method (1) can insure the precise result when the iteration is in the optimal research range, but might trap in local extrema. Method (2) can balance the above problems. Method (3) has a relation with the iteration time, and gradually decreases the *Step*, so decreases the various quantities of optimized variables in each iteraton time. Using the adaptive step, we can select the *Step* more randomly which can guarantee the fast convergence, the result's precision and stability

For example, using fixed *Step* = 2 and adaptive step ($\beta$ = 1.2) of method (3), we optimize a nonlinear function with many local extreme points by AFSA and improved AFSA, the function $f(x, y) = x\sin(2\pi x) - y\sin(2\pi y) + 1$, and $x, y \in [-1, 2]$.The comparison result is shown in Fig. 12.1. With the adaptive

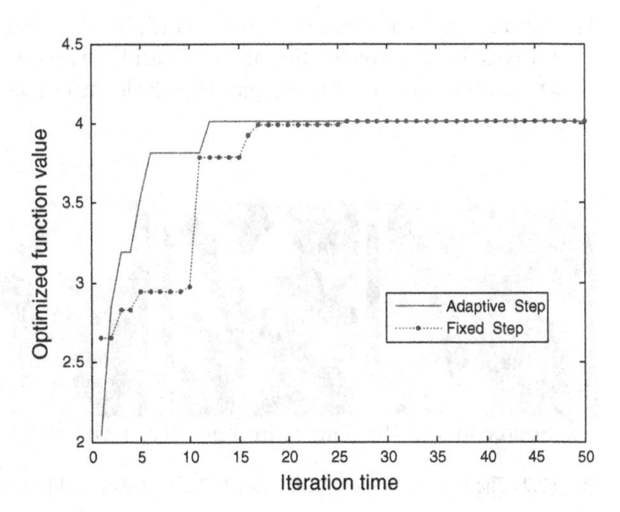


**Fig. 12.1** Convergence speed with adaptive step and fixed step

step, convergence speed is faster and the result is more stable especially when the optimized variables are many

## 12.3 Image Segmentation by Improved AFSA

### *12.3.1 The Multi-Threshold Image Segmentation*

The image segmentation is key pre-step for the object identification; there are many conventional segmentation methods such as the maximum variance between clusters method, maximum entropy (ME) threshold method, minimum error threshold method, etc. The principle of image segmentation based on ME [14] is: the image gray level $\in[0,l-1]$,$t$ is the threshold for the image segmentation, $P_t = \sum_{i=0}^{t} p_i$ and $p_i$ is the probability of$i$gray level of image. Select the optimal$t$; make the image entropy $H(t) = -\sum_{i=0}^{t} \frac{p_i}{p_t} \ln \frac{p_i}{p_t} - \sum_{i=t+1}^{l-1} \frac{p_i}{1-p_i} \ln \frac{p_i}{1-p_t}$ be maximal. Extending to multi-threshold, the problem is: select multi-threshold, make the image entropy $H(t_1, t_2, \ldots, t_k) = -\sum_{i=0}^{t_1} \frac{p_i}{p_{t_0}} \ln \frac{p_i}{p_{t_0}} - \sum_{i=t_1+1}^{t_2} \frac{p_i}{p_{t_1}} \ln \frac{p_i}{p_{t_1}} - \ldots - \sum_{i=t_i}^{i=l-1} \frac{p_i}{p_{t_k}} \ln \frac{p_i}{p_{t_k}}$ be maximal. We optimize and select 2, 3 thresholds with improved AFSA for image segmentation.

### *12.3.2 The Parameter Selection and Results*

The parameter selection: fish number is 15, *try*_number is 3, iteration is 50, *Step* is 10, *Visual* is 10, crowded factor $\delta$ is 0.8 and the result is shown in the Fig. 12.2.

(a) Source image (b). Single threshold (c) Two thresholds (d) Three thresholds

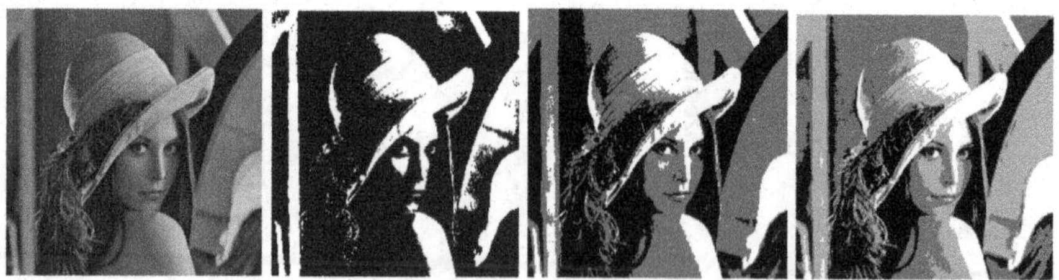

(a) Source image (b) Single threshold (c) Two thresholds (d) Three thresholds

**Fig. 12.2** The image segmentation with 1, 2 and 3 thresholds respectively

**Table 12.1** The comparison of GA, PSO and improved AFSA

| Method | 2-threshold | ME | Time | 3-threshold | ME | Time |
|---|---|---|---|---|---|---|
| GA | 82,151 | 12.613 | 17.24 s | 58,115,162 | 15.571 | 20.91 s |
| PSO | 80,151 | 12.613 | 8.67 s | 59,109,162 | 15.590 | 9.96 s |
| AFSA | 80,151 | 12.613 | 4.19 s | 60,109,162 | 15.590 | 6.9 s |

We also use the GA and PSO for the same problem, the GA parameter selection: population size is 20, iteration number is 100, 0 is 0.8 and$P_m$ is 0.05, the PSO parameter selection: the number of particles is 20, the iteration number is 100, $W_{max}=0.9$, $W_{min}=0.4$, learning factor $c1 = c\ 1 = 2$, the result is shown in Table 12.1.

## 12.4 Discussion and Conclusion

### *12.4.1 Discussion*

From the principle of the optimization algorithm, the AFSA is a neighborhood search, with the aid of heuristic search strategy, and has the ability of global optimization. The AFSA has no special requirements to the object function and initial values, so the initial value can be set with stochastic values or fixed values, and other parameters can be set in a wide range. In brief, the algorithm has strong adaptability.

In Table 12.1, with the optimized threshold results of image segmentation by GA, PSO and improved AFSA, we can see that the properties of AFSA are better than PSO, and the properties of PSO are better than GA. But how to compare them fairly is a problem; the AFSA has low computing complexity, faster convergence speed and greater precision. Future work needs to be a consideration of establishing the theory standard and adjusting method of other parameters such as adaptive *Visual*, crowded factor effect etc. Fish fast recognition and decision mechanism can be considered into the algorithm and some other improved ideas can be seen in [10, 13].

### *12.4.2 Conclusion*

We propose three adaptive step methods in the AFSA and analyze the properties of improved method, and then we use it in the multi-threshold image segmentation, and compare it with the GA and PSO methods. The results show that the AFSA can obtain better performance than the GA and PSO. Some improved methods and ideas are proposed, the applied performance can be better, the AFSA can obtain the optimal or sub-optimal results on theory, and it is believed that the AFSA can be used in some other complicated optimization applications.

**Acknowledgements** This work is supported by the National Natural Scientific Foundation of China (No. 60672036, No. 60672037), the Natural Science Foundation of Shandong Province of China (No. Y2006G06), and the Catalan Government (Generalitat de Catalunya, Spain) under grant SGR2005-00690.

## References

1. Goldberg DE (1989) Genetic algorithms in search, optimization and machine learning. Addison-Wesley Longman
2. Eberhart R, Shi Y, Kennedy J (2001) Swarm intelligence. Morgan Kaufmann, San Francisco
3. Dorigo M, Stzle T (2004) Ant colony optimization. MIT Press, Cambridge
4. Pham DT, Ghanbarzadeh A, Koc E (2005) The bees algorithm. Technical note, Manufacturing Egnineering Centre, Cardiff University, UK
5. Li XL (2003) A new intelligent optimization-artificial fish swarm algorithm. Doctor thesis, Zhejiang University of Zhejiang, China
6. Jiang MY, Yuan DF (2005) Wavelet threshold optimization with artificial fish swarm algorithm. In: Proc. of IEEE International Conference on Neural Networks and Brain, Beijing China, pp 569–572
7. Xiao JM, Zheng XM, Wang XH (2006) A modified artificial fish-swarm algorithm. In: Proc. of IEEE the 6$^{th}$ World Congress on Intelligent Control and Automation, Dalian China, pp 3456–3460
8. Zhang MF, Cheng S, Li FC (2006) Evolving neural network classifiers and feature subset using artificial fish swarm. In: Proc. of IEEE International Conference on Mechatronics and Automation, Luoyang China, pp 1598–1602
9. Shan XJ, Jiang MY (2006) The routing optimization based on improved artificial fish swarm algorithm. In: Proc. of IEEE the 6th World Congress on Intelligent Control and Automation, Dalian China, pp 3658–3662
10. Jiang MY, Wang Y, Pfletschinger S, Lagunas MA (2007) Optimal multiuser detection with artificial fish swarm algorithm. In: Proc. of International Conference on Intelligent Computing (ICIC 2007), CCIS 2, Springer-Verlag Berlin Heidelberg, pp 1084–1093
11. Yu Y, Tian YF, Yin ZF (2005) Multiuser detector based on adaptive artificial fish school algorithm. In: Proc. of IEEE International symposium on communications and information technologym, pp 1480–1484
12. Jiang MY, Wang Y, Rubio F (2007) Spread Spectrum Code Estimation by Artificial Fish Swarm Algorithm. In: Proc. of IEEE International Symposium on Intelligent Signal Processing (WISP'2007), Alcalá de Henares, Spain
13. Jiang MY, Yuan DF (2006) Artificial fish swarm algorithm and its applications. In: Proc. of International Conference on Sensing, Computing and Automation, Chongqing China, pp 1782–1787
14. Zhang YJ (2001) Image segmentation. Science press, Beijing

# Chapter 13
# Propagation of Elastic-Plastic Waves in Bars

**Olga Martin**

**Abstract** In this paper we study the impact loading of semi-infinite prismatic bars, using a numerical algorithm based on the finite element method (FEM) and finite-differences method (FDM). A detailed analysis of the calculus errors is made and a theorem concerning the approximations is proved. The numerical example confirms the efficiency of the proposed method.

## 13.1 Introduction

The action of loads suddenly applied to a body is not propagated instantaneously but is transmitted from one particle to another in a wave-like manner. The strength problems for various machines and structures subjected to impact (or explosions) can be studied only if a clear understanding is gained of the laws of propagation of elastic-plastic deformation preceding failure. Also, dynamic problems may be of importance in the analysis of high-speed technological metal-forming processes.

Kachanov [6], Rakhmatulin [11], and Karman and Duwez [7] have published the studies regarding the propagation of elastic-plastic compression (extension) waves in a bar. Then, through Kubenko [8], Bajons [2], and Cristescu [4] were studied various generalizations of these problems. Detailed references may be found in the books [4, 8].

## 13.2 Problem Formulation

Let us consider the problem of waves propagation in a semi-infinite prismatic bar, whose axis coincide with the $x$-axis. We consider the following basic assumptions.

O. Martin (✉)
University "Politehnica" of Bucharest, 313, Splaiul Independentei, Romania
e-mail: omartin_ro@yahoo.com

N. Mastorakis et al. (eds.), *Proceedings of the European Computing Conference*, Lecture Notes in Electrical Engineering 28,
DOI 10.1007/978-0-387-85437-3_13, 

- During deformation the cross-sections of the bar remain plane and normal to the $x-$axis.
- Deformations are small and, hence, changes in the dimensions of a bar can be neglected.
- The inertia forces corresponding to the motion of particles of a bar in transverse directions (due to the contraction or expansion of the section) can be neglected.
- The effect of the rate of strain on the curve representing the relation between stress $\sigma = \sigma_x$ and $\varepsilon = \varepsilon_x$ can be neglected. In a dynamic problem, the rates of strain are large and their effect on the stress – strain curve may be appreciable. In this paper, we adopt a relation $\sigma = \sigma(\varepsilon)$ that is a first approximation and it refers to a certain average strain rate over the given range.

Let us consider a bar of the cross-section $A$ and the density is $\rho$. The displacement of the point $x$ at the time $t$ is denoted by $u(x, t)$ and the strain is

$$\varepsilon = \frac{\partial u(x,t)}{\partial x} = u_x \text{ and } \dot{u} = \frac{\partial u(x,t)}{\partial t}$$

Suppose that the bar is in the rest at $t = 0$ and for the end $x = 0$ some disturbances are given at time $t \geq 0$. For instance, a velocity

$$\dot{u}\big|_{x=0} = \frac{\partial u(x,t)}{\partial t}\big|_{x=0} = V(t), t \in [0, T].$$

The total energy at the time $t$ is the sum

$$W_{total} = K + P \tag{13.1}$$

where

$$K = \frac{\rho A}{2} \int_0^L \dot{u}^2 dx \tag{13.2}$$

is the kinetic energy,

$$P = A \int_0^L W dx$$

is the potential energy and $W$ is the strain energy of a unit volume.

The strain energy $W$ has the following properties:

- the derivatives $\frac{d^2 W}{du_x^2}$ are positive and bounded;
- $\frac{d^2 W}{du_x^2}$ is proportionally with the square of wave velocity;

- $\frac{dW}{du_x} = \sigma(u_x)$, where $\sigma(u_x)$ is the stress relative to the references configuration (Piola-Kirchhoff stress).

  Then, the principle of energy conservation:

$$\dot{W}_{total} = 0 \tag{13.3}$$

will lead to the equation of motion. Its solution belongs to the measurable set $D = H^0([0, L]) \times [0, T]$, where $H^0([0, L]) = L_2([0, L])$ is a Hilbert space with the scalar product

$$(u(t), s(t)) = \int_0^L u(x, t)s(x, t)dx \tag{13.4}$$

and the norm

$$\|u(t)\|^2 = \int_0^L u^2(x, t)dx \tag{13.5}$$

for any $t \in [0,T]$. Then our problem (13.3) becomes

$$\rho A(\ddot{u}, \dot{u}) - A(\sigma(u_x), \dot{u}_x)) = 0 \tag{13.6}$$

with the condition at the end $x = 0$:

$$\dot{u}(0, t) = V(t), \forall t \in [0, T] \tag{13.7}$$

In order to get the solution of the problem (13.6)–(13.7), we consider on the $x$ axis a points system: $0 = x_0 < x_1 < \ldots < x_N$, with the constant step $h = x_{k+1} - x_k$, $k = 0, 1, 2, \ldots, N$.

We seek an approximate solution of the form

$$U(x, t) = \sum_{k=1}^{N} \alpha_k(t)\phi_k(x) \tag{13.8}$$

where $\phi_k(x)$ are the coordinate functions:

$$\phi_k(x) = \begin{cases} \frac{x - x_{k-1}}{h}, x \in [x_{k-1}, x_k] \\ \frac{x_{k+1} - x}{h}, x \in [x_k, x_{k+1}] \\ 0, \quad x \notin [x_{k-1}, x_{k+1}] \end{cases}$$

with $\phi_k(x_k) = 1$. Let us now consider a further points system for the interval $[0, T]$: $0 = t_0 < t_1 <\ldots.< T_R = T$, with the constant step $\tau = t_l - t_{l-1}$, $l = 1, 2, \ldots,$ $R$.. Using the notations:

$$\alpha_k^l = \alpha_k(l\tau), u^l = u(x, l\tau) \tag{13.9}$$

we define the progressive and regressive differences

$$\nabla_t u^l = \frac{u^{l+1} - u^l}{\tau}, \Delta_t u^l = \frac{u^l - u^{l-1}}{\tau} \tag{13.10}$$

Then, the numerical derivatives of first and second order with respect to $t$, become

$$\begin{aligned} \delta_t u^l &= \frac{1}{2}(\nabla_t + \Delta_t)u^l = \frac{u^{l+1} - u^{l-1}}{2\tau} \\ \delta_t^2 u^l &= \nabla_t \Delta_t u^l = \frac{u^{l+1} - 2u^l + u^{l-1}}{\tau^2} \end{aligned} \tag{13.11}$$

Replacing (13.9) and (13.11) in the equation (13.6), this becomes

$$A\rho\left(\delta_t^2 U^l, \delta_t U^l\right) + A\left(\sigma(U_x^l), \delta_t U_x^l\right) = 0 \tag{13.12}$$

where

$$\begin{aligned} U^l &= U(x, l\tau) = \sum_{k=1}^{N} \alpha_k^l \phi_k(x) \\ U_x^l &= \frac{\partial U(x, l\tau)}{\partial x} = \sum_{k=1}^{N} \alpha_k^l \frac{d\phi_k(x)}{dx} \\ \delta_t U_x^l &= \sum_{k=1}^{N} \delta_t \alpha_k^l \frac{d\phi_k(x)}{dx} \end{aligned} \tag{13.13}$$

We now define the mass matrix

$$m_{kl} = \rho A \int_0^L \phi_k(x)\phi_i(x)dx = \rho A(\phi_k(x), \phi_i(x)) \tag{13.14}$$

and rewrite the equation (13.12) in the form

$$\sum_{k,i=1}^{N} m_{ki}\, \delta_t^2 \alpha_k^l\, \delta_t \alpha_i^l + A\left(\sigma\left(\sum_{k=1}^{N} \alpha_k^l \frac{d\phi_k(x)}{dx}\right), \sum_{i=1}^{N} \delta_t \alpha_i^l \frac{d\phi_i(x)}{dx}\right) = 0$$

$l = 1, 2, \ldots, R-1$.

Since $\delta_t \alpha_i^l$ are arbitrary, we obtain for a fixed $t_l$ the following system with unknowns $\alpha_i^l$

$$\sum_{k=1}^{N} m_{ki}\delta_t^2\alpha_k^l + A\left(\sigma\left(\sum_{k=1}^{N}\alpha_k^l\frac{d\phi_k}{dx}\right), \frac{d\phi_i}{dx}\right) = 0, i = 1, 2, \ldots, N \tag{13.15}$$

The second term of the equation (13.15) defines the rigidity of the model. For instance, in the case of the linear theory we have

$$\sigma(u_x^l) = Eu_x^l = E\sum_{k=1}^{N}\alpha_k^l\frac{d\phi_k(x)}{dx} \tag{13.16}$$

where $E$ is Young's modulus and

$$AE\left(\sum_{k=1}^{N}\alpha_k^l\frac{d\phi_k}{dx}, \frac{d\phi_i}{dx}\right) = \sum_{k=1}^{N}\gamma_{ki}\alpha_k^l \tag{13.17}$$

Here $\gamma_{ki}$ are the elements of the stiffness matrix from the finite element theory. In our case, the constitutive equation is nonlinear. Such equation will be presented in the numerical example.

## 13.3 Approximation Study

Let us now consider the derivable function $\sigma(u_x)$ with $c_0^2 = \inf|\sigma'(u_x)|$, $c_1^2 = \sup|\sigma'(u_x)|$. Then, for two gradients $u_x$ and $w_x$, where $w_x$ there is in a neighborhood of $u_x$, we have

$$\sigma(u_x) = \sigma(w_x) \pm \sigma'(w_x + \theta(u_x - w_x))(u_x - w_x), \theta \in (0, 1) \tag{13.18}$$

This relation leads to the following inequalities

$$\begin{aligned} &c_0^2(u_x - w_x, V_x) \leq (\sigma(u_x) - \sigma(w_x), V_x) \leq \\ &\leq c_1^2(u_x - w_x, V_x), \forall V \in D. \end{aligned} \tag{13.19}$$

In view of the finite differences theory, if $u(x,t)$ has the derivatives of the fourth order with respect to $t$ bounded on $D$, we have for $t_l = l\tau$ that

$$\delta_t^2 u(x, l\tau) = \frac{\partial^2 u(x, l\tau)}{\partial t^2} + \omega^l(x, t) \cdot \tau \tag{13.20}$$

Let us now consider the exact solution $u$ of the problem (13.6)–(13.7) and the numerical solution $U$ defined by (13.7). By virtue of (13.12) and (13.20), we get

$$\begin{aligned} &\rho(\delta_t^2 U^l, V) + (\sigma(U_x^l), V_x) = 0 \quad &(a) \\ &\rho(\delta_t^2 u^l, V) + (\sigma(u_x^l), V_x) = \rho(\omega^l \tau, V) \quad &(b) \end{aligned} \tag{13.21}$$

where $V = \delta_t U^l$. We now introduce an interpolation polynomial of the solution $u$ for a fixed $t_l$

$$\tilde{u}^l = \tilde{u}(x_k, t_l) = u(x_k, t_l) = u^l, k = 0, 1, .., N \tag{13.22}$$

and its interpolation error

$$\tilde{e}^l = \tilde{u}^l - u^l \tag{13.23}$$

We define for any fixed $t_l$

$$\varepsilon^l = \tilde{u}^l - U^l \tag{13.24}$$

and therefore, the finite element error is of the form

$$e^l = u^l - U^l = \varepsilon^l - \tilde{e}^l \tag{13.25}$$

Using the interpolation theorem of the finite element theory for the functions, which belong to the $C^k$ functions space, [12], we remind the following properties: there are the constants $\mu$and $\mu_1$ independently of step $h$, such that

$$\|\tilde{e}^l\| \le \mu h^2, \quad \|\tilde{e}_x^l\| \le \mu_1 h \tag{13.26}$$

**Lemma** *Taking the relations (13.18)–(13.21) into account, there are for any $t_l$, the constant $\mu$, independently of $h$, such that*

$$c_0^2 \|\varepsilon_x^l\|^2 + \rho(\delta_t^2 e^l, \varepsilon^l) \le \frac{c_1^2}{4\mu} \|\tilde{e}_x^l\|^2 + c_1^2 \mu \|\varepsilon_x^l\|^2 + \rho(\omega^l \tau, \varepsilon^l) \tag{13.27}$$

Let us consider (13.21$a$), where $u^l$ is replaced by $\tilde{u}^l$ an $V = \varepsilon^l = \tilde{u}^l - U^l$

$$\rho(\delta_t^2 \tilde{u}^l, \tilde{u}^l - U^l) + (\sigma(\tilde{u}_x^l), \tilde{u}_x^l - U_x^l) = 0 \tag{13.28}$$

*Subtracting* (13.28) *and* (13.21$a$), *where* $V = \tilde{u}^l - U^l$ *and using* (13.19), *we get*

$$\begin{aligned}&\rho\left(\delta_t^2(\tilde{u}^l - U^l), \tilde{u}^l - U^l\right) + c_0^2\|\tilde{u}_x^l - U_x^l\|^2 \leq \\ &\leq \rho\left(\delta_t^2(\tilde{u}^l - U^l), \tilde{u}^l - U^l\right) + \left(\sigma(\tilde{u}_x^l) - \sigma(U_x^l), \tilde{u}_x^l - U_x^l\right) \leq \\ &\leq \rho\left(\delta_t^2(\tilde{u}^l - U^l), \tilde{u}^l - U^l\right) + c_1^2\|\tilde{u}_x^l - U_x^l\|^2\end{aligned} \tag{13.29}$$

*With the help of the inequality*

$$(a,b) \leq \frac{1}{4\mu}\|a\|^2 + \mu\|b\|^2, \mu > 0 \tag{13.30}$$

*we obtain by the addition and subtraction of $u^l$ in the first term and of the stress $\sigma(u_x^l)$ in the second term the following inequalities*

$$\begin{aligned}\rho\left(\delta_t^2\varepsilon^l, \varepsilon^l\right) + c_0^2\|\varepsilon_x^l\|^2 &= \rho\left(\delta_t^2(\tilde{u}^l - U^l), \tilde{u}^l - U^l\right) + c_0^2\|\tilde{u}_x^l - U_x^l\|^2 \leq \\ &\leq \rho\left(\delta_t^2(\tilde{u}^l - u^l + u^l - U^l), \tilde{u}^l - U^l\right) + \\ &+ \left(\sigma(\tilde{u}_x^l) - \sigma(u_x^l) + \sigma(u_x^l) - \sigma(U_x^l), \tilde{u}_x^l - U_x^l\right) \leq \\ &\leq \rho\left(\delta_t^2(\tilde{u}^l - u^l), \tilde{u}^l - U^l\right) + \left(\sigma(\tilde{u}_x^l) - \sigma(u_x^l), \tilde{u}_x^l - U_x^l\right) + \\ &+ \rho\left(\omega^l\tau, \tilde{u}^l - U^l\right) \leq \rho\left(\delta_t^2\tilde{e}^l, \varepsilon^l\right) + c_1^2\left(\tilde{e}_x^l, \varepsilon_x^l\right) + +\rho\left(\omega^l\tau, \varepsilon^l\right) \leq \\ &\leq \rho\left(\delta_t^2\tilde{e}^l, \varepsilon^l\right) + \frac{c_1^2}{4\mu}\|\tilde{e}_x^l\|^2 + c_1^2\mu\|\varepsilon_x^l\|^2 + \\ &+ \rho\left(\omega^l\tau, \varepsilon^l\right) \leq \rho\left(\delta_t^2\tilde{e}^l, \varepsilon^l\right) + \frac{c_1^2}{4\mu}\|\tilde{e}_x^l\|^2 + c_1^2\mu\|\varepsilon_x^l\|^2 + \rho\left(\omega^l\tau, \varepsilon^l\right)\end{aligned} \tag{13.31}$$

*Finally, we obtain*

$$c_0^2\|\varepsilon_x^l\|^2 + \rho\left(\delta_t^2 e^l, \varepsilon^l\right) \leq \frac{c_1^2}{4\mu}\|\tilde{e}_x^l\|^2 + c_1^2\mu\|\varepsilon_x^l\|^2 + \rho\left(\omega^l\tau, \varepsilon^l\right) \tag{13.32}$$

**Theorem** *If* (13.27) *exists for a fixed $t_l$, then the finite element error verifies the inequality*

$$\|e^l\| \leq Ch \tag{13.33}$$

*Subtracting the relations* (13.21), *where $V = \varepsilon^l$ and using* (13.19) *and* (13.30), *we get*

$$\begin{aligned}&\rho\left(\delta_t^2 e^l, \varepsilon^l\right) \leq \left(\omega^l\tau, \varepsilon^l\right) - c_0^2\left(\varepsilon_x - \tilde{e}_x^l, \varepsilon_x^l\right) \leq \\ &\leq \left(\omega^l\tau, \varepsilon^l\right) + c_0^2\left(\tilde{e}_x^l, \varepsilon_x^l\right) \leq \left(\omega^l\tau, \varepsilon^l\right) + \frac{c_0^2}{4\mu}\|\tilde{e}_x^l\|^2 + c_0^2\mu\|\varepsilon^l\|^2\end{aligned} \tag{13.34}$$

*Subtracting* (13.27) *and* (13.34), *we have*

$$c_0^2\|\varepsilon_x^l\|^2 \leq \frac{c_1^2 - c_0^2}{4\mu}\|\tilde{e}_x^l\|^2 + (c_1^2 - c_0^2)\mu\|\varepsilon_x^l\|^2$$

*Consequently,*

$$(c_0^2 - c_1^2\mu + c_0^2\mu)\|\varepsilon_x^l\|^2 \leq ah^2$$

*If* $\mu \in (0, 1)$, *we obtain*

$$\|\varepsilon_x^l\|^2 \leq C_1 h^2 \tag{13.35}$$

*Using the definition* (13.25) of $e^l$, *we get*

$$\|e_x^l\| \leq \|\varepsilon_x^l\| + \|\tilde{e}_x^l\| \leq C_2 h + \mu_1 h = C_3 h$$

*and in accordance with the Friedrichs's inequality, we finally have*

$$\|e^l\| \leq k\|e_x^l\| \leq Ch \tag{13.36}$$

## 13.4 Numerical Example

We study the waves propagation in a semi-infinite bar $x \geq 0$, which is at rest at $t = 0$. Let us consider that at $t > 0$ the velocity $V$ of the point $x = 0$ is of the form

$$\dot{U}(0,t) = \begin{cases} 2\,V(t/t_0), t < t_0/2 \\ V, \quad t \geq t_0/2 \end{cases}, V = 2000\,cm/\sec, t_0 = 5 \cdot 10^{-6}\sec \tag{13.37}$$

The density $\rho = 0.00785$ kg/cm$^3$ and the constitutive equation is of the form, [7]:

$$\sigma = \sigma_c + K \cdot \left(|\varepsilon|^{0.4} - |\varepsilon_c|^{0.4}\right) \tag{13.38}$$

where $\sigma_c = 500\text{daN/cm}^2, E = 10^6\text{daN/cm}^2, K = 5000\text{daN/cm}^2$.

The step $h = 0.05$ cm and the time step $\tau = h/c_0 = 4.43 \cdot 10^{-6}$, where $c_0 = \sqrt{E/\rho}$ is the velocity of the elastic wave and $\varepsilon_c = \sigma_c/E$.

The displacements $U$ of the bar points will be calculated by finite element – finite differences methods for the time intervals $[i\tau, (i+1)\tau]$, $i = 0, 1, 2, \ldots$

Step 1. $0 \leq t \leq \tau$

$$U_{11} = U(x_1, \tau) = V\tau^2/t_0 = 7.85 \cdot 10^{-3}\text{cm}$$

From the definition (13.7), we observe that

$$U(x_k, t_l) = \alpha_k^l \quad U(x_1, t_1) = \alpha_1^1 \tag{13.39}$$

Step 2. $\tau \leq t \leq 2\tau$

$$U_{12} = \alpha_1^2 = U(x_1, 2\tau) = V \cdot (2\tau)$$
$$U_{22} = \alpha_2^2 = U(x_2, 2\tau) = U_{11}, x_2 = 2h$$

Evidently, the displacements $U_{ik} = 0$, if $k<i$, this assumption been in good agreement with the experimental observations.

Step 3. $2\tau \leq t \leq 3\tau$

$$U_{33} = U_{11} = \alpha_3^3, U_{13} = \alpha_1^3 = V(3\tau)$$

By using the system (13.15) and the definition (13.12), we can obtain $U(x_2, t_3) = \alpha_2^3$ from the equation

$$\begin{aligned} &\rho\left[\delta_t^2\alpha_1^2 \int_{x_1}^{x_2} \phi_2\phi_1 dx + \delta_t^2\alpha_2^2 \int_{x_1}^{x_3} \phi_2\phi_2 dx + \delta_t^2\alpha_3^2 \int_{x_2}^{x_3} \phi_2\phi_3 dx\right] = \\ &= -K\left|\frac{U_{22} - U_{12}}{h}\right|^{0.4} + K\left|\frac{-U_{22}}{h}\right|^{0.4} \end{aligned} \tag{13.40}$$

Hence we find $\alpha_2^3 = U_{23}$.

Step 4. $3\tau \leq t \leq 4\tau$

$$U_{14} = \alpha_1^4 = V \cdot (4\tau) \text{ and } U_{44} = \alpha_4^4 = U_{11}$$

The system (13.15) becomes

$$\begin{aligned} &\rho\left[\delta_t^2\alpha_1^3 \int_{x_1}^{x_2} \phi_2\phi_1 dx + \delta_t^2\alpha_2^3 \int_{x_1}^{x_3} \phi_2\phi_2 dx + \delta_t^2\alpha_3^3 \int_{x_2}^{x_3} \phi_2\phi_3 dx\right] = \\ &= -\left(\sigma\left(\sum_{k=1}^{3} \alpha_k^3 \frac{d\phi_k}{dx}\right), \frac{d\phi_2}{dx}\right) \end{aligned}$$

where, for instance

$$\left(\sigma\left(\sum_{k=1}^{3} \alpha_k^3 \frac{d\phi_k}{dx}\right), \frac{d\phi_2}{dx}\right) = 5000\left[\left|\frac{U_{23} - U_{13}}{h}\right|^{0.4} - \left|\frac{U_{33} - U_{23}}{h}\right|^{0.4}\right]$$

Finally, we find $U_{24}$ and $U_{34}$ from the iterative matrix equation

$$\begin{bmatrix} 4 & 1 \\ 1 & 4 \end{bmatrix} \begin{bmatrix} U_{24} \\ U_{34} \end{bmatrix} - 2 \begin{bmatrix} 4 & 1 \\ 1 & 4 \end{bmatrix} \begin{bmatrix} U_{23} \\ U_{33} \end{bmatrix} + \begin{bmatrix} 4 & 1 \\ 1 & 4 \end{bmatrix} \begin{bmatrix} U_{22} \\ 0 \end{bmatrix} = \\ = - \begin{bmatrix} \left( \sigma \left( \sum_{k=1}^{3} \alpha_k^3 \frac{d\phi_k}{dx} \right), \frac{d\phi_2}{dx} \right) + U_{14} - 2U_{13} + U_{12} \\ \left( \sigma \left( \sum_{k=1}^{3} \alpha_k^3 \frac{d\phi_k}{dx} \right), \frac{d\phi_3}{dx} \right) + U_{44} \end{bmatrix}$$

or

$$D \cdot U^{(4)} = -B + 2D \cdot U^{(3)} - D \cdot U^{(2)} \tag{13.41}$$

where

$$D = \begin{bmatrix} 4 & 1 \\ 1 & 4 \end{bmatrix}$$

*Step* $n+1$. $n\tau \leq t \leq (n+1)\tau$

The matrix equation (13.41) becomes

$$D \cdot U^{(n+1)} = -B + 2D \cdot U^{(n)} - D \cdot U^{(n-1)} \tag{13.42}$$

where

$$D = \begin{bmatrix} 4 & 1 & 0 & 0 & \cdots & 0 \\ 1 & 4 & 1 & 0 & \cdots & 0 \\ \vdots & \vdots & \vdots & \vdots & \vdots & \vdots \\ 0 & 0 & \cdots & 1 & 4 & 1 \\ 0 & 0 & \cdots & 0 & 1 & 4 \end{bmatrix}.$$

The values of $U_{k,l} = U(x_k, t_l)$, for $k = 0, 1,.., 4$ and $l = 0, 1,\ldots,4$, are

$$U = \begin{bmatrix} 0.00785 & 0.018 & 0.027 & 0.035 & 0.044 \\ 0 & 0.00785 & 0.018 & 0.027 & 0.035 \\ 0 & 0 & 0.00785 & 0.014 & 0.02 \\ 0 & 0 & 0 & 0.00785 & 0.014 \\ 0 & 0 & 0 & 0 & 0.00785 \end{bmatrix}$$

In the Fig. 13.1 are presented the values of the displacements $U(x_i, t_l)$, that correspond to the sections $x = x_k$, $k = 1, 2,\ldots, 5$ for the time: $t_l$, $l = 1, 2, 3, 4$.

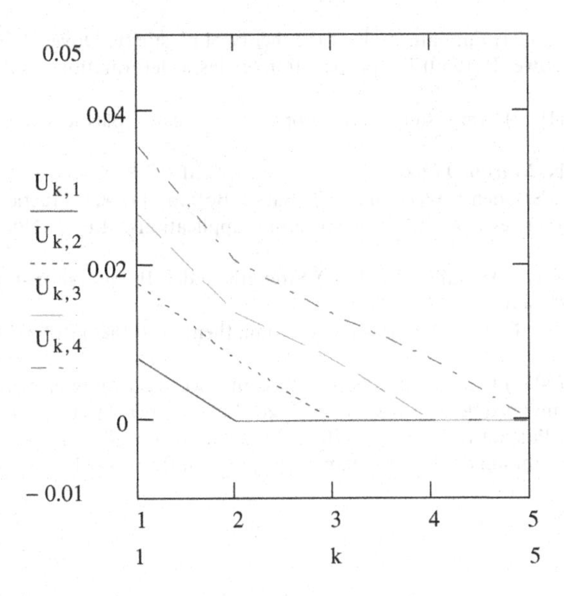


**Fig. 13.1** Values of the displacements $U(x_i, t_l)$

## 13.5 Conclusions

The action of loads suddenly applied to a structure is not propagated instantaneously but it is transmitted from one particle to another in a wave-like manner. In the numerical example we approach the case of large plastic strains, disregarding the elastic strains. In this case, the waves' velocity is variable, the larger strain propagating with a lower velocity; as time goes on, the waves "diffuse". Unlike the characteristics method, the effects of this phenomenon and the interactions of the direct and reflected waves are automatically determined with the help of the numerical method presented in the paper.

The study of the errors that correspond to the numerical solution shows that these are of the first order with respect to the spatial step *h*. In a future paper we will apply the finite element-finite differences method to the unloading phenomenon of the elastic-plastic bars.

## References

1. Babuska I, Szabo B (1991) Finite element analysis. John Wiley and Sons, New York
2. Bajons P (1980) The problem of resonance for longitudinal small amplitude excitation of an elastic-plastic bar. Acta Mecanica 37:199–215
3. Benitez F, Lu L, Rosakis A (1993) A boundary element formulation based on the three-dimensional elasto-static fundamental solution. Int J Numer Meth Eng 36:3097–3130
4. Cristescu N (2007) Dynamic plasticity. World Scientific
5. Hongwei Ma (1999) Experimental studies on dynamic plastic buckling of circular cylindrical shells under axial impact. Acta Mecanica 15:275–282

6. Kachanov L (2004) Fundamentals of the theory of plasticity. Dover Publications
7. Karman T, Duwez P (1950) The propagation of plastic deformation in solids. J Appl Phys 21:987–994
8. Kubenko V (1972) Symposium on the propagation of elasic-plastic waves. Int Appl Mech 8(5)
9. Mastorakis N, Martin O (2007) Finite elements. WSEAS Press
10. Proudhomme S, Oden J, Westermann T, Bass J, Botkin M (2003) Practical methods for a posteriori error estimation in engineering applications. Int J Numer Meth Eng 56:1193–1199
11. Rakhmatulin K, Demianov Y (1961) Strength under Intense short-time loads. MIR Publishers, Moscow
12. Reddy D, Oden J (1976) An introduction to mathematical theory of MEF. J. Wiley and Sons
13. Schreyer H (2005) Consistent diagonal mass matrices and finite element equations for one-dimensional problems. Int J Numer Meth Eng 12:1171–1184
14. Voyiadjis G, Palazotto A, Gao X (2002) Modeling of metallic materials at high strain rates with continuum damage mechanics. Appl Mech Rev 55:481–493

# Chapter 14
# Shahyat Algorithm as a Clustering Method

**Omid Khayat and Hamid Reza Shahdoosti**

**Abstract** This paper presents a new method of clustering and classification. Here is discussed a new method of clustering based on defining the heat function. Shahyat algorithm can be implemented either in a supervised or unsupervised manner. We consider each pattern as a heat source; the points of feature space are affected by the heat of all patterns. Therefore, the closer the patterns, the higher the heat, and we can determine compact regions of feature space noticing the heat of space. We calculate the heat of patterns, and choose cluster centers based on their heat. Finally, we solve the iris clustering problem and present several specifications of shahyat algorithm.

## 14.1 Introduction

There are several methods of pattern clustering and classification. Clustering and classification are among the most important topics in artificial intelligence, signal processing, image processing, pattern recognition, etc.

Several popular approaches are known for clustering and classification such as k-means [5], nearest neighbor [3] and fuzzy c-means [1].

Shahyat algorithm presents a new method for clustering and classification. We have a large number of patterns and want to cluster them into an identical number. Shahyat algorithm is based on the definition of a Heat Function. We consider each pattern in the N-dimensional feature space as a heat source. Each pattern affects the space totally, that makes every point of N-dimensional feature space be affected by all patterns. This definition helps us to discern the concentration of patterns.

Because of the high concentration in clusters, heat in the clusters is higher than out of clusters, and this fact is used to determine and confine the clusters. By calculating the heat of patterns, we can find cluster centers. Each pattern belongs to nearest center. Therefore the clusters are being determined.

O. Khayat (✉)
Department of Biomedical Engineering, Amirkabir University, Iran

N. Mastorakis et al. (eds.), *Proceedings of the European Computing Conference*, Lecture Notes in Electrical Engineering 28,
DOI 10.1007/978-0-387-85437-3_14, 

## 14.2 Description of Shahyat Algorithm

We consider k patterns in the N-dimensional feature space as below:

$$pattern = \{X_1, X_2, ..., X_k\} \; X_i = (d_1, d_2, .., d_n)(i = 1, 2, ..., k) \tag{14.1}$$

In which $X_i$ $(i = 1, 2, ..., k)$ is the $i$'th pattern and $d_j$ $(j = 1, 2, ..., k)$ is the $j$'th feature of patterns. We want to classify k patterns into C clusters. First, we map each pattern as a point in the N-dimensional feature space. Assume that the N-dimensional space is orthogonal. The difference between two patterns is the Euclidean distance between them. We want to assign patterns with lower distance to the same cluster; since the patterns with low distance cause concentration, we need to determine high and low concentration of patterns in the space. We consider each pattern as a heat source to determine concentration by heat effectiveness. Patterns (heat sources) affect the whole space. It is assumed that the interaction of near points (in a cluster) is considerable. On the other hand it is negligible when the points are far from each other. This means that the effect decreases when the distance increases. Heat function is defined as below:

$$H(X_0|X_j) = F(DISTANCE\ BETWEEN\ X_0\ \&\ X_j)(j = 1, 2, ..., k) \tag{14.2}$$

where $H(X_0|X_j)$ is the heat of arbitrary point$X_0$ (which is affected by pattern $X_j$), which depends on the distance between $X_0$ and pattern $X_j$. $F$ is a function of distance that should satisfy 3 conditions:

1. $F(0) < \infty$
2. $F(\infty) \to 0$
3. $if\ d < d' \to F(d) > F(d')$

The first condition points to the heat limitation in patterns, the second condition expresses that the heat is negligible at far, and the third shows that the function is decreasing by distance.

By increasing the distance from the patterns, heat decreases by a decreasing function (for example exponentially function). It is obvious that the intensity of heat could be explanatory of high or low concentration of patterns. The heat can be calculated for each arbitrary point $X_0$ affected by K patterns as below:

$$H(X_0) = \sum_{j=1}^{k} H(X_0|X_j) \tag{14.3}$$

Here, $H(X_0|X_j)$ is the heat of point $X_0$ affected by pattern $X_j$ and $H(X_0)$ is the heat of arbitrary point $X_0$ totally (affected by all patterns). First, it is needed to define a decreasing function and use it as $F$.

We can determine the decreasing function $F$ by considering that the effect of each pattern on patterns in other clusters should be negligible. By having the necessary decrease in each direction we can determine the type of function (exponential or any other nonlinear decreasing function) and the coefficients in the function. For example when clusters are close, we consider $F$ high decreasing function to have low heat between clusters (relative to in the clusters).

After determining the decreasing function $F$, we calculate the heat of all patterns. Then sort them according to their heat.

Let the sorted collection be $\{X_1, X_2, ..., X_k\}$ where $X_1$ has the maximum heat and $X_k$ has the minimum heat among the pattern collection.

Now we determine the straight line connecting patterns $X_j$ and $X_{j+1}$ $(j = 1, 2, ..., k - 1)$, and calculate the Heat function on points on the path from $X_1$ to $X_k$. Then we plot the heat as a function of the patterns met on the path.

Now we have a continuous diagram (if we define $F$ to be a continuous) or a discrete diagram (if we define $F$ to be a discrete function). By traversing the path of $X_1$ to $X_k$, we move in the clusters and between clusters. Out of clusters, the heat is low. Therefore it's needed to find the minimums of heat function (on the path). The minimums should satisfy a condition as below:

$$dH(l)/dl = 0 \rightarrow \\ l = l_1, l_2, \ldots, l_T | \left[ (H(X_{previous}) - H(l_i)) \geq H(X_k) \right]_{(i=1,2,\ldots,T)} \tag{14.4}$$

Whereby $dH(l)/dl$ is the differential of heat function relative to the path, $X_{previous}$ is $l_i$'s previous pattern, and $l_i (i = 1, 2, \ldots, T)$ are the roots of the derivative.

If the heat difference between minimum and previous pattern is higher than $H(X_k)$, the first pattern after minimum (on the path) is being considered as a candidate for the center of the cluster (assumed that minimums occur out of clusters).

Furthermore, we can find the candidate centers without calculating the minima of function (without differentiating the Heat function relative to path) as below.

### *14.2.1 Finding the Candidate Centers Without Differentiating from Heat Function*

The heat of points in the whole space can be calculated. We traverse the path of $X_1$ to $X_k$ (after $X_j$, we go to $X_{j+1}$), considering the heat of path ,where the gradient of heat on a line (the line between two patterns) becomes higher than $H(X_k)$, first next pattern is a candidate center. Because it is first pattern after a minimum .By traversing the path of $X_1$ to $X_k$, we find all patterns that can be a candidate center.

Assumed that we have$T$ candidate centers and one main center $X_1$, totally $T-1$ candidate center. Considering the C centers are needed, 3 cases may occur:

- First case: $C > T-1$, it means that the number of candidate centers $(T-1)$ is less than the needed number of centers $C$, so, we should select $C-(T-1)$ minima among other minimum. Now we extenuate $H(X_k)$ (it results the number considered center increases) till the number of candidate center be equal to or larger than needed center. Therefore we have $C : T-1$.
- Second case: $C = T-1$, it means that needed number of centers is equal to the number of candidate centers ,so we choose the $T-1$ candidate centers as the main centers. Third case: $C-1$, this case often occurs, it means the number of candidate centers is more than the number of needed centers. We should select $C$ center from $T-1$ candidate centers by the following method.

### 14.2.2 Selecting C Main Centers from $T+1$ Candidate Centers

First we calculate the distance between every two centers among $T-1$ candidate centers. Therefore, we have $\begin{bmatrix} T+1 \\ 2 \end{bmatrix}$ calculated distances. Assume that the shortest calculated distance (they belong to the same cluster) is between $X_m$ and $X_n$, and $H(X_m) \geq H(X_n)$, Shahyat algorithm removes $X_n$ from the candidate center list (it puts $X_n$ in the cluster with $X_m$ as the center).

Now we have $T$ candidate centers, we find the shortest distance (after removing $X_n$) and assume that the shortest distance is between $X_p$ and $X_q$, $(H(X_p) \geq H(X_q))$ therefore remove the $X_q$. This procedure resume until the number of candidate centers becomes equal to the needed number of centers.

When the number of candidate centers becomes equal to $C$, each pattern belongs to the nearest center. The clustering is finished and the clusters are determined.

## 14.3 Solving a Clustering Problem by Shahyat Algorithm

We have 40 patterns $X_j = (d_1, d_2)_{(j=1,2,\ldots,40)}$ in a 2-dimensional feature space $(d_1, d_2)$, and want to assign them to 3 clusters. The patterns are shown in Fig. 14.1, in which the horizontal axis represents feature $d_1$ and vertical axis represents feature $d_2$. Now we calculate the heat of patterns based on Shahyat algorithm. Assumed that our arbitrary decreasing function is an exponentially decreasing function, as below:

$$H(X_0) = \sum_{j=1}^{k} e^{Fune(\|X_0 - X_J\|)} \tag{14.5}$$

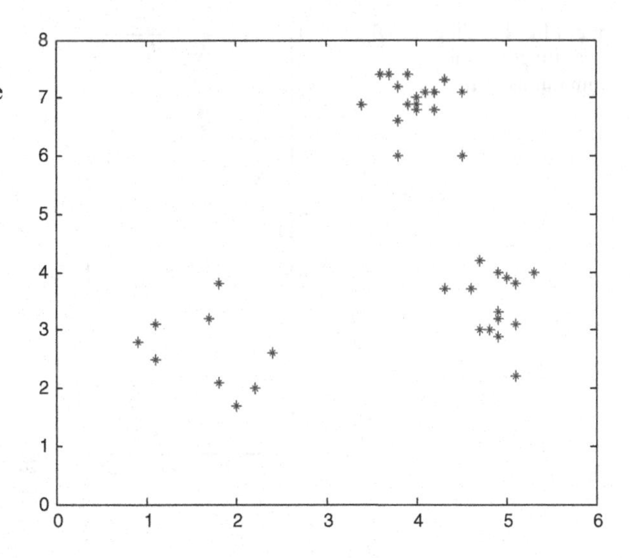


**Fig. 14.1** Separation of 40 patterns in the 2-dimensional feature space

where $H(X_0)$ is the heat of the arbitrary point $X_0$ affected by 40 pattern $\{X_1, X_2, \ldots, X_{40}\}$, and $Func(\|X_0 - X_j\|)$ is the function that correlates with the distance between point $X_0$ and point $X_J$ in the feature space. We consider $Func(\|X_0 - X_j\|)$ as below:

$$Func(\|X_0 - X_j\|) = \mu_1(d_{1X_0} - d_{1X_J})^2 + \mu_2(d_{2X_0} - d_{2X_J})^2 \tag{14.6}$$

where $X_0 = (d_{1_{X_0}}, d_{2_{X_0}}), X_j = (d_{1_{X_j}}, d_{2_{X_j}})$, and $\mu_1, \mu_2$ are the coefficients that determine the scale of the heat decrease in each direction. Noticing the distance between patterns in a cluster and between clusters at each direction we assume $\mu_1 = \mu_2 = 1$ (the problem give us this information).

So the Heat function for arbitrary point $X_0$ is:

$$H(X_0) = \sum_{j-1}^{k} e^{-\left[(d_{1X_0} - d1_{X_J})^2 + (d_{2X_0} - d_{2X_J})^2\right]} \tag{14.7}$$

According to Shahyat algorithm, we calculate the heat of patterns and then sort them from maximum to minimum. The diagram of Heat function versus path (from $X_1$ with maximum heat to $X_40$ with minimum heat) is shown in Fig. 14.2.

The diagram has 5 minima that heat difference between the minimum and previous pattern is higher than $H(X_k)$. In Fig. 14.3 the first pattern after each minimum is determined.

The candidate center collection is $\{X_1, X_{16}, X_{29}, X_{31}, X_{36}, X_{37}\}$.

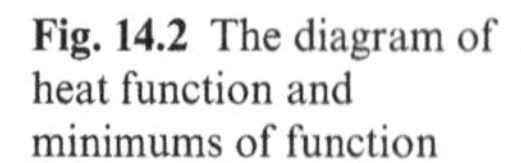

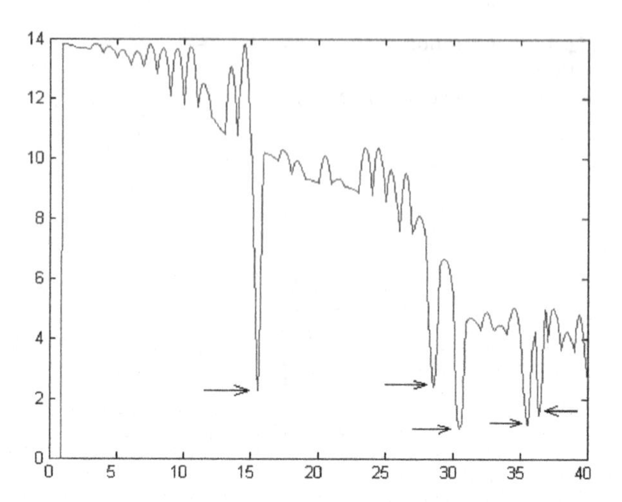


**Fig. 14.2** The diagram of heat function and minimums of function

The shortest distance is the distance between $X_1$ and $X_{2}9$ (the distance is 1.04) so Shahyat algorithm removes $X_{2}9$ from candidate center collection (because $H(X_1) \geq X(X_{29})$), after removing $X_{29}$ from the candidate center collection, the shortest distance is between $X_{16}$ and $X_{36}$ (it is 1.25).

Hence, the algorithm removes $X_{36}$ from candidate center collection (because $H(X_{16}) \geq H(X_{36})$).

Now we have 4 centers and need to remove one center, the shortest distance is between $X_{31}$ and $X_{37}$, so we remove $X_{37}$ (because $H(X_{3}1) \geq H(X_{37})$). Now we have 3 centers, therefore, we consider them as the main centers. The centers are shown in Fig. 14.4.

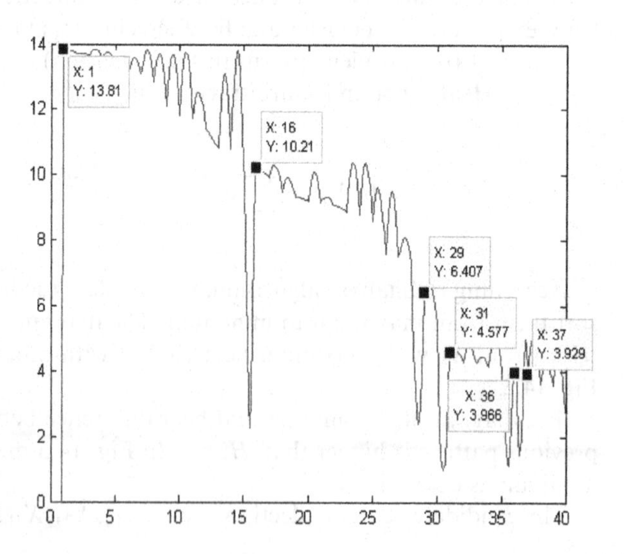


**Fig. 14.3** First pattern after each minimum is determined

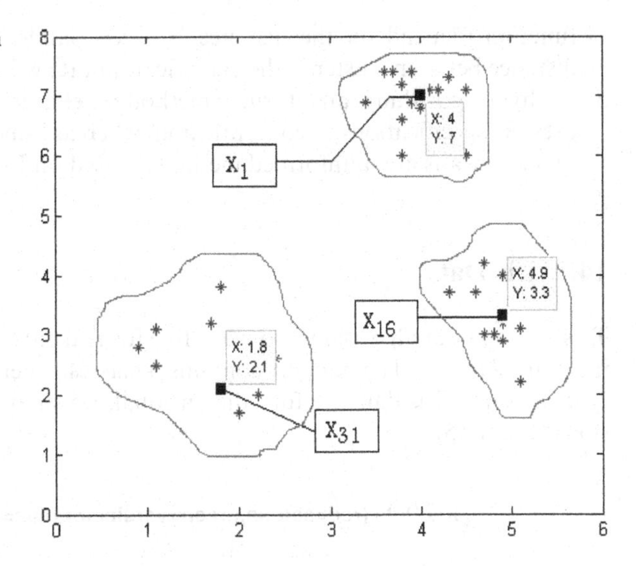


**Fig. 14.4** Clustered pattern

Now the centers are determined. Each pattern belongs to the nearest center. Now clusters are specific.

## 14.4 Specifications of Shahyat Algorithm

1. The algorithm is complete; this means that under each condition the algorithm can cluster and does not depend on special conditions. Because of in the clusters, the patterns are near together, and concentration in the clusters is higher than out of clusters, hence the heat in the clusters is more than out of clusters. So by defining an appropriate heat function the algorithm (based on heat in the clusters and out of clusters) can separate the clusters.
2. The maximum complexity of solving each problem by Shahyat algorithm is known, which depends on number of patterns, number of cluster that we want, distribution of patterns in the feature space and dimension of the patterns. It seems Shahyat algorithm after choosing candidate centers, works like nearest neighbor. But solving the classification problems by Shahyat algorithm has less complexity, necessary memory and needed processing relative to solving by k-means, nearest neighbor, fuzzy c-means.
3. running the algorithm once, we can classify the patterns into an arbitrary number of clusters, and it is not needed to run the algorithm again when we want to change the number of clusters. But k-means should process the algorithm procedures again if the number of clusters (that we want) changes.
4. For a certain problem, the algorithm has the ability to reduce the error (if error exists) by changing the type of Heat function and the coefficients in function. For example by varying the coefficients $\mu_1, \mu_2$ in exponential

function (depend on the distance between patterns in a cluster and the distance between clusters), the classification rate will change.

5. Shahyat algorithm is an informed method for clustering and classification. It uses the specifications of concentration, whereas k-means, nearest neighbor, fuzzy c-means are uninformed methods (based on Euclidean distance).

## 14.5 Iris Data

To show that our algorithm also work fine for real data sets, we use the famous iris plant data set. The data set contains 3 classes, where each class refers to a type of plants. Iris data are four dimensional; we show it in every two dimensions (Fig. 14.5).

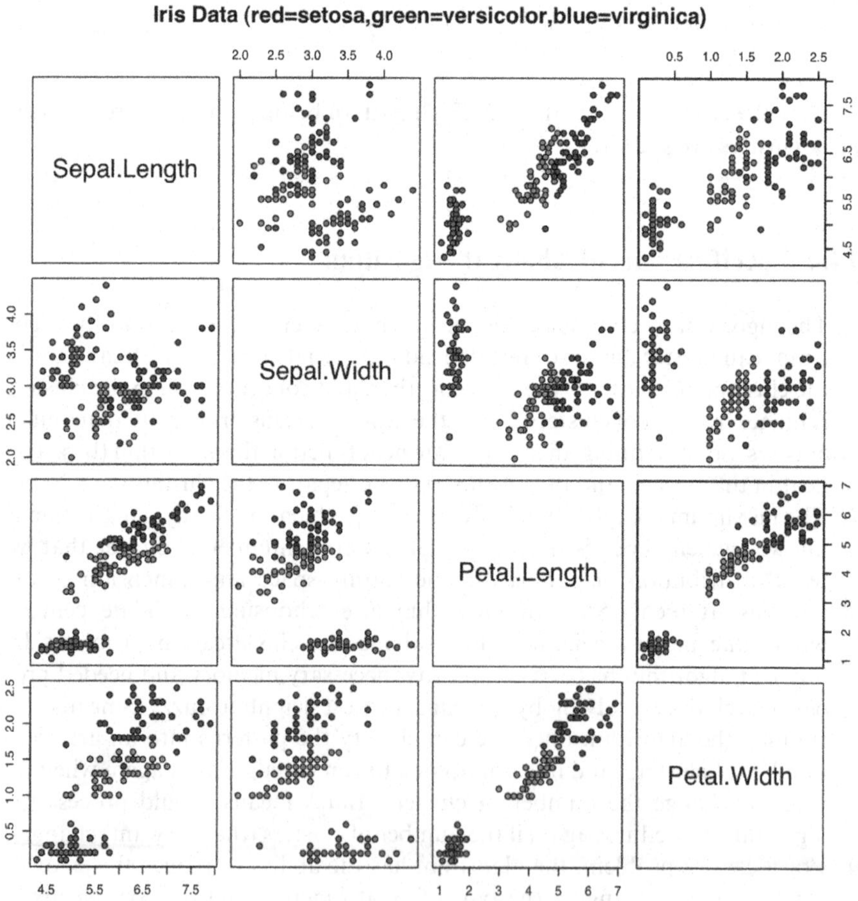

**Fig. 14.5** Iris data

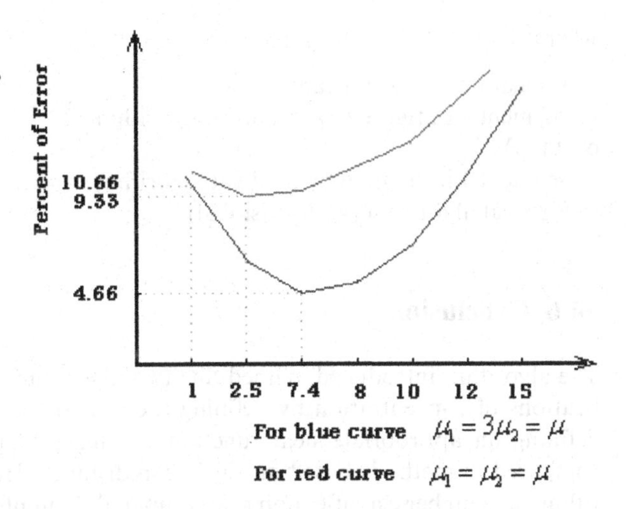


**Fig. 14.6** Percent of error relative to variation of $\mu_1, \mu_2$, segment 1 (*blue curve*) and segment 2 (*red curve*)

Now we consider segments 1 and 2 in Fig. 14.5.

We cluster them by Shahyat algorithm. Assume that the decreasing function $F$ of Shahyat algorithm is exponential with coefficients $\mu_1, \mu_2$. The diagram of clustering error relative to gradient of $\mu$ for two segments is shown in Fig. 14.6.

As we see, the error of clustering for segment 1 is lower than segment 2 (when we consider it only 2-dimensional). By considering iris data in 4-dimensional, Shahyat algorithm cluster iris data with classification rate 96%.

Most works report a classification rate that does not exceed 90%. K-means and fuzzy c-means, for example both have a classification rate of 89.3% on the iris data set. Alternative fuzzy c-means (AFCM) (Wu and Yang, 2002), reaches a classification rate of 91.3%.

The block diagram of classification rate is shown in Fig. 14.7.

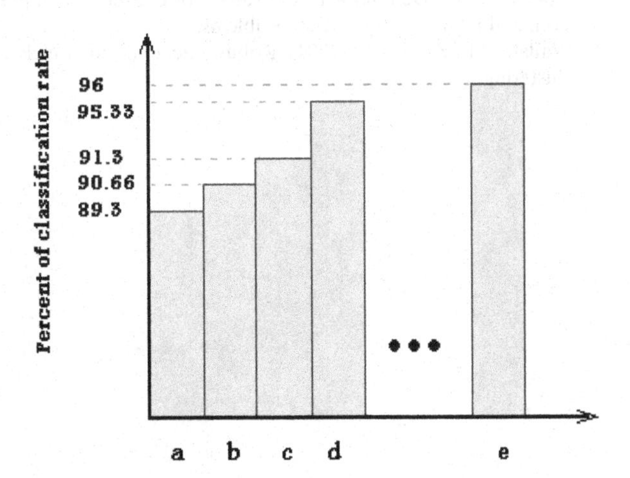


**Fig. 14.7** Block diagram of classification rate for iris data clustering

where:

a: k-means & fuzzy c-means
b: segment 2 in figure 6 by Shahyat algorithm
c: AFCM
d: segment 1 in figure 6 by Shahyat algorithm
e: Shahyat algorithm (4-dimensional)

## 14.6 Conclusion

The algorithm introduced, named Shahyat algorithm, makes use of the specifications of concentration by defining the heat function. This means that by defining an appropriate heat function, we have a larger classification rate compared to methods based on Euclidean distance. Hence, the Shahyat algorithm has a higher classification rate compared to uninformed methods that are only based on Euclidean distance.

## References

1. Alanzado AC (2003) Noise clustering using fuzzy c-means and mixture distribution model
2. Ding C, He X (2004) K-nearest-neighbor consistency in data clustering: incorporating local information into global optimization
3. Eick CF, Zeidat N, Vilalta R (2005) Using representative-based clustering for nearest neighbor dataset editing
4. Grabusts P, Borisov A (2002) Using grid-clustering methods in data classification
5. Kanungo T, Mount DM, Netanyahu NS, Piatko CD (2003) A local search approximation algorithm for k-means clustering
6. Osipova N (2004) Classification and clustering methods development and implementation for unstructured documents collections. Department of Programming Technology, Faculty of Applied Mathematics and Control Processes, St. Petersburg State University
7. Roze K (1998) Deterministic annealing for clustering compression, classification, regression, and related optimization problems.
8. Wallace M, Kollias S (2004) Robust, generalized, quick and efficient agglomerative clustering

# Chapter 15
# A Modified Electromagnetism-Like Algorithm Based on a Pattern Search Method

**Ana Maria A.C. Rocha and Edite M.G.P. Fernandes**

**Abstract** The Electromagnetism-like (EM) algorithm, developed by Birbil and Fang (J Global Optim 25(3):263–282, 2003) is a population-based stochastic global optimization algorithm that uses an attraction-repulsion mechanism to move sample points towards optimality. A typical EM algorithm for solving continuous bound constrained optimization problems performs a local search in order to gather information for a point, in the population. Here, we propose a new local search procedure based on the original pattern search method of Hooke and Jeeves, which is simple to implement and does not require any derivative information. The proposed method is applied to different test problems from the literature and compared with the original EM algorithm.

## 15.1 Introduction

Many real life global optimization problems that arise in areas such as physics, chemistry, and molecular biology involve multi-modal and non-differentiable nonlinear functions of many variables that are difficult to handle by conventional gradient-based algorithms. As a result, many researchers have devoted themselves to finding reliable stochastic global optimization methods that do not require any derivative computation. Recently, Birbil and Fang proposed the electromagnetism-like (EM) algorithm that is a population-based stochastic search method for global optimization [1, 2]. This algorithm simulates the electromagnetism theory of physics by considering each point in the population as an electrical charge. The method uses an attraction-repulsion mechanism to move a population of points towards optimality. The original algorithm incorporates a simple random local search procedure that is applied coordinate by coordinate to one point only or to all points in the population [2]. In an attempt to improve the accuracy of the results and to accelerate convergence we propose

A.M.A.C. Rocha (✉)
Minho University, Campus de Gualtar, 4710-057 Braga, Portugal
e-mail: arocha@dps.uminho.pt

N. Mastorakis et al. (eds.), *Proceedings of the European Computing Conference*, Lecture Notes in Electrical Engineering 28,
DOI 10.1007/978-0-387-85437-3_15, 

a modification to the original EM algorithm by replacing the random local search with a pattern search method [6] with guaranteed convergence.

The method is to work on nonlinear optimization problems with box constraints in the following form:

$$\min f(x), \text{subject to} x \in \Omega \tag{15.1}$$

where $f : \Re^n \rightarrow \Re$ is a nonlinear function and $\Omega = \{x \in \Re^n : l \leq x \leq u\}$ is a bounded feasible region.

The paper is organized as follows. Section 15.2 briefly introduces the original EM algorithm and Section 15.3 is devoted to a description of the main ideas concerning the Hooke and Jeeves pattern search method. Section 15.4 contains the numerical results and some conclusions are drawn in Section 15.5.

## 15.2 Electromagnetism-Like Algorithm

The EM algorithm starts with a population of randomly generated points from the feasible region. Analogous to electromagnetism, each point is a charged particle that is released to the space. The charge of each point is related to the objective function value and determines the magnitude of attraction of the point over the population. The better the objective function value, the higher the magnitude of attraction. The charges are used to find a direction for each point to move in subsequent iterations.

The regions that have higher attraction will signal other points to move towards them. In addition, a repulsion mechanism is also introduced to explore new regions for even better solutions. The following notation is used: $x^i \in \Re^n$ denotes the $i$th point of a population; $x^{best}$ is the point that has the least objective function value; $x_k^i \in \Re (k = 1, \ldots, n)$, is the $k$th coordinate of the point $x^i$ of the population; $m$ is the number of points in the population; MaxIt is the maximum number of EM iterations; LSIt denotes the maximum number of local search iterations; and $\delta$ is a local search parameter, $\delta \in [0,1]$.

The EM algorithm comprises four main procedures.

**Algorithm EM** ($m$, MaxIt, LSIt, $\delta$)
*Initialize*()
iteration $\leftarrow$ 1
**while** termination criteria are not satisfied **do**
  *Local*(LSIt, $\delta$)
  $F \leftarrow$ *CalcF*()
  *Move*($F$)
  iteration $\leftarrow$ iteration + 1
**end while**

Details of each procedure follow. *Initialize* is a procedure that aims to randomly generate $m$ points from the feasible region. Each coordinate of a point $x_k^i (k = 1, \ldots, n)$ is assumed to be uniformly distributed between the corresponding upper and lower bounds, i.e., $x_k^i = l_k + \lambda(u_k - l_k)$ where $\lambda \sim U(0, 1)$. After computing the objective function value for all the points in the population, the procedure identifies the best point, $x^{best}$, which is the point with the best function value.

The *Local* procedure performs a local refinement and can be applied to one point or to all points in the population. The local search presented in [2] is a random line search algorithm that is applied coordinate by coordinate only to the best point in the population. First, the procedure computes the maximum feasible step length, $Length = \delta(max_k(u_k - l_k))$, based on $\delta$. This quantity is used to guarantee that the local search generates feasible points. Second, the best point is assigned to a temporary point $y$ to store the initial information. Next, for each coordinate $k$, a random number $\lambda$ between zero and one is selected as a step length and the point $y_k$ is moved along that direction, $y_k = y_k + \lambda\ Length$. If an improvement is observed, within LSIt iterations, the best point is replaced by $y$ and the search along that coordinate ends.

The *Calc F* procedure aims to compute the total force exerted on a point via other points. First a charge-like value, $q^i$, that determines the power of attraction or repulsion for the point $x^i$, is assigned. The charge of the point is calculated according to the relative efficiency of the objective function values, i.e.,

$$q^i = \exp\left(-n\frac{(f(x^i) - f_{best})}{\sum_{j=i}^{m}(f(x^i) - f_{best})}\right), i = 1, \ldots, m \tag{15.2}$$

Hence, the points that have better objective function values possess higher charges. The total force vector $F^i$ exerted on each point is calculated by adding the individual component forces, $F_j^i$, between any pair of points $x^i$ and $x^j$,

$$F^i = \sum_{j \neq i}^{m} F_j^i = \begin{cases} (x^j - x^i)q^i q^j/(\|x^j - x^i\|^3), & \text{if } f(x^j) < f(x^i) \\ (x^i - X^j)q^i q^j/(\|x^j - x^i\|^3), & \text{if } f(x^j) \geq f(x^i) \end{cases} \tag{15.3}$$

$i = 1, \ldots, m.$

Finally, the *Move* procedure uses the total force vector $F^i$, to move the point $x^i$ in the direction of the force by a random step length,

$$x_k^i = x_k^i + \lambda\frac{F^i}{\|F^i\|}(RNG), i = 1, 2, \ldots, m \text{ and } i \neq best \tag{15.4}$$

where $RNG$ is a vector that contains the allowed range of movement towards the lower bound $l_k$, or the upper bound $u_k$, for each coordinate $k$. The random step length $\lambda$ is assumed to be uniformly distributed between 0 and 1. Note that feasibility is maintained by using the normalized force exerted on each point. The best point, $x^{best}$, is not moved and is carried out to the subsequent iteration.

## 15.3 Hooke and Jeeves Pattern Search Method

In this section, we describe our modification to the original EM algorithm. In this algorithm, the *Local* procedure is based on a random line search method [2]. Here a new *Local* procedure based on the Hooke and Jeeves (HJ) pattern search algorithm is proposed. This is a derivative-free method that searches in the neighbourhood of a point $x^i$ for a better approximation via exploratory and pattern moves [4, 6]. To reduce the number of function evaluations, the HJ pattern search algorithm is applied to the current best point only. This algorithm is a variant of the coordinate search, in the sense that incorporates a pattern move to accelerate the progress of the algorithm, by exploiting information obtained from the search in previous successful iterations. The exploratory move carries out a coordinate search (a search along the coordinate axes) about the best point, with a step length $\delta$. If a new trial point, $y$, with a better function value than $x^{best}$ is encountered, the iteration is successful. Otherwise, the iteration is unsuccessful and $\delta$ should be reduced. If the previous iteration was successful, the vector $y - x^{best}$ defines a promising direction and a pattern move is then implemented, meaning that the exploratory move is carried out about the trial point $y + (y - x^{best})$, rather than about the current point $y$. Then, if the coordinate search is successful, the returned point is accepted as the new point; otherwise, the pattern move is rejected and the method reduces to coordinate search about $y$. Please see [4] for details. To ensure feasibility in the HJ pattern search algorithm an exact penalty strategy is used. This technique considers solving

$$\min F(x) \equiv \begin{cases} f(x) & \text{if} x \in \Omega \\ \infty & \text{otherwise} \end{cases} \tag{15.5}$$

rather than problem (15.1). This means that any infeasible trial point is rejected, since the objective function value is $\infty$.

## 15.4 Numerical Results

Computational tests were performed on a PC with a 3 GHz Pentium IV microprocessor and 1 Gb of memory. We compare the original EM algorithm, as described in Section 15.2, with the herein proposed EM algorithm modified with the HJ *Local* procedure, described in Section 15.3. We use a collection of 18 test functions [2, 3, 5] (see Table 15.1).

**Table 15.1** Test functions and the corresponding parameters used by EM

| Test function | $n$ | Box constraints | $f_{global}$ | $m$ | MaxIt | LSIt | $\delta$ |
|---|---|---|---|---|---|---|---|
| Shekel5 | 4 | $[0,10]^4$ | −10.153200 | 40 | 150 | 10 | 1.00E-03 |
| Shekel7 | 4 | $[0, 10]^4$ | −10.402941 | 40 | 150 | 10 | 1.00E-03 |
| Shekel10 | 4 | $[0, 10]^4$ | −10.536410 | 40 | 150 | 10 | 1.00E-03 |
| Hartman3 | 3 | $[0,1]^3$ | −3.862782 | 30 | 75 | 10 | 1.00E-03 |
| Hartman6 | 6 | $[0,1]^6$ | −3.322368 | 30 | 75 | 10 | 1.00E-03 |
| Goldstein-price | 2 | $[-2,2]^2$ | 3.000000 | 20 | 50 | 10 | 1.00E-03 |
| Branin | 2 | $[-5,10]\times[0,15]$ | 0.397887 | 20 | 50 | 10 | 1.00E-03 |
| Six-hump camel | 2 | $[-3,3]\times[-2,2]$ | −1.031628 | 20 | 50 | 10 | 1.00E-03 |
| Shubert | 2 | $[-10, 10]^2$ | −186.730909 | 20 | 50 | 10 | 1.00E-03 |
| Griewank | 2 | $[-100, 100]^2$ | 0.000000 | 30 | 100 | 20 | 1.00E-03 |
| Himmelblau | 2 | $[-6, 6]^2$ | 0.000000 | 10 | 50 | 5 | 1.00E-03 |
| Sine envelope | 2 | $[-0.5, 0.5]^2$ | 0.000000 | 20 | 75 | 10 | 5.00E-04 |
| Bohachevsky | 2 | $[-10, 10]^2$ | 0.000000 | 20 | 75 | 20 | 1.00E-03 |
| Easom | 2 | $[-10, 10]^2$ | −1.000000 | 20 | 50 | 10 | 1.00E-03 |
| Hump | 2 | $[-5, 5]^2$ | 0.000000 | 20 | 50 | 10 | 1.00E-03 |
| Spherical | 2 | $[-100, 100]^2$ | 0.000000 | 30 | 75 | 20 | 1.00E-03 |
| Three-hump | 2 | $[-5, 5]^2$ | 0.000000 | 20 | 50 | 10 | 1.00E-03 |
| Zakharov4 | 4 | $[-5, 10]^4$ | 0.000000 | 30 | 75 | 20 | 1.00E-03 |

The first four columns of the table refer to the name of the function, the dimension of the problem, $n$, the default box constraints, and the known global optimum, $f_{global}$. The last four columns list the parameters used by EM for each function.

The results obtained by the original EM algorithm are shown in Table 15.2 and the results of the EM algorithm modified with the HJ *Local* procedure are presented in Table 15.3. We use average results for comparison, over 25 runs. Tables 15.2 and 15.3 report average number of function evaluations, $Ev_{avg}$, the average best function values, $f_{avg}$, and the best function value, $f_{best} = \min(f^i_{best}, i = 1, \ldots, nruns)$, over $nruns = 25$ runs. Values of the mean absolute error

$$\text{MAP} = \frac{|f_{avg} - f_{global}|}{n} \tag{15.6}$$

and the standard deviation

$$\text{SD} = \sqrt{\frac{\sum_{i=1}^{nruns} \left(f^i_{best} - f^i_{avg}\right)^2}{nruns}} \tag{15.7}$$

are also listed for each problem.

The termination criteria and the used parameters are the ones proposed in [2]. Thus, both algorithms stop when the number of iterations exceeds MaxIt unless the relative error in the best objective function value, with respect to

**Table 15.2** Results of original EM with *Local* procedure applied to the best point

| Test function | $Ev_{avg}$ | $f_{avg}$ | $f_{best}$ | MAE | SD |
|---|---|---|---|---|---|
| Shekel5* | 1865 | −10.152770 | −10.153163 | 0.000108 | 0.000270 |
| Shekel7 | 1480 | −10.402471 | −10.402884 | 0.000117 | 0.000486 |
| Shekel10 | 1486 | −10.535939 | −10.536256 | 0.000118 | 0.000243 |
| Hartman3 | 1260 | −3.862487 | −3.862722 | 0.000098 | 0.000111 |
| Hartman6 | 1850 | −3.322267 | −3.322365 | 0.000017 | 0.000091 |
| Goldstein Price | 488 | 3.000141 | 3.000001 | 0.000071 | 0.000077 |
| Branin | 512 | 0.397906 | 0.397888 | 0.000009 | 0.000012 |
| Six Hump Camel | 291 | −1.031599 | −1.031628 | 0.000015 | 0.000026 |
| Shubert | 357 | −186.723068 | −186.730906 | 0.003920 | 0.005750 |
| Griewank** | 1582 | 0.000062 | 0.000008 | 0.000031 | 0.000049 |
| Himmelblau | 347 | 0.000047 | 0.000008 | 0.000023 | 0.000023 |
| Sine envelope | 518 | 0.000044 | 0.000002 | 0.000022 | 0.000027 |
| Bohachevsky | 793 | 0.000063 | 0.000001 | 0.000032 | 0.000041 |
| Easom | 489 | −0.999966 | −1.000000 | 0.000017 | 0.000029 |
| Hump | 287 | 0.000044 | 0.000001 | 0.000022 | 0.000027 |
| Spherical | 903 | 0.000033 | 0.000001 | 0.000017 | 0.000030 |
| Three-hump | 382 | 0.000041 | 0.000001 | 0.000021 | 0.000030 |
| Zakharov4 | 1621 | 0.000068 | 0.000014 | 0.000017 | 0.000022 |

$f_{global}$, is less than 0.01%. In the HJ algorithm, the factor used to reduce $\delta$, whenever an unsuccessful iteration is found, is 0.1 and the minimum step length allowed was $1 \times 10^{-8}$. In the tables, $*$ means that 4 (in Table 15.1) and 5 (in Table 15.2) runs of Shekel5 did not converge and $**$ means that 3 runs of Griewank did not converge.

**Table 15.3** Results of EM with HJ *Local* procedure applied to the best point

| Test function | $Ev_{avg}$ | $f_{avg}$ | $f_{best}$ | MAE | SD |
|---|---|---|---|---|---|
| Shekel5* | 1880 | −10.153169 | −10.153186 | 0.000008 | 0.000023 |
| Shekel7 | 1677 | −10.402855 | −10.402934 | 0.000022 | 0.000205 |
| Shekel10 | 1679 | −10.536320 | −10.536405 | 0.000022 | 0.000177 |
| Hartman3 | 1482 | −3.862541 | −3.862775 | 0.000080 | 0.000106 |
| Hartman6 | 2378 | −3.322308 | −3.322361 | 0.000010 | 0.000075 |
| Goldstein price | 405 | 3.000036 | 3.000001 | 0.000018 | 0.000054 |
| Branin | 372 | 0.397900 | 0.397888 | 0.000006 | 0.000012 |
| Six Hump camel | 261 | −1.031588 | −1.031628 | 0.000020 | 0.000025 |
| Shubert | 641 | −186.714293 | −186.730827 | 0.008308 | 0.067471 |
| Griewank** | 2095 | 0.000018 | 0.000000 | 0.000009 | 0.000017 |
| Himmelblau | 294 | 0.000018 | 0.000000 | 0.000009 | 0.000021 |
| Sine envelope | 423 | 0.000025 | 0.000000 | 0.000012 | 0.000030 |
| Bohachevsky | 1892 | 0.000018 | 0.000000 | 0.000009 | 0.000011 |
| Easom | 476 | −0.999965 | −1.000000 | 0.000018 | 0.000033 |
| Hump | 315 | 0.000023 | 0.000000 | 0.000012 | 0.000032 |
| Spherical | 514 | 0.000031 | 0.000000 | 0.000015 | 0.000032 |
| Three-hump | 337 | 0.000027 | 0.000001 | 0.000013 | 0.000021 |
| Zakharov4 | 1430 | 0.000013 | 0.000001 | 0.000003 | 0.000027 |

The results obtained with the EM algorithm modified with the HJ *Local* procedure are better than the ones produced by the algorithm of Section 15.2, as far as the accuracy of the results is concerned. The proposed algorithm achieves in general the lowest numerical errors (MAE) and lowest standard deviations (SD) for 25 runs. However, Table 15.3 reveals in some cases larger number of function evaluations.

## 15.5 Conclusions

We have studied the Electromagnetism-like algorithm and implemented a new *Local* procedure based on a pattern search method. This new algorithm was applied to different test problems from the literature and compared with the original EM. The preliminary results seem promising. Future developments will focus on extending the numerical experiments to other test functions with larger dimensions and many local minimizers in the feasible region. The purpose here is to analyze the pattern search ability to drive the best point towards the global minimizer instead of to a non-global one.

## References

1. Birbil SI (2002) Stochastic global optimization techniques. PhD Thesis, North Carolina State University
2. Birbil SI, Fang S (2003) An electromagnetism-like mechanism for global optimization. J Global Optim 25(3):263–282
3. Dixon LCW, Szegö GP (1978) The global optimization problem: an introduction. In: Dixon LCW, Szegö GP (eds) Towards Global Optimisation 2. North-Holland, Amsterdam, pp 1–15
4. Hooke R, Jeeves TA (1961) Direct search solution of numerical and statistical problems. J Assoc Comput Mach 8:212–229
5. Huyer W, Neumaier A (1999) Global optimization by multilevel coordinate search. J Global Optim 14:331–355
6. Lewis RM, Torczon V (1999) Pattern search algorithms for bound constrained minimization. SIAM J Optim 9(4):1082–1099

# Chapter 16
# Latency Hiding and Adaptability for Parallel Iterative Algorithms

**Javier Roca, José Antonio Álvarez, and Jose Jesús Fernández**

**Abstract** Switching from the MPI process model to a threaded model in the parallel environment via user level threads takes advantage of the existing concurrence in applications. Major aims for this work are to face the computational demands of scientific applications by exploiting concurrency, minimizing latency due to message passing, and to present and evaluate strategies to implement adaptability when imbalance occurs.

## 16.1 Introduction

In contrast to the legacy and structured programming model used in standard MPI based applications, multithreaded programming [1] provides a way to divide a program into different pieces that can be executed concurrently. Moreover using user level threads [2] in places where concurrence can be exploited would allow us to achieve latency hiding, better scalability, skills to avoid overhead on processors due to faster context switching and abilities to migrate threads for load balancing purposes. On the other hand, data locality is an important issue that ties up a large amount of scientific applications. Image processing such as filtering or 3D tomographic reconstruction algorithms [3], are examples. In the design of a parallel strategy for this kind of application, data distributions have to be devised so that the locality is preserved as much as possible. Otherwise, the application will exhibit poor performance due to excessive communications. Data distributions that preserve locality are specially important in systems where the workload is dynamically reassigned.

Our application, *BICAV* [4], addresses the tomographic reconstruction problem using iterative methods. These methods are far superior to the standard

J. Roca (✉)
Departamento de Arquitectura de Computadores y Electrónica, Universidad de Almería, Almeria, Spain
e-mail: jroca@ace.ual.es

N. Mastorakis et al. (eds.), *Proceedings of the European Computing Conference*, Lecture Notes in Electrical Engineering 28,
DOI 10.1007/978-0-387-85437-3_16, 

technique, weighted back projection, in terms of reconstruction quality [4]. However, they are much more computationally demanding.

In this work, we present and evaluate a parallel implementation of iterative reconstruction methods using a multithreaded approach with AMPI [5]. This framework embeds MPI processes into user level threads, a more efficient technique in aspects like context switching, migration of tasks to other processors. The aim is to face the computational demands of these methods by exploiting concurrency and minimizing latency due to message passing. Evaluating the efficiency achieved when applying adaptability techniques, in scenarios where load imbalance – either intrinsic or extrinsic – arises, is another major objective. For this purpose, a particular implementation of the 3D Jacobi algorithm [6], a classical iterative method for solving partial differential equations that exhibits strong locality properties alike – when data is distributed along only one dimension – to those presented by *BICAV*. Jacobi is, therefore, used to test the load balancing strategy implemented, that seizes data locality properties inherent to aforementioned applications. Experiments performed for Jacobi show gains and key spots to work on when using load balancing strategy with our tomographic applications

This paper is organized as follows. Section 16.2 analyzes the iterative reconstruction techniques and its data dependences. Section 16.3 describes the adaptability issues aroused when load balancing is applied. Section 16.4 analyzes how the threaded implementation behaves when compared with its MPI version, and also describes and compares the load balancing strategies. Finally, in the last section, the conclusions are exposed.

## 16.2 Parallel Iterative Reconstruction Methods

Series expansion reconstruction methods assume that the 3D object, or function $f$, can be approximated by a linear combination of a finite set of known and fixed basis functions, with density $x_j$. The aim is to estimate the unknowns, $x_j$. These methods are based on an image formation model where the measurements depend linearly on the object in such a way that $y_i = \sum_{j=1}^{J} l_{i,j} \cdot x_j$, where $yi$ denotes the$i$th measurement of $f$ and $l_{i,j}$ the value of the $i$th projection of the $j$th basis function. Under those assumptions, the image reconstruction problem can be modeled as the inverse problem of estimating the $x_j$'s from the $y_i$'s by solving the system of linear equations aforementioned.

Assuming that the whole set of equations in the linear system may be subdivided into $B$ blocks, a generalized version of component averaging methods, *BICAV*, can be described. The processing of all the equations in one of the blocks produces a new estimate. All blocks are processed in one iteration of the algorithm. This technique produces iterates which converge to a weighted least squares solution of the system.

A volume can be considered made up of 2D slices. When voxel basis functions are used, the slices are independent. However, the use of the spherically symmetric volume elements (blobs) [7], makes slices interdependent because of blob's overlapping nature.

We have implemented the parallel iterative reconstruction method adapting legacy MPI code into a multithreaded environment by means of Adaptive MPI (AMPI). The block iterative version of the component averaging methods, *BICAV*, has been parallelized following the Single Program Multiple Data (SPMD) approach [4]. The volume is decomposed into slabs of slices that will be distributed across the nodes (for MPI) or threads (for AMPI). The threads can, thereby, process their own data subdomain, however, the interdependence among neighbor slices due to the blob extension makes necessary the inclusion of redundant slices into the slabs. In addition, there must be a proper exchange of information between neighbor nodes to compute the forward-projection or the error back-projection for a given slab and to keep redundant slices updated. These communication points constitute synchronization points in every pass of the algorithm in which the nodes must wait for the neighbors. The amount of communications is proportional to the number of blocks and iterations. Reconstruction yields better results as the number of blocks is increased

In any parallelization project where communication between nodes is involved, latency hiding becomes an issue. That term stands for overlapping communication and computation so as to keep the processor busy while waiting for the communications to be completed. Our improvement proposal achieves latency hiding via user level threads.

## 16.3 Adaptability

In general, load balancing strategies are based on the distinction of heavily versus lightly loaded processors at the moment in which load balancing decision is taken by the system. The main issue consists then of migrating tasks among processors so that the nodes in the system get an even workload. Nevertheless, in general, such strategies do not take into consideration data locality, and they may assign pieces of work that share data to different processors. Consequently, an unnecessary increment of communications flow would appear.

The strategy for locality-preserving load balancing (RakeLB) was implemented as a centralized load balancer. For RakeLB, two processors are neighbours (adjacent) if both have at least a thread with data in common. RakeLB has been implemented taking advantage of the migration facilities, also, provided by AMPI. Then the migration of workload among nodes is carried out in terms of threads. Employing user level threads make load balancing issues easier in SPMD applications. These user level threads own an universal ID, the rank, this rank establishes an order relationship between them. When computation is ignited, each thread picks up a chunk of data to work on; therefore, threads with consecutive ranks will need, at some point in time, to communicate with each other. Therefore, thread migration decisions should conserve as much as possible the established thread order. A FIFO class structure keeps threads ordered during migrations. When threads migration is advised, those with the minimal data locality restrictions in the node (the ones occupying the end of the stack) will be moved. RakeLB behaviour has been contrasted with a, standard centralized, dynamic load balancing strategy, like Greedy [8].

## 16.4 Results

The platform used to evaluate our algorithms was our research group cluster, VERMEER, which owns as a front-end a dual Pentium IV XEON 3.2 Ghz, 512 KB Cache node with Gentoo Linux, Kernel 2.4.32 SMP and two Pentium IV XEON 3.06 Ghz, 512 KB L2 Cache per computing node with 2 GB of ECC DDR SDRAM, connected via 1 Gb Ethernet.

This work pursues the study of the gain obtained by the multithreaded implementation of *BICAV*, which overlaps computation/communication. Comparison to its MPI version, where latency hiding is achieved using advanced programming techniques, to interleave computations and communication, was also an aim. Thereby, scaling tests were carried out varying the number of threads/processor and the number of processors, for both versions AMPI and MPI. As a concluding phase, the load balancing strategy developed (RakeLB), is evaluated and compared to Greedy, a standard strategy. The initial set of experiments was intended to quantify the influence of the number of blocks over the global performance. As the number of blocks increases, the convergence of the algorithm is faster, but the amount of communications also increases. This test was performed on the Vermeer cluster using up to 32 physical processors.

Figure 16.1 shows the speedup of the MPI version and the AMPI version with 128 threads, for two different volumes (256 and 512), using several numbers of blocks $K$. It can be observed that below a threshold of K=64 both versions seem to behave similarly, showing slight improvement in AMPI. But above that threshold, and as K is increased, AMPI behaves better than MPI especially for more than 16 processors. AMPI is getting benefits from the hidden concurrency by means of an intelligent overlapping of running thread communication and sleeping threads' computation still to be done. Speedup curves in Fig. 16.1 clearly show that the application implemented with MPI loses the linear speedup when using more than eight processors. However, AMPI keeps speedup almost linear up to 32 processors. Therefore, thread switching succeeds in maintaining an optimal speedup.

In the subsequent sets of experiments, the 3D version of the Jacobi algorithm was used because it exhibits stronger data dependence and, hence, emphasizes the influence of data locality preservation on the performance. As the bigger

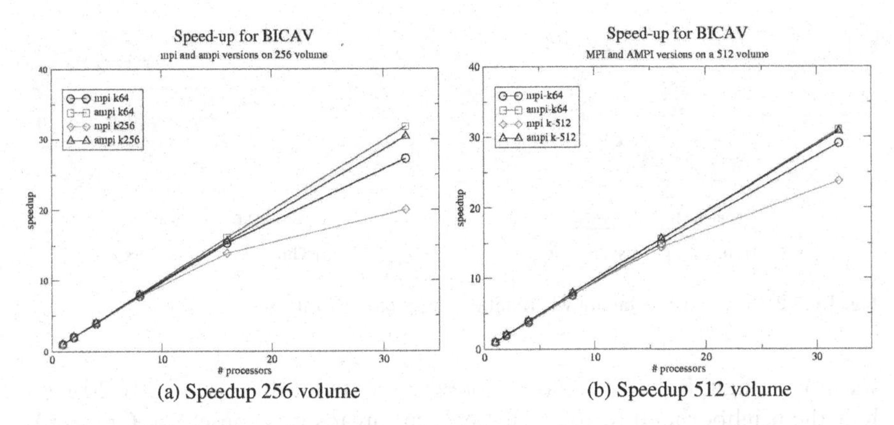


(a) Speedup 256 volume (b) Speedup 512 volume

**Fig. 16.1** Speedup in Vermeer cluster for volumes of dimensions 256 and 512

number of threads are created, the higher accuracy we obtain, but also computational load increases. Tests carried out were characterized by Jacobi performing its work along ten iterations and the load balancer strategy being invoked three times. In order to simulate imbalanced scenarios, processors were loaded with random workload (background load). At the initial step, threads from our application were evenly distributed among nodes.

A formal prove of imbalance was carried out by employing as a metric the standard deviation of the whole system load, normalized to the average load, The initial imbalance value reflected by was over 0.5. After load balancing, both strategies achieved a similar $\sigma$ value (0.051 for GreedyLB and 0.045 for RakeLB), but there's a key point not stated yet, *data dependency*.

After load balancing, a complete homogeneous load distribution is achieved (almost zero). Although load balancing strategies get an almost alike load distribution, it is key to note that these distributions are completely different in terms of data locality, an aspect that turns out to be an issue for performance when data locality is a significant characteristic for applications.

Figure 16.2 intends to show that although each load balancer did a reasonable load distribution to respond to the imbalanced situation, for this application that

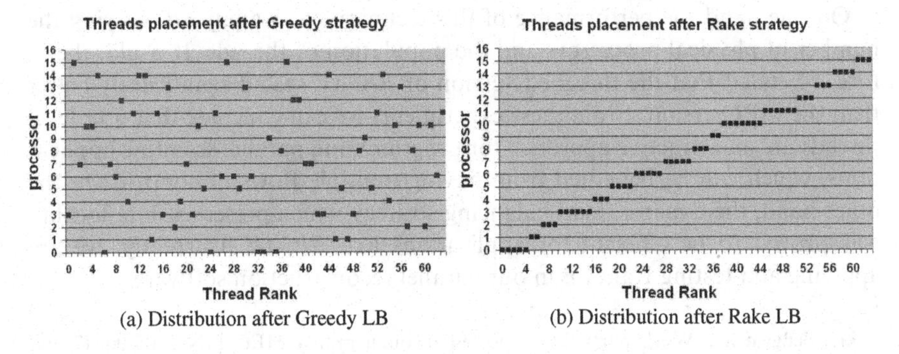


(a) Distribution after Greedy LB (b) Distribution after Rake LB

**Fig. 16.2** Thread placement, before and after the first balancer call

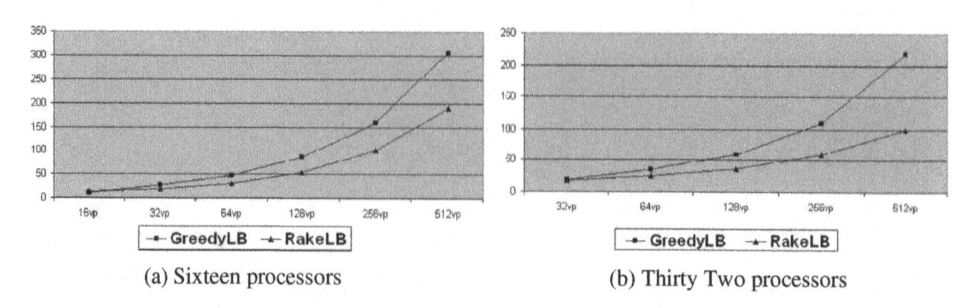


(a) Sixteen processors (b) Thirty Two processors

**Fig. 16.3** Performance of Jacobi when data locality based strategies are used

was not enough. RakeLB algorithm preserved data locality, see Fig. 16.2(b), on how the neighborhood relationship between threads was conserved. GreedyLB strategy Fig. 16.2(a) balanced load without paying attention to locality issues, and not preserve the original ordering among threads when they migrate to other nodes. As stated before, the version of Jacobi used is characterized by a proportional growth of computation as the number of threads is increased. As seen in Fig. 16.3, scaling for the application being balanced with RakeLB experiments is a better behaviour than its counterpart strategy. In every test, keeping the maximum number of neighboring threads in the same processor as RakeLB does, leads to a performance improvement in the application. This performance improvement begins to be much more significant for Jacobi when balanced with the RakeLB strategy in contrast with the cases where Jacobi was balanced with Greedy strategy.

## 16.5 Discussion and Conclusions

Through experiments, using AMPI, we illustrated how one of our scientific iterative applications was directly benefited as the cpu and walltime gap per iteration was almost zero, due to an optimal concurrence exploitation for the threaded version, in contrast to the MPI version which did not respond so well.

On one hand the performance of this technique was analyzed, varying the number of physical processors and the population of threads. It could, therefore, be stated that the threaded version of *BICAV* scaled significantly better than the MPI version. From these experiments we can conclude that user level threads are an attractive approach for programming parallel scientific applications, which can be benefited from having multiple flows of control. On the other hand, the dynamic load balancing strategy which preserves data locality was proved to be efficient for applications like *BICAV*. Currently, we are applying and testing RakeLB in our parallel reconstruction software.

**Acknowledgements** Work partially supported through grants MEC-TIN 2005-00447 and JA-P06-TIC-01426.

## References

1. Pancake CM (1993) Multithreaded languages for scientific and technical computing. In: Proceedings of the IEEE 81(2)
2. Price GW, Lowenthal DK (2003) A comparative analysis of fine-grain threads packages. J Parallel Distr Comput, College Station, Texas, 63:1050–1063
3. Kak AC, Slaney M (2001) Principles of computerized tomographic imaging. SIAM Society for Industrial and Applied Mathematics
4. Fernández J, Lawrence AF, Roca J, García I, Ellisman MH, Carazo JM (2002) High performance electron tomography of complex biological specimens. J Struct Biol 138:6–20
5. Huang Ch, Zheng G, Kumar S, Kale LV (2006) Performance evaluation of Adaptive MPI. In: Proceedings of ACM SIGPLAN Symposium on Principles and Practice of Parallel Programming
6. Press W, Teukolsky S, Vetterling W, Flannery B (1992) Numerical recipes in C. The art of scientific computing. Cambridge University Press
7. Matej S, Lewitt R, Herman G (1996) Practical considerations for 3-D image reconstruction using spherically symmetric volume elements. IEEE Trans Med Imag 15:68–78
8. Aggarwal G, Motwani R Zhu (2006) The load rebalancing problem. J Algorithm 60:42–59

## References

# Chapter 17
# Clustering the Linearly Inseparable Clusters

**Hamid Reza Shahdoosti and Omid Khayat**

**Abstract** In this paper a new method for clustering is introduced that works fine for linearly inseparable clusters. This algorithm uses the of patterns' concentration specification. We consider each pattern as a heat source to determine concentration by heat effect. Because of high concentration in clusters, heat in the clusters is higher than out of them, and this fact is used to determine and bound the clusters. CLIC algorithm can be implemented for any linearly inseparable cluster. The classification rate of this algorithm is higher than previously known methods (k-means, fuzzy c-means, etc). Finally, we cluster real iris data by using CLIC algorithm.

## 17.1 Introduction

There are several methods of pattern clustering and classification. Clustering and classification are among the most important topics in artificial intelligence, such as signal processing, image processing, and pattern recognition. Several popular methods are known for clustering and classification such as k-means, nearest neighbor and fuzzy c-means that are based on Euclidean distance. CLIC algorithm presents a new method for clustering and classification that is based on defining the heat function. We have a large number of patterns and want to cluster them into a desired number of clusters. We consider each pattern as a heat source in the N-dimensional feature space. Each pattern affects the whole N-dimensional feature space; this which causes the points in the N-dimensional feature space to be affected by the all patterns. CLIC algorithm finds the hyperplanes in the N-dimensional feature space that separate the clusters, based on the heat intensity. If several linearly inseparable clusters exist, the algorithm enables us to separate them two by two. The classification rate of this algorithm is higher than the classification rate of previously known methods, such as

H.R. Shahdoosti (✉)
Department of Biomedical Engineering, Amirkabir University, Iran

N. Mastorakis et al. (eds.), *Proceedings of the European Computing Conference*, Lecture Notes in Electrical Engineering 28,
DOI 10.1007/978-0-387-85437-3_17, 

k-means, nearest neighbor, fuzzy c-means. In the following sections the algorithm is explained in greater detail.

## 17.2 Description of the Algorithm

We have two clusters in the N-dimensional feature space, which are not linearly separable. Consider the following clusters:

$$X = \{x_1, x_2, \ldots, x_n\} Y = \{y_1, y_2, \ldots, y_n\}$$
$$x_i, y_i = (d_1, d_2, \ldots, d_N)_{(I=1,2,\ldots,n)} \tag{17.1}$$

where $X$ and $Y$ are collections of patterns for first and second clusters respectively, and $d_j$(j = 1…N) is the *j*th feature of patterns. Since the close patterns cause concentration, it is necessary to determine high and low concentration of patterns in the space. We consider each pattern as a heat source to determine concentration by heat effect. Patterns (heat sources) affect the whole space. The interaction of near patterns (in a cluster) is considerable. On the other hand, it is negligible when the patterns are far from each other. This means the effect decreases when the distance increases. By considering this effect as the heat, we define heat function as below:

$$H(X_0|X_j) = F(DISTANCE\ BETWEEN\ X_0\ \&\ X_j),\ (j = 1, 2, \ldots, k) \tag{17.2}$$

where $H(X_0|X_j)$ is the heat of arbitrary point $X_0$ affected by pattern $X_j$, which depends on the distance between point $X_0$ and pattern $X_j$. *F*is a function of distance that satisfies 3 conditions:

1. $F(0) < \infty$
2. $F(\infty) \to 0$
3. $if\ d < d' \to F(d) > F(d')$

The first condition points to the heat limitation in patterns, the second condition expresses that the heat is negligible for high distances, and the third shows that the function is decreasing by distance.

By increasing the distance from each pattern, heat effectiveness of pattern decreases by a decreasing function (for example exponential function). It is obvious that intensity of heat could be explanatory of high or low concentration of patterns. The heat can be calculated for each arbitrary point $X_0$ affected by K patterns as below:

$$H(X_0) = \sum_{j=1}^{k} H(X_0|X_j) \tag{17.3}$$

Here, $H(X_0|X_j)$ is the heat of point $X_0$ affected by pattern $X_j$, and $H(X_0)$ is the total heat of arbitrary point $X_0$ (affected by all patterns).

First, it is needed to define a decreasing function and use it as $F$. Depending on conditions, the decreasing function $F$ can be determined. We can determine the decreasing function $F$, by considering the effect of each pattern on patterns in other clusters should be negligible. By having the necessary decrease in each direction we can determine the type of function (exponential or any other nonlinear decreasing function) and the coefficients of the function. For example, when clusters are close, we consider $F$ high decreasing function to have low heat between clusters (versus in the clusters).

Now we want to find the HYPER-PLANE in the N-dimensional space, which separates the clusters. A hyper-plane in N-dimensional space is defined as below:

$$a_1 d_1 + a_2 d_2 + \ldots + a_N d_N = a_0 \tag{17.4}$$

In which $d_i$ are coordination axis, $a_i$ are coefficients, and $a_0$ is equation's constant. By varying the coefficients, the equation can embrace all hyper-planes in the space. Since the coefficients have no limitation and they are unlimited we can replace them with $\tan \theta_i$, in which $-\pi/2 < \theta_i < \pi/2$:

$$\alpha_i \approx \tan(\theta_i), -\infty < a_i < \infty \rightarrow -\pi/2 < \theta_i < \pi/2 \tag{17.5}$$

By replacing $a_i$ with $\tan \theta_i$ we have:

$$\begin{aligned} &\tan \theta_1 . d_1 + \tan \theta_2 . d_2 + \ldots + \tan \theta_N . d_N \\ &= \sum_{i=1}^{N} \tan \theta_i . d_i = a_0, -\pi/2 < \theta_i < \pi/2 \end{aligned} \tag{17.6}$$

By discriminating the $\theta_i$, the hyper-plane rotates step by step, and the step size affects the accuracy and complexity of problem.

Considering the fact that the heat of each point is affected by all patterns, we can calculate the heat of hyper-plane as below:

$$H(hyper \quad plane) = \int_{d_{1\min}}^{d_{1\max}} \int_{d_{2\min}}^{d_{2\max}} \cdots \int_{d_{N\min}}^{d_{N\max}} \left( \sum_{j=1}^{n} H(ds|X_j) \right) ds \tag{17.7}$$

where $H(hyper \quad plane)$ is the heat of hyper-plane,$ds$ is the unit of surface on the hyper-plane that its heat depends on distance between center of unit with all patterns. $d_{i_{\min}}$ and $d_{i_{\max}}$ determine the limitations of integral that are being determined based on the intensity of heat decrease around the clusters . The goal is finding the best hyper-plane which can separate the clusters. Each state of hyper plane has certain $\theta_1, \theta_2, \theta_3, \theta_4$. By varying the angles with an identified step ($\Delta\theta_i$), the state of hyper plane varies. At each state of hyper-plane (with an identified angles), we move the hyper-plane in the direction of hyper plane's normal vector to cross the clusters. Then we plot the diagram of heat of hyper plane (at each move) versus the number of patterns distanced by hyper-plane. Some diagrams of hyper-plane's heat (relative to the number of distanced patterns), have two main maxima (main maximum occurs when hyper plane travels through each cluster center) and a main minimum (main minimum occurs when hyper plane traverses between clusters) between them. Of course, the diagram may have one main maximum (has no main minimum) when hyper-plane traverses two cluster's center together (almost). Now we consider all diagrams have a main minimum, and find the diagram that has lowest main minimum. Assume that the hyper plane with lowest minimum is as below:

$$\tan\theta_1.d_1 + \tan\theta_2.d_2 + \ldots + \tan\theta_N.d_N = a_0 \qquad (17.8)$$

Whereby $\theta_{i_{(i=1,2,\ldots,N)}}$ and $a_0$ are certain constants. The hyper- plane with the lowest heat (that stands between centers) is shown in equation above. The hyper-plane separates two clusters. If we assume the number of pattern assigned to first cluster is $n'_1$ and the number of pattern assigned to second cluster is $n'_2$, 3 cases may occur:

- First case: $n_1, n_2$ are not identified, so consider the hyper-plane mentioned as a cluster separator, that results two clusters are specific.
- Second case: $n_1, n_2$ are identified, and $n'_1 = n_1$, $n'_2 = n_2$. Similar to first state we consider the mentioned hyper plane as a cluster separator, that results two clusters are specific.
- Third case: $n_1, n_2$ are identified, $n'_1 \neq n_1$ and $n'_2 \neq n_2$. Assuming $n'_1 < n_1$, we need to choose the difference number of patterns from second cluster and assign them to first cluster. We consider heat distribution of the mentioned hyper-plane (the hyper plane with lowest heat), assume the point with highest heat on the hyper plane is $S = (d_1, d_2, \ldots, d_N)$ and the necessary number of patterns is $n_1 - n'_1 = n''$, now find the $n''$ nearest patterns to point $S$ and assign them to first cluster. So the number of assigned patterns to first cluster (by algorithm) becomes equal to the real number of patterns (also for second cluster).

## 17.3 Iris Data

To show that our algorithm also work fine in real data set, we use the famous Iris plant data set. The data set contains three classes of 50 instances each, where each class refers to a type of Iris plant: Setosa, Versicolor, and Virginica. The attributes consist of the Sepal Length, Sepal Width, Petal Length and Petal Width. One class (red patterns) is linearly separable from the other two; the

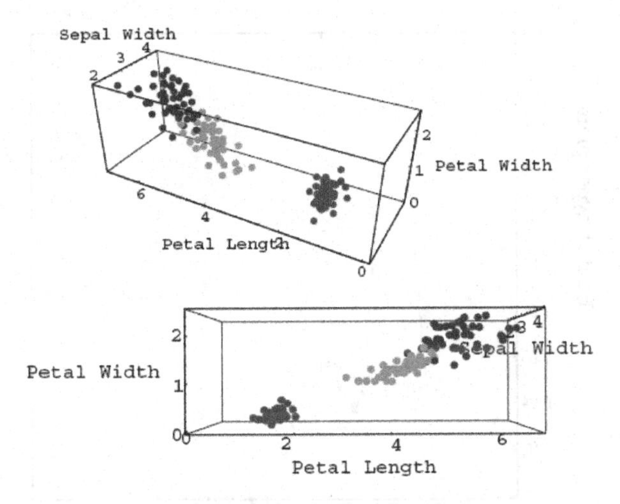


**Fig. 17.1** Iris data

latter are not linearly separable from each other (green and blue patterns). Iris data is shown in Fig. 17.1. We want to separate them by algorithm.

First we need to define a decreasing function and use it as $F$. An appropriate function is exponential function. So we can calculate the heat of arbitrary point $X_0$ affected by pattern $X_j$ as below:

$$H(X_0|X_j) = Exp\left(-\sum_{k=1}^{N} \mu_k (d_{k_{x_0}} - d_{k_{xj}})\right) \tag{17.9}$$

$$x_0 = (d_{1x_0}, d_{2x_0}, \ldots, d_{Nx_0}) x_j = (d_{1x_j}, d_{2x_j}, \ldots, d_{Nx_j})$$

That $\mu_{i_{(i=1,2,\ldots,N)}}$ is coefficient of distance at direction $d_i$, and $X_0$ is an arbitrary point and $X_j$ is a pattern.

Now consider two linearly inseparable clusters that are shown in Fig. 17.1 with blue and green patterns. Noticing the average distance between patterns in the same clusters and distance between clusters, we put $\mu_1 = \mu_2 = \mu_3 = \mu_4 = 5$ (the coefficient $\mu = 5$ is an appropriate coefficient). Now we consider the hyper plane in 4-dimensional coordination (Table 17.1).

By varying the angle $\theta_{i_{(i=1,2,\ldots,N)}}$ (with an identified gradient $\Delta\theta_i$) with certain $\mu_{i_{(i=1,2,\ldots,N)}}$ the hyper plane with minimum heat belongs to diagram Fig. 17.2.

**Table 17.1** The value of classification rate relative to coefficients $\mu_1 = \mu_2 = \mu_3 = \mu_4 = \mu$ and gradient of angle ($\Delta\theta_1 = \Delta\theta_2 = \Delta\theta_3 = \Delta\theta_4 = \Delta\theta$)

| | $\Delta\theta = 30^0$ | $\Delta\theta = 15^0$ | $\Delta\theta = 10^0$ | $\Delta\theta = 5^0$ | $\Delta\theta = 2^0$ | $\Delta\theta = 1^0$ |
|---|---|---|---|---|---|---|
| $\mu = 0.3$ | 73.33% | 80% | 85.33% | 89.33% | 90.66% | 90.66% |
| $\mu = 1$ | 74.66% | 90.66% | 90.66% | 93.33% | 96% | 96% |
| $\mu = 5$ | 82.66% | 93.33% | 97.33% | 98.66% | 98.66% | 98.66% |
| $\mu = 20$ | 80% | 90.66% | 93.33% | 96% | 97.33% | 97.33% |

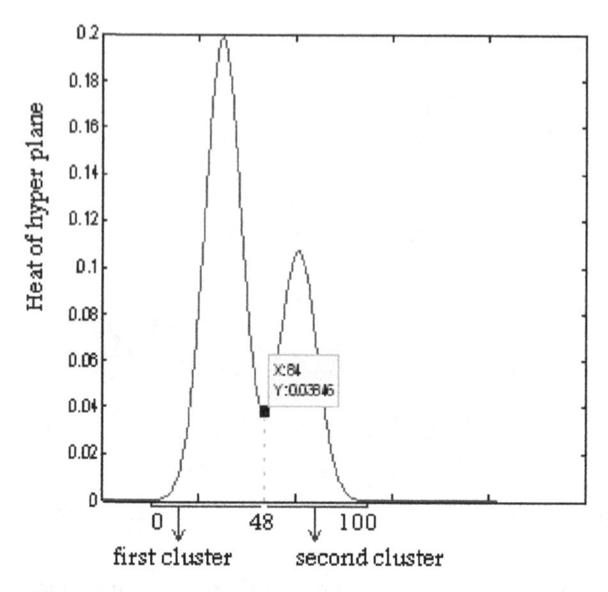


**Fig. 17.2** The diagram of heat of hyper plane versus number of patterns distanced $\theta_1 = -30^0, \theta_2 = 25^0, \theta_3 = 45^0, \theta_4 = 45^0$, $\mu_1 = \mu_2 = \mu_3 = \mu_4 = 5$, $a_0 = 4.25$ $\Delta\theta = 5^0$

The number of assigned pattern to first cluster by the hyper plane with minimum heat is 48 patterns in which 47 patterns belong to the first cluster. We find two nearest patterns (the real number of first cluster is 50) to the point with maximum heat on the hyper plane. Two found patterns belong to the first cluster, so we have:

$$\textit{classification rate} = \frac{\textit{true assigned patterns}}{\textit{Total of patterns}} = 98.66\% \qquad (17.10)$$

## 17.4 Conclusion

Noticing the result of clustering based on CLIC algorithm, we find CLIC algorithm as a strong and useful algorithm. The methods only based on Euclidean distance like k-means, nearest neighbor, fuzzy c-means are uninformed methods which need high memory capacity. CLIC algorithm is based on defining the heat function; it can use program knowledge. The algorithm introduced (CLIC) is an informed algorithm. That result is a high classification rate more than the classification rate of methods only based on Euclidean distance (fuzzy c-means, nearest neighbor, etc).

## References

1. Osipova N (2004). Classification and clustering methods development and implementation for unstructured documents collections. Department of Programming Technology, Faculty of Applied Mathematics and Control Processes, St. Petersburg State University
2. Rose K (1998) Deterministic annealing for clustering, compression, classification, regression, and related optimization problems
3. Wallace M, Kollias S (2004) Robust, generalized, quick and efficient agglomerative clustering
4. Wolkenhauer Olaf Fuzzy classification, the iris-and-admission data sets

# Chapter 18
# A Top-Down Minimization Algorithm for Non-Deterministic Systems

**Adrian Zafiu and Stefan Holban**

**Abstract** Nowadays, the decisional diagrams principle is generally adopted as the base technique for decisional systems analysis. The proposed technique is based on different principles and strategies. This technique opens the way for more native interpretations of logic comprised by a decisional system, free by constriction of hardware implementation possibilities.

## 18.1 Introduction

There are no methods that can process the minimization into a polynomial time. There are many algorithms or methods approaching the problematic of minimization of multi-valued or binary decisional systems, like graphics systems (Veitch-Karnaugh diagrams), Quine McCluskey [1] method, Espresso [2, 3] and SIS [4]. The SIS software also minimizes multi-valued specifications, but in a binary way, coding the multi-valued variables. This method is a pseudo multi-valued one. The MVSIS [5, 6] application minimizes multi-valued systems in a natural manner. BOOM [7] and BOOM-II [8] tools perform a heuristic implementation with outstanding results regarding the execution time, but applicable only for binary specifications. The Discrimination method [9] is the last known exact method for the multi-valued minimization.

The presented algorithm use a different principle named discrimination [9]. According to this principle we search the set of difference between groups of input combinations with the same output.

A. Zafiu (✉)
Department of Electronics, Communications and Computers, University of Pitesti, 110040 Pitesti Str., Targul din Vale, nr.1, Arges, Romania

N. Mastorakis et al. (eds.), *Proceedings of the European Computing Conference*, Lecture Notes in Electrical Engineering 28,
DOI 10.1007/978-0-387-85437-3_18, 

## 18.2 Method Presentation

The minimization algorithm presented here requires an input table with disjointed rows. The selected minimal set of implicants is processed next to eliminate any potential redundancies. The method consists of three main steps: pre-processing, generating the complete set of implicant vectors and selecting a non-redundant solution.

**The pre-processing**: The given specification must be pre-processed to obtain an equivalent table appropriate for the next step. In this stage, there is eliminated implicit non-determinism, composed vectors are translated into primes vectors and, if it is possible, a superficial compression is performed.

**Disjoint rows**: Firstly, any pair of rows, which have implicit non-determinism, is decomposed. The results of decomposition transform the implicit non-determinism into an explicit one.

Let us consider the next function with default output $f: Z_3 \times Z_3 \times Z_3 \times Z_3 - Z_2 \times Z_3$:

**Table 18.1** $f: Z_3 \times Z_3 \times Z_3 \times Z_3 - Z_2 \times Z_3$

| # | $i_1$ | $i_2$ | $i_3$ | $i_4$ | / | $o_1$ | $o_2$ |
|---|---|---|---|---|---|---|---|
| 1 | 1 | – | 2 | – | / | 1 | 0 |
| 2 | – | 2 | – | 0 | / | 1 | 1 |
| 3 | 2 | – | 2 | 2 | / | – | 2 |
| 4 | | | | | / | 0 | – |

Rows 1 and 2 have an implicit non-determinism because the input combination (1 2 2 0), common for both rows, has output values zero and one for second output variable. The decomposition is applied for each pair of rows, initial or resulted rows, with implicit non-determinism. The result of decomposition is a set of rows with explicit non-determinism (row one of the result):

**Table 18.2** Decomposition result for rows 1 and 2

| # | $i_1$ | $i_2$ | $i_3$ | $i_4$ | $o_1$ | $o_2$ |
|---|---|---|---|---|---|---|
| 1 | 1 | 2 | 2 | 0 | 1 | 0,1 |
| 2 | 1 | 0,1 | 2 | 1,2 | 1 | 0 |
| 3 | 1 | 0,1 | 2 | 0 | 1 | 0 |
| 4 | 1 | 2 | 2 | 1,2 | 1 | 0 |
| 5 | 0,2 | 2 | 0,1 | 0 | 1 | 1 |
| 6 | 0,2 | 2 | 2 | 0 | 1 | 1 |
| 7 | 1 | 2 | 0,1 | 0 | 1 | 1 |

Split composed rows into primaries rows.

**Light compression**: The last operation of the pre-processing consists in a light compression of rows. Row 0 2 - 0 / 1 1, for example, replaces rows 10, 12 and 14. This process avoid redundant computations without denaturize the results.

By considering all these operations, the specification presented in Table 18.2 becomes:

**Table 18.3** Prime input vectors

| # | Row | # | Row | # | Row |
|---|---|---|---|---|---|
| 1 | 1220/10,1 | 6 | 1020/10 | 11 | 2200/11 |
| 2 | 1021/10 | 7 | 1120/10 | 12 | 0210/11 |
| 3 | 1121/10 | 8 | 1221/10 | 13 | 2210/11 |
| 4 | 1022/10 | 9 | 1222/10 | 14 | 0220/11 |
| 5 | 1122/10 | 10 | 0200/11 | 15 | 2220/11 |

**The implicants computation**: This stage of processing generates a complete set of implicant vectors having a minimum number of literals (in two steps). The algorithm generates implicants with only one non-DC position and, if there remain uncovered rows, it continues to find implicants with two non-DC positions, and so on. By using this principle, is obvious that only necessary implicants with a minimum number of non-DC positions form the results.

**The generator**: The generator uses packages which split the search space into disjunctive zones. The candidates comprised into a package have at least a common component that separates this candidate from the candidates comprised into other packages.

Let's consider a function defined on the domain $Z_3 \times Z_3 \times Z_3$. The generator creates a candidate vector for each value of each input subdomain. The first line of Table 18.4 contains the first package.

**Table 18.4** Prime input vectors

| Pack | 1 | 2 | 3 | 4 | 5 | 6 | 7 | 8 | 9 |
|---|---|---|---|---|---|---|---|---|---|
| 1 | 0-- | 1-- | 2-- | -0- | -1- | -2- | --0 | --1 | --2 |
| 1.1 | 00- | 01- | 02- | 0-0 | 0-1 | 0-2 | | | |
| 1.2 | 10- | 11- | 12- | 1-0 | 1-1 | 1-2 | | | |

The second package (1.1) is generated from the package one by preserving its elements. The first candidate is combined with candidates from four to nine of the first package. Items two and three cannot be combined with the first candidate, because the first position must have the same specified value.

**The attribute computation**: For each candidate, a set of attributes is computed, in order to characterize the candidate status. The set of attributes has a part commonly used for all output functions of each row and a part which is specific to each output function and to each output row.

The attribute $<C, I, O_f>$ of a row and of an output function consists from a set $C$ of covered rows, a set $I$ of intersected rows and a set $O\_f$ of possible function's output values (Fig. 18.1).

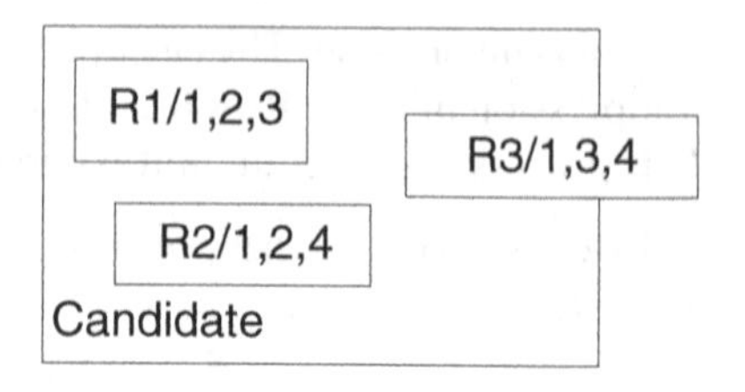


Fig. 18.1 Attribute's interpretation

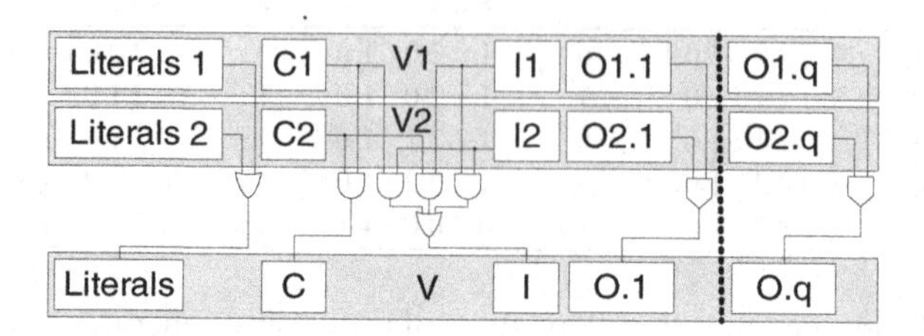


**Fig. 18.2** Attribute's computation

If a candidate covers two rows R1 and R2 and intersect row R3, like in Fig. 18.2, it is obvious that the only possible output value is one. The output cannot be two because R3 do not have this output value on the common zone, it cannot be three because R2 do not have this output value, it cannot be four because R1 do not have this output value and it cannot be five because R1 and R2 do not have this output value.

The candidates attribute for a function with $n$ outputs is $<C, I, <O_{f_1}, O_{f_2}, O_{f_3}, \ldots, O_{f_n}>>$.

Starting from two candidates $V_1<C_1, I_1, O_{1f}>$ and $V_2<C_2, I_2, O_{2f}>$, the attributes of new candidate $V<C, I, O_f>$ are performed using the following formulas:

- $C = C_1 \cap C_2$
- $I = (C_1 \cap I_2) \cup (I_1 \cap C_2) \cup (I_1 \cap I_2)$

The last attribute $O_f$ is performed starting from $O_{1f}$ or $O_{2f}$. The algoritm uses this attribute to compute appropiate output values for the current candidate. An output value is acceptable if it belongs to any covered or intersected row. The used structure is a table that keeps in a descendance order the number of incidences of each output value for covered or intersected rows.

By using this supplementary structure which keeps the set of attributes computed for each candidate, the algorithm avoids redundant computations. In a complex system like the one presented, this approach increases with a factor the used memory which in turn exponentially decreases the computation.

**The candidate's classification**: Each candidate is classifiable according to it's attributes. The classification is performed after each computed "child" (component of C). If set $C$ is void, then the candidate is not viable and the algorithm deletes it out of the package. This candidate cannot cover any row and, even by combining it with other candidate, in no way will cover any row. If a candidate with a non void $C$ has available outputs for a function ($O_f$ is not void), it is an implicant for that function and its copy is kept in the set of results. In the package, the attribute $O_f$ is set to "null", and the algorithm will no longer compute the attribute for this function for any child of $C$. The algorithm keeps candidates if its sets $C$ are not void and all sets $O_f$ are different from the null set.

**Overview of generation of the implicants**: All previous presented methods are parts of the general algorithm. After the generation of all implicants with the same number of literals, a general structure that keeps state of the algorithm is updated.

**Definition**: A function for a row is covered if all its output values are covered.

**Definition**: A row is covered if all its functions are covered.

According to previous definitions, this stage of algorithm is ended when all rows are covered.

Another approach could be performed if the definition of a covered function is changed, i.e. a function for a row is covered if at least one output value of the function is covered. The second approach not preserves all initial specifications, but, in many cases, generates a satisfactory solution.

The structure keeping the algorithm status could be updated after the generation of all implicants with the same number of literals, or after each implicant generation. In the second case, the algorithm has a higher convergence, but not all implicants with the same number of literals are generated. Such an approach leads to a heuristic method. All these observations are easily to be implemented and have impact in algorithm at the level of candidate classification only.

## 18.3 Conclusions

The presented method minimizes multiple output multivalent decisional systems, deterministic or nondeterministic, completely or incompletely specified, with correlated or non-correlated outputs. The classical methods of multivalent minimization use the reduction of decision diagrams to the computation of a minimal decisional tree for the given specification. If the given specification is incompletely specified, then the results of this approach include all the input space. The presented method uses only a part of the non-specified space. The scope of this method is not to include unspecified combinations of inputs. The method finds the elements that discriminate input combinations which generate different outputs [9]. The binary specifications become a particular case of the covered class of problems.

The algorithm facilitates the parallel and network minimization. The network minimization automatically establishes an order for the function computation which gives remarkable results even for some classical problems (ex. BCD 7 segments) [10]. The network minimization is applicable on the intrinsic multivalued variable or on the valences of outputs (each output valence corresponds to a subsystem witch decides whether or not if the value is delivered to the output). The optimal solution is evaluated after its minimum number of literals. There exist many equivalent solutions with the same number of literals. The method computes and gives as the result only the first of them. If there exists some supplementary criteria that could make the difference between two equivalent solutions, both having a minimum number of literals (one is better, or another is not acceptable), the algorithm could be adjusted to include the new rule and to reduce the searched space of solutions.

## References

1. Quine WV (1952) The problem of simplifying truth functions. Am Math Mon 59(8):521–531
2. Brayton RK, Hachtel GD, McMullen C, Sangiovanni-Vincentelli AL (1984) Logic minimization algorithms for VLSI synthesis. Kluwer Academic Publisher, Boston
3. Hachtel GD, Somenzi F (1984) Logic synthesis and verification algorithms. Kluwer Academic Publisher, Boston
4. Sentovich EM, et al. (1992) SIS: A system for sequential circuit synthesis. University of California, Berkeley, CA 94720, Tech. Rep. UCB/ERL M92/41 Electronics Research Lab
5. Gao M, Jiang J-H, Jiang Y, Li Y, Sinha S, Bryton R (2001) MVSIS. In: the Note of the International Workshop on Logic Synthesis, Tahoe City
6. Chai D, Jiang J-H, Jiang Y, Li Y, Mischenko A, Brayton R (2003) MVSIS 2.0 User's manual. University of Berkeley CA 94720
7. Fišer P, Hlavička J (2003) BOOM – A heuristic boolean minimizer. Comput Inform 22(1):19–51
8. Fišer P, Kubatova H (2006) Flexible two-level boolean minimizer BOOM-II and its applications. In: Proceedings of the 9th EUROMICRO Conference on Digital System Design, pp 369–376
9. Ştefănescu I (2005) DISCRIMINATION: A new principle in the efficient minimization of the binary and multiple-valued functions. In: 1st Int Conf on Electronics, Computers and Artificial Intelligence, ECAI 2005, University of Piteşti, Romania
10. Ştefănescu I, Zafiu A (2007) An efficient network strategy in deep minimizations of deterministic and nondeterministic multivalued decisional systems. In: 2nd Int Conf on Electronics, Computers and Artificial Intelligence ECAI 2007, University of Piteşti, Romania

# Part III
# Natural Phenomena and Computer Models

This part will discuss the issues and the problems associated with the simulation of natural phenomena. This is a difficult area of research since it generally involves complex databases and in many instances time variant phenomena. The computational loads can become very high as one considers the physics or the mathematical modeling of structures. Most items in, for example, nature, wind, clouds, fire and comets have not been displayed realistically in computer graphics. This lack stems from a few different problems, all of which are significant. The first is the fact that realistic descriptions require large amounts of storage and consequently large compute time. Nature is able to create diverse detail at the most minute levels within an object of very large scale. The second problem is that of diversity of design within a given framework. Humans typically become tired after the first few iterations of such a design process, with a resulting degradation in the subsequent models. Here we present several results for a multi-layer finite volume solution of wind induced basin flow, NASIR heat generation and transfer solver for RCC dams, water hammer modeling using 2nd order Godunov finite volume method, simulation of wind pressure on two tandem tanks and the tropical cyclone forecaster integrated with case-based reasoning. However, it appears that natural phenomena will require more research into the fundamental way things happen in nature, and in terms of computer graphics that still require new modeling techniques.

# Chapter 19
# Multi-Layer Finite Volume Solution of Wind Induced Basin Flow

**Saeed-Reza Sabbagh-Yazdi, Hamed Arabi, and Nikos E. Mastorakis**

**Abstract** In this paper, a multi-layers numerical model is introduced for modeling shallow water flows. The model numerically solves equations of continuity and motions in order to compute water depth and velocity patterns in each layer. The model can consider the elevation variation of upper layers and bed and wall geometric complexities and resistances. It can also consider the effect of wind on surface layer. The governing equations are discretized using cell vertex finite volume method in triangular unstructured meshes. For stabilizing the explicit solution process, artificial viscosity formulations are adopted for the unstructured meshes in such a way that preserves the accuracy of the numerical results. The accuracy of the results of present multi-layer flow solver is assessed by simulating wind induced flow in a circular basin and by a comparison of the computed results with the results of previous research works.

## 19.1 Introduction

Using the hydrostatic pressure assumption and integration of the incompressible continuity and momentum equations ends up with two-dimensional depth averaged flow equations known as shallow water equations (SWE).

In some cases of the real world lake flow problems (i.e. shallow lakes with considerable bed roughness which are subject to wind on water surface) vertical variations in horizontal velocity components are significant. In such cases, although the vertical momentum is mostly due to hydrostatic pressure, consideration of vertical gradients of the horizontal momentum is essential, particularly for modeling the times variation of some parameters like heat, salt and density and other materials which may be solved or suspended in the water. The

S.-R. Sabbagh-Yazdi (✉)
Associate Professor of K.N. Toosi University of Technology, Department of Civil Engineering, 1346 Valiasr St. 19697 Tehran, Iran
e-mail: SYazdi@kntu.ac.ir

N. Mastorakis et al. (eds.), *Proceedings of the European Computing Conference*, Lecture Notes in Electrical Engineering 28,
DOI 10.1007/978-0-387-85437-3_19, 

multi-layer solution of SWE is one of the techniques which can be used for numerical solution of such cases.

Some of the works on numerical multi-layer simulation of the lakes flow could be referred to Ezar & Mellor in 1993 which used $k-\varepsilon$ turbulence model in multi-layer form for analyzing the lakes flow [1]. Mellor et al. in 1998 analyzed the two dimensional model in sigma coordinate considering the effects of advection and diffusion, density and pressure gradients considered [2].

The interaction of wind effects and geometrical features of the lakes (i.e. bathymetry and coastal irregularities) plays an important role in formation of wind induced flow patterns. Therefore, development of a finite volume SWE solver with an algorithm suitable for unstructured meshes is considered in this work.

In this work, the equations of continuity and motions (in horizontal plane) are chosen as the mathematical model in which the pressure distribution in flow is hydrostatic and the effect of momentum in vertical direction is negligible. Because of the characteristics of the application problem desired in this work (topography, bed slop and boundary fractures), the sigma coordinate multi-layer system is used for numerical modeling. In order to reduce the computational workload, a reconstruction free cell-vertex finite volume algorithm is utilized in the present multi-layer SWE solver.

## 19.2 Governing Equations

With the assumption of water incompressibility the mathematical multi-layer model contains equations of continuity and motions. The continuity equation for the $k$th layer can be written in vector form as follows:

$$\frac{\partial h_k}{\partial t}+\frac{\partial h_k u_k}{\partial x}+\frac{\partial h_k v_k}{\partial y}=\pm(q_z)_k \tag{19.1}$$

where $t$ is time, $x$ and $y$ are Cartesian coordinates, $h$ is water depth, $u$ and $v$ are velocity components in $x$ and $y$ directions. In this equation, the first term shows the changing of flow mass and the second and third terms show the transported mass of flow in $x$ and $y$ directions. $q_z$ is the net balance of water volume exchange in vertical direction between the $k$th layer and its neighboring media.

The horizontal $x$ and $y$ directions momentum equations for the $k$th layer could be written as follows:

$$\frac{\partial(h_k v_k)}{\partial t}+\frac{\partial(h_k v_k u_k)}{\partial x}+\frac{\partial\left(h_k v_k^2\right)}{\partial x}+g h_k\frac{\partial\zeta}{\partial y}=\frac{\tau_{by}}{\rho}+\left[\frac{\partial}{\partial x}h_k\left(\left(T_{yx}\right)_k\right)+\frac{\partial}{\partial y}h_k\left(\left(T_{yy}\right)_k\right)\right] \tag{19.2}$$

$$\frac{\partial(h_k v_k)}{\partial t}+\frac{\partial(h_k v_k u_k)}{\partial x}+\frac{\partial(h_k v_k^2)}{\partial y}+gh_k\frac{\partial\zeta}{\partial y}=\frac{\tau_{ky}}{\rho}+\left[\frac{\partial}{\partial x}h_k((T_{yx})_k)+\frac{\partial}{\partial y}h_k((T_{yy})_k)\right] \tag{19.3}$$

Here, $\zeta$ is depth of each layer from bed (water level in each layer), $\partial(h_k u_k)/\partial t, \partial(h_k v_k)/\partial t$ are momentums changing in time of $\Delta t$ and $\partial(h_k u_k^2)/\partial x, \partial(h_k u_k v_k)/\partial y, \partial(h_k v_k u_k)/\partial x, \partial(h_k v_k^2)/\partial y$ show the momentum effect on flow in $x$ and $y$ directions. $gh_k\partial\zeta/\partial x$,$gh_k\partial\zeta/\partial y$ cite the changing of bed slop and their effect of on flow in $x$ and $y$ directions, and also the effect of upper layer weight on desired layer [3]. $\tau_{kx}, \tau_{ky}$ representing the $x$ and $y$ directions' global stresses of the $k$th layer, including the effect wind stresses $\tau_{tx}, \tau_{ty}$ on the upper layer, $\tau_{bx}, \tau_{by}$ representing the effect of the bed roughness in the lower layer [4].

It is worth noting that, negligible shear stress between the layers may be assumed, and, the effects of wind and bed roughness can be interpolated between the momentum equations of all the layers.

The wind stresses that should be added to the above mentioned $x$ and $y$ directions' global stresses of the top layer are defined as,

$$\tau_{tx}=\rho_a C_\omega W_x|W_{10}| \tag{19.4}$$

$$\tau_{ty}=\rho_a C_\omega W_y|W_{10}| \tag{19.5}$$

where, $W_{10}$ is the wind speed at 10 meters above the water surface and $C_w$ is the drag coefficient which can be obtained from following graph (Fig. 19.1).

The global friction due to bed roughness can be attained from the following relations:

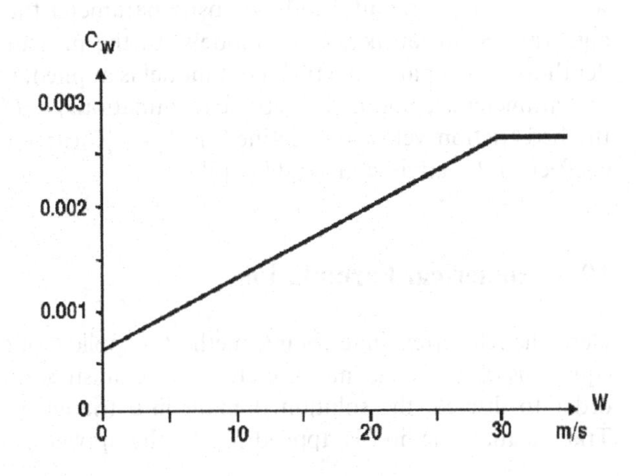


**Fig. 19.1** Relationship between wind speed and drag coefficient [5]

$$\tau_{bx} = C_f u \sqrt{(u^2 + v^2)_k} \tag{19.6}$$

$$\tau_{by} = C_f v \sqrt{(u^2 + v^2)_k} \tag{19.7}$$

where, $C_f$ is the effective global dissipative coefficient.

In multi-layer cases, in which shear stress between the layers are assumed to be negligible (due to uniformity of the fluid properties), the global effect of bed friction that must be added to all the layers may be calculated using following coefficient of friction,

$$C_f = \left[\frac{1}{K}\log(\frac{30Z_b}{k_s})\right]^{-2} \tag{19.8}$$

where, $k_s$ is the bed friction coefficient, $k$ is Von-Karman coefficient which equals to 0.4 and $Z_b$ is distance from bed.

The effect of global dissipation in $x$ and $y$ directions are computed using the following formulation [6].

$$T_{xx} = \frac{\partial}{\partial x}\left(2v_t h \frac{\partial u}{\partial x}\right) \tag{19.9}$$

$$T_{xy} = \frac{\partial}{\partial y}\left(v_t h\left(\frac{\partial u}{\partial y} + \frac{\partial v}{\partial x}\right)\right) \tag{19.10}$$

$$T_{yx} = \frac{\partial}{\partial x}\left(v_t h\left(\frac{\partial v}{\partial x} + \frac{\partial u}{\partial y}\right)\right) \tag{19.11}$$

$$T_{yy} = \frac{\partial}{\partial y}\left(2v_t h \frac{\partial v}{\partial y}\right) \tag{19.12}$$

where $v_t$ is the horizontal eddy viscosity parameter that can be computed with algebraic, Smagronsiky, $k - \varepsilon$ models. In the present work, the widely used depth-averaged parabolic turbulent model is applied, in which the eddy viscosity parameter is computed algebraic formulation$\nu_t = \theta h U_*$. In this formulation the bed friction velocity is defined as $U_* = [C_f(u^2 + v^2)]^{0.5}$ and the empirical coefficient $\theta$ is advised around 0.1 [7].

## 19.3 Numerical Formulation

Here, the cell vertex finite volume method is applied for converting the governing equations of the layers into discrete form on unstructured triangular meshes. In order to discrete the solution domain into triangular subdomains, Deluaney Triangulation method is applied [8]. In the applied solution algorithm, at the outset, the computational control volumes are formed by gathering the triangles on every node. Then, the governing equations are integrated over each control volume. Application of the Green's theorem to the integrated equations results in:

$$\int_{\Omega} \frac{dW}{dt} d\Omega + \oint_{\Gamma} ((E\Delta y - F\Delta x) + (G\Delta y - H\Delta x)) = \int_{\Omega} S d\Omega \tag{19.13}$$

where,

$$W = \begin{pmatrix} h_k \\ h_k u_k \\ h_k v_k \end{pmatrix}, E = \begin{pmatrix} h_k u_k \\ h_k u_k^2 \\ h_k u_k v_k \end{pmatrix}, F = \begin{pmatrix} h_k v_k \\ h_k u_k v_k \\ h_k v_k^2 \end{pmatrix},$$
$$G = \begin{pmatrix} 0 \\ h_k \nu_{Th} \frac{\partial u_k}{\partial x} \\ h_k \nu_{Th} \frac{\partial v_k}{\partial x} \end{pmatrix}, H = \begin{pmatrix} 0 \\ h_k \nu_{Th} \frac{\partial u_k}{\partial y} \\ h_k \nu_{Th} \frac{\partial v_k}{\partial y} \end{pmatrix}, S = \begin{pmatrix} q_{zk} \\ -g h_k \frac{\partial \xi}{\partial x} - \frac{\tau_{kx}}{\rho_w} \\ -g h_k \frac{\partial \xi}{\partial y} - \frac{\tau_{ky}}{\rho_w} \end{pmatrix}$$

And $\Omega$ and $\Gamma$ are the area and perimeter of the control volume formed by gathering triangular cells sharing a computational node, respectively.

In order to solve the value of the unknown vector $W$ at the central node of the control volumes, the above integral equation can be transformed to the following algebraic formulation.

$$W_i^{n+1} = W_i^n - \frac{\Delta_t}{\Omega_i} \left[ \sum_{i=1}^{m} ((\bar{E}\Delta y - \bar{F}\Delta x) + (\overline{G}\Delta y - \overline{H}\Delta x)) \right]^n + S^n \tag{19.14}$$

where, $W_i^{n+1}$ is the value of $W_i^n$ to be computed after $\Delta t$ in the $i$th node. The parameters $\bar{E}$, $\bar{F}$, $\overline{G}$ and $\overline{H}$ are the averaged values of the fluxes at boundary edges of the control volume $\Omega$ [8].

## 19.4 Artificial Viscosity

In the explicit solution of the convection dominated problems (i.e. flow with negligible friction and turbulent effects) lack of some enough physical dissipation may give rise to some numerical oscillations. This problem, which endangers the stability of the computations, usually appears near regions with high gradient of flow parameters. For damping out these numerical noises artificial dissipation terms can be added to the derived formulations. The formulation of these additional terms should suit unstructured nature of the mesh data. The dissipation terms must damp out the numerical errors without degradation of the accuracy. Although for the flows with gradual variations independent variables the fourth order term (Biharmonic operator) produces enough dissipations, near the high gradient region it is necessary to add second order operators (Laplacian operator) to the dissipation formulation. In order to prevent unwanted dissipation in the smooth flow regions, the Laplacian operator can be multiplied by a water elevation switch. Hence, the artificial dissipation operators are introduced as,

$$\left(\nabla^4 W_i\right)_k = \varepsilon_4 \sum_{j=1}^{Ne} \left[_{ij}\left(\nabla^2 W_j - \nabla^2 W_i\right)\right]_k$$

$$and \quad \left(\nabla^2 W_i\right)_k = \varepsilon_2 \sum_{j=1}^{Ne} \left(W_j - W_i\right)_k \tag{19.15}$$

Where λ is scaling factor, $\varepsilon_4$ and $\varepsilon_2$ are the artificial dissipation coefficients ($1256 < \varepsilon_4 < 3/256$ and $0.2 < \varepsilon_2 < 0.3$), which must be tuned to minimum value required for the stability of computations. The scaling factor $_{ij}$ is computed using maximum nodal values of Eigen values of Jacobian matrix at the edges connected to the centre node of the control volume.  is evaluating as follow.

$$_k = \left[\left|\vec{U}.\hat{n}\right| + \sqrt{U^2 + C^2(\Delta x^2 + \Delta y^2)}\right]_k \quad where \quad C = \sqrt{gh_k} \tag{19.16}$$

Where, $C$ is celerity, $\vec{U}$ is averaged computed velocity, and $\hat{n}$ is normal vector at of edges. Note that, the computations of artificial dissipation operators are performed over the edges connected to the central node of the control volume [8].

## 19.5 Boundary Condition

All the inflow and outflow volumes may be considered as source and sinks of the continuity equation at desired nodal points.

Free-slip velocity condition at coastal walls can be imposed where no flow passes through the flow boundaries. These boundaries can be used to reduce the computational domain assuming that the thickness of near wall boundary layer is considerably small in comparison with the horizontal dimensions of the problem. At these boundaries the component of the velocities normal are set to zero. Therefore tangential computed velocities are kept using free slip condition at wall boundaries [8].

## 19.6 Verification Test

Numerical solution of flow in a circular basin is chosen as a test case to evaluate the model performance in simulating the effect of wind flow in a shallow lake.

This test is a circular basin that Krunburg in 1992 design it as shallow basin case. The circular basin has a radius of 192 meter and the depth of 0.6 meter.

A steady wind directly passes from east to west of this circle and its speed is 10 meter per second. The drag coefficient is nearly equal to 0.002 [9].

In present work, the water depth is divided into three layers as surface, middle and bottom layer that it is shown in Fig. 19.2.

As there is no internal or external flow through the basin, thereafter just wind stress effects are considered in this case. Still water is assumed as the initial condition. Then, a constant wind with speed of 10 meter per second considered to blow 10 meters over the basin. This west to east direction wind formed two circulating flows in the basin. In the following figures the computed flow pattern at each layer are plotted in the form of stream lines (Fig. 19.3) and velocity vectors (Fig. 19.4).

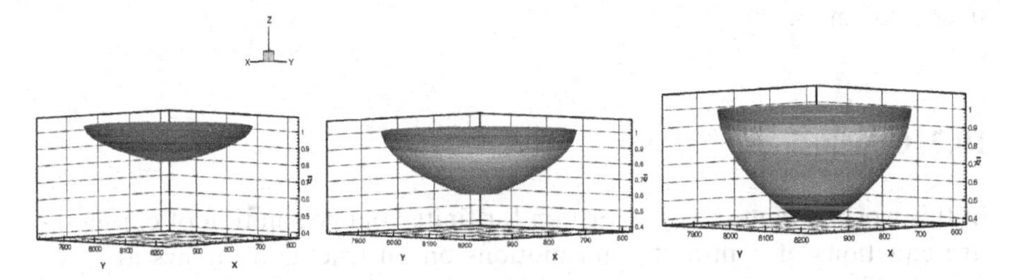

Fig. 19.2 Topology of the three layers of the basin

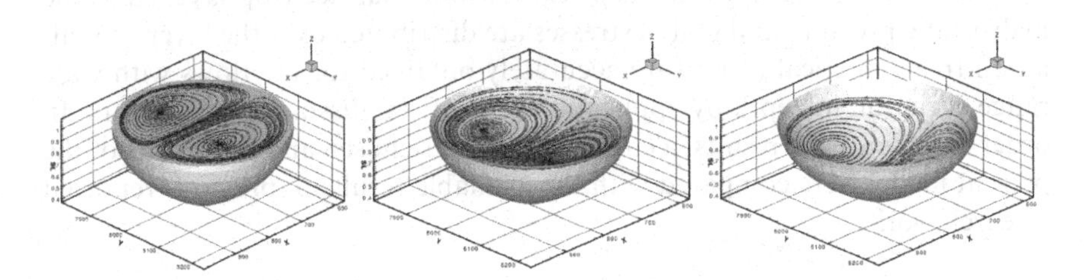

Fig. 19.3 Computed stream lines on surfaces of three layers

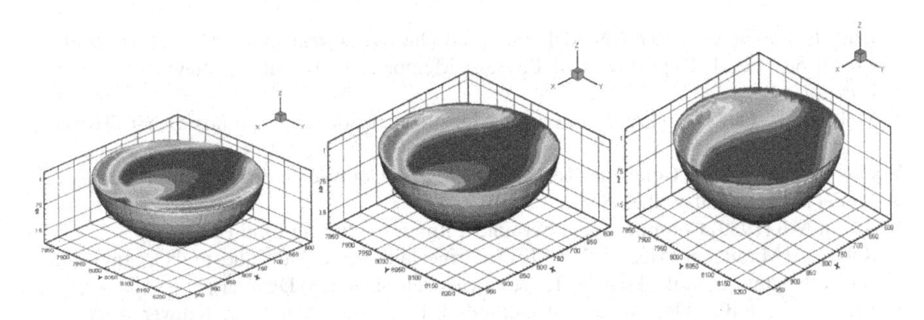

Fig. 19.4 Computed velocity contours on surfaces of three layers

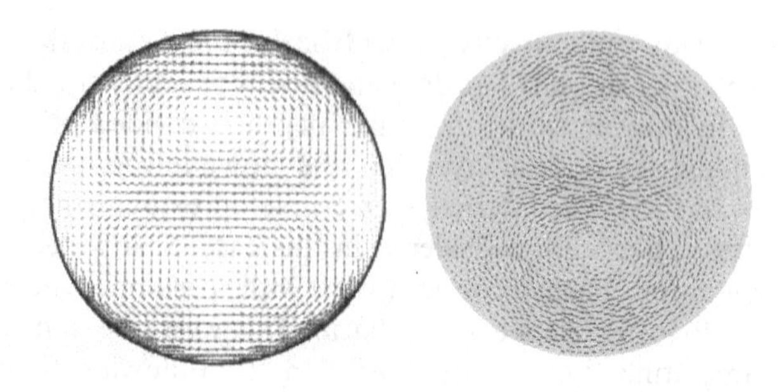

**Fig. 19.5** Comparison between velocity vectors of reference paper (*up*) and results of the present model (*down*)

Figure 19.5 presents the results of the present model on unstructured mesh which is very similar to the results reported by previous workers on structured mesh [9].

## 19.7 Conclusions

In this work, a multi-layer algorithm for finite volume solution of conservative equations of continuity and motions on unstructured meshes are presented. The multi-layer equations are solved on triangular unstructured meshes using cell vertex finite volume method. In this algorithm, the effects of wind stress are successfully imposed on water surface (top layer) and the hydrostatic pressure and global stresses are distributed over the layers. Oscillation free numerical results are accurately obtained for the flows with wide range variation in depth by application of artificial dissipation operators. In order to validate the results, a circular basin was chosen and the results of the present model were compared with the available results of previous reported research works.

## References


1. Ezar T, Mellor GL (1993) Model simulated changes in transport. Meridional heat & Costal Sea Level, Department of Physics, Memorial University of New Found land, Canada
2. Mellor GL, Oey LY (1998) American Metrological Society. J Atmos Ocean Technol 15:11–22
3. User Documentation Coherens, Release 8.4 (1999) A Coupled Hydro dynamical Ecological Model for Regional & Shelf Seas
4. Abbot MB (1980) Computational hyraulic. Pitman Publishing Limited
5. Reference Manual of MIKE 3/21 FM, Estuarine and Coastal Hdraulics and Oceangraphy, Hydrodynamic Module, DHI Water & Enviromental, Agem, Denmark
6. Vreugdenhil CB (1994) Numerical methods for shallow-water flow. Kluwer Academic Publishers

7. Sabbagh-Yazdi SR, MohamadZadeh M (2004) Finite volume solution of two-dimensional convection dominated sub-critical free surface flow using unstructured triangular meshes. Int J Civil Eng 2(2):78–91
8. Weatherill NP, Hassan O, Marcum DL (1993) Calculation of steady compressible flow-fields with the finite elements method. 31 Aerospace Sciences Meeting and Exhibit, AIAA-93-0341
9. Kranenburg C (1992) Wind-driven chaotic advection in a shallow mode lake. J Hydr Res 30(1):29–47

# Chapter 20
# NASIR Heat Generation and Transfer Solver for RCC Dams

**Saeed-Reza Sabbagh-Yazdi, Ali-Reza Bagheri, and Nikos E. Mastorakis**

**Abstract** In this paper, the heat transfer module of NASIR (Numerical Analyzer for Scientific and Industrial Requirements), software which solves the temperature field on unstructured finite volumes, is introduced. In this software, the transient PDE for heat transfer in solid media is coupled to a suitable concrete heat generation algebraic relation. The discrete form of the heat transfer equations is derived by multiplying the governing equation by a piece wise linear test function and integrating over a sub-domain around any computational node. The solution domain is divided into hybrid structured/unstructured triangular elements. The triangular elements in the structured part of the mesh can be activated for simulating gradual movement of the top boundary domain due to advancing the concrete lifts. The accuracy of the developed model is assessed by comparison of the results with available analytical solutions and experimental measurements of two-dimensional heat generation and transfer in a square domain. The computer model is utilized to simulate the transient temperature field in a typical RCC dam.

## 20.1 Introduction

The heat exchange in RCC dams principally takes place in the transverse direction of the dam axis. Therefore, two-dimensional models are generally adopted for temperature studies [1, 2]. The temperature analysis of a concrete structure is associated with temperature dependent heat sources of cement hydrations all over the domain. However, the heat generation rate of the cement contents with gradually increasing section, depends on the temperature of the surrounding concrete and the heat conduction properties of the aggregates; the computation of the temperature profiles is not an easy task.

S.-R. Sabbagh-Yazdi (✉)
Department of Civil Engineering, K.N. Toosi University of Technology, 1346 Valiasr St. 19697 Tehran, Iran
e-mail: SYazdi@kntu.ac.ir

N. Mastorakis et al. (eds.), *Proceedings of the European Computing Conference*, Lecture Notes in Electrical Engineering 28,
DOI 10.1007/978-0-387-85437-3_20, © Springer Science+Business Media, LLC 2009

In this paper, first the mathematical model of the phenomena including the transient partial differential equation of heat transfer and associating boundary conditions in conjunction with an algebraic formula for the source term of the heat generation rate is described. Then, the numerical techniques employed for deriving the algebraic formula suitable for numerical solution on a triangular mesh is discussed. For the present solution algorithm, the domain discretization technique using hybrid unstructured/structured triangular mesh, taking into account the layer-based movement of the top surface boundary of the concrete structure is introduced. The accuracy of the solution of the spatial derivative part of the governing equation is verified using the analytical solution of heat transfer in a square domain. Then, the quality of the formula adopted for the source term of cement heat generation rate is assessed by the use of available experimental measurements on a concrete block. Finally, the developed model is applied for prediction of the transient temperature field in a typical RCC dam section during the construction and after the completion.

## 20.2 Governing Equations

Assuming isotropic thermal properties for the solid materials, the familiar two-dimensional equation defining heat generation and transfer is of the form,

$$\alpha \nabla^2 T + \frac{\alpha}{\kappa} \dot{Q} = \dot{T} \tag{20.1}$$

where, the parameters are $T(^oC)$ temperature, $K(W/m^o c)$ heat conduction coefficient and $\dot{Q}(KJ/m^3\,h)$ rate of heat generation per unit volume and the thermal diffusion is defined as $\alpha = \kappa/\rho C$ with $\rho(Kg/m^3)$ being the density and $C(KJ/Kg^o c)$ the specific heat of concrete. The parameter $\dot{Q}$, which acts as a source term, is the rate of heat generation due to cement hydration in the concrete field.

The natural boundary condition on concrete external surface is taken as,

$$\kappa(\nabla T.\hat{n}) + q = 0 \tag{20.2}$$

where, $q$ is the rate of heat exchange per unit volume of concrete surface with surrounding ambient and $\hat{n}$ is the unit vector normal to the boundary surface. The rate of heat exchange $q$ is taken into consideration through three mechanisms, $q_c$ (convection), $q_r$ (long wave radiation of concrete to surroundings) and $q_s$ (solar radiation absorption). Hence, the total rate of heat exchange can be defined as, $q = \pm q_c + q_r - q_s$ [1, 2].

Note that, $q_c$ is through air (or water) movement and depends on the difference between temperature of the concrete surface, $T_s$, with surrounding ambient temperature, $T_{air}$, and is given by $q_c = h_c(T_s - T_{air})$. Where, $h_c = h_n + h_f$ is coefficient of thermal convection compromising from the natural convection coefficient, $h_n = 6(W/m^{2\circ}C)h_n = 6(W/m^{2\circ}C)$, and forced convection coefficient, $h_f = 3.7\ V$, where $V(m/s)$ is the wind speed.

For long wave radiation a similar relationship is used as $q_r = h_r(T_s - T_{air})$. Where, $h_r$ is the coefficient of thermal radiation.

Short wave exchange rate $q_s$ is given by $q_s = \alpha I_n$. Where $\alpha$ is surface absorption coefficient and the parameter $I_n$ is incident normal solar radiation.

One of the main factors influencing temperature profiles in concrete sections is the amount and rate of heat generated by the cementitious materials present in the concrete [3]. Due to the considerable influence of concrete temperature on the hydration rate of cementation materials, it is necessary to take this effect into account in the simulation method [1, 4, 5].

The effect of temperature history of each point on its rate of heat generation was taken into account in this research by the concepts of maturity functions $f(t)$ and equivalent time, $t_e$. Maturity functions define the temperature dependence of cement hydration reactions. Two commonly used such functions are named after Rastrup [2] and Arrhenius [1, 5] and are given as:

$$\text{(Rastrup)}\ f(T) = k2^{\left(\frac{T}{10}\right)} \tag{20.3}$$

$$\text{(Arrhenius)}\ f(t) = k\exp\left\{\frac{-E}{R(273+T)}\right\} \tag{20.4}$$

$K =$ Constant factor
$T =$ Concrete temperature in
$E =$ Activation Energy $(J/mol)$
$E = 33500$ for $T \geq 20^\circ C$
$E = 33500 + 1470$ (20-$T$) for $T < 20^\circ C$
$R =$ Molar gas constant $= 8.314\ J/^\circ Cmol$

As seen above, an increase in concrete temperature, $T$, results in an increase in the maturity function, $f(t)$. Using such functions, the relative speed of reaction at any temperature, $T$, in comparison with that at a reference temperature, $T_r$, can be obtained. The relative speeds of reaction, $H(T)$, based on Rastrup and Arrhenins functions are derived as

$$\text{(Rastrup)}\ H(T) = 2^{0.1(T-T_r)} \tag{20.5}$$

$$\text{(Arrhenius)}\ H(T) = \exp\left\{\frac{E}{R}\left(\frac{1}{273+T_r} - \frac{1}{273+T}\right)\right\} \tag{20.6}$$

Having obtained the relative speed of reaction at various temperatures to that at a reference temperature $T_r$, the equivalent time $(t_e)$ at this reference temperature for each point in concrete with its particular temperature profile $T(t)$, can be calculated:

$$t_e = \int H(T)dt \tag{20.7}$$

By obtaining the equivalent time, $t_e$, for each point in concrete, its rate of heat generation can be derived from the base heat generation curve of the concrete under a reference temperature, $T_r$. The basic heat generation curve is generally derived for the concrete mix under study by experimental determination.

In this work, suitable functions are fitted to the experimental data and the heat generation function $Q(t_e)$ is obtained. Different concretes have different heat generation characteristics. Therefore, there have been a number of functions proposed for defining this parameter.

Two of the common ones are proposed by Rastrup [2] and Gotfredson [5]:

$$(\text{Rastrup})\ Q(t_e) = A + E\ \exp\left\{-b[t_e]^n\right\} \tag{20.8}$$

$$(\text{Gotfredson})\ Q(t_e) = A(1 - e^{\frac{-t_e}{B}}) \tag{20.9}$$

where, $A, E, B$ and $b$ are constants determined by regression analysis of experimental results.

Based on the derivations given above the rate of heat generation for any point in concrete can now be determined at any equivalent time, $t_e$, using the following procedure:

$$\dot{Q}(t_e) = \left(\frac{dQ(t_e)}{d(t_e)}\right)\left(\frac{d(t_e)}{dt}\right) \tag{20.10}$$

Remember that from equation (20.7) we have,

$$\frac{d(t_e)}{dt} = H(t) \tag{20.11}$$

Choosing Rastrups relative rate of reaction for $H(T)$, i.e: equation (20.5) we get:

$$\frac{d(t_e)}{dt} = 2^{0.1(T-T_r)} \tag{20.12}$$

For the term $dQ(t_e)/d(t_e)$, if we choose Rastrups function for basic heat generation curve $Q(t_e)$, we will get:

$$\frac{dQ(t_e)}{d(t_e)} = nbE\left\{(t_e)^{-n-1}\exp[-b(t_e)^{-n}]\right\} \tag{20.13}$$

Combining equations (20.12) and (20.13) for obtaining $\dot{Q}(t_e)$ we have:

$$\dot{Q}(t_e) = nbE(t_e)^{-n-1}\exp\{-b(t_e)^{-n}\}2^{0.1(T-T_r)} \tag{20.14}$$

However, the program enables the user to choose any other combination of maturity function and basic heat generation functions for the derivation of $\dot{Q}(t_e)$.

## 20.3 Numerical Solutions

Here, a fast and accurate numerical technique is introduced which efficiently enables the solution of the temperature field in two-dimensional domains with complex and moving boundaries by the use of mixed structured/unstructured triangular meshes [6].

First the governing equation for heat generation and transfer is written in the following two-dimensional form,

$$(i = 1, 2)\frac{\partial T}{\partial t} + \frac{\partial F_i^T}{\partial x_i} = S \tag{20.15}$$

where $T$ (temperature) is the unknown parameter and $S_n$ is the heat source, and temperature flux in $i$ direction is defined as $F_i^T = \alpha \partial T / \partial x_i$.

The governing equation is multiplied by a piece wise linear test function on triangular element meshes, and then, it is integrated over all triangles surrounded every computational nodes. By application of Guass divergence theorem and using the property of the test function the boundary integral terms can be omitted. After some manipulations, the resulted equation can be written as,

$$(i = 1, 2)\int_{\Omega}\left(\frac{\partial T_n}{\partial t}\right)\phi d\Omega = \int_{\Omega}\left(F_i^d\frac{\partial \phi}{\partial x_i}\right)_n d\Omega + \int_{\Omega} S_n \phi d\Omega \tag{20.16}$$

The procedure of deriving algebraic equation will end up with an efficient explicit numerical algorithm by using of following formula within each sub-domain $\Omega$ consisting the triangles associated with every node $n$ [6],

$$(i = 1, 2)T_n^{t+\Delta t} = T_n^t + \Delta t\left[S_n - \frac{3}{2\Omega_n}\left(\sum_{k=1}^{N} F_i^d \Delta l_i\right)_n\right] \tag{20.17}$$

where, $\Delta l_i$ is the $i$ direction component of normal vector of $m$th edge of each triangle. The area of sub-domain, $\Omega$ can be computed by summation of the area of the triangles $\Lambda$ associated with node $n$, using $\Lambda = \int_\Lambda x_i (d\Lambda)_i \approx \sum_k^3 [\bar{x}_i \delta \ell_i]_k$.

By application of the Guass divergence theorem, the piece wise constant temperature flux in $i$direction, $F_i^d$, at each triangular element can be calculated using following algebraic formula,

$$F_i^d = \frac{1}{\Lambda}\sum_{m=1}^{3}(\bar{T}\Delta l_i)_m \qquad (20.18)$$

where, $\bar{T}$ is the average temperature of the edge.

Using meshes with various sub-domain sizes of sub-domains, to maintain the stability of the explicit time stepping the minimum time step of the domain of interest should be as,

$$\Delta t = k\left(\frac{\Omega_n}{\alpha_n}\right)_{\min} \qquad (20.19)$$

Remember, the heat source for each node n in concrete body is defined by $S_n = \alpha_n \dot{Q}(t_e)_n / \kappa_n$.

If $a$ and $\kappa$ at node $n$ are considered independent of time and temperature and maturity of the concrete, then for determination of the heat source in every time step, only the value of heat generation rate $\dot{Q}(t_e)$ at nodes located in concrete body must be updated for the nodes using equation (20.3). It should be noted that the value of $\dot{Q}(t_e)$ in each node is a function of $t_e$ for that node. Therefore, $t_e$ should be updated at each time step using following formula,

$$t_e = \sum_{t=t_0}^{t=t_0+N\Delta t}(2^{0.1(T-T_r)})\Delta t \qquad (20.20)$$

Two types of boundary conditions are usually applied in this numerical modeling. The essential and natural boundary conditions are used for temperature at boundary nodes and temperature diffusive flux (gradients) at boundary elements, respectively [7–9].

For the boundaries where the natural boundary conditions are to be applied, by the use of equation (20.2) and the temperature at the nodes at the vicinity of the boundary, the temperature at boundary nodes can be computed and imposed.

## 20.4 Domain Discretization

The solution comprises of the concrete dam body and the foundation (Fig. 20.1). Considering the regular geometry of dam cross-section and in order to facilitate the movement of the upper boundary of concrete as layered sequences of construction progresses, the use of structured mesh [10] is considered for dam body. Numbering the nodes and elements from the base layer to the top layer of the dam, provides the ability to ignore the desired upper layers of the dam.

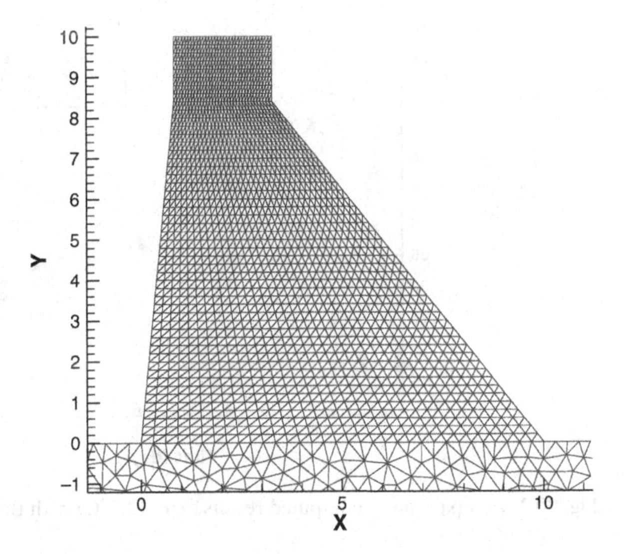


**Fig. 20.1** Hybrid structured/unstructured triangular mesh of dam and foundation

Since the age of concrete varies from layer to layer in the body of RCC dam, the use of regular mesh spacing allows moving the top boundary by gradually increasing the number of layers with constant thickness. The structured mesh is therefore quite appropriate because of offering control on the layer thickness and sequence of construction.

The dam foundation, where the moving boundary is not required, is discretized using unstructured mesh. This will facilitate consideration of irregularities in foundations. The irregular triangular mesh was produced using Deluaney Triangulation technique [10]. The use of unstructured mesh for the foundation of the dam has the advantage of allowing finer mesh size near high heat temperature gradient zones (i.e. the dam body). This technique increases the speed and accuracy of the computations.

## 20.5 Verification of Solution

In order to evaluate the accuracy of the source term representing the rate of the heat generation, a set of experimental measurements on a concrete cube with 60 (cm) dimensions caste by 450 kilogram per cubic meter is used. The concrete block was insolated all over the faces. The properties of concrete is considered as $\alpha_c = 0.0038\ (m^2/h)$ and $\kappa = 9(KJ/m^o ch)$. The averages of measured temperature at three points of this concrete block (one point at the center and two points near the faces) reported by previous workers [11].

A two dimensional rectangular triangular mesh with 6 (cm) spacing is utilized for numerical simulations of this case. The results of the computer model with the 13.5°C concrete placing temperature present reasonable agreements with the experimental data for the permanently isolated concrete block (Fig. 20.2).

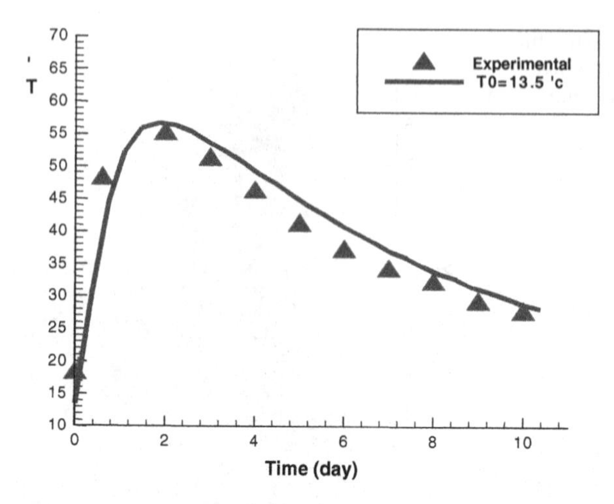


**Fig. 20.2** Comparison of computed results (T0=13.5°C) with the experimental data

At the next stage, the accuracy of the solution of spatial derivative terms is investigated by comparison of the results of the numerical solver with the analytical solution of the following two-dimensional boundary value problem with a constant source term as,

$$\Omega = \{0 < (x_1, x_2) < 1\} \frac{\partial^2 T}{\partial x_i^2} = 1 \tag{20.21}$$

The boundary conditions are considered $T = 0$ at $x_1 = 1$, $x_2 = 1$ and $\partial T/\partial n = 0$ at $x_1 = 0$, $x_2 = 0$.

The numerical modeling of the case is performed on 20 × 20 grid for triangular meshing. In Fig. 20.3, the results of numerical computations are compared with the analytical solution of the case [8].

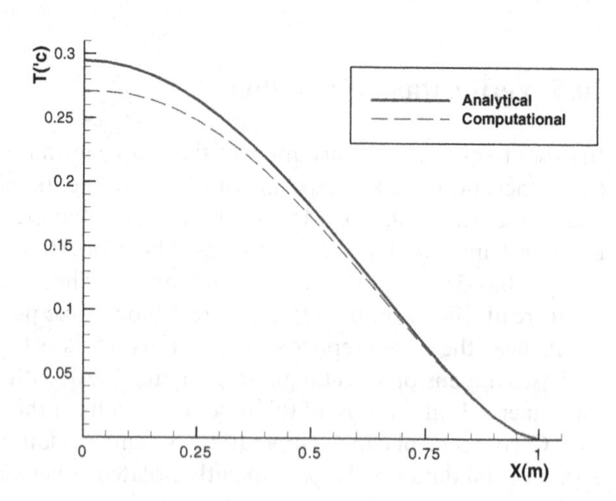


**Fig. 20.3** Comparison of computed and analytical results on diagonal line of square

## 20.6 Applications to a Typical Case

In order to present the ability of the model to deal with real world cases, it is applied to a typical gravity dam section. Both the base and the height of the section are considered as 10 (*m*) while the dam crest width is considered as 2.5 (*m*). Upstream and downstream slops are designed as 0.1:1.0 and 0.8:1.0, respectively. The radius of the half circle shape far field boundary was considered 1.5 times of the dam base dimension.

The regular mesh of the RCC dam body is generated with 0.2 (*m*) vertical spacing. The horizontal spacing at the dam base is considered equal to 0.3 (*m*) and reduced in upper layers proportional to the dam width. The horizontal mesh spacing at the dam base was taken as the finest part of the unstructured triangular mesh of the foundation part of the domain, and the triangles' sizes increase far from the dam base center.

The following assumptions were made for the concrete placing program. The layers are constructed every 24 hours at a thickness of 0.4 (*m*). The cement content of the concrete is 150 ($kg/m^3$). The concrete placing temperature is considered to be 25°*C*. The ambient temperature variation is assumed to vary between 20 and 30°*C*. The properties of RRC dam are considered as $\alpha = 0.0038$ ($m^2/h$) and $\kappa = 9(KJ/m^o ch)$. Rock foundation is considered to be quartzite with material properties of $\alpha = 0.009$ ($m^2/h$) and $\kappa = 6(KJ/m^o ch)$.

The results of the computational model for some of the stages of layer-based construction of a 10 (*m*) height RCC dam are shown in the form of temperature contour maps at Fig. 20.4.

## 20.7 Conclusions

A module of NASIR software which solves the equation of heat transfer considering the heat generation rate (due to cement hydration) is introduced in this paper. The governing equation is discetized on hybrid structured-unstructured mesh of triangular element. The resulted algorithm provides light explicit computation of time dependent problems.

The numerical model was verified in two stages. Firstly, the adopted formulation for the term representing heat generation rate of the concrete is evaluated by the comparison of the computed results with the available experimental measurements on a concrete block. Secondly, the accuracy of the solution of the heat transfer terms was assessed using a boundary value problem and its analytical solution. The results of the developed model present reasonable agreements to the analytical solution and experimental measurements.

By application of the structured mesh for the concrete dam part of the domain, an efficient modeling of the layered based construction of concrete structures (RCC Dam) is achieved. Gradual movement of the top boundary of the structure during the construction period is simulated by deactivating the

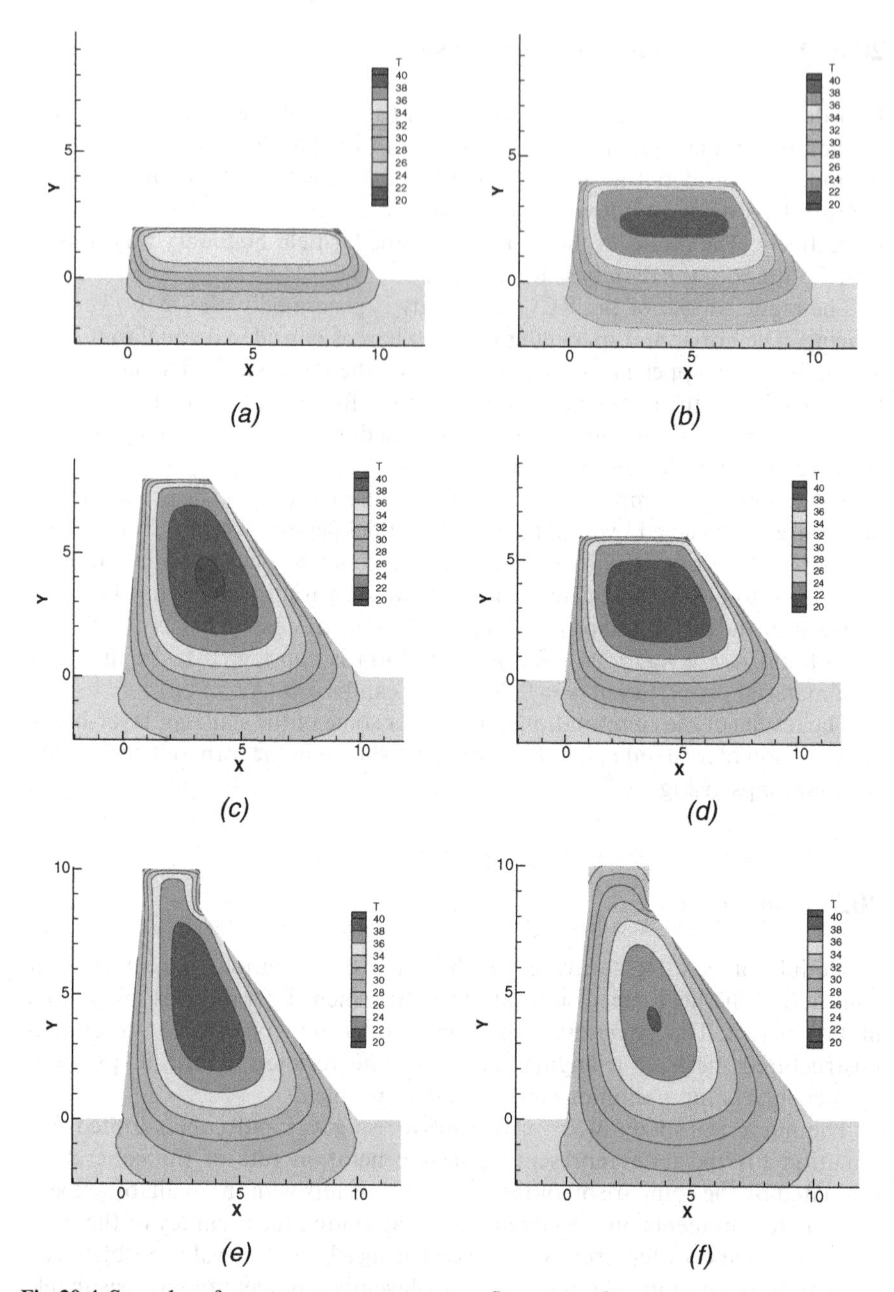


**Fig. 20.4** Snap shots from temperature contours at five stages of layered concrete placing of a 10 m height RCC dam without form works during 25 days and finally 10 days after dam completion

upper part of the dam, in which no concrete is placed. For discetetization of the dam foundation, unstructured mesh of triangles is considered, which facilitates considering irregularities in geometrical features and material properties of the natural foundation. The developed model was applied to a typical RCC dam section and the results of the temperature fields obtained showed the general pattern expected for such structures.

## References


1. Tatro SB, Shrade EK (1992) Thermal analysis for RCC, a practical approach. In: Proceeding of the conference: Roller compacted concrete 3, A.S.C.E., New York
2. Hansen KD, Reinhardt WG (1991) Roller compacted concrete dams. Mac Grow Hill Company, New York
3. Gotfredson HH, Idorn GM (1986) Curing technology at the force Bridge Denmark. ACI S.P. 95:17–33
4. Bagheri AR (1990) Early age thermal effects in conventional and micro-silica concrete linings. PhD Thesis, University of Newcastle, UK
5. Wilson EL (1968) Determination of temperatures within mass concrete structure. Report No. 68-17, Department of Civil Eng., University of California, Berkeley
6. Sabbagh Yazdi SR, Bagheri AR (2001) Thermal numerical simulation of laminar construction of RCC dams. Moving Boundaries VI, WIT Press, pp 183–192
7. Patankar SV (1980) Numerical heat transfer and fluid flow. McGraw Hill
8. Reddy JN (1984) An introduction to the Finite Element Method. McGraw-Hill, Mathematics and Statistics Series
9. Anderson DA, Tannehill JC, Pletcher RH (1984) Computational fluid mechanic and heat transfer. Cambridge Hemispher Press
10. Thompson JF, Soni BK, Weatherill NP (1999) Hand book of grid generation. CRC Press.
11. Branco FA, Mendes PA, Mirambell E (1992) Heat of hydration effects in concrete structures. ACI Mater J, 89(2):139–145
12. Lachemi M, Aitcin PC (1997) Influence of ambient and fresh concrete temperatures on maximum temperature and thermal gradient in a high performance concrete. ACI Mater J 94(2):102–110
13. Sykes LA (1990) Development of a two-dimensional navier-stokes algorithm for unstructured triangular grids. ARA Report 80

# Chapter 21
# Water Hammer Modeling Using 2nd Order Godunov Finite Volume Method

**Saeed-Reza Sabbagh-Yazdi, Ali Abbasi, and Nikos E. Mastorakis**

**Abstract** Analyzing and interpreting the water-hammer (unsteady flow) phenomena in a pipeline are not easy tasks. For complicated cases, the governing partial differential equations can only be solved numerically. Various numerical approaches have been introduced for pipeline transient calculations. In this paper, second-order Finite Volume (FV) Godunov type scheme is applied for water hammer problems and the results are analyzed. The developed one-dimensional model is based on Reimann solution. The model is applied to two classic problems (systems consisting of a reservoir, a pipe and a valve). The stability and accuracy of the developed second-order Godunov scheme is examined for Courant number less than or equal to unity. The pressure waves computed by present second order Godunuve finite volume method are in close agreement with the analytical solution for the water hammer problem in a frictionless pipe.

## 21.1 Introduction

The wave resulting from instantaneous and complete valve-closure in pressurized pipe systems propagates upstream and/or downstream in the hydraulic system. The pressure (head) and velocity of the flow (waves) are important parameters in the design of pipeline systems. Thus accurate modeling of water hammer events (hydraulic transient) is vital for proper design and safe operation of pressurized pipeline systems. The design of pipeline systems, and the prediction of water quality impacts, requires efficient mathematical models capable of accurately solving water hammer problems [1].

Various numerical approaches have been introduced for pipeline transient calculation. They include the method of characteristics (MOC), finite difference

S.-R. Sabbagh-Yazdi (✉)
Department of Civil Engineering, K.N. Toosi University of Technology, 1346 Valiasr St. 19697 Tehran, Iran
e-mail: SYazdi@kntu.ac.ir

N. Mastorakis et al. (eds.), *Proceedings of the European Computing Conference*, Lecture Notes in Electrical Engineering 28,
DOI 10.1007/978-0-387-85437-3_21, 

(FD), wave plan (WP), finite volume (FV), and finite element (FE). Among these methods, the MOC proved to be the most popular among water hammer experts. The MOC approach transforms the water hammer partial differential equations into ordinary differential equations along characteristic lines. The integration of these ordinary differential equations from one time step to the next requires that the value of the head and flow at the foot of each characteristic line be known. This requirement can be met by one of two approaches: (i) use the MOC-grid scheme; or (ii) use the fixed-grid MOC scheme and employ interpolation in pipes. However, unwanted numerical damping may associate with application of MOC when the Courant number is less than one. Therefore, application of MOC may end up with accuracy degradation due to the problem of making the Courant number exactly equal to one in all the grid points along the pipelines. [2].

Results of solving the water hammer equations by the explicit and implicit FD schemes show that these second-order FD schemes produce better results than the first-order MOC [1, 2].

Finite element methods (FE) are noted for their ability to: (i) use unstructured grids (meshes), (ii) provide convergence and accurate results, and (iii) provide results in any point of problem domain. Jovic (1995) used the combined method of MOC and FE for water hammer modeling in a classic system (a system consisting of a reservoir, a pipe, and a valve) [3].

FV methods are widely used in the solutions of hyperbolic systems, such as gas dynamics and shallow water waves. FV methods are noted for their ability to: (i) conserve mass and momentum, (ii) provide sharp resolution of discontinuities without spurious oscillations, and (iii) use unstructured grid (mesh). The first application of the FV method to water hammer problems was a first-order scheme which was highly similar to MOC with linear space-line interpolation [4]. Application of second order Godunov scheme for the FV solution of continuity and momentum equations, in which the convective term is ignored, produced accurate results for very low Mach numbers [5].

The objectives of this article are: (1) to implement the Godunov type solution for water hammer and (2) to investigate the accuracy of this method for providing a solution to the water hammer problem in a smooth pipe in which there is no physical friction for damping out the numerical errors and oscillations.

## 21.2 Governing Equations

Unsteady closed conduit flow is often represented by a set of 1D hyperbolic partial differential equations [6]:

$$\text{Continuity equation}\ \frac{\partial H}{\partial t} + V\frac{\partial H}{\partial x} + \frac{a^2}{g}\frac{\partial V}{\partial x} + V\sin\theta = 0 \tag{21.1}$$

$$\text{Momentum equation}\ \frac{\partial V}{\partial t} + V\frac{\partial V}{\partial x} + g\frac{\partial H}{\partial x} + J = 0 \tag{21.2}$$

where $t$ = time; $x$ = distance along the pipe centerline; $H = H(x,t)$ = piezometric head; $V = V(x,t)$ = instantaneous average fluid velocity; $g$ = gravitational acceleration; $a$ = wavespeed; $\theta$ = the pipe slope; and $J$ = friction force at the pipe wall.

The nonlinear convective terms $V^{\partial H}/\partial x$ and $V^{\partial V}/\partial x$ are included in Eqs. (21.1) and (21.2). These terms, although small for the majority of water hammer problems, are not neglected in this paper. Maintaining the convective terms in the governing equations makes the scheme applicable to a wide range of transient flow problems.

## 21.3 Formulation of Finite Volume Schemes for Water Hammer

The computational grid involves the discretization of the $x$ axis into reaches each of which has a length $\Delta x$ and the $t$ axis into intervals each of which has a duration $\Delta t$. Node $(i,n)$ denotes the point with coordinate $x = [i - (1/2)]\Delta x$ and $t = n\Delta t$. A quantity with a subscript $i$ and a superscript $n$ signifies that this quantity is evaluated at node $(i,n)$. The $i$th control volume is centered at node $i$ and extends from $i$–1/2 to $i+1/2$. That is, the $i$th control volume is defined by the interval $[(i\text{–}1)\Delta x, i\Delta x]$. The boundary between control volume $i$ and control volume $i+1$ has a coordinate $i\Delta x$ and is referred to either as a control surface or a cell interface. Quantities at a cell interface are identified by subscript such as $i$–1/2 and $i+1/2$.

The Riemann-based FV solution of Eqs. (21.3) and (21.4) in the $i$th control volume entails the following steps: (1) the governing equations are rewritten in control volume form; (2) the fluxes at a control surface are approximated using the exact solution of the Riemann problems; and (3) a time integration to advance the solution from $n$ to $n+1$.

Equations (21.1) and (21.2) can be rewritten in a matrix form as follows:

$$\frac{\partial u}{\partial t} + A\frac{\partial u}{\partial x} = s \tag{21.3}$$

Where

$$u = \begin{pmatrix} H \\ V \end{pmatrix}; \ A = \begin{pmatrix} V & a^2/g \\ g & V \end{pmatrix}; \text{ and } s = \begin{pmatrix} -\sin\theta \\ -J \end{pmatrix}.$$

For hyperbolic systems in non-conservative form, Eq. (21.3) can be approximated as follows [7]:

$$\frac{\partial \boldsymbol{u}}{\partial t} + \frac{\partial \boldsymbol{f}(\boldsymbol{u})}{\partial x} = \boldsymbol{s}(\boldsymbol{u}) \tag{21.4}$$

where $f(\boldsymbol{u}) = \overline{A}\boldsymbol{u}$; $\overline{A} = \begin{pmatrix} \overline{V} & a^2/g \\ g & \overline{V} \end{pmatrix}$; and $\overline{V}$ = mean value of $V$ to be specified later. Setting $\overline{V} = 0$, the scheme reverts to the classical water hammer case where the convective terms are neglected.

The mass and momentum equations for control volume $i$ is obtained by integration Eq. (21.4) with respect to $x$ from control surface $i$–1/2 to control surface $i+1/2$. The results is:

$$\frac{d}{dt}\int_{i-1/2}^{i+1/2} \boldsymbol{u}dx + \boldsymbol{f}_{i+1/2} - \boldsymbol{f}_{i-1/2} = \int_{i-1/2}^{i+1/2} sdx \tag{21.5}$$

Equation (21.5) is the statement of laws of mass and momentum conservation for the $i$th control volume. Let $\boldsymbol{U}_i$ = mean value of $\boldsymbol{u}$ in the interval $[i-1/2, i+1/2]$. Equation (21.5) becomes

$$\frac{d\boldsymbol{U}}{dt} = \frac{\boldsymbol{f}_{i-1/2} - \boldsymbol{f}_{i+1/2}}{\Delta x} + \frac{1}{\Delta x}\int_{i-1/2}^{i+1/2} sdx \tag{21.6}$$

The fluxes at cell interfaces can be determined from the Godunov schemes that requires the exact solution of the Riemann problem. Godunov schemes are conservative, explicit, and efficient. The formulation of a Godunov scheme for the mass and momentum flux $\boldsymbol{f}_{i+1/2}$ in Eq. (21.6) for all $i$ and for $t \in [t^n, t^{n+1}]$requires the exact solution of the following Riemann problem:

$$\frac{\partial \boldsymbol{u}}{\partial t} + \frac{\partial \boldsymbol{f}(\boldsymbol{u})}{\partial x} = 0 \quad \text{and} \quad u^n(x) = \begin{cases} \boldsymbol{U}_L^n & for \quad x < x_{i+1/2} \\ \boldsymbol{U}_R^n & for \quad x > x_{i+1/2} \end{cases} \tag{21.7}$$

where $U_L^n$ = average value of $\boldsymbol{u}$ to the left of interface $i+1/2$ at $n$; and $U_R^n$ = average value of $\boldsymbol{u}$ to the right of interface $i+1/2$ at $n$. The exact solution of Eq. (21.7) at $i+1/2$ for all internal nodes $i$ and for $t \in [t^n, t^{n+1}]$ is as follows:

$$\boldsymbol{u}_{i+1/2}(t) = \begin{pmatrix} H_{i+1/2} \\ V_{i+1/2} \end{pmatrix} = \frac{1}{2}\begin{pmatrix} (H_L^n + H_R^n) + \frac{a}{g}(V_L^n - V_R^n) \\ (V_L^n + V_R^n) + \frac{g}{a}(H_L^n - H_R^n) \end{pmatrix} = \boldsymbol{BU}_L^n + \boldsymbol{CU}_R^n \tag{21.8}$$

where

$$\boldsymbol{B} = \frac{1}{2}\begin{pmatrix} 1 & a/g \\ g/a & 1 \end{pmatrix}; \quad \text{and} \quad \boldsymbol{C} = \frac{1}{2}\begin{pmatrix} 1 & -a/g \\ -g/a & 1 \end{pmatrix}.$$

Using Eq. (21.8), the mass and momentum fluxes at $i+1/2$ for all internal nodes and for $t \in [t^n, t^{n+1}]$ are as follows:

$$\boldsymbol{f}_{i+1/2} = \bar{\boldsymbol{A}}_{i+1/2}\boldsymbol{u}_{i+1/2} = \bar{\boldsymbol{A}}_{i+1/2}\boldsymbol{BU}_L^n + \bar{\boldsymbol{A}}_{i+1/2}\boldsymbol{CU}_R^n \tag{21.9}$$

The evaluation of the right-hand side of Eq. (21.9) requires that $\overline{\boldsymbol{A}}_{i+1/2}$ ,$\boldsymbol{U}_L^n$ , and $\boldsymbol{U}_R^n$ are approximated. To estimate $\overline{\boldsymbol{A}}_{i+1/2}$ , the entry associated with the advective terms, $\overline{V}_{i+1/2}$ , needs to be approximated. Setting $\overline{V} = 0$ is equivalent

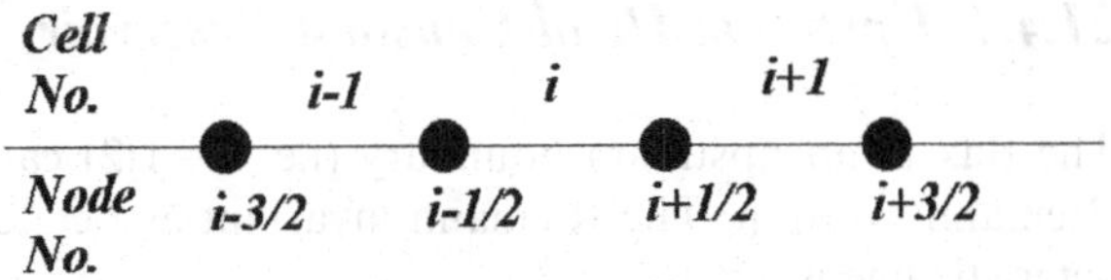

Fig. 21.1 Finite volume grids system [7]

*Cell No.* *i-1* *i* *i+1*

*Node No.* *i-3/2* *i-1/2* *i+1/2* *i+3/2*

to neglecting the advective terms from the governing equations. In general, an arithmetic mean be used to evaluate $\overline{V}_{i+1/2}$ [i.e.,$\overline{V}_{i+1/2} = 0.5(V_i^n + V_{i+1}^n)$] (Fig. 21.1).

In general, the numerical dissipation in first-order scheme is more than in second-order scheme. Limiters increase the order of accuracy of a scheme while ensuring that the results are free of spurious oscillations [4].

Using MINMOD limiter, an approximation for $\boldsymbol{U}_L^n$ and $\boldsymbol{U}_R^n$ that is second order in space and time is obtained as follow:

$$\boldsymbol{\sigma}_{j-1}^n = (\boldsymbol{U}_j^n - \boldsymbol{U}_{j-1}^n)/\Delta x \quad \text{and} \quad \sigma_j^n = (\boldsymbol{U}_{j+1}^n - \boldsymbol{U}_j^n)/\Delta x \tag{21.10}$$

$$MINMOD(\sigma_j^n, \sigma_{j-1}^n) = \begin{cases} \sigma_j^n & if \ \left|\sigma_j^n\right| < \left|\sigma_{j-1}^n\right| \ \text{ and } \ \sigma_j^n.\sigma_{j-1}^n > 0 \\ \sigma_{j-1}^n & if \ \left|\sigma_j^n\right| > \left|\sigma_{j-1}^n\right| \ \text{ and } \ \sigma_j^n.\sigma_{j-1}^n > 0 \\ 0 & if \ \ \sigma_j^n.\sigma_{j-1}^n < 0 \end{cases} \tag{21.11}$$

$$\boldsymbol{U}_{i-(1/2)^+}^n = \boldsymbol{U}_i^n - 0.5\Delta x MINMOD(\sigma_j^n, \sigma_{j-1}^n) \tag{21.12}$$

$$\boldsymbol{U}_{i-(1/2)^-}^n = \boldsymbol{U}_i^n + 0.5\Delta x MINMOD(\sigma_j^n, \sigma_{j-1}^n) \tag{21.13}$$

$$\boldsymbol{U}_{i+(1/2)^-}^{n*} = \boldsymbol{U}_{i+(1/2)^-}^n + \frac{1}{2}\frac{\Delta t}{\Delta x}[f(\boldsymbol{U}_{i-(1/2)^+}^n) - f(\boldsymbol{U}_{i+(1/2)^-}^n)] \tag{21.14}$$

$$\boldsymbol{U}_{i-(1/2)^+}^{n*} = \boldsymbol{U}_{i-(1/2)^+}^n + \frac{1}{2}\frac{\Delta t}{\Delta x}[f(\boldsymbol{U}_{i-(1/2)^+}^n) - f(\boldsymbol{U}_{i+(1/2)^-}^n)] \tag{21.15}$$

$$\boldsymbol{U}_R^n = \boldsymbol{U}_{i+(1/2)^+}^{n*} \quad \text{and} \quad \boldsymbol{U}_L^n = \boldsymbol{U}_{i+(1/2)^-}^{n*} \tag{21.16}$$

Inserting Eq. (21.16) into Eq. (21.9) can give Godunov second-order scheme for the water hammer.

## 21.4 Boundary Conditions

The implementation of boundary conditions is a important step in solving partial differential equations. The boundary conditions in this model are [4]:

### 21.4.1 Upstream Head-Constant Reservoir

The flux at an upstream boundary (i.e., i = 1/2) can be determined from the Riemann solution. The Riemann invariant associated with the negative characteristic line is:

$$H_{1/2} - \frac{a}{g} V_{1/2} = H_1^n - \frac{a}{g} V_1^n \tag{21.17}$$

Coupling this Riemann invariant with a head-flow boundary relation determines:

$$V_{1/2}^{n+1} = V_{1\frac{1}{2}}^n + \frac{g}{a}(H_{1/2} - H_{1\frac{1}{2}}^n) \tag{21.18}$$

For an upstream reservoir where $H_{1/2}^n = H_{res}$ , the flux at the upstream boundary is:

$$f_{1/2} = \begin{bmatrix} \overline{V}_{1/2} H_{res} + \frac{a^2}{g}(V_1^n + \frac{g}{a}(H_{res} - H_1^n)) \\ g H_{res} + \overline{V}_{1/2}(V_1^n + \frac{g}{a}(H_{res} - H_1^n)) \end{bmatrix} \tag{21.19}$$

### 21.4.2 Fully Closed Downstream Valve

The flux at a downstream boundary can be determined from the Riemann solution. The Riemann invariant associated with the positive characteristic line is:

$$H_{Nx+1/2} + \frac{a}{g} V_{Nx+1/2} = H_{Nx}^n + \frac{a}{g} V_{Nx}^n \tag{21.20}$$

Downstream boundary condition is valve closure in $T_c$. Head-flow boundary relation determines:

$$V_{Nx+1/2}^{n+1} = V_{steady}(1 - \frac{t}{T_c}) 0 \leq t \leq T_c \tag{21.21}$$

$$V_{Nx+1/2}^{n+1} = 0 t > T_c \tag{21.22}$$

$$H_{Nx+1/2}^n - H_{Nx-1/2}^n - \frac{a}{g}(V_{Nx+1/2}^{n+1} - V_{Nx-1/2}^n) = 0 \tag{21.23}$$

As a result, the flux at the boundary is determined as follows:

$$f_{Nx+1/2} = \begin{bmatrix} \overline{V}_{Nx+1/2}(H_{Nx} + \frac{a}{g} V_{Nx}^n - \frac{a}{g} V_{Nx+1/2}) + \frac{a}{g} V_{Nx+1/2} \\ g(H_{Nx} + \frac{a}{g} V_{Nx}^n - \frac{a}{g} V_{Nx+1/2}) + \overline{V}_{Nx+1/2} V_{Nx+1/2} \end{bmatrix} \tag{21.24}$$

## 21.5 Time Integration

The previous section provided a second-order scheme for the flux terms. In order to advance the solution from $n$ to $n+1$, Eq. (21.6) needs to be integrated with respect to time. In the absence of friction, the time integration is exact and leads to the following [4]:

$$U_i^{n+1} = U_i^n - \frac{\Delta x}{\Delta t}(f_{i+1/2}^n - f_{i-1/2}^n) \tag{21.25}$$

In the presence of friction, a second order Runge-Kutta solution is used and results in the following explicit procedure:

$$\overline{U}_i^{n+1} = U_i^n - \frac{\Delta x}{\Delta t}(f_{i+1/2}^n - f_{i-1/2}^n) \tag{21.26}$$

$$\overline{\overline{U}}_i^{n+1} = \overline{U}_i^{n+1} + \frac{\Delta t}{2} s(\overline{U}_i^{n+1}) \tag{21.27}$$

$$U_i^{n+1} = \overline{U}_i^{n+1} + \Delta t . s(\overline{\overline{U}}_i^{n+1}) \tag{21.28}$$

The time step should satisfy the Courant-Friedrichs-Lewy (CFL) condition for the convective part:

$$Cr = \frac{a.\Delta t}{\Delta x} \leq 1 \tag{21.29}$$

Although another stability condition should be used for the updating of a source term, it is found that the CFL condition is sufficient for the cases where the magnitude of the source term is small.

## 21.6 Numerical Results and Discussion

The objective of this section is to compare the accuracy and efficiency of the Godunov scheme in solving water hammer problems. Analytical solution data [8], and MOC results [9] are used to investigate the accuracy of the proposal model (Godunov scheme). The geometrical and hydraulic parameters for this test case are given in Table 21.1. This test case consists of a simple reservoir-pipe-valve configuration.

**Table 21.1** Properties for the test case 1

| Pipe diameter (m) | 0.5 |
|---|---|
| Pipe length (m) | 1000 |
| D.W friction factor | 0.00 |
| Unsteady friction factor | 0.00 |
| Wave speed(m/s) | 1000 |
| Reservoir head-upstream (m) | 0 |
| Initial mean velocity(m/s) | 1.02 |
| Cause of transients | Downstream instantaneous fully valve closure |

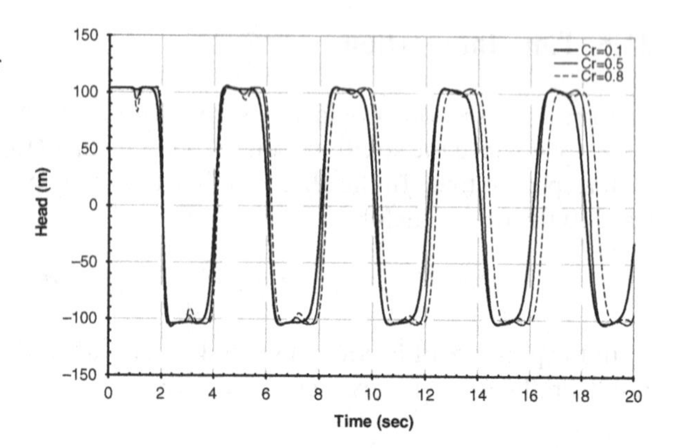


**Fig. 21.2** Pressure head traces at valve for various Cr No. (FVM)

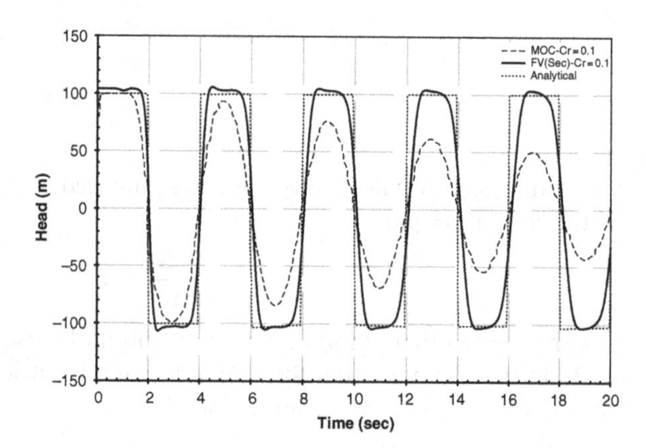


**Fig. 21.3** Hydraulic head at the valve for MOC and FVM and analytical solution

As can be seen in Fig. 21.2, there are minor numerical dissipations produced by the Godunov scheme for Courant number less than one. It is worth noting that the scheme produces pressure waves very similar to the analytical solution when Cr = 1.0.

Figure 21.3 shows the analytical and computed solutions for the variations in hydraulic head at the valve as a function of time. As expected, the head traces results by both schemes (MOC and FVM) exhibit numerical dissipation, but the numerical dissipation in FVM is less than the MOC, particularly for small Courant numbers (i.e. 0.1).

## 21.7 Conclusions

In this paper, nonlinear convective terms are included in governing equations and explicit finite volume formulations are derived using the second-order Godunov-type scheme and the model is applied for solution of the water hammer problem in smooth pipes. The results of this scheme are compared with the analytical solution from different research groups and with numerical data produced by a MOC model. The results are as follows.

1. Finite volume formulation conserves mass and momentum and produces sharp shock fronts.
2. Omission of damping effect of friction due to pipe roughness does not produce any numerical instability which may degrade the solution results.
3. The results of the explicit Godunov finite volume scheme is in close agreement with the analytical solution for the Courant numbers less than one.
4. Numerical dissipation of the explicit Godunov finite volume scheme is considerably less than MOC, particularly for the Courant numbers less than one.

## References


1. Wylie EB, Streeter VL (1993) Fluid transients in system. Prentice-Hall, Engle-wood
2. Lai C (1989) Comprehensive method of characteristic models for flow simulation. J Hydraul Eng 114(9):1074–1097
3. Jovic V (1995) Finite element and method of characteristics applied to water hammer modeling. Int J Eng Model 8(3–4):21–28
4. Guinot V (2000) Riemann solvers for water-hammer simulations by Godunov method. Int J Numer Meth Eng 49:851–870
5. Zhao M, Ghidaoui MS (2004) Godunov-type solution for water hammer flows. J Hydraul Eng 130(4):341–348
6. Chaudhry MH (1979) Applied hydraulic transients. Van Nostrand Reinhold
7. Toro EF (1997) Reimann solvers and numerical methods for fluid dynamics. Springer, Berlin
8. Sharp BB (1981) Water-hammer problems and solutions, University of Melborn, Edward Arnold Ltd.
9. Turani M (2004) Water hammer analysis with MOC for Karoun IV Penstocks (in Farsi). M.Sc thesis, K.N.Toosi University of Technology

# Chapter 22
# Simulation of Wind Pressure on Two Tandem Tanks

**Saeed-Reza Sabbagh-Yazdi, Farzad Meysami-Azad, and Nikos E. Mastorakis**

**Abstract** In this work, the set of two dimensional for the incompressible fluid is combined with an subgrid scale (SGS) turbulence viscosity turbulence mode. The SGS turbulence viscosity model is applied to simulate the changes in the pressure distributions in two storage tanks in tandem arrangement due to the wind flow at supercritical Reynolds number ($1.43\times10^8$). These two storage tanks are located in the KAZERUN power station, which are under construction in the southern part of IRAN. NASIR 2D air flow-solver is applied to investigate the wind pressure on the structural surfaces of these tanks. This Numerical model solves the governing equations for wind flow on unstructured finite volumes. The equation of continuity is simultaneously solved with the two equations of motion in a coupled manner for the steady state problems by application of the pseudo-compressibility technique. The discrete form of the two-dimensional flow equations are formulated using the Galerkin Finite Volume Method for the unstructured mesh of triangular cells. Using unstructured meshes provides the merit of accurate geometrical modeling of the curved boundaries of the tanks.

## 22.1 Introduction

The interaction of neighboring tanks may considerably change the pressure filed on a storage tank. Moreover, wind flow through particular arrangement of storage tanks may produce unexpected pressure fields, which may cause disastrous structural loading condition. Therefore, modeling of final design is recommended by most of the codes of practice to evaluate actual pressure loads on them.

The availability of high performance digital computers and development of efficient numerical models techniques have accelerated the use of Computational

S.-R. Sabbagh-Yazdi (✉)
Department of Civil Engineering, K.N. Toosi University of Technology, 1346 Valiasr St. 19697 Tehran, Iran
e-mail: SYazdi@kntu.ac.ir

N. Mastorakis et al. (eds.), *Proceedings of the European Computing Conference*, Lecture Notes in Electrical Engineering 28,
DOI 10.1007/978-0-387-85437-3_22, 

Fluid Dynamics. The control over properties and behavior of fluid flow and relative parameters are the advantages offered by CFD which make it suitable for the simulation of the applied problems. Consequently, the computer simulation of complicated flow cases has become one of the challenging areas of the research works. In this respect, many attempts have been made to develop several efficient and accurate numerical methods suitable for the complex solution domain.

In the present paper, a computation of pressure distribution of two circular tanks with different diameters in tandem arrangement at supercritical Reynolds number ($1.43 \times 10^8$) is performed using NASIR (Numerical Analyzer for Scientific and Industrial Requirements) finite volume flow-solver and the results are discussed.

## 22.2 Governing Equations

The Navier-Stokes equations for an incompressible fluid combined with a subgrid scale (SGS) turbulence viscosity model are used for the large eddy simulation (LES) of the flow around storage tanks. The non-dimensional form of the governing equations in Cartesian coordinates can be written as:

$$\frac{\partial W}{\partial t} + \left(\frac{\partial F^c}{\partial x} + \frac{\partial G^c}{\partial y}\right) + \left(\frac{\partial F^c}{\partial x} + \frac{\partial G^c}{\partial y}\right) = 0 \tag{22.1}$$

where,

$$W = \begin{pmatrix} \frac{p/\rho_0}{\beta^2} \\ u \\ v \end{pmatrix}, F^c = \begin{pmatrix} u \\ u^2 + p/\rho_0 \\ uv \end{pmatrix}, G^c = \begin{pmatrix} v \\ uv \\ v^2 + p/\rho_0 \end{pmatrix},$$

$$F^v = \begin{pmatrix} 0 \\ \nu_T \frac{\partial u}{\partial x} \\ \nu_T \frac{\partial v}{\partial x} \end{pmatrix}, G^v = \begin{pmatrix} 0 \\ \nu_T \frac{\partial u}{\partial y} \\ \nu_T \frac{\partial v}{\partial y} \end{pmatrix}$$

$W$ represents the conserved variables while, $F^c, G^c$ are the components of convective flux vector and $F^v, G^v$ are the components of viscous flux vector of $W$ in non-dimensional coordinates x and y, respectively, $u$ and $v$ the components of velocity, $p$ pressure are three dependent variables. $\nu_T$ is the summation of kinematic viscosity $\nu$ and eddy viscosity $\nu_t$.

The variables of above equations are converted to non-dimensional form by dividing $x$ and $y$ by $L$, a reference length $u$ and $v$ by $U_o$, upstream wind velocity, and p by $\rho_o U_o^2$.

The parameter $\beta$ is introduced using the analogy to the speed of sound in the equation of state of compressible flow. The application of this pseudo-compressible transient term converts the elliptic system of incompressible flow equations into a set of hyperbolic type equations [1]. Ideally, the value of the pseudo-compressibility is to be chosen so that the speed of the introduced waves approaches that of the incompressible flow. This, however, introduces a problem of contaminating the accuracy of the numerical algorithm, as well as affecting the stability property. On the other hand, if the pseudo-compressibility parameter is chosen such that these waves travel too slowly, then the variation of the pressure field accompanying these waves is very slow. Therefore, a method of controlling the speed of pressure waves is a key to the success of this approach. The theory for the method of pseudo-compressibility technique is presented in the literature [2].

Some algorithms have used constant value of pseudo-compressibility parameter and some workers have developed sophisticated algorithms for solving mixed incompressible and compressible problems [3]. However, the value of the parameter may be considered as a function of local velocity using following formula proposed [4]

$$\beta^2 = Maximum(\beta^2_{\min}\ or\ C|U^2|)$$

In order to prevent numerical difficulties in the region of very small velocities (i.e., in the vicinity of stagnation pints), the parameter $\beta^2_{\min}$ is considered in the range of 0.1-0.3, and optimum $C$ is suggested between 1 and 5 [5].

The method of the pseudo-compressibility can also be used to solve unsteady problems. For this purpose, consider additional transient term. Before advancing in time, the pressure must be iterated until a divergence free velocity field is obtained within a desired accuracy. The approach in solving a time-accurate problem has absorbed considerable attention [6]. In the present paper, the primary interest is in developing a method of obtaining steady-state solutions.

## 22.3 Numerical Method

The governing equations can be changed to a discrete form for the unstructured meshes by the application of the Galekin Finite Volume Method. This method ends up with the following 2D formulation:

$$\frac{\Delta W}{\Delta t} = \frac{(W_j^{n+1} - W_j^n)}{\Delta t} = -\frac{P}{\Omega}\left[\sum_{k=1}^{Nedge}(F^c\Delta y - G^c\Delta x)\right] - \frac{P}{A}\left[\frac{3}{2}\sum_{i=1}^{Ncell}(F^v\Delta y - G^v\Delta x)\right] \quad (22.2)$$

where, $W_i$ represents conserved variables at the center of control volume $\Omega_i$ (Fig. 22.1).

Here, $F^c, G^c$ are the mean values of convective fluxes at the control volume boundary faces and $F^v, G^v$ are the mean values of viscous fluxes which are computed at each triangle. Superscripts $n$ and $n+1$ show nth and the $(n+1)$th computational steps. $\Delta t$ is the computational step (proportional to the

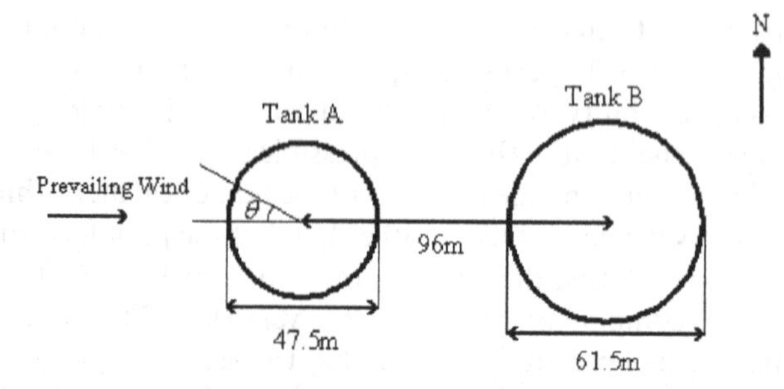


**Fig. 22.1** Schematic diagram of the arrangement of two tanks in KAZERUN power plant

minimum mesh spacing) applied between time stages $n$ and $n+1$. In present study, a three-stage Runge-Kutta scheme is used for stabilizing the computational process by damping high frequency errors, which this in turn, relaxes CFL condition.

In this study, the Smagorinsky model is used for the subgrid scale (SGS) turbulence viscosity. Eddy viscosity $\nu_t = \nu_{SGS}$ is computed as follows [7]:

$$\nu_{SGS} = (C_s\Delta)^2[1/2\bar{s}_{ij}\bar{s}_{ij}]^{1/2} \tag{22.3}$$

$$\approx \frac{1}{\Delta}\sum_{1}^{3}[\bar{v}\Delta y - \bar{u}\Delta x]\,\bar{s}_{ij} = \frac{\partial \bar{u}_i}{\partial x_j} + \frac{\partial \bar{u}_j}{\partial x_i} \tag{22.4}$$

where, $i, j = 1,2$ are for the two-dimensional computation in this paper. The subgrid scale model is used for definition of $\nu_{SGS}$, where $\Delta$ is the area of a triangular cell and the $C_s = 0.15$ are used. In equation 22.4, $\bar{u}, \bar{v}$ are mean values of velocity in each edge of the triangular element. $\Delta x, \Delta y$ for edge k of control volume $\Omega$ (Fig. 22.1) are computed as fallow:

$$\Delta y_k = y_{n_2} - y_{n_1}, \Delta x_k = x_{n_2} - x_{n_1} \tag{22.5}$$

In order to damp unwanted numerical oscillations associated with the explicit solution of the above algebraic equation a fourth order (Bi-Harmonic) numerical dissipation term is added to the convective, $C(W_i)$ and viscous, $D(W_i)$ terms. Where;

$$C(W_i) = \sum_{k=1}^{N_{edges}}[F^c\Delta y - G^c\Delta x], D(W_i) = \sum_{k=1}^{Ncell}[F^v\Delta y - G^v\Delta x]$$

The numerical dissipation term, is formed by using the Laplacian operator as follow;

$$\nabla^4 W_i = \varepsilon_4 \sum_{j=1}^{Ne}\lambda_{ij}(\nabla^2 W_j - \nabla^2 W_i) \tag{22.6}$$

The Laplacian operator at every node *i*, is computed using the variables W at two end nodes of all $N_{edge}$ edges (meeting node *i*).

$$\nabla^2 W_i = \sum_{j=1}^{Ne} (W_j - W_i) \tag{22.7}$$

In equation 22.6, $\lambda_i$, the scaling factors of the edges associated with the end nodes *i* of the edge *k*. This formulation is adopted using the local maximum value of the spectral radii Jacobian matrix of the governing equations and the size of the mesh spacing as [6]:

$$\lambda_i = \sum_{k=1}^{Ne} \left| (u.\Delta y - v.\Delta x)_k \right| + \sum_{k=1}^{Ne} \sqrt{(u.\Delta y - v.\Delta x)_k^2 + (\Delta x^2 + \Delta y^2)_k} \tag{22.8}$$

## 22.4 Application of the Model

According to API 650 storage tanks should remain stable during hurricanes with the speed of 100 m/h (44 m/s) [8]. for air we have $\rho = 1.23\text{kg/m}^3$, $\mu = 1.795 \times 10^{-5}$ So Reynolds number for tank A is computed as;

$$\text{Re} = \frac{\rho v D}{\mu} = \frac{1.23 \times 44 \times 47.5}{1.795 \times 10^{-5}} 1.43 \times 10^8$$

The performance of the this flow-solver is examined by solving turbulence flow around two storage tanks in the KAZERUN power station located in the southern part of IRAN. The schematic diagram (Fig. 22.1) shows the arrangement of two tanks in the KAZERUN site.

In order to assess the changes of pressure distribution when two tanks are in tandem arrangement the flow-solver is applied to solve the turbulence flow on three different meshes; first tank A, second tank B and finally two tanks with each other (Fig. 22.2).

In this work, No-slipping condition is considered at the solid wall nodes by setting zero normal and tangential components of computed velocities at wall nodes. At inflow boundaries unit free stream velocity and at outflow boundaries unit pressure is imposed. The free stream flow parameters (outflow pressure and inflow velocity) are set at every computational node as initial conditions.

The results on the tank A at supercritical Reynolds number ($1.43 \times 10^8$) is plotted in terms of velocity vectors respectively (Fig. 22.3).

For the case of two storage tanks with tandem arrangement velocity, vectors are presented in (Fig. 22.4). In this case, two tanks behave as a single lengthened

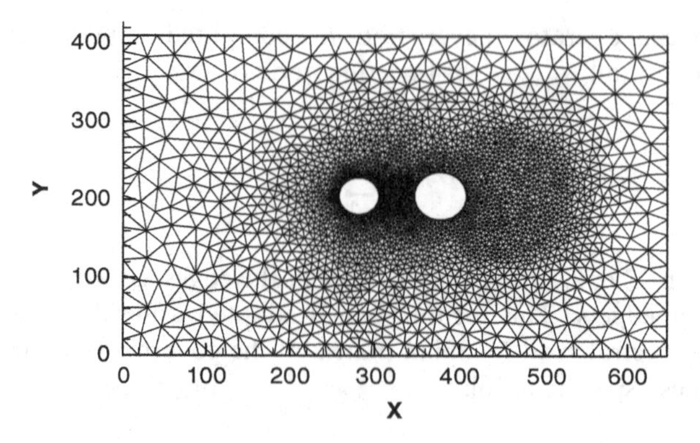


Fig. 22.2 Mesh for two tanks in tandem arrangement

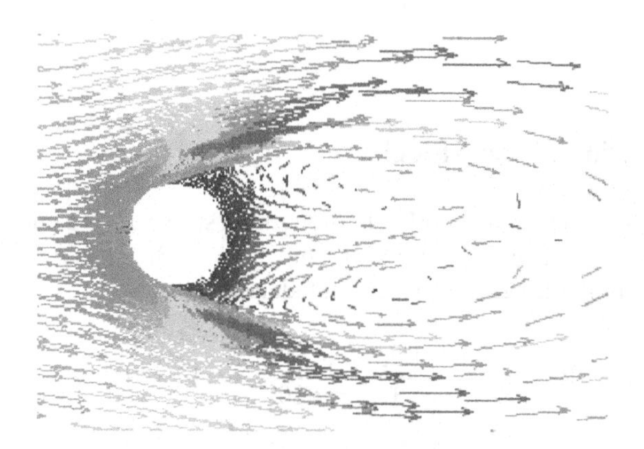

Fig. 22.3 Velocity vectors around a single tank (A) at supercritical Reynolds number ($4.52\times10^5$)

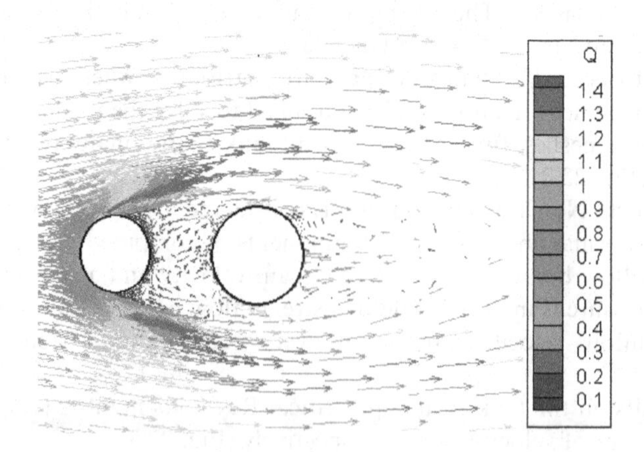


Fig. 22.4 Velocity vectors around two tandem tanks at supercritical Reynolds number ($4.52\times10^5$)

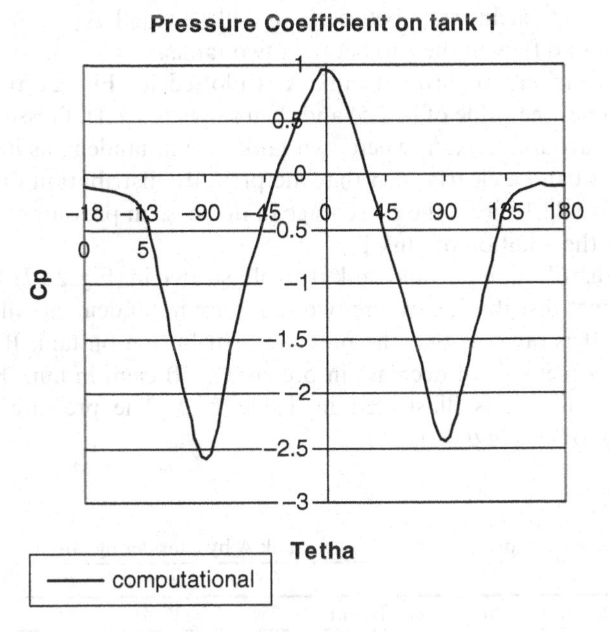


**Fig. 22.5** Pressure distributions around tank A (without tank B)

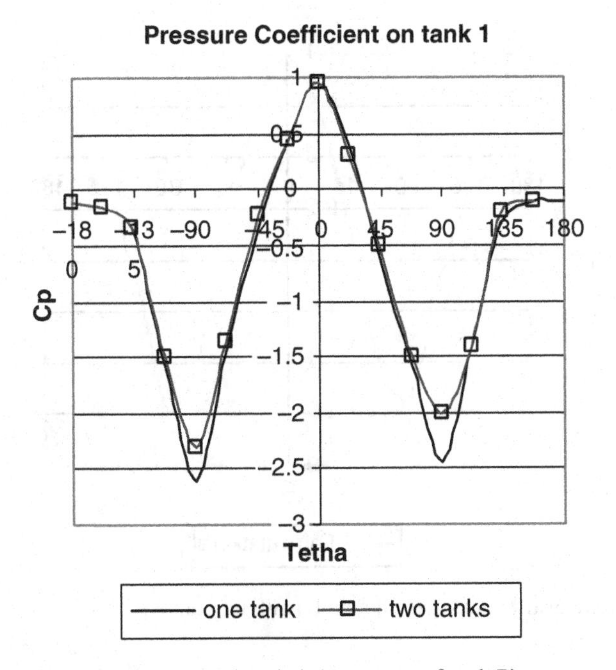


**Fig. 22.6** Pressure distribution around tank A (up-stream of tank B)

body because the spacing ratio between two tanks is small. As can be seen, there is almost no fluid flow in the gap between two tanks.

Pressure distributions around tank A is plotted in (Fig. 22.5). Minimum pressure reaches the value of (−2.5) and then raises to (+1). Pressure distribution changes around tank A, where two tanks are in tandem, as illustrated in (Fig. 22.6), it can be clearly seen that the pressure distribution on tank A is reduced. Table 22.1 shows the percentage of decrease in pressure coefficient in tank A after the addition of tank B.

Pressure distributions around tank B are illustrated in (Fig. 22.7). Changes in tank B pressure distribution, where two tanks are in tandem, are illustrated in (Fig. 22.8),. It is obvious that the pressure distribution on tank B is reduced highly. The percentage of decrease in pressure coefficient in tank B, after the addition of tank A, is illustrated in Table 22.2. The pressure coefficient decreases up to 90% in $\theta = 0$.

**Table 22.1** Decrease in pressure coefficient on tank A by considering tank B at downstream

| $\theta$ | $\theta = 0$ | $\theta = 90$ | $\theta = 180$ |
|---|---|---|---|
| Decrease in pressure coefficient | 2% | 17% | 0% |

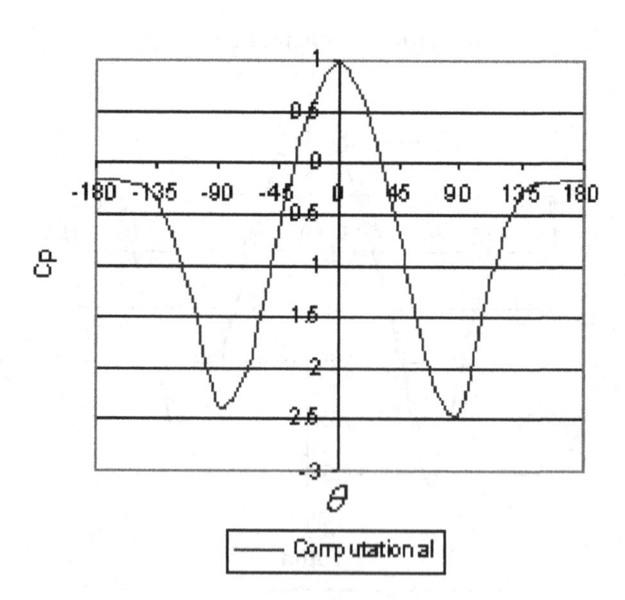


**Fig. 22.7** Pressure distributions around tank B (without tank A)

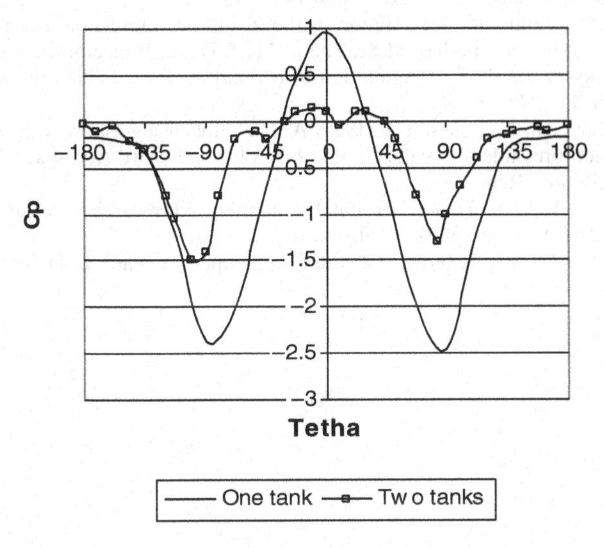


Fig. 22.8 Pressure distribution around tank B (down-stream of tank A)

Table 22.2 Decrease in pressure coefficient on tank B by considering tank A at upstream

| $\theta$ | $\theta = 0$ | $\theta = 90$ | $\theta = 180$ |
|---|---|---|---|
| Decrease in pressure coefficient | 90% | 53% | 50% |

## 22.5 Discussion

For the present case of two tanks in tandem arrangement, the downstream tank is submerged in the wake of the upstream one. The computed results show that the interference will be greatly affected by the wake behavior of the upstream tank. When the spacing ratio is small, two tanks behave as a single lengthened body. The downstream tank is wrapped in the shear layers separated from the upstream tank. In the present case of tandem arrangement, zero computed pressure is detected on a wide area in the front part of the downstream tank.

## References

1. Chorin A (1967) A numerical method for solving incompressible viscous flow problems. J Comput Phys 2:12–26
2. Chang JL, Kwak D (1984) On the method of pseudo-compressibility for numerically solving incompressible flow. AIAA 84-0252, 22nd Aerospace Science Meeting and Exhibition, Reno

3. Turkel E (1986) Preconditioning methods for solving the incompressible and low speed compressible equations. ICASE Report 86-14
4. Dreyer J (1990) Finite volume solution to the steady incompressible Euler equation on unstructured triangular meshes. M.Sc. Thesis, MAE Dept., Princeton University
5. Rizzi A, Eriksson L (1985) Computation of inviscid incompressible flow with rotation. J Fluid Mech 153:275–312
6. Belov A, Martinelli L, Jameson A (1995) A new implicit algorithm with multi-grid for unsteady incompressible flow calculations. AIAA 95-0049, 33rd Aerospace Science Meeting and Exhibition, Reno
7. Yu D, Kareem A (1996) Two-dimensional simulation of flow around rectangular prisms. J Wind Eng Ind Aerodynam 62:131–161
8. Lieb JM (2003) API 650 External pressure design appendix. Tank Industry Consultants

# Chapter 23
# Tropical Cyclone Forecaster Integrated with Case-Based Reasoning

**James N.K. Liu, Simon C.K. Shiu, and Jane You**

**Abstract** One of the major challenges for predicting tropical cyclone intensity is that we lack the understanding of coupling relationships of physical processes governing tropical cyclone intensification. This paper presents a Java-based case-based reasoning model to assist tropical cyclone forecasters to determine the intensity change of the tropical cyclone. Cases are constructed by using the data mining algorithms to uncover the hidden relationships between physical processes and tropical cyclone intensity. We specify the domain data, definitions of features from the data, tool for data exploration, and architecture of case-based reasoning model. Preliminary results are found to be useful to forecasters when faced with some unusual problem and under different weather situations.

## 23.1 Introduction

Tropical cyclone (TC) forecasting is especially difficult because cyclones are highly chaotic events and do not consistently take the same obvious form or follow obviously established patterns. Cyclones, however, may offer certain consistent information that is not apparent and this information may be available for classification and further use if suitable techniques can be developed and applied.

The intensity of a TC is one of the important parameters for estimating the possible impact by the TC. The prediction of TC intensity, which is defined as the maximum sustained one-minute surface winds, remains a challenge task in all tropical cyclone basins. It can be estimated by associating certain cloud features with conventional estimates of its strength. Procedures have been designed to improve both the reliability and the

J.N.K. Liu (✉)
Department of Computing, Hong Kong Polytechnic University, Hung Hom, Kowloon, Hong Kong
e-mail: csnkliu@comp.polyu.edu.hk

N. Mastorakis et al. (eds.), *Proceedings of the European Computing Conference*, Lecture Notes in Electrical Engineering 28,
DOI 10.1007/978-0-387-85437-3_23, 

consistency of intensity estimates made from satellite imagery [2]. These procedures rely on the interpretive skill of the meteorologists who manually view the satellite images and make a subjective assessment of the features that may be involved. The image is matched with a number of possible pattern types: curved band pattern, shear pattern, eye pattern, etc [5]. Each TC goes through a lifecycle that may be classified into eight categories (T-numbers) by its appearance in visible satellite imageries and the use of a decision tree. However, due to the subjective analysis technique and the rapid development of TCs in severe weather conditions, such prediction always leads to inaccurate results [11].

We have applied data mining techniques to meteorological data lately and uncovered interesting inter-transactional association rules for weather prediction [4]. This represents a step forward for the association of explanation with the weather phenomena. The aim of this study is to develop some means of extracting cases based on the previous meteorological information and important signals and implements a system for forecasting the intensity of a TC. We note that Case-Based Reasoning (CBR) is a problem solving methodology [1]. It is based on similar past problems and solutions to propose a solution to a current problem. The CBR system is a computer program which simulates human recognition process. Learning capability is an important feature of a CBR system. The accuracy of the system increases with the number of cases in the case library. The paper will illustrate the use of case-based reasoning to assist forecasters in making a better decision.

In Section 23.2, problem formation and data sets for the experiments are described. In Section 23.3, the development of CBR intensity prediction model is given. The system design and architecture are presented in Section 23.4 and the proposed CBR system is evaluated in Section 23.5. Conclusions will follow in the last section.

## 23.2 Problem Formulation

We consider incorporate with related synoptic information to form cases which can be referenced to improve forecast and enable reasoning support. It is particularly useful when satellite data indicate a loosely organized developing system, surface observations of an overall circulation (e.g. Westerly winds on the equatorward side) can help considerably with the satellite interpretation [7].

With reference to [8], we propose an extensive view of the forecasting framework as shown in Fig. 23.1 – a forecaster's subjective judgments regarding the similarity and usefulness of past cases to a current case (case retrieval cycles shown in Fig. 23.1 will provide an explanation as to what, why, when and how particular cases have been selected as best analogs). The use

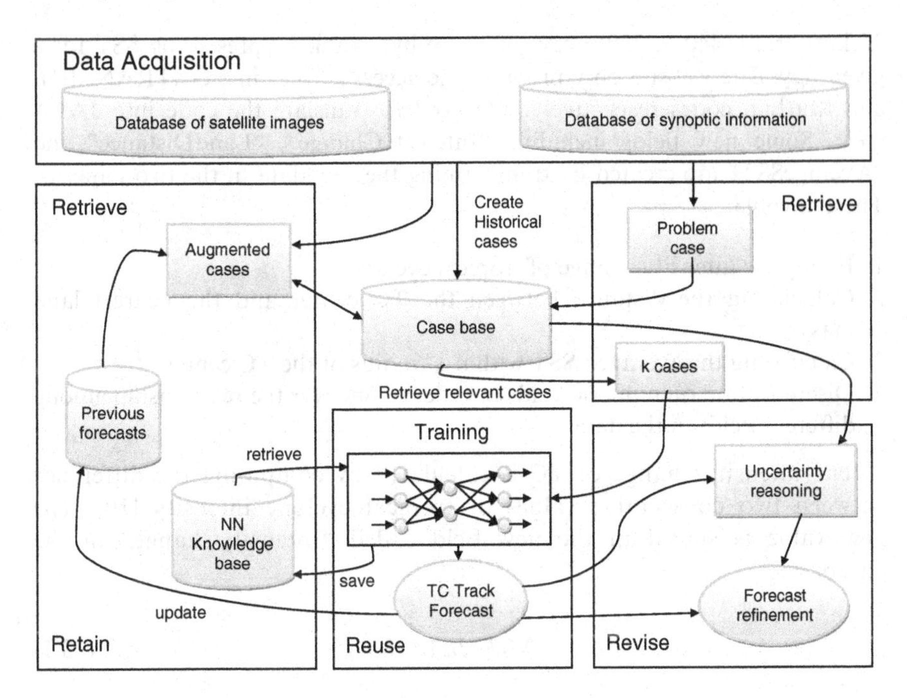


**Fig. 23.1** An integrated forecasting framework

of outranking relations in selecting the best analogs suggests a means for capturing expert knowledge. Past cases are selected based on their degree of match and usefulness to the current situation. This is done by a series of partial matching of past cases with the current case, and by ranking across case dimensions until a smaller set of matching and useful cases are retrieved (e.g. [9]). Past decisions, actions or consequences associated with the best analogs may also be reused or adapted to justify decisions or actions for the current case. The proposed approach therefore enhances learning from experience.

As a preliminary study, we have used two sets of data which can be downloaded from the Web. One of the datasets is the best track (BT) data from the NOAA National Hurricane Center (NHC) (http://www.nhc.noaa.-gov/pastall.shtml). The "best track" means that the data is a comprehensive TC track analysis after considering all available observations and expert interpretation. Another dataset is the optimally interpolated (OI) sea surface temperature (SST) product derived from TRMM/TMI (Tropical Rainfall Measuring Mission/TRMM Microwave Imager) satellite observations. The OI TMI data which could be downloaded at http://www.ssmi.com/sst/microwave_oi_sst_browse.html is available from January 1998 to the current time. The files are in binary format consisting of a $0.25 \times 0.25$ degree (about

25 km) grid (1440 × 720 array) of single byte values representing SST for a given day. The website only provides the access of files in FORTRAN, IDL and MatLab codes respectively. We need to translate the code into JAVA code. Some new fields including “IntensityChange”, “LandDistance” and “AverageSST” are created by transforming the raw data in the two datasets. These involve:

1. Filling the intensity change of tropical cyclone
2. Calculating the distance between the TC center and the nearest land mass
3. Calculating the averaged SST within 1° radius of the TC center
4. Discretization on numeric fields in order to uncover the relationship among different fields in the data

The intensity change of TC is calculated by computing the difference between two consecutive values of tropical cyclone intensity [10]. The new value is stored in the new field called “IntensityChange” in the database:

$$\Delta I_t = I_{t+1} - I_t \tag{23.1}$$

where $\Delta I_t = TC$ intensity change at time t; $I_t = TC$ intensity at time t

The binary file of the OI TMI SST products contains a byte array which has 720 rows and 1440 columns. The center of the first cell of the 1440 column and 720 row map is at 0.125E longitude and -89.875 latitude, and that for the second cell is 0.375E longitude and –89.875 latitude.

The distance between the center of the TC and the nearest land mass is calculated using the land marks (Byte value: 255) in the OI TMI SST products of Remote Sensing System. Great-circle distance is used to calculate the shortest distance between two points on the surface of the earth (Assume the radius of the earth is 6371 km). The value is stored in the new field called “LandDistance” in the database. We compute the sum of the SST within 1° radius of the tropical cyclone center (*) including (X) as:

$$SST(^{\circ}C) = (bytevalue^{*}0.15) - 3.0) \tag{23.2}$$

The averaged SST is computed by dividing the result by number of (*) including (X). The value is stored in the new field called “AverageSST” in the database. The database contains two tables storing the raw data extracted from the datasets, and transformed data for finding association rules for subsequent experiments.

## 23.3 Development of CBR Intensity Prediction Model

Apriori algorithm is applied to the data mining process. An ACSII file is generated by Clementine[1] for our case converter to convert all association rules into cases.

### 23.3.1 Case Representation

We represent the case by a 60-bits array. The first 57 bits represent the antecedent of an association rule. The last 3 bits represent the consequent of the association rule. For example, the cumbersome "00000 00000 00100 00000 00010 00000 00000 00000 00000 00000 00000 00100" is used to represent the rule: Wind_4 and AverageSST_High · IntensityChange_High. If the TC belongs to hurricane category 1 and the averaged sea surface temperature is higher than 28.8°C, then the TC will intensify in the next 6 hours.

### 23.3.2 Building a Case Store

We extract TC data from datasets NHC best track data and OI TMI SST product and have them stored in our TC database. The system stores cases in a list in the main memory after loading the case file. The case can be retrieved and compared efficiently with a new case because there is no I/O after the loading process. Apriori algorithm is applied to generate association rules and stores in an ASCII file. We extract the consequent and antecedents of each case from the case store.

### 23.3.3 Case Retrieval

To address the case retrieved from the case store, an appropriate similarity measure is needed whose purposes are (1) retrieved cases that can adapt easily to the current problem (2) retrieved cases that can solve the current problem.

When a tropical cyclone arises, we would like to retrieve all cases in the case store that match the weather situation at the time. The matching process can be as simple as just checking whether all the case feature values are identical. But this is too raw and not suitable for practical applications. We measure the corresponding similarity between TC information including latitude, translation speed, intensity, center pressure, land distance and averaged sea surface temperature for the prediction of change in the next six hours.

---

[1] A datamining toolkit developed by Integral Solutions Ltd: http://www.isl.co.hk/

After the computation of the similarities, we rely on the selection of some threshold (e.g. 0.6 found from experimental testing) to confirm the similar cases. If the similarity is greater than this threshold value, the selected case is identified as one of the best cases. There are three consequents (1) Intensifying Group (2) Stable Group and (3) Weakening Group. Each group has a weight which is the sum of the similarities in the best cases. The consequent of the new case depends on the highest value of three weights.

## 23.4 Results and Discussion

The TCs from 1998 to 2002 in the North Atlantic basin are used for data mining. There were more than 200 association rules found by using the Apriori algorithm. The minimum antecedent support and minimum rule confidence were 1 and 60% respectively. As shown in Table 23.1, we carry out tests with data for all TCs within the chosen period, TCs within the TC seasons, the exclusion of tropical depression and storm, the use of new binning features (i.e. High, Med, Low) respectively.

The number of correct predictions increases if the actual intensity change of TC is the same with our system's prediction. In result three, the prediction accuracy is about 58%, meaning that our CBR system could predict 58 events correctly out of 100 tropical cyclone events on average.

We note that the association rules for TC, which is weakening in the next six hours, are less than the rules for TC, which is intensifying in the next six hours. This may be because the data is not being distributed normally. More data are biased to being stable and intensifying. The experimental result improves if we filter the tropical depression, tropical storm category A and tropical storm category B. The result of all the TC intensity is less than 64 knots, which is too weak and will not be considered. The accuracy also increases if we filter the non-TC seasonal data, or adopt the binning process. Nevertheless, the model appears to not be doing very well in cases of TC fast intensification, re-intensification and fast weakening. The processing speed of our predicting module is very fast. It depends on the number of cases in the case stores. The time used to predict more than 500 new TC events is less than 1 second including loading the case store under a Pentium 4 class computer with 1 GB main memory.

**Table 23.1** TC prediction results

| | Training data TCs / Events | Testing data TCs / Events | Correct predict. | Predict Acc. (%) |
|---|---|---|---|---|
| Exp. Result 1 | 68 / 2202 | 16 / 550 | 309 | 56.18 |
| Exp. Result 2 | 68 / 2179 | 13 / 474 | 249 | 52.53 |
| Exp. Result 3 | 46 / 1840 | 11 / 448 | 259 | 57.81 |
| Exp. Result 4 | 68 / 2202 | 16 / 550 | 331 | 60.18 |

## 23.5 Conclusion and Future Work

Tropical cyclone prediction is a very complex process requiring analysis of an enormous amount of possibly low quality meteorological data [6]. This study uses CBR to predict the intensity change of the TC. The prediction process is fast because all cases are stored in the main memory. One of the limitations here is that the TC predictor can only predict the intensity change in the next six hours. The prediction accuracy of our CBR model may be bound by the data mining process.

For future work, the TC predictor can be extended to provide a short-range forecast. More attributes which may be extracted from other datasets can be added to the existing database in order to get better association rules with higher confidences. The accuracy will be improved if the association rules are more reliable [3]. In addition, further work on the fuzziness of those TC parameters may improve the results of prediction (e.g. [4, 9]). Predicting the moving direction of TCs is also important. This is the subject of a future paper.

**Acknowledgments** The authors would like to acknowledge the partial support of CRG grant G-U186 of The Hong Kong Polytechnic University.

## References

1. Aamodt A, Plaza E (1994) Case-based reasoning: foundational issues, methodological variations, and system approaches. AICom – Artificial Intelligence Communications, IOS Press, 7(1):39–59
2. Dvorak VF (1975) Tropical cyclone intensity analysis and forecasting from satellite imagery. Mon Wea Rev 103:420–430
3. Feng L, Dillon T, Liu J (2001) Inter-transactional association rules for prediction and their application to studying meteorological data. Data Knowl Eng, Elsevier, 37:85–115
4. Hansen B, Riordan D (1998) Fuzzy case-based prediction of ceiling and visibility. In: Pro. of 1st Conf. on Artificial Intelligence, American Meteorological Society, pp 118–123
5. Lee RST, Liu JNK (2000) Tropical cyclone identification and tracking system using integrated neural oscillatory elastic graph matching and hybrid RBF network track mining techniques. IEEE Trans NN 11(3):680–689
6. Li B, Liu J, Dai H (1998) Forecasting from low quality data with applications in weather forecasting. Int J Comp Inform 22(3):351–358
7. McGregor JL, Walsh KJ, Katzfey JJ (2000) Climate simulations for Tasmania. In: Pro. of 4th Int. Conf. on Southern Hemisphere Meteorological and Oceanography, American Meteorological Society, pp 514–515
8. Pal Sankar K, Shiu Simon CK (2004) Foundations of soft case-based reasoning. Wiley
9. San Pedro J, Burstein F (2003) A framework for case-based fuzzy multi-criteria decision support for tropical cyclone forecasting. In: Pro. of 36th Hawaii Int. Conf., System Sciences
10. Tang J, Yang R, Kafatos M (2005) Data mining for tropical cyclone intensity prediction. In: Pro. of 6th Conf. on Coastal Atmospheric and Oceanic Prediction and Processes, Session 7.5
11. TC Formation Regions, NOAA, http://www.srh.noaa.gov/jetstream/tropics/ tc_basins.htm

# Part IV
# Knowledge Engineering, Decision Rules and Data Bases

In this part papers from the areas of knowledge engineering, decision rules and databases are unified. One of the major problems in information processing is automated knowledge transfer and utilization, which is a key factor in productivity improvement. Knowledge engineering is a field of science in which the primary goal is to develop techniques and methodologies for achieving automated knowledge transfer and utilization. A decision rule is either a function that maps from the current state to the agent's decision or choice or a mapping from the expressed preferences of each of a group of agents to a group decision. The former is more relevant to decision theory and dynamic optimization whereas the latter is relevant to game theory. Knowledge engineering and decision rules are major tools for dealing with large databases. Here we present some results about the understanding of knowledge reuse in an on-line environment, handling contradictions in knowledge discovery applications, cooperative distributed query-processing approaches, estimating join and projection selectivity factors, connection between the decision rules and metamodel to organize information, a computerized solution for the financial diagnose of the Small and Medium Enterprises (SMEs), tracking deadlocks and phantoms in databases, concurrency control for multilevel secure distributed real-time databases and a new framework for non-deterministic multi-valued system minimization.

# Chapter 24
# Toward an Understanding of Knowledge Reuse in an On-Line Environment

**Hsieh-Hua Yang, Jui-Chen Yu, and Hung-Jen Yang**

**Abstract** For the past ten years, a growing amount of research has shown the knowledge reuse's role in knowledge management. This paper follows the steps and tries to reveal a model to explain the relation among intention and meta-cognition. In this study, first setup theory model via literature review and an investigating process was conducted by using an on-line questionnaire with internal-reliability alpha value of 0.93. In analyzing the collected data, a confirmatory factor analysis was performed on the data using the LISREL program. The measurement models were evaluated using maximum likelihood estimation. Convergent and discriminates validity of the items and scales were tested with CFA. Based upon 781 participants' responses, the proposed theory model was proved at well fit conditions. It is concluded that knowledge reuse behavior is affected by both intention and meta-cognitive intention of knowledge reuse.

## 24.1 Introduction

Knowledge creation is often viewed as somehow more important, more difficult to manage, and less amenable to information technology support. However, the effective reuse of knowledge is arguably a more frequent organizational concern and one that is clearly related to organizational effectiveness.

Despite its importance, knowledge reuse is something that we know relatively little about. Although knowledge reuse has been observed and researched under many different names in many different settings, findings about knowledge reuse have remained relatively dispersed and un-integrated. One possible explanation is that knowledge reuse is seen as a unitary phenomenon – pretty much the same regardless of who does it, how, and why.

H.-H. Yang (✉)
Department of Health Care Administration, Chang Jung Christian University,
396 Chang Jung Rd., Sec. 1, Kway Jen, Tainan, Taiwan, R.O.C
e-mail: yangsnow@mail.cjcu.edu.tw

N. Mastorakis et al. (eds.), *Proceedings of the European Computing Conference*, Lecture Notes in Electrical Engineering 28,
DOI 10.1007/978-0-387-85437-3_24, 

### *24.1.1 Statement of Research Problem*

Knowledge reuse is a basic component of knowledge management systems. While earlier research interest on knowledge management tended to revolve around using information and communication technologies and procedures, attention is increasingly shifted to investigate the role of contextual factors and psychological factors in promoting or inhibiting knowledge reuse. As the importance of knowledge management grows rapidly in most environments, however, few prior or related empirical studies of factors promoting or inhibiting knowledge reuse in an on-line environment context exist. The problem of this study was what are those factors that affect individual knowledge reuse propensity within an on-learning environment.

### *24.1.2 Purposes of the Study*

The purpose of this study was to identify factors that affect individual knowledge reuse propensity based upon both contextual factors and individual psychological factors. Specifically, research was designed to answer the following questions:

1. Is individuals' knowledge reuse propensity influenced by their perception of self-interest in an on-line environment?
2. Is individuals' knowledge reuse propensity influenced by their perception of altruism in an on-line environment?
3. Is individuals' knowledge reuse propensity influenced by their perception of self-esteem in an on-line environment?
4. Is individuals' knowledge reuse propensity influenced by their perception of ownership belief in an on-line environment?

## 24.2 Literature Review

### *24.2.1 Knowledge Reuse as a Process*

The knowledge reuse process can be described in terms of the following stages: capturing or documenting knowledge, packaging knowledge for reuse, distributing or disseminating knowledge (providing people with access to it), and reusing knowledge [1].

Capturing and documenting knowledge can occur in at least four ways.

First, documenting can be a largely passive by-product of the work process, as when virtual teams or communities of practice automatically generate archives of their informal electronic communication that can later be searched.

Second, documenting knowledge for potential reuse can occur within a structure such as that provided by facilitators using brainstorming techniques, perhaps mediated by the use of electronic meeting systems [2].

Third, documenting can involve creating structured records as part of a deliberate, before-the-fact knowledge reuse strategy.

Fourth, documentation can involve a deliberate, after-the-fact strategy of filtering, indexing, packaging, and sanitizing knowledge for later reuse, as in the creation of learning histories, expert help files, or the creation of a data warehouse.

Packaging knowledge is the process of culling, cleaning and polishing, structuring, formatting, or indexing documents against a classification scheme. Among the activities involved in knowledge packaging are: authoring knowledge content, codifying knowledge into "knowledge objects" by adding context [3], developing local knowledge into "boundary objects" by deleting context [4], filtering and pruning content, and developing classification schemes [5].

### 24.2.2 Meta-Cognitive Knowledge

The term "meta-cognition" is most often associated with John Flavell [6]. According to Flavell [6, 7], meta-cognition consists of both meta-cognitive knowledge and meta-cognitive experiences or regulation. Meta-cognitive knowledge refers to acquired knowledge about cognitive processes, knowledge that can be used to control cognitive processes. Flavell further divides meta-cognitive knowledge into three categories: knowledge of person variables, task variables and strategy variables.

Stated very briefly, knowledge of person variables refers to general knowledge about how human beings learn and process information, as well as individual knowledge of one's own learning processes. One may be aware that a study session will be more productive if he/she works in the quiet library rather than at home where there are many distractions. Knowledge of task variables includes knowledge about the nature of the task as well as the type of processing demands that it will place upon the individual. For example, you may be aware that it will take more time for you to read and comprehend a science text than it would for you to read and comprehend a novel. Knowledge about strategy variables include knowledge about both cognitive and meta-cognitive strategies, as well as conditional knowledge about when and where it is appropriate to use such strategies.

### 24.2.3 Meta-Cognitive Regulation

Meta-cognitive experiences involve the use of meta-cognitive strategies or meta-cognitive regulation [8]. Meta-cognitive strategies are sequential processes that one uses to control cognitive activities, and to ensure that a cognitive goal (e.g., understanding an on-line text) has been met. These processes help to regulate and oversee learning, and consist of planning and monitoring cognitive activities, as well as checking the outcomes of those activities.

For example, after reading a paragraph in an on-line platform a learner may question herself about the concepts discussed in the paragraph. Her cognitive goal is to understand the text. Self-questioning is a common meta-cognitive comprehension monitoring strategy. If she finds that she cannot answer her own questions, or that she does not understand the material discussed, she must then determine what needs to be done to ensure that she meets the cognitive goal of understanding the text. She may decide to go back and re-read the paragraph with the goal of being able to answer the questions she had generated. If, after re-reading through the text she can now answer the questions, she may determine that she understands the material. Thus, the meta-cognitive strategy of self-questioning is used to ensure that the cognitive goal of comprehension is met. Information technology may play an assistance role in helping learners with searching, indexing, storing, or marking.

### *24.2.4 Meta-Cognitive Strategies*

Most definitions of meta-cognition include both knowledge and strategy components; however, there are a number of problems associated with using such definitions. One major issue involves separating what is cognitive from what is meta-cognitive.

Flavell himself acknowledges that meta-cognitive knowledge may not be different from cognitive knowledge [6]. The distinction lies in how the information is used.

Recall that meta-cognition is referred to as "thinking about thinking" and involves overseeing whether a cognitive goal has been met. This should be the defining criterion for determining what is meta-cognitive. Cognitive strategies are used to help an individual achieve a particular goal while meta-cognitive strategies are used to ensure that the goal has been reached. Meta-cognitive experiences usually precede or follow a cognitive activity. They often occur when cognitions fail, such as the recognition that one did not understand what one just read. Such an impasse is believed to activate meta-cognitive processes as the learner attempts to rectify the situation [9].

Meta-cognitive and cognitive strategies may overlap in that the same strategy, such as questioning, could be regarded as either a cognitive or a meta-cognitive strategy depending on what the purpose for using that strategy may be. For example, you may use a self-questioning strategy while reading as a means of obtaining knowledge (cognitive), or as a way of monitoring what you have read (meta-cognitive). Because cognitive and meta-cognitive strategies are closely intertwined and dependent upon each other, any attempt to examine one without acknowledging the other would not provide an adequate picture.

Knowledge is considered to be meta-cognitive if it is actively used in a strategic manner to ensure that a goal is met.

Meta-cognition or the ability to control one's cognitive processes (self-regulation) has been linked to knowledge reuse. By applying information technology in an on-line environment, knowledge reuse behavior would be affected. Meta-cognition is executive processes that control other cognitive components as well as receive feedback from these components. It is responsible for figuring out how to do a particular task or set of tasks, and then making sure that the task or set of tasks are done correctly. Theses executive processes involve knowledge reuse behavior.

Meta-cognition enables learners to benefit from instruction and influences the use and maintenance of cognitive strategies. Meta-cognitive Strategies for Successful Knowledge Learning are:

**Consciousness:**

- Consciously identify what you already know
- Define the learning goal
- Consider your personal resources (e.g. textbooks, access to the library, access to a computer workstation or a quiet study area)
- Consider the task requirements (essay test, multiple choice, etc.)
- Determine how your performance will be evaluated
- Consider your motivation level
- Determine your level of anxiety

**Planning:**

- Estimate the time required to complete the task
- Plan study time into your schedule and set priorities
- Make a checklist of what needs to happen when
- Organize materials
- Take the necessary steps to learn by using strategies like outlining, mnemonics, diagramming, etc.

**Monitoring and Reflection:**

- Reflect on the learning process, keeping track of what works and what doesn't work for you
- Monitor your own learning by questioning and self-testing
- Provide your own feedback
- Keep concentration and motivation high

As knowledge workers become more skilled at using meta-cognitive strategies, they gain confidence and become more independent as learners. Independence leads to ownership as worker's realize they can pursue their own intellectual needs and discover a world of information at their fingertips.

The task of technology for knowledge management is to acknowledge, cultivate, exploit and enhance the meta-cognitive capabilities of all knowledge workers.

## 24.3 Theory Frame

Based upon the theory of information reuse [10, 11] and meta-cognition theory [12, 13], a multilevel model was constructed to simultaneously investigate these factors which in both theories could impact individuals' knowledge reuse intentions.

Information reuse theory identified social and individual antecedents of knowledge reuse, and at the same time it distinguishes information product from expertise to explain that people reuse knowledge has a different motivation.

Meta-cognition theory helps us to understand that knowledge reuse is conducted according to the learner's evaluation via strategy, reflection, and self-regulation.

### *24.3.1 Intention of Knowledge Reuse*

Intention of knowledge reuse relates to personal perception of easy use. According to the technology acceptance model, TAM, ease of use is a factor which contributes to technology attitude [14]. Under TAM, ease of knowledge access

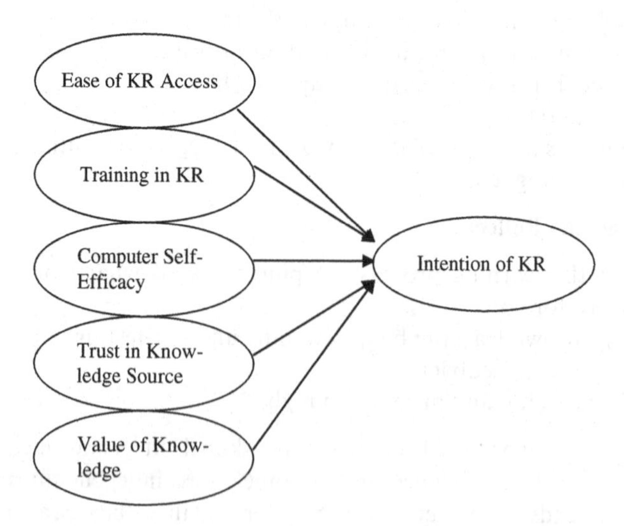


**Fig. 24.1** Theory model of intention of knowledge reuse

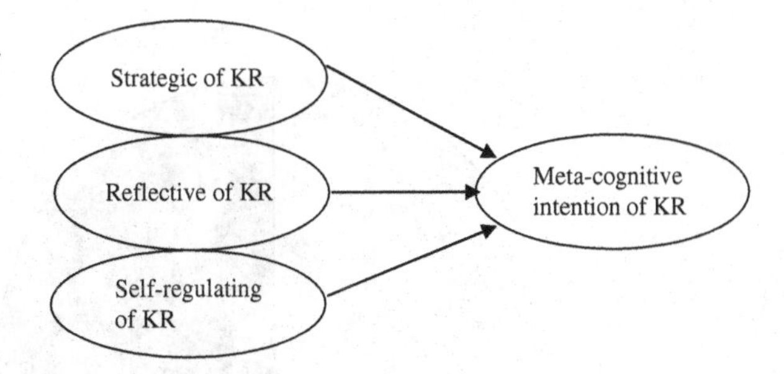


**Fig. 24.2** Theory model of meta-cognitive intention of knowledge reuse

can be seen as an indicator of ease of use, and thus, as a predictor of the user's intention to use, and their ultimate use, of the knowledge system.

*Hypothesis 1: Ease of knowledge access will be positively related to the intention of knowledge reuse.*

One way the knowledge workers could raise their expectancy perceptions is through adequate training in the activities they are expected to perform. With advanced training, knowledge workers are more likely to achieve the outcome, or get better results, with the same level of effort.

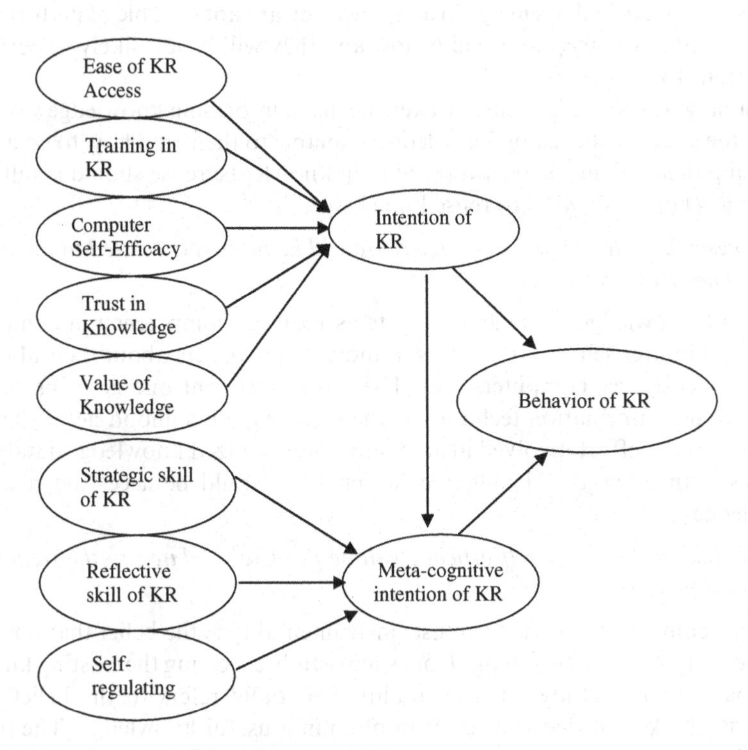


**Fig. 24.3** Theory model of knowledge reuse behavior

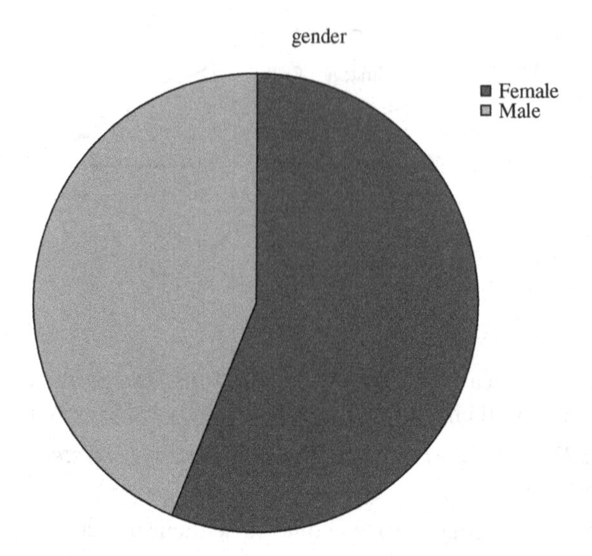


**Fig. 24.4** Gender information of respondents

If people are asked to engage in activities they are not capable of performing, their expectancy perceptions will be low and they will be less likely to perform the action [15].

Training in knowledge reuse is a key means of informing knowledge workers about the sources of existing knowledge available to them and how to go about accessing them. Thus, increased training in knowledge reuse should result in a higher level of motivation to reuse knowledge.

*Hypothesis 2: Training in knowledge reuse will be positively related to the intention of knowledge reuse.*

Use of knowledge management systems requires competency in computer usage. Computer self-efficacy refers to individuals' beliefs about their abilities to competently use computers [16]. The more confident one is in his or her ability to use information technology, the greater the likelihood he or she will believe that the effort involved in accessing computerized knowledge databases will result in a successful outcome, which here would be accessing relevant knowledge.

*Hypothesis 3: Computer self-efficacy will be positively related to the frequency of knowledge reuse.*

In the context of knowledge reuse, instrumentality is the belief that one will get relevant, reusable knowledge from successfully accessing the existing knowledge base. In this study, instrumentality essentially refers to the belief that access to the knowledge will result in obtaining useful knowledge. The more an individual trusts that the knowledge in the database of existing knowledge is timely and accurate, the greater the belief that accessing the database will result in the reward of a useful piece of knowledge. Trust in the system can be viewed as reflecting the perceived usefulness of that system, and therefore, under the TAM, trust would be positively associated with the use of the system. Thus, there should be a positive relationship between trust in the knowledge and the intention of knowledge reuse.

*Hypothesis 4: Trust in knowledge source will be positively related to the intention of knowledge reuse.*

Whereas instrumentality refers to the belief that one will get something for achieving the outcome, another key aspect that should be considered is the values of what one will receive for achieving the outcome. The greater the valence is, the higher the motivation for engaging in the behavior will be. Therefore, we expect there to be a positive relationship between the perceived value of the knowledge received through knowledge reuse and the intention of knowledge reuse.

*Hypothesis 5: Value of knowledge will be positively related to the intention of knowledge reuse.*

### *24.3.2 Meta-Cognitive Intention of Knowledge Reuse*

Knowing how to learn, and knowing which strategies work best, are valuable skills that differentiate expert learners from novice learners. Meta-cognition is a critical ingredient to successful learning knowledge. It consists of two basic processes occurring simultaneously: monitoring your progress as you learn, and making changes and adapting your strategies if you perceive that you are not doing so well.

Meta-cognitive skills include taking conscious control of learning, planning and selecting strategies, monitoring the progress of learning, correcting errors, analyzing the effectiveness of learning strategies, and changing learning behaviors and strategies when necessary [17]. Strategic, reflective, and self-regulating are three components of meta-cognitive skills. These skills become factors which affect the meta-cognitive intention of knowledge reuse. Thus, there exists a relationship between strategic and meta-cognitive intention of knowledge reuse.

*Hypothesis 6: Strategic skill will be positively related to the intention of meta-cognitive intention of knowledge reuse.*

*Hypothesis 7: Reflective skill will be positively related to the intention of meta-cognitive intention of knowledge reuse.*

*Hypothesis 8: Self-regulating skill will be positively related to the intention of meta-cognitive intention of knowledge reuse.*

### *24.3.3 Behavior of Knowledge Reuse*

The behavior of knowledge reuse is to conduct capturing and packaging knowledge for reuse [1]. The behavior is affected by the intention of knowledge reuse and meta-cognitive.

The frequency of knowledge behavior would be positively affected by one's intention of reusing knowledge. Whenever a person intended to reuse certain knowledge, he/she would re-think what is pursued. At this moment, the meta-cognition process would begin. Thus, it is proposed that reuse intention affects reuse behavior. The reuse intention also affects meta-cognitive intention. The behavior is also affected by the intention of meta-cognition.

*Hypothesis 9: Knowledge reuse intention will be positively related to the reuse behavior.*

*Hypothesis 10: Knowledge reuse intention will be positively related to the meta-cognitive intention.*

*Hypothesis 11: The meta-cognitive intention of knowledge reuse will be positively related to the reuse behavior.*

## 24.4 Research Method

### *24.4.1 Data Collection*

Data to test the model and hypotheses were drawn from a cross-sectional field study of students in the National Kaohsiung Normal University in Taiwan. The knowledge reuse behavior data were acquired from an online-learning platform that students used for academic learning activity. There were a total of 781 subjects in this study.

### *24.4.2 Measurement of Constructs*

Constructs were measured by using multiple-item scales; a pilot study of 80 subjects was conducted to ensure the reliability of the questionnaire. The overall internal consistence value is 0.93 and the subdomain value is from 0.78 to 0.85. There were twenty two items in the questionnaire. All items use seven – point Likert scales anchored between "strongly disagree" and "strongly agree". There were three levels of factors. The detail of the structure was listed in Table 24.1.

### *24.4.3 Scale Validation Method*

In analyzing the collected data, a confirmatory factor analysis was performed on the data using the LISREL program. The measurement models were

**Table 24.1** Research variables lists according to factor and level

| Levels | Factors | Variables |
|---|---|---|
| 1 | Ease of KR access | EA1 |
| Intention of KR | | EA2 |
| | Training in KR | Train1 |
| | | Train2 |
| | Computer self-efficacy | CSE1 |
| | | CSE2 |
| | Trust in knowledge | Tru1 |
| | | Tru2 |
| | Value of knowledge | Value1 |
| | | Value2 |
| 1 | Strategic of KR | ST1 |
| Meta-cognitive | | ST2 |
| intention of KR | Reflective of KR | Ref1 |
| | | Ref2 |
| | Self-regulating of KR | Reg1 |
| | | Reg2 |
| 2 | Intention of KR | Inten1 |
| | | Inten2 |
| | | Inten3 |
| 2 | Meta-cognitive intention of KR | Meta1 |
| | | Meta2 |
| | | Meta3 |
| 3 | Behavior of KR | BKR1 |
| | | BKR2 |
| | | BKR3 |

evaluated using maximum likelihood estimation. Convergent and discriminates validity of the items and scales were tested with CFA.

## 24.5 Data Analysis and Results

### *24.5.1 Samples*

Seven hundred and eighty one participants completed the survey via an online website. There were 438 female and 343 male. The percentage was listed in Table 24.2.

**Table 24.2** Gender distribution of participants

| | | Frequency | Percent | Valid percent |
|---|---|---|---|---|
| Valid | Female | 438 | 56.1 | 56.1 |
| | Male | 343 | 43.9 | 43.9 |
| | Total | 781 | 100.0 | 100.0 |

### 24.5.2 Measurement Equations

The LISREL estimates results were listed in Tables 24.3, 24.4, 24.5, 24.6, and 24.7. In Table 24.3, the factor of intent1 would significantly contribute to intention of knowledge reuse and predict at 41%.

The factor of inten2 would significantly contribute to the intention of knowledge reuse and predict at 76%. The factor of intent3 would significantly contribute to intention of knowledge reuse and predict at 23%.

In Table 24.4, the factor of meta1 would significantly contribute to intention of knowledge reuse and predict at 26%.

**Table 24.3** Measurement equations of intention of knowledge reuse

| | | |
|---|---|---|
| inten1 = 0.65*IKR, | Errorvar. = 0.61, | $R^2$ = 0.41 |
| | (0.025) | |
| | 24.55 | |
| inten2 = 0.91*IKR, | Errorvar. = 0.27, | $R^2$ = 0.76 |
| (0.032) | (0.019) | |
| 28.63 | 14.37 | |
| inten3 = 0.52*IKR, | Errorvar. = 0.92, | $R^2$ = 0.23 |
| (0.022) | (0.051) | |
| 23.46 | 18.00 | |

**Table 24.4** Measurement equations of meta-cognitive intention of knowledge reuse

| | | |
|---|---|---|
| meta1 = 0.55*Metacog, | Errorvar. = 0.83, | $R^2$ = 0.26 |
| | (0.034) | |
| | 24.11 | |
| meta2 = 0.91*Metacog, | Errorvar. = 0.21, | $R^2$ = 0.80 |
| (0.031) | (0.016) | |
| 29.37 | 13.19 | |
| meta3 = 0.95*Metacog, | Errorvar. = 0.12, | $R^2$ = 0.89 |
| (0.031) | (0.012) | |
| 30.27 | 9.44 | |

**Table 24.5** Measurement equation of behavior of knowledge reuse

| | | |
|---|---|---|
| bkr1 = 0.79*BKR, | Errorvar. = 0.55, | $R^2$ = 0.53 |
| | (0.026) | |
| | 20.98 | |
| bkr2 = 0.94*BKR, | Errorvar. = 0.15, | $R^2$ = 0.85 |
| (0.023) | (0.016) | |
| 41.55 | 9.58 | |
| bkr3 = 0.93*BKR, | Errorvar. = 0.16, | $R^2$ = 0.84 |
| (0.013) | (0.012) | |
| 72.10 | 13.95 | |

**Table 24.6** Measurement equations the first level factors of intention of knowledge reuse

| Equation | Errorvar. | $R^2$ |
|---|---|---|
| ea1 = 0.66*EaseAcce, | Errorvar. = 0.59, | $R^2$ = 0.42 |
| (0.032) | (0.031) | |
| 20.41 | 18.99 | |
| ea2 = 0.88*EaseAcce, | Errorvar. = 0.32, | $R^2$ = 0.71 |
| (0.030) | (0.025) | |
| 29.09 | 12.86 | |
| train1 = 0.65*Training, | Errorvar. = 0.67, | $R^2$ = 0.38 |
| (0.027) | (0.023) | |
| 23.63 | 29.76 | |
| train2 = 0.84*Training, | Errorvar. = 0.34, | $R^2$ = 0.67 |
| (0.032) | (0.023) | |
| 26.36 | 14.93 | |
| cse1 = 1.05*CSE, | Errorvar. = 0.26, | $R^2$ = 0.81 |
| (0.034) | (0.037) | |
| 30.4 | 66.93 | |
| cse2 = 0.82*CSE, | Errorvar. = 0.44, | $R^2$ = 0.60 |
| (0.033) | (0.031) | |
| 24.64 | 14.04 | |
| tru1 = 0.96*Trust, | Errorvar. = 0.26, | $R^2$ = 0.78 |
| (0.034) | (0.035) | |
| 28.3 | 77.56 | |
| tru2 = 0.97*Trust, | Errorvar. = 0.27, | $R^2$ = 0.78 |
| (0.034) | (0.034) | |
| 28.3 | 87.88 | |
| value1 = 0.99*Value, | Errorvar. = 0.37, | $R^2$ = 0.73 |
| | (0.030) | |
| | 12.43 | |
| value2 = 0.78*Value, | Errorvar. = 0.29, | $R^2$ = 0.68 |
| (0.031) | (0.020) | |
| 24.69 | 14.00 | |

**Table 24.7** Measurement equations the first level factors of meta-cognitive intention of knowledge reuse

| Equation | Errorvar. | $R^2$ |
|---|---|---|
| st1 = 1.05*Strategi, | Errorvar. = 0.35, | $R^2$ = 0.76 |
| (0.026) | (0.020) | |
| 40.83 | 17.51 | |
| st2 = 0.63*Strategi, | Errorvar. = 0.71, | $R^2$ = 0.36 |
| (0.032) | (0.037) | |
| 19.98 | 19.40 | |
| ref1 = 1.09*Reflecti, | Errorvar. = 0.20, | $R^2$ = 0.86 |
| (0.025) | (0.022) | |
| 43.12 | 8.93 | |
| ref2 = 0.41*Reflecti, | Errorvar. = 0.51 , | $R^2$ = 0.25 |
| (0.026) | (0.027) | |
| 15.51 | 18.84 | |
| reg1 = 0.51*Self_reg, | Errorvar. = 0.89, | $R^2$ = 0.23 |
| (0.023) | (0.026) | |
| 22.53 | 34.00 | |
| reg2 = 0.66*Self_reg, | Errorvar. = 0.63, | $R^2$ = 0.41 |
| (0.020) | (0.028) | |
| 33.13 | 22.65 | |

The factor of meta2 would significantly contribute to intention of knowledge reuse and predict at 80%. The factor of meta3 would significantly contribute to intention of knowledge reuse and predict at 89%.

In Table 24.5, the factor of bkr1 would significantly contribute to intention of knowledge reuse and predict at fifty-three percent. The factor of bkr2 would significantly contribute to intention of knowledge reuse and predict at eighty-five percent. The factor of bkr3 would significantly contribute to intention of knowledge reuse and predict at 84%.

In Table 24.6, the first level factors on measuring knowledge reuse intention were listed and the contribution of factors is between 38 and 81%.

In Table 24.7, the first level factors on measuring meta-cognitive knowledge reuse intention were listed and the contribution of factors is between 23 and 86%.

All factors in the models were significantly contributing in predicting. These findings demonstrate that the relationships in these models exist.

### *24.5.3 Structural Equations*

In Table 24.8, the structural equations of knowledge reuse behavior are listed. The contributions of equations are sixty-nine, ninety-seven, and eighty-three percent.

In Table 24.9, the reduced form equations of knowledge reuse behavior model were listed. For model IKR, there are five factors. For model meta-cognitive, there are seven factors. For model behavior, there also are seven factors. For model value, there is one factor.

**Table 24.8** Structural equations of knowledge reuse behavior

| |
|---|
| IKR = 0.039*Value + 3.52*EaseAcce – 2.82*Training + 0.25*CSE + 0.41*Trust, Errorvar. = 0.31, $R^2$ = 0.69 |
| Metacog = 0.60*IKR + 0.68*Self_reg – 3.75*Reflecti + 3.00*Strategi, Errorvar. = 0.027, $R^2$ = 0.97 |
| BKR = – 0.031*IKR + 0.95*Metacog, Errorvar. = 0.15, $R^2$ = 0.85 |
| Value = 0.91*Strategi, Errorvar. = 0.17 , $R^2$ = 0.83 |

**Table 24.9** The reduced form equations of knowledge reuse behavior model

| |
|---|
| IKR = 3.52*EaseAcce – 2.82*Training + 0.25*CSE + 0.41*Trust + 0.036*Strategi, Errorvar. = 0.32, $R^2$ = 0.68 |
| Metacog = 2.12*EaseAcce + 0.68*Self_reg – 3.75*Reflecti – 1.70*Training + 0.15*CSE + 0.24*Trust + 3.02*Strateg, Errorvar. = 0.14, $R^2$ = 0.86 |
| BKR = 1.90*EaseAcce + 0.65*Self_reg – 3.56*Reflecti – 1.53*Training + 0.13*CSE + 0.22*Trust + 2.86*Strategi, Errorvar. = 0.27, $R^2$ = 0.73 |
| Value = 0.91*Strategi, Errorvar. = 0.17, $R^2$ = 0.83 |

### 24.5.4 Conclusion

Based on the LIRREL estimate measurement equations, it is concluded that all eleven hypothesis held true. This proves that all of the separate relations exist. According to all three structural equations, theory model as a whole also exists.

Based upon these findings, knowledge reuse behavior is affected by both the intention and meta-cognitive intention of knowledge reuse and the meta-cognitive intention is affected by intention of knowledge reuse.

## References


1. Alavi M (2000) Managing organizational knowledge.Pinnaflex Educational Resources, OH
2. Dean DL, Lee JD, Pendergast MO, Hickey AM, Nunamaker JF (1997–1998) Enabling the effective involvement of multiple users: methods and tools for collaborative software engineering. J Manage Inform Syst 14(3):179–222
3. Roth G, Kleiner A (1998) Developing organizational memory through learning histories. Organ Dyn 27:43–59
4. Andreoli JM, Borghoff UM, Pareschi R (1996) The constraint-based knowledge broker model – semantics, implementation and analysis. J Symb Comput 21:635–667
5. Manville B (1999) A complex adaptive approach to KM: reflections on the ease of MeKinsey & Company, Inc.. Knowledge Management Review, pp 26–31
6. Flavell JH (1979) Metacognition and cognitive monitoring: A new area of cognitive-developmental inquiry. American Psychologist 34
7. Flavell JH (1987) Speculations about the nature and development of metacognition. Hillside, Lawrence Erlbaum Associates, New Jersey
8. Brown AL (1987) Metacognition, executive control, self-regulation, and other more mysterious mechanisms. In: Weinert FE, Kluwe RH (eds) Metacognition, motivation, and understanding. Hillsdale, Lawrence Erlbaum Associates, New Jersey, pp 65–116
9. Roberts MJ, Erdos G (1993) Strategy selection and metacognition. Educ Psychol 13:259–266
10. Watson S, Hewett K (2006) A multi-theoretical model of knowledge transfer in organizations: Determinants of knowledge contribution and knowledge reuse. J Manage Stud 43:141–173
11. Lee HJ, Ahn HJ, Kim JW, Park SJ (2006) Capturing and reusing knowledge in engineering change management: A case of automobile development. Inform Syst Front 8:375–394
12. Call J (2003) On linking comparative metacognition and theory of mind. Behav Brain Sci 26:341–342
13. Borkowski JG (1996) Metacognition – Theory or chapter heading. Learn Individ Differ 8:391–402
14. Davis FD, Bagozzi RP, Warshaw PR (1989) User acceptance of computer technology: a comparison of two theoretical models. Manage Sci 35:982–1033
15. Pinder CC (1997) Work Motivation in Organizational Behavior. Upper Saddle River, NJ, Prentice-Hall
16. Compeau DR, Higgins CA (1995) Computer self-efficacy: development of a measure and initial test. MIS Quarterly 19:189–211
17. Ridley DS, Schutz PA, Glanz RS, Weinstein CE (1992) Self-regulated learning: the interactive influence of metacognitive awareness and goal-setting. J Exp Educ 60:293–306

# Chapter 25
# Handling Contradictions in Knowledge Discovery Applications

**David Anderson**

**Abstract** There have been a number of very credible attempts to devise paraconsistent logics to deal with the problems caused by the unavoidability of contradictions in knowledge bases and elsewhere. This paper suggests a set of principles for generating logical operator semantics which are broadly drawn from classical logic. These lead directly to a paraconsistent logic, $LM_4$, which significantly outperforms other systems and solves the problem for all practical purposes without giving rise to the difficulties inherent in other attempted solutions.

## 25.1 What's Wrong with Contradictions?

Let |- be a relation of logical consequence. We may say that |- is *explosive* iff for arbitrary formulae q and r, (q& $\neg q$} |- r. Classical logic, intuitionistic logic, and most other standard logics are explosive but ordinary human reasoning is not. The explosive character of standard logic has long been a cause of concern among mathematicians and logicians whose early attempts to address the difficulty are summed up by Quine thus:

> Consider… the crisis which was precipitated in the foundations of mathematics, at the turn of the century, by the discovery of Russell's paradox and the other antinomies of set theory. These contradictions had to be obviated by ad hoc devices; our mathematical myth-making had become deliberate and evident to all. [1]

It is in very much the same spirit of dissatisfaction with attempts to avoid rather than to solve the problem which inconsistency poses, that Graham Priest attempts to incorporate, in a controlled way, paradoxes into the very heart of a system of reasoning by formulating a logic (LP) in which the relation of logical

D. Anderson (✉)
School of Computing, University of Portsmouth, Portsmouth PO1 2EG, UK
e-mail: david.anderson@port.ac.uk

N. Mastorakis et al. (eds.), *Proceedings of the European Computing Conference*, Lecture Notes in Electrical Engineering 28,
DOI 10.1007/978-0-387-85437-3_25, 

consequence are not explosive. Logics displaying this feature are said to be paraconsistent and there are quite a few variants available.

## 25.2 Correcting Classical Logic's Mistake?

According to Priest, classical logic is founded, at least in part on an error:

> Classical logic errs in assuming that no sentence can be both true and false. We wish to correct this assumption. If a sentence is both true and false, let us call it "paradoxical"(p). If it is true but not false we will call it "true only" (t) and similarly for "false only" (f). [2]

We are further told that:

> A sentence is true if its negation is false, hence the negation of a true and false sentence is false and true, i.e., paradoxical. The negation of a true only sentence is false only ... the negation of a false only sentence is true only. [2]

On the basis of these assertions, and more in the same vein, Priest presents us with Tables 25.1 and 25.2 for negation, conjunction and disjunction.

**Table 25.1** Negation and conjunction tables in LP

| ¬ | | & | t | p | f |
|---|---|---|---|---|---|
| t | f | t | t | p | f |
| p | p | p | p | p | f |
| f | t | F | f | p | f |

**Table 25.2** Disjunction and material implication tables in LP

| V | t | p | f | => | t | p | f |
|---|---|---|---|---|---|---|---|
| t | t | t | t | t | t | p | f |
| p | t | p | p | p | t | p | p |
| f | t | p | f | f | t | t | t |

In LP both t and p are designated values because both are the values of "true" sentences. The intended effect of these innovations is to render invalid *ex contradictione quodlibet* [ q & ¬q |- r ] but it is not possible to remove single theorems from logic and unfortunately for LP a number of valuable forms of reasoning get lost in the process. Other significant casualties include:

| | | | |
|---|---|---|---|
| q, ¬q V r | \|- | r | *Disjunctive Syllogism* |
| q => (r & ¬ r) | \|- | ¬ q | *Reductio Ad Absurdum* |
| q => r, r => s | \|- | q => s | *Transitivity* |
| q, q => r | \|- | r | *Modus Ponens* |
| q => r, ¬r | \|- | ¬ q | *Modus Tollens* |

It does not require a great deal of specialised interest in formal logic to realise that the loss of so many common forms of reasoning represents a significant inhibition on the power and attractiveness of the resultant system. Indeed, Priest admits that "We cannot give these inferences up without crippling classical reasoning" [4] but putting a brave face on matters suggests that we should regard the inference forms above not as invalid (which in reality they are) but as "quasi-valid". "Quasi" because they are truth-preserving provided that there are no paradoxical sentences involved. We are encouraged to accept the following methodological maxim:

MM Unless we have specific grounds for believing that paradoxical sentences are occurring in our argument, we can allow ourselves to use both valid and quasi-valid inferences. [4]

The methodological maxim is an admission that LP cannot realistically be used as the logic of first choice nor, consequently, as the basis of an inference engine for normal use by an autonomous machine.

Discomfort with endorsing a rejection of the principle of bivalence on which LP, and many other multi-valued systems, depends may go some way to explaining why the development of paraconsistent logics as a response to the problem of inconsistency has been less than universally embraced. The idea of a sentence being literally "true and false" strikes at the heart of the very notions of truth and falsity and it would be a rash logician indeed who did not hope for an alternative approach. Perhaps if the deductive power of LP were greater it would be enough to sweeten the pill, but as things stand the temptation is hardly overwhelming.

## 25.3 A Different Approach – Knowledge Bases and Presence Values

Perhaps we would profit from approaching this problem from a slightly different angle. Anyone who has recently considered seriously building an autonomous intelligent machine will have turned their minds to the problems caused by the inevitability of inconsistency getting into the machine's knowledge base. One approach following David Lewis [4] might be to keep a multitude of small knowledge bases each relatively independent from the others in hopes that when contradiction does arise it might have only limited impact and when discovered might be relatively easy to correct. This, in essence, is the approach Lenat and Guha adopt in CYC:

There is no need – and probably not even any possibility – of achieving a global consistent unification of... [a] very large KB... How should the system cope with inconsistency? View the knowledge space, and hence the KB, not as one rigid body, but rather as a set of independently supported buttes... [As] with large office buildings, independent supports should make it easier for the whole structure to weather tremors such as local anomalies ... [I]nferring an inconsistency is only slightly more serious than the usual sort of "dead end" a searcher runs into... [5]

Copeland [6] is surely right to be uneasy about such a cavalier approach, for if the information contained in the buttes is available to other buttes (as is the case with CYC) then the effect of contradiction will spread. However if the buttes are completely independent then it is difficult to see how they can in any real sense act as part of a larger whole.

All is not lost. Whatever reservations logicians might have about saying that a sentence can literally be true and false there can be no such objections raised to saying that a given knowledge base K contains both a proposition q and its negation ¬q. After all, this is just another way of saying that K contains a contradiction. It is perfectly possible in the face of this information about K to say q is actually true (or false) and that ¬q (or q) ideally should be removed from K. Of course, we could follow Priest and say q is paradoxical and K merely reflects this fact or we need not actually make any truth claims at all. Certainly if K contains q but does not contain ¬q, any system relying solely on K would assert q if interrogated, but that does not settle the truth value of q. As system builders we need not to takes sides concerning the truth value of the potential constituent elements of knowledge bases although we do have an interest in what may be called the *presence* values of those elements.

Thus, let L be a propositional language whose set of propositional variables is P.

For any given propositional variable q of P, let M(q) be a mapping of q onto {in,out}.

A knowledge base K, is specified by the mapping of each of the propositional variables comprising P onto {in, out}. Suppose L has just three propositional variables q, r, s and

$$M(q) = >\{in\}, M(r) = >\{in\}, \text{and } M(s) = >\{out\} \text{ then K will be } \{q, r\}$$

That is to say, for any proposition q, which may be expressed in our propositional language then either q is present in a given knowledge base K or it is not. We may think of this as the principle of presence-bivalence.

Reasoning, from the view of programs using knowledge bases, can be thought of as a matter of expanding (and contracting[1]) knowledge bases. To regulate this process let us permit the expansion of the knowledge base in accordance with the following seven principles:

- Negation

1. Any knowledge base for which $M(q) = \{in\}$ may be extended so that $M(\neg\neg q) = \{in\}$ and vice versa.

[1] FN>1 I will not deal here with the issue of contracting knowledge bases as it has little bearing on the main issue I want to address.

- Conjunction

2. Any knowledge base for which M(q) = {in} and M(r) = {in} may be extended so as M(q&r) = {in}
3. Any knowledge base for which M(q) = {out} or M(r) = {out} may be extended so as M(¬(q&r)) = {in}

- Disjunction

4. Any knowledge base for which M(q) = {in} or M(r) = {in} may be extended so as M(qVr) = {in}
5. Any knowledge base for which M(q) = {out} and M(r) = {out} may be extended so as M(¬(qVr)) = {in}

- Material implication

6. Any knowledge base for which M(q) = {out} and M(¬q) = {in} or for which M(r) = {in} and M(¬r) = {out} may be extended so as M(q=>r) = {in}
7. Any knowledge base for which [M(r) = {out} or M(¬r) = {in}] and¬[M(q) = {out} and M(¬q) = {in}] may be extended so as M(¬(q=>r)) = {in}

   Now, let v(q):=> {a,d} (i.e., v is a presence valuation of the propositional variables) according to the following rules
   R1. If M(q) =>{in} and M(¬q) => {out} then v(q) = a (assert)
   R2. If M(q) =>{out} and M(¬q) => {in} then v(q) = d (deny)

Applying the seven principles of expansion and the valuation rules yields the presence-tables (Tables 25.3 and 25.4).

Let $v^+$ be the natural extension of v to all the sentences of L using the tables above.

If E is a set of sentences of L, we define:

$$E| - \text{q iff there is no v}(q) = \text{d, but for all r} \in \text{E}, \text{v}^+(\text{r}) = \text{a}| - \text{q}$$

$$\text{iff for allv}, \text{v}^+(q) <> \text{d}$$

**Table 25.3** Two-valued presence tables for negation and conjunction

| ¬ | | & | a | d |
|---|---|---|---|---|
| a | d | a | a | d |
| d | a | d | d | d |

**Table 25.4** Two-valued presence tables for disjunction and material implication

| V | a | d | => | a | d |
|---|---|---|---|---|---|
| a | a | a | a | a | D |
| d | a | d | d | a | a |

Note that if one wants to read assert as "true" and deny as "false" these are the familiar tables of classical logic and if "a" is designated as above we have created a perfectly standard classical system.

So far, so good. However, R1 and R2 do not cater for all the situations which might arise. Specifically we have not considered the following situations:

Case 3: M(q) = {out} and M(¬q) = {out}
Case 4: M(q) = {in} and M(¬q) = {in}.

If K contains q and also contains ¬q (Case 4) a system relying solely on K would equivocate with respect to q. If K does not contain q and also does not contain ¬q (Case 3) a system relying solely on K may be said to be indifferent with respect to q.

Thus, let v(q):=> {a, d, e, i} (i.e., v is a presence valuation of the propositional variables) according to the following rules

R1.If M(q) =>{in} and M(¬q) => {out} then v(q) = a (assert)
R2.If M(q) =>{out} and M(¬q) => {in} then v(q) = d (deny)
R3.If M(q) =>{in} and M(¬q) => {in} then v(q) = e (equivocate)
R4.If M(q) =>{out} and M(¬q) => {out} then v(q) = i (indifferent)

This yields Tables (25.5 and 25.6). In this case let us designate both a and e, so let $v^+$ be the natural extension of v to all the sentences of L using these tables.

If E is a set of sentences of L, we define;

$$E| - \text{q iff there is no } v(q) = d, \text{ but for all } r \in E,$$
$$v^{+}(r) = a \text{ or } e| - q \text{ iff for all } v, v^{+}(q) < > d$$

Table 25.5 Four-valued presence tables for negation and conjunction

| ¬ | | & | a | d | E | i |
|---|---|---|---|---|---|---|
| A | d | a | a | d | E | i |
| D | a | d | d | d | d | d |
| E | e | e | e | d | e | d |
| I | i | i | i | d | d | i |

Table 25.6 Four-valued presence tables for disjunction and material implication

| V | A | d | e | i | => | a | d | e | i |
|---|---|---|---|---|---|---|---|---|---|
| a | a | a | a | a | a | a | d | d | d |
| d | a | d | e | i | d | a | a | a | a |
| e | a | e | e | a | e | a | d | d | d |
| i | a | i | a | i | i | a | d | d | d |

The features of this system make it a very powerful tool. Like LP, $LM_4$ is not explosive as *Ex Contradictione Quodlibet*[(q &¬q} |- r] is $LM_4$ invalid as is *Disjunctive Syllogism* [q, ¬q V r |- r] but the following familiar forms are preserved (a longer list of results is included as an appendix):

q => (r & ¬ r) |- ¬ q *Reductio Ad Absurdum*
q => r, r => s |- q => s *Transitivity*
q, q => r |- r *Modus Ponens*
q => r, ¬r |- ¬ q *Modus Tollens*
q & r |- q | r *And Elimination*
q, r |- q & r *And Introduction*
q |- q V r *Or Introduction*
q V r, q =>s, r =>s |- s *Or Elimination*
q => r |- ¬q V r *Material Conditional(Or)*
q => r |- ¬(q & ¬r) *Material Conditional(And)*
q => (r => s) |- (q & r) => s *Exportation*
q |- q V r *Or Introduction*
q V r, q =>s, r =>s |- s *Or Elimination*

Some important and familiar logical equivalences still hold:

q = ¬¬q *Double Negation*
q = q & q *Idempotentency(And)*
q = q V q *Idempotentency(Or)*
q & r = r & q *Commutativity (And)*
q V r = r V q *Commutativity (Or)*
q & (r & s) = (q & r) & s *Associativity (And)*
q V (r V s) = (q V r) V s *Associativity (Or)*
¬(q & r) = ¬q V ¬r *De Morgan's Law (And)*
¬(q V r) = ¬q & ¬r *De Morgan's Law (Or)*
q => r = ¬r => ¬q *Contraposition*

## 25.4 Conclusion

The improved deductive power of $LM_4$ relative to paraconsistent systems like LP is readily apparent and is sufficient to enable us to turn on its head the methodological maxim which Priest is forced to adopt. Hence we may stick with $LM_4$ unless we can be completely assured (in some particular context perhaps) that contradictions cannot arise in our knowledge bases. Furthermore, by confining itself to talking about presence values $LM_4$ permits the parties for whom it is a genuine concern to differ on the question of truth while still using the system to good effect.

The operator semantics of $LM_4$ arise naturally as the working out of classical precepts in a more complicated environment. In the bivalent situation these precepts produce what we might call $LM_2$ and this is effectively classical logic[2]

[2] $LM_3$ is the subject of a separate paper.

[10]. In the situation for which $LM_4$ was designed they yield an effective and productive system which should prove useful for A.I. system builders. While $LM_4$ cannot cure all of the problems faced, for example by CYC, it does provide a safe mechanism for dealing with contradictions either as an alternative to using buttes or within buttes.

## Appendix: Indicative Results for $LM_4$

The test results presented in Table 5.7 are drawn from two sets of tests: the first was a group standard inference forms, the second were representations of the sequents and exercises which appear in the first few chapters of E.J. Lemmon's, Beginning Logic [7]. The results are presented here without pruning.

- Column 2: Indicates if a logical relation of entailment exists from the premise(s) to the conclusion. Where invalid, I have indicated the first position on the presence table where an invalidity was found.
- Column 3: Indicates if a logical relation of entailment exists from the conclusion to the premise(s). Where invalid, I have indicated the first position on the presence table where an invalidity was found.
- Column 4: "=" indicates the relation of logical equivalence exists between the premise(s) and the conclusion.
- Column 5: The representation of the sequent which was tested.

NB: Although it does not occur in any of the sequents tested below, it should be possible to get a sequent for which the premise(s) entails the conclusion and vice versa but where the premise(s) and conclusion have distinct presence tables. Such a situation could not occur in classical logic ($LM_2$).

**Table 25.7** Indicative results for $LM_4$

| Ex | | | | |
|---|---|---|---|---|
| Contradictione Quodlibet | Invalid when q=e r=a | Invalid when q=a r=e | Dif | (q & ¬q) \|- ¬r |
| Disjunctive Syllogism | Invalid when q=e r=e | Invalid when q=d r=a | Dif | ((q) & ¬q V r) \|- r |
| Conditional Proof | Invalid when q=a r=e | Invalid when q=d r=a | Dif | ((q) & r) \|- q => r |
| And Elimination | Valid | Invalid when q=a r=e | Dif | (q & r) \|- q |
| And Introduction | Valid | Valid | = | ((q) & r) \|- q & r |
| Associative Laws (&) | Valid | Valid | = | (q & (r & s)) \|- (q & r) & s |
| Associative Laws (V) | Valid | Valid | = | (q V (r V s)) \|- (q V r) V s |

**Table 25.7** (continued)

| Ex Contradictione Quodlibet | Invalid when q = e r = a | Invalid when q = a r = e | Dif | (q & ¬q) \|- ¬r |
|---|---|---|---|---|
| Commutative Laws (&) | Valid | Valid | = | (q & r) \|- r & q |
| Commutative Laws (V) | Valid | Valid | = | (q V r) \|- r V q |
| Contraposition | Valid | Valid | = | (q => r) \|- ¬r => ¬q |
| Double Negation | Valid | Valid | = | (q) \|- ¬¬q |
| Distributive Laws (&) | Valid | Valid | = | (q & (r V s)) \|- (q & r) V (q & s) |
| Distributive Laws (V) | Valid | Valid | = | (q V (r & s)) \|- (q V r) & (q V s) |
| Idempotentency (&) | Valid | Valid | = | (q) \|- q & q |
| Idempotentency (V) | Valid | Valid | = | (q) \|- q V q |
| Material Conditional (1) | Valid | Invalid when q = a r = e | Dif | (q => r) \|- ¬q V r |
| Material Conditional (2) | Valid | Invalid when q = a r = e | Dif | (q => r) \|- ¬(q & ¬r) |
| De Morgan's Laws (&) | Valid | Valid | = | (¬(q & r)) \|- ¬q V ¬r |
| De Morgan's Laws (V) | Valid | Valid | = | (¬(q V r)) \|- ¬q & ¬r |
| Modus Ponens | Valid | Invalid when q = a r = e | Dif | ((q) & q => r) \|- r |
| Modus Tollens | Valid | Invalid when q = d r = a | Dif | ((q => r) & ¬r) \|- ¬q |
| Or Elimination | Valid | Invalid when q = a r = a s = e | Dif | (((q V r) & q => s) & r => s) \|- s |
| Or Introduction | Valid | Invalid when q = d r = a | Dif | (q) \|- q V r |
| Reductio Ad Absurdum | Valid | Invalid when q = e r = a | Dif | (q => (r & ¬r)) \|- ¬q |
| Transitivity | Valid | Invalid when q = a r = e s = a | Dif | ((q => r) & r => s) \|- q => s |
| Exportation | Valid | Invalid when q = e r = i s = d | Dif | (q => (r => s)) \|- (q & r) => s |
| Lemmon 001 | Valid | Invalid when q = a r = e | Dif | ((q => r) & q) \|- r |
| Lemmon 002 | Valid | Invalid when q = a r = a | Dif | ((¬r => (¬q => r)) & ¬r) \|- ¬q => r |
| Lemmon 003 | Valid | Invalid when q = a r = a s = e | Dif | (((q => r) & r => s) & q) \|- s |
| Lemmon 004 | Valid | | Dif | |

**Table 25.7** (continued)

<table>
<tr><th>Ex Contradictione Quodlibet</th><th>Invalid when q=e r=a</th><th>Invalid when q=a r=e</th><th>Dif</th><th>(q & ¬q) |- ¬r</th></tr>
<tr><td></td><td></td><td>Invalid when q=a r=a s=e</td><td></td><td>(((q => (r => s)) & q => r) & q) |- s</td></tr>
<tr><td>Lemmon 005</td><td>Valid</td><td>Invalid when q=d r=a</td><td>Dif</td><td>((q => r) & ¬r) |- ¬q</td></tr>
<tr><td>Lemmon 006</td><td>Valid</td><td>Invalid when q=a r=e s=a</td><td>Dif</td><td>(((q => (r => s)) & q) & ¬s) |- ¬r</td></tr>
<tr><td>Lemmon 007</td><td>Valid</td><td>Invalid when q=d r=e</td><td>Dif</td><td>((q => ¬r) & r) |- ¬q</td></tr>
<tr><td>Lemmon 008</td><td>Valid</td><td>Invalid when q=a r=a</td><td>Dif</td><td>((¬q => r) & ¬r) |- q</td></tr>
<tr><td>Lemmon 009</td><td>Valid</td><td>Valid</td><td>=</td><td>(q => r) |- ¬r => ¬q</td></tr>
<tr><td>Lemmon 010</td><td>Valid</td><td>Valid</td><td>=</td><td>(q => (r => s)) |- r => (q => s)</td></tr>
<tr><td>Lemmon 011</td><td>Valid</td><td>Invalid when q=a r=e s=d</td><td>Dif</td><td>(r => s) |- (¬r => ¬q) => (q => s)</td></tr>
<tr><td>Lemmon 012</td><td>Valid</td><td>Valid</td><td>=</td><td>((q) & r) |- q & r</td></tr>
<tr><td>Lemmon 013</td><td>Invalid when q=e r=i s=d</td><td>Valid</td><td>Dif</td><td>((q & r) => s) |- q => (r => s)</td></tr>
<tr><td>Lemmon 014</td><td>Valid</td><td>Invalid when q=a r=e</td><td>Dif</td><td>(q & r) |- q</td></tr>
<tr><td>Lemmon 015</td><td>Valid</td><td>Invalid when q=d r=a</td><td>Dif</td><td>(q & r) |- r</td></tr>
<tr><td>Lemmon 016</td><td>Valid</td><td>Invalid when q=e r=i s=d</td><td>Dif</td><td>(q => (r => s)) |- (q & r) => s</td></tr>
<tr><td>Lemmon 017</td><td>Valid</td><td>Valid</td><td>=</td><td>(q & r) |- r & q</td></tr>
<tr><td>Lemmon 018</td><td>Invalid when q=e r=a s=a</td><td>Invalid when q=d r=a s=d</td><td>Dif</td><td>(r => s) |- (q & r) => (q & s)</td></tr>
<tr><td>Lemmon 019</td><td>Valid</td><td>Valid</td><td>=</td><td>(q V r) |- r V q</td></tr>
<tr><td>Lemmon 020</td><td>Invalid when q=e r=e s=d</td><td>Invalid when q=a r=a s=d</td><td>Dif</td><td>(r => s) |- (q V r) => (q V s)</td></tr>
<tr><td>Lemmon 021</td><td>Valid</td><td>Valid</td><td>=</td><td>(s V (q V r)) |- q V (s V r)</td></tr>
<tr><td>Lemmon 022</td><td>Valid</td><td>Invalid when q=e r=a</td><td>Dif</td><td>((q => r) & q => ¬r) |- ¬q</td></tr>
<tr><td>Lemmon 023</td><td>Valid</td><td>Invalid when q=e</td><td>Dif</td><td>(q => ¬q) |- ¬q</td></tr>
<tr><td>Lemmon 024</td><td>Valid</td><td>Valid</td><td>=</td><td>(q=r) |- r=q</td></tr>
<tr><td>Lemmon 025</td><td>Valid</td><td>Invalid when q=a r=e</td><td>Dif</td><td>((q) & q=r) |- r</td></tr>
<tr><td>Lemmon 026</td><td>Valid</td><td>Invalid when q=a r=e s=a</td><td>Dif</td><td>((q=r) & r=s) |- q=s</td></tr>
</table>

**Table 25.7** (continued)

<table>
<tr><th>Ex Contradictione Quodlibet</th><th>Invalid when q=e r=a</th><th>Invalid when q=a r=e</th><th>Dif</th><th>(q & ¬q) |- ¬r</th></tr>
<tr><td>Lemmon 027</td><td>Valid</td><td>Invalid when q=e r=a</td><td>Dif</td><td>((q & r)=q) |- q => r</td></tr>
<tr><td>Lemmon 028</td><td>Valid</td><td>Invalid when q=a r=e</td><td>Dif</td><td>(q & (q=r)) |- q & r</td></tr>
<tr><td>Lemmon 029</td><td>Valid</td><td>Valid</td><td>=</td><td>(q) |- q</td></tr>
<tr><td>Lemmon 030a</td><td>Invalid when q=e r=i s=d</td><td>Valid</td><td>Dif</td><td>((q & r) => s) |- q => (r => s)</td></tr>
<tr><td>Lemmon 030b</td><td>Valid</td><td>Invalid when q=e r=i s=d</td><td>Dif</td><td>(q => (r => s)) |- (q & r) => s</td></tr>
<tr><td>Lemmon 031a</td><td>Valid</td><td>Valid</td><td>=</td><td>(q & (q V r)) |- q</td></tr>
<tr><td>Lemmon 031b</td><td>Valid</td><td>Valid</td><td>=</td><td>(q) |- q & (q V r)</td></tr>
<tr><td>Lemmon 032a</td><td>Valid</td><td>Valid</td><td>=</td><td>(q V (q & r)) |- q</td></tr>
<tr><td>Lemmon 032b</td><td>Valid</td><td>Valid</td><td>=</td><td>(q) |- q V (q & r)</td></tr>
<tr><td>Lemmon 033a</td><td>Valid</td><td>Valid</td><td>=</td><td>(q V q) |- q</td></tr>
<tr><td>Lemmon 033b</td><td>Valid</td><td>Valid</td><td>=</td><td>(q) |- q V q</td></tr>
<tr><td>Lemmon 034a</td><td>Invalid when q=e r=a</td><td>Invalid when q=d r=e</td><td>Dif</td><td>((q) & ¬(q & r)) |- ¬r</td></tr>
<tr><td>Lemmon 034b</td><td>Invalid when q=d r=e</td><td>Invalid when q=e r=a</td><td>Dif</td><td>(¬r) |- q & (¬(q & r))</td></tr>
<tr><td>Lemmon 035a</td><td>Valid</td><td>Invalid when q=a r=e</td><td>Dif</td><td>(q => r) |- ¬(q & ¬r)</td></tr>
<tr><td>Lemmon 035b</td><td>Invalid when q=a r=e</td><td>Valid</td><td>Dif</td><td>(¬(q & ¬r)) |- q => r</td></tr>
<tr><td>Lemmon 036a</td><td>Valid</td><td>Valid</td><td>=</td><td>(q V r) |- ¬(¬q & ¬r)</td></tr>
<tr><td>Lemmon 036b</td><td>Valid</td><td>Valid</td><td>=</td><td>(¬(¬q & ¬r)) |- q V r</td></tr>
<tr><td>Lemmon 038</td><td>Invalid when q=e</td><td></td><td>Dif</td><td>|- q => q</td></tr>
<tr><td>Lemmon 039</td><td>Invalid when q=e</td><td></td><td>Dif</td><td>|- q => ¬¬q</td></tr>
<tr><td>Lemmon 040</td><td>Invalid when q=e</td><td></td><td>Dif</td><td>|- (¬¬q) => q</td></tr>
<tr><td>Lemmon 041</td><td>Invalid when q=d r=a</td><td></td><td>Dif</td><td>|- q & (r => q)</td></tr>
<tr><td>Lemmon 042</td><td>Valid</td><td></td><td>Dif</td><td>|- (q => r) => (¬r => ¬q)</td></tr>
<tr><td>Lemmon 043</td><td>Valid</td><td></td><td>Dif</td><td>|- (q => (r => s)) => ((q => r) => (q => s))</td></tr>
<tr><td>Lemmon 044</td><td>Valid</td><td></td><td>Dif</td><td>|- q V ¬q</td></tr>
<tr><td>Lemmon 045</td><td>Invalid when q=e r=i</td><td>Valid</td><td>Dif</td><td>(q) |- (q & r) V (q & ¬r)</td></tr>
<tr><td>Lemmon 046</td><td>Invalid when q=d r=a</td><td>Valid</td><td>Dif</td><td>(q => r) |- q & (r=q)</td></tr>
<tr><td>Lemmon 047a</td><td>Valid</td><td>Invalid when q=e r=a</td><td>Dif</td><td>((q & r)=q) |- q => r</td></tr>
<tr><td>Lemmon 047b</td><td></td><td>Valid</td><td>Dif</td><td>(q => r) |- (q & r)=q</td></tr>
</table>

**Table 25.7** (continued)

<table>
<tr><td>Ex Contradictione Quodlibet</td><td>Invalid when<br>q=e r=a</td><td>Invalid when<br>q=a r=e</td><td>Dif</td><td>(q & ¬q) |- ¬r</td></tr>
<tr><td></td><td>Invalid when<br>q=e r=a</td><td></td><td></td><td></td></tr>
<tr><td>Lemmon 048</td><td>Invalid when<br>q=a r=e</td><td>Valid</td><td>Dif</td><td>(¬q V r) |- q => r</td></tr>
<tr><td>Lemmon 050</td><td>Invalid when<br>q=e r=a</td><td>Invalid when<br>q=d r=e</td><td>Dif</td><td>(q) |- r => q</td></tr>
<tr><td>Lemmon 051</td><td>Invalid when<br>q=e r=e</td><td>Invalid when<br>q=a r=a</td><td>Dif</td><td>(¬q) |- q => r</td></tr>
<tr><td>Lemmon 053</td><td>Invalid when<br>q=d r=e</td><td>Invalid when<br>q=a r=a</td><td>Dif</td><td>((¬r) & q V r) |- q</td></tr>
<tr><td>Lemmon 054</td><td>Invalid when<br>q=e r=e</td><td></td><td>Dif</td><td>|- (q => r) V (r => q)</td></tr>
<tr><td>Lemmon 055</td><td>Valid</td><td>Invalid when<br>q=d r=a</td><td>Dif</td><td>((q => r) & ¬r) |- ¬q</td></tr>
<tr><td>Lemmon Exercise 1(a)</td><td>Valid</td><td>Invalid when<br>q=a r=e</td><td>Dif</td><td>((q => (q => r)) & q) |- r</td></tr>
<tr><td>Lemmon Exercise 1(b)</td><td>Valid</td><td>Invalid when<br>q=d r=a<br>s=a</td><td>Dif</td><td>(((r => (q => s)) & ¬s) & r) |- ¬q</td></tr>
<tr><td>Lemmon Exercise 1(c)</td><td>Valid</td><td>Invalid when<br>q=a r=e</td><td>Dif</td><td>((q => ¬¬r) & q) |- r</td></tr>
<tr><td>Lemmon Exercise 1(d)</td><td>Valid</td><td>Invalid when<br>q=a r=e</td><td>Dif</td><td>((¬¬r => q) & ¬q) |- ¬r</td></tr>
<tr><td>Lemmon Exercise 1(e)</td><td>Valid</td><td>Invalid when<br>q=a r=e</td><td>Dif</td><td>((¬q => ¬r) & r) |- q</td></tr>
<tr><td>Lemmon Exercise 1(f)</td><td>Valid</td><td>Valid</td><td>=</td><td>(q => ¬r) |- r => ¬q</td></tr>
<tr><td>Lemmon Exercise 1(g)</td><td>Valid</td><td>Valid</td><td>=</td><td>(¬q => r) |- ¬r => q</td></tr>
<tr><td>Lemmon Exercise 1(h)</td><td>Valid</td><td>Valid</td><td>=</td><td>(¬q => ¬r) |- r => q</td></tr>
<tr><td>Lemmon Exercise 1(i)</td><td>Valid</td><td>Invalid when<br>q=a r=e<br>s=a</td><td>Dif</td><td>((q => r) & r => s) |- q => s</td></tr>
<tr><td>Lemmon Exercise 1(j)</td><td>Valid</td><td>Invalid when<br>q=a r=e<br>s=d</td><td>Dif</td><td>(q => (r => s)) |- (q => r) => (q => s)</td></tr>
<tr><td>Lemmon Exercise 1(l)</td><td>Valid</td><td>Invalid when<br>q=a r=e<br>s=a</td><td>Dif</td><td>(q => r) |- (r => s) => (q => s)</td></tr>
<tr><td>Lemmon Exercise 1(m)</td><td>Valid</td><td>Invalid when<br>q=d r=a</td><td>Dif</td><td>(q) |- (q => r) => r</td></tr>
<tr><td>Lemmon Exercise 1(n)</td><td>Valid</td><td>Invalid when<br>q=d r=a<br>s=a</td><td>Dif</td><td>(q) |- (¬(r => s) => ¬q) => (¬s => ¬r)</td></tr>
<tr><td>Lemmon Exercise 1 p.27-8 (a)</td><td>Invalid when<br>q=a r=e</td><td>Invalid when<br>q=d r=e</td><td>Dif</td><td>(q) |- r => (q & r)</td></tr>
</table>

**Table 25.7** (continued)

| Ex Contradictione Quodlibet | Invalid when q = e r = a | Invalid when q = a r = e | Dif | $(q \& \neg q) \vdash \neg r$ |
|---|---|---|---|---|
| Lemmon Exercise 1 p.27-8 (b) | Valid | Valid | = | $(q \& (r \& s)) \vdash r \& (q \& s)$ |
| Lemmon Exercise 1 p.27-8 (c) | Valid | Valid | = | $((q => r) \& (q => s)) \vdash q => (s \& r)$ |
| Lemmon Exercise 1 p.27-8 (d) | Valid | Invalid when q = a r = e | Dif | $(r) \vdash q \lor r$ |
| Lemmon Exercise 1 p.27-8 (e) | Valid | Invalid when q = a r = e | Dif | $(q \& r) \vdash q \lor r$ |
| Lemmon Exercise 1 p.27-8 (f) | Valid | Valid | = | $((q => s) \& (r => s)) \vdash (q \lor r) => s$ |
| Lemmon Exercise 1 p.27-8 (j) | Valid | Invalid when q = e | Dif | $(\neg q => q) \vdash q$ |
| Lemmon Exercise 1 p.33 (a) | Valid | Invalid when q = a r = e | Dif | $((r) \& q = r) \vdash q$ |
| Lemmon Exercise 1 p.33 (b) | Valid | Valid | = | $((q => r) \& r => q) \vdash q = r$ |
| Lemmon Exercise 1 p.33 (c) | Valid | Valid | = | $(q = r) \vdash \neg q = \neg r$ |
| Lemmon Exercise 1 p.33 (d) | Valid | Valid | = | $(\neg q = \neg r) \vdash q = r$ |
| Lemmon Exercise 1 p.33 (e) | Invalid when q = a r = e | Invalid when q = d r = a | Dif | $((q \lor r) = q) \vdash q => r$ |
| Lemmon Exercise 1 p.33 (f) | Valid | Invalid when q = a r = a s = a | Dif | $((q = \neg r) \& r = \neg s) \vdash q = s$ |
| Lemmon Exercise 1 p.41 (a) | Valid | Valid | = | $(q \lor r) \vdash q \lor r$ |
| Lemmon Exercise 1 p.41 (b-1) | Valid | Valid | = | $(q \& q) \vdash q$ |
| Lemmon Exercise 1 p.41 (b-2) | Valid | Valid | = | $(q) \vdash q \& q$ |
| Lemmon Exercise 1 p.41 (c-1) | Valid | Valid | = | $(q \& (r \lor s)) \vdash (q \& r) \lor (q \& s)$ |
| Lemmon Exercise 1 p.41 (c-2) | Valid | Valid | = | $((q \& r) \lor (q \& s)) \vdash q \& (r \lor s)$ |
| Lemmon Exercise 1 p.41 (d-1) | Valid | Valid | = | $(q \lor (r \& s)) \vdash (q \lor r) \& (q \lor s)$ |
| Lemmon Exercise 1 p.41 (d-2) | Valid | Valid | = | $((q \lor r) \& (q \lor s)) \vdash q \lor (r \& s)$ |
| Lemmon Exercise 1 p.41 (e-1) | Valid | Invalid when q = e r = i | Dif | $(q \& r) \vdash \neg(q => \neg r)$ |
| Lemmon Exercise 1 p.41 (e-2) | Invalid when q = e r = i | Valid | Dif | $(\neg(q => \neg r)) \vdash q \& r$ |
| Lemmon Exercise 1 p.41 (f-1) | Valid | Invalid when q = a r = e | Dif | $(\neg(q \lor r)) \vdash \neg q \lor \neg r$ |
| Lemmon Exercise 1 p.41 (f-2) | Invalid when q = a r = e | Valid | Dif | $(\neg q \lor \neg r) \vdash \neg(q \lor r)$ |

Table 25.7 (continued)

| Ex Contradictione Quodlibet | Invalid when q = e r = a | Invalid when q = a r = e | Dif | (q & ¬q) \|- ¬r |
|---|---|---|---|---|
| Lemmon Exercise 1 p.41 (g-1) | Valid | Valid | = | (¬q & ¬r) \|- ¬(q V r) |
| Lemmon Exercise 1 p.41 (g-2) | Valid | Valid | = | (¬(q V r)) \|- ¬q & ¬r |
| Lemmon Exercise 1 p.41 (h-1) | Valid | Valid | = | (q & r) \|- ¬(¬q V ¬r) |
| Lemmon Exercise 1 p.41 (h-2) | Valid | Valid | = | (¬(¬q V ¬r)) \|- q & r |
| Lemmon Exercise 1 p.41 (i-1) | Valid | Invalid when q = a r = e | Dif | (q => r) \|- ¬q V r |
| Lemmon Exercise 1 p.41 (j-1) | Valid | Invalid when q = d r = e | Dif | (¬q => r) \|- q V r |
| Lemmon Exercise 1 p.41 (j-2) | Invalid when q = d r = e | Valid | Dif | (q V r) \|- ¬q => r |
| Lemmon Exercise 1 p.62 (a) | Valid | | Dif | \|- (r => s) => ((q => r) => (q => r)) |
| Lemmon Exercise 1 p.62 (b) | Invalid when q = a r = e | | Dif | \|- q => (r => (q & r)) |
| Lemmon Exercise 1 p.62 (c) | Valid | | Dif | \|- (q => s) => ((r => s) => ((q V r) => (s))) |
| Lemmon Exercise 1 p.62 (d) | Valid | | Dif | \|- ((q => r) & ¬r) => ¬q |
| Lemmon Exercise 1 p.62 (e) | Valid | | Dif | \|- (¬q => q) => q |
| Lemmon Exercise 2 p.63 (a) | Valid | Invalid when q = a r = e | Dif | (((q => r) => q) & q => r) \|- q |

Secondly, while not being a relevance logic, $LM_4$ does stop a number of the sequents which would concern Relevance Logicians like Lemmon 050 (q |- r => q).

## References


1. Quine WVO (1948) On what there is. The Review of Metaphysics 2:21–28, Reprint in: From a logical Point of View, Cambridge, Harvard University Press (1980), p 18
2. Priest G (1979) The logic of paradox. J Philos Logic, p 226
3. Priest G (1979) The logic of paradox. Journal Philos Logic, p 235
4. Lewis D (1982) A logic for equivocators. Noûs 16(3):431–441
5. Lenat DB, Feigenbaum EA (1991) On the thresholds of knowledge. Artif Int 47:217–222
6. Copeland BJ (1993) Artificial intelligence: A philosophical introduction. Blackwell, Oxford, p 120
7. Lemmon EJ (1992) Beginning Logic. Hackett, pp 1–83

# Chapter 26
# A Cooperative Distributed Query-Processing Approach

**Saeed K. Rahimi and Frank S. Haug**

**Abstract** Although Distributed Database Management Systems (DDBMS) have been studied for many years there has been little or no emphasis that these studies have put on dynamic systems where database systems can join and depart a group of co-operating DBMS systems. We introduce a new approach to distributed query processing that allows local systems to freely join and leave a distributed system. We achieve this by eliminating the tight synchronization of the global data dictionary and local data dictionaries.

## 26.1 Introduction

A centralized database management system (DBMS) is a complex software program that allows an enterprise to control its data on a single machine. The recent trend in mergers and acquisitions has resulted in organizations owning more that one DBMS, not all of which have come from the same vendor. It is not too uncommon for an enterprise to own a mainframe database management system from IBM DB2 and a few other departmental databases such as Oracle and Microsoft SQL Server. To control the information in such an environment one uses a Distributed Database Management System (DDBMS) (See Fig. 26.1) [1].

A client computer in such a system issues an SQL request to access information that is distributed across the computers in the network. The fact that required data items are distributed and stored on different computers is hidden from the user by the DDBMS system. In other words, the users of distributed data issue their queries as if the system only had one database, was controlled by one database management system (DBMS) and resided on one computer. In order for this to work, a DDBMS system must provide for a set of transparencies.

---

S.K. Rahimi (✉)
Opus College of Business, University of St. Thomas, 2115 Summit Ave., St. Paul, MN 55105, USA

N. Mastorakis et al. (eds.), *Proceedings of the European Computing Conference*, Lecture Notes in Electrical Engineering 28,
DOI 10.1007/978-0-387-85437-3_26, 

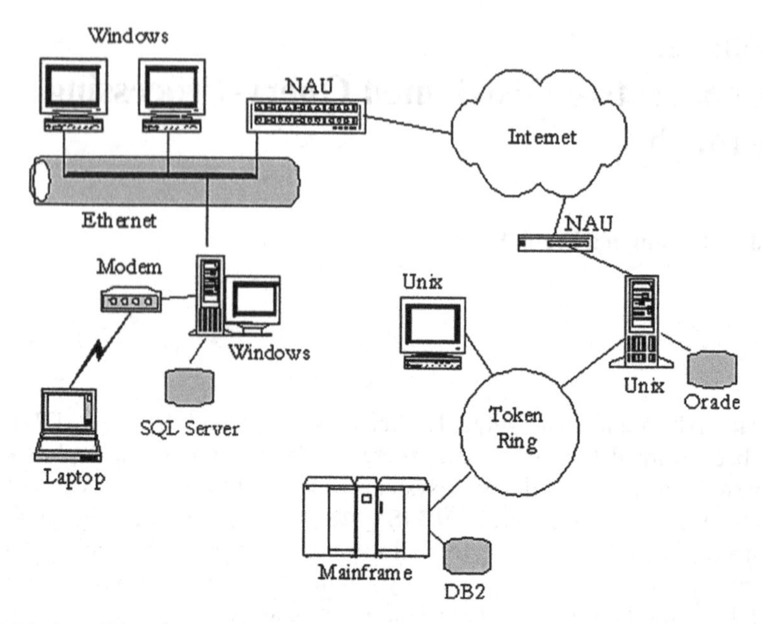


**Fig. 26.1** A traditional DDBMS

### *26.1.1 Transparency Types*

There are many transparency types that a distributed database system should support. The more transparencies that a DDBMS system supports, the easier is the system to use.

1. DBMS transparency – the fact that data are controlled by one or more Local Database Management Systems (LDBMS) from different manufacturer is hidden from the user.
2. Location transparency – the fact that data may be stored anywhere in the system and it is not necessarily local is hidden from the user.
3. Distribution transparency – the fact that database tables might be fragmented is hidden from the user.
4. Replication transparency – the fact that there might be more than one copy of a table or a fragment of a table is hidden from the user.
5. Operating system transparency – the fact that different computers in the system are controlled by different operating systems is hidden from the user.
6. Network/communication transparency – the fact that different computers of the system use different communication protocols to talk to each other is hidden from the user.
7. Hardware/architecture transparency – the fact that different computers may have different architecture (32 bit vs. 64 bit machine, for example) is hidden from the user.

## 26.2 Traditional DDBMS

The key components of a DDBMS system that implements these transparencies are the Global Data Dictionary (GDD) and Global Execution Manager (GEM). The GDD is a data model representing the definition and deployment details for all the global tables and columns [2]. When the user of a distributed system issues a query to access distributed data, the GEM uses the GDD to validate the definition for all the global tables and columns referenced by the query and then fetches the deployment details from GDD. The GEM will then use these details to create a distributed execution plan for accessing each physical table distributed across the network that implements each global table (or each fragment of the global table) required for the query [3, 4].

Although sounding simple and straightforward, the success, efficiency and validity of a given plan depends heavily upon the accuracy and freshness of the deployment details stored in the GDD. As individual LDBMS become active or inactive in the environment, failure to synchronize the deployment details carefully can cause the GEM to build an inaccurate plan even when an alternate LDBMS is available to complete the plan. Even when recovery techniques are used to build alternate plans in the face of failures, this can be an inefficient and complicated process.

## 26.3 Co-operative and Dynamic DDBMS

Our proposal for processing queries in a distributed system is different from traditional approaches that use a GDD and GEM. These approaches depend heavily on the synchronicity of the GDD and the actual data deployment. This synchronicity is hard to maintain and it is time-consuming. We propose an approach to distributed query processing in dynamic environments in which databases can freely join and leave the servers in an Intranet. Although our approach uses the concept of a GDD, our GDD only stores the global table and column definitions. Our technique for building the distributed execution plan requires the GDD to be synchronized with the global table and column definitions but not the actual local data deployment details. This approach lends itself nicely to peer-to-peer systems [5], development, testing, and research environments where tables' fragmentation, replication and allocation change constantly. It also allows for multiple, alternative deployments to coexist within the same system seamlessly.

### *26.3.1 Architecture*

Our architecture consists of four programmatic components, two metadata schemas, and the local data stores themselves.

#### 26.3.1.1 Client

Clients are user-written components that can be implemented in J2EE [6], Dot Net, Web Services [7], or similar-minded approaches. The client library/framework can be quite minimal. The clients are small, lightweight, and multi-threaded. Placing too much logic or communication overhead in the client library, would make it impractical or impossible for users to create the programs they want.

#### 26.3.1.2 Distributed Query Service Provider

The DQSP is deployed as a Web service and registered dynamically via UDDI. The DQSP is also a reasonably lightweight component. The DQSP can be deployed as several instances within the intranet, and can be pooled within an application server (like the Dot Net Framework's managed assemblies or a J2EE application server like JBOSS [8]). The DQSP is merely an encapsulation and decoupling of the communication complexities and potentially complex threading required to process queries and their results correctly and efficiently.

#### 26.3.1.3 Data Dictionary Service Provider

The DDSP contains the Global "Interface" Schema (GIS). The GIS replaces the traditional GDD and contains the global table names, global column names, and global data types/nullability details. The definition of this schema is the only thing that needs to be synchronized system-wide. The deployment and implementation details are not located here; those details are only located with each LQSP.

This component is also responsible for connecting each DQSP with the appropriate set of LQSP components. This can be a trivial task and can be done by requiring the LQSP components to be registered in such a way that the DDSP can find them. This is done via some form of messaging, perhaps using some standard approach (like WSDL and UDDI). This information can be stored in a database or just cached in memory.

#### 26.3.1.4 Local Query Service Provider (LQSP)

This component is responsible for the local query processing and actually connecting to the LDBMS. Each LQSP would use a schema that connects the GIS with the Database, Location, Fragmentation, and Replication transparency details for all pieces located at this component. This schema is called the Local Deployment Schema (LDS).

## 26.3.2 Distributed Execution Plan Strategy

The execution plans devised in our approach are no different than those in a traditional system, the only difference is how we actually build and execute them. In general, a distributed execution plan is formed by first translating the global query from SQL into a fully qualified relational algebra expression. Next, we replace each global table with the set of all distributed tables (or fragments) that represent it and the joins and unions required to reconstruct the global tables from the fragments. The operations can be rearranged to ensure that each operation can be executed within a single LQSP if possible.

At this point, we have a plan suitable for execution, but not an optimal plan. If we look at the relational algebra expression, it is easy to see how things can be rearranged to optimize the query. By "pushing" the projection and selection operations "down" (closer to the actual tables) and rearranging the order in which things are joined we can improve the performance by reducing the number and size of the rows produced by each intermediate step and reduce the number of times that results need to be combined from multiple LQSP sites. Lastly, when the results from more than one LQSP do need to be combined (using union, join, or semi-join operations) we can use the general guiding principle of sending the smaller result set to the LQSP with the larger result set. This minimizes the communication cost. There can be more complex strategies used for special cases but this overall approach is sufficient for most simple queries.

## 26.3.3 Communication Overview

Figure 26.2 depicts the overall communication of our system components.

1. A Client initializes the library, which looks up the DQSP to use.
2. A Client sends a query to the DQSP as raw SQL text.
3. The DQSP parses the SQL and builds a simple XML [9] version of the query.
4. The DQSP sends the XML query to the DDSP.
5. The DDSP disambiguates and validates the query. It then builds a list of candidate LQSP's that could potentially fulfill this query.
6. The DDSP returns the candidate list and fully qualified query to the DQSP. The DQSP sends this query request to each candidate LQSP.
7. Each candidate LQSP will look at the request and look at their LDS determining what (if any) part it could answer along with a "selfranking" indicating some measure of the "coverage" or "efficiency" with which it can do things. For example, if a LQSP had an instance of all tables versus only some tables versus only some fragment(s) of some table(s).
8. The system needs to elect the actual LQSP from the candidate LQSP in the list. This is controlled and coordinated by the DQSP by evaluating the "self-ranking" information from each responding LQSP.

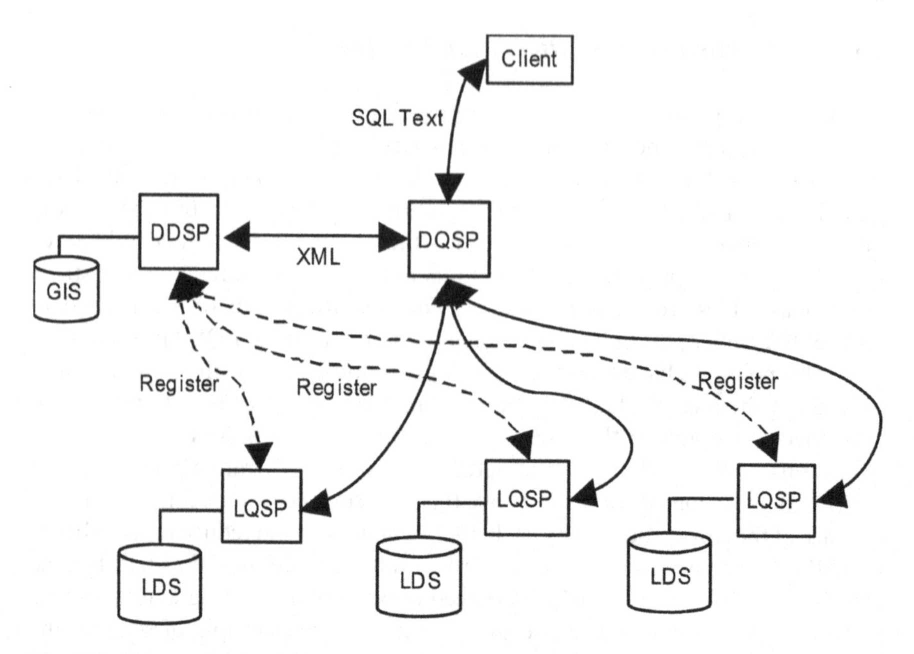


**Fig. 26.2** Communication in a P2P DDBMS

9. DQSP uses a timeout-based approach and waits for the "best" answer available and then closes the election once a "good enough" group of candidates has responded.
10. Now the DQSP can tell the elected LQSP to perform the processing. The DQSP sends the list of elected LQSP along with the self-ranking information returned by each to each elected LQSP. Each LQSP can then use the same information and algorithm to build their portion of the execution plan. This means that each LQSP determines its local plan, determines which component it should send its results to, and determines which components it should expect to receive results from. When results are received, they are merged and passed to the appropriate LQSP or cached for retrieval by the client. The LQSP containing the final results will message the DQSP when they are ready.
11. The DQSP receives notification of the query completion status (success or failure) and messages the client with the result.
12. The client receives notification and can then send additional messages to retrieve the result in multiple chunks or in its entirety.

#### 26.3.3.1 DQSP Process

1. A DQSP initializes, which looks up the DDSP to use.
2. The DQSP receives a SQL query from a client.
3. The DQSP parses the SQL into simple XML.

4. The DQSP sends the XML query to the DDSP.
5. The DQSP receives acknowledgement that the SQL is valid along with a LQSP candidate list and fully qualified query, or an error if there was a problem with the query (in which case it then sends the error back to the client.)
6. The DQSP sends query request messages to the LQSP candidate list.
7. The DQSP receives responses from the candidates and determines if it has enough "good enough" candidates responding within the timeout. If it does not then it sends the error back to the client.
8. The DQSP sends an "OK" to all the elected candidates and processes the results (merging if necessary).
9. The DQSP returns the results to the client.

#### 26.3.3.2 DDSP Process

1. A DDSP initializes, which connects to the GIS and registers the DDSP for use.
2. The DQSP registers and un-registers LQSP components as needed.
3. The DDSP receives the simple XML query. It validates this against the GIS, and makes it fully qualified. If it is not valid, or if it is ambiguous, it returns an error to the client.
4. The DDSP looks at the registered LQSP and the fully qualified query. It builds a list of candidate LQSP and returns the fully qualified XML query along with the list of candidate LQSP to the DQSP.

#### 26.3.3.3 LQSP Process

1. A LQSP initializes, which looks up the DDSP to use.
2. The LQSP registers itself with the DDSP.
3. The LQSP receives global query (in XML) from DQSP. It uses the statistics and deployment information in the LDS to determine what type of coverage it can provide for this query. As part of this process, it builds up a local execution plan.
4. The LQSP sends its coverage and related statistics (such as number of rows, number of columns, types of fragmentation it has, etc.) to the DQSP.
5. The LQSP receives the notification that it has been elected to process a query. It processes the local plan for the query and sends its results to DQSP or other LQSP's as per plan.

### *26.3.4 Query Processing Example*

Consider a simple global schema containing two tables. The Emp table is vertically fragmented to separate sensitive salary details from the less sensitive

**Global Table**: Emp(Eno, Name, Title, Salary, Dno)
**Fragment**: Emp_VF1(Eno, Salary, Dno)
**Fragment**: Emp_VF2(Eno, Name, Title)
**Global Table**: Dept (Dname, Dno, Budget, Loc)
**Fragment**: Dept_HF1 (Dname, Dno, Budget, Loc)
**Fragment**: Dept_HF2 (Dname, Dno, Budget, Loc)

**Fig. 26.3** Example fragmentation

employee name and job title columns while the Dept table is horizontally fragmented by location (See Fig. 26.3). For example, suppose Dept_HF_1 were defined to contain all rows with Loc in ('MN', 'IA', 'WI') while Dept_HF2 will contain all rows with any other value for Loc.

For this example, there are three LQSP sites (See Fig. 26.4). The Emp_VF1 fragment containing the salary information is deployed within the LQSP at Site-A. The LQSP at Site-B contains the other vertical fragment (Emp_VF2) and one of the horizontal fragments (Dept_HF1) while Site-C contains the other horizontal fragment (Dept_HF2).

Suppose a client submitted the SQL query shown in Fig. 26.5 column 1. The client would send this SQL string to the DQSP. The DQSP would then translate this query into relational algebra, encode it in XML and pass it to the DDSP for verification and disambiguation. The DDSP would verify that all the global tables and columns referenced in the query were valid and generates the query in column.

Using the current list of registered LQSP components, DDSP would build a list of candidate sites (here Site-A, Site-B, and Site-C). The fully qualified query and candidate site list would be returned to the DQSP. The DQSP would send the fully qualified query to all candidate sites and await their responses.

Upon receiving the query, each LQSP would then form a local plan and determine how well it can satisfy the query. This self-ranking would be based upon the fragmentation plan and actual fragments deployed at each LQSP site.

**Fig. 26.4** Example deployment

Site A stores Emp_VF1
Site B stores Emp_VF2 and Dept_HF1
Site C stores Dept_HF2

| Original Global Query | Fully Qualified Global Query |
|---|---|
| Select Name, Salary, Budget<br>From Emp, Dept<br>Where Emp.Dno = Dept.Dno<br>AND Title = 'Engineer'; | Select T1.Name, T1.Salary, T2.Budget<br>From Emp T1, Dept T2<br>Where T1.Dno = T2.Dno<br>AND T1.Title = 'Engineer'; |

**Fig. 26.5** Example global query

Each LQSP would also use locally gathered and cached statistics to estimate the size (number of rows and row width) for each result produced at that LQSP.

In this example, the local execution plan built by the LQSP at Site-A would return the entire fragment. The coverage information would indicate that the Emp_VF1 fragment was deployed here by including a unique plan id (a GUID) and bit mask flag for the Emp table. The size estimate would use the size of the columns and locally gathered statistics to indicate the expected size for the result. In this case it is merely the total number of rows and sum of the column widths.

The local execution plan built by the LQSP at Site-B would use two separate commands and produce two separate results. The first command would select only rows with Title = 'Engineer' from the Emp_VF2 fragment and project the Eno and Name columns as the final result. The second command would project the Dno and Budget columns from the Dept_HF1 fragment as the final result. The coverage information returned would indicate that the EMP_VF2 fragment was deployed here by including the same plan GUID as Site-A, and a complimentary mask value. Similarly, the second command would indicate a different plan GUID for the Dept fragmentation and a different bit mask value for the Dept_HF1 horizontal fragment deployed here. The size estimate for each result would be included in the self-ranking response from Site-B. In this case, the first result would use locally cached selectivity statistics to estimate the number of rows matching 'Engineer' and the size of the Eno and Name columns. The results from the second command would use the number of rows statistic for the Dept_HF1 fragment and the column size for the Dno and Budget columns.

At Site-C, the LQSP would build a plan containing a single command, similar to the second command at Site-B. In other words, it would project the Dno and Budget columns from the Dept_HF2 fragment. The coverage information from this command would use the same plan GUID as the second command at Site-B, and a complimentary bit mask value. The size estimate would use the number of rows in the Dept_HF2 fragment and the size of the Dno and Budget columns.

When different fragments from the same fragmentation plan for the same table are deployed at different LQPS sites, the LDS must be populated with the deployment information required. In other words, the LDS at each site contains the fragmentation plan criteria, a plan GUID unique to this plan, a bit mask value representing the fragment deployed here, and a bit mask value representing the reconstructed table. In our example, the Emp coverage information at Site-A could have a bit mask value of 0×01, at Site-B a value of 0×02 and both sites would know that the reconstructed table requires a bit mask of 0×03. The GUID stored for the Emp fragmentation plan at each site would be the same value. Similarly, the Dept fragmentation plan deployed at Site-B and Site-C would use a different GUID than the Emp fragmentation plan and bit mask values for each fragment and the reconstructed Dept table based upon its plan.

All of the coverage and size estimate information from each LQSP is returned to the DQSP. The DQSP uses the actual response along with the coverage and size details it contains to elect the official list of sites which are to be used. In this case, Site-A, Site-B, and Site-C are all available and useful. Therefore, the DQSP elects them and sends another message to each elected LQSP. This message would include the list of all elected sites as well as the coverage and size statistics that each elected site had returned prior to the election.

Using this information, the local plan at Site-A would know that it must join with results from the first command at Site-B in order to reconstruct the Emp table. Similarly, Site-B and Site-C know that the results of the second command at Site-B and the results of the command at Site-C must be unioned together to reconstruct the Dept table. All three sites know that the reconstructed Emp table and Dept table need to be joined and that the Name, Salary, and Budget columns need to be projected from this result in order to satisfy the original query.

Since the elected LQSP at each site has the same algorithm, same fully qualified query, and same coverage and size statistics, they can each come to the same decision with respect to where results (both the intermediate and final) are sent. For our example, suppose the first result from Site-B is smaller than the result for Site-A's local plan. Site-B would then know to send its result to Site-A, and Site-A would know to expect it. Similarly, suppose that the result from Site-C is smaller than the second result at Site- B. Both sites would expect the results from Site-C to be sent to Site-B for the union operation. Lastly, using the size of the results for the two Emp fragments about to be joined, all sites can estimate the size of the result, it will be the same number of rows as the smaller of the two fragments (in other words the same number of rows as the first result at Site-B) and the size of the Dno, Name, and Salary columns. The size of the union would be the sum of the size for the second result at Site-B and the result at Site-C. Once again, all sites can determine this for themselves and come to the same decision as to where these two results will be joined to produce the final result. In this case, suppose the Emp result is smaller, Site-A would then send its result to Site-B. The result from Site-A would be joined with the result at Site-B and the final result would be ready for the client to process. The LQSP at Site-B would then notify the DQSP that the results are available.

As the client requests rows to be returned from the DQSP, the DQSP can connect to this LQSP and retrieve the rows. Although it is possible to return all the rows in a single message, it is most likely more efficient to have the number of rows returned be a configurable or parameterized value less than this. The client could use the DQSP as a proxy or connect directly to the LQSP containing the results depending upon the physical network topography, security concerns and installation decisions chosen for a given organization.

## 26.4 Conclusion

Using the GIS and LDS to decouple the deployment details from the global table and column definitions; we can eliminate the need to keep a single monolithic and synchronized GDD. This separation also reduces the amount of information that needs to be sent to the DDSP when a given LQSP needs to register with the DDBMS. By eliminating the synchronization of deployment details and reducing the size and complexity of information necessary to register a LQSP with the system, it becomes practical for dynamic LQSP registration. Returning the self-ranking coverage and size statistics from each candidate LQSP allows us to build the execution plan in a dynamic and distributed fashion. A byproduct of the separation of deployment details is the ability to support multiple deployment plans for the same global table. Although each LQSP providing fragments for the same table in the same query must use the same plan, the system can choose to use any plan for a given table provided that the LQSP required for that plan all respond to the election process in a timely fashion.

Unlike a traditional DDBMS, the there is no bottleneck at the GEM/DEM for query processing because the majority of the work is done in parallel by the LQSP. In a traditional DDBMS, the synchronization of the GDD can cause deadlock and live lock concerns within the GDD (query processing requiring read access while LDBMS activation and deactivation requires write access). In our approach, this is no longer required. The only time we need to update the GIS is when there is either a new global construct or a change to the interface of an existing global construct (for example, adding a new global column to an existing global table, etc.) This is unavoidable, but it is also an event that happens far less frequently than the activation/deactivation of an LQSP or a new deployment plan.

Although other variations could be implemented, such as allowing the LQSP to communicate while deciding where to send results, this would increase the communication complexity and cost with only minor potential for improved performance for most typical queries.

## References

1. Özsu MT, Valduriez P (1999) Principles of distributed database systems. Second edn. Prentice Hall, New Jersey
2. Larson J, Rahimi S (1984) Tutorial: Distributed database management. Computer Society Press, Los Alamitos
3. Kossmann D (2000) The state of the art in distributed query processing. ACM comput surv 32(4):422–469
4. Ceri S, Pelagatti G (1984) Distributed databases – Principles and systems. McGraw Hill Inc., New York, NY
5. Ripeanu M (2001) Peer-to-peer architecture case study: Gnutella network. In: Proceedings of international conference on peer-to-peer computing, Linkopings, Sweden, pp 99–100
6. Singh I, Stearns B, Johnson M, and the Enterprise Team (2002) Designing Web Services with the J2EE(TM) 1.4 Platform: JAX-RPC, SOAP, and XML Technologies (The Java Series). Addison-Wesley
7. Singh I, Brydon S, Murray G, Ramachandran V, Violleau T, Sterns B (2004) Designing Web Services with the J2EE(TM) 1.4 Platform: JAX-RPC, SOAP, and XML Technologies. Addison-Wesley, Boston
8. Fleury M, Reverbel F (2003) The JBoss Extensible Server. In: Proceedings of the 2003 ACM/IFIP/USENIX international middleware conference, pp 344–373
9. Extensible Markup Language (XML), May 2007, http://www.w3.org/XML/

# Chapter 27
# Estimating Join and Projection Selectivity Factors

**Carlo dell' Aquila, Ezio Lefons, and Filippo Tangorra**

**Abstract** The selectivity factor of relational operations is a critical parameter for determining the cost function of query processing. Good estimates of these parameters allow the optimizers to choose the least expensive path in the query execution. A method for estimating the join and projection selectivity factors based on the orthogonal polynomial series is presented. Experimental results on real data are also reported which show the good performance of the approach.

## 27.1 Introduction

With the growing number of large data warehouses for decision support applications, to efficiently execute data queries is becoming increasingly important. The use of a data profile yields advantages not only in establishing efficient strategies to minimize access to large data sets in the traditional database field, but also can be used in activities such as data warehouses in which data are continually collected from different source databases, in data mining applications that require the access of large data sets for complex exploratory activity, and in networking management in which decisions are needed on how to move data to minimize the data traffic. Data profiles maintain statistics about data as the shape form of data distribution, data cardinality, the number of distinct values, and so on, in order to determine accurate estimates of the query counts and selectivity factors. Using this information, the query optimizer parses and analyzes the query, constructs possible access paths, and estimates the cost of each path in order to determine the least expensive one.

In addition, data profiles are also important in data mining applications and data analysis activity, where the user explores various hypotheses on data and may prefer an approximated result to the query in real time, rather than waiting for the exact one. This research field has conduced to the design of systems

C. dell' Aquila (✉)
Dipartimento di Informatica, Università di Bari, Via E. Orabona 4, I-70125 Bari, Italy
e-mail: dellaquila@di.uniba.it

N. Mastorakis et al. (eds.), *Proceedings of the European Computing Conference*, Lecture Notes in Electrical Engineering 28,
DOI 10.1007/978-0-387-85437-3_27, 

which involve the estimates of summary data and the execution of aggregate queries with approximate answers derived by the data profile.

To obtain reasonable cost estimates of query execution, the query optimizers need estimates of sizes of the final and intermediate relations involved, based on the selectivity factors [2]. The selectivity factor corresponds to the fraction of tuples which satisfy the query.

The methods to estimate the actual query selectivity factors are an important research topic in traditional databases. Various methods, such as parametric [4, 14], sampling [3, 13], histogram [10], wavelet [18], and discrete cosine transform [11], etc., have been proposed.

We use the analytic approach based on the approximation of the actual multivariate data distribution of attributes by series of orthogonal polynomials [12]. The method is a special case of the least squares approximation by orthonormal functions and summarizes all the information on the data distribution by few data—the computed coefficients of the polynomial series, or the so-called *Canonical Coefficients* (*CC*) of the data. The coefficients *CC* allow the selectivity factors and main data statistics to be easily derived and efficiently computed with no further access to the data warehouse.

According to the good performance in estimating the selectivity factor of the restriction operation in multidimensional environment [12], we have applied the analytical approach to estimate the selectivity factor of join and projection operations. Relative to these relational operations, the knowledge of (counts of) distincts of database attributes plays a crucial role in improving the performance of the method in estimating the selectivity factors. The analytic method, opportunely adapted, permits to enrich the descriptive contents of data profile with few metadata relative to the distribution of distances of unique values, in the attribute value ranges.

Here, we describe the application of the analytic method to determine the selectivity factor of join and projection operations using a new method to estimate the attribute distincts.

## 27.2 Related Work

In this Section, we briefly review methods used in estimating the selectivity factor respectively for join and projection relational operations.

Here, "join" refers to the *equijoin* relational operation. When considering the join T between relations R and S, the canonical formula giving the cardinality of join is $card(\mathrm{T}) = j\rho \times card(\mathrm{R}) \times card(\mathrm{S})$ where $j\rho$ is the join selectivity factor and represents the fraction of tuples of the Cartesian product of R and S which satisfy the join condition.

There are several methods for estimating join cardinality [1, 8, 9, 11, 16, 17]. Some methods are based on integrity constraints of the relational schema and provide limits for the join cardinality estimation (*e.g.*, $card(\mathrm{T}) \bullet card(\mathrm{R} \times \mathrm{S})$). These methods do not provide accurate estimations. Other methods assume arbitrarily that join attributes are uniformly distributed and that the number of distinct values of attribute T.X is *min*(*dist*(R.X), *dist*(S.Y)) [15]. Indeed, it is very rare that attributes are uniformly distributed in real cases and the uniformity assumption leads to pessimistic results when data attributes accord to other distributions such as Normal, Gamma, Zipf, etc.

Other methods have been proposed. In the worst case assumption method, join cardinality is estimated as *card*(R) × *card*(S). The worst case divided 2 method minimizes errors produced by the worst case method. Methods of perfect knowledge does not produce errors but all the occurrences of attributes values must be stored, so it is very expensive in storage and time consuming for updating [1]. Piece-wise uniform method has been proposed and used in [16] to estimate the frequency distribution of join attributes by equal-width histograms. In this approach, the attribute domain is divided into intervals and the number of tuples holding values which fall into each interval is stored. Attribute distribution is approximated by a piece-wise linear curve which interpolates some points by a relative frequency histogram, which is then used to estimate the size of relational query and parameters of intermediate results. Join range is divided into a number of intervals whose scale is function of the number of distinct values within the join range. To determine the number of distinct values, the (number of) distinct values of joining attributes should be known, but no proposed method estimates them. So, a way to obtain this is to sort join attribute values and to eliminate duplicates. However, this method is impractical due to the high cost of processing and maintenance after updates. The approach presented here estimates the resulting size of the relational join by evaluating the distinct values and using the estimation of actual distribution of join attributes.

The selectivity of projection is the ratio of reduction of cardinalities. In fact, notice that the projection removes not only attributes but also the resulting duplicate tuples.

In general, the approach for estimating the size of a projection consists of the ability of data statistical profile to fit the actual multidimensional data distribution in order to derive estimates of the marginal distribution of the projected attributes.

In [15] is described an approach based on the multidimensional histogram divided into equal-sized buckets, assuming uniformity within the buckets. Parametric methods described in [6] assume that attributes are uniformly distributed and independent each other. Under these assumptions, authors studied the probability distributions of the sizes of the projections testing various hypotheses. The assumption of independence and/or uniformity of attributes result inadequate to correctly represent many actual database instances. However, they simplify and speed the computations of parameters for run-time query optimizer in evaluating access plans. The model presented in [5] assumes that values of attributes determine a time dependent active domain. Others approaches are based on fast ways to estimate distinct values [7, 8].

The approach described here estimates the resulting size of the relational projection by evaluating distinct values of active domains of the attributes involved in the operation.

## 27.3 Data Distribution Approximation by Orthogonal Series

Let R be a relation of cardinality $N$ and let X be an attribute of R. We assume $dom(\mathrm{X}) = [\mathrm{a}, \mathrm{b}]$ and let $\mathrm{x}_1, \mathrm{x}_2, \ldots, \mathrm{x}_N$ be the occurrences of X in R.

The probability density function $g(\mathrm{x})$ of the attribute X is approximated as

$$g(\mathrm{x}) = \frac{1}{\mathrm{b}-\mathrm{a}} \sum_{i=0}^{n} (2i+1) c_i P_i(\mathrm{x}) \tag{27.1}$$

for x∈X and the opportune $n$. For each $i = 0, 1, \ldots, n$, $P_i(\mathrm{x})$ is the Legendre orthogonal polynomial of degree $i$, and coefficient $c_i$ is the mean value of $P_i(\mathrm{x})$ on the instances of X. That is,

$$c_i = \frac{1}{N} \sum_{\mathrm{x} \in \mathrm{X}} P_i(\mathrm{x}) \tag{27.2}$$

The $c_0, c_1, \ldots, c_n$'s are computed with simple recursive formulæ [12] and they are the so-called *Canonical Coefficients* of X.

The approximation of the cumulative distribution function $G(\mathrm{x})$ of $g(\mathrm{x})$ is defined as

$$G(\mathrm{x}) = \int_{\mathrm{a}}^{\mathrm{x}} g(y)dy = \frac{1}{\mathrm{b}-\mathrm{a}} \sum_{i=0}^{n} c_i (P_{i+1}(\mathrm{x}) - P_{i-1}(\mathrm{x})) \tag{27.3}$$

Let $\mathrm{I} = [\mathrm{x}_1, \mathrm{x}_2] \subseteq [\mathrm{a}, \mathrm{b}]$ be a generic query-range of X. We denote with $count(\mathrm{x}; \mathrm{I})$ or $N \times percent(\mathrm{x}; \mathrm{I})$ the number of tuples of R.X whose x-values belong to the interval I. Then, $count(\mathrm{x}; \mathrm{I})$ can be approximated by $N \times (G(\mathrm{x}_2) - G(\mathrm{x}_1))$.

## 27.4 Join Selectivity Estimation

In this Section, we present how to estimate the cardinality of join operation between relations R and S on the respective attributes X and Y using the analytic model described in the Section 27.3.

Let $\mathrm{T} = R \bowtie_{\mathrm{X}=\mathrm{Y}} \mathrm{S}$. In order to estimate $card(\mathrm{T}) = j\rho \times card(\mathrm{R}) \times card(\mathrm{S})$ or equivalently $j\rho$, let $x_1, x_2, \ldots, x_{dist(\mathrm{X})}$ and $y_1, y_2, \ldots, y_{dist(\mathrm{Y})}$ be the ordered distinct values of attributes X and Y, respectively. Then,

$$card(\mathrm{T}) = \sum_{i=1}^{dist(\mathrm{X})} \sum_{j=1}^{dist(\mathrm{Y})} [C_{\mathrm{X}}(x_i) \times C_{\mathrm{Y}}(y_j)]_{x_i = y_j} \tag{27.4}$$

where $C_X(x_i)$ ($C_Y(y_j)$ respectively) denotes the number of the tuples of R (resp., S) such that X = $x_i$ (resp.,Y = $y_j$). In general, the estimation of $C_X(x_i)$ and $C_Y(y_j)$ is obtained without scans of R and S, but only by using the aggregate function *percent* (*cf.*, Section 27.3), as

$$C_x(X_i) \cong count(x; lx_i) = card(\mathrm{R}) \times percent(x; lx_i)$$

$$c_y(Y_j) \cong count(y; ly_j) = card(\mathrm{S}) \times percent(y; ly_j)$$

where $\mathrm{I}x_i = [(x_i+x_{i-1})/2, (x_i+x_{i+1})/2]$ and $\mathrm{I}y_j = [(y_j+y_{j-1})/2, (y_j+y_{j+1})/2]$. So, the selectivity factor is approximated as follows.

$$j\rho \cong \sum_{i=1}^{dist(\mathrm{X})} \sum_{j=1}^{dist(\mathrm{Y})} \left[percent(x; \mathrm{I}_{x_i}) \times percent(y; \mathrm{I}_{y_j})\right]_{x_i=y_j} \tag{27.5}$$

The *percent* value depends on the distinct values of attributes X and Y, which are often unknown. Many researchers suppose that differences (distances) between two adjacent domain values are approximately equal [17].

The approach described here does not make hypotheses about distinct values spacing distribution or their joining equivalence. Distinct values are evaluated using the canonical coefficients which are easily computed and updated whenever a new distinct value is inserted into the database attribute.

Let $\{x_1, x_2, \ldots, x_d\}$ be the distinct values of attribute X. The canonical coefficients $(d_i)_{1 \leq i \leq n}$ up to degree *n*, that contain information about how distinct values are spaced, are computed as follows.

$$d_i = \frac{1}{d \times dns} \sum_{j=1}^{d} \sum_{k=1}^{dns} P_i(x_j - \frac{\delta}{2} + \delta \times rand(0,1)); i = 0, 1, \ldots, n \tag{27.6}$$

where we suppose that, for each $x_j$, there are distributed a sufficiently high number *dns* of random values in the interval $[x_j - \delta/2, x_j + \delta/2]$ , with $\delta = (\mathrm{b} - \mathrm{a}/card(\mathrm{X})$ . We assume that *x* is an approximation of the distinct value $x_i$, if it is verified the condition $count(x; \mathrm{I}) \bullet dns$ in the interval I or, equivalently, $|count(\mathrm{x}; \mathrm{I}) - dns| \bullet \varepsilon$ for $\varepsilon = k \times dns$ $(0 < k < 1)$. It follows the algorithm that computes the approximation of distinct values and their number based on the knowledge of canonical coefficient $d_i$.

*Algorithm* 27.1: determination of distinct values.

$d$:= 1; $x(d)$:= a; $x_1$:= a + $\delta/2$; $x_2$:= $x_1$ + $\delta$; *eps*:= 0.05 × *dns*;
**while** ($x_2 <$ b) **do begin**

I:= $[x_1, x_2]$; *occ*:= $count(x; \mathrm{I})$;

**if** $eps \geq |dns - occ|$ **then begin**

$d := d + 1;$

$x(d) := (x_1 + x_2)/2$ **end**

$x_1 := x_2; x_2 := x_1 + \delta$

**end**

$d := d + 1; x(d) := \mathrm{b}$

When Algorithm 27.1 terminates, the value $d$ approximates the number of distinct values of X in [a, b], and the vector $x[1..d]$ contains the approximation of distinct values.

As in computing canonical coefficients up to degree $n$ in $\mathrm{I}_j = [x_j - \delta/2, x_j + \delta/2]$ we have distributed *dns* occurrences, then it is expected that $count(x; \mathrm{I}_j) \cong dns$. So, Algorithm 27.1 analyses interval [a, b] searching for sub-interval I which satisfies the condition of the required estimation accuracy at the given confidence level, defined empirically as the fraction 0.05 of the *dns* quantity in the interval I of amplitude δ.

Series of canonical coefficients $(d_i)_{0 \le i \le n}$ for distincts is different from series computed to approximate the distribution of attribute X. In particular, aggregate function *count* in Algorithm 27.1 is calculated using $(d_i)_{0 \le i \le n}$.

Assuming that sets $X = \{x_i\}_{0 \le i \le dist(\mathrm{X})}$ and $Y = (y_j)_{0 \le j \le dist(\mathrm{Y})}$ estimated by Algorithm 27.1 are denoted with $\overline{X} = \{\overline{x}_i\}_{0 \le i \le dist(\mathrm{X})}$ and $\overline{Y} = (\overline{y}_j)_{0 \le j \le dist(\mathrm{Y})}$, to compute join selectivity we use, instead of applying (27.5), the following

$$j\rho \cong \left[ \sum_{i=1}^{dist(\mathrm{X})} \sum_{j=1}^{dist(\mathrm{Y})} percent(x; \overline{\mathrm{I}}_{\overline{x}_i}) \times percent(y; \overline{\mathrm{I}}_{\overline{y}_j}) \right]_{\overline{x}_i \cong \overline{y}_j} \qquad (27.7)$$

We assume the equivalence $\overline{x}_i \cong \overline{y}_j$ if $|\overline{x}_i - \overline{y}_j|$ is lower than the mean distance of adjacent approximate distincts respectively of X and Y.

We have performed experiments on a real case considering the relation R(A, B, C, D, E, F) having $m = 104{,}828$ tuples.

The values of attributes A and B of relation R are integer (with equally-spaced distinct values). The values of attributes C, D, E and F are real numbers.

Table 27.1 reports the join selectivity estimates by comparing our approach, or the CC method, and the histogram approach, or the EW method. In the CC approach, no hypotheses have been made on the number of distinct values and about their spacing in the join range [a, b]. In this case, distinct values for the joining attributes X and Y have been estimated using the method discussed in Section 27.2. According to empirical tests, the distinct values of the attributes are approximated using $dns = 50$, $\varepsilon = 0.05 \times dns$ and approximation degree $n = 27$, while the join selectivity factor has been estimated using formula (27.7).

**Table 27.1** Mean percentage errors of join selectivity estimates

| | | | | | *Mean Percentage Relative Errors* | |
|---|---|---|---|---|---|---|
| X⋈Y | *d* | $dc_c$ | $de_w$ | $nc_c$ | CC | EW |
| A⋈A | 16 | 15 | 18 | 8 | 1.04 | 3.12 |
| | | | | 13 | 0.81 | 2.80 |
| | | | | 27 | 0.93 | 2.95 |
| B⋈B | 16 | 15 | 18 | 8 | 1.98 | 0.42 |
| | | | | 13 | 1.70 | 0.04 |
| | | | | 27 | 1.58 | 0.08 |
| C⋈C | 80 | 179 | 85 | 8 | 22.09 | 695.94 |
| | | | | 13 | 10.15 | 821.94 |
| | | | | 27 | 3.09 | 888.55 |
| D⋈D | 286 | 285 | 298 | 8 | 0.90 | 13.85 |
| | | | | 13 | 8.23 | 4.62 |
| | | | | 27 | 5.26 | 1.36 |
| E⋈E | 253 | 225 | 285 | 8 | 13.87 | 34.85 |
| | | | | 13 | 8.97 | 14.82 |
| | | | | 27 | 5.01 | 2.11 |
| F⋈F | 175 | 174 | 184 | 8 | 12.34 | 5586.70 |
| | | | | 13 | 5.84 | 5586.70 |
| | | | | 27 | 4.49 | 5586.70 |

In the EW approach, the number of distinct values for each attribute has been estimated considering the join range divided into equal-width intervals and values in each interval have been considered to be join-equivalent [10].

The entries in Table 27.1 indicate the join attributes, the actual number of distinct values *d* on join range [a, b], the number of distinct values $d_{cc}$ estimated using Algorithm 27.1, and the number of distinct values $d_{ew}$ estimated by the EW method. The mean percentage error is measured using the metric

$$M = |j\rho - \overline{j\rho}|/j\rho$$

where $j\rho$ and $\overline{j\rho}$ are, respectively, the join selectivity factor and its estimation relative to methods CC and EW.

For each method, when estimating the cumulative distribution function by canonical coefficients, we have used approximation degrees 5 to 33. However, for the sake of readability, here we only report the mean percentage error measured for $n_{cc} = 8, 13, 27$ using metric $M$.

The errors obtained using the CC estimates of join selectivity are significantly reduced with respect to the EW ones. In particular, the results of EW are pessimistic for the attributes whose distinct values are not equally-spaced. For these attributes, the mean percentage error is generally lower for CC method than the EW one.

## 27.5 Projection Selectivity Estimation

The model of Gardy-Puech [6], based on the attribute independence and uniform distribution assumptions, is frequently used to estimate the size of projections. Using probabilistic arguments, they derive formulas to compute the $l$ tuples of projection $_Y(R)$ from the $m$ tuples of a relation R(X), with $Y \subseteq X$, as the expected value of all possible randomly generated projections. This value is computed as

$$l = \frac{d_x}{d_y}\left[1 - \left(1 - \frac{d_y}{d_x}\right)^m\right] \tag{27.8}$$

where $d_x$ and $d_y$ are the product of distinct value numbers respectively of attributes X and Y.

We have used the Gardy-Puech formula, but we estimate the distinct values with canonical coefficients using the effective distinct values in the domain (*i.e.*, the *active* domain). In fact, canonical coefficients allow to estimate current distincts for they are updated when a new/old occurrence is added/removed in/from database. For example, if values $(d_i)_{0 \le i \le n}$, *dns*, and $\delta$ are stored in the statistical profile at a certain time and the occurrence $x'$ arrives for attribute X in database, then Algorithm 27.2 updates canonical coefficients $d_i$ if it establishes that $x'$ is a new distinct value.

*Algorithm* 27.2: update of canonical coefficient $d_i$ for distinct values, $0 \le i \le n$.

$I = [x' - \delta/2, x' \delta/2]$,

**if** $| d \times dns \times percent(x; I) - dns | \le eps$

**then** $x'$ already exists in database

**else** $x'$ is a new distinct value hence **begin**

$$d_i := \frac{d_i \times (d \times dns) + \sum_{k=1}^{dns} P_i(x' - \frac{\delta}{2} + \delta \times rand(0,1))}{(d+1) \times dns}; i = 0, \ldots, n;$$

$d := d + 1$ **end** (27.9)

Using the same data and assumptions of the previous Section, Fig. 27.1a and b report the performance of the projection selectivity respectively in the bidimensional and tridimensional cases, by comparing the method of Gardy-Puech (GP) to the analytic method (CC).

The errors obtained in applying the analytic method estimates to projection selectivity are in many cases significantly lower than those by GP method. We observe that the analytic method improves sensibly its performance if the projection is performed on two attributes respect to the case of three attributes. This result offers a relevant advantage when considering that plans of

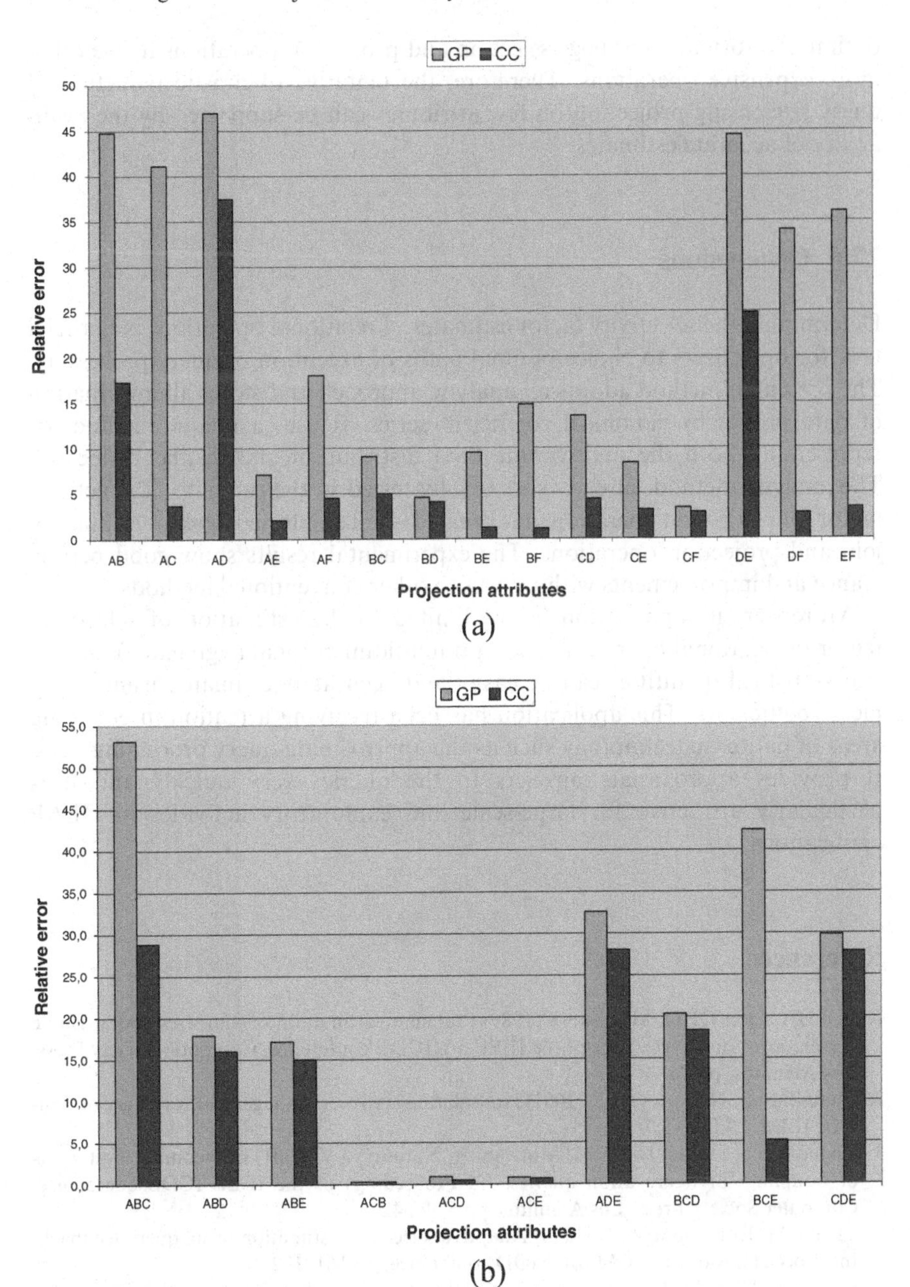


**Fig. 27.1** Comparison of the performance of GP and CC methods for estimating projection selectivity respectively in the **(a)** bidimensional and **(b)** tridimensional cases

optimizers normally privilege selection and projection operations to the other more expensive operations. Therefore, the tentative of drastic reduction of query sizes using projection on few attributes can be supported by the availability of accurate estimates.

## 27.6 Conclusions

Determining the selectivity factor estimates of relational operations is a crucial task for optimizers to choice optimal paths of execution of query processing. The presented method adopts an analytic approach and stores all information of data profile by canonical coefficient series. It uses a unique method for representing both the multivariate data distribution and distinct value set. The analytic method, already successfully tested in the estimate of selectivity factor for restriction operation, has been adapted and tested here for estimating join and projection operations. The experimental results show good performance and improvements with respect to other conventional methods.

Moreover, its application is not limited to the estimation of selectivity factor of relational operations. Several multidimensional aggregate functions and statistical quantities can be easily and accurately estimated using canonical coefficients. This application has been receiving attention in emerging areas of database technology such as the approximate query processing field. It provides approximate answers to the queries very quickly and it is particularly attractive for large-scale and exploratory activities in OLAP applications.

## References

1. Bell DA, Ling DHO, McClean S (1989) Pragmatic estimation of join sizes and attribute correlations. In: Proceedings of the IEEE 5th ICDE conference, Computer Society Press, Los Alamitos, pp 76–84
2. Chaudhuri S, Narasayya VR (2001) Automating statistics management for query optimizers. IEEE TKDE 13:7–20
3. Chaudhuri S, Das G, Datar M, Motwani R, Narasayya V (2001) Overcoming limitations of sampling for aggregation queries. In: Proceedings of the IEEE ICDE conference, Computer Society Press, Los Alamitos, pp 534–542
4. Chen CM, Roussopoulos N (1994) Adaptive selectivity estimation using query feedback. In: Proceedings of the ACM SIGMOD conference, pp 161–172
5. Ciaccia P, Maio D (1995) Domains and active domains: what this distinction implies for the estimation of projection sizes in relational databases. IEEE TKDE 4:641–654
6. Gardy D, Puech C (1984) On the sizes of projections a generating functions approach. Inf Syst 9:231–235
7. Gibbons PB (2001) Distinct sampling for highly-accurate answers to distinct values queries and event reports. In: Proceedings of the 27th VLDB conference, Roma, Italy, pp 541–550

8. Haas PJ, Naughton JF, Seshadri S, Stokes L (1995) Sampling-based estimation of the number of distinct values of an attribute. In: Proceedings of the 21th VLDB conference, Morgan Kaufmann Publishers, San Francisco, CA, pp 311–322
9. Ioannidis YE, Christodoulakis S (1993) Optimal histograms for limiting worst-case error propagation in the size of the join results. ACM TODS 18:709–748
10. Ioannidis YE, Poosala V (1995) Balancing histogram optimality and practicality for query result size estimation. In: Proceedings of the ACM SIGMOD conference, pp 233–244
11. Jiang Z, Luo C, Hou W-C, Yan F, Zhu Q (2006) Estimating aggregate join queries over data streams using discrete cosine transform. In: Proceedings of the DEXA conference, pp 182–192
12. Lefons E, Merico A, Tangorra F (1995) Analytical profile estimation in database systems. Inf Syst 20:1–20
13. Ling Y, Sun W (1995) A comprehensive evaluation of sampling-based size estimation. In: Proceedings of the IEEE 11th ICDE conference, pp 532–539
14. Ling Y, Sun W (1999) A hybrid estimator for selectivity estimation. IEEE TKDE 11:338–354
15. Merrett TH, Otoo E (1979) Distribution models of relations. In: Proceedings of the 5th VLDB conference, pp 418–425
16. Mullin JK (1993) Estimating the size of a relational join. Inf Syst 18:189–196
17. Sun W, Ling Y, Rishe N, Deng Y (1993) An instant and accurate size estimation method for joins and selection in retrieval-intensive environment. In: Proceedings of the ACM SIGMOD conference, pp 79–88
18. Vitter JS, Wang M (1999) Approximate computation of multidimensional aggregates of sparse data using wavelets. In: Proceedings of the ACM SIGMOD conference, pp 193–204

# Chapter 28
# Decision Rules: A Metamodel to Organize Information

**Sabina-Cristiana Mihalache**

**Abstract** This paper tries to introduce the idea that the form of specialized knowledge relating to decision models, rules, and strategies may be a way to organize information by attaching meaning to the existing formalized entities at the computer-based information system level. We realize an exemplification by using a rule in order to attach a meaning to an existing class from the domain model in the form of a new property. The paper presents some technologies and tools that we used for the exemplified decision problem, along with some remarks and conclusions. We titled the article "thinking in decision rules" starting from the assumption that the way we think when we make decisions should be the metamodel to organize information. When we ask to obtain information we do not ask for specific data, but we ask on concepts that are usually interrelated in IF. . .THEN. . .ELSE rules.

## 28.1 Introduction

Considering decisions to be a matter of problem-solving transforms data and information models into problems' models. Information that is necessary in making decisions is a metamodel to organize information in a way that proves to be useful for decision-making. Problems' models organize data that is necessary in evaluating decision alternatives and optimize the value for each alternatives conforming to a set of criteria. Information models organize data conforming to an accepted common sense for data in a way that maps reality through representing knowledge.

The literature concerning computer-based decision models and decision-problems search to provide general problem-solving models applicable in all sorts of situations. Either in the form of analytical models, data mining models,

S.-C. Mihalache (✉)
Business Information Systems Department, Alexandru Ioan Cuza University of Ia i, Carol I Blvd, no. 700505, Ia i, Romania
e-mail: sabina.mihalache@gmail.com

N. Mastorakis et al. (eds.), *Proceedings of the European Computing Conference*, Lecture Notes in Electrical Engineering 28,
DOI 10.1007/978-0-387-85437-3_28, 

or expert systems models they all embed a model for solving a problem and transforms from what intended to be general into a specifically solution. All of these models prove to be static, non-adaptive, and incapable to react when the conditions change or interact with a decision-maker. The idea to realize decision models is good and necessary, but the technology is not sufficient. We try to argue in this article that the whole idea of realizing decision models should be used in improving information provided in decision-making and all that is necessary is to create an environment, a metamodel, to organize information so that the decision-makers can specify the decision-rules that constrain a problem.

## 28.2 Problem Formulation

The real problem in specifying decision-models consists in representing knowledge as control constructs. It is ordinarily known that in order to make a decision there is necessary information and knowledge application. Information technologies and technology providers have tried always either to offer possibilities to embed decision problems simulations either to provide a way to have easy access on aggregated data, usually named information. Decision support systems and Business Intelligence technology's providers have always tried to animate the idea that analyzing data means structuring problems and therefore information and knowledge is disseminated to the user. But knowledge in its pure sense is a pure form of what proved to be true, either changed conditions. What is true has always an unambiguous value and not a fuzzy value or a probable value derived from some comportment presented for datasets.

What we try to say is that the philosophy presented for some technology trends is good, but the infrastructure or the metamodel for decision modeling is not unified, probably because there are many different approaches in analyzing decision problems. In the following we will present two problems: (1) the output for a decision-model: a result of an act or information that can transform in action? (2) the ambiguous ideology of knowledge representation.

### *28.2.1 The Output for a Decision-Model*

Making decisions concretizes in the phase of implementation, in actually performing an action. We have a philosophy concerning decision-making steps provided by H.Simon which is unbeatable because it is always true: we always make decisions by choosing from some alternatives one alternative that satisfies best a set of criteria preferences. Modeling decisions doesn't mean that it has to be realized in the same way: implementing the alternatives, the criteria and the algorithm that optimizes the criteria. We think in a specialized way so it cannot be provided a general model for decisions by implementing models as alternatives and choosing from the respective alternatives. The output for a decision-model should have to be information with potential in activating an action and not the action itself. The information provided should refer to possible effects for a decision and not the actual way for acting because often the decision-maker will not trust the model: the respective model hides knowledge that is implemented in the form of control constructs. Usually the decision-maker doesn't know why the implemented model provided a solution because he/she and the model do not share the same meanings and amount of knowledge.

During the time we have experienced different forms of technology: from data processing (which is actually the "intelligent" form for computer-based decision models) to knowledge repository, data mining and knowledge discovery, knowledge-based systems or even the buzz form business intelligence.

It is not actually the fault of the technology; we consider that the fault belongs to the ideology in using these technologies. We should not believe that processing data means providing intelligent data or that if we deposit knowledge we actually do. It cannot be the success of one technology because we had so many examples that failed in the past. All that ideology of decision-support systems failed because the implemented models didn't prove to be interactive: they had a major problem in interacting with the user because humans and computers don't understand symbols in the same way. So the technology providers embedded all kinds of data structures and representation formats in order to get more close to the user. But, at the moment, the computer-based models and the user cannot constitute both in a system as they cannot be integrated.

The output for a decision model should be information capable to activate action. The respective action should be the user's desire, wish, or belief and the information must refer to that. It doesn't make any sense to try to emulate all kinds of desires, wishes, and beliefs because this is obviously unattainable. The implemented models must provide a way to the user to specify his/her beliefs or wishes and in that way the computer-based models are able to transmit actionable information.

## 28.2.2 The Ideology of Knowledge Representation

The ideology of representing knowledge belongs to Artificial Intelligence and it had interesting approaches during the time. The knowledge level was firstly proposed by A. Newell who considered that in order to develop intelligent models there is a need to represent a knowledge level which is above the physical level. This idea led to studies being conducted in developing inference engines but also provided a basis for data modeling. The knowledge representation principle consider that differentiating knowledge by structuring it conforming to different roles that pieces of knowledge may have could provide easier access to data.

Knowledge representation principle offered different paradigms for programming languages: frame-based or object-oriented. Whatever be the paradigm its philosophy, tools, languages and data structures try to virtually connect data structures to knowledge from a reality known as discourse universe. At the physical level it is all about data structures and algorithms. Developing models transforms in implementing computer-based applications in a programming language. The Artificial Intelligence critics said always that implementing models by specifying all the existential conditions doesn't mean Artificial Intelligence: it is all about data structures, procedures and control structures.

On the other side, knowledge is pretty difficult to define. Knowledge is something that is always true and as Einstein said: the truth is what passes the test of experience. Representing the world and its existence also has connections in philosophy by the approaches concerning ontology.

The literature from the knowledge-based systems technology field provided different approaches. The most pragmatic that we consider are inference engines and ontologies. All the principles and philosophy that other technologies promote are good but these two technologies are more oriented on knowledge and representing concepts. The expert systems technology, the fuzzy systems technology, and neural networks technology are more oriented to problem solving. This aspect of problem solving was conducted to consider only some sort of problems that were being satisfactory for Artificial Intelligence: diagnosis, monitoring, explanation. In fact what proves to be important in developing these models isn't the knowledge that they embed but the algorithms used to manipulate that knowledge. All the know-how used on knowledge is embedded in a hidden form of procedural knowledge like: forward chaining and backward chaining, fuzzy algorithms, genetic algorithms, and stored procedures.

So, in order to actually implement declarative knowledge we must have access to larger represented datasets, existential knowledge that always is true like ontologies. The technology to be used must have implemented general forms of inference on the concepts from ontologies. All the conditions, constraints, and controls on concepts must be edited by the user, by editing the metamodel of inference.

So we can formulate another principle for knowledge representation: knowledge engineering must provide the metamodel used in inferring and instantiating knowledge. Instantiating knowledge from the values specified for control data structures must be realized by the users of the metamodel: inferring knowledge doesn't belong to knowledge engineering, but the metamodel does.

The specialized knowledge belongs to users and the user location represents the place where knowledge application is performed. General knowledge and its forms of theories, axioms, and ontology are very little supported by technologies and usually are confounded with specialized knowledge. Even a mathematical model starts from a set of assumptions in order for a problem to have a solution. The mathematical models aren't a form of general knowledge; they are a form of procedural knowledge.

We ask what is the metamodel for decision problems. Everyone makes decisions in terms of IF...THEN...ELSE rules. The metamodel of IF...THEN...ELSE rule must be implemented in technology and the metamodel editing must be realized by the user.

## 28.3 Problem Solution

The main solution that we provide in assuring the generality for decision-models is considering the metamodel being one of importance and not the specific conditions that characterize a certain problem.

From the informatics point of view we propose the following technologies: relational databases (to organize and store data), ontology (to organize meanings), inference engines (to organize constraints, conditions, rules defined by the user).

Inference engines are a technology that uses production rules, meaning that they provide an environment to infer derived knowledge from facts. We imagined a scenario in which the user can edit the decision rules used for inferring facts from a decisional problem or situation. She/he uses an environment that permits editing, using and questioning the individuals represented in ontology. The individuals are the instances from a database or another source for data written in a format that an ontology editor may recognize.

Supposing that at one moment the user say: <<“IF property a for class A is greater than 2 and property b of class B is greater than 3 THEN C”; “IF C and property c of class D is greater than 5 THEN E”>>; the user must have functionalities provided by technology in order to specify this and actually infer the new facts (new properties or new classes). After doing so, maybe she/he wants to query the ontology which individuals respect her/his constraints. By implementing the metamodel without specifying the control on data, the answer provided will be semantically enriched if the rules edited by the user are fired on ontology’s individuals.

The decision model remains valid even if in the near future the constraints actually change and the user writes his/her new rules of thinking without changing the metamodel.

From our experience in modeling decisions during the PhD program and two research grant programs that we manage we observed that all the literature related to decision modeling refers to analytical models usually based on mathematics or expert systems models based on the so-called knowledge captured from experts. In fact, in practice these models are not used. Every idea, philosophy or technology proves its usefulness by using it in practice. But we cannot say that the analytical models or expert systems models are used in practice because these models have an embedded form of knowledge hidden in some sort of technology. In our opinion, for the decision-makers to use some models it is necessary that they can specify the constraints, the inferring chains of rules without knowing the intrinsic part of the metamodel.

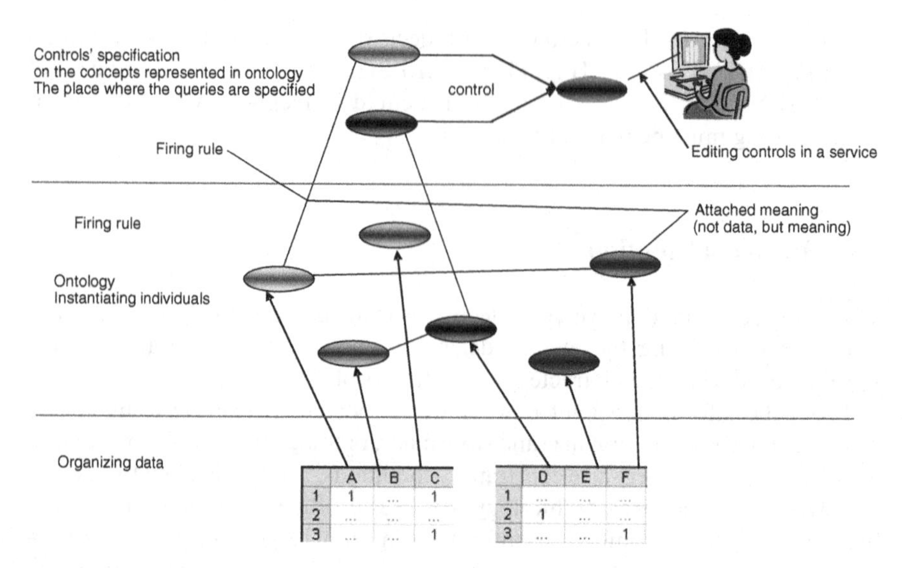


**Fig. 28.1** The places where the control on concepts is specified

The actual infrastructure to organize data and have access on information is presented in Fig. 28.1.

The metamodel must be represented on the concepts and the metamodel editing must be performed by the user.

A basic question is why do we use the term concept? A concept is better suited to a class of objects that contain instances; it describes an entity, some would say. Not exactly: our opinion is that a concept is a combination of instances that characterizes different entities. And they should be different in order to provide a special meaning in the sense of a concept, otherwise is a class or a table or an entity.

The expert system model technology represented a good solution for combining concepts in IF…THEN…ELSE rules which represented the key of this technology and the best suited way to represent the specialized knowledge provided by an expert. The major disadvantage was the fact that an instance from the database source corresponds to a single concept. The mapping between the concepts and data was at the instance level which implies a redundancy of concepts because the concepts aren't organized as data. Organizing concepts otherwise than in the form of IF…THEN…ELSE rules must provide a way to represent information, to attach meaning in a way that humans do. Humans recognize, classify and evaluate messages in order to perceive and to attach meaning. This would have to be a method to organize concepts to represent some sort of meaning and this kind of method uses semantic technologies models.

### *28.3.1 An Example to Justify the Solution*

In the following we will present a problem and a possible conceptualization to this problem to sustain our ideas.

In order to provide information there must be a mapping between different data sources stored usually in relational databases. Considering the case of a simple decision anyone can assume that the data sources are heterogeneous. If we take as example the case of deciding if an asset is impaired, data stored in the organization's database aren't sufficient. We must take into account the market value for the asset or the so-called capitalized value or utility value. Assuming that for finding this value someone would choose the Internet, the data found for the respective value would have to be saved in a file format that another program should "understand", which for an application means read and perform a predefined function, calculus or execute a procedure on that data. For that data to be combined with the data derived from a database they must share a common ontology. Usually for data extracted from a database it is recommended the RDF [13] format, which is a description in the form of triples, each consisting of a subject, a predicate and an object. We exemplified this problem by using an external table in PostgreSQL which stores data resulted from a query performed on a databases, which contains more tables and by using an .xml file as we would save the data from a website.

To transform the two different formats in which we have stored data about assets we used D2RQ [1] which is a declarative language to describe mappings between relational database schemata and OWL/RDFS ontologies. The mappings allow RDF applications to access the content of huge, non-RDF databases using Semantic Web query languages like SPARQL. The mapping is actually quite simple to make and implies two command lines in the MS-DOS prompt.

After that, we combined the two files simply by eliminating the redundant triples.

The standard language recommended to build ontology is OWL [14]. There are a lot of ontology editors but the most known tool is Protégé Editor which is open-source (there is a big community who use this tool to develop ontologies and to build plugins). In order to transform the file from RDF format to the OWL format we used SWOOP [6] which is another open-source tool that supports the RDF format but it doesn't have so many plugins as Protégé has. We simply saved the file in the OWL format.

Finally we were able to see the database relational schema as ontology. We can observe that the ontology refers to a super class called THING, which belongs to Protégé, and that we have a single class whose name corresponds to the table's name from relational database schema; the class has properties whose values are the fields' names and individuals whose names corresponds to the table's attributes values. We can say in this moment that we managed to combine two different data sources that share the ontology defined by the table source data and that this data are the resulted stored in one table from a query addressed to an initial multiple-tables database which had to respect some integrity constraints.

In this moment it is good to ask ourselves if we have represented concepts. And a good answer is: not so much according to our above definition. We have data stored in individuals that have properties that belong to a class so we may say that the model could present the capacity to recognize and classify, but not to evaluate because the data represented has no attached meaning.

In order to evaluate this meaning for a computer W3C recommends SWRL Semantic Web Rule Language which permits editing the rules of existence for the data to be recognized in ontology. SWRL permits only editing the rules and not firing on them (capability that would permit adding a property value to an individual and in this way reaching the model's semantics).

Under the umbrella of the rule engines are a lot of tools but because Protégé is Java-based the Java engine recommended is Jess. The users' community developed also a JessTab Plugin which is available with the last version of Protégé. Jess is a rule engine that derives from CLIPS and uses Horn clauses to represent the rules. The users' community also uses bridges to map SWRL Rules into Jess rules but because this is not a trivial task we managed to do it without using a bridge. So we wanted to attach the meaning that for an asset to have the capitalized value under the net accounting value is to be impaired.

Once the relevant OWL concepts and SWRL rule have been represented in Jess, the Jess execution engine can perform inference. As rules fire, new Jess facts are inserted into the fact base. Those facts are then used in further inference. When the inference process completes, these facts can then be transformed into OWL knowledge, a process that is the inverse of the mapping mechanism.

After firing the presented rule the facts stored in Jess are the same but they have an additional slot named "impairment" if the condition specified in the rule is true. In our case, only one fact presents this aspect and we present the fact in Fig. 28.2. There exists a slot named "depreciere" (impairment) that we defined in OWL ontology as a property of vocab0:mijlocfix_nrinventar (fixed_asset) with the accepted values "da" (yes) and "nu" (no). The slot does not belong to the ontology provided by relational database schemata; it was defined by us and its value is attached to the individuals only if the rule proves to be true.

What we deliberately omitted to say is that we used the D2RQ tool especially because it provides a way to visualize the individuals from ontology in a browser, so in order for a user to extract useful information it is sufficient to access the n3 format of the ontology in a browser. The D2R Server address is http://localhost/2020 and it permits viewing data in n3 and in the RDF format

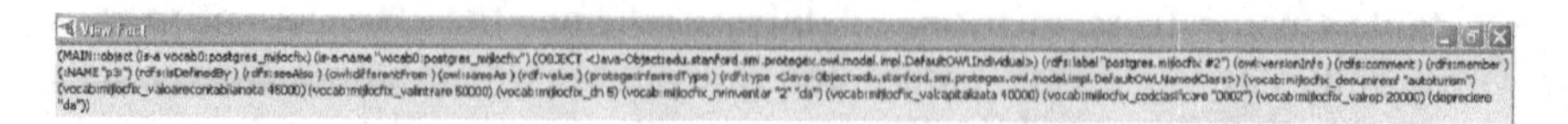

**Fig. 28.2** The fact that contains the additional slot "depreciere" (impairment)

and more important querying the ontology by using SPARQL. Protégé doesn't permit saving the OWL format back into n3 format once the OWL ontology has individuals defined by using rules.

## 28.4 Conclusions

Better decisions means improving the information provided. We tried in this paper to outline that the knowledge of acting from the decision models must be implemented to improve information, to actually provide better information to the user. Using the decision models for the problem-solving task proved to be a failure in the past. It resulted in very static models, non-adaptive ones, with no utility for the user because they captured a kind of model that imposes performing an action by the decision maker in the form of transferring knowledge from the model to the human being.

We do not try to argue that this is the best model that we want to recommend but in the moment the technology and the technology's producers aren't too concept-oriented in the business domain although there is a clear wish to solve the problem of interoperability. Only for our little example we used no more than six tools to manage to transmit a simple message and we are not saying that this is a sure way to represent meaning. The tools that we used have a lot of facilities that we possibly omitted to consider.

We provided this model oriented somehow on decisions due to the last standardization efforts in the field of information integration and due to the fact that we study decisions models as a special concerning in this moment.

We considered all the possible applications in practice: decision-support systems (this is a field where information is considered to be some aggregated data and not meanings organized in ontologies), information retrieval, knowledge engineering, knowledge-based systems project and management.

## References


1. Bizer C (2006) D2RQ V0.5 – Treating non-RDF databases as virtual RDF graphs. At http://sites.wiwiss.fu-berlin.de/suhl/bizer/D2RQ/, accessed 30 June 2007
2. Chandrasekaran B (1990) Design problem solving: A task analysis. Am Assoc Artif Intell 11(4):59–71, http://www.cse.ohio-state.edu/~chandra/ai-mag-design-ps.pdf
3. Chandrasekaran B (1988) From numbers to symbols to knowledge structures: AI perspectives on the classification task. IEEE 18(3):415–425, http://www.cse.ohio-state.edu/~chandra/From-numbers-symbols-knowledge. pdf
4. Chen H, et al. (2006) Toward a semantic web of relational databases: a practical semantic toolkit and an in-use case from traditional chinese medicine. In: ISWC 2006 Conference, http://iswc2006.semanticweb.org/items/ Chen2006kx.pdf
5. Fayyad U, Shapiro G, Smyth P (1996) From data mining to knowledge discovery in databases. American Association for Artificial Intelligence, Menlo Park, CA, www.aaai.org/AITopics/assets/PDF/AIMag17-03-2-article.pdf

6. Google (2007) Semantic Web Ontology Editor. At http://code.google.com/ p/swoop/ accessed 30 June 2007
7. Musen M (1998) Modern architectures for intelligent systems: reusable ontologies and problem-solving methods. In: AMIA Annual Symposium, Orlando, FL, http://smi.stanford.edu/smi-web/reports/SMI-98-0734.pdf
8. Newell A, Simon H (1963) GPS, A program that simulates human thought. In: Feigenbaum EA, Feldman J (eds) Computers and thought. (1995) American Association for Artificial Intelligence Press Edition, Menlo Park, CA, pp 279–293, http://www.cog.jhu.edu/faculty/smolensky/050.326-626/Foundations%20 Readings%20PDFs/Newell&Simon-1963-GPS.pdf
9. Protégé (2007) The Protégé Ontology editor and knowledge acquisition system. Stanford University, at http://protege.stanford.edu/, accessed 30 June 2007
10. Schreiber G, et al. (2000) Knowledge engineering and management – The CommonKADS Methodology. MIT Press, London, England, pp 403–418
11. Studer R, Benjamins VR, Fensel D (1998) Knowledge Engineering: Principles and Methods. Data & Knowledge Engineering 25(1–2):161–197 http://hcs.science.uva.nl/usr/richard/postscripts/dke.ps
12. Tang Z, MacLennan J (2005) Data mining with SQL server 2005. Wiley Publishing Inc, Indianapolis, Indiana
13. W3C, Resource Description Framework (RDF). At http://www.w3.org/RDF/ accessed 30 June 2007
14. W3C, Web Ontology Language (OWL). At http://www.w3.org/TR/owl-features/ accessed 30 June 2007

# Chapter 29
# A Computerized Solution for the Financial Diagnose of the SMEs

**Vasile Daniel Păvăloaia**

**Abstract** In the now-a-days business environment, when efficiency and profitability are two major aspects to be taken into account by any enterprise, including the small and medium sized enterprises (SMEs), interested in obtaining better economic results, the use of computerized tools for performing the financial diagnose on the activity of an enterprise is a must. From the multitude of information technologies that can be used in order to design such a tool, we choose Exsys Corvid; this software product is very reliable to use, allows integration with other programs (database query) and has its own browser to run the application. One of the paper's major aims is to illustrate a manner of computerizing the field of financial diagnosis for the case of a Romanian enterprise. Thus, the current study will be conducted by implementing into a real application the theory (from the specialized literature) and expertise (collected from experts) related to the field of financial diagnose and will utilize, for exemplification purposes, the financial ratios that are used in the process of a financial diagnose. The computerized solution can be easily implemented for the case of the SMEs, as well as for any other enterprise, as long as the bankruptcy prediction model used is the proper one.

## 29.1 Introduction

The financial phenomenons are characterized as being restless and they modify its structure under the impulse of the factors that generate the dynamics of social life. In light of this information, a system based on financial ratios should be extremely flexible and sensitive in order to appropriately observe and reflect the reality. As technology progresses rapidly, specific tools must be used for managing such a system that is made up of financial ratios, in order to speed up

V.D. Păvăloaia (✉)
Business Information Systems Department, Alexandru Ioan Cuza University of Iasi, Bdul Carol I nr.22, Iasi 700505, Romania
e-mail: danpav@uaic.ro

N. Mastorakis et al. (eds.), *Proceedings of the European Computing Conference*, Lecture Notes in Electrical Engineering 28,
DOI 10.1007/978-0-387-85437-3_29, 

the time consumed by the ratios commensuration. At the same time, it is required to improve the quality of decision in the field of financial diagnose [1]. This task is only accomplished by human experts that acquire knowledge throughout many years of practice. Therefore, we intend to capture such knowledge and implement it into the computerized solution.

From a theoretical point of view, the causes and factors that lead an economic entity to bankruptcy are not only of a financial nature, they can conduct to a low profitableness and solvency of the enterprise. The main information obtained from the literature, related to the field of financial diagnose, is that an enterprise will encounter major difficulties in fulfilling its relationship with the contractor's if it doesn't have a minimal rate of profitableness and a satisfactory level of liquidity.

Therefore, in order to ensure its survival (if not its growth), the enterprise must conserve its financial autonomy; otherwise, the investments that require supplemental financial resources cannot be fulfilled. The lack of profitableness implies an insufficient level of auto-finance capacity with respect to the required investment rate. On the other side, the lack of profitableness obliges the enterprise to become indebted as a low level of profitableness doesn't allow it to attract new capital from investors. This measure will lead to an increase of the enterprise expenditures, affecting its solvency and furthermore, leading to incapacity in terms of payment.

As a result, one may notice the importance of maintaining an economic equilibrium for the enterprise functionality which, mainly, resides in performing a profitable activity concurrent with monetary equilibrium – all of those actions are translated into a satisfactory level of liquidity. Having in mind the importance of profitableness -liquidity correlation required for the persistence of an enterprise, the specialized literature value the "health state of an enterprise" by including them into four groups, as illustrated in Fig. 29.1.

As an interpretation of the information included in Fig. 29.1, the enterprise that suffers from "transient malady" has a certain level of profitableness but encounters some liquidity problems. It is the case of first-years' enterprises, in full growth, whose investments in capital assets and working capital are higher than their auto-finance capacity. These economic entities are less exposed to the risk of bankruptcy if they can overcome the liquidity issues.

Whether they are young or mature, the enterprises that belong to the third category "chronicle malady" have profitableness issues related to either the difficulties in maintaining the market share or the high weight of fixed expenses

| Ratios / values | | *Profitableness* | |
|---|---|---|---|
| | | *positive* | *negative* |
| *Liquidity* | *positive* | 1 „in shape" | 3 „chronicle malady" |
| | *negative* | 2 „transient malady" | 4 „imminent ending" |

**Fig. 29.1** Profitableness, liquidity and the financial health of the enterprise

within the quantum of the total manufacturing expenses. When the total collected income covers the payment expenses and the investments are kept within a normal level, the liquidity ratio can be rated as satisfactory, allowing the enterprise to develop a profitableness activity.

An economic entity that is included in the fourth category "imminent ending" cumulates an insufficient profitableness and a total lack of liquidity ratio. Its survival can only be attained after profound reorganization, accompanied by a reinforcement of the incoming financial resources.

As a consequence, in the Romanian economic system, in order to perform an enterprise diagnose, it is required to test the liquidity rate and the profitableness of the economic entity. In the following chapters of the paper we will try to formulate and solve this matter.

## 29.2 Problem Formulation

As stated in earlier paragraphs of the paper, in order to perform an enterprise diagnose, it is required to test the liquidity rate and the profitableness of the economic entity.

### *29.2.1 Testing the Enterprise Liquidity Ratio*

In order to test the enterprise liquidity level there, a ratio system illustrated in Table 29.1 is used; the best rate of the liquidity ratios will be attained when an enterprise fulfills all the conditions imposed by the ratios (Table 29.2).

### *29.2.2 Testing the Enterprise's Profitableness*

Enterprise's economic and financial performances can be measured throughout N note that is obtained by averaging five ratios and the value of each of ratio is determined by using the following pattern (see Eq. 29.1):

**Table 29.1** The system of Liquidity ratios

| Ratios name | Formulas | Explanations |
|---|---|---|
| General liquidity rate (RLG) | = Ac/Dts | Ac = Current assets; Dts = Short term liabilities; |
| Low liquidity ratio (RLR) | = (Ac-St)/Dts | St = Reserves; |
| Immediate liquidity ratio(RLI) | = Disp/Dts | Disp = Liquidness; FTE = Cash flows; A = Anuity; |
| The rate of covering the payment engagements (Ro) | = FTE/ (A + $\text{Div}_{\text{plătite}}$ + Inv) | $\text{Div}_{\text{plătite}}$ = Paid dividends; Inv = Investment payments; |

Table 29.2 The interpretation of the ratios values

| Liquidity ratios | Obtained values | |
|---|---|---|
| | Favourable | Unfavourable |
| RLG | RLG >= 1 | RLG < 1 |
| RLR | RLR >= 0.8 | RLR < 0.8 |
| RLI | RLI >= 0.2 | RLI < 0.2 |
| Ro | Ro >= 1 | Ro < 1 |

$$\frac{\text{Value of } R_i \text{ ratio}}{\text{Default value of } R_i \text{ ratio}} \tag{29.1}$$

The default value of $R_i$ ratio is determined for a population obtained from the default enterprise and turns out to be the median of the enterprises that belongs to the same sector of activity. The general model of a score function that gives a mark for each particular case, is as follows (see Eq. 29.2):

$$N = a_1R_1 + a_2R_2 + a_3R_3 + a_4R_4 + a_5R_5 \tag{29.2}$$

The enterprise that has a perfect state according with a default state has a mark N computed as follows (see Eq. 29.3):

$$N = a_1 + a_2 + a_3 + a_4 + a_5 \tag{29.3}$$

Building and applying such a model may look like a simple task but its relevance was acknowledged by the empiric tests which were conducted in order to verify the statistic method.

$$Z = 0.24X_1 + 0.22X_2 + 0.16X_3 - 0.87X_4 - 0.10X_5 \tag{29.4}$$

The Conan-Holder model had been statistically tested and it is associated to a bankruptcy probability in a score function [2]. The functions suggested for the industrial (see Eq. 29.4) and manufacturing(see Eq. 29.5) sectors have the following pattern:

$$Z = 0.0197X_3 + 0.0136X_2 + 0.0341X_6 + 0.0185X_7 - 0.0158X_8 - 0.01222 \tag{29.5}$$

The purports [3] of the parameters used in the above models are explained in Table 29.3.

The obtained score value can be explained by using the interpretations showed in Table 29.4.

In order to test enterprise's liquidity and profitableness ratios, the input data will be taken over from a spreadsheet file that has the structure shown in Fig. 29.2.

**Table 29.3** The parameters used in the Conan-Holder model

| Parameter | Formulas | Explanations |
|---|---|---|
| $X_1$ | = EBE/Dt | EBE = Gross operating excess; Dt = Total liabilities; Cperm = Permanent capital; At = Total asset; Disp = Money liquidity Chfin = Financial expense; CA = Turnover (except VAT); Chpers = Employment expenses; Vad = Added value; Autofin = Auto-financing; Bt = Total Balance sheet; NFR = Operating capital |
| $X_2$ | = Cperm/At | |
| $X_3$ | = Disp/At | |
| $X_4$ | = Chfin/CA | |
| $X_5$ | = Chpers/Vad | |
| $X_6$ | = Autofin/Bt | |
| $X_7$ | = EBE/Bt | |
| $X_8$ | = NFR/CA | |

**Table 29.4** Score functions value interpretation for different activity sectors

| Value interpretation | Manufacture | Commerce |
|---|---|---|
| Good shape | Z >= 4 | Z >= 20 |
| Danger | Z < 4 | Z < 20 |

| | A | B | C | D | E | F | G |
|---|---|---|---|---|---|---|---|
| 4 | | | | | | | |
| 5 | Parameter | Ac | Dts | St | FTE | A | Divp |
| 6 | Value | 73492284 | 79840528 | 40522862 | 100000 | 20000 | 132450 |
| 7 | | | | | | | |
| 8 | Parameter | Inv | Disp | Rec | At | Rnet | Cpr |
| 9 | Value | 500000 | 6166480 | 17057763 | 1064528882 | 5213587 | 25669712 |
| 10 | | | | | | | |
| 11 | Parameter | d | CA | EBE | Dt | CP | Chfin |
| 12 | Value | 10,5 | 172907428 | 9050306 | 942642 | 26612354 | 10851112 |
| 13 | | | | | | | |
| 14 | Parameter | Chpers | Vad | AF | Bt | NFR | |
| 15 | Value | 10995878 | 20466383 | 12762482 | 2129057764 | 73492284 | |

**Fig. 29.2** The spreadsheet file containing the parameter's values

### *29.2.3 Inferring the Recommendations*

If the enterprise is diagnosed as "in shape", the user of the computerized solution will be prompted to choose whether he/she wants to find out about the recommendations and solutions he/she has in order to maintain and improve the enterprise performances. In this case, the ratios whose values are near to the normal limit are tested in order to identify the actions that can be undertaken so that the ratios don't run over their normal limits.

When the enterprise is diagnosed with any of the other states ("transient malady", "chronicle malady" or "imminent ending") the situation is considered to be critical. In this case, the user will be prompted to find out the causes that led to this dangerous situation. This action allows the analyst to identify the threats and weaknesses that generated the abnormal functionality of the enterprise.

## 29.3 Problem Solution

Exsys Corvid provides a powerful environment for developing Web-enabled Knowledge Automation Systems for a wide range of decision-making problems. This software allows the logical rules and procedural steps used by an expert to reach a decision, to be efficiently described in a manner that is easy to read, understand and maintain. Corvid converts logic to a form that the underlying Runtime Inference Engine can process in order to emulate the questions, process, and recommendations of the expert in an interactive session, allowing the end-users to interact with the system as if they were talking to the expert, to produce situation-specific recommendations and advice on a wide range of subjects [5].

The computerized solution, implemented with Exsys Corvid, took advantage of all the above features of the software that, throughout its rule base, queries an external spreadsheet file by using a friendly interface. Thus, the main aim of the solution is to help the financial analyst run a financial diagnose on an enterprise only knowing the values of certain financial parameters (shown in Fig. 29.1). The computerized solution has two goals:

1. To establish, based on the input information, the financial state of the enterprise (one of the four category illustrated in Fig. 29.1)
2. To reveal the causes that generated the particular financial state of the enterprise

All the information included in Tables 29.1to 29.5 was implemented into Corvid under the form of variables, logic and command blocks, all of them forming the rule base of the system.

### *29.3.1 The System's Variables*

All the elements that are used in the formulas presented in Tables 29.1 and 29.2 were implemented as Corvid numeric type variables. Their values were obtained by querying the spreadsheet file shown in Fig. 29.2. From the specialized literature and also from the domain expertise, all the possible recommendations were implemented as Corvid confidence type variables. Some information was required from the user and thus, several variables were inputted as static list type variables. The system has a total of sixty four variables of the three types mentioned above.

When working with numeric type variables, for best results, it is required an integration of the expert system with an external database that contains the elements that the ratios system uses. This action can be best accomplished by using a spreadsheet file as an external database. Among the reasons that brought us to the above statement we only stress two of them: the ease of working with this type of database and also the availability of the spreadsheet programs among computer users. In this manner, the "raw" data provided by the spreadsheet file would be processed by expert systems prototype.

### 29.3.2 Logic Blocks

The reasoning base of the system is formed, in Corvid, by the logic blocks. A logic block will contain a certain combination of variables and it must obey the IF-THEN-ELSE reasoning control. For instance, in Fig. 29.3, we illustrate the logic block that contains the reasoning attached to the profitableness ratio specific to the commerce activity. On the first line it is defined that the logic block applies only when the sector of activity is commerce; on the second line the formula is implemented – the values for the parameters are taken from the external spreadsheet file; the third line contains the recommendation: good level of profitableness. The fourth line contains the specifications for the case when the profitableness has an inadequate value (line five).

In the same manner were implemented all the required blocks containing the reasoning system. Another logic block, very important, that contains the reasoning for all the Liquidity ratios (illustrated in Table 29.1) can be viewed in Fig. 29.4.

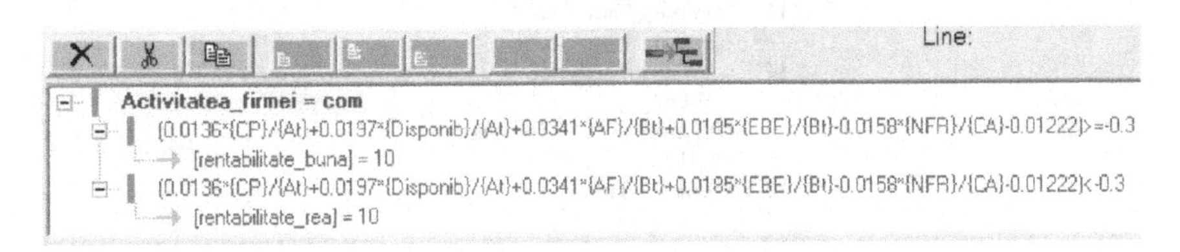


**Fig. 29.3** The logic block containing the profitableness ratio specific to the commerce sector

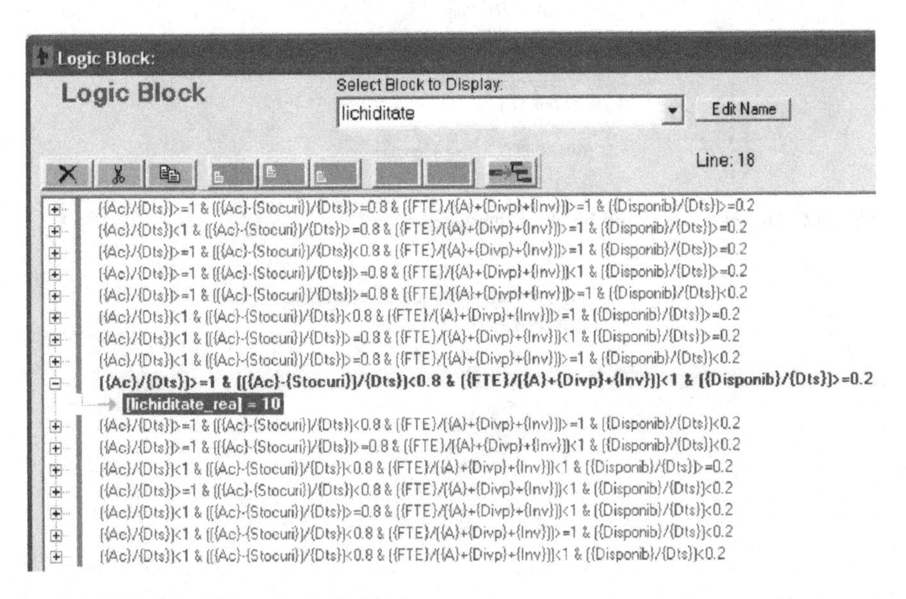


**Fig. 29.4** The Liquidity ratios logic block

### 29.3.3 Command Blocks

The Corvid inference engine runs the logic blocks in the order specified in the command block and infer the recommendations, namely the financial diagnose. Corvid allows both the backward and forward chaining when it comes to setting up the inference strategy.

For exemplification reason, in Fig. 29.5, it is illustrated the main command block of the system that sets up the strategy applied for the functionality of the system.

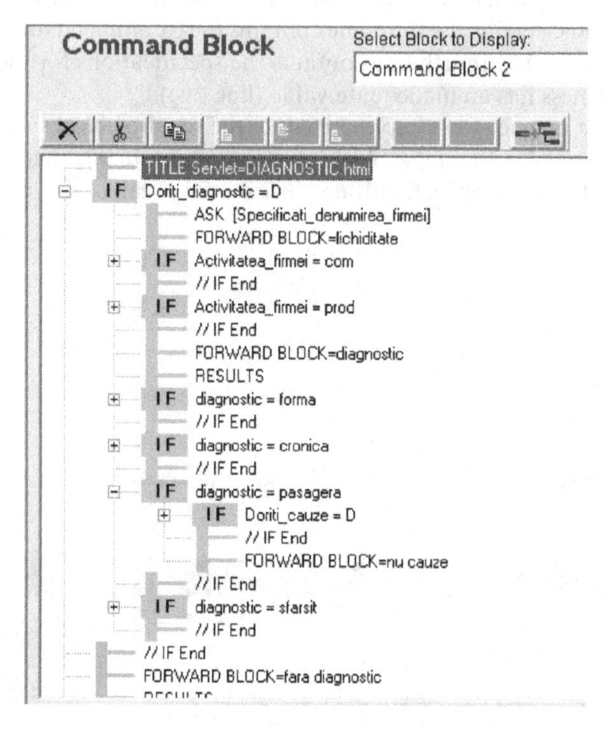


**Fig. 29.5** The main command block of the system

**Fig. 29.6** The main screen of the system

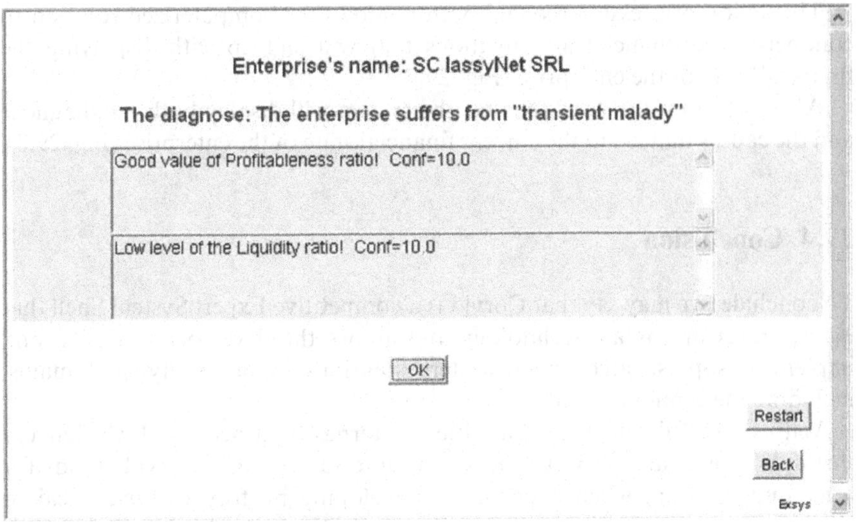

**Fig. 29.7** The screen containing the financial state of the diagnosed enterprise

### *29.3.4 Testing the Computerized Solution and Displaying the Results*

Once launched, the system displays the main screen (Fig. 29.6) containing the name of the system and the OK button.

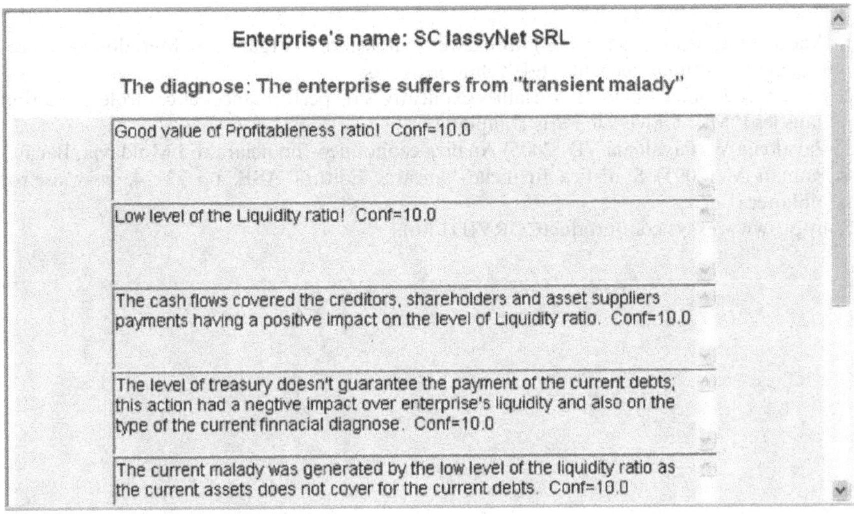

**Fig. 29.8** The results screen containing both the financial diagnose and the causes

The action of pressing the OK button allows the computerized solution to commence the financial investigations that will end up with displaying the financial state of the enterprise (Fig. 29.7).

Also, based on the user's choice, the system will display both the diagnose and the causes that led to the current financial state of the enterprise (Fig. 29.8).

## 29.4 Conclusion

To conclude, we may say that Corvid is a competitive Expert System Shell that incorporates up-to-date technology and allows the developer to design and implement sophisticated expert system solutions for a variety of domains, including the economic field.

With powerful integration abilities (external data access), Java delivery through applets and servlets and strong interface design, Corvid stands for a leading software when it comes to developing prototypes for a local or Web environment. Therefore, one of the future directions for the development of the current research would be to incorporate it into a web page and make it available throughout the Internet. In this way, the current solution could be widely used by a variety of firms to test their financial performances.

The field of financial analysis and diagnose can be very easily computerized with the Expert system technology due to its abilities to interpret, analyze and infer recommendation and not just determine the value for a certain ratio.

## References

1. Andone I, Pavaloaia VD (2004) Modelarea cunoasterii in organizatii. Metodologie obiectual pentru solu ii inteligente, Ed. Tehnopress, Iasi
2. Conan D, Holder M (1979) Variables explicatives de performances et controle de gestion dans les P.M.I. Universite Paris Dauphine
3. Pavaloaia W, Pavaloaia VD (2005) Analiza economico-financiara. Ed.Moldavia, Bacău
4. Roman M (2005) Statistica financiar-bancara. Editura ASE, pp 23–24, www.ase.ro/biblioteca
5. http://www.exsys.com/productCORVID1.html

# Chapter 30
# DELP System: Tracking Deadlocks and Phantoms in Databaseas

**Cristea Boboila, Simona Boboila, and Constantin Lupsoiu**

**Abstract** Database systems are the core of applications from various fields. In many of these, the well-functioning of databases can be extremely important, even critical. Deadlocks and phantoms are some of the problems that may appear in database systems, leading to information loss, with a highly detrimental impact. This paper presents DELP (DEadLocks and Phantoms), a tracking system used to study different database access scenarios in which deadlocks and phantoms may appear.

## 30.1 Introduction

A database transaction represents a sequence of operations executed as a whole resulting in an interaction with a database management system. Considered independently from other transactions, a transaction should demonstrate coherence and reliability in all situations. On the other hand, when several transactions are running simultaneously, issues like deadlocks [1, 5, 6, 7] or phantoms [8, 9] are likely to appear.

In databases, deadlock refers to the situation when two processes are waiting to access a record, but they are blocked by each other; each of them waits for the other to release the lock on some record. The phantom phenomenon may occur when one transaction establishes a set of records as the result of a query and other concurrent transactions insert or delete records which would have been part of that set [11].

This paper presents the DELP (**DE**adLocks and **P**hantoms) system, a tool for analyzing how deadlocks and phantoms appear in a database management system. We present the system anatomy as well as the experimental procedure, and give a comprehensive description of current experimental results.

C. Boboila (✉)
Faculty of Mathematics and Computer Science, University of Craiova, Romania
e-mail: boboila@central.ucv.ro

N. Mastorakis et al. (eds.), *Proceedings of the European Computing Conference*, Lecture Notes in Electrical Engineering 28,
DOI 10.1007/978-0-387-85437-3_30, 

The rest of the paper is structured as follows: Section 30.1 presents related work in the area. Section 30.2 describes the system's structure and implementation aspects. Section 30.3 synthesizes the results obtained while carrying out experiments at different isolation levels. Section 30.4 concludes the paper.

## 30.2 Related Work

Database management systems have benefited from extended attention in the current literature. Our work adds to the research effort of tracing and preventing integrity constraints violations in transactional systems [2, 10, 13]. Moreover, several studies enlarge upon the effect of queries and updates during the interaction with the database system [3, 4, 14].

Deadlocks are among the main causes that prevent a correct functioning of systems in which several entities have to act concurrently. In a transactional system, simultaneous actions coming from many users may lead to deadlocks. A constant interest in recent research has been given to methods of database functionality optimization by preventing deadlocks [1, 5, 6, 7], as well as to methods of accurate informational data gathering by eliminating the risk of phantom occurrences [8, 9].

In contrast with most of the current research, where theoretical surveys are presented, our paper targets a more practical perspective, and describes an environment that uses actual database technologies to trace deadlocks and phantoms.

## 30.3 System Description

The DELP framework and the interaction flow between components are presented in Fig. 30.1. The *Caller Module* communicates with the *DB Access Module* by invoking the procedure for database access.

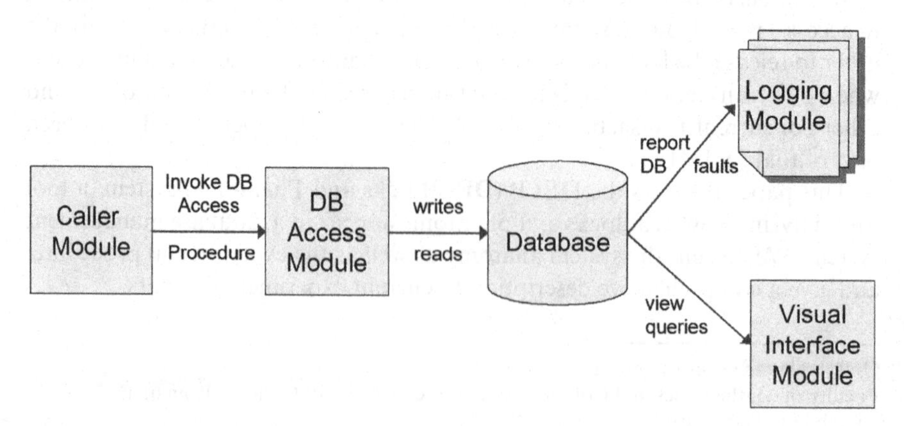


**Fig. 30.1** System anatomy

Next, the DB Access Module directly interacts with the database, performing writes (updating and inserting) or reads (selecting). The *Visual Interface Module* delivers the state of the database to the user, upon user interrogation. Furthermore, the *Logging Module* records a series of statistics about the number of deadlocks, lost updates and phantoms that have occurred during the interaction with the database.

The Caller Module has been designed as a Java Client, which ensures that our implementation is cross-platform and provides a flexible database access mechanism. The Caller Module invokes the DB Access Module by directly calling the access procedure to the database.

The DB Access Module runs the SQL scripts which perform reads or writes on the database. The SQL scripts may be provided by the user as an input, or may reside in the shared space of the DELP system. In order to intercept the Java Client's requests, the database management system runs the Sybase server [12].

## 30.4 Experiments and Results

We have run a series of experiments on a prototype database storing information about an on-line store. The e-store database contains prices and in-stock quantity associated with each item as well as other data about the products.

During our experiments, we have focused on how the state of the database changes when several transactions attempt to update and query it simultaneously. The DB Access Module runs the SQL scripts that interact with the database. For efficiency purposes, we have implemented the SQL scripts as stored procedures which reside on the server side (the Sybase server).

In order to achieve transaction consistency, choice of which lock-based concurrency control mechanism is used by the database management system is crucial. At SERIALIZABLE isolation level, the transactions are isolated, and occur as if they were running one after the other. This way, the locking mechanism prevents consistency violations from happening. Due to the locking control, transactions may block each other and end in deadlock. One of the concurrent transactions must be aborted and rolled back. Our experiments in Section 30.4.1 study an illustrative scenario of deadlock occurrence in a database management system.

The next isolation level that we focus on is REPEATABLE READS, where the read locks are placed only on the records returned by a SELECT statement, and which are present in the database at a specific time. The read locks are not on the whole range requested with the SELECT statement, thus allowing new records that match the WHERE condition to be inserted in the database and generate phantoms (see Section 30.4.2).

In order to analyze the behavior of concurrent transactions, we have run multiple transactions in parallel during our experiments. The following results are obtained with 20 transactions running concurrently in our testing environment.

### *30.4.1 Deadlock Occurrences*

The stored procedure updates the price of one of the products in the e-store. Each manufacturer or brand has a specific policy of price update. For example, the manufacturer may decide to increase the price of its products when the number of items on stock falls below a certain threshold.

Therefore, our stored procedure must first count the number of items from the e-store with a read operation. Next, the same stored procedure modifies the price according to the policy. The following example presents the update procedure for a pottery manufacturer:

```
CREATE PROC UpdatePrice AS
DECLARE @count FLOAT
BEGIN TRANSACTION
SELECT @count =
(SELECT COUNT(*) FROM Products
WHERE Category = 'pottery')
IF @count   < 1000 THEN
UPDATE Products SET price = 1.05*price
WHERE Category = 'pottery'
END IF
COMMIT TRANSACTION
```

The update action (database write), taken alone, poses no problems to the database management system. In the case presented previously, the write is preceded by a read (the select operation), and therefore the time interval that may interpose between select and update can be a source of deadlocks.

Since we have set the isolation level to be SERIALIZABLE, any of the 20 transactions can obtain the read locks. The conflict appears when one of the transactions wants to get the write lock to do the update. Neither of the transactions can obtain the write lock, because other transactions already hold read locks on the same database records, due to the COUNT statement. Since read locks are long at this level of isolation, none of the transactions will release the read lock until commit. Therefore, a deadlock appears and one of the transactions will eventually abort and give up the read lock.

### *30.4.2 Phantom Occurrences*

The next experiment studies database faults that may appear at REPEATABLE READS isolation level. The DB Access Module runs two different stored procedures that are executed independently from concurrent transactions. One of the stored procedures performs the price update:

Table 30.1 Deadlocks and phantoms

| Procedure Name | Isolation Level | Fault Type | Number of Faults |
|---|---|---|---|
| UpdatePrice | SERIALIZABLE | Deadlock | 34 |
| SetPrice/ReadPrice | REPEATABLE READS | Phantom | 15 |

```
CREATE PROC SetPrice AS
BEGIN TRANSACTION
UPDATE Products SET price = 1.05*price
WHERE Category = 'pottery'
COMMIT TRANSACTION
```

The second stored procedure carries out successive reads on the same data, in order to observe whether phantoms have appeared:

```
CREATE PROC ReadPrice AS
BEGIN TRANSACTION
SELECT price FROM Products WHERE Category = 'pottery'
SELECT price FROM Products WHERE Category = 'pottery'
COMMIT TRANSACTION
```

In this case, we observe that our database is prone to phantoms, because the update can take place in-between the database reads.

Table 30.1 is a synthesis of the number of deadlocks and phantoms that occurred in the system when 20 transactions were running concurrently at the two isolation levels discussed.

## 30.5 Conclusions

The large applicability of database management systems in several fields that deal with information storage and processing has favored the perception of databases as one of the most important and demanding topics in today's applications and research. In this context, the well-functioning of the database system is imperative for accurate data.

This paper presents DELP system, a monitoring tool for database performance in different scenarios where concurrent transactions are accessing the database. Using DELP, the user may test the transaction in an experimental environment, and determine whether their interaction is detrimental, before running them in the actual storing environment. We are presenting a comprehensive experimental study, which points out the importance of choosing wisely the isolation levels for SQL databases. A thorough analysis of the situations in which deadlocks and phantoms appear can lead to significantly better results in terms of information management in database systems. The current work focuses on tracing deadlocks and phantoms in the database management system. Future extensions include methods to automatically detect types of faults that may appear at other SQL isolation levels (e.g. lost updates, etc.).

## References

1. Chen X, Davare A, Hsieh H, Sangiovanni-Vincentelli A, Watanabe Y (2005) Simulation based deadlock analysis for system level designs. In: 42nd Design Automation Conference, Anaheim, CA
2. Decker H (2002) Translating advanced integrity checking technology to SQL. Database integrity: Challenges and solutions. Idea Group Publishing, Hershey, PA
3. Elkan C (1990) Independence of logic database queries and update. In: 9th symposium on principles of database systems, ACM SIGACT-SIGMOD-SIGART symposium on Principles of database systems, pp 154–160
4. Godfrey P, Gryz J, Zuzarte C (2001) Exploiting constraint-like data characterizations in query optimization. In: ACM SIGMOD international conference on management of data, pp 582–592
5. Kaveh N, Emmerich W (2001) Deadlock detection in distributed object systems. In: Joint proceedings of the 8th European software engineering conference and the 9th ACM SIGSOFT symposium on the foundations of software engineering, Vienna, Austria, pp 44–51
6. Kobayashi N (2000) Type systems for concurrent processes: From deadlock-freedom to livelock-freedom, time-boundedness. Springer, Berlin Heidelberg, vol 1872/2000, p 365
7. Krivokapic N, Kemper A, Gudes E (1999) Deadlock detection in distributed database systems: A new algorithm and a comparative performance analysis. VLDB J 8(2):79–100
8. Rakow T, Gu J, Neuhold E (1990) Serializability in object – oriented database systems. In: 6th international conference on data engineering, pp 112–120
9. Reimer M (1983) Solving the phantom problem by predicative optimistic concurrency control. In: 9th international conference on very large data bases, pp 81–88
10. Ross KA, Srivastava D, Sudarshan S (1996) Materialized view maintenance and integrity constraint checking: trading space for time. In: ACM SIGMOD international conference on management of data, pp 447–458
11. Silberschatz A, Korth HF, Sudarshan S (2005) Database system concepts, 5th edn. McGraw-Hill, Boston, MA
12. Sybase SQL Server$^{TM}$ Reference Manual. http://download.sbase.com/ pdfdocs/ srg1100e/sqlref.pdf. Accessed: March 15th, 2007
13. Vielle L, Bayer P, Küchenhoff V, Lefebvre A (1999) 0Checking integrity and materializing views by update propagation in the EKS system. In: Materialized views: techniques, implementations, and applications, MIT Press, Cambridge, MA
14. Wetzel G, Toni F (1998) Semantic query optimization through abduction and constraint handling. In: 3rd international conference on flexible query answering systems, pp 366–381

# Chapter 31
# Concurrency Control for Multilevel Secure Distributed Real-Time Databases

**Valuchandhar Vadivelu, and S. Albert Rabara**

**Abstract** The technology evaluation of database management system has integrated with significant improvement in distributed computing. The requirements for multilevel security have been shown to conflict with database integrity and consistency. Concurrency control ensures that individual users meet consistent states of the databases with security. Concurrency control is an integral part of the database systems to manage the concurrent execution of operations by different transactions in the same data or distributed data with consistency. In this paper, it is proposed a model to enhance the performance of concurrent transactions for multilevel security for distributed database. This model reduces the data access time and wait time for every transaction monitored by the sub-query analyzer. The simulation study reveals the effective enhancement of performance for concurrent transactions with different levels of security.

## 31.1 Introduction

The importance of databases in modern businesses, public and private organizations, banks, educational institutions and in general day-to-day applications is already huge and still growing. Many critical applications require databases. These databases contain data of different degrees of importance and confidentiality, and are accessed by a wide variety of users. Integrity violations for a database can have a serious impact on business processes and disclosure of confidential data. Traditional security provides techniques and strategies to handle such problems with respect to database servers in a non-distributed environment. In a global enterprise the database access is required round the clock. The Database Management Systems (DBMS) has coincided with significant developments today in distributed computing and

V. Vadivelu, (✉)
Department of Computer Science, St. Joseph's College (Autonomous),
Tiruchirappalli– 620 002, Tamilnadu, India

N. Mastorakis et al. (eds.), *Proceedings of the European Computing Conference*, Lecture Notes in Electrical Engineering 28,
DOI 10.1007/978-0-387-85437-3_31, 

concurrent access technologies and this becomes the dominant data management tools for highly data intensive applications. A distributed database is a collection of multiple logically interrelated databases distributed over a computer network. A distributed DBMS is defined as the software system that permits the management of the distributed database and makes the distribution transparent to the users [1]

In this scenario, Data Base Management Systems (DBMS) are designed to meet the requirements of performance, availability, reliability, security and concurrency. The recent rapid proliferation of Web-based applications and information systems has further increased the risk exposure of databases in a distributed environment and hence data protection is more crucial than ever. The data needs to be protected not only from external threats but also from insider threats [2]. The solution to data security is classified into three major categories: (1) the protection of data against unauthorized disclosure, (2) prevention of unauthorized and improper modification and (3) prevention of denial of services. In addition to the different levels of security, the DBMS is to provide concurrent access of highly available data in the presence of large and diverse user populations. Therefore it is obvious that multilevel security must be provided to the DBMS mainly on distributed environment. Concurrency control is an integral part of the database systems. It is used to manage the concurrent execution of operations by different transactions on the same data with consistency. Several methods have been proposed to provide secure concurrency control to achieve correctness and at reduced cost of high security level transactions. One of the most important issues for concurrency control in the MLS database system is the cover channel problem [3]. It naturally comes due to the contention for the shared data items by transactions executing at different security levels. The most common instances of totally ordered security levels are the Top-Secret(TS), Secret(S), and Unclassified(U) security levels encountered in the military and government sectors. In this paper, it is proposed a model to enhance the performance of concurrency for the Multilevel Secure Distributed Database System. This model allows users to access a database concurrently from geographically dispersed locations through use of a concurrency control locking algorithm.

The paper is organized as follows: The review of literature on MLS distributed database and related issues on concurrency control algorithms are presented in Section 31.2 Section 31.3 presents the proposed multilevel secure distributed database model. Section 31.4 presents lock based protocol for the proposed concurrency model. Section 31.5 concludes the paper.

## 31.2 Review of Literature

The operations of the database can be performed in the form of transactions. Several units of works that form a single logical unit of work can be called a transaction. This can be performed under the supervision of the transaction manager. A valid transaction must be satisfied the Atomicity, consistency, Isolation, Durability properties [3]. The transaction manager can allow two or more transactions to access the same data called concurrent transactions. The uncontrolled concurrent transaction leads to an inconsistent database and violate the isolation property. Hence, the transaction manager needs to control the interactions between the transactions. To achieve this task, the transaction manager uses several concurrency control mechanisms, such as locks, timestamps, etc.

James M. Slack [3] presented two security mechanisms: the first mechanism is based on a one-way protected group. A one-way protected group is a set of one-way protected objects. Each one-way protected object in the group will accept messages only from a distinguished object in the group called the interface object. A one-way protected group supports data integrity and access integrity. The second mechanism is a two-way protected group. It is an extension of a one-way protected group. Each object in the group could only send the messages of that group. This model fails to address how security and integrity policies are implemented with protected groups.

Keishi Tajimia [4] has developed a technique to statically detect the security flaws in OODBSs. He designed a framework to describe the security requirements and developed an algorithm to determine the security flaws. In this technique a user can bypass the encapsulation and abuse the primitive operations inside the functions and also the properties of the aggregate are not given.

Linda M. Null et al. [5] have defined security policy for object-oriented data model. This security policy addresses mandatory as well as discretionary access controls. This model is based on the classes which derive security classification constraints from their instances and logical instances. The implementation and suitability of these policies in the object-oriented environment is not presented in their proposal.

Sushil Jajodia et al. [6] have proposed a database security model for mandatory access control that details with the Object-Oriented Data model. This includes a message filtering algorithm that protects the illegal flow of information among objects of various security levels. Finally a set of principles is defined to design and implement security policies in Object-Oriented Database Management Systems. This model fails to address information flow are rendered explicitly.

Roshan. K. Thomas et al. [7] provide a kernelized architecture for multilevel secure Object-Oriented Database Management Systems (DBMS's) which support write-up. However, supporting write-up operations in object-oriented systems is complicated by the fact that such operations are no longer primitive, but can be arbitrarily complex and therefore can take arbitrary amounts of processing time. This architecture supports Remote Procedure Call (RPC) based on write-up operations. Dealing with the timing of such write-up operations consequently holds less time.

Bertino et al. [8] proposed the practical relevance of nested transactions and for the theoretical issues related to the development of suitable locking mechanisms and serializability theories for nested transaction. In particular, the interactions between parent and child transactions in the same nested transactions can be executed concurrently. This requires revise and extend primitives and the locking protocol, however, as mentioned. Serializability theory for nested transactions is substantially more complex.

Lin et al. [9] a simplified simulation model is used to compare the performance of basic timestamp, multi-version timestamp, and two-phase locking algorithms. It does not include different data distributions (partitioned, replicated, etc.), and simplified communication delay by combining CPU processing time, communication delays, and I/O processing time for each transaction.

Navdeep Karur et al. [10] have presented a simulation model of a multilevel secure distributed database system using secure concurrency algorithm. It addresses the performance price paid for maintaining security in a MLS/DDBMS, but performance of higher security level transactions in a replicated database has not been studied.

Having studied the above literature, it has been identified that most of the research efforts in the area of secure concurrency control are focused on centralized databases. Concurrent access of distributed database is mandatory for applications like banking, financial, enterprises, industry and institutes etc. Concurrent Transactions within a distributed database management system face several restrictions. The proposed model is designed to perform efficient, secured concurrent transactions based on a sub-query analyzer.

## 31.3 Computing Complexity

The data stored in a database should be secured from the unauthorized users. The data retrieval time is minimized by the Lock Manager (LM). This manager handles the locking mechanism for distributed sites. All transactions are sorted through the security manager (SM). The security manager (SM) handles the authentication of the user using with proper verification. The different query levels proposed in the model are View level ($SL(D_u)$), Secret level ($SL(D_s)$) and Top secret level ($SL(D_t)$). If the user is permitted access to the view level, the query is limited with the view level transaction ($SL(Q)<SL(D_u)$). Otherwise, the query is rolled back to Transaction Level security. Transaction level is also classified into secret and top secret respectively upon the user classifications.

The query optimizer further divides the query into various levels for distributed access and creates a new optimized query according to the data distribution. The Transaction Manager (TM) pools the query in the transaction queue and allows the transaction to be executed according to the load balance of the transaction concurrency. The network traffic is also considered by the Transaction Manager (TM). The Transaction Concurrency Manager (TCM) again analyzes the query and arranges the query for various levels of concurrency. The waiting timestamp is used for each transaction. The transactions are executed without concurrency when the waiting timestamp is expired. So the infinity waiting time of

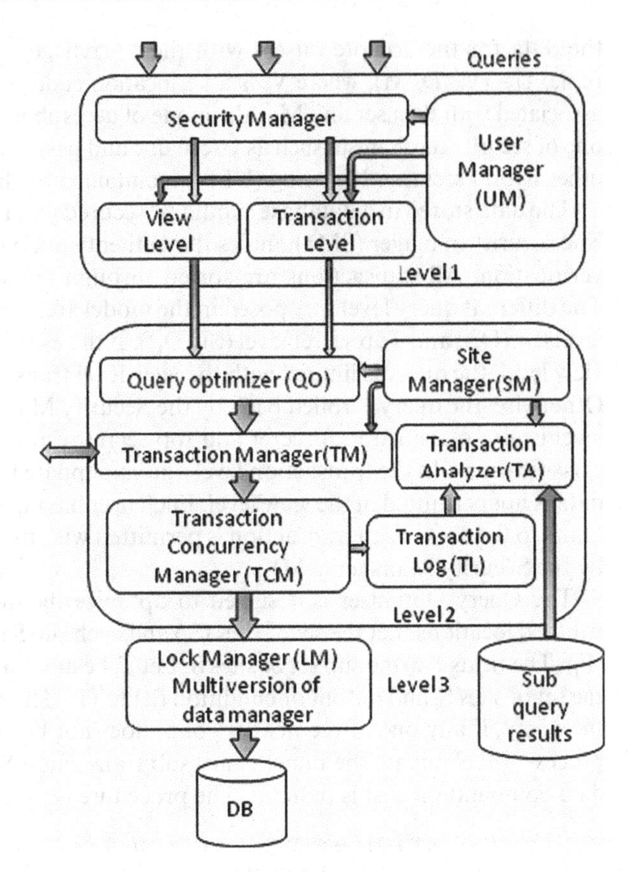


**Fig. 31.1** Proposed model

transaction is avoided. The waiting timestamp for each transaction is proportional to the security levels. The lock manager transmits the lock signal to the entire distributed database sites. It allows the transaction to update the data only when there is no objection from any other sites. Otherwise, it rolls back the transaction. If there is no objection from all other lock managers then it locks the data. The transaction is rolled back by the lock manager if the transaction exceeds the timestamp limit. Thus the proposed model felicitates to access data only after passing through multiple security levels and also allowed to access concurrently without any access conflict. The proposed model is depicted in Fig. 31.1.

### *31.3.1 Secure Concurrency Control Protocol*

The various modules of the secure concurrency control protocol including the concurrency components are illustrated as follows:

The various modules proposed in the MLS framework including concurrency components are briefly presented here. The User Module (UM) handles the data for the accepted users with their privileges. It is defined as a three tuple. U = ($V_s$, D, M), where $V_s$ is a verification code set, D is the data which is associated with the user and M is the mode of accessibility. The set $V_s$ may be any one of security mechanism such as Username and password, IP based security and other hybrid security. User Log (UL) is maintained by the User Module (UM).

The data stored in a database should be secured from the unauthorized users. The security manager (SM) handles the authentication of the user using proper verification. All transactions are sorted through the security manager (SM). The different query levels proposed in the model are View level ($SL(D_u)$), Secret level ($SL(D_s)$) and Top secret level ($SL(D_t)$). If the user is permitted access to the view level, the query is limited with the view level transaction ($SL(Q)<SL(D_u)$). Otherwise, the query is rolled back by the Security Manager (SM). Transaction level is also classified into secret and top secret respectively based on the user classifications. In the transaction level one can update the data, but updation of data is not permitted in the view level. Each user having their own level of access limits to the data. Each transaction is permitted with the access limits controlled by the Security Manager (SM).

The Query Optimizer is designed to optimize the quires received from distributed locations. Let the set of sites ($S_i$) and each site S is having the set of fields ($F_j$). The fields F is the sub set of sites S. Let $T_i$ be any transaction which involves the set of sites $S_i$ and output or condition fields $TF_i$. The site $S_i$ is eliminated from the query, if any one of the fields $F_j$ of $S_i$ does not belongs to the set $TF_i$. This process will eliminate the unnecessary sub transaction $ST_i$. So the network and data computation cost is reduced. The procedure is presented below.

```
void queryOptimizer(transaction T)
  TF0=getOutputFields(T)
  TF0=getComputationalFields(T)//add to output fields
  S0=getDistributedDatabase(T)
  F00=getFields(S) //two dimensional table match the each row
  F[x]0 with TF0
If F[x]0 is not match with any TF0
          eliminate S from T
```

The Site Manager (SMR) looks after all the distributed data which is created or configured for the transaction. This helps to secure the data from the unauthorized network access to the services. The site manager sends the signal to its own network boundary about the various locations of which data distributed and its configuration. This helps to attach and detach of sites over the network. The site manager has the details about the sites ($S_i$) and the associated fields ($F_i$). A transaction $T_i$ is rolled back, if the transaction wants to access field $TF_i$ in site $S_i$, which is not belongs to $S_i$. The Site Manager (SMR) handles the distributed sites and their locations. It also handles the frequency of updating the records. So it assigns a timestamp for each site. This timestamp is used for the transaction analyzer for flush the Sub-query Results (SQR).

The Transaction Manager (TM) further divides the query into various levels for distributed access and creates a new optimized query according to the data distribution. The Transaction Manager (TM) pools the transactions $T_i$ in the transaction queue and allows the transaction to be executed according to the load balance of the transaction concurrency. The network traffic is also considered by the Transaction Manager (TM). Let $T_i$ be any transaction and $ST_{ij}$ be any sub transaction which is derived from the parent transaction $T_i$. We can define dispatcher as a four tuple $T_i = \{ST, TS, DB, CL\}$. Let ST be the sub transaction, TS be the timestamp, DB be the database and CL be the listing port for the sub transaction in the network. The transaction $T_i$ is rolled back, if its all sub transactions $S_i$ is not finished with in the timestamp order. The transaction manager (TM) saves the sub-query result in the disk until it receives the updated message from the database. So before sending the Sub Transaction (ST) it checks its own Sub-query Results (SQR). For that reason we can avoid repeated transaction for the same data. The Transaction Dispatcher (TD) sends the sub-query to various sites. After completing that query it assembles the result and produces the result. The dispatcher using the timestamp mechanism for getting the results back.

The Transaction Analyzer (TA) analyzes the pooled transactions arrived from various distributed locations using the transaction log and creates the Frequently Accessed Fields (FAF) table. It also set the expiry time for each record in Frequently Accessed Fields (FAF) table. Using the timestamp, the FAF table entry is flushed. Let $T_i$ be any transaction, $TF_i$ be a set of fields which is going to be access by the transaction $T_i$, $FAF_i$ be any entry in FAF table, $FAFT_i$ be the timestamp for record $FAF_i$, $FAFS_i$ be the current state of the Sub-query Results (SQR). The sub transaction $ST_i$ for any transaction $T_i$ is allowed to access the other site, if there is no entry in FAF table or the current state of $FAFS_i$ is cleared. The Sub-query Results (SQR) is updated, if or the current state of $FAFS_i$ is cleared and the sub transaction $ST_i$ for any transaction $T_i$ is allowed to access the other site. Transaction Analyzer (TA) flushes the Sub-query Results (SQR) when it receives the updated message from any other site. The Sub-query Results (SQR) is only for the transactions which are frequently involved. SQR proposed in this model reduce the unnecessary data access time and wait time for every transaction.

The Transaction Concurrency Manager (TCM) analyzes the concurrent transactions $T_i$ and arranges the transaction for various levels of concurrency $TC_i$. The waiting timestamp is used for each transaction. The transactions are executed without concurrency when the waiting timestamp is expired. Thus the infinite waiting time of transaction is avoided. The waiting timestamp for each transaction is proportional to the security levels.

Let $T_1,T_2,...T_n$ be n transactions which can be executed concurrently. Let $U_i$ be the unit of a transaction for any transaction $T_i$. The schedule S is created for the transactions $T_1,T_2,...T_n$. The schedule S can be defined as a two tuple. $S = (U , T)$. where U is a unit and T is a transaction. Let Unit Tree (UT) be the executed unit of transaction which are waiting for commit. The node for the unit tree is constructed when each unit transaction U is executed by the concurrency manager. The Transactions ($T_i$) are associated with the one-to-one mapping to the unit tree UT. The state for the Unit Tree (UT) is marked as roll back tree when the unit transaction of Schedule is failed. The Transaction associated with $UT_i$ will not be executed in the Schedule if the Unit Tree $UT_i$is set by the Concurrency Manager (CM). The final commit is only after finishing the entire schedule. The commit is only for the Unit Tree which are not marked as a roll back tree by the Concurrency Manager (CM). The According to this new concurrency algorithm, If the schedule S is failed then there is no need to roll back the entire transaction.

The Lock Manager (LM) transmits the lock signal to the entire distributed database sites. It allows the transaction to update the data only when it passes all the clearance from all security levels. Otherwise, it rolls back the transaction. If there is no objection from all other lock managers then it locks the data. The transaction is rolled back by the lock manager if the transaction exceeds the timestamp limit. The locking mechanism follows a **non cyclic** tree structure. If here any cyclic tree is formed by the transaction $T_i$ then the $T_i$ is rolled back. Hence the deadlock is prevented. Thus the proposed model felicitates to access data only after passing through multiple security levels and also allowed to access concurrently without any access conflict.

### *31.3.2 Concurrency Control Algorithm*

Distributed database management system (DDBMS) allows users to access a database concurrently from geographically dispersed locations interconnected by a network. Concurrent accesses to the database have to be synchronized in order to maintain data consistency and to ensure correctness. This is achieved through use of distributed locking protocol which is applied in this proposed model. The efficient method of implementing concurrency is depicted in Fig. 31.2 and presented in this section.

Let $T_u$ denotes the unclassified security level transaction. Let $T_s$ denotes the secret security level transaction and $T_{ts}$ denotes the top secret security level transaction, i.e. $SL(T_u) \leq SL(T_s) SL(T_{ts})$.

Let r(x) and w(x) be the read and write of data item x respectively.

Based upon the above assumption, the following conflicts may occur:

1. (Read-down conflict among different levels): Read-down conflict occurs between $SL(T_s) \leq SL(T_{ts})$*'s* read operation, *r[x]*, and $SL(T_{ts})$*'s* write operation, *w[x]*.
2. (Read-write conflict at same level): Read-write conflict occurs between $SL(T_s)_i$*'s* read operation, *r[x]*, and $SL(T_s)_j$*'s* write operation, *w[x]*. Where $SL(T_s)i, SL(T_s)_j$ and *x* are at the same security level.
3. (Write-write conflict at same level): Write-write conflict occurs between $SL(T_{ts})_i$*'s* write operation, *w[x]*, and $SL(T_{ts})_j$*'s* write operation, *w[x]*. Where $SL(T_{ts})i, SL(T_{ts})_j$ and *x* are at the same security level.

Every transaction in this security model must obtain a read lock before reading a data item and a write lock before writing a data item. The security model allows a transaction to issue read-equal, read-down and write-equal operations. This is sufficient to prove that security is not violated through data access. The execution of a distributed transaction T is divided into sub-transactions $T_i$, where $i = 1$ to n. A sub-transaction $T_i$ is sent to the node $N_i$ where the data is available and executed under the local security and concurrency transaction manager. If a sub-transaction fails, then the parent transaction is rolled-back and restarts after some delay to avoid repeating the restart. The TLM (Transaction Lock Manager) determines at which node data items requested by a transaction are located. If the data is available in the parent node $N_i$, it is accessed in the same node $N_i$ otherwise, if there is no local copy and multiple copies exist at more than one node, then one copy is randomly selected and locks other copies of the same data. It creates one sub-transaction for each node $N_i$ that needs to be visited and acts as the coordinator in the distributed two-phase commit process. Even though the dispatches of sub-transactions of a transaction appear sequential, they are dispatches concurrently. Parent transactions originate from a fixed number of terminals and their number in the system is the sum of terminals connected to each node.

The CTM (Concurrency Transaction Manager) coordinates concurrency control activities with other nodes. In the case of data replication, it implements a read-one-write-all policy for read requests. For a write request, it consults all nodes that hold a copy of the desired data item. A DM (Data Manager) at every node contains information about data distribution and replication.

Let V1 be the set of transactions $V1 = \{T1, T2, \ldots, Tn\}$ and V2 be the set of Distributaded Databases $V2 = \{DB1, DB2, \ldots, DBn\}$. The set of all links $E = \{e1, e2, \ldots e_m\}$ connecting from $T_i$ where $T_i \in V1$ to DBj where $DB_j \in V2$ denotes the transaction $T_i$ can access the data on the distributed database $DB_j$ as illustrated in the Fig. 31.3.

A bipartite graph is a triple $G = (V1, V2, E)$ where V1 and V2 are two disjoint sets of vertices, respectively the top and bottom vertices, and $E \subseteq V1 \times V2$ is the set of edges.

The difference with classical graphs lies in the fact that edges exist only between top vertices and bottom vertices.

Two degree distributions can naturally be associated to such a graph, namely the **top degree distribution**($V1_k$)

$$V1_k = \frac{|\{t \in V1 : d(t) = k\}|}{|V1|} \tag{31.1}$$

Where d(t) denotes degree of a vertex t and the **bottom degree distribution**($V2_k$),

$$V2_k = \frac{|\{t \in V2 : d(t) = k\}|}{|V2|} \tag{31.2}$$

**Table 31.1** Access permissions in the proposed model

| Transactions | Data Items $SL(x_{ul})$ | $SL(y_{sl})$ | $SL(z_{ts})$ |
|---|---|---|---|
| $SL(T_{ts})$ | r[x] | r[y] | r[z],w[z] |
| $SL(T_{sl})$ | r[x] | r[y],w[y] | – |
| $SL(T_{ul})$ | r[x] | – | – |

A transaction cannot request additional locks once it has issued an unlock action. It holds on to all its locks (read or write) until it completes. A top secret security level transaction must release its read lock on a low data item when an unclassified security level transaction requests a write lock on the same data item and the aborted top secret security level transaction is restarted after some delay. Thus multiple transactions are performed simultaneously with minimum cost.

Table 31.1 shows the access permissions in the proposed model.

***Concurrency Control Algorithm***

```
void concurrency(transaction T)
U0=divide the transaction into various unit of transaction
Analyze the transactions, which are in the queue
Create the Schedule for the various units of transaction
While(schedule finish)
  U= next unit to be executed
  UT=the Unit Tree associated with U
  If (UT is not marked as roll back tree)
Execute the unit (U) in the schedule
If(U is success)
  Place U in the Unit Tree (UT)
else
  mark the Unit Tree (UT) as roll back tree
while (all unit tree UT in CM is traversed)
if (UT is not marked as roll back tree)
  commit the Unit Tree UT
else
reject the transaction Ti which is associated
  with Unit Tree UT
```

## 31.4 Simulation Results

The protocols for evaluating the performance of concurrency control are tabulated. This evaluation is based on the performance presented in [11]. The aim of this experiment is to test the transaction performance with the proposed

**Table 31.2** Simulation parameters

| Parameter | Value |
|---|---|
| NumDBS | 5 |
| No. Query / sec | 31 |
| No. CPU | 5 |
| Disk for each site | 5 |
| Log disk | 1 |
| Concurrent Transaction / sec | 11 |
| Write ratio | 6 |
| Read ratio | 25 |
| Waiting time out transaction | 4 |
| Execution time | 0.0715 s |
| Schedule Execution time | 0.2015 s |
| Transaction size | 10 records |
| CPU time / unit | 0.0119 s |
| Network delay | 0.01 s |
| Time for optimize | 0.002 |
| Time to partition | 0.014 |
| Unit / transaction | 6 |

concurrency algorithm. The model is simulated in a real time environment presented in Table 31.2.

The performance study is carried out with varying time factor. The number of transactions is directly proportional to the increased level of security with varying time factor. Let NT be the number of transactions and TT be the time to finish the transactions then,

$$NT \,\alpha\, 3.525\, TT \tag{31.3}$$

The time for executing the transaction is directly proportional to the security level also. Let SL be the security level and TT be the time to finish the transaction then,

$$TT \,\alpha\, 1.1928\, SL \tag{31.4}$$

The simulation results graphically represented in Figs. 31.3, 31.4, and 31.5 show that the performance of concurrent transactions increased when the transactions arrival rate increased. The performance of the transaction is high when the number of distributed database is increased.

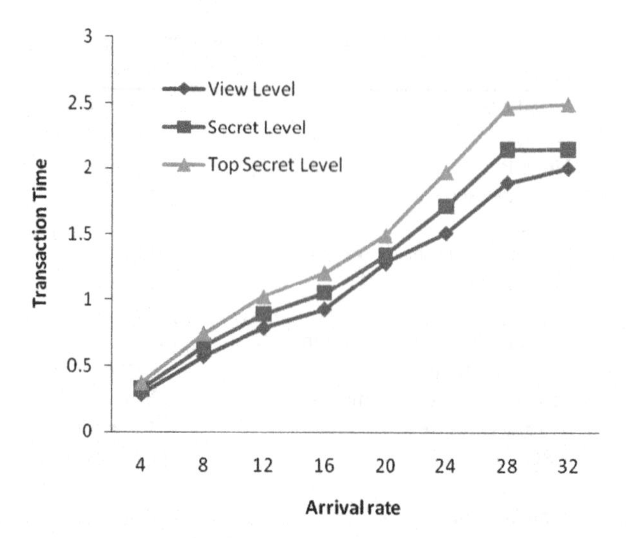


**Fig. 31.3** Transaction time vs arival rate

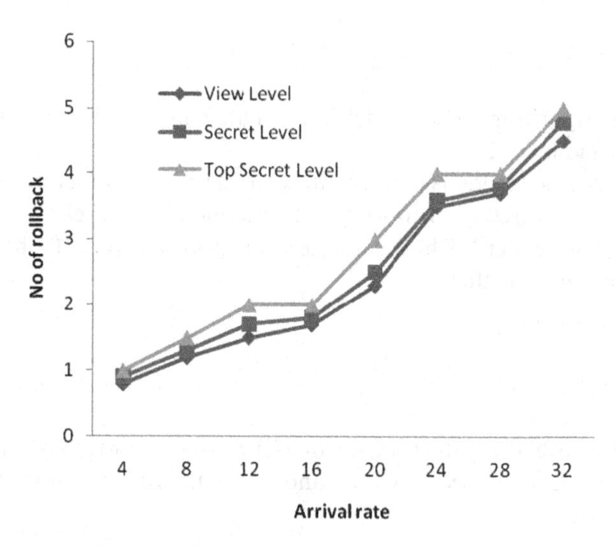


**Fig. 31.4** No of rollback vs arival rate

## 31.5 Conclusion

Distributed database systems today play a reality. Several organizations are now deploying distributed database systems. Security is a serious concern a while accessing data. Different security levels need to be integrated into the database to avoid access conflict. We have proposed a new model for Multilevel Secure Real-time Database Systems (MLSDD) and deployed concurrency control for the

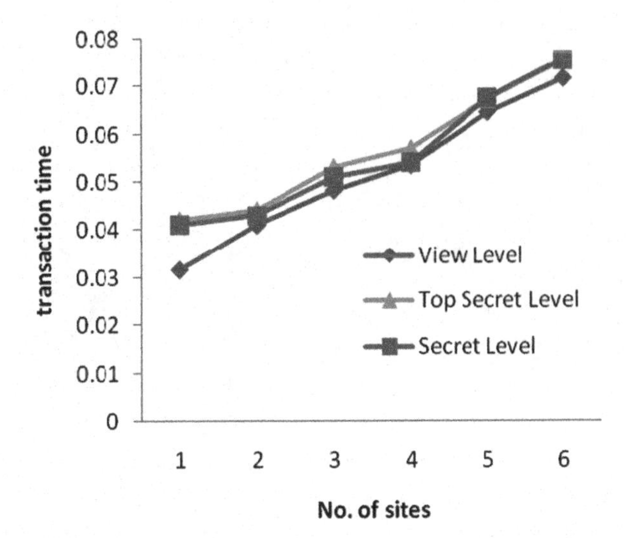


**Fig. 31.5** Transaction time vs no. of sites

secured access of data. The simulation study reveals that the performance of concurrent transactions is enhanced with multiple levels of security over the distributed database. This model can be applied for any real-time environment such as, corporate, financial enterprises, academic institute etc., performing day-to-day transactions in a distributed environment, cutting edge to different levels of security.

## References

1. Bertino E, Sandhu R (2005) Database security – concepts, approches, and challenges. IEEE Trans Dependable and Secure Comput 2(1):2–19
2. Tamper Ozsu M, Valduriez Patrick (1996) Distributed and parallel database systems. ACM Computing Surveys 28(1)
3. Slack James M (1993) Security in an object-oriented databases. Proceedings of the of ACM Transactions, pp 155–159
4. Tajimia Keishi (1996) Static detection of security flaws in OODBMS. RIFMS, Kyoto University Japan. In: ACM SIGMOD international conference on management of data. SIGMOD96, 25(2):341–352
5. Null Linda M (1992) The DIAMOND security policy for object-oriented databases. In: ACM annual Conference on Communicaions, pp 49–56
6. Jajodia S, Kogan B, Sandhu Ravi S A Multilevel Secure Object-Oriented Data Model. In: Proceedings of ACM SIGMOD, pp 596-616.
7. Thomas Roshan K, Sandhu Ravi S (1993) A kernelized architecture for multilevel secure object-oriented databases supporting write-up. J Comput Secur
8. Bertino E, Catania B, Ferrari E (2001) A nested transaction model for multilevel secure database management systems. ACM Trans Inf Sys Secur 4(4):321–370
9. Lin W, Nolte J (1983) Basic timestamp, multiple version timestamp, and two-phase locking. In: Proceedings of 9th international conference on VLDB Conference, Florence, Italy
10. Kaur N, Sarje AK, Misra M (2004) Performance evaluation of concurrency control algorithm for multillevel secure distributed databases. IEEE Compter Society
11. Xeoing M et al. (2002) Mirror: A state consciious concurrency control protocol for replicated real time database. J Inf Sys 27:277–297

# Chapter 32
# A New Framework for Nondeterministic Multi-valued System Minimization

**Adrian Zafiu and Ion Stefanescu**

**Abstract** The main goal of this paper is to present a new minimization method for the nondeterministic networks. The motivation of this article is given by the recently developed minimization system named COMIN. This method comprises a deep analysis of the deterministic and nondeterministic multi-valued networks, complete or incomplete specified networks and of the possible adopted strategies during the combinational minimization.

## 32.1 Introduction

Multi-valued logic, as a tool of expressing and optimizing the input–output digital correspondences, expressing the decisional parts of digital systems and algorithms, becomes of a major importance in the field of software minimization, as well in the field of new developments of digital circuits, like current and optical implementation of gates and interconnection means. The field of interest is logical design and optimization of hardware and software decisional systems. A natural way of expressing a function of a decisional system is a multi-valued or a binary value table. Thus, the optimization kernel of a decisional system, either an algorithm logical structure, or a digital circuit structure, becomes the minimization of the input- output correspondence, represented by a value table. Presented method use a different principle named discrimination [1]. According to this principle we search the set of difference between groups of input combinations with the same output.

A. Zafiu (✉)
Department of Electronics, Communications and Computers, University of Pitesti, 110040 Pitesti Str. Targul din Vale, nr.1, Arges, Romania
e-mail: adrian_zafiu@yahoo.com, adrian.zafiu@upit.ro

N. Mastorakis et al. (eds.), *Proceedings of the European Computing Conference*, Lecture Notes in Electrical Engineering 28,
DOI 10.1007/978-0-387-85437-3_32, 

## 32.2 Minimization Tools

Here is a decisional part of a given algorithm:

```
int function(int x1,int x2,int x3){
  if (x1 = = 1 && x2= = 3 && x3= = 1) return 0;
  if (x1 = = 2 && x2= = 1 && x3= = 4) return 3;
  if (x1 = = 3) if(x2= = 5 && x3= = 7) return 9;
                else     if(x2= = 7 && x3= = 2) return 7;
  if (x1 = = 4 && x2= = 2 && x3= = 2) return 0;
  if (x1 = = 5) if(x2= = 7) if(x3= = 2) return 7;
                                  elseif (x3= = 3) return 0;}
```

The above algorithm could be written as a multi-valued input-output correspondence table and the result of reconverting the optimized table into an algorithm is:

```
int function(int x1,int x2,int x3){
      if(x1= = 3) return 9;
      if(x1= = 4) return 3;
      if(x2= = 7 && x3= = 2) return 7;
      return 0;}
```

There are known minimization tools, like ESSPRESO, SIS, MVSIS, MVSIS-BLIF, and BALM (the later one, for some steps of automata minimization [2]). As far as we know at this moment, the main reference in the field of multi-valued decision table minimization is the program MVSIS 3.0 [3, 4], but in many cases the results not represents an optimum, especially in the field of network minimization, as well as in the case of pure multi-valued input-output deterministic and nondeterministic correspondences, and in the case of correlated outputs.

The principles and programs recently developed at the University of Pitesti [1,5, 6, 7–9] solve the most of the mentioned problems, offering a multi-valued minimization technique, based on a different principle of minimization – the discrimination principle [5, 6, 7] – and on an original generative minimization method [8], as well on an original network strategy [6].

The nondeterministic systems are minimized in two ways:

- full (n/n) non-determinism, by considering all initial options (i.e. preserving in the minimized solution all values of an output variable corresponding to a given input combination in the non-minimized table – which represents a nondeterministic correspondence, having a deterministic semantics),
- partial (k/n) non-determinism, by selecting an optimum (from the minimization point of view) subset out of the set of values of an output variable corresponding to a given input specification (having real nondeterministic semantics).

There is an option of minimizing system containing correlated outputs as well a selection of a default output value for deterministic and nondeterministic tables.

## 32.3 The Proposed Principles of Minimization

The basic principles of the minimization techniques used in our works are mentioned:

**Discrimination principle** [1, 7]: according to this principle, there are preserved – in the input part of each row of a value table – the minimum number of specified positions, so that the row is still different (discriminated) from all other tables rows having different output values.

**Preserving the specification integrity**: the result of a minimization (the minimized table) preserves all the specifications of the initial input-output correspondence (of the initial table). Thus, when the specifications of an initial row are applied to the simplified table, then there are obtained the same output values as in the initial deterministic or nondeterministic non-minimized table;

**Minimization on output value groups:** the results of a multi-valued decision system produce decisions for delivering each output value. Thus, the minimized outputs are, in fact, the value outputs.

**Deterministic meaning of the non-univocities:** each nondeterministic table having implicit non-univocities is converted into a table having explicit non-univocities only, and then expanded on the values of its output variables.

**Deterministic semantics of non-univocities**: An explicit non-univocity is considered as a set of simultaneous values corresponding to a given input specification row, if not otherwise specified (so called "non-determinism with deterministic semantics", or "full non-determinism", denoted by n/n) – the reason is to preserve all initial intentions of the table designer.

**Parallel or network minimization of multiple outputs tables:** tables with multiple outputs could be minimized in two ways:

- parallel: each output is minimized separately, independent of the other outputs;
- network: the output variables can be connected as inputs to other outputs, resulting simpler minimized results, due to supplementary information offered by these interconnections;
- network minimization oriented on the values of output variables:

**Preserving the correlations among different outputs, or groups of outputs in a nondeterministic network table:** in case of a nondeterministic table having declared correlated outputs, the minimization must consider the aspect and produce a minimized table preserving all correlations implemented into the non-minimized table.

**A single version of minimized solution:** the minimization result presents one version of solutions out of a class of equivalent minimized solutions.

## 32.4 Network Minimization of Nondeterministic Tables with Correlated Outputs

The network strategy includes an option selecting a minimization preserving all value correlations between different output variables, as they are manifested in the initial non-minimized table. The option has a meaning in the case of non-deterministic tables having more rows with the same input specification. In the rest, the correlations between outputs are automatically preserved during the minimization, regardless of the network or of the parallel type of minimization. If the correlation option is not activated, the network minimization could produce simpler results, because there are fewer constraints, but the correlations among the outputs are not, generally, preserved. For example, a part of a nondeterministic table, having two rows with the same input specification and two outputs with different non-univoque specifications, is given in the next table:

**Table 32.1**

| x1 | x2 | q1 | q2 |
|---|---|---|---|
| 2 | 0 | 1 | (1,2) |
| 2 | 0 | (1,2,3) | 3 |

As a result of minimization, there are two solutions, depending on the initial intentions of the designer:

- outputs q1 and q2 are independent, not informationally correlated, in which case the above table could be condensed, as in Table 32.1, implying nine possible (q1 q2) output pairs corresponding to input (2 0), namely: (1 1), (1 2), (1 3), (2 1), (2 2), (2 3), (3 1), (3,2), (3 3) (see Figs. 32.1 and 32.2)

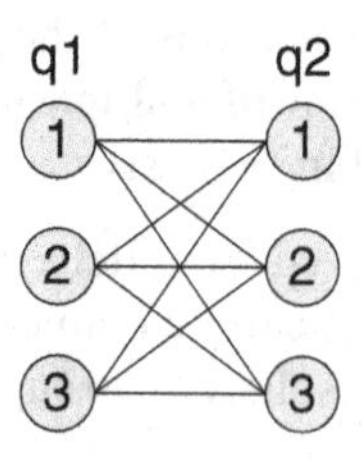


**Fig. 32.1** Output interconnections in a non-correlated minimization

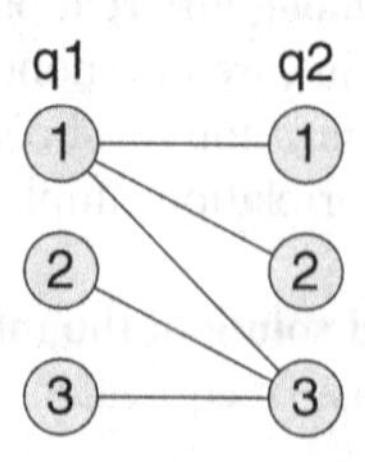


**Fig. 32.2** Selective output interconnections in a correlated minimization

**Table 32.2**

| x1 | x2 | q1 | q2 |
|---|---|---|---|
| 2 | 0 | (1,2,3) | (1,2,3) |

**Table 32.3**

| x | y | z | / | a | y | / | B |
|---|---|---|---|---|---|---|---|
| 0 | – | – | / | 1 | 1 | / | {3,4} |
| 0 | – | 1 | / | {1,2} | | / | 2 |
| 1 | 0 | – | / | 2 | | | |
| | | | / | 0 | | | |

**Table 32.4**

| x | Y | b | / | a | y | / | b |
|---|---|---|---|---|---|---|---|
| 0 | – | – | / | 1 | 1 | / | {3,4} |
| 0 | – | 3 | / | {1,2} | | / | 2 |
| 1 | 0 | – | / | 2 | | | |
| | | | / | 0 | | | |

- outputs q1 and q2 are informationally correlated, which means the only allowed output combinations are those directly resulting out of Table 32.1, and no more, namely: (1 1), (1 2), (1 3), (2 3), (3 3).

Results with uncorrelated and correlated outputs, full non-determinism (n/n): In the case of correlated outputs, variable „a" contains, as feedback, output „b", thus satisfying the correlation between „a" and „b". There is also noticed the uncorrelated outputs results are generally simpler than correlated outputs results.

## 32.5 Conclusions

Based on a different minimization principle – the discrimination [1, 7] – and on an appropriate original multi-valued minimization algorithm [8], there is presented here an efficient technique to obtain powerful minimizations of binary and multi-valued decisional systems, by using a strategy of network expansion and processing of the output nodes.

The discrimination does not need to consider explicitly the unspecified input combinations ("don't care").

The network structuring on output values is a natural one in the multi-valued case, where the minimization and the results must be expressed on values of each multi-valued output. The network strategy is recommended even in single multi-valued outputs, by expanding the output variables on their values.

The nondeterministic input-output correspondences could be treated either integrally, by considering all nondeterministic specifications (full non-determinism), or by selecting a part of the nondeterministic specifications (partial non-determinism), in order to get a simpler solution.

Due to cross-correlations appearing among different value outputs in a network minimization, all the interdependencies between outputs, settled by an initial statement of a nondeterministic decisional system (by the initial table), are automatically considered (if so specified), by means of a special option.

There is also implemented an option to select, if possible, a default value, taking account of nondeterministic constraints.

As presented in [8], the minimization method and algorithms do not depend exclusively on the number of input variables; the inner structure of the table decides the upper limits.

## References

1. Ştefănescu I (2005) Discrimination: A new principle in the efficient minimization of the binary and multiple-valued functions. In: 1st international conference on electronics, computers and artificial intelligence, ECAI 2005, University of Piteşti, Romania
2. Brayton Robert K[1], Jiang Jie-Hong R[1], Mishchenko Alan[1], Villa Tiziano[2], Yevtushenko Nina[3] BALM User's manual. [1]University of California, Berkeley, [2]University of Udine, Italy, [3]Tomsk State University, Russia
3. Gao M, Jiang J-H, Jiang Y, Li Y, Sinha S, Bryton R (2001) MVSIS. In: the Note of the international workshop on logic synthesis, Tahoe City, CA
4. Chai D, Jiang J-H, Jiang Y, Li Y, Mischenko A, Brayton R (2003) MVSIS 2.0 User's manual. University of Berkeley, Berkeley, CA 94720
5. Zafiu A, Ştefănescu I (2005) Minimization aspects of hardware and software systems – an extension of the discrimination method. In: 1st international conference on electronics, computers and artificial intelligence, ECAI 2005, University of Piteşti
6. Stefanescu I, Zafiu A (2007) An efficient network strategy in deep minimizations of deterministic and nondeterministic multi-valued decisional systems. In: 2nd internation conference on electronics, computers and artificial intelligence ECAI 2007, University of Pitesti, Romania
7. Stefanescu I (2006) Proiectarea logica a sistemelor decizionale hardware so software – Aspecte de baza. (Logic Design of Hardware and Software Decisional Systems – Basic Aspects – (in romanian)), Matrix Rom, Bucuresti, p 630
8. Zafiu A (2007) An efficient generative algorithm developed for the minimization of the multi-valued specifications. In: 2nd international conference on electronics, computers and artificial intelligence, ECAI 2007, University of Pitesti, Romania
9. Zafiu A, Stefanescu I, Franti E (2007) A new method for combinational minimization of multi-valued decisional system specifications. Services and Software Architectures, Bucharest, Romanian Academy, pp 27–37

# Part V
# Web-Based Applications

Web-based applications provide users with the opportunity to save time and money, and improve interaction with clients, suppliers and business partners. In this part are presented many Web application frameworks which facilitate rapid application development by allowing the programmer to define a high-level description of the program. In addition, there is potential for the development of applications on Internet Operating Systems, although currently there are not many viable platforms that fit this model. The papers cover different aspects: thread pool-based improvement of the mean-value analysis algorithm, security enhancements for web-based applications, description and discovery of Web applications in grid, user's preferences about personal data, peak analysis for efficient string sorting, parametric studies for the AEC domain using InteliGrid platform and an auction-based resource allocation strategy with a proportional share model for economic-based grid systems.

# Chapter 33
# Thread Pool-Based Improvement of the Mean-Value Analysis Algorithm

**Ágnes Bogárdi-Mészöly, Takeshi Hashimoto, Tihamér Levendovszky, and Hassan Charaf**

**Abstract** The performance of information systems is an important consideration. With the help of proper performance models and evaluation algorithms, performance metrics can be predicted accurately. The goal of our work is to improve the Mean-Value Analysis (MVA) evaluation algorithm based on the investigation of thread pools. In our work, the performance metrics of multi-tier information systems are predicted with the help of a queueing model and the improved MVA. The proposed evaluation algorithm has been implemented, the inputs have been estimated based on one measurement, and the model has been evaluated to predict performance metrics. In addition, ASP.NET web applications have been tested with concurrent clients in order to validate the proposed algorithm in different versions of ASP.NET environments.

## 33.1 Introduction

One of the most important factors of information systems is performance, because they face a large number of users, and they must provide high-availability services with low response time, while they guarantee a certain throughput level. With the help of a properly designed performance model and an appropriate evaluation algorithm, the performance metrics of a system can be determined at early stages of the development process.

Several methods have been proposed in the past few years to address this goal. A group of them are based on queuing networks or extended versions of queueing networks [7, 11, 15]. Another group uses Petri-nets or generalized stochastic Petri-nets [1, 8]. The third group of the approaches uses a stochastic extension of process algebras, like TIPP (Time Processes and Performability

Á. Bogárdi-Mészöly (✉)
Department of Automation and Applied Informatics, Budapest University of Technology and Economics Goldmann György tér 3. IV. em., 1111, , Budapest, Hungary

N. Mastorakis et al. (eds.), *Proceedings of the European Computing Conference*, Lecture Notes in Electrical Engineering 28,
DOI 10.1007/978-0-387-85437-3_33, 

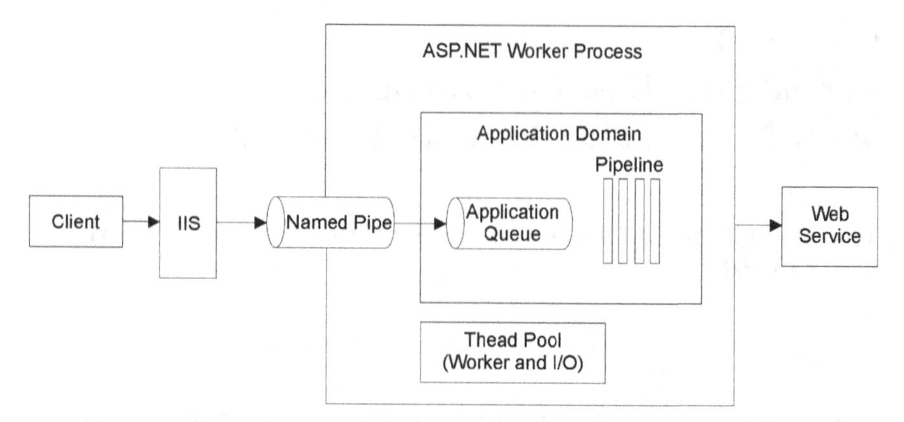


**Fig. 33.1** The architecture of ASP.NET environment

Evaluation) [6], EMPA (Extended Markovian Process Algebra) [2], and PEPA (Performance Evaluation Process Algebra) [5].

Performance metrics depend on many factors. Several papers have investigated how various configurable parameters affect the performance of web-based information systems. Statistical methods and hypothesis tests are used to retrieve factors influencing the performance. One approach [14] applies analysis of variance, another [3, 4] performs an independence test.

Today one of the most prominent technologies of information systems is Microsoft .NET. The application server has several settings which can affect the performance [10]. The request of the client goes through several subsystems before it is served as depicted in Fig. 33.1.

From the IIS (Internet Information Services), the accepted HTTP (Hypertext Transfer Protocol) connections are placed into a named pipe. This is a global queue between IIS and ASP.NET, where requests are posted from native code to the managed thread pool. From the named pipe, the requests are placed into an application queue, also known as a virtual directory queue. There is one queue for each virtual directory. The global queue is managed by the process that runs ASP.NET; its limit is set by the *requestQueueLimit* property. The application queue limit is configured by the *appRequestQueueLimit*property. When the limit is reached, the requests are rejected. When an application pool receives requests faster than it can handle them, the unprocessed requests might consume all of the memory, slowing the server and preventing other application pools from processing requests.

The *maxWorkerThreads* attribute means the maximum number of worker threads; the *maxIOThreads* parameter is the maximum number of I/O threads in the .NET thread pool. These attributes are automatically multiplied by the number of available Central Processing Units (CPUs). The *minFreeThreads*attribute limits the number of concurrent requests, because all incoming requests will be queued if the number of available threads in the thread pool falls below the value for this setting. The *minLocalRequestFree-Threads* parameter is similar to *minFreeThreads*, but it is related to requests from localhost (for example a local web service call). These two attributes can be used to prevent deadlocks by ensuring that a thread is available to handle callbacks from pending asynchronous requests.

In our previous work [3, 4], the effects of the limits of the thread and the queue size have been statistically analyzed, and six influencing parameters of the performance have been found: *maxWorkerThreads*, *maxIOThreads*, *minFreeThreads*, *minLocalRequestFreeThreads*, *requestQueueLimit*, and *appRequestQueueLimit*, which is proven by a statistical method, namely, the chi square test of independence.

The paper is organized as follows. Section 33.2 covers backgrounds and related work. Section 33.3 describes our improved MVA algorithm. Section 33.4 presents the experimental configurations, the performance prediction with the proposed algorithm and its validation in different versions of ASP.NET environments. Finally, Section 33.5 reports conclusions and future work.

## 33.2 Backgrounds and Related Work

Queueing theory [7] is one of the key analytical modeling techniques used for information system performance analysis. Queueing networks and their extensions (such as queueing Petri nets [9]) are proposed to model web-based information systems [11, 12, 15].

In [15], a queueing model is presented for multi-tier information systems, which are modeled as a network of $M$queues: $Q_1,\ldots,Q_M$ illustrated in Fig. 33.2. Each queue represents an application tier. A request can take multiple visits to each queue during its overall execution, thus, there are transitions from each queue to its successor and its predecessor as well. Namely, a request from queue $Q_m$ either returns to $Q_{m-1}$ with a certain probability $p_m$, or proceeds to $Q_{m+1}$ with the probability $1-p_m$. There are only two exceptions: the last queue $Q_M$, where all the requests return to the previous queue ($p_M = 1$) and the first queue $Q_1$, where the transition to the preceding queue denotes the completion of a request. $S_m$ denotes the service time of a request at $Q_m$ ($1 \leq m \leq M$).

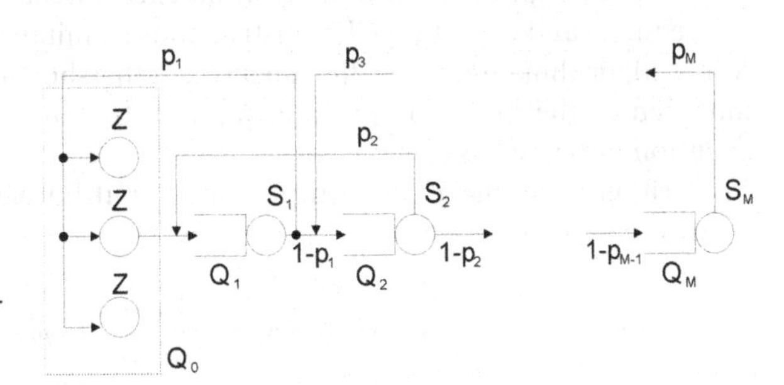


**Fig. 33.2** Modeling a multi-tier information system using a queueing network

Internet workloads are usually session-based. The model can handle session-based workloads as an infinite server queueing system $Q_0$ that feeds the network of queues and forms the closed queueing network depicted in Fig. 33.2. Each active session is in accordance with occupying one server in $Q_0$. The time spent at $Q_0$ corresponds to the user think time $Z$.

The model can be evaluated for a given number of concurrent sessions $N$. A session in the model corresponds to a customer in the evaluation algorithm. The MVA algorithm for closed queueing networks [7, 13] iteratively computes the average response time and the throughput performance metrics. The input parameters of the algorithm are $N$ the number of customers, $M$ the number of tiers, $Z$ the average user think time, $V_m$ the visit number and $S_m$ the average service time for $Q_m$ $(1 \leq m \leq M)$. The output parameters are $\tau$the throughput, $R$ the response time, $R_m$ the response time for $Q_m$ and $L_m$ the average length of $Q_m$. The initialization is $L_m = 0$ $(1 \leq m \leq M)$.

The algorithm introduces the customers into the queueing network one by one $(1 \leq n \leq N)$. The steps of an iteration are as follows:

$$R_m = V_m \cdot S_m \cdot (1 + L_m), \tag{33.1}$$

$$\tau = \frac{n}{Z + \sum_{m=1}^{M} R_m}, \tag{33.2}$$

$$L_m = \tau \cdot R_m. \tag{33.3}$$

The cycle terminates when all the customers have been entered.

## 33.3 Improving MVA Based on Thread Pool Investigation

The MVA algorithm can be effectively improved to model the behavior of the thread pool. Consider that the actual request contains CPU as well as I/O calls. In case of multiple threads, I/O calls do not block the processing, because the execution can continue on other non-blocked threads. This enables handling I/O requests and executing CPU instructions simultaneously. Therefore, in the MVA algorithm, updating the queue length (shown in Eq. 33.3) must be modified in the input, output and input/output tiers; in the CPU tiers this equation preserves its original form.

**Proposition 1** In the input, output and input/output tiers Eq. 33.3 is the following:

$$L_{I/O} = \tau \cdot R_{I/O} - \frac{S_{CPU}}{S_{I/O}} \cdot L_{CPU}, \tag{33.4}$$

*where the CPU index means a CPU tier index and the I/O index corresponds to an I/O tier index from* $1 \leq m \leq M$*.Additionally,* $L_{CPU}$ *is from the previous iteration step, namely,* $L_{CPU}$ *is updated only after this step. After Eq.33.4, one more checking step is necessary:*

$$if \, L_{I/O} < 0 \; then \; L_{I/O} = 0. \tag{33.5}$$

*Proof* Consider that the CPU queue and the I/O queue contain some requests respectively. If a new request arrives, it must wait until each previous request has been handled in the CPU queue. It takes $L_{CPU} \cdot S_{CPU}$ time unit to handle each CPU request. During this time the requests residing in the I/O queue can be handled simultaneously, because the CPU and the I/O devices can work in parallel. If an I/O call occurs, the CPU has to wait until the I/O device handles the request. In case of multiple threads the CPU does not have to wait, because the execution can continue on another thread as depicted in Fig. 33.3.

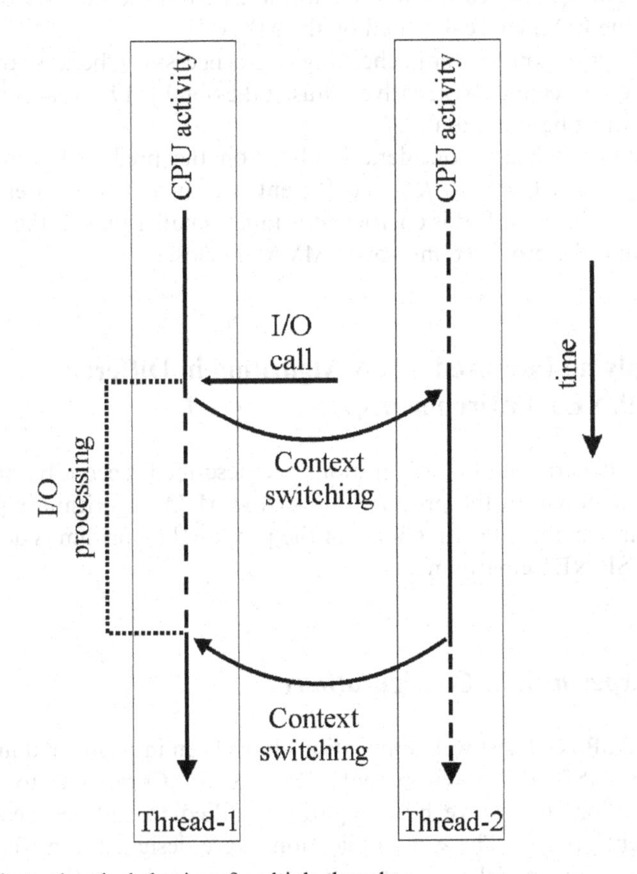


**Fig. 33.3** Illustrating the behavior of multiple threads

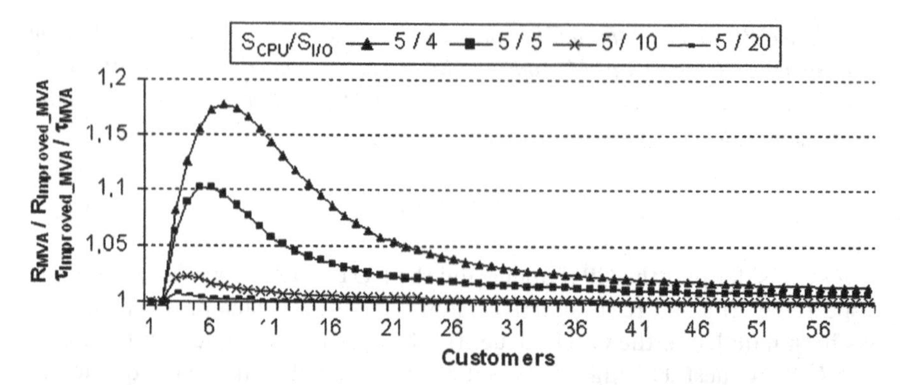


**Fig. 33.4** The effect of the proposed coefficient

Namely, during this time interval $S_{CPU}/S_{I/O} \cdot L_{CPU}$ I/O requests can be handled, which affects that the number of requests waiting in the I/O queue decreases. This difference should be subtracted from the number of requests waiting in the I/O queue calculated by the MVA.

After this equation one more checking step is necessary, because the average length of a queue cannot be negative. Thus, if the obtained $L_{I/O}$ value is less than 0, the $L_{I/O}$ must be equal to 0. □

This extension has a considerable effect on the predicted response time and throughput if the $S_{CPU}/S_{I/O}$ coefficient is near to 1 or greater than 1 as illustrated in Fig. 33.4. If this coefficient is much smaller than 1, the outputs of the MVA and the proposed improved MVA are closer.

## 33.4 Applying Improved MVA Algorithm in Different ASP.NET Environments

Firstly, our experimental configurations are described. Secondly, our performance prediction with the proposed improved MVA algorithm is presented. Finally, our experimental validation of the proposed algorithm is described in different ASP.NET environments.

### *33.4.1 Experimental Configurations*

Three-tier ASP.NET test web applications have been implemented in different versions of ASP.NET environments (Fig. 33.5). Compared to a typical web application, they have been slightly modified to suit the needs of the measurement process. The web applications were designed in a way that the input values of the model parameters can be determined from the results of one

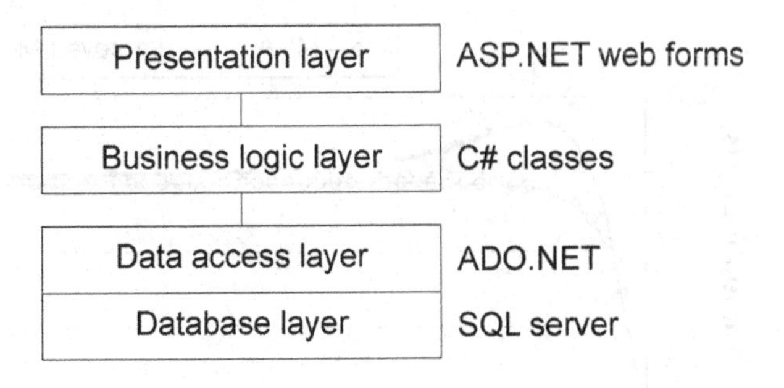


**Fig. 33.5** The test web application architecture

measurement. Each page and class belonging to the presentation, business logic or database was measured separately.

The web server of our test web application was Internet Information Services (IIS) 6.0. The server runs on a 2.8 GHz Intel Pentium 4 processor with Hyper-Threading technology enabled. It had 1 GB of system memory; the operating system was Windows Server 2003 with Service Pack 1. There were two environments. In the first environment, the application server was ASP.NET 1.1 runtime environment, the database management system was Microsoft SQL Server 2000 with Service Pack 3. In the second environment, ASP.NET 2.0 and Microsoft SQL Server 2005 with Service Pack 1 were used.

During our measurements the limits of the thread type (*maxWorkerThreads*, *maxIOThreads*, *minFreeThreads*, *minLocalRequestFreeThreads*,) and the queue size (*requestQueueLimit*, *appRequestQueue-Limit*) were held on their default settings. Hyper-Threading technology is enabled on the processor of the server. Thus, it seems to have two processors, although physically there is only one processor. Hence the first two parameters (*maxWorkerThreads* and *maxIOThreads*), automatically multiplied by two.

The clients ran on another PC on a Windows XP Professional computer with Service Pack 2. The supporting hardware was a 3 GHz Intel Pentium 4 processor with Hyper-Threading technology enabled, and it also had 1 GB system memory. The connection between the computers was provided by a 100 Mb/s network. The emulation of the browsing clients and the measuring of the response time were performed by JMeter, which is an open-source load tester. Tests can be created on a graphical interface. Virtual users send a list of HTTP requests to the web server concurrently. Each virtual user inserts an exponentially distributed think time between its requests with a mean of 4 seconds. With the help of JMeter, the measurement process can be easily automated.

### *33.4.2 Performance Prediction with the Improved MVA Algorithm*

It is worth decomposing a web application to multiple tiers, and modeling it according to Fig. 33.2, because the service time of each tier can be very different.

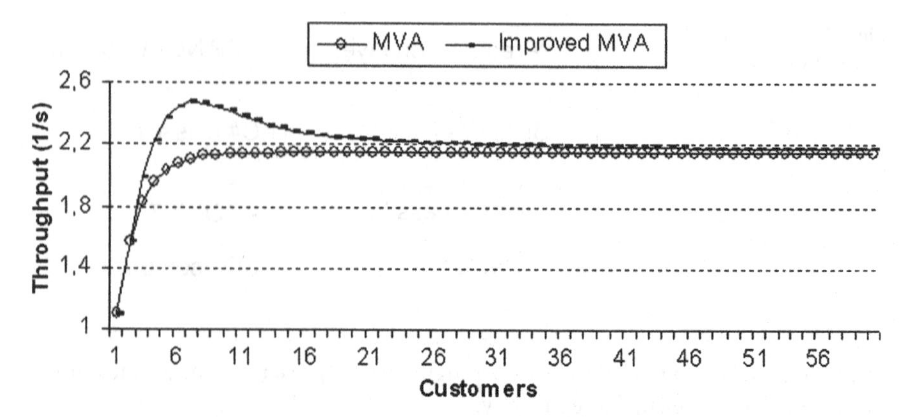


**Fig. 33.6** The predicted throughput with MVA and with extended MVA in the first environment

The presentation tier (UI) is an output tier and the database tier (DB) corresponds to an input/output tier. Thus, the proposed improvement must be applied in these tiers. The business logic layer (BLL) corresponds to a CPU tier.

Firstly, the MVA and the proposed improved MVA algorithm for closed queueing networks have been implemented with the help of MATLAB. Secondly, the input values of the algorithms have been estimated in both environments from one-one measurement. The inputs of the script are the number of tiers $M$, the maximum number of customers (simultaneous browser connections), the average user think time $Z$, the average service times $S_m$ and the visit numbers $V_m$ for $Q_m$ $(1 \leq m \leq M)$.

During the measurement the number of tiers was constant (three). The maximum number of customers means that the load was characterized as follows: we started from one simultaneous browser connection, then we continued with two, until the maximum number (arbitrary, in this case it is set to 60) had been reached. Since JMeter can measure only the response time of the three tiers without user think time between the subrequests, in the algorithms the average user think time between the requests is 0.

Finally, the model has been evaluated. The script computes the response time and the throughput up to a maximum number of customers in both environments (depicted in Figs. 33.6 and 33.7).

In these cases the proposed improvement has a considerable effect on the predicted response time and throughput in both environments, because the $S_{BLL}/S_{UI}$ coefficient is near to 1. Thus, high curvatures can be observed in the response time and in the throughput as well.

### *33.4.3 Experimental Validation*

The proposed extended MVA algorithm has been validated in the two ASP.NET environments. Typical web applications (Fig. 33.5) have been tested with

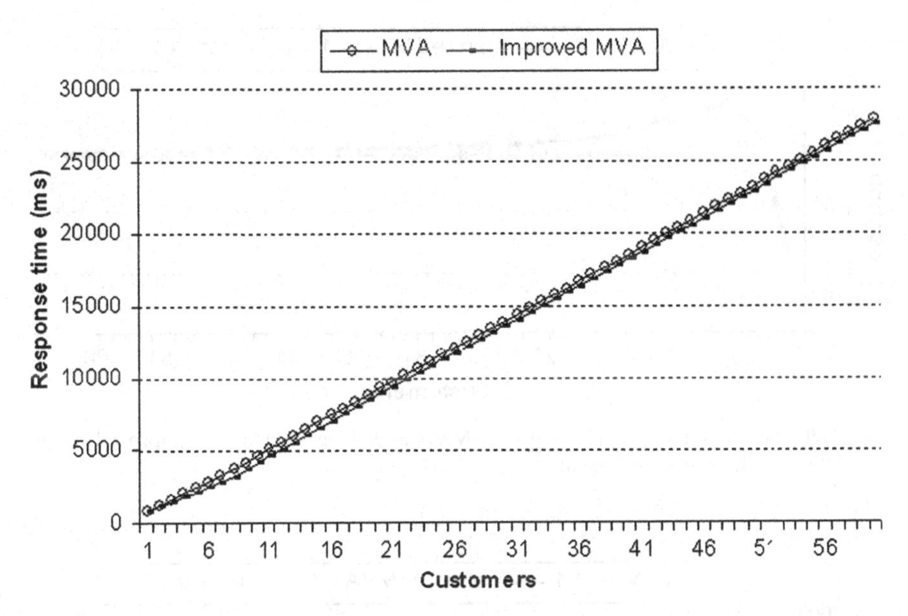


Fig. 33.7 The predicted response time with MVA and with extended MVA in the first environment

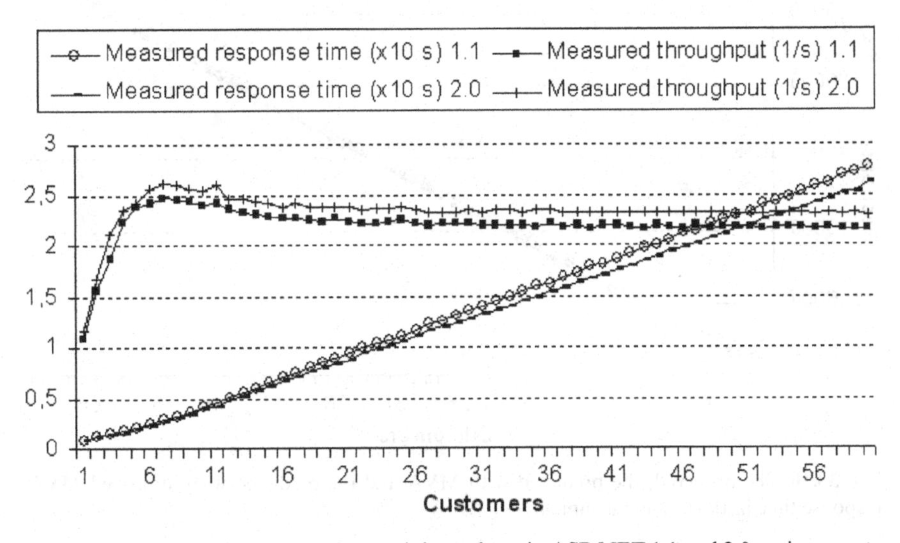


Fig. 33.8 The observed response time and throughput in ASP.NET 1.1 and 2.0 environments

concurrent user sessions, comparing the observed and predicted values in order to validate the proposed algorithm in both environments.

While the number of simultaneous browser connections varied, the average response time and throughput were measured in the two environments. As illustrated in Fig. 33.8, the new version of ASP.NET runtime environment

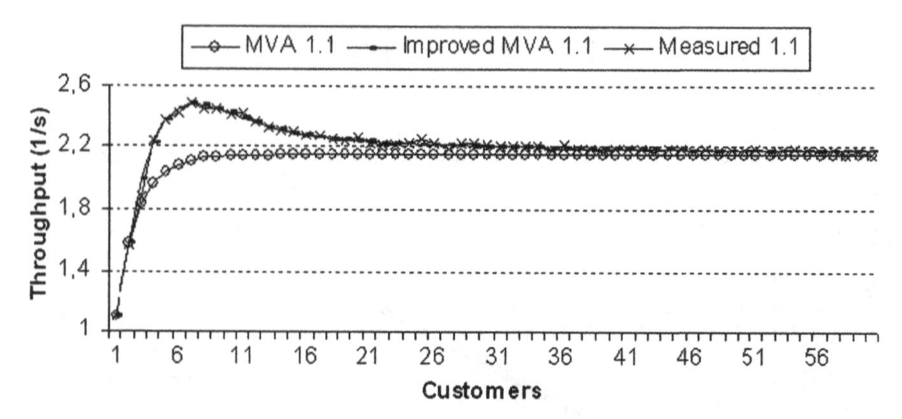


**Fig. 33.9** The observed, the predicted with MVA and the predicted with improved MVA throughput in the first environment

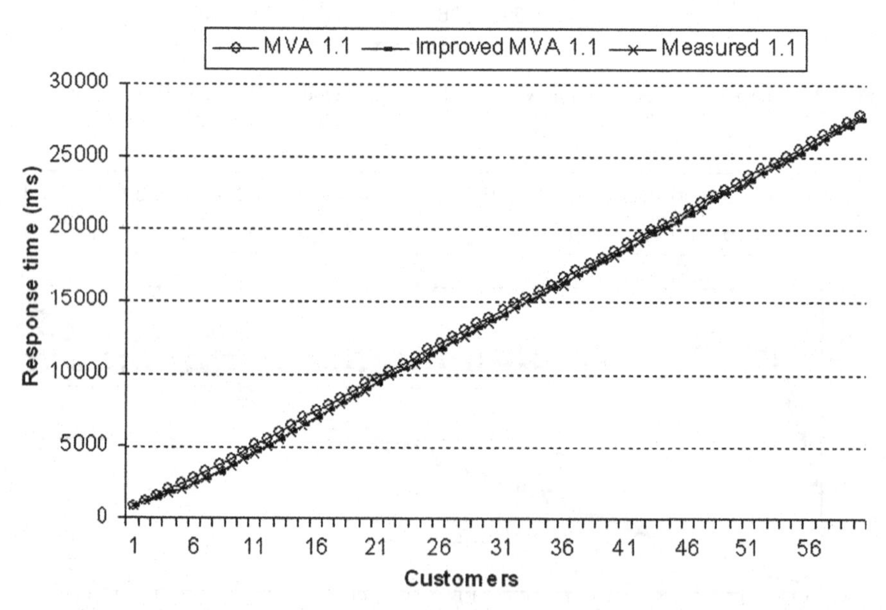


**Fig. 33.10** The observed, the predicted with MVA and the predicted with improved MVA response time in the first environment

and SQL Server database management system are slightly faster than the older version, but the character of the curve shapes are very similar. The results correspond to the common shape of the response time and throughput performance metrics. Increasing the number of concurrent clients, the response time grows, since the queue limits are large enough, thus, all of the requests can be successfully served.

Our proposed algorithm was experimentally validated by comparing the observed, the predicted values with MVA, and the predicted values with our improved MVA in the two environments. Our results (depicted in Figs. 33.9, 33.10, 33.11, 33.12) have shown that the output of the proposed algorithm approximates the measured values much better than the MVA considering the shape of the curve and the values as well in both environments. Thus, the

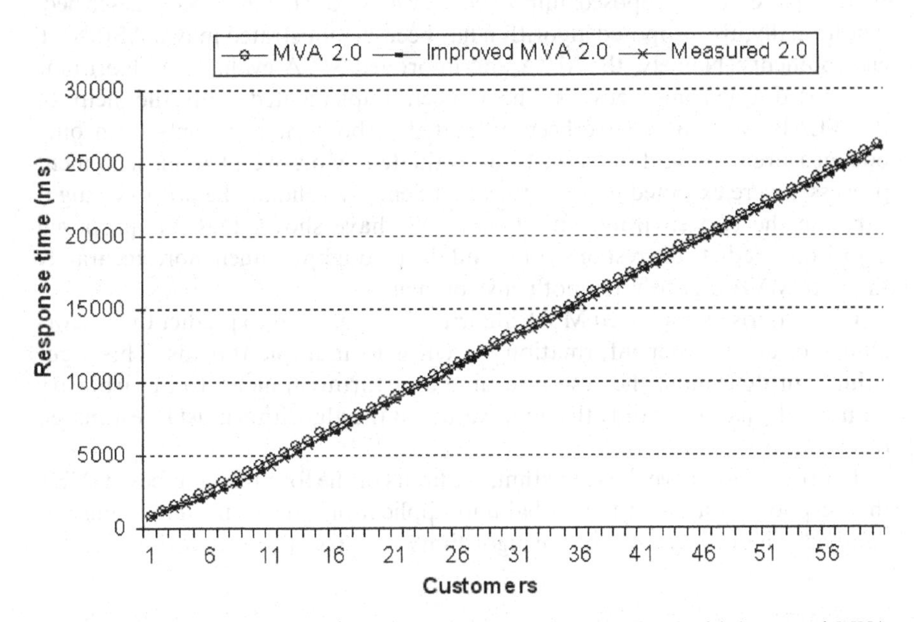


**Fig. 33.11** The observed, the predicted with MVA and the predicted with improved MVA response time in the second environment

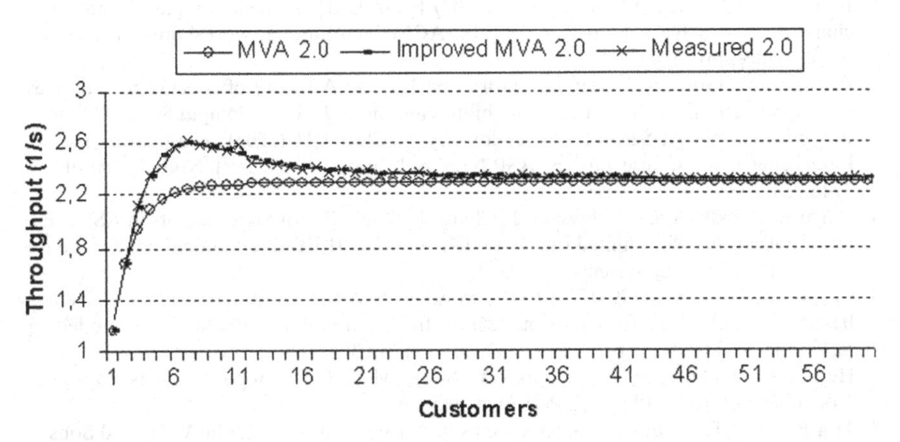


**Fig. 33.12** The observed, the predicted with MVA and the predicted with improved MVA throughput in the second environment

proposed algorithm predicts the response time and throughput much more accurately than the MVA algorithm in both environments.

## 33.5 Conclusions and Future Work

In this paper our proposed improved MVA algorithm has been described. Additionally, the proposed algorithm has been demonstrated in two ASP.NET environments. Namely, the MVA and improved MVA evaluation algorithms for closed queueing networks have been implemented with the help of MATLAB, the inputs have been estimated in both environments from one-one measurement, and the model was evaluated. Moreover, two measurement processes were executed in order to experimentally validate the proposed algorithm in the two environments. Our results have shown that our proposed algorithm predicts the response time and the throughput much more accurately than the MVA algorithm in both environments.

Our proposed improved MVA algorithm can be used to predict the performance of a multi-tier information system with multiple threads. This algorithm can determine the response time and throughput up to an arbitrary number of customers; only the input values of the algorithm must be estimated or measured.

In order to improve the algorithm, the limits of the four thread types in .NET thread pool, along with the global and application queue size limits must be handled. These extensions of the algorithm are parts of future work.

## References


1. Bernardi S, Donatelli S, Merseguer J (2002) From UML sequence diagrams and statecharts to analysable petri net models. In: ACM international workshop software and performance, pp 35–45
2. Bernardo M, Gorrieri R (1998) A tutorial on EMPA: A theory of concurrent processes with nondeterminism, priorities, probabilities and time. J Theor Comput Sci 202:11–54
3. Bogárdi-Mészöly Á, Szitás Z, Levendovszky T, Charaf H (2005) Investigating factors influencing the response time in ASP.NET web applications. Lect Notes Comput Sci 3746:223–233
4. Bogárdi-Mészöly Á, Levendovszky T, Charaf H (2006) Performance factors in ASP.NET web applications with limited queue models. In: 10th IEEE international conference on intelligent engineering systems, pp 253–257
5. Gilmore S, Hillston J (1994) The PEPA workbench: A tool to support a process algebra-based approach to performance modelling. In: 7th international conference modelling techniques and tools for performance evaluation, pp 353–368
6. Herzog U, Klehmet U, Mertsiotakis V, Siegle M (2000) Compositional performance modelling with the TIPPtool. J Perf Eval 39:5–35
7. Jain R (1991) The art of computer systems performance analysis. John Wiley and Sons
8. King P, Pooley R (1999) Derivation of petri net performance models from UML specifications of communication software. In: 25th UK performance engineering workshop

9. Kounev S, Buchmann A (2003) Performance modelling of distributed e-business applications using queuing petri nets. In: IEEE international symposium on performance analysis of systems and software
10. Meier JD, Vasireddy S, Babbar A, Mackman A (2004) Improving .NET application performance and scalability (patters & practices). Microsoft Corporation
11. Menascé DA, Almeida V (2001) Capacity planning for web services: Metrics, models, and methods. Prentice Hall PTR, Upper Saddle River, NJ
12. Smith CU, Williams LG (2000) Building responsive and scalable web applications. Computer measurement group conference, pp 127–138
13. Reiser M, Lavenberg SS (1980) Mean-value analysis of closed multichain queuing networks. J Assoc Comput Mach 27:313–322
14. Sopitkamol M, Menascé DA (2005) A method for evaluating the impact of software configuration parameters on e-commerce sites. In: ACM 5th international workshop on software and performance, pp 53–64
15. Urgaonkar B (2005) Dynamic resource management in internet hosting platforms. Dissertation, University of Massachusetts Amhersto

# Chapter 34
# Security Enhancements for Web-Based Applications

**Shivanand B. Hiremath and Sameer S. Saigaonkar**

**Abstract** Web applications use inputs from Hypertext Transfer Protocol (HTTP) requests sent by the users to determine the response. Attackers can tamper with any part of the HTTP request, including the Uniform Resource Locator (URL), query string, headers, cookies, form fields and hidden fields and attempt to bypass the application's security mechanisms. Common input-tampering attacks include forced browsing, command insertions, cross-site scripting, buffer overflows, format string attacks, Structured Query Language (SQL) injection, cookie poisoning and hidden field manipulation. In this paper, we have proposed an algorithm to detect hidden fields, the form fields and URL parameters manipulation and the algorithm is implemented in Java Server Pages (JSPTM). Security aspects are illustrated with fine remarks.

## 34.1 Introduction

Building Web-based applications is challenging from a security perspective as HTTP transactions are connectionless, one-time transmissions. A Web transaction consists of a pair of connections – one to fetch the web form that collects the transaction data and another to transmit the collected data to the server [4]. For applications that lead a user through a series of input forms, the application must temporarily store field data entered on previous pages.

Developers have two choices for storing this "state" information: on the server side or on the client side. Most developers find it easy to store this "state" information on the client side, which is sent back with each transaction. State information can be stored in a browser in two ways:

- HTML form "hidden" fields
- Browser "cookies"

S.B. Hiremath (✉)
National Institute of Industrial Engineering, Mumbai, India

N. Mastorakis et al. (eds.), *Proceedings of the European Computing Conference*, Lecture Notes in Electrical Engineering 28,
DOI 10.1007/978-0-387-85437-3_34, 

### 34.1.1 Hidden Fields

The browser doesn't display hidden fields embedded in HTML and hence are called "hidden". They are an easy way for preserving state information in a web application. As the user proceeds through each input screen, you can use hidden fields to store the information already collected so it is sent back to the server as part of the next transaction.

However, the hidden variable is visible in the raw HTML that the server sends to the browser. To see the hidden variables, a user just has to select "view source" from the browser menu [9].

Example: log in and change address

The following is a part of the ASP™ script of "change address" application that allows a user update their mailing address.

The application displays two pages:

1. A form for mailing address information
2. A page indicating the result of the transaction

```
...
<input type="hidden" name="userid" value="<%=userid%>">
<input type="hidden" name="form_expires"
  value="<%=form_expires%>">
<input type="submit" name="chaddr" value="Change address">
...
```

(We are omitting a user login page, plus the script.)

When we run the above script in the browser and view the HTML source code we see the values stored in "hidden" fields to store information from the login process:

```
...
<input type="hidden" name="userid" value="nitie">
<input type="hidden" name="form_expires"
  value="20070301:12:45:20">
...
```

When we fill in some information and press the submit button, a confirmation screen uses the value of the userID hidden field from the HTML form.

A hacker can use the browser's "Save as" feature to save the HTML of the change address form to his/her computer. The complete HTML, including values in the hidden fields are saved.

Then he/she can open the HTML file with a text editor and change the userID field and save the file. And then he/she can alter the other hidden fields as well.

Thus he/she can open the file in his/her web browser and submit the form. As the application is trustworthy with respect to its contents of the hidden fields, any user knowledgeable enough to save and load the HTML is able to alter the userID field and make changes freely to the addresses of any user they like.

Experienced web programmers believe that this type of tampering can be easily prevented by checking the HTTP_REFERER variable.

Most browsers send an HTTP header named "HTTP_REFERER". It contains the URL of the page the user viewed before the current one. If the hacker saves the form to his/her computer and resubmits it, HTTP_REFERER is blank or contains a different URL.

This is not a safe method to validate a form against tampering. Just like form fields, the value of HTTP_REFERER is set by the web browser. A user with only a little knowledge can write a script to spoof this header along with the contents of the form.

Checking HTTP_REFERER will catch trivial attempts to tamper with forms, but cannot be relied on for serious web applications [8].

### *34.1.2 Cookies*

A "cookie" is a mechanism developed by the Netscape Corporation to make up for the stateless nature of the HTTP protocol [9]. Normally, each time a browser requests the URL of a page from a Web server the request is treated as a completely new interaction. Although this makes the Web more efficient, this stateless behavior makes it difficult to create things like shopping carts that must remember the user's actions over an extended period of time. Cookies solve this problem. A cookie is a small piece of information, often no more than a short session identifier that the HTTP server sends to the browser when the browser connects for the first time [9].

Because cookies are not part of the standard HTTP specification, only some browsers support them: currently Microsoft Internet Explorer 3.0 and higher, and Netscape Navigator 2.0 and higher [9].Cookies are small chunks of data stored by the browser and they are of two types:

- Transient (or session) cookies: Ones without an expiry date.
- Persistent cookies: Ones with an expiry date set in the future.

Persistent cookies are stored by the web browser on disk. MS Internet Explorer, for example, stores each persistent cookie as a unique file in the user's "Application Data\cookies" folder. Netscape and Mozilla-based browsers store all cookies in one file named "cookies.txt". Since they are stored in files, persistent cookies can be altered by the user using a text editor.

Transient cookies are usually kept only in memory and erased when the browser window closes, making them more difficult to alter. However, several HTTP proxies are available that allow users to view and modify transient cookies, hence they must not be considered secure.

## 34.2 Building Blocks

If all session input from a web browser can be altered and is untrustworthy, how can a web application detect tampering?

### *34.2.1 Digest Algorithms*

Message digests, like checksums, determine a mathematical "fingerprint" of the characters in a string. The fingerprint can be compared with a saved or known good value to detect tampering [3]. Digest algorithms create a unique "signature" string for any given input data, such that it is practically impossible to alter the original data and still produce the same signature. Digests are used in SSL browser connections, Virtual Private Networks (VPNs) and Public Key Infrastructure (PKI) systems to "sign" data [8].

The same technique can be used in a web application to "sign" hidden fields. The most common digest algorithm is Message Digest 5 (MD5), but many other algorithms exist such as SHA1 and RIPEMD [8].

### *34.2.2 Digests and HTML Forms*

We can use MD5 or other digest algorithm to generate a fingerprint of hidden field data in an HTML form. We can simply concatenate the values of all the hidden fields into a string, run it through a digest algorithm to get a fingerprint and send it out in another hidden field. When the form is submitted by the user, the fingerprint of hidden field data is generated again and compared to the original fingerprint.

If the hacker comes to know that Message Digest 5 algorithm is in operational, it is possible for him/her to alter the hidden fields and generate a new fingerprint from the form data.

Fortunately, Digest algorithms are designed to make it impossible to determine the contents of a message from its fingerprint. We can use this property to add a secret component to the data before generating fingerprint so that the user never gets to know. The user may be able to compute the fingerprint of the "hidden" form fields, but without knowing the secret component, he/she cannot compute the correct fingerprint.

By adding a secret component to a digest, we are actually constructing what cryptographers call a "message authentication code" (MAC). The standard way to do this is via a function called HMAC.

### *34.2.3 The HMAC Standard*

The standard way to use a digest for message authentication is HMAC (RFC 2104) [8]. HMAC is a generic term for an algorithm that uses a keyless hash function and a cryptographic key to produce a keyed hash function [5]. This algorithm folds the text to be authenticated with two keys and three iterations through a digest such as MD5 or SHA1.

HMAC functions are either part of or an available addition to every programming language [8]:

```
Perl    Digest::HMAC  http://search.cpan.org/
PHP     mhash()       http://mhash.sourceforge.net/
Java    KeyGenerator  http://java.sun.com/products/jce/
MS.Net  HMACSHA1      Dot Net Framework 1.1
```

Example: a tamper-resistant form

The following is a part of JSP script of "change address" application again, but this time we add a MD5 HMAC "signature" to the form.

```
....
<input type="hidden" name="userid" value="<%=userid%>">
<input type="hidden" name="form_expires" value="<%=form_
  expires%>">
<input type="hidden" name="signature" value="<%=
  signature%>">
<input type="submit" name="chaddr" value="Change address">
...
```

When we run the above script in browser and view the HTML source code we see that the form has a new "hidden" field named "signature" to store a MD5 HMAC.

```
...
<input type="hidden" name="userid" value="nitie">
<input type="hidden" name="form_expires"
  value="20070 301:12:45:20">
<input type="hidden" name="signature"
  value="YJSG2af XQRSsvLdDXJpjFaxLLYo">
...
```

The fingerprint was generated using the values of the three hidden fields, plus a secret key stored only on the server.

When we submit the form, the contents of the three hidden fields are combined and an HMAC using the secret key is generated. If it doesn't match the "signature" field sent with the original form, the fields have been tampered with.

Now try saving the form to the computer and edit the "userID" field again. This time the tampering is detected. This technique can also be used with web applications that store the actual data on the server and use a client-side cookie to store a session key.

## 34.3 Proposed Approach

The user's selection on html pages is stored as a form field and is sent to the application as HTTP request (GET/POST). These form fields can be hidden or visible. The form fields of the URL can be easily modified by the hackers. Thus, in this manner, the attacker can gain an unauthorized access to the data.

The web application can be made secured by securing hidden parameters, form input parameters and URL parameters. For each page (whether it has a form element or not), we store all the hidden parameters and their possible set of values in the session as a hash table. In the case of a hidden parameter having more than one permissible value, all permissible values are stored in the session as a hash table. Similarly, form input parameters which are not hidden (e.g. list of values in combo box) are also stored in session as a hash table. To prevent manipulation of parameters in URL, the URL is secured by generating a key (e.g. hash value) corresponding to the URL to guarantee its integrity. Now when a request is made, the controller first checks the request for integrity of the fields and the URL and then performs necessary actions.

The following functions are implemented and used in web pages to enhance security of web-based applications.

1. checkSubmit
   This function is called first in each page to ensure that the web page is being called only through the request controller.
2. secureHiddenParameters
   For each page (whether it has a form element or not), we store all the hidden parameters and their possible set of values in the session as a hash table. In case of a hidden parameter having more than one permissible value, all permissible values are stored in the session as a hash table. Then call to this function is made by passing this hash table.
3. setParameterValues
   The form input parameters which are not hidden (e.g. list of values in combo box) are also stored in session as a hash table. Then call to this function is made by passing

   - Name of input parameter to be secured and
   - Value/list of values for that parameter.

4. allowNullValuesForParameter
   If a particular form element is disabled, the value returned by the selected parameter is null, hence this function is called, only if the disabled condition for that form element is true.

5. generateKeyForURL

To prevent manipulation of parameters in URL, the URL is secured by generating a key (e.g. hash value) corresponding to the URL to guarantee its integrity. Now when a request is made, the controller first checks the request for integrity of the fields and the URL and then performs necessary actions.

Each and every link generated in the web page must be secured using this function. The link which is generated using the following must be secured.

- href
- action = (for form)
- window. (window.open or window.location)
- <iframe src =
- Request?
- Response.

If the URL is being generated in a JavaScript function by concatenating the request parameters (JavaScript variables), the part of the URL except for the JavaScript variables must be secured. In such cases the parameters whose value is assigned using JavaScript are not secured.

## 34.4 Implementation

Model-View-Controller (MVC) is the most popular and implemented architecture in industry. All of the competing web technologies like JSP™, ASP.Net™ or PHP™ have provided an implementation of MVC. MVC organizes an interactive application into three separate modules: one for the application model with its data representation and business logic, the second for views that provide data presentation and user input, and the third for a controller to dispatch requests and control flow [2].

The following code (due to space constraints we have given pseudo-code) demonstrates how web security can be enforced in J2EE™.

Let ReqController be the controller servlet of the MVC Model in the web tier and Web Security be the class which contains functions for web security.

ReqController class is responsible for processing all requests received from the users and generating necessary response to the requests.

Process Logic for detection of form fields, URL parameters, and Hidden fields is shown in Fig. 34.1.

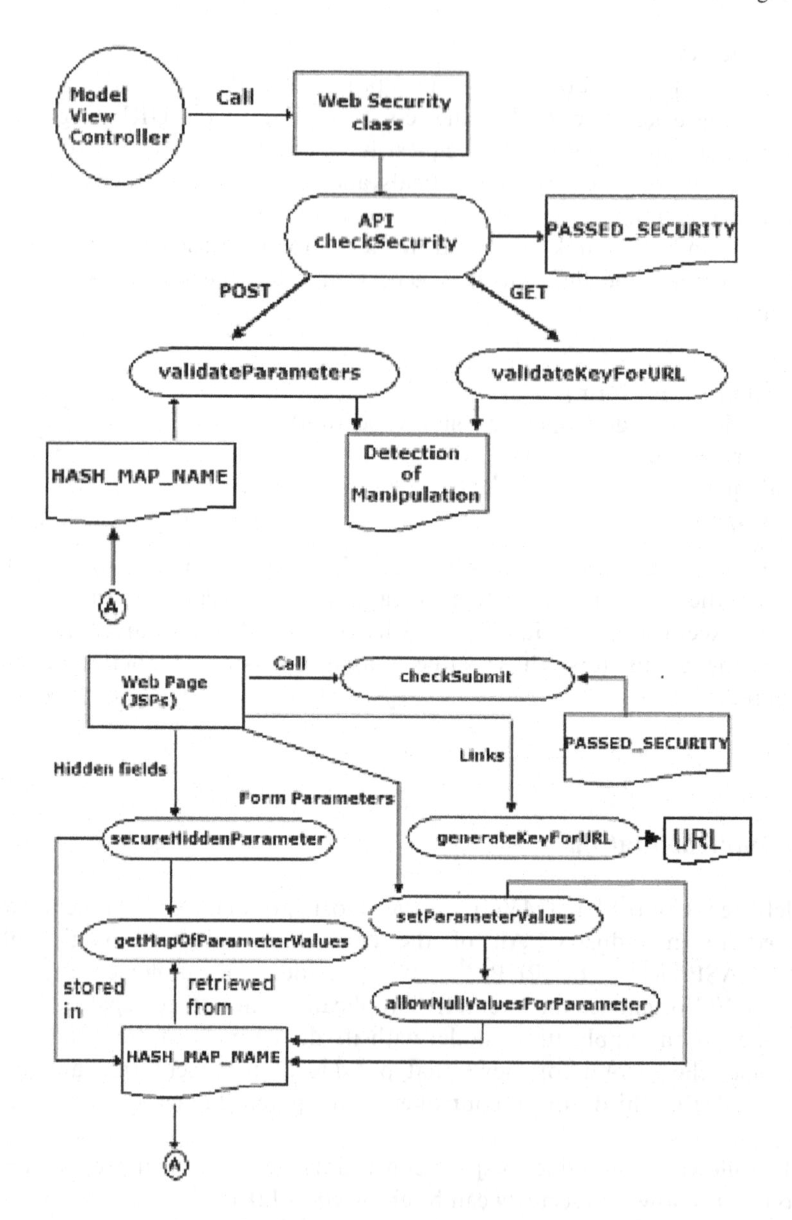


**Fig. 34.1** Process logic

Pseudo-code:

```
public class ReqController extends HttpServlet
{
        ...
        public void doGet(HttpServletRequest request,
 HttpServletResponse response) throws ServletException,
 IOException
        {
...
WebSecurity.checkSecurity(request,response);
...
  }
        public void doPost(HttpServletRequest request,
 HttpServletResponse response) throws ServletException,
 IOException
        {
                ...
        WebSecurity.checkSecurity(request,response);
                ...
        }
  ...
}
```

WebSecurity class contains all the functions for enforcing Web Security.

```
// This API will be called from the ReqController //
 Servlet.
//It will call all security related APIs.
void check Security(HttpServlet, HttpServletRequest,
 HttpServletResponse)
{
Set PASSED_SECURITY = true in request.
if (method = "POST")
{
Get HASH_MAP_NAME from the HttpServletRequest as
 w_mapName
Call validateParameters(request,mapName);
}
else if(method = "GET")
{
Call validateKeyForURL(request,servlet);
}
}
 // This API will be called from all JSPs.
```

```
// PASSED_SECURITY attribute in Request is set
//to true if web page is being called through the
//controller.
void checkSubmit(HttpServletRequest, HttpServletResponse,
ServletContext)
{
If ( RemoteHostAddress == LocalHostAddress OR RemoteHostAd-
dress == "127.0.0.1")
Then Ignore
Else If request.getAttribute(PASSED_SECURITY) == null
Then throw Exception("This jsp has bypassed Security
Checks.");
}

//This API will be called from all JSPs to secure
//hidden parameters.
//The paramater a_paramValues should be as
//follows:
// e.g. "a:1,b:2,c:3:4:5,d:4:6"
void secureHiddenParameter(HttpServletRequest, Http
ServletResponse, ServletContext, String paramValues)
{
Get HASH_MAP_NAME from the HttpServletRequest as mapName
Get HashMap corresponding to mapName from the Session as map
map = getMapOfParameterValues(map,paramValues);
Set HashMap corresponding to mapName in the Session.
}

//called from secureHiddenParameter
HashMap getMapOfParameterValues(HashMap map,String param-
Values)
{
Separate the parameters and their corresponding values from
paramValues
Put (parameter, values) pairs in hashmap map
return map;
}

//This API will be called from all JSPs to secure
//input parameters.
void
setParameterValues(HttpServletRequest,HttpServlet
Response,ServletContext,String paramName,Object
param Values)
{
```

```
Get HASH_MAP_NAME from the HttpServletRequest as mapName
Get HashMap corresponding to mapName from the Session as map
Get ArrayList of values corresponding to paramName from the
HashMap
as valueList
Add paramValues to the ArrayList valueList
Put valueList coressponding to paramName in HashMap map.
Set HashMap corresponding to mapName in the Session.
}

void allowNullValuesForParameter(HttpServletRequest,
Http ServletResponse, ServletContext, String paramName)
{
WebSecurity.setParameterValues(a_request,a_respon-
se,a_application,a_paramName,"Dummy");
}

void validateParameters(HttpServletRequest,String
mapName)
{
Get HashMap corresponding to mapName from the Session as map
For each parameter param in hash map
{
Get ArrayList of values corresponding to param from the
HashMap as valueList
Get ArrayList of values corresponding to param from the
request as reqValues
For i = 1 to no_Of_Values_In_Request
{
For j = 1 to size_of_valueList
{
        If (w_valueList[j].equals (w_reqValues [i]))
{
w_matched++;
break;
}
}
}
if (w_matched==0)
{
throw Exception("The parameter has value which is not in
the allowed set of values.");
}
}
}
```

```
void validateKeyForURL(HttpServletRequest a_request,
HttpServlet a_servlet)
{
String w_secretParameterName = "key";
String w_secretParamValue = "RandomString";
String w_keyStr[] = a_request.getParameterValues(w_secret-
ParameterName);
if (w_keyStr == null || w_keyStr.length == 0)
{
throw Exception ("Keys Manipulated");
}
String w_queryStr = a_request.getQueryString();
if (w_queryStr==null)
{
throw Exception ("No query String");
}

String w_qryStr = w_queryStr.substring(0,w_queryStr.last-
IndexOf("&"+w_secretParameterName+"="));

Find digest of (w_qryStr+w_secretParamValue) using "MD5"
message digest algorithm as w_NewKey.
if(!w_NewKey.equals(w_keyStr[w_keyStr.length-1])) //Pick
the last one
{
throw new Exception ("Query String Manipulated");
}
}

// This API will be called from all JSPs.
String generateKeyForURL(HttpServletRequest a_request,
HttpServletResponse a_response,ServletContext a_applica-
tion,String a_url)
{
if (a_url.indexOf("? ")==-1) //there is no ? in the url
{
return a_url;
}
String w_secretParameterName = "key";
String w_secretParamValue = "RandomString";
String w_temp = a_url.substring(0,a_url.indexOf("? "));
String w_params = a_url.substring(a_url.indexOf("? ")+1,
a_url.length());
Find digest of (w_params+w_secretParamValue) using "MD5"
message digest algorithm as w_Key.
```

```
return
w_temp+"?"+w_params+"&"+w_secretParameterName+"="+
w_Key;
}
```

## 34.5 Conclusions

The proposed approach can detect tampering of form fields, URL parameters and hidden fields, but it is not foolproof. If GET method is used in request then the validateKeyForURL procedure is not secure enough. The weakness of this method lies in the "secret key" stored on the web server. If the server is broken into or someone manages to view the source code of the application, the method is no longer secure. However, assuming that the web server is reasonably secure from break-ins and that we change the key regularly, the approach is secure enough.

## References

1. Schneier B (1996) Applied cryptography. John Wiley, New York
2. Singh Inderjeet, Stearns Beth, Johnson Mark, and the Enterprise Team (2002) Designing Enterprise Applications with the J2EE™ Platform. 2nd edn. Addison-Wesley, Michigan
3. Graff MG, van Wyk KR (2002) Secure coding. O'Reilly & Associates, California
4. Smith RE (1999) Internet cryptography. Addison-Wesley, Michigan
5. Venkatramanayya S, Bishop M (2006) Introduction to computer security. Addison–Wesley, Michigan
6. CERT advisory CA-2000-02. http://www.cert.org/advisories/CA-2000-02.html
7. Hidden form field vulnerability. White papers (InfoSec Labs), http://www.infoseclabs.com/mschff/mschff.htm
8. Preventing HTML form tampering. http://advosys.ca/papers/form-tampering.html
9. World Wide Web security FAQ. http://www.w3.org/Security/Faq/www-security-faq.html

# Chapter 35
# Description and Discovery of Web Applications in Grid

**Serena Pastore**

**Abstract** Most of the shared distributed services on the Internet are developed as modular web applications, based on infrastructures that should provide several functionalities in order to guarantee the effective management and use of resources. Thus an adequate resources' representation is required for discovery, composition and security goals. This paper starts from the experiences gathered in the broad context of the GRID-IT project, aiming at porting in grid astrophysical applications, to report about the efforts and solutions proposed in describing a web-based application resource and its relationships with discovery. Starting from a grid-based data model (GLUE schema), the descriptions involve the Web Services Description Language (WSDL), its relationships with a web service-based data structure (UDDI schema) to a semantic annotation language for service capabilities (OWL-S or WSDL-S) in order to enhance the discovery task. The steps needed to process the web application in order to provide such information are provided by using existing open-source software tools.

## 35.1 Introduction

The Internet functions as a global network of shared distributed resources and services. Most of the services were developed as web applications by using several paradigms and programming languages (i.e. Java, PHP, etc.). The underlying distributed infrastructures that should provide main functionalities to guarantee the effective management and use of the distributed, networked resources are implemented as a grid system [1]. However, the need of specific resources' description following a detail data model that affects the management, discovery, composition and security tasks is required. This paper starts from these issues to describe experiences made in testing the solutions for

S. Pastore (✉)
INAF – Astronomical Observatory of Padova, Vicolo Osservatorio
5 – 35122 – Padova, Italy

N. Mastorakis et al. (eds.), *Proceedings of the European Computing Conference*, Lecture Notes in Electrical Engineering 28,
DOI 10.1007/978-0-387-85437-3_35, © Springer Science+Business Media, LLC 2009

working with web applications in a grid system [2]. The work has arisen from a broad framework, the GRID-IT project (http://www.grid.it) which aimed to establish and test the Italian grid infrastructure. In specific, the WP10 project group, conducted by INAF (http://www.inaf.it), had the task of studying the porting in grid astrophysical applications. The nature of such applications involves web services technologies [1] for its interoperability features, grid technologies for the exploitation of computing and storage facilities, and a secure environment (i.e. like the grid ones) to satisfy local policies. These are some of the practices that the astronomical community grouped in the IVOA alliance (http://www.ivoa.net) proposes in the optics of having a unified framework of resources managed by each site. The paper faces grid resource description challenges, referring to specific web application resources made up of web services deployed in the grid distributed system provided by the EGEE-II (http://www.egee-eu.org) infrastructure. In such a framework the lack of adequate software resource description can affect several aspects, including the discovery and use of such resources. This study provides the "application layer point of view", since an adequate description funded on the fact that the software resource is poorly described in EGEE-compliant grid system affects both the application development and the resource provider approach in the discovery task. Starting from a specific grid data model (i.e. the GLUE schema, http://glueschema.forge.cnaf..infn.it), the description of such resources involves the WSDL language [1], its relationship with a web service-based data structure (UDDI schema, http://www.uddi.org), to semantic annotation languages for service capabilities to enhance discovery. The solutions proposed are based on the adoption of existing grid and web services standards through available open-source software implementations that may be customized.

## 35.2 The Project Scenario and Resource's Provider Challenges

The EGEE-II infrastructure as the evolution of European grid projects in the last years represents an ideal grid platform for running applications. In Italy the access to the infrastructure is directed by the INFN (http://www.infn.it) which coordinates its use by providing a set of main grid services. The other institutions structured in Virtual Organizations (VOs) [1] contribute their resources by connecting to the grid of one or more sites, made up of physical machines which provide the logical functionalities. EGEE-like architecture [3] implemented through the gLite toolkit (http://glite.cern.web.ch) distinguishes logical nodes that, with customized specific software, perform the grid core services and the grid resources: the workload and data management, the information system, computing resources (CE), storage resources (SE) and processing nodes (WN). Thus, there is both a centralized management of core functionalities, needed for the entire system to work and a decentralized (VO or site-based) management of resources (through the CE, WN and SE nodes). A software resource provider is realized by the hosting environment (i.e. an application server like the Apache Tomcat, http://tomcat.apache.org for Java-based applications). However, their use is related to the effective integration of the software components within the grid core functionalities. In an ideal environment a consumer (i.e. human, application or agent) should be able to automatically match its requirements with the provider's opportunity which offers a network-oriented interface by which the client invokes the service. A user/client takes advantage of the grid power if there is an automatic match between the requirements and the best resources available at that particular time.

## *35.2.1 Grid Services, Grid Resources and Web Services-Based Application*

Grid services (i.e. discovery) provide uniform interfaces to the types of system elements (i.e. grid resources). According to grid definitions [1], resources are heterogeneous, geographically distributed, connected by a wide-area network, and accessible on-demand. However in the EGEE model [3], grid resources focus on hardware features of each VO's site. The data model adopted for resources is in fact the GLUE schema, an abstract model available for both LDAP and XML schemas, which describes the attributes and the values of each of the grid sites (Fig. 35.1). It distinguishes core entities (*site* and *service*), computing resource (*cluster, CE* and *subcluster*) and storage resource (*SE, storage area*) elements. A site collects the resources owned and managed by the same organization and it is made up of clusters (an aggregation entity that represents a complex computing resource). In this context, a software resource is described by the "*Software*" entity as a feature of the *subcluster* element, which in turn provides details of the machines (*clusters*) that offer job execution environments. The software entity describes what software packages are available in the worker nodes part of the *subcluster*. Its attributes consist essentially

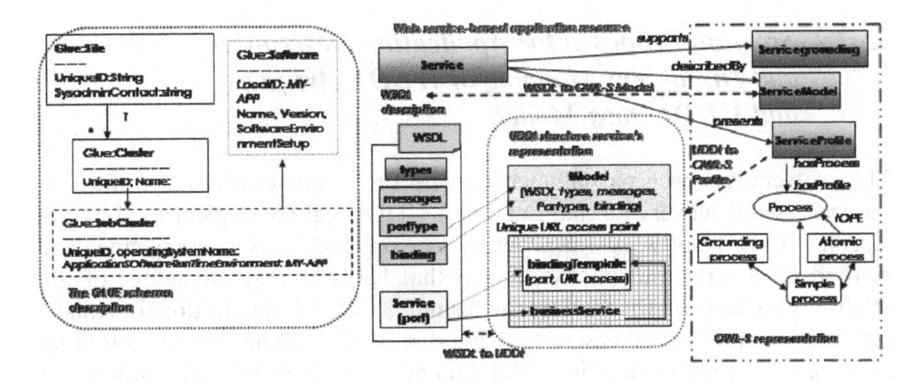


**Fig. 35.1** GLUE, WSDL, UDDI and OWL-S representation of web application

of the name, the version and some values about the runtime environment. On the other hand, the service element, which was introduced later, is not specific to the web service software and refers to the site element features. All grid resources are indexed by the grid information system (IS) [3] that is in charge of collecting and monitoring resources. Initially implemented as an LDAP directory access system (through the BDII system) [3], it is now going to be substituted with the R-GMA system (http://www.r-gma.org) that bases relational database store entities information into tables and uses SQL commands instead of LDAP tools for searching tasks. However, since it is based on the same data model, software discovery is made through the attributes associated to the "software" element and searching tools are thus limited to a few keywords (i.e. name and version). This becomes a problem for web service-based applications in which descriptions are rich in information and can be used in the discovery task. Software resources, like java web service applications, in fact provide a rich set of information useful for sharing, publishing, discovering and invoking them across the web in a secure way. A web service may be considered as the new breed of web applications, opposed to packaged based products. Layered upon an XML-based stack [1], it is described by specific languages that describe its capabilities and exchanges messages to be processed according to the detail specifications (i.e. SOAP protocol). A SOAP-based web service, for example, exposes on the network a WSDL interface that separates the abstract functionalities from the details of how and where the service is offered. According to the version 1.x of the specification, the service in the WSDL document is described (Fig. 35.1) within the <service> element as a collection of endpoints (<port>). Each endpoint (port) defines a combination of binding (<binding>) which specifies a protocol and data format and a network address. A portType collects the operations supported that are specified in terms of input/output messages (<message>) and data types (<types>) exchanged. Unfortunately all of this information cannot be used in the grid discovery task with the current core functionalities.

### *35.2.2 Solution Proposed for Application Description and Discovery: Mapping of WSDL Artifacts and UDDI Data Model*

The problem has been partially solved with the installation of a specific web service-based discovery mechanism and a UDDI registry implemented by the Apache jUDDI software (http://ws.apache.org/juddi) as a complementary method [4] to grid information system that helps to take advantage of the relationships between the UDDI schema and the WSDL description for enhancing the web application description and discovery. This has been followed by an integration procedure. The UDDI data model (Fig. 35.1) describes both the provider and the services it exposes via a number of bindings to different protocols and at different physical locations (network access) through the core entities (*businessEntity, businessService* and *bindingTemplate*). Moreover it defines the *tModels* element that, used as reference to other elements, serves as a place holder for concepts to identify property namespaces and categorization schemes for classification.

UDDI mainly uses property-based lookup and taxonomic categorization. The properties (identifiers) with names and values attached to the data structures are supported inside the *identifierBag* element and are represented by *keyedReference* elements. In a similar way, categories which are applied to classify objects are supported by the *categoryBag* and *keyedReference* elements. Main category schemes refer to industry, services or products and geographical locations, but others are supported by *tModels*. This structure essentially contains a *tModelkey* as the unique identifiers, a name and a description, an *overviewDoc* element as URL to document describing *tModel*, and *identifierBag* and *categoryBag* structures. WSDL and UDDI constructs together allow for a full description of web service application and for realizing the main operations (find, invoke and publish). While the WSDL document describes the network interface, the *tModel* document provides metadata descriptions and pointers to specifications that describe the implementation. Given this provision, WSDL documents tie into the UDDI data structure in a couple of pieces and this mapping allows for the storage of WSDL information into the UDDI registry. For each WSDL document two *tModels* (one for the portType and one for the binding) are created and one *businessService* with one *bindingTemplate*. *tModel*, as an abstract structure, does not specify service or port elements. It is actually the *bindingTemplate* that is created for each unique URL access point used by the service that carries out this function. Libraries and APIs allow for finding web services through such registry (by using WSDL metadata), but it is done so in a disjointed manner in respect to standard grid mechanisms. A first integration test has been done for including such information in the LDAP grid schema [4] thanks to the RFC 4403 (http://www.ietf.org) that maps UDDI entities into the LDAP schema. By using the XML version of the GLUE schema, it probably will be possible to integrate the different schemas and to use the standard grid tools to find a specific WSDL or UDDI structure. Now improvement in the grid middleware has introduced the Service Discovery package [3] as a front-end to several discovery methods. This has led to further work on the UDDI solution and the adoption of richer resource descriptions. In this optics, a web services semantic description gives the possibility to enhance the automation.

## 35.3 A Semantic Description for Resources

Semantics that help to describe the resources in an unambiguously computer-machine manner and are able to promote more refined search results compared to syntactic matching, is based on a slew of technologies and infrastructures based on XML, in part developed within the W3C consortium (http://www.w3.org). Resources are defined as classes and subclasses of objects (defined within a domain vocabulary or ontology) with relations about them (expressed with properties), and a set of inference rules. Its application in the web service area has been taken on by several projects aiming at providing languages or specific data models. Semantic web service is a research topic of the DAML program (http://www.daml.org) and it has its result on the OWL-S (OWL-based web service ontology) ontology language, now a W3C standard, as a technology for building semantic web services by integrating business logic with semantic web's meaningful content. The language provides a set of constructs for describing the properties and capabilities of web services. On the other hand, the W3C consortium hosts an activity that takes account of all of the semantic efforts (i.e. METEOR-S (http://lsdis.cs.uga.edu/projects/meteor-s/) and WSMO (http://www.wsmo.org) projects). It proposes the WSDL-S as a mechanism to associate semantic annotations with the WSDL-based version 2.0 (but also with WSDL 1.1) descriptions, and as a method to assess annotations to services. Finally, the Semantic Web Services Framework, which consists of a language and ontology has been proposed to the W3C by the SWSI initiative (http://www.swsi.org). Even if there are a lot of initiatives in dealing with semantic web services, the work has focused on the OWL-S language since the availability of open source software to work which was developed by the Software Agent group of the Carnegie Mellon University (http://www.cs.cmu.edu). It is likely that the WSDL-S language will be more useful for the studied web applications, but at present there are too few tools to work with.

### 35.3.1 Web Application Description Enrichment with OWL-S Language and its Application to Discovery Task

According to OWL-S, for each distinct service composing the web application there is an instance in which the class Service, to whom it associates three properties and thus three XML-based descriptions (*ServiceProfile, ServiceModel and ServiceGrounding*), provides all the information needed by a client or an agent to interact with (Fig. 35.2).

A service can be described by at most one model; however it is associated with having only one grounding. The Profile (*presents*) defines the service's capabilities in terms of the processes performed and thus gives information to the discovery task, while the Model (*describedBy*), which describes how it is used and the Grounding (*supports*) which offers the details about transport protocols and data formats are useful to make use of once found. When the web application is semantically described, its discovery needs a customized registry enabled to parse semantic annotation when required by a client. All the steps shown in Fig. 35.2 are needed to enrich the grid application description, process and use it through a discovery in the grid, which are then realized by tools and

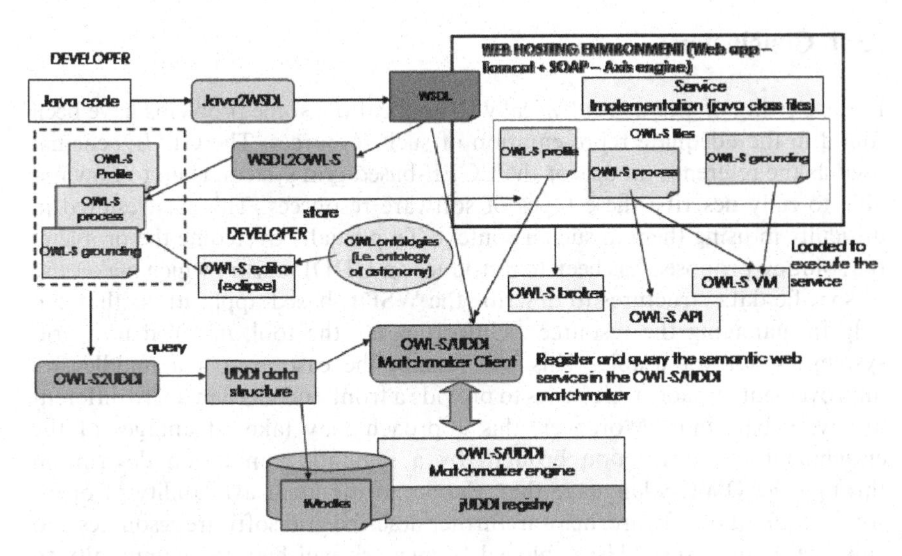


**Fig. 35.2** An overall representation of all the modules and tools needed to create, deploy and use a semantic-enhanced web service application

components provided by the Carnegie Mellon software. Starting on the development phase when the Java source code that is parsed by the Java2WSDL conversion, the WSDL file is exported as a network interface inside a grid machine, thus accessible by all grid consumers which are authorized to access the web service framework where the service implementation (java class files) are deployed. Moreover the UDDI registry is also deployed inside a grid node. At this point the semantic annotation of the service description is done through the WSDL2OWL-S tool which creates the needed OWL-S files. This step, however requires a further elaboration to fill out missing information and if necessary to validate it. In this phase for example, other ontologies (i.e. the astronomical ontology) are applied in order to better define the application context. All files are hosted in a web/grid accessible location. The realization of an OWL-S-enabled registry is provided by the OWL-S/UDDI matchmaker, a software that consists of an engine enhancing the capabilities of a jUDDI-based registry in order to understand and process semantic information related to service description and the client component (java library). This is then used to publish and query information on the registry. The approach takes advantage of both data model relationships and of the *tModel* concept as the way to store ontology information by realizing the mapping of OWL-S constructs with WSDL artifacts and UDDI data. Specifically, OWL-S Model refers to WSDL artifacts, and OWL-S Profile to UDDI structures. The conversion between the OWL-S files and the UDDI data are made by the OWL-S2UDDI tool, while *tModels* representing OWL-S structures are stored in the relational database of the registry.

## 35.4 Conclusions

In the porting of grid astrophysical web application, some problems have been found in the adequate representation of such resources. The GLUE schema, used as the reference model for the EGEE-based grid system, is up to now not able to fully describe these types of software resources. This has resulted in difficulty in using them in such a context. To partially overcome the problem, the solution proposed has been to introduce an UDDI registry which makes use of specific data structures to describe the WSDL-based applications that can help in managing the resources. Unfortunately the tool, installed in a grid system, is not yet usable. This will remain the case even if a middleware improvement is made, which tries to provide a front-end mechanism to different discovery backends. Moreover, this approach may take advantages of the enrichment of information brought by a semantic annotation description through the OWL-S language that, thanks to the great availability of open-source tools and software, helps to further describe grid software resources and enhance its discovery. The deployed framework will help the community to learn and test grids, web services and semantic technologies in order to better describe, use and manage web applications.

## References

1. Sotomayor B, Childers L (2006) Globus Toolkit 4. Elsevier, Amsterdam
2. Volpato A, Taffoni G, Pastore S et al (2005) Astronomical database related applications in the Grid.it project. In: ASP Conference Series, Vol XXX. pp P1-3-5
3. EGEE JRA1 Team (2004) EGEE Middleware architecture. EU deliverable DJRA1.4, EGEE-DJRA1.1-594698-v1.0
4. Pastore S (2006) The service discovery methods issue. J Network Comp Appl, Elsevier Science, to be published

# Chapter 36
# An Approach to Treat the User's Preferences About Personal Data

**Sérgio D. Zorzo, Reginaldo A. Gotardo, Paulo R. M. Cereda, Bruno Y. L. Kimura, Ricardo A. Rios, and Robson E. Grande**

**Abstract** The concern with users' privacy in the Internet is an argued subject in some areas of knowledge. Mechanisms and actions exist to promote guarantees of user's privacy; the majority of them, however, present limitations in the privacy control. This paper describes a procedure for the user's privacy preferences treatment in web systems based on a contextualization model. It offers a personalization choice to the user guaranteeing that his/her privacy preferences are respected. The procedure is applied in a case study and the observed results are presented, evidencing that the privacy personalization improves the user trust in web transactions.

## 36.1 Introduction

Services personalization has a great importance in Web enterprises, because it is a kind of direct marketing which is based on users' profiles, directs contents and advertisement. According to research conducted by Kobsa [2], websites that offered personalized services obtained a 47% raise in the number of new clients.

Direct marketing is substituting the traditional mass marketing strategies. However, to apply personalization, data collection about users is necessary. This brings into consideration a discussion about privacy.

When a process of data collecting is started, the user needs to be informed about it (and agree to it) as well as being informed of how the data will be treated.

According to Teltzrow [5], 64% of Web users gave up accessing some website, or buying something online, because they didn't know how the information they gave to the system would be used.

Data collection can occur in implicit or explicit ways. Explicit collection happens when an individual sends information to a certain destiny, trusting that the information will be stored and conveniently treated. Implicit collection

S.D. Zorzo (✉)
Computer Science Department, Federal University of São Carlos, Rod.Washington Luis, km 235 – 13565-905, São Carlos – SP – Brasil
e-mail: zorzo@dc.ufscar.br

N. Mastorakis et al. (eds.), *Proceedings of the European Computing Conference*, Lecture Notes in Electrical Engineering 28,
DOI 10.1007/978-0-387-85437-3_36, 

happens without an individual noticing it and generally consists of gathering data based on a user's interactions with the system.

Implicit collection allows the creation of profiles based on users' interests, mainly inferred from his/her browsing patterns. Therefore, explicit and implicit collections differ from each other in relation to the authenticity of the information. When information is collected in an explicit mode, its authenticity is the user's responsibility. However, when the information is collected in an implicit mode, its authenticity can be questionable, since the user is unaware of the whole collecting process.

In many situations the user's confidence in web systems depends on the guarantee of privacy offered by these systems. User's suspicion about non-authorized spreading or bad use of his/her information leads him/her to give up accessing certain services.

This paper is not related to the legal restrictions applied by some legislative or normative sector. The proposed approach aims to respect the user's choice about his/her information treatment, congregating the main characteristics of legislations, norms and principles about existing data collection.

Privacy protection techniques are used to increase the user's control about his/her information in Web browsing. The creation of pseudonyms, use of privacy contracts and the introduction of anonymity for browsing are examples of these techniques. Each one of them approaches in different forms the user's privacy, as much for implicit collection as for explicit collection, allowing or prohibiting data acquisition for personalization application.

Based on the existing techniques, the information spreading about websites privacy practices is the one that better enables the user to decide about his/her preferences. The user acquires knowledge about offered privacy options, deciding (or not) to subscribe to the service. However, this technique uses static privacy policies, which make possible only one between two choices: to access or not to access a Web service.

Aiming at increasing the user control over information, this paper proposes a model inspired by the contextualization method created by Kobsa and Teltzrow [3], which implements a scheme for different possible choices for privacy treatment combined with personalization methods.

In Section 36.2 some Web privacy concepts are presented, highlighting the relevant ideas and mechanisms related to the work described in this paper. In Section 36.3, the user-controlled privacy model is described, and in Section 36.4 a case study conducting using a website is presented. The results and conclusions are presented in Section 36.5.

## 36.2 Current Web Privacy Treatment

Web Privacy is, in a brief form, the user's capacity in controlling his/her data. It can be related with users' information maintenance right in a secure way, without the possibility to identify them while browsing Internet.

In this paper the privacy treatment will be done according to the main agent. Thus, the forms of Web privacy treatment, as such by the user, or information owner, as by the server, or information collector, will be related. To the user, tools and techniques for privacy protection are available. The server is in charge of policies and how to collect and notify the user, allowing or disallowing the control by him/her.

There are techniques and mechanisms that offer ways to control improper information collection to the user like anonymity, pseudonyms, privacy agents and filters. Also, the privacy can be managed in the own content server for several techniques that can be synthesized in privacy policies [1, 4], negotiation protocols (P3P) [1], privacy seals [4], or privacy preferences contextualization model [3, 5].

## 36.3 Web Privacy Controlled by User

When browsing a certain website, the user must be aware about the data collection done. The website must show explicitly, through its privacy policies, all the pertinent information about this collection. The user must read the privacy policy and, if he/she agrees to it, continue his/her browsing, or, if not, must leave the website.

This is an approach that competes with the user browsing; therefore the user needs to observe all the privacy policies to verify which treatment is given to his/her privacy in the website. This can lead to the desistance of privacy policies reading of the accessed Websites, being the user confidence based only on the Website popularity. In this form, the user's privacy is harmed, because there's no concern with the website privacy practices, but with the image that the website itself represents.

This paper proposes a negotiation between user and website, in which the user defines his/her privacy profile. At the moment where the user has control of his/her privacy, defining it through his/her preferences, the website will explicitly collect only the allowed data.

According to Kobsa, the privacy is related to the control that a user has about his/her information. In this way, transactions that provide trustworthiness to the user relating to his/her data use, will provide better commercial results and they will prevent failures in the business model, since the user has the fundamental role in the website–costumer relationship [2].

Web Privacy Controlled by User, presented in this paper, proposes alterations to Kobsa and Teltzrow model, treating the privacy preferences in a centered way, and not contextualized. In this form, the data collection has only the global context, that can be configured according to the choices of each user.

Additionally, the concept of "privacy status" is presented, where the user can visualize his/her privacy preferences for a certain website, as well as modifying these preferences. It is a guide who presents to the user, briefly, details about the used mechanisms for data collection. This mechanism demonstrates the continuous concern of the system in dealing with the user's privacy and adjusting such treatment to his/her choices.

Through “privacy status”, the user is aware about the mechanisms that are being used for data collection and how they are being used in an accessed page. By the “What is this?” guide, the user obtains clear information that explains the website’s intention in offering such functionalities. This guide also helps users with information about the used collection mechanisms.

Another modifier is the maintenance of the user model, which keeps his/her privacy preferences in his/her computer using Cookies, making it possible to keep an access session and offer personalized services.

The Web Privacy Controlled by User Approach modifies some of the functionalities of Kobsa and Teltzow proposal [3, 5]. Through this approach, the user obtains direct access to his/her privacy preferences and this negotiation influences website behavior. The useful benefits of the mechanisms continue presented to the user, as previously, and by “privacy status”, available on the interface, the user can view and modify his/her privacy preferences.

The privacy policies that describe the objectives and used mechanisms of collection, also present the benefits related to each mechanism and type of data collected. In this form the user can control the data collection and he/she can verify the influence of this collection and his/her control on this.

Currently, it is predominated the use of P3P for Website privacy policy spreading. However, besides presenting low use, it also has limitations, as not to present the benefits of data collection and to request that the user takes decisions previously, without relating them to the context of use.

The use of Privacy Seals collaborates with the ratification of practices described in the Website Privacy Policy. However, the companies that supply the Privacy Seals must be trustworthy and well known by the users.

The used mechanisms by the user (privacy agents and filters, pseudonyms and anonymity) harm the offer of personalized services by websites. Such mechanisms do not supply ways of data collection negotiation between websites and the users, differently to the Approach proposed in this paper.

In the Kobsa and Teltzrow model [3, 5] the reading of the privacy policies becomes easier; therefore it transforms into lesser parts the reading of extensive policies. However, the developer’s work becomes more onerous, because he must treat each aspect of the business model considering the privacy question.

It is important to say that this Approach can be used with other presented mechanisms.

## 36.4 Case Study

The Case’s Study Approach was done using the Federal University of São Carlos’ Graduate Course website, in the period of 4 November to 5 December 2006. In this website, the interested candidates in the Course get information about it, about the Institution and the Faculty. The interested parties can use the on-line registration. The registration process requires that personal information and digital documents be sent. The Approach evaluation was exactly during the period of on-line registration, because of the great access flow to the website.

For the creation of the user's privacy preferences two mechanisms had been used: Cookies and ClickStream. Allied to them it was implemented the anonymity controlled by the website and of user choice.

A Cookie is a set of information changed between the user's browser and the web server. This information is placed in an archive of text that can be stored in the user's computer or only in the computer's memory, during the browsing.

ClickStream is a technique to analyze the user's browsing information in the websites. Through the analysis of the user's clicks sequence it is possible to infer information about pages preferences or the user's profile. ClickStream data can be collected using information generated by the same page, information of the Internet Service Provider (ISP) or using the servers' logs.

ClickStream acts with some data mining technique to achieve the inference of information, being also used for the generation of statistical information. When the user disables this option, his/her requests of pages won't be collected.

Every user has a sessionId for his/her identification. In this case, Session Cookies have been used and this has been excluded after the user browsing has been finished. The information that had been recorded in the user's machine is related to his/her privacy preferences about the website and they are only recorded with the permission of each user.

In this case, when the user accesses the website again, it is identified the existence of a Cookie with the determined user's privacy preferences and, thus, the standard preferences are changed by the chosen ones.

One clear privacy policy was constructed, guaranteeing trustworthiness for the users. Allied to this policy was created a page with auxiliary information that explain all the used mechanisms.

The identification anonymity was offered to the users. So, when they are using this option, the website doesn't collect information that can identify the users, such as their IP address.

This mechanism, normally supplied available tools to the users, was implemented in the same website, what it demonstrates, again, the concern with the preferences of users' privacy.

In the Website Privacy Policy declaration the necessary forms for alteration of the user's privacy preferences are available. The user can observe his/her privacy status and, if necessary, modify it. As already displayed, the website's behavior is modified according to the alterations that have been made, differently to the other privacy proposals.

Through the guide "what is this?" the user can get information about the collection mechanisms used in the website and the functioning of these.

In this case study it was observed that 79% of the users who had read the privacy policy – acquiring knowledge themselves of the data collection mechanisms used – had supplied their personal data to the website for registration in the Graduate Program Course. It demonstrates that the really interested users in the offered Course had searched information related to the security of their data and their privacy.

The website structure contemplates information pages for visualization by the users and pages to send personal information for registration, as well as personal document copy.

This service type is highly directed to the specific public and to only one transaction object, having no generic focus, as e-commerce applications.

The measurement about privacy preferences, in this case, is related only to the interest in the Course and the registration for the selective process, different to conventional e-commerce that offers several products and services to the users.

## 36.5 Conclusions

This paper contributes to the area of Personal Data Protection by proposing an approach for Web privacy preferences and discussing its implementation.

It presents quantitative data gathered during approximately one month, with constant monitoring. The results give an insight as to the user's behavior related to the website privacy treatment. They allowed to conclude that a great amount of users read the privacy policy and were aware of the proposed approach, trusting the system and supplying personal information.

The applicability of the approach, however, depends on many factors, such as the context and the current laws, which influence its implementation.

It is a fact, however, that common sense tells that websites should be concerned about users' right to privacy, and respect their preferences related to the private characteristic of the information which websites collect. This can be accomplished by means of providing a personalization procedure.

The Approach to Web Privacy Preferences implementation results has proved to be an important tool for users interested in controlling their preferences, assuring their privacy when using the Web for private transactions.

## References

1. Cranor LF (2002) Web privacy with P3P – The platform for privacy preferences. O'Reilly, Sebastopol, California
2. Kobsa A (2003) Tailoring privacy to user's needs. Lect Notes Comp Sci 2109(1):303
3. Kobsa A, Teltzrow M (2004) Contextualized communication of privacy practices and personalization benefits: Impacts on users' data sharing and purchase behavior. In: Martin D, Serjantov A (eds) Privacy Enhancing Technologies, Fourth International Workshop, PET 2004, Toronto, Canada
4. Mai B, Menon N, Sarkar S (2006) Online privacy at a premium. In: XXXVI Hawaii International Conference on Systems Sciences
5. Teltzrow M, Kobsa A (2004) Communication of privacy and personalization in e-business. In: Proceedings of the Workshop "WHOLES: A Multiple View of Individual Privacy in a Networked World", Stockholm, Sweden

# Chapter 37
# Ssort: Peak Analysis for Efficient String Sorting

**Sunil Mallya and Kavi Mahesh**

**Abstract** With the growth of the Internet, processing large amounts of textual data has become a common requirement in practical applications. This paper presents an efficient algorithm, Ssort that is specialized for sorting large numbers of strings. It is an improvement over Quicksort and its variants such as Meansort. It has an excellent average case behavior. Moreover, its worst case is highly unlikely to occur in practice.

## 37.1 Introduction

In today's world, due to the advent of the Internet and large-scale text databases, it has become increasingly important to have a good text processing toolkit. String sorting is an essential part of such a text processing toolkit. Unfortunately, currently available sorting algorithms are not ideal for sorting long lists of strings and not much work has been done in this domain. Most sorting algorithms available for sorting numbers are extended to sort strings with a suboptimal performance and thus there is a need for a specialized algorithm for sorting strings.

Divide and conquer techniques for sorting algorithms have been known for their superiority in sorting by comparison. Of all the algorithms of this paradigm, Quicksort by Hoare [2] is the most popular algorithm due to its efficiency parameters. One of the drawbacks of Quicksort is the worst case behavior of $O(n^2)$ comparisons, in the standard version of Quicksort. This case will occur when the file is sorted or reverse sorted. The Meansort algorithm by Motzkin [1], which is a modification of Quicksort algorithm calculates the arithmetic mean and then partitions the array. This improves the Quicksort algorithm because arithmetic mean provides for a better pivot resulting in two almost equal partitions. Meansort has not been extended to strings. Meansort algorithm cannot be applied directly to string sorting without significant changes.

S. Mallya (✉)
Department of Computer Science and Engineering, PES Institute of Technology, Bangalore 560085, India
e-mail: mallya16@gmail.com

N. Mastorakis et al. (eds.), *Proceedings of the European Computing Conference*, Lecture Notes in Electrical Engineering 28,
DOI 10.1007/978-0-387-85437-3_37, 

Meansort will fail to achieve good results when the input is concentrated in a particular area. For example,

E.g. 1: If we take the input to contain 10 strings starting with the letter`a': a1, a2...a10 and 100 strings starting with the letter`z': z1, z2...z100,

The arithmetic mean calculated would be 'x', resulting in a left partition having 10 elements and a right partition having 100 elements. This is highly undesirable because such repeated bad partitions can result in $O(n^2)$ growth. Drawbacks of Meansort algorithm are overcome in our algorithm by a method called Artificial Pivot Induction in which the mean is shifted by a small constant factor, the API factor. This method follows the principle of Locality of Reference and tries to find a better pivot in the vicinity of the current mean. This will work for some cases, but for the example described earlier we need a better method to tackle the problem. This is accomplished by our algorithm through a method called Peak-Analysis. In Peak-Analysis, as illustrated in more detail below, we choose strings that belong to the maximum character count or 75% of the maximum count and the arithmetic mean of this reduced set is taken. This ensures that the mean is influenced only by input strings that begin with the character having the maximum count, thus ensuring a better partition.

The algorithm described in this paper has three main parts, namely, Mean Partitioning, Artificial Pivot Induction and Peak-Analysis. The first part of the algorithm partitions the input based on the arithmetic mean of some prefix of the strings to be sorted. Artificial Pivot Induction tries to induct a pivot artificially when the first part of the algorithm fails to achieve a good partition. The technique of Peak-Analysis computes a better pivot string when the initial part of the algorithm fails to get a good pivot by finding the mean. The algorithm tries to get a partition ratio around one, which is defined to be the ratio of number of elements in the left sub-array to the number of elements in the right sub-array, i.e., the array is divided into equal sub-arrays. Finally the overall algorithm achieves an efficiency of $\Omega(n\log_2 n)$ for the best and $\theta(n\log_2 n)$ average cases. As shown later, this algorithm has a lesser number of partitioning steps. This algorithm is unlikely to encounter an $O(n^2)$ growth for practical inputs. The superiority of this algorithm is that it has the same time complexity and space complexity for a given data set and does not depend on the ordering of the input elements.

## 37.2 Algorithm Description

During the first pass, the mean of the first characters of all the strings is calculated. This is our initial compare string based on which we partition the strings. We keep a String Count Pointer which keeps the index of character under comparison. Three cases can occur: the partition is good and is accepted, one of the partitions is empty thus indicating that all the strings have the same character as the mean that was used for comparison, or an uneven (or skewed) partition indicating that there is concentration of input in a certain area as illustrated in the example with a's and z's shown earlier. In the second case, we attempt to increment the String Count Pointer by 1 character and repeat the partitioning procedure. When the third case occurs, we can artificially induct a pivot and then check if this has achieved a better partition ratio. If yes, we partition and sort, else Peak-Analysis is done to break the input concentration that is described as a "Peak". Artificial Pivot Induction works well for cases such as the one in the following example:

E.g. 2: Consider 10 strings a1, a2, a3, c1, c2, c3, f1, g1, z1, z2.

The mean calculated would be 'g' and results in a partition ratio of 4:1. Now, because the left array is heavy, we can reduce the mean artificially by an API factor of 2 and get the new mean to be 'e'. The new partition ratio is 3:2, which is an improvement over the earlier method. After sufficient experiments and analysis 2 was found to be a good API factor and catered well to the principle of Locality of Reference. Artificial Pivot Induction works well when the mean is affected by a few larger valued strings or very low valued strings, as was the case in example Eg2. But Artificial Pivot Induction fails when there is a Peak. In a Peak, the Peak elements are the ones that begin with a character that has the maximum count (or has 75% or more of the maximum count). In Peak-Analysis, a mean compare string that is the pivot is computed based only on these Peak elements. This ensures that the mean compare string will partition the elements with an acceptable partition ratio since the new mean that is calculated is not influenced by other strings that are not present in the Peak. The Peak-analysis is done only if the sub-array has greater than 1,000 elements since Peak-Analysis has an overhead which makes it profitable only for a large number of elements. The sort is finished when there are no more sub-arrays. To better understand Peak-Analysis, let us consider an example:

E.g. 3: consider 1000 strings a1, z1, z2...z999

The mean calculated would be 'y', which would achieve a partition ratio of 1:999 and Artificial Pivot Induction would not help either. In such cases, Peak-Analysis will choose 999 strings (z1, z2...z999) that constitute the Peak and the new mean string calculated based only on these strings will be 'z540', thus achieving a good partition ratio. The cost of the Peak-Analysis has been determined to be O(n). Most of the parameters such as the Artificial Pivot Induction factor and the maximum count for peak element selection can be set according to the nature of input dataset.

## 37.3 Algorithm

This section provides the detailed Ssort algorithm in the form pseudo-code. Since the algorithm is fairly complex, it is essential to include the entire algorithm in the paper.

```
BeginProcedure avg_sort (string_array,n,str_count_
  pointer,compare_string)
/* input_array is array of pointers to the string array
  to be sorted */
```

```
if (n=1)
  input_array[counter] ←string_array[0]
  counter ←count +1
  return
endif
for(i←0 to n)
  alpha_count[string_array [i][0]]← alpha_count[string_
    array [i][0]] +1
/* determine the count of first characters of the strings */
  sum ← sum +string_array [i][str_count_pointer]
/* sum of characters pointed by current string count pointer */
endfor
avg = sum / n
compare_string [str_count_pointer] ←avg
compare_string [str_count_pointer +1] ←'\0'
do {
for (i←0 to n)
if (strncmp(string_array[i],compare_string,str_count_
  pointer)  <=0)
        if ( !strncmp(string_array[ i ],compare_string,
          str_count_pointer) )
                  if(!strcmp(string_array[ i ],
                              compare_string))
                              dupe_count ←dupe_count +1
/* this is done to identify duplicate strings and to make the
  sort stable by restoring the position of the duplicate
  elements */
        endif
      endif

  if(dupe_count  <2)
                              left_array[l_c] ←string_
                                array[i]
      l_c ← l_c +1
        else
                              right_array[r_c] ←string_
                                array[i]
             r_c ← r_c +1
  endif
else
right_array[r_c] ←string_array[i]
r_c ← r_c +1
endif
endfor
```

```
restore_flag ←0
if (r_c =0)
  if ( api_flag =1 )
    compare_string[str_count_pointer]
←compare_string[str_count_pointer]-api_factor
  restore_flag ←1
    /* this is used when artificial pivot induction results in
      a new mean which is beyond the boundary of strings under
      comparison. */
  endif
else
  str_count_pointer ←str_count_pointer +1
  no_partion_flag ←1
/* all strings have the same characters pointed by the current
String count pointer as the mean calculated, so increment the
pointer and partition based on the next character. So skip the
artificial pivot induction and peak analysis code*/
endif
if (l_c=0 and api_flag =1)
compare_string[str_count_pointer ]
  ←compare_string[str_count_pointer ] +api_factor
restore_flag ←1
endif
if(no_partiton_flag !=1)
if(restore_flag !=1)
  partition_ratio←l_c /(float)r_c
  partition_count ←partition_count +1
  if (partition_ratio> 2and partition_count   <2 and n>100)
/* Artificial Pivot Induction:
    left array has more elements so decrease the mean */
compare_string[str_count_pointer ]
              ←compare_string[str_count_pointer ] -api_
                factor
  api_flag ←1
  endif
  else if (partition_ratio   <0.3 and partition_count   <2
    and n>100)
/* Artificial Pivot Induction:
  left array has lesser elements so increase the mean */
compare_string[str_count_pointer ]
              ←compare_string[str_count_pointer ] +api_
                factor
api_flag ←1
  else
    api_flag ←0
```

```
  endif
  endif
} while ((api_flag =1 and partition_count   <2) or restore_
  flag =1)
if ( partition_count >=2 andapi_flag=0 and n>1000)
/* Peak-Analysis */
big ←alpha_count[33]
big_index ←33
for(index ←34 to 123)
                if(big <=alpha_count[index])
                              big ←alpha_count[index]
                              big_index ←index
                  endif
            endfor
/* determine the largest count and the corresponding strings
that begin with it and put them in a new array called the peak
_array*/
  index ←big_index
  while (index <123 && alpha_count[index] >= (0.75*big))
        alpha_array[index] ←1
        index ← index +1
  endwhile
  index ←big_index
  while(index >33 && alpha_count[index] >=(0.75*big))
        alpha_array[index] ←1
        index ←index -1
  endwhile
  for(index ←33 to 123)
    if(alpha_array[index] =1)
          peak_array[peak_index] ←index
          peak_index ←peak_index +1
  endfor
  for(peak_index2 ←0 to n)
      for(peak_index ←0 to peak_index)
if(string_array[peak_index2][0] =peak_array[peak_
  index1])
          peak_string_array[peak_n] ←string_array[peak_
  index2]
          peak_n ← peak_n + 1
        endfor
      endfor
callprocedurepeak_avg_find(peak_string_array,peak_n,
peak_compare_string)
wcs ←-1
  do {
```

```
        l_c ←0; r_c←0; wcs ←wcs+1;
        if(wcs !=0)
  temp_len ←temp_len + 1
  /* peak compare string can result in r_c =0, to avoid this
    anomaly we partition based on each character if wcs is set
    to any value other than ' 0' */
          endif
          for (i←0 to n)
              if(wcs =0)
                    temp_len ←strlen(string_array[i])
              endif
  c_count ←c_count +1
              if ( strncmp(arr[i],
  peak_compare_string,temp_len)   <=0)
                        left_array[l_c] ←string_array[i]
                        l_c ←l_c +1
                else
                    right_array[r_c] ←string_array[i]
                    r_c ←r_c +1
                endif
              endfor
              if(wcs_flag=0)
       temp_len ←0; wcs_flag ←1
          endif
    } while(r_c =0 or l_c =0)
  endif
  endif /* for no_partition flag */
  if(l_c !=0)avg_sort(left_array,l_c, str_count_pointer,
    compare_string)
  if(r_c !=0 ) avg_sort(right_array,r_c, str_count_pointer,
    compare_string)
  endprocedure avg_sort
  Begin procedure
  peak_avg_find (peak_string_array [][40], peak_n, peak_
    compare_string)
    for ( x ←0 to max_str_len )
        peak_string_sum[x] ←peak_string_avg[x] ←0
    endfor
    for ( x ←0 to   <peak_n )
          for(y ←0 to peak_string_array[x][y] !=' \0' )
  peak_string_sum[y] ←peak_string_sum[y] + peak_string_
    array[x][y]
          endfor
    endfor
    for(x=0 to y)
```

```
      peak_string_avg[x] ←peak_string_sum[x] /peak_n
      if(peak_string_avg[x]  <' !' and peak_string_avg[x]
        !=' \0' )
peak_string_avg[x]←33
/* ignore null character * /
      endif
      if(peak_string_avg[x] >' z' )
peak_string_avg[x] ←122
      endif
  endfor
peak_string_avg[x] ←' \0'
strcpy(peak_compare_string,peak_string_avg)
return peak_compare_string
endprocedure peak_avg_find
```

## 37.4 Result Analysis

The WordNet [5] lexical database [4] was used as the dataset to be sorted. It had a set of 116,938 Strings which was replicated as lowercase and uppercase words to expand the dataset to 233,876 strings. Different test datasets were generated from the original dataset by randomizing the data. Randomizing was done by repeatedly applying a card shuffling algorithm. The standard version of Quicksort was used where in the first element of the sub-array was chosen as the pivot. As the data in Table 37.1 indicates, fewer number of partition steps were required by Ssort to sort the input data. Quicksort only showed improvements for totally random data. It is evident from Table 37.2 that the number of bad partitions, which are partitions that had a ratio greater than 2 or lesser than 0.5, were found to be far lesser than in the case of Quicksort. In the best case for Quicksort, the number of bad partitions was still found to be twice as that of Ssort. The largest bad partition was found to be 121 times than that of Ssort. Even in the best case of Quicksort it was 1.3 times than that in Ssort as indicated in the Table 37.2. Time analysis showed that for a randomization factor 1 that is for highly shuffled data the performance of Quicksort was better. But for any randomization factor lesser than 0.75, Ssort was found to be far better.

**Table 37.1** Comparison of Ssort and Quicksort

| Randomizing factor* | Total partitions | | Time in seconds | |
|---|---|---|---|---|
| | Quicksort | Ssort | Quicksort | Ssort |
| 0 | 378343 | 233696 | 196.99 | 1.01 |
| 0.1 | 331633 | 233696 | 80.33 | 1.01 |
| 0.5 | 233876 | 233696 | 3.30 | 1.01 |
| 1 | 233876 | 233696 | 0.41 | 1.01 |

* Randomizing factor 0 indicates sorted input.

**Table 37.2** Partition analysis Ssort and Quicksort

| Randomizing factor* | Number of bad partitions | | Largest bad partition | |
|---|---|---|---|---|
| | Quicksort | Ssort | Quicksort | Ssort |
| 0 | 263343 | 49145 | 108557 | 897 |
| 0.1 | 263343 | 49145 | 108557 | 897 |
| 0.5 | 109084 | 49145 | 1169 | 897 |
| 1 | 76714 | 49145 | 1169 | 897 |

* Randomizing factor 0 indicates sorted input.

Quicksort varied from 197 seconds to 0.41 seconds to sort from the worst case to the best case, whereas Ssort was consistent and did not depend on the order of input elements and was measured to sort in 1.01 seconds in all cases. Time was measured on an Intel Core Duo T2050, 1 GB RAM machine running on Fedora Core 5 Linux Operating System. This worst case behavior of Quicksort makes it unreliable to be used for real life applications of Internet or text processing where there is a time constraint. It is undesirable for the user to wait for long periods (e.g., for a web page to refresh) merely because the data involved was a worst case for Quicksort.

It may be noted that fully randomized data is unlikely to occur in practical cases of string sorting. Because of the way in which large collections of strings are put together in any practical application, it is highly likely that there will be pockets of locally sorted data (albeit partially sorted), thereby indicating that the worst case for Ssort of fully randomly ordered strings where Quicksort is much better than Ssort is unlikely to occur in real applications.

As can be seen from the Table 37.2, the largest bad partition in Ssort was still of size 897. This is because Peak-Analysis is not done for sub-arrays having less than 1,000 strings. Due to the overhead of Peak-Analysis it becomes desirable to use it only for sub-arrays which have greater than 1,000 strings. The worst case for Ssort will occur where the algorithm goes to peak-analysis every time as it recursively partitions the dataset. But since peak-analysis and partitioning on mean are mutually exclusive in the algorithm, the algorithm is highly unlikely to suffer from $O(n^2)$ growth.

## 37.5 Further Work

Though for practical input data we claim that the worst case for Ssort is very unlikely to occur, Peak-Analysis can fail when there is a "Peak" within a Peak as illustrated next. Ignoring the fact that Peak-Analysis is done for partitions with more than 1,000 elements for the present, let us consider that Peak-Analysis is done for a Peak array whose contents are aa, aza, azza, azzza, azzzza. The mean compare string calculated would be 'au' which would result in a partition ratio of 1:4. Again for the next iteration, the mean compare string would be 'azu' and the partition ratio would be 1:3. This keeps recurring, resulting in a $O(n^2)$

growth. But this kind of input is highly unlikely to be encountered in practical cases. This is the worst case for Ssort and further work on Ssort can be done to achieve a better Peak-Analysis algorithm so that such worst case occurrences can be eliminated.

## 37.6 Conclusion

Experimental results show that Ssort is an improvement over Quicksort in terms of achieving better partitions. It also is consistent, that is, its performance is not dependent on the input ordering, a behavior which is not seen in Quicksort. Quicksort has a varied performance and its time efficiency can degrade to $O(n^2)$. The occurrence of an $O(n^2)$ worst case for Ssort is very remote because the Artificial Pivot Induction and Peak-Analysis parts of the algorithm seldom allow for a bad partition. It is a stable sort and is best suited for real time processing where the sorting has to be done in a fixed amount of time and any $O(n^2)$ growths cannot be tolerated. This is especially significant in today's world where processing textual data in large quantities is an essential requirement in many practical applications.

## References

1. Motzkin D (1983) Meansort. Commun ACM 26(4):250–251
2. Hoare CAR (1961) Quicksort (Algorithm 64). Commun ACM, 4(7):321–322
3. Morato J et al (2004) WordNet applications. In: GWC 2004, pp 270–278
4. Fellbaum C (1999) WordNet: an electronic lexical database. MIT Press, Cambridge
5. http://wordnet.princeton.edu

# Chapter 38
# Parametric Studies for the AEC Domain Using InteliGrid Platform

**Matevž Dolenc and Robert Klinc**

**Abstract** Parametric studies play a vital role in science and engineering in general. They usually consist of a number of repetitive, independent calculations that are therefore easily parallelized. Scientists and engineers involved in this type of work need a computing environment that delivers large amounts of computational power over a long period of time. Such environment is provided by the InteliGrid project testbed. The paper introduces the InteliGrid platform and provides a detailed description of the demonstrated IDA parametric studies and its execution on the InteliGrid testbed as well as benchmark analysis of the benefits provided by the deployed Condor high-throughput computing system.

## 38.1 Introduction

Parametric studies in science and engineering usually consist of a number of repetitive, independent calculations that are therefore easily parallelized. Scientists and engineers involved in this type of work need a computing environment that delivers large amounts of computational power over a long period of time. Such environment is called a High-Throughput Computing (HTC) environment. The European Union (EU) project InteliGrid (2004–2007) [4] combined and extended the state-of-the-art research and technologies in the areas of semantic interoperability, virtual organisations and grid technology to provide diverse engineering industries with a platform prototype with flexible, secure, robust, interoperable, pay-per-demand access to information, communication and processing infrastructure. As one of the important requirements for the InteliGrid platform recognised by the end-users is the ability to perform complex parametric studies.

M. Dolenc (✉)
Faculty of Civil and Geodetic Engineering, University of Ljubljana, Jamova 2, 1000 Ljubljana, Slovenia
e-mail: mdolenc@itc.fgg.uni-lj.si

N. Mastorakis et al. (eds.), *Proceedings of the European Computing Conference*, Lecture Notes in Electrical Engineering 28,
DOI 10.1007/978-0-387-85437-3_38, 

## 38.2 InteliGrid Architecture Components

InteliGrid delivers a generic grid-based integration and web semantic-based interoperability platform for creating and managing networked virtual organisations. The following common characteristics can be defined for all InteliGrid services as main components that can be deployed either at some workstation or on the grid: (1) services are modular components that can be semantically described, registered, discovered and finally used by clients, (2) services can participate in specific business workflows, where the order in which they are sent and received affects the outcome of the operations by a service, (3) services may be completely self-contained, or they may depend on the availability of other services, (4) services are able to advertise details such as their capabilities, interfaces and supported communication protocols according to pre-defined concepts and ontologies, and (5) all capabilities provided by services as well as communication and data channels among them and clients are protected by security mechanisms.

There are three main kinds of components on the InteliGrid platform. Firstly, there are business applications that are the consumers of other components. Most services are accessed through portal, but access through other desktop software is possible as well. Secondly, InteliGrid platform identifies services that comply with WSRF [12] services or pure web services extended with GSI [6], and talk to each other as well as to the client applications. They can be further divided into (a) interoperability services, (b) business or domain specific services, and (c) middleware services that offer traditional grid middleware functionality. All these resources are available through a well-defined secure protocols and grid-enabled generic core services for remote data access.

As the detailed discussion of the InteliGrid architecture is out of the scope of this paper, only a short overview of the components enabling the HTC environment is presented (see Fig. 38.1):

- *Portal framework*. The InteliGrid Portal is based on the GridSphere portal framework [8] which provides an open-source portlet based Web portal. GridSphere enables developers to quickly develop and package third-party portlet web applications that can be run and administered within the GridSphere portlet container.
- *Grid authorization service*. Build on GAS [5] that has been designed as a trusted single logical point for defining security policies and enables the implementation of various advanced virtual organisation security scenarios.
- *Resource management*. It is built on a GRMS service [7] which allows developers to build and deploy resource management systems for large scale distributed computing infrastructures.
- *Job management*. Condor [2] is a batch queuing system for managing compute-intensive jobs. It performs it by providing an HTC environment that efficiently harnesses and makes the best use of all available resources (Fig. 38.2) while providing a high throughput for jobs.

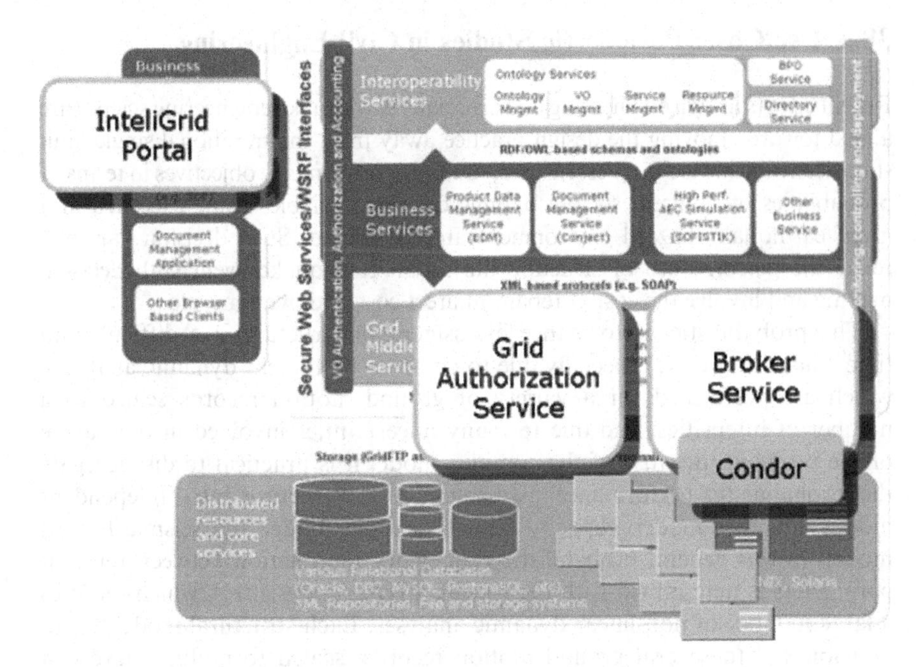


Fig. 38.1 InteliGrid architecture components supporting HTC environment

OpenSees Client 1.0

Tasks list:

| No. | Input File | Job Id | Job Status | Remove | Stdout | Stderr |
|---|---|---|---|---|---|---|
| 1 | C:\seesClient\inputfile.11.tcl | krim.fgg.uni-lj.si.4633.0 | DONE (0) | del | view | view |
| 2 | C:\seesClient.3.old\seesClient\inputfile.11.tcl | krim.fgg.uni-lj.si.4634.0 | DONE (0) | del | view | view |
| 3 | C:\seesClient.3.old\seesClient\inputfile.110.tcl | krim.fgg.uni-lj.si.4635.0 | RUNNING | del | view | view |
| 4 | C:\seesClient.3.old\seesClient\inputfile.111.tcl | krim.fgg.uni-lj.si.4636.0 | QUEUED-ACTIVE | del | view | view |
| 5 | C:\seesClient.3.old\seesClient\inputfile.112.tcl | krim.fgg.uni-lj.si.4637.0 | QUEUED-ACTIVE | del | view | view |
| 6 | C:\seesClient.3.old\seesClient\inputfile.113.tcl | krim.fgg.uni-lj.si.4638.0 | QUEUED-ACTIVE | del | view | view |
| 7 | C:\seesClient.3.old\seesClient\inputfile.114.tcl | krim.fgg.uni-lj.si.4639.0 | QUEUED-ACTIVE | del | view | view |
| 8 | C:\seesClient.3.old\seesClient\inputfile.115.tcl | krim.fgg.uni-lj.si.4640.0 | QUEUED-ACTIVE | del | view | view |
| 9 | C:\seesClient.3.old\seesClient\inputfile.116.tcl | krim.fgg.uni-lj.si.4641.0 | QUEUED-ACTIVE | del | view | view |
| 10 | C:\seesClient.3.old\seesClient\inputfile.117.tcl | krim.fgg.uni-lj.si.4642.0 | QUEUED-ACTIVE | del | view | view |
| 11 | C:\seesClient.3.old\seesClient\inputfile.118.tcl | krim.fgg.uni-lj.si.4643.0 | QUEUED-ACTIVE | del | view | view |
| 12 | C:\seesClient.3.old\seesClient\inputfile.119.tcl | krim.fgg.uni-lj.si.4644.0 | QUEUED-ACTIVE | del | view | view |

Project: Open... Save as... Add Files... Add Directory... Clear All

Submit: Run Job Cancel Job Get Results... Reset

Status: SUBMITTED

Configuration: Edit...

Misc: About... Quit

Fig. 38.2 Condor pool job monitor

## 38.3 Use Case: Parametric Studies in Civil Engineering

Recent efforts in civil engineering and especially earthquake engineering science are aimed towards moving the design practice away from the traditional single limit state deterministic techniques. Instead, multiple performance objectives in terms of probabilities of different socio-economic decision variables being exceeded in a given earthquake hazard environment are considered. Such decision variables relate the performance of structure during an earthquake event with decision making and involve the cost of repair, indirect cost of down time, etc.

The probabilistic performance assessment of structure is a difficult and time consuming task, especially due to complex non-linear dynamic analyses, which are performed for a variety of ground motion records scaled to a number of intensities, and due to many uncertainties involved in the nature of the problem. Because of the complex model it is practical to disaggregate the probabilistic performance assessment of a structure into independent models like loss model, capacity model, demand model and seismic hazard model. In this general probabilistic framework the relation between relevant parameters is determined with the so called IDA analysis [11], which requires a large number of non-linear dynamic analyses. Each structural model has to be subjected to several ground motion records, scaled to multiple levels of intensity in order to obtain enough data about the behaviour of the structure. It is obvious, that the number of analyses and therefore the required computer time are significant.

The bridge (see Fig. 38.3) was analysed using the IDA analysis to determine the relation between the engineering demand parameter and seismic intensity measure. OpenSees [9] computer code was used to model the bridge. Several steps were taken during the analysis:

1. *Numerical modelling*. The numerical modelling of the structure that was the subject of investigation was performed. Since the near-collapse region was investigated, all possible failure mechanisms had to be accurately modelled.
2. *Earthquake records*. Selection of earthquake records (this is typically selected from an Internet database such as the ESD [1]) was chosen. Due to the characteristics of the database's user interface, the download of earthquake records cannot be automated after the selection has been made. The ground motions are normalised.
3. *Creation of job files*. Job files for Condor, automated with Matlab, were prepared. The performance of the structure was evaluated using IDA analysis, a dynamic equivalent of the static pushover analysis. Following the procedure, the structural model was subjected to several ground motion records, each scaled to multiple levels of intensity. A nonlinear dynamic analysis was performed at each step. Several engineering demand parameters were monitored.
4. *Running the analyses*. This step was performed five times with a different number of available computing resources to determine the benefits of using

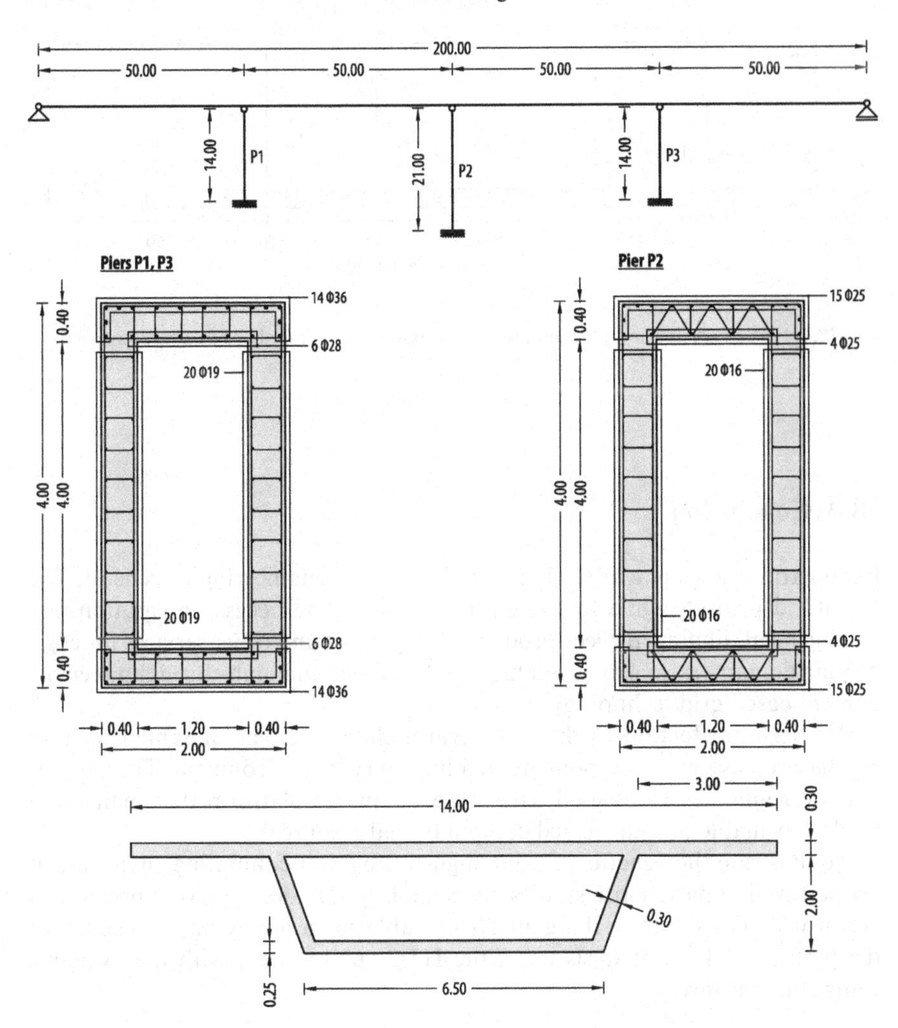


Fig. 38.3 Test structure characteristics: the box section deck (pinned at the abutments), the concrete piers (rectangular hollow cross section, designed according to the Eurocode standards), the piers (side piers – 14 m high, central pier – 21 m high)

the HTC environment for performing parametric studies. Figure 38.4 summarises calculation times for performing the analyses and clearly shows the advantages of parallelization of parametric studies.

5. *Post analysis.* After all load cases had been evaluated, the acquired data was analysed. The goal was to provide probability of exceeding a chosen performance level, such as the collapse prevention damage state. Cornell's approach [3] to assess the mean annual probability of exceeding a limit state was used in this example.

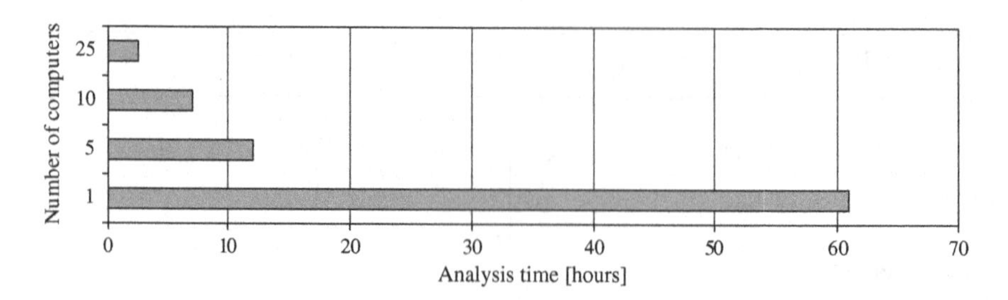


**Fig. 38.4** IDA analysis computational times compared to the number of computers used

## 38.4 Conclusion

Parametric studies play a vital role in science and engineering in general. The InteliGrid project, while focusing on providing data access, integration and management, lifetime project/product management and other services for engineering domains, also provides a high-performance and high-throughput environment based grid technology.

A brief introduction to the InteliGrid architecture was presented with an emphasis on essential components enabling an HTC environment. The Condor batch queuing system is used within the InteliGrid platform that manages a pool of available non-dedicated computational resources.

To illustrate the benefits of using high-throughput computing environment for performing parametric studies an exemplary IDA analysis of a bridge was performed. The benchmarking analysis established a nearly linear speedup of the performed IDA analysis using the HTC environment with the available computing resources.

## References

1. Ambraseys N, Smit P, Sigbjornsson R, Suhadolc P, Margaris B (2002) Internet site for European Strong-Motion Data European Commission. Research Directorate General, Environment and Climate Programme
2. Condor Project. www.cs.wisc.edu/condor/
3. Cornell CA, Fatemeh J, Hamburger RO, Foutch DA (2002) Probabilistic basis for 2000 SAC federal emergency management agency steel moment frame guidelines. J Struct Eng 128(4):526–533
4. Dolenc M, Turk Z, Katranuschkov P, Kurowski K, Hannus M (2005) Towards grid enabled engineering collaboration environment. In: Topping BHV (ed) Proceedings of the Tenth International Conference on Civil, Structural and Environmental Engineering Computing, Civil-Comp Press, Edinburgh
5. Grid authorization service (GAS). www.gridlab.org/gas

6. Grid security infrastructure (GSI). www.globus.org/security/overview.html
7. GridLab resource management system. http://www.gridlab.org/grms
8. GridSphere portal framework. http://www.gridsphere.org
9. Mazzoni S, McKenna F, Scott MH, Fenves GL, Jeremic B (2006) Open system for earthquake engineering simulation (OpenSees) [Command Language Manual].University of California, Berkeley, PEER
10. Pinto AV (1996) Pseudodynamic and shaking table tests on R. C. Bridges (Report No. 5). The European Laboratory for Structural Assessment, Ispra
11. Vamavatsikos D, Cornell CA (2002) Incremental Dynamic Analysis. Earthquake Eng Struct Dyn 31(3):491–514
12. WS-Resource Framework (WSRF). www.oasis-open.org/

# Chapter 39
# An Auction-Based Resource Allocation Strategy with a Proportional Share Model for Economic-Based Grid Systems

**Kuan-Chou Lai, Chao-Chin Wu, and Jhen-Wei Huang**

**Abstract** Due to the complexity caused by distributed computing and resource heterogeneity, the resource management and application scheduling in grid systems are complex. This paper proposes a flexible auction-based resource allocation strategy with a proportional share model for economic-based Grid systems. Instead of maximizing resource utility, this proposed strategy aims to complete jobs by meeting their deadline and budget requirements with the lowest resource costs. The experimental results show that the proposed strategy could complete jobs before the deadline with lower budgets.

## 39.1 Introduction

Grid systems have been adopted as the next generation platform for many scientific applications. The Grid system consists of geographically distributed computer systems which make use of the Internet to connect dynamic, shared resources and services [4]. In order to share, integrate and manage these resources and services, Grid systems adopt resource allocation and management strategies to enhance the system throughput and utilization. In Grid systems, the producers (i.e., resource owners) and consumers (i.e., resource users) have different purposes and targets respectively from system-centric and user-centric viewpoints. System-centric approaches aim to optimize system performance; however, user-centric approaches, which are common to conform to the economic principle, aim to maximize utility based on users' QoS requirements [2].

In an economic-based Grid system, resource allocation and scheduling decisions are made according to the user requirements at runtime. The economic model charges users for shared resources and services that they consume based on the costs which are defined for running applications. A user adjusts the

K.-C. Lai (✉)
Department of Computer and Information Science, National Taichung University, Taiwan, R.O.C.

N. Mastorakis et al. (eds.), *Proceedings of the European Computing Conference*, Lecture Notes in Electrical Engineering 28,
DOI 10.1007/978-0-387-85437-3_39, 

budget to acquire resources and services to compete with other users by his demand.

In this study, we integrate the proportional share model with the first-price sealed-bid auction model [2] to develop a new resource allocation strategy which aims to complete jobs by meeting their deadline and budget requirements with the lowest resource costs. The experimental results show that the proposed strategy could complete jobs before the deadline with lower budgets. The rest of this paper is organized as follows. Section 39.2 introduces related works. Section 39.3 presents the proposed algorithm. Experimental results and performance analyses are provided in Section 39.4. Concluding remarks are then offered.

## 39.2 Related Works

The economic-based Grid system must contain tools for users and resource providers to convey their needs and goals. In general, there are many resource allocation approaches developed for economic-based Grid systems; these approaches apply some parameters (e.g., deadline, budget and resource cost) to make decisions dynamically at runtime. Economic models could be divided into two categories [1, 2]: the non-price-based-model and the price-based-model. In the non-price-based-model, users share the resources together to meet their requirements; however, in the price-based-model, there are trading systems for users to bid for resources and services. The price-based-model includes the commodity markets model, posted price model, bargaining model, tendering/contract-net model and auction model. In the auction model, a resource accepts the bids from many users to acquire it; then, the user making the highest bidding wins the auction. There are several resource allocation algorithms, including the deadline and budget constrained cost-time optimization algorithm [2, 4, 7], Grid classified optimization scheduling algorithm [5], the Analytic Hierarchy Process and History Based Policy algorithm [8], and the Time Optimization algorithm with Auction-Based Proportional Share model [9].

In general, the Time Optimization algorithm with Auction-Based Proportional Share model [9] outperforms other proposed ones. However, it doesn't support the flexible bidding time period, and then it can't adjust the priority order to bid in real time. In addition to the lack of the flexibility of the bidding time period, this proposed algorithm [9] conservatively decides the initial bidding costs. The user broker uses the resource cost in the worst case scenario as the initial bidding cost to prevent the job's execution time from exceeding the deadline. But, when the difference of the resource performance between the worst case and the average case is too big, it would cause the abnormal situation that the initial bidding cost excessively exceeds the expected bidding cost, and then wastes the budget.

## 39.3 Proposed Algorithm

In this study, we propose a Flexible Auction-based Resource Allocation (FARA) strategy with a proportional share model for Bag-of-Tasks (BoT) applications. This strategy aims to use lower budgets to complete jobs before their deadlines. The user broker retrieves useable resource in the local Virtual Organization (VO) from the Grid Information System (GIS). The operating steps of our proposed algorithm are listed as follows:

1. Resources register the machine name, the resource capability and the price to the GIS.
2. Each user has a broker which is responsible for scheduling; each user submits tasks, deadline and budget to the broker, and the user broker makes the deployment of communication and the lower level actions.
3. The user broker searches information in the GIS to select appropriate resources for the job after receiving the user's demand.
4. The user broker sends tasks to the selected resource and continues to monitor the status for increasing or reducing the budgets in order to complete jobs in each time period, P. The user broker then pays the total cost to the resource provider.

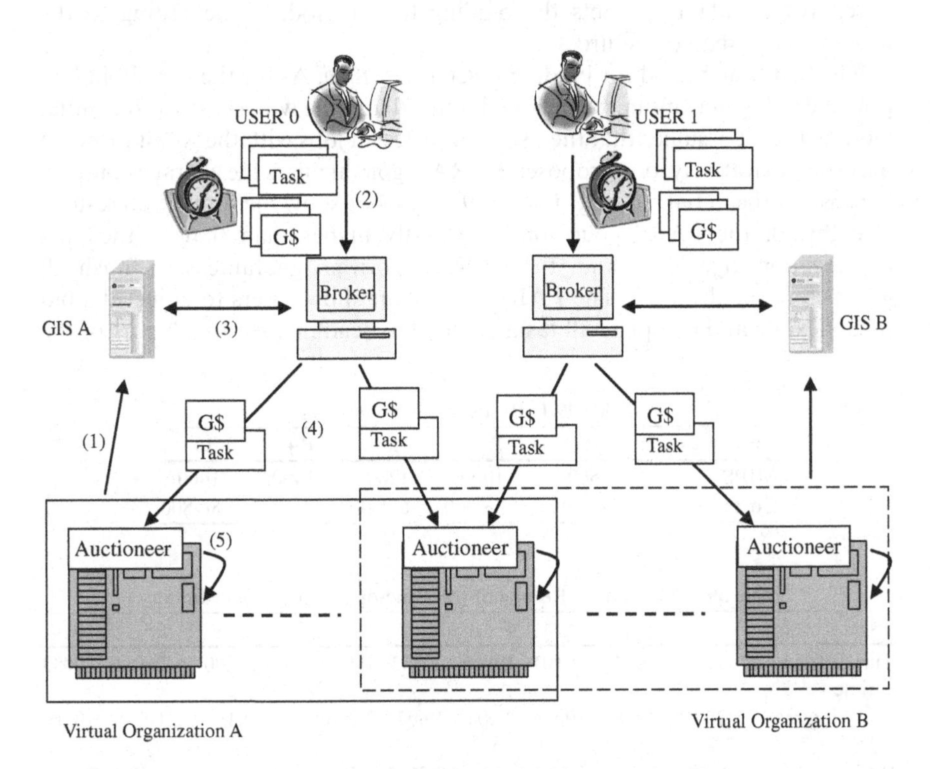


**Fig. 39.1** Framework of the proposed FARA algorithm

5. The auctioneer makes the resource work. If the Auctioneer receives the new price sent from the broker in time period, P, it will reallocate the proportion of Grid resources to each user in the next time period.

The framework of the Flexible Auction-based Resource Allocation FARA) strategy with a proportional share model is shown in Fig. 39.1.

## 39.4 Experimental Results

The simulation environment is written by JAVA language. The resource characteristics are shown in Table 39.1. There are five resources (R1, R2, R3, R4 and R5) and 10 users. The number of jobs for each user is 100. All jobs are generated randomly with the job lengths between 9,000–11,000 million instructions (MI). The deadline of each user is 1,000. In this study, we adopt the ratio of the performance divided by the cost (P/C) as the comparison index for the effectiveness of resource allocation algorithms. That is, the higher the P/C is, the more jobs are completed per unit.

We assume that user *i* updates the bid to resume participating the auction every 100 seconds in the time optimization scheduling [9]; however, our proposed FARA algorithm sets the bidding time period, $P_i$, according to the proportion of shared resources.

The first time period for bidding is set as *est_time*. As for the initial bidding price step, the time optimization scheduling [9] uses the worst case for the initial bidding (i.e., it assumes that the user completes all jobs with the worst resource only); comparatively, our proposed FARA algorithm uses the average computing cost for the initial bidding. Tables 39.2 and 39.3 show the simulation results. The P/C of the FARA algorithm is slightly higher than that of the time optimization algorithm; and, the FARA algorithm guarantees to finish all jobs before the deadline. The FARA algorithm allows users to adjust the bid price flexibly and complete all tasks within the deadline. Although part of the

**Table 39.1** Resource characteristics

| Resource name | R1 | R2 | R3 | R4 | R5 |
|---|---|---|---|---|---|
| MIPS | 500 | 1000 | 1200 | 1400 | 1600 |
| Cost range | 1~100 | 5~500 | 6~600 | 7~700 | 8~800 |

**Table 39.2** Simulation Results of the Time-optimization Scheduling

| User | 1 | 2 | 3 | 4 | 5 | 6 | 7 | 8 | 9 | 10 |
|---|---|---|---|---|---|---|---|---|---|---|
| Turnaround time | 19962 | 20044 | 20101 | 19966 | 20062 | 19998 | 20012 | 20022 | 20036 | 20033 |
| Finishing rate [%] | 20.04 | 19.93 | 19.86 | 20.04 | 19.92 | 20.00 | 19.99 | 19.97 | 19.95 | 19.96 |
| P/C | 0.50 | 0.50 | 0.50 | 0.50 | 0.50 | 0.50 | 0.50 | 0.50 | 0.50 | 0.50 |

**Table 39.3** Simulation Results of the FARA algorithm

| User | 1 | 2 | 3 | 4 | 5 | 6 | 7 | 8 | 9 | 10 |
|---|---|---|---|---|---|---|---|---|---|---|
| Turnaround time | 3915 | 3920 | 3919 | 3933 | 3928 | 3930 | 3939 | 3924 | 3936 | 3942 |
| Finishing rate [%] | 100 | 100 | 100 | 100 | 100 | 100 | 100 | 100 | 100 | 100 |
| P/C | 0.94 | 0.93 | 0.94 | 0.94 | 0.94 | 0.94 | 0.94 | 0.94 | 0.95 | 0.95 |

tasks would be assigned more expensive resources, the FARA algorithm still guarantees that all the jobs are completed before the deadline.

## 39.5 Concluding Remarks

This study proposes a flexible job allocation strategy for economic-based grids to complete jobs before the deadline with lower budgets. This study improves the time-optimization algorithm in the initial auction, and user brokers could flexibly adjust the bidding time and prices to complete jobs. Since the FARA algorithm allows users to adjust the bid price flexibly and thus complete all tasks within the deadline, the simulation results shows that the P/C of the FARA algorithm is higher than that of the time optimization algorithm. Moreover, the FARA algorithm guarantees to finish all jobs before the deadline. The experimental results show that the proposed strategy could outperform the Time Optimization algorithm with Auction-Based Proportional Share model [9].

This study only focuses on CPU resources; however, other resources (such as memory and network bandwidth) could be considered to make the resource allocation more comprehensive.

**Acknowledgement** This work was partially supported by the National Science Council of the Republic of China under the contract numbers: NSC 95-2221-E-142-004- and NSC 95-2622-E-142-001-CC3.

## References

1. Buyya R, Stockinger H, Giddy J, Abramson D (2001) Economic models for management of resources in peer-to-peer and Grid computing. In: SPIE International Conference on Commercial Applications for High-Performance Computing, Denver, CO
2. Buyya R, Abramson D, Venugopal S (2005) The Grid economy. Proc IEEE 93(3):698–714
3. Chun B, Culler D (2000) Market-based proportional resource sharing for clusters. Technical Report CSD-1092, University of California, Berkeley, CA
4. Foster I, Kesselman C (1999) The Grid: Blueprint for a future computing infrastructure. Morgan Kaufmann, San Francisco, CA
5. Li Kenli, Tang Xiaoyong, Zhaohuan (2005) Grid classified optimization scheduling algorithm under the limitation of cost and time. In: Proceedings of the second International Conference on Embedded Software and Systems

6. Pourebrahimi B, Bertels K, Kandru GM, Vassiliadis S (2006) Market-Based Resource Allocation in Grids. In: Second IEEE International Conference on e-Science and Grid Computing (e-Science'06), p 80
7. Rajkumar B, Manzur M (2002) A deadline and budget constrained cost-time optimisation algorithm for scheduling task farming applications on global grids. Technical Report, Monash University, Australia
8. Sai Rahul Reddy P (2006) Market Economy Based Resource Allocation in Grids. Master Thesis, Indian Institute of Technology, Kharagpur, India
9. Sulistio A, Buyya R (2005) A time optimization algorithm for scheduling bag-of-task applications in auction-based proportional share systems. In: Proceedings of the 17th International Symposium on Computer Architecture and High Performance Computing (SBAC-PAD), Rio de Janeiro, Brazil, pp 235–242
10. William V (1961) Counterspeculation, auctions, and competitive sealed tenders. J Finance 16(1):8–37

# Part VI
# Advances in Computer Science and Applications

In this part we selected some advanced topics in the computer science and its applications. The papers include contributions from a wide range of disciplines – numerical analysis, discrete mathematics and system theory, natural sciences, engineering, information science, electronics, and naturally, computer science itself. The results are addressed not only to the specialist in a particular field but also to the general scientific audience. The contributions include propagation in cylindrical inset dielectric guide structures, damage in nonlinear mesoscopic materials, diagnosis of robotized assembling systems, open architecture systems for MERO walking robots control, mass transfer analysis in the case of the EFG method, information system architecture for data warehousing, dimensional dynamics identification of reconfigurable machine tools, compaction-induced deformation on flexible substrates, robust and accurate visual odometry by stereo, financial auditing in an e-business environment, reputed authenticated routing ad-hoc networks protocol and quantum mechanical analysis of nanowire FETs.

# Chapter 40
# Propagation in Cylindrical Inset Dielectric Guide Structures

**Ioannis O. Vardiambasis**

**Abstract** The inset dielectric guide (IDG) is just a rectangular groove filled with dielectric, which retains many of the advantages of dielectric guides, without fabrication and loss problems. Quasi-planar IDGs are attractive for applications in microwave and millimeter-wave frequency ranges, as they can be used as low loss and low cost transmission lines, directional couplers, nonreciprocal phase shifters, antennas and other devices. In this paper we address circular IDG-type structures, investigating hybrid wave propagation in single microslot lines loaded with semi-circular dielectric cylinders. The analysis is based on systems of singular integral-integrodifferential equations of the first kind. The proposed algorithms are rapidly convergent, yielding very accurate results for the propagation constants and for the modal equivalent magnetic currents of the principal as well as the higher-order modes.

## 40.1 Introduction

Grooves filled with dielectric (Inset Dielectric Guides, IDGs) have been proposed as low loss transmission lines, as directional couplers and bandpass filters, as well as a basis for planar antenna arrays [1, 2, 3, 4, 5, 6].

Although rectangular IDGs have been extensively studied in the past using semi-analytical methods (like the transverse resonance diffraction method [1, 2, 3, 4, 5] and the effective dielectric constant method [6]), circular ones have not yet received appropriate attention [7] due to the complexity of the pertinent boundary value problems (mixed circular-plane boundaries) [8, 9]. In this paper we address the latter configurations, combining cylindrical IDG-type structures

I.O. Vardiambasis (✉)
Microwave Communications and Electromagnetic Applications Laboratory, Division of Telecommunications, Department of Electronics, Technological Educational Institute (T.E.I.) of Crete – Chania Branch, Romanou 3, Chalepa, 73133 Chania Crete, Greece
e-mail: ivardia@chania.teicrete.gr

N. Mastorakis et al. (eds.), *Proceedings of the European Computing Conference*, Lecture Notes in Electrical Engineering 28,
DOI 10.1007/978-0-387-85437-3_40, 

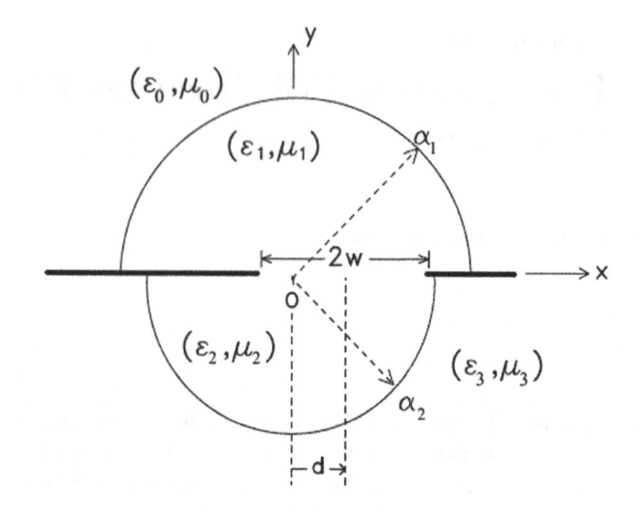


**Fig. 40.1** Geometry of a slotline loaded with two semi-clinders

with microslot lines, which are widely used in modern microwave integrated circuitry. Hybrid propagating modes' characteristics are investigated in single microslots on a ground plane loaded with open or shielded dielectric semi-circular cylinders, as shown in Fig. 40.1.

The analysis, briefly discussed in Section 40.2, is based on exact systems of coupled singular integral-integrodifferential equations of the first kind, which are efficiently discretized using the methods explored in [8, 9]. More details on the formulation and the method of solution may be found in [9] dealing with 3D scattering by these same structures. The proposed algorithms yield very accurate results for the propagation constant as well as for the modal electric currents of all modes, as will be seen in Section 40.3 where such typical results are presented for several cases.

## 40.2 Formulation

Figure 40.1 shows a microslot line of width 2w, covered by two semi-circular dielectric cylinders of radii $\alpha_1$ and $\alpha_2$ (regions 1 and 2).

Each of the regions 0 ($r>\alpha_1$, $\varphi>0$) and 3 ($r>\alpha_2$, $\varphi<0$) may be either air or perfect electric conductor (PEC). Each of the regions is characterized by its dielectric permittivity $\varepsilon_s$ and its magnetic permeability $\mu_s$ (s = 0,1,2,3). For simplicity in notation the semi-cylinders are taken to be coaxial with their common axis displaced from the slot axis by d.

Assuming propagation in the $-\hat{z}$ direction, the slotline's equivalent surface magnetic current densities $M_z(x)\cdot e^{j(\omega t+\beta z)}$ and $M_x(x)\cdot e^{j(\omega t+\beta z)}$ satisfy the following homogeneous system of coupled singular integral-integrodifferential equations [9]:

$$\sum_{s=1}^{2}\left\{-\frac{k_{cs}^{2}}{2\omega\mu_{s}}\int_{C}M_{z}(x)\cdot H_{0}^{(2)}(k_{cs}|x-x'|)dx-\right.$$
$$-\frac{j\beta}{2\omega\mu_{s}}\frac{d}{dx'}\int_{C}M_{x}(x)\cdot H_{0}^{(2)}(k_{cs}|x-x'|)dx+$$
$$+\sum_{n=0}^{\infty}\in_{n}J_{n}(k_{cs}x')\left[2B_{n}^{M}\int_{C}M_{z}(x)\cdot J_{n}(k_{cs}x)dx-\right. \tag{40.1}$$
$$\left.\left.-\frac{1}{k_{cs}}\int_{C}M_{x}(x)\left(j\beta B_{n}^{M}\cdot h_{n}^{-}(k_{cs}x)+\omega\varepsilon_{s}A_{n}^{M}\cdot h_{n}^{+}(k_{cs}x)\right)dx\right]\right\}=0(x'\in C)$$

$$\sum_{s=1}^{2}\left\{-\frac{j\beta}{2\omega\mu_{s}}\frac{d}{dx'}\int_{C}M_{z}(x)\cdot H_{0}^{(2)}(k_{cs}|x-x'|)dx-\right.$$
$$-\frac{1}{2\omega\mu_{s}}\left[\frac{d^{2}}{dx'^{2}}+k_{s}^{2}\right]\int_{C}M_{x}(x)\cdot H_{0}^{(2)}(k_{cs}|x-x'|)dx+$$
$$+\frac{1}{k_{cs}}\sum_{n=0}^{\infty}\in_{n}\left[\left(j\beta B_{n}^{M}h_{n}^{-}(k_{cs}x')-\omega\varepsilon_{s}B_{n}^{J}h_{n}^{+}(k_{cs}x')\right)\cdot\int_{C}M_{z}(x)\cdot J_{n}(k_{cs}x)dx+\right. \tag{40.2}$$
$$+\frac{1}{2k_{cs}}\int_{C}M_{x}(x)\left[\beta h_{n}^{-}(k_{cs}x')\left(\beta B_{n}^{M}h_{n}^{-}(k_{cs}x)-j\omega\varepsilon_{s}A_{n}^{M}h_{n}^{+}(k_{cs}x)\right)+\right.$$
$$\left.\left.\left.+\omega\varepsilon_{s}h_{n}^{+}(k_{cs}x')\left(\omega\varepsilon_{s}A_{n}^{J}h_{n}^{+}(k_{cs}x)+j\beta B_{n}^{J}h_{n}^{-}(k_{cs}x)\right)\right]dx\right]\right\}=0(x'\in C)$$

where $\beta$ is the unknown propagation constant, $H_0^{(2)}(.)$ is the Hankel function of second kind and of zero order, $J_n(\cdot)$ is the Bessel function of order n, C denotes the slot interval $d-w\leq x'\leq d+w$, $k_{cs}^2=k_s^2-\beta^2$, $k_s^2=\omega^2\varepsilon_s\mu_s$, $\in_n=2-\delta_{n0}$ is the Neumann's factor ($\delta_{nm}$ is the Kronecker delta), $h_n^{\pm}(\cdot)=J_{n-1}(\cdot)\pm J_{n+1}(\cdot)$, whereas $A_n^M\equiv A_n^M(s,s)$, $B_n^M\equiv B_n^M(s,s)$, $A_n^J\equiv A_n^J(s,s)$, $B_n^J\equiv B_n^J(s,s)$ are specified in the Appendix of [9]. Note that by s = 1, 2 we denote the regions (semi-cylindrical or half plane) from both sides of the microslotline.

The solution of the system (40.1) and (40.2) is carried out on a moment method basis using expansions of the magnetic current distributions in terms of first and second kind Chebyshev polynomials. Following singular integral equation techniques in the way outlined in [9], we arrive at the infinite system of linear algebraic equations:

$$\sum_{N=0}^{\infty}\left[a_N R_{MN}^{zz}+b_N R_{MN}^{zx}\right]=0,\sum_{N=0}^{\infty}\left[a_N R_{MN}^{xz}+b_N R_{MN}^{xx}\right]=0;\quad M=0,1,\ldots,\infty \tag{40.3}$$

The matrix elements $R_{MN}^{uv}$ $(u, v \equiv z, x)$ assume rapidly converging analytical expressions given in [9]. Non-zero solutions of (40.3), for the expansion constants $a_N$ and $b_N$ of the slotline's equivalent surface magnetic currents, require vanishing of its determinant. This yields to the dispersion equation of the structure. The propagation constants and equivalent current distributions of the principal as well as of the higher order modes can be efficiently obtained with very high accuracy for any combination of the physical and geometrical parameters of the configuration.

## 40.3 Numerical Results

In single-slot configurations loaded with open dielectric semi-cylinders there are always two (for single load) or three (for double load) principal modes, which are never cutoff in analogy with [8]. These include: (a) the slot mode (S-mode), and (b) the $HE_{11}$ modes from the regions 1 and 2.

In the special case $(\varepsilon_0 \to \infty, \varepsilon_3 \to \infty, \alpha_1 = \alpha_2)$ all principal as well as all higher order modes coincide with the corresponding modes of the circularly shielded microslot lines studied in [10]. This coincidence of results may be used as a validity check for the developed here numerical codes.

In Fig. 40.2, where some typical results are presented in case of a symmetrical slot (d = 0), there are mode transformation phenomena that take place between the two principal modes as well as between the 4th, 5th and 6th modes. Such

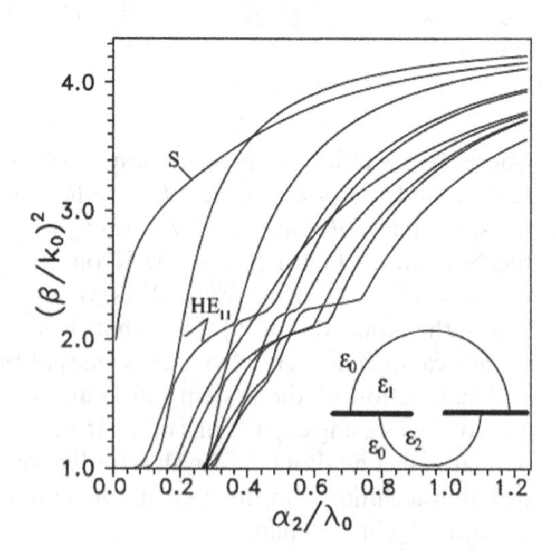


**Fig. 40.2** Effective dielectric constant $\varepsilon_{eff} = (\beta/k_0)^2$ versus normalized frequency for the first 10 propagating modes of the structure of Fig. 40.1 with $2w = 0.5\alpha_2$, $d = 0$, $\varepsilon_1 = 2.32\varepsilon_0$, $\varepsilon_2 = 4.34\varepsilon_0$, $\varepsilon_3 = \varepsilon_0$, $\mu_i = \mu_0$, $\alpha_1 = 2\alpha_2$

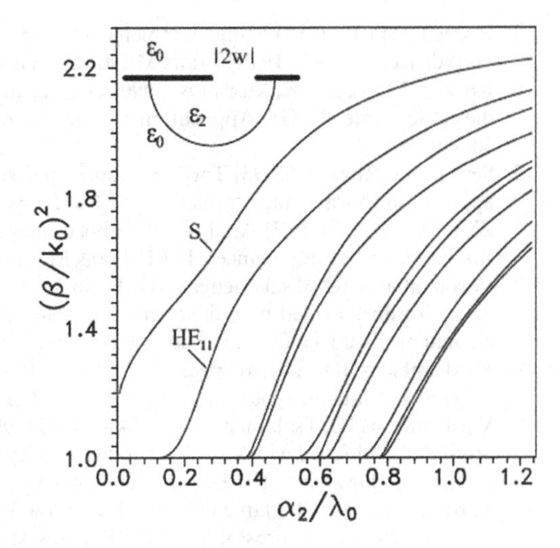


**Fig. 40.3** $\varepsilon_{eff}$ vs $\alpha_2/\lambda_0$ for the first 10 propagating modes of the structure of Fig. 40.1 with $2w = 0.5\alpha_2$, $d = 0.5\alpha_2$, $\varepsilon_1 = \varepsilon_3 = \varepsilon_0$, $\varepsilon_2 = 2.32\varepsilon_0$, $\mu_i = \mu_0$

phenomena do not occur in case d $\neq$ 0, as illustrated in Fig. 40.3, where the S-mode has the highest $\varepsilon_{eff}$ over the whole frequency range.

## 40.4 Conclusions

Generalized cylindrical IDG structures, or microslot lines loaded by one or two semicircular dielectric cylinders, could be useful in microwave applications. The analysis of this structure was based on exact singular integral equation techniques. The algorithms that have been developed converge rapidly and can be used to obtain the propagation constant and the modal currents of any mode with high accuracy.

**Acknowledgments** This work is co-funded by the European Social Fund (EKT) with 75% and National Resources (Greek Ministry of National Education and Religious Affairs) with 25% under the ΕΠΕΑΕΚ project "Archimedes – Support of Research Groups in TEI of Crete – 2.2.-7 – Smart antenna study and design using techniques of computational electromagnetics and pilot development and operation of a digital audio broadcasting station at Chania (SMART-DAB)".

Also, the author would like to thank Prof. John L. Tsalamengas and Ass. Prof. George Fikioris, of the National Technical University of Athens, for their helpful discussions and insightful comments.

## References

1. Pennock SR, Izzat N, Rozzi T (1992) Very wideband operation of twin-layer inset dielectric guide. IEEE Trans Microwave Theory Techn 40:1910–1917
2. Rozzi T, Hedges SJ (1987) Rigorous analysis and network modeling of the inset dielectric guide. IEEE Trans Microwave Theory Techn 35:823–833

3. Rozzi T, Ma L (1988) Mode completeness, normalization, and Green's function of the inset dielectric guide. IEEE Trans Microwave Theory Techn 36:542–551
4. Rozzi T, Ma L, Leo R, Morini A (1990) Equivalent network of transverse dipoles on inset dielectric guide (IDG): Application to linear arrays. IEEE Trans Antennas Propagat 38:380–385
5. Sewell PD, Rozzi T (1993) The continuous spectrum of the inset dielectric guide and its application to waveguide transitions. IEEE Trans Microwave Theory Techn 41:282–289
6. Zhou W, Itoh T (1982) Analysis of trapped image guides using effective dielectric constant and surface impedances. IEEE Trans Microwave Theory Techn 30:2163–2166
7. Vardiambasis IO, Tsalamengas JL, Fikioris JG (1998) Hybrid wave propagation in microslot lines loaded with dielectric semi-cylinders. In: 1998 URSI International Symposium on Electromagnetic Theory Proceedings 1:163–165
8. Vardiambasis IO, Tsalamengas JL, Fikioris JG (1997) Hybrid wave propagation in generalised Goubau-type striplines. IEE Proc Microwaves Antennas Propag 144:167–171
9. Vardiambasis IO, Tsalamengas JL, Fikioris JG (1998) Plane wave scattering by slots on a ground plane loaded with semicircular dielectric cylinders in case of oblique incidence and arbitrary polarization. IEEE Trans Antennas Propagat 46:1571–1579
10. Vardiambasis IO, Tsalamengas JL, Fikioris JG (1995) Hybrid wave propagation in circularly shielded microslot lines. IEEE Trans Microwave Theory Techn 43:1960–1966

# Chapter 41
# On the Damage in Nonlinear Mesoscopic Materials

**Veturia Chiroiu, Dan Dumitriu, Ana Maria Mitu, Daniel Baldovin, and Cornel Secara**

**Abstract** This paper is devoted to the analysis of the damage in nonlinear mesoscopic materials, which are aggregates of grains which act as rigid vibrating units, while the contacts between them – the bond system – constitute a set of interfaces that control the behaviour of the material. The interfaces are mesoscopic, with a typical size of one micrometer. A constitutive micropolar model for material interfaces is presented. The model is based on damage coupled to plastic or viscoplastic slip, stick and dilatation (separation), and it is able to describe the succesive degradation and failure of an interface.

## 41.1 Introduction

Damaged materials form a class of unusual elastic materials that show extreme nonlinearity, hysteresis and discrete memory. This class includes pearlitic steel, fiber-reinforced metal matrix composites, cement, concrete, ceramics, rocks, sand, soil etc. Materials whose elastic properties are determined by such a bond system are called nonlinear mesoscopic materials.

The bond systems consist of a fabric of defects (cracks, voids) that participate in the elastic response of the material. The grains have a random position and the intergranular interfaces contain cavities, microvoids, and defects. The mesoscopic materials cannot be described by the traditional theory of elasticity [1, 2]. The degradation and deterioration on the microstructural level may be viewed as the successive evolution with load and time of microdefects in the shape of microvoids and microcracks, which are part of the microstructure heterogeneity. For example, residual microstresses after the manufacturing processes, effects form the large precipitates present in grey-cast iron are initiators of microcracks [3]. This paper is devoted to the analysis of the damage in nonlinear mesoscopic materials, which are aggregates of grains which act as rigid vibrating units, while the contacts between them – the bond system – constitute a set of interfaces that

V. Chiroiu (✉)
Institute of Solid Mechanics, Ctin Mille 15, Bucharest 010141, Romania

N. Mastorakis et al. (eds.), *Proceedings of the European Computing Conference*, Lecture Notes in Electrical Engineering 28,
DOI 10.1007/978-0-387-85437-3_41, 

control the behaviour of the material. The interfaces are mesoscopic, with a typical size of 1 $\mu$m [4–6]. At the mesoscopic scale, the micropolar continuum mechanics possesses a great potential to describe the long-range interactions among the particles in materials [7–11].

## 41.2 Basic Micropolar Equations

Throughout this paper a rectangular coordinate system $(x_1, x_2, x_3)$ is employed and repeated indices denote summation over the range (1,2,3). Balance of momentum equation is written as $t_{ji,i} + \rho(f_i - \dot{v}_i) = 0$, where $t_{ij}$ is the stress tensor, $f_i$ the body force vector, $v_i$ the velocity vector, $\rho$ the density, a superposed dot indicates material differentiation, i.e. $\dot{v}_i = v_{i,t} + v_k v_{i,k}$ and a comma represents partial differentiation with respect to the variable. The balance of moment of momentum equation is $m_{ji,j} + e_{ijk}t_{jk} + \rho(l_i + j\dot{\nu}_i) = 0$, where $m_{ij}$ is the couple stress tensor, $l_i$ is the body couple vector, $\nu_i$ is the microgyration vector which is also the material derivative of the microrotation vector $\phi_i$, $j$ is the micro-inertia, and $e_{ijk}$ is the permutation tensor. The constitutive law can be written as

$$t_{ij} = \lambda\varepsilon_{kk}\delta_{ij} + (\mu + \kappa)\varepsilon_{ij} + \mu\varepsilon_{ji},$$

$$m_{ij} = \alpha x_{kk}\delta_{ij} + \beta x_{ji} + \gamma x_{ij}, x_{ij} = \phi_{j,i},$$

where $\delta_{ij}$ is the Kronecker delta and the microstrain tensor $\varepsilon_{ij}$ links the displacement vector $u_i$ to the microrotation vector through the relations $\varepsilon_{ij} = u_{j,i} + e_{jik}\phi_k$. The kinematic characteristics of the strain are given by $\varepsilon_{ij}$ and $x_{ij}$. Six micropolar elastic moduli are introduced: $\lambda$ and $\mu$ are the Lame constants from classical elasticity, and $\kappa$, $\alpha$, $\beta$ and $\gamma$ are additional constants in micropolar theory [7].

The above formulation is complete since there are 33 equations in the 33 variables $t_{ij}, m_{ij}, \varepsilon_{ij}, \phi_i$ and $u_i$. As boundary conditions, we can prescribe the surface tractions and couples: $t_{(n)i} = t_{ji}n_j, m_{(n)i} = m_{ji}n_j$ on $S$, where $t_{(n)i}$ and $m_{(n)i}$ are the surface stress and couple vectors, respectively, the subscript $(n)$ referring to the bounding surface $S$ whose exterior unit normal vector is $n_j$. It is also possible to prescribe the displacement $u_i$ and the microrotation $\phi_i$ on $S$: $\phi_i = \phi_i^0$, $u_i = u_i^0$ on $S$, or to construct mixed boundary conditions in displacements and microrotations over a portion of the surface $S_1$ and tractions and couples over the remainder surface $S_2, S = S_1 \cup S_2, S_1 \cap S_2 = 0$.

## 41.3 Proposed Model for the Interface

Let us consider an interface in the shape of a band of width $\delta$, in a local coordinate system, along which the micro constituents A and B are not perfectly bonded, because of the slip and dilatation which may occur along the interfaces (Fig. 41.1).Let $q_i$ and $s_i$

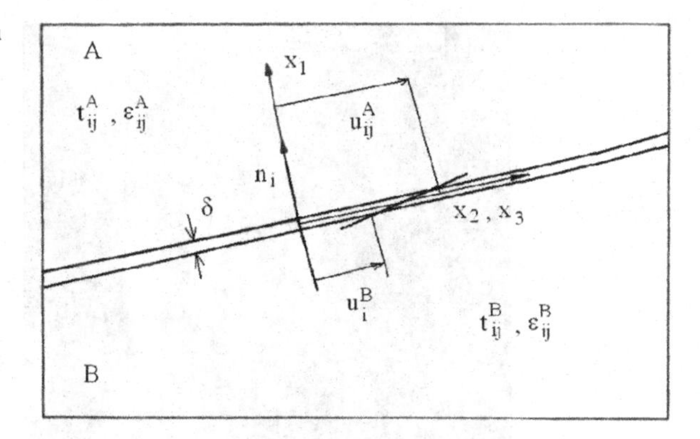


**Fig. 41.1** Interface between the bodies A and B

be the traction vector and respectively the couple vector acting on the body B defined by the normal $n_i = (1,0,0), q_i = t_{ij}n_j = t_{i1}, s_i = m_{ij}n_j = m_{i1}, i = 1,2,3$, where $q_i$ and $s_i$ are continuous along the interface: $[q_i] = 0, [s_i] = 0, i = 1,2,3$ (the bracket indicates a jump across the interface $[q_i] = q_i^A - q_i^B$).

It results that the components $t_{11}, t_{12}, t_{13}$ and $m_{11}, m_{12}, m_{13}$ are always continuous and these unique values may be associated with the interface. For the remaining components $t_{22}, t_{23}, t_{33}$ and $m_{22}, m_{23}, m_{33}$ it is not possible to associate unique values with the interface. So, in a manner similar to [3], the independent uniquely defined state variables at the interface are

$$t_{ij} = \begin{bmatrix} t_{11} & t_{12} & t_{13} \\ t_{12} & a & a \\ t_{13} & a & a \end{bmatrix}, m_{ij} = \begin{bmatrix} m_{11} & m_{12} & m_{13} \\ m_{12} & b & b \\ m_{13} & b & b \end{bmatrix}, \varepsilon_{ij} = \begin{bmatrix} c & c & c \\ c & \varepsilon_{22} & \varepsilon_{23} \\ c & \varepsilon_{23} & \varepsilon_{33} \end{bmatrix},$$

where *a,b,c* are non-unique variables. For variables *a,b*, termodynamically associated stresses and couple-stresses that are energy conjugates to $\varepsilon_{22}, \varepsilon_{23}, \varepsilon_{33}$, are introduced. For variable $c$, thermodynamic strains and micro-strains that are energy conjugates to $t_{11}, t_{12}, t_{13}, m_{11}, m_{12}, m_{13}$, are introduced. It follows

$$\tilde{t}_{ij} = \begin{bmatrix} \tilde{a} & \tilde{a} & \tilde{a} \\ \tilde{a} & \tilde{t}_{22} & \tilde{t}_{23} \\ \tilde{a} & \tilde{t}_{23} & \tilde{t}_{33} \end{bmatrix}, \tilde{m}_{ij} = \begin{bmatrix} \tilde{b} & \tilde{b} & \tilde{b} \\ \tilde{b} & \tilde{m}_{22} & \tilde{m}_{23} \\ \tilde{b} & \tilde{m}_{23} & \tilde{m}_{33} \end{bmatrix}, \tilde{\varepsilon}_{ij} = \begin{bmatrix} \tilde{\varepsilon}_{11} & \tilde{\varepsilon}_{12} & \tilde{\varepsilon}_{13} \\ \tilde{\varepsilon}_{12} & \tilde{c} & \tilde{c} \\ \tilde{\varepsilon}_{13} & \tilde{c} & \tilde{c} \end{bmatrix}.$$

In order to define these thermodynamic variables, we consider a relative motion in the interface to couple kinetically the motion with damage. We introduce the jump of displacements $u_i$ and the microrotation $\phi_i$ across the interface, defined as $u_i^0 = u_i^A - u_i^B, \phi_i^0 = \phi_i^A - \phi_i^B$ and $\varepsilon_{1j} = u_{j,1} + e_{j1k}\phi_k, \chi_{1j} = \phi_{j,1}$.

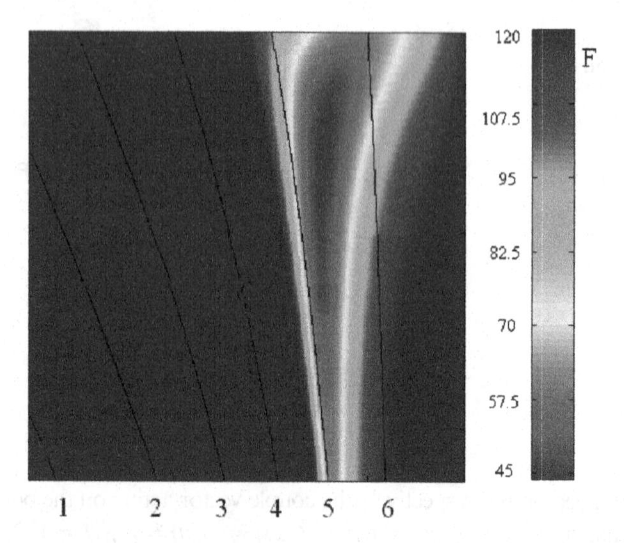


**Fig. 41.2** The field of forces showing the successive localization and shear band development

## 41.4 Degradation and Failure of the Interface

To analyze the successive degradation and failure of an interface as shown in Figs. 41.1 and 41.2, we consider that this interface is subjected to monotonic loading ([12–16]). The states one to six in Figure 41.2 refer to the corresponding field of forces which show the successive localization and shear band development in the interface.

## 41.5 Conclusions

The propagation of the interface damage into materials A and B is schematically illustrated in Fig. 41.3. The damage of the interface affects also the state of integrity of the adjacent bodies. The lines represent the directions in which the cracks initiated into the interface move outside the interface into bodies A and B.

The failure of the interface generates also an avalanche of plastic minicraters into the material as shown in Fig. 41.4. These minicraters are observed

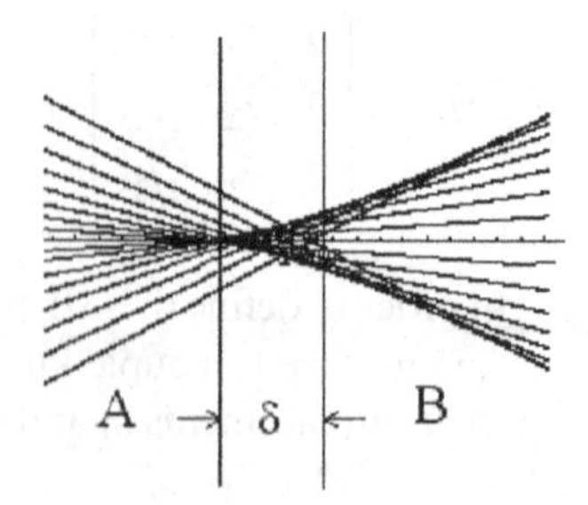


**Fig. 41.3** The propagation of cracks into the bodies A and B

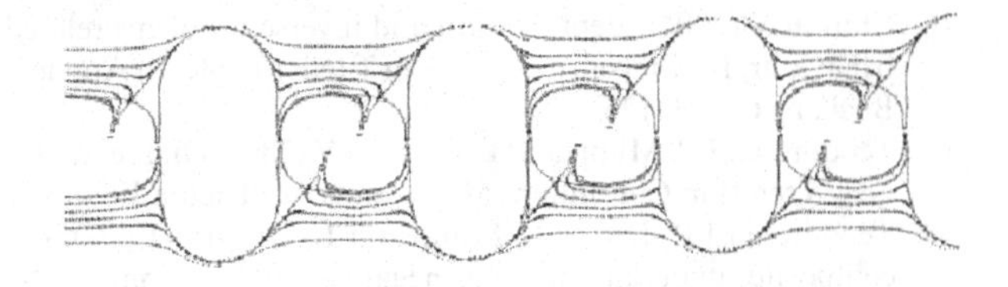

**Fig. 41.4** Intragranular minicraters of damage

macroscopically along the grain boundaries at ferrous alloys such as steel with a pearlitic ministructure. They can be attributed to the failure in the sense of nucleation, growth and coalescence [3, 17, 18].

**Acknowledgment** The authors acknowledge the financial support of the CEEX postdoctoral grant 1531/2006. Best thanks are presented also to Professor *Pier Paolo Delsanto* and to Dr. *Marco Scalerandi* (Dipartimento di Fisica, Politecnico di Torino) for the important suggestions and discussions in the field of this topic.

## References

1. Guyer RA, McCall KR, Boitnott GN (1995) Hysteresis, discrete memory, and nonlinear wave propagation in rock. Phys Rev Lett 74:3491–3494
2. Guyer RA, Johnson PA (1999), Nonlinear mesoscopic elasticity: Evidence for a new class of materials. Physics Today
3. Cannmo P (1997) A damage based interface model: Application to the degradation and failure of a polycrystalline microstructure. Chalmers Univ. of Techn., Div. of Solid Mechanics, Göteborg, Sweden, pp 1–64
4. Kachanov LM (1958) Time of the rupture process under creep conditions. Izv Akad, SSSR, Otd. Tech Nauk 8:26–31
5. Lemaitre J (1985) A continuous damage mechanics model for ductile fracture. ASME, J Eng Mater Technol 107:83–89
6. Lemaitre J (1990) Micromechanics of crack initiation. Int J Fracture 42:87–99
7. Eringen AC (1966) Linear theory of micropolar elasticity. J Math Mech 15:909–924
8. Eringen AC (1968) Theory of micropolar elasticity. In: Liebowitz R (ed) Fracture, vol 2. Academic Press, New York, pp 621–729
9. Mindlin RD (1964) Microstructure in linear elasticity. Arch Rat Mech Anal 16:51–78
10. Kröner E (1963) On the physical reality of torque stresses in continuum mechanics. Int J Engng Sci 1:261–278
11. Berglund K (1982) Structural models of micropolar media. In: Mechanics of Micropolar Media, World Scientific, Singapore
12. Gauthier RD (1962) Experimental investigations on micropolar media. In: Brulin O, Hsieh RKT (eds) Mechanics of Micropolar Media. World Scientific, Singapore, pp 395–463
13. Lakes RS (1986) Experimental microelasticity of two porous solids. Int J Solids, Struct 22:55–63
14. Badea T, Chiroiu C, Soare M, Chiroiu V (2000) The inverse problem of micro-defects identification. Rev Roum Sci Techn 45(1):31–39
15. Badea T, Nicolescu M (2004) A Preisach model of hysteretic behavior of nonlinear elastic materials. In: Chiroiu V, Sireteanu T (eds) Topics in Applied Mechanics. Ed. Academiei, Bucharest, 1:1–25

16. Chiroiu V (2004) Identification and inverse problems related to material properties and behaviour. In: Chiroiu V, Sireteanu T (eds) Topics in Applied Mechanics. Ed. Academiei, Bucharest, 1:83–126
17. Teodorescu PP, Munteanu L, Chiroiu V (2005) On the wave propagation in chiral media. New Trends in Continuum Mechanics, Ed. Thetha Foundation, Bucharest, pp 303–310
18. Teodorescu PP, Badea T, Munteanu L, Onişoru J (2005) On the wave propagation in composite materials with a negative stiffness phase. New Trends in Continuum Mechanics, Ed. Thetha Foundation, Bucharest, pp 295–302

# Chapter 42
# Diagnosis of Robotized Assembling Systems

**C. Ciufudean, R. Serban, C. Filote, and B. Satco**

**Abstract** In this paper we propose a Petri net within a diagnosis system for robot construction design. We assimilate the robot construction process with an assembly process that composes parts and/or subassemblies into a unique product. We assume that the assembly supervisor (AS) system is distributed, and it solves several local AS̀s attached to the nodes of the Petri net model of the assembly process. The research issues that we address in this work include the modelling of assembly process, the determination of cost-effective assembly plans for efficient building, and the real time adoption of a plan to a given product to be assembled. This work extends the known assembly Petri nets to a powerful framework enabling to derive the diagnosis of assembly processes whose path may vary, and the objective function is maximized. The presented approach can be used to evaluate transient and steady state performances of alternative design based on a construction example.

## 42.1 Introduction

In this paper we focus on the diagnosis of asynchronous systems. Typical examples are robot-assembled systems, such as shown in Fig. 42.1, where the supervisor system is distributed, and it involves several local supervisors, attached to some nodes of the construction network. Each local supervisor has only a partial view of the overall system. The different local supervisors have their own local time, but they are not synchronized. Alarms are reported to the global supervisor, and this supervisor performs the diagnosis. We notice that events may be correctly ordered by local supervisors, but communicating observations via network causes a loss of synchronization, and results in a non-deterministic supervisor [1–6]. Alternative process plans can be exploited in real time to react to machine

C. Ciufudean (✉)
Faculty of Electrical Engineering and Computer Science, University "Stefan cel Mare", Suceava, Romania
e-mail: calin@eed.usv.ro

N. Mastorakis et al. (eds.), *Proceedings of the European Computing Conference*, Lecture Notes in Electrical Engineering 28,
DOI 10.1007/978-0-387-85437-3_42, 

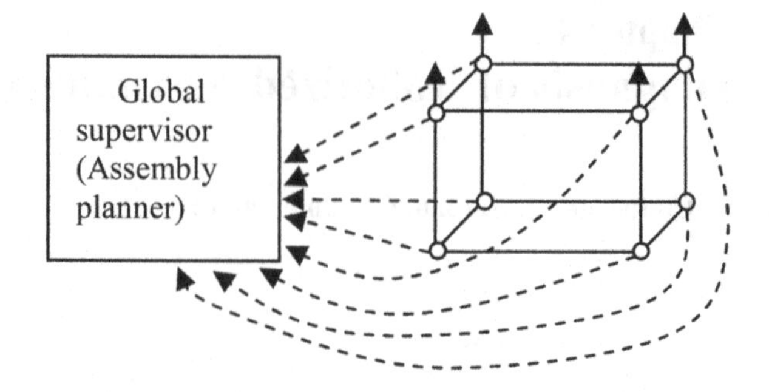


Fig. 42.1 Supervision of a construction system

failures, in order to avoid having bottleneck machines, and to enable adaptive production planning of failure-prone construction systems. The work proposes an adaptive process-planning scheme that can manage process changes and adapt the process to specific assembly conditions. In order to solve this problem, the paper proposes a methodology for the design and implementation of an adaptive assembly planner based on assembly Petri nets (APN).

The advantages of using APN's include the following [7, 8]: allowing the dynamic behaviour to be visualized; representing both the assembly process and system resources in a single presentation for diagnosis and easy control implementation; allowing a linear programming formulation to find optimal assembly plans. The proposed planner integrates, as shown in Fig. 42.1, an assembly system, and this principia scheme is represented in Fig. 42.2. The input to the system consists of raw materials; the output is the finite product (construction) and what remain to be dumped, secondary raw materials, s.a. During assembly execution, observations made by local supervisors are transferred directly to the global supervisor (assembly planner – see Fig. 42.1). They are used to update the predicted values of each component and respective assembly costs. Hence, the assembly system adapts the predictive process plan to the new data and generates an adapted plan that may lead to a new termination goal. The

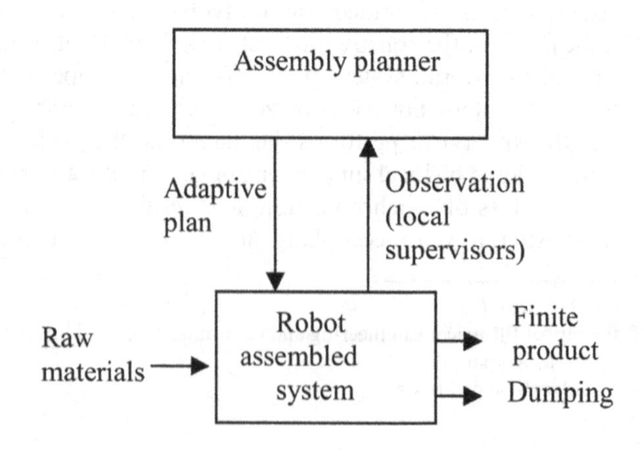


Fig. 42.2 Adaptive robot assembled system

objectives of this paper are to present a method for developing an adaptive planner and to illustrate assembly process planning via a specific design with execution success rate and respective costs. Section 42.2 describes the proposed robot assembled Petri net model and planning algorithm, respectively the assembly supervisor Petri nets (ASPN). Section 42.3 deals with an important problem concerning the Petri nets modelling process, respectively the size of the Petri net model. This paper presents an approach that decomposes the ASPN into several basic configurations (e.g., subnets), in order to determine the state estimation function (Se-function) that describes the Petri net state-space size. This approach allows the designer to choose a proper ASPN model for a complex robot assembled system. Section 42.4 presents a design and implementation methodology for a robot assembled system.

## 42.2 Modelling the Robot Assembled Process: Assembly Supervisor Petri Nets

This paper extends the known assembled Petri nets (APN) into assembly supervisor Petri nets (ASPN). Thus ASPNs can accurately describe the robot construction topology, mating relations and precedence relations. In an ASPN a transition (assembly process) and a place (a product, or its subassembly) are associated with utility information (cost/benefit). Each transition is also associated with pre-firing and post-firing values. The pre-firing value is a decisional value which indicates the priority level for a transition to fire [8], respectively its associated assembly operation to perform. The post-firing value represents a probability that indicates the success rate of its assembly operation, which is updated based on the observations received from the local supervisors. The ASPN can estimate the assembly performance, e.g. net profit, and also decides the best actions among various corrective actions, in order to maximize the profit. An assembly supervisor Petri net is defined as: ASPN = (P, T, W, Mo, $f_1$, $f_2$, $v_d$, $v_p$), where P and T are finite sets of places and transitions, respectively; W$\subseteq$ (PXT) U (TXP) is a set of directed arcs; $M_0$: P$\rightarrow$N is the initial marking, where N is the set of nonnegative numbers. The set of input (output) transitions of a place p $\in$ P is denoted by $^0p$ ($p^0$). The set of input (output) places of a transition t $\in$ T is denoted by $^0t$ ($t^0$) [5, 6]. We also have:

- $f_1$: P$\rightarrow$$R^+$ is the resource utility function assigned to a place, where R is the set of nonnegative real numbers;
- $f_2$: T$\rightarrow$$R^+$ is the cost function assigned to a transition, where $R^+$ is the set of nonnegative real numbers;
- $v_d$: T$\rightarrow$N is a decisional value assigned to a transition. This value is assigned according to a planning algorithm. Value $v_d$ is used to decide the firing priority of the transitions;
- $v_p$: T$\rightarrow$ [0,1] is a probability value associated with a transition, that is updated according to the sensing result of the corresponding assembly

operation. The value $v_p(t)$ represents the success rate of an assembly operation. The value 1 - $v_p(t)$ represents the failure rate.

We notice that in an ASPN model, a place with multiple output transitions represents a subassembly with multiple ways to be assembled. These different assembly choices should determine a common set of assembled parts. Multiple output transitions from a place form a Logic-OR relation and multiple output places from a transition form a Logic-AND relation. Both place and transition utility functions are used to generate an optimal assembly plan. The decision and probability value of transitions are used to execute, respectively to adapt the construction plan. ASPN defined in such a way belongs to the class of free-choice Petri nets [9]. That means that the global ASPN model has the structure given in Fig. 42.3, where $P_{si}$, i = 1,. . .,n are the partial subassemblies of the construction, and $P_f$ represents the final product. Obviously, a construction planning schedule is to determine the best order of assembly operations, i.e. transition firings. In order to reject the uncertainty of the assembly operations, different-level priorities are assigned to different assembly alternatives for all the subassemblies [10, 11].

In an ASPN, a place with multiple transitions implies assembly methods, each of them having its own value ($v_w$). Introducing the $v_d$ (decisional value) to each transition enables an easy determination of assembly order. For example let place p having k output transitions: $t_1$, $t_2$, . . ., $t_k$ and their values $v_{wi}$, i = 1,. . .,k. The priority levels to transitions $t_i$ are assigned accordingly to the following relation:

$$v_d(t_j) = i \tag{42.1}$$

Where $v_{wi}$ is the ith smallest among $v_{wi}$, i = 1,. . .,k. When an assembly operation fails (e.g. the ASPN diagnosis is revealing a bottleneck or a too expensive way of firing transitions), the assembly planner selects a transition with the next largest $v_d$ value, and so on. The $v_p$ values assigned to transitions are designed to adapt the ASPN for the maximum expected assembled value. Initially, all $v_p$ are set accordingly to assembly planner (designer) experience. During execution, for each assembly operation, the number of successes are recorded and different $v_{pi}$, i = 1,. . .,k are re-adjusted with an exponential rate $e^{-N/N_s \cdot t_i}$ where N is the number of transitions fired for a subassembly, and $N_s$ is

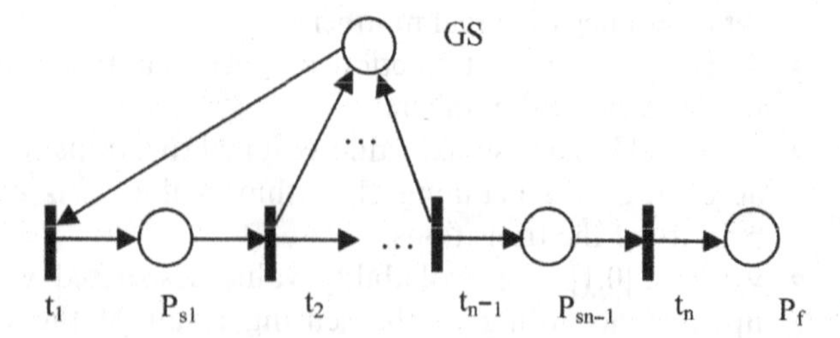


**Fig. 42.3** Global ASPN structure

the number of successes. We notice that $v_d$ is assigned based on both $v_p$ and $v_w$ for each transition.

## 42.3 SE – Functions for Basic Configurations of ASPN

According to the assumptions made in the previous section, we define a generally firing policy for the ASPN:

$$a_{it} = \min_{Tj/m_{it}=0} v_{p_j}, \forall i, j \in \{T\} \tag{42.2}$$

where $a_{it}$ is the success rate of firing transition i at time t; $m_{it}$ is the marking of place $p_i$ at time t; $T_j$ is the transition which follows the place $p_i$ with marking $m_{it}$; $v_{pj}$ is the value $v_p$ assigned to transition $T_j$.

We determine the size of the state-space for the underlying system, irrespective of its dimension, by decomposing it into several basic configurations, e.g. subnets (SN) that can form complex (large) ASPN, while combined interconnections for mechanisms of execution are presented here. ASPN can model the execution of sequential, parallel and choice operations. This approach allows the designer to choose a properly ASPN model for a complex robot assembled system. Figure 42.4 illustrates two subnets in series: $SN_1$, respectively $SN_2$.

The possible distribution of r tokens among the places of the subnets $SN_1$, and $SN_2$ is due to the marking repartition among the respective places, separated by transitions $T_i$, i = 1,2,3, with $a_{it}$ success rate of firing. The interconnection Se–function is given by the next relation [12, 15]:

$$Se_{series} = \sum_{i=0}^{r} Se_{a_{1t}}(i)\, Se_{a_{2t}}(r-i) \tag{42.3}$$

where $Se_{a_{nt}}(i)$ is the kth partition of r, is an n-vector of non-negative integers that sum to r, i = 1,2, at time t. For n subnets, the Se-function is:

$$Se_{series} = \sum_{i=0}^{r} Se_{a_{1t}}(i).Se_{a_{2t}}(i-1)\ldots Se_{a_{i-1t}}(r-i+1).Se_{a_{it}}(r-i) \tag{42.4}$$

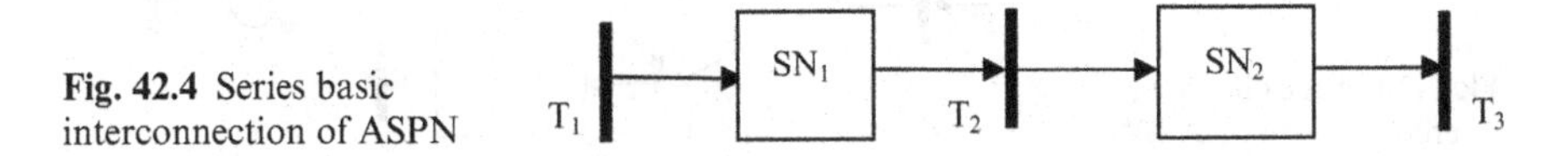


Fig. 42.4 Series basic interconnection of ASPN

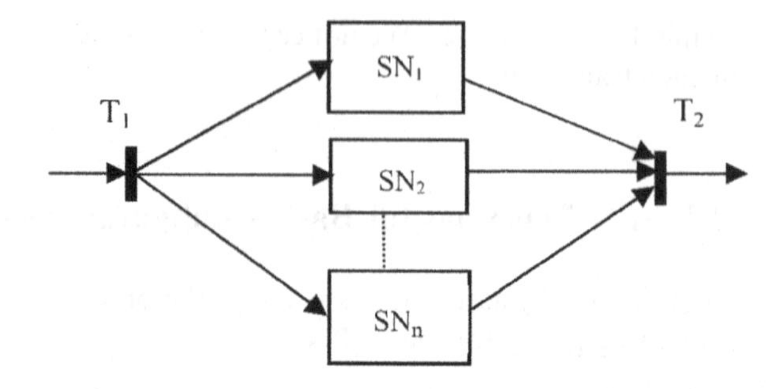


**Fig. 42.5** Parallel basic interconnection of ASPN

Figure 42.5 depicts the parallel execution of the operations. For every token entering through $T_1$, there is one in each subnet. The Se-function is given by the next relation [11, 13]:

$$Se_{parallel} = \prod_{i=1}^{n} Se_{a_{it}}(r) \tag{42.5}$$

The choice among subnets is represented in Fig. 42.6. We consider three configurations: $SN_1$, respectively SN2 and, as a group, the places $P_1$ and $P_2$. Having r tokens among $P_1$ and $P_2$ generates $C_r^2 = r + 1$ states. For n subnets, we have:

$$Se_{choise} = \sum_{i=0}^{n} Se_{1_{a1t}}(i).Se_{2_{a2t}}(i-1)\ldots Se_{a_{i-1t}}(r-i+1).Se_{a_{it}}(r-i).(i+1) \tag{42.6}$$

ASPN of systems containing unreliable components may include models for the failure and repair of these components. Figure 42.7 depicts a common model for these operations. Under normal functioning, $T_2$ fires instead of $T_{fail}$, leaving $SN_2$ out of consideration. The choice between $T_2$ and $T_{fail}$ allows $SN_2$ to use all r tokens. Thus, $SN_1$ and $SN_2$ are, in series, producing the following Se-function:

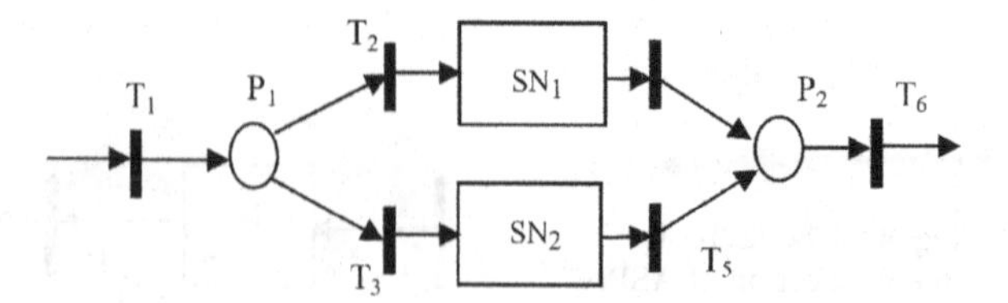


**Fig. 42.6** Free choice basic interconnection of ASPN

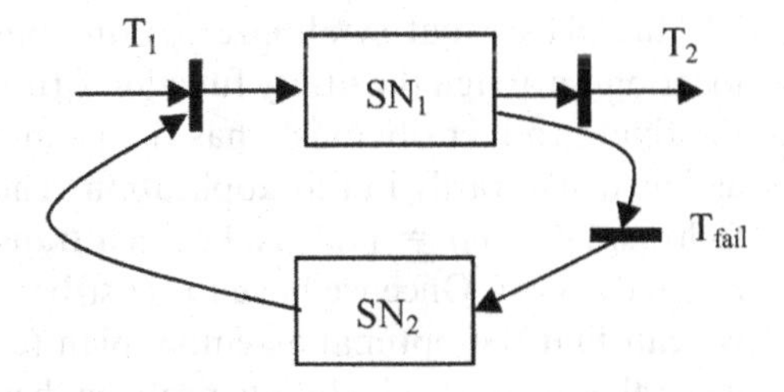


**Fig. 42.7** Failure/repair interconnection of ASPN

## 42.4 Design and Implementation Method

Using the proposed ASPN and algorithms, we have the following design and implementation steps for a construction system: (a) Construct an ASPN given the product information and the properly Se-function; (b) Associate all the data with places and transitions in the ASPN; (c) Run the ASPN based on the construction resources. In order to understand these steps we implement an assembly (construction) system as a plausible example. We have raw material type A,B,C,D,E (e.g. parts A,B,C,D,E). The possible way to assembly these parts to obtain the final subassembly F is depicted in Fig. 42.8. We noted with

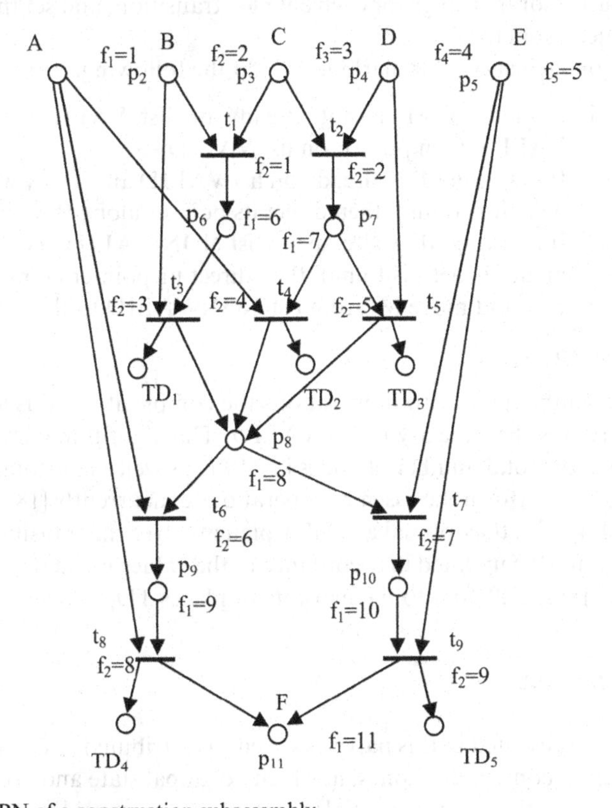


**Fig. 42.8** ASPN of a construction subassembly

TD the places that symbolize the dumping materials. In Fig. 42.8, for each location we assign the utility function $f_1(p_1 \div p_{11}) = (1,2,3,\ldots,11)$. For example, the final product obviously has the greatest value, and the other values were assigned arbitrarily in this application. The assembly cost in this example is as follows: $f_2(t_1 \div t_9) = (1,2,\ldots,9)$. Each transition has allocated in Fig. 42.8 the respective cost. Once we have the cost/benefits values of places and transitions we can find the optimal assembly plan (e.g. the optimal way in ASPN). The general job-shop scheduling problem has been shown to be NP-complete. Therefore, we resort to the heuristic search algorithm to solve this problem. We use a heuristic best-first search procedure known as $A^*$ algorithm [10].

This algorithm is the following one:

Step 1. Place initial marking $M_0$ on the list VALID

Step 2. If VALID is empty terminate with failure.

Step 3. Choose a marking M from the list VALID with maximal cost f(M) and move it from the list VALID to the list NON-VALID.

Step 4. If M is the final marking, construct the searched way from the initial marking to the final marking, and finish.

Step 5. Calculate $v_d(t=j)$ (see Section 42.2 – relation (42.1)) and generate the successor markings for each enabled transition, and set the way from successors to M.

Step 6. For each successor marking M', do the following:

- 6.1. If marking M' is not already on list VALID or list NON-VALID, then put M' on list VALID;
- 6.2. If marking M' is already on list VALID and a way with a higher benefit is found, then direct its pointer along the current way;
- 6.3. If marking M' is already on list NON-VALID and a way with a higher benefit is found, then direct its pointer along the current way and move M' from list NON-VALID to the list VALID.

Step 7. Go to step 2.

For an assembly plan with n operations the complexity of this algorithm is O(bn), where b is the capacity of list VALID. The algorithm complexity also depends upon the total number of nodes in ASPN as well as the total number of raw materials to perform the assembly operations concurrently [18–21]. For the example in Fig. 42.8 the optimal assembly plan involves the transition ($t_2$, $t_4$, $t_7$, $t_9$). This way may be updated in accordance to the values $v_{di}$, and $v_{pj}$, where $i = 1,\ldots,11$, and $j=1,\ldots,9$, and to the marking of places $TD_k$, where $k=1,\ldots,5$.

## 42.5 Conclusion

The approach presented in this paper is suited to distributed and asynchronous systems, such as construction ones, in which no global state and no global time is available, and therefore a partial order model is considered. This work proposes a methodology for design and implementation of adaptive assembly systems. In order to model the planning problem, assembly supervisor Petri nets (ASPNs) are introduced, with two functions: one attached to places and the other to transitions. The first is a resource utility function, which represents the value of a subassembly or a part to be used, and the second function represents the cost of performing a particular assembly operation. A bottom-up state-space size estimation technique for ASPN has been described. The estimation relies on the computation of state estimating functions (Se-functions). A lot of model analysis and control algorithms are based on the model state-space, and there it is proven that they are affected by large state-space sizes. The benefits of this approach include the simple representation of Se-functions that facilitates automation, and the possibility to interject hand-computed results into the estimation. Errors in the estimation may result from changes in the ASPN model to permit analysis in the Se functions.

Further research will include the colored Petri nets into the ASPNs, and the calculus of their Se-functions. To incorporate the uncertainty caused by different assembly conditions and the quality of resources, a probability value is assigned to each transition. Probability values can be updated during process execution. Future research will also focus on adapting interpreted Petri nets to the presented approach.

## References


1. Sampath M, Sengupta R, Lafortune S, Sinnamohideen K, Tewneketzis D (1995) Diagnosability of discrete event systems. IEEE Trans on Autom Contr 40(9):1555–1575
2. Jalcin A, Boucher ThO (1999) An architecture for flexible manufacturing cells with alternate machining and alternate sequencing. IEEE Trans Robot Autom 15(6):1126–1130
3. Liu Z, Towsley D (1994) Stochastic scheduling in tree-networks. Adv Appl Probability 20:222–241
4. Lahmar M, Ergan H, Benjafaar S (2003) Resequencing and feature assignment on an automated assembly line. IEEE Trans on Robot and Autom 19(1):89–102
5. Viswandham N, Narahari Y, Johnson TL (1990) Deadlock prevention and deadlock avoidance in flexible manufacturing systems using petri net models. IEEE Trans Robot Autom 6(6):438–456
6. Nebel B, Koehler J (1995) Plan reuse versus plan generation – A theoretical and empirical analysis. Artif Intell 76:427–454
7. Suzuki T, Kanehara T, Inaba A, Okuma S (1993) On algebraic and graph structural properties of assembly petri net. In: Atlanta GA (ed) Proceeding of the IEEE International Conference on Robotics and Automation, pp 507–514
8. Zussman E, Zhou MCh (1999) A methodology for modelling and adaptive planning of disassembly processes. IEEE Trans on Robot and Autom 15(1):190–193
9. Zussman E, Zhou MCh (2000) Design and implementation of an adaptive process planner for disassembly processes. IEEE Trans on Robot and Autom 16(2):171–179
10. Rogers RV, Preston White K Jr (1991) Algebraic mathematical programming and network models of the deterministic job-shop scheduling problem. IEEE Trans Syst Man and Cybern 21(3):350–361

11. Xie X (2002) Fluid stochastic event graph for evaluation and optimization of discrete event systems with failures. IEEE Trans on Robot and Autom 18(3):360–367
12. Watson JF, Desrochers AA (1992) Methods for estimating state-space size of Petri nets. In: Proceeding of the IEEE International Conference on Robotics and Automation, Nice, France, pp 1031–1036
13. Ciufudean C (2001) Petri nets in railway traffic systems. Advances in electrical and computer engineering, University "Stefan cel Mare" of Suceava, vol 1(18), no 1(15):21–26
14. Ciufudean C (2002) Discrete event systems for modelling the railway traffic. Matrix Rom Publishing House, Bucuresti, Romania
15. Ciufudean C, Larionescu AB (2002) Safety criteria for production lines modeled with Petri nets. Adv Electrical and Comput Eng, vol 2(9), no 2(18):15–20
16. Ciufudean C (2007) Discrete event systems – Applied themes. Matrix Rom Publishing House, Bucuresti, Romania
17. Ciufudean C, Graur A, Filote C, Turcu C, Popa V (2006) Diagnosis of complex systems using ant colony decision petri nets. In: First European Conference on Availability, Reliability and Diagnosis (ARES 2006), Wien, Austria, pp 567–573, ISBN 0-7695-2567-9
18. Ciufudean C, Filote C, Popescu D (2006) Worflows in constructions modelled with stochastic artificial Petri nets. In: Proceedings of the 23rd International Symposium on Automation and Robotics in Construction (ISARC 2006), Tokyo, Japan, pp 773–778, ISBN 4-9902717-1-8
19. Ciufudean C, Satco B, Filote C (2007) Reliability Markov chains for security data transmitter analysis. In: Proceedings of the Second International Conference on Availability, Reliability and Security (ARES 2007), Vienna University of Technology, Austria, pp 886–892, ISBN 0-7695-2775-2
20. Ciufudean C, Satco B, Nitu C (2007) Markov decision models for immune mechanism analysis. WSEAS Trans Inf Sci Appl 4(1):23–29, ISSN 1790-0832
21. Ciufudean C, Graur A, Filote C, Turcu C (2007) A new formalism for failure diagnosis: Ant colony decision Petri nets. J Software (JSW) 2(1):39–46, ISSN: 1796-217X

# Chapter 43
# Open Architecture Systems for MERO Walking Robots Control

**Luige Vladareanu, Radu Munteanu, Adrian Curaj, and Ion N. Ion**

**Abstract** The paper presents an Open Architecture system for the MERO walking modular robot, which is a robot on 3×6 degrees of freedom, autonomous, being able to interact with the environment. Starting from purpose kinematics and kinetostatics analysis the mathematic model of the inverted kinematics is determined for controlling the main trajectory that defines legs' support and weigh centre positions. Related to this there is presented an Open Architecture system for the MERO robot position control in Cartesian coordinates through real time processing of the Jacobean matrix obtained out of the forward kinematics using the Denevit-Hartenberg method and calculating the Jacobean inverted matrix for feedback. The obtained results prove a significant reduction of the execution time for the real time control of MERO robot's position in Cartesian coordinates and increased flexibility.

## 43.1 Introduction

A robot can be viewed as a serial link manipulator where the links sequence is connected by an actuated joints mathematical relation which ensures coordinate transformation from one axis to the other. In general, due to having six degrees of freedom, the mathematical analysis for the manipulator becomes very complicated. There are two dominant coordinates systems: Cartesian coordinates and joint coordinates. Joint coordinates represent angles between links and link extensions. They form the coordinates where the robot links are moving with direct control by the actuators. The position and orientation of each segment of the linkage structure can be described using the Denavit-Hartenberg [D-H] transformation [1, 2].

**Kinematics modeling.** Kinematics modeling of the legs' mechanisms related to the main system attached to the MERO modular walking robot is presented

L. Vladareanu (✉)
Institute of Solid Mechanics, Romanian Academy, C-tin Mille 15, Bucharest, Romania
e-mail: luigiv@arexim.ro

N. Mastorakis et al. (eds.), *Proceedings of the European Computing Conference*, Lecture Notes in Electrical Engineering 28,
DOI 10.1007/978-0-387-85437-3_43, 

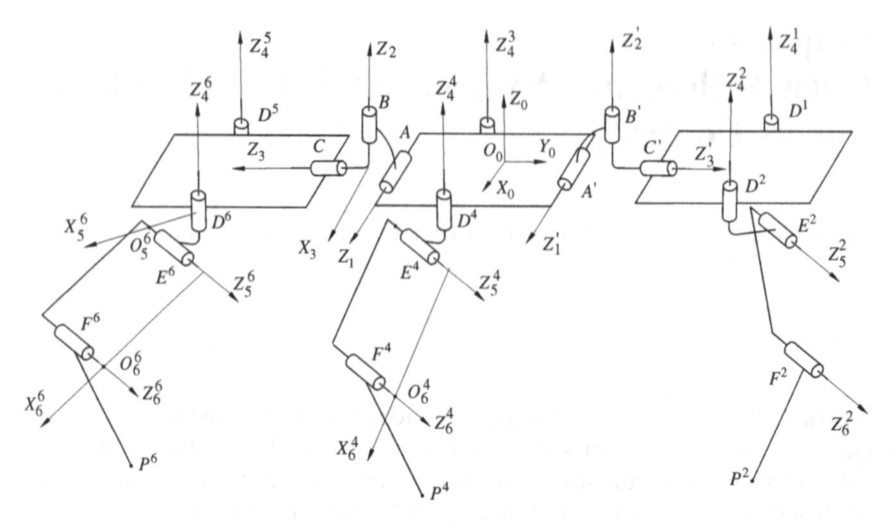


**Fig. 43.1** The H-D axis system attached to the leg of MERO modular walking robot

in Fig. 43.1. There are three modules with two legs each, connected by three joint angles. Each leg is composed of a mechanism with three elements connected by three rotation actuators. The **D-H** matrix method in homogenous coordinates is used in order to determine the position of the legs against the platform. Following the analysis of the **D-H** links system for the robot leg mechanism we have found a quite particular position [3], characterized by the following values: $\alpha_1 = 900$, $s_3 = 0$, $\alpha_2 = 0$, $a_1 = 0$, $\alpha_3 = 0$, $a_2 = 11 = 250$ mm, $s_1 = 30$ mm, $a_3 = 12 = 250$ mm, $s_2 = 0$. This led to decreasing the number of parameters of the transformation matrix $A_i$ from six to four.

The work presents both the transforming connections of the point's coordinates – from system $(\mathbf{i+1})$ to the system $(\mathbf{i})$, that are realized by a transformation matrix $\mathbf{A_i}$ – as well as the transforming relations of the $P_i$ point's coordinates from the system attached to the element (3) of the leg in the platform system. Considering as already known the legs coordinates $X_{OiP}$, $Y_{OiP}$, $Z_{OiP}$, the mathematic model with variable parameters of the leg joint angles $\theta_{ij}$ ($j = 1,6$, $i = 4,6$) has been determined by inverted kinematics analysis. Thus, knowing the trajectory that the legs should describe, the parameters necessary to the control system in real time of the robot can be established. Using direct kinematics analysis are calculated $X_{OiP}$, $Y_{OiP}$, $Z_{OiP}$, and by inverted kinematics analysis $\theta_{4i}$, $\theta_{5i}$, $\theta_{6i}$. The relative speed is calculated by derivation of movement equation and the relative accelerations by double derivation.

In the general case of the walking robots with 2n legs – the "myriapod" robots having "n" connected modules – the legs kinematics related to the environment system have been determined by the same algorithm.

## 43.2 Robot Position Control

The control using forward kinematics consists of transforming the actual joint coordinates, resulting from transducers, to Cartesian coordinates and compared with the desired Cartesian coordinates. The resulting error to the reference position in Cartesian coordinated is transmitted to every axis processing system. Using the inverting Jacobean matrix the angular errors will be generated on every motor control axis [3, 4]. There were identified five main processes for robot control: process 1-current position determined by the matrix in $X_C$ Cartesian coordinates; process 2-error position resulting in $\delta X_C$ Cartesian coordinates; process 3-Jacobean matrix setting; process 4-obtaining the triangular Jacobean matrix; and process 5- processing joint errors $\boldsymbol{\delta\theta}_{\mathbf{i}\ 1,2,3}$, i = 1–6, by inverse substitution.

The robot position control based on the **D-H** transformation is illustrated in Fig. 43.2. The robot joint angles, $\theta_c$, are transformed in $X_c$ – Cartesian coordinates with **D-H** transformation, where a matrix results from $X_C$ and $\delta\theta$, as below, with $\theta_j$ -joint angular $_j$ -offset distance, $a_j$ – link length, $\alpha_j$ – twist.

Position and orientation of the end effectors with respect to the base coordinate frame is given by $X_C = A_1 * A_2 * A_3$. Position error $\Delta X$ is obtained as a difference between desired and current position. There is difficulty in controlling robot trajectory, if the desired conditions are specified using position difference $\Delta X$ with continuous measurement of $\boldsymbol{\delta\theta}_{1,2,3}$, current position. The relation between the end-effector's position and orientation considered in Cartesian coordinates given at a certain moment and the robot joint angles $\boldsymbol{\delta\theta}_{1,2,3}$, is : $x_i = f_i(\theta)$, where $\boldsymbol{\theta}$ is the vector representing the freedom degrees of robot axis. By derivation we will obtain: $\boldsymbol{\delta}\ {}^3\mathbf{X}_3 = J(\theta)*\delta\theta_{1,2,3}$, where $\boldsymbol{\delta}\ {}^3\mathbf{X}_3$ represents differential linear and angular changes in the end effectors at the currently values of $\mathbf{X}_3$ and $\boldsymbol{\delta}\ \boldsymbol{\theta}_{1,2,3}$ representing the differential change of the set of joint angles. $J(\theta)$ is the Jacobean matrix in which the elements $\mathbf{a}_{ij}$ satisfy the relation: $\mathbf{a}_{ij} = \boldsymbol{\delta}\mathbf{f}_{i-1} / \boldsymbol{\delta\theta}_{j-1}$, where i, j correspond to

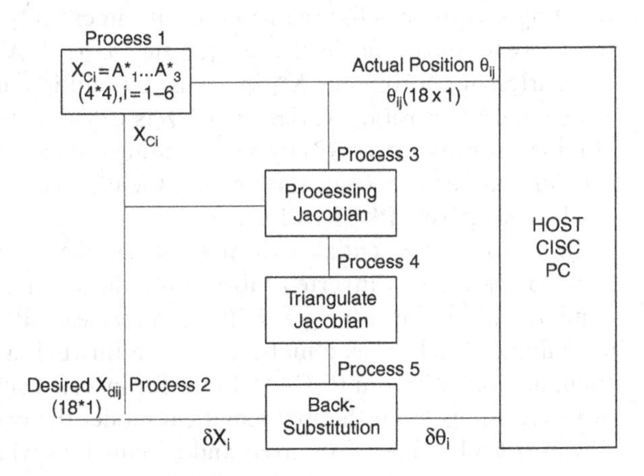


**Fig. 43.2** The robot position control based on the **D-H** transformation

the dimensions ofx respectively **θ**. The inverse Jacobean transforms the Cartesian position $\delta^3 X_3$ respectively **ΔX** into joint angular errors ($\Delta\theta$): $\boldsymbol{\delta\theta}_{1,2,,3} = \mathbf{J}^{-1}(\boldsymbol{\theta})^*\delta^3\mathbf{X}_3$.

The Jacobean calculation consists in consecutive multiplication of the A manipulator matrix. The Gaussian elimination method provides an efficient implementation of the matrix inversion. The method consists of reducing the **J** matrix to the upper triangulate form and finding errors in **Δθ** joint coordinates using back-substitution. The joint angular errors **Δθ** can be used directly as control signals for robot motors.

**The Multiprocessor System PLC.** As a result of the studies, for the position control of the MERO walking robot a real time control system with open architecture OAH has been conceived. The system has four main components: (1) the programmable logical controller (PLC0), (2) PC system with open architecture (PC-OPEN), (3) multiprocessor system PLC (SM-PLC) and (4) the position control system ($CP^k{}_i$, i = 1–3, k = 1–6). The system ensures the implementation of the algorithm for robot position control in Cartesian coordinates through real time processing of the Jacobean matrix obtained by direct kinematics of the robot, respectively of the invert Jacobean matrix for feedback [4]. Joint angular errors for actuators on each robot freedom axis corresponding to the robot's new desired position generated in Cartesian coordinates are obtained by processing the Jacobean and inverted substitution.

**The Multiprocessor System PLC (SM-PLC)** and PLC0 has the role to send in real time, through the ARCNET fast communication network, the angular reference positions for the position regulator. For feedback, the current values $\boldsymbol{\theta}^k{}_{C\ i}$, (i = 1–3, k = 1–6), received from the position transducers ($TP_{ki}$, i = 1–3, k = 1–6) are transmitted by PLC0 through ARCNET. Five processes for implementing the mathematical model of robot control have been identified. In the active topology for process (1), each PLC generates an ascendant data flux from PLC0 to PLC5.

By calculating the transformed matrix $^{\mathbf{i}}\mathbf{A}^{\mathbf{k}}{}_{\mathbf{j}}$ for the leg k (k = 1–6), from the axis i to the axis j, we obtain the coordination matrix in axis j, and finally the resulting coordinates for the robot environment $\mathbf{X}^{\mathbf{k}}{}_{\mathbf{C}} = {}^1A^k{}_3 = A_1{}^*A_2{}^*A_3$. In the active topology for the process (2), the matrix $^{\mathbf{i-1}}\mathbf{A}^{\mathbf{k}}{}_{\mathbf{i}}$ is stored for each PLC, the Cartesian coordinates $\mathbf{X}^{\mathbf{k}}{}_{\mathbf{i}}$ in i axis, by multiplying with $^{\mathbf{1}}\mathbf{A}^{\mathbf{k}}{}_{\mathbf{j-1}}$, are determined and the position variation $\boldsymbol{\delta}\mathbf{X}^{\mathbf{k}}{}_{\mathbf{C}}$ is calculated. The Jacobean matrix is obtained during process (3) by an ascending data flow, correlated with process (1). During the matrix calculation $^{\mathbf{1}}\mathbf{A}^{\mathbf{k}}{}_{\mathbf{j}}$, the PLC0 is assigned to $J(\theta_1)$, matrix of (3×1), respectively PLC5 to $J(\theta_3)$.

In a process sequential with process (4) takes place the active process (5) which determines by inverted substitution the $\boldsymbol{\delta\theta}_i$ angular error value in accordance to the relation: $\boldsymbol{\delta\theta}_{1,2,3} = \mathbf{J}^{-1}(\boldsymbol{\theta})^*\delta^3\mathbf{X}_3$. To each PLC there has been allotted a column of the Jacobean matrix, the data flux going from PLC0 to PLC5 for triangulation and from PLC5 to PLC0 for inverted substitution. With the help of the relations from the mathematical model the execution program for the PLC0-PLC5 has been conceived and executed, in which each central unit has

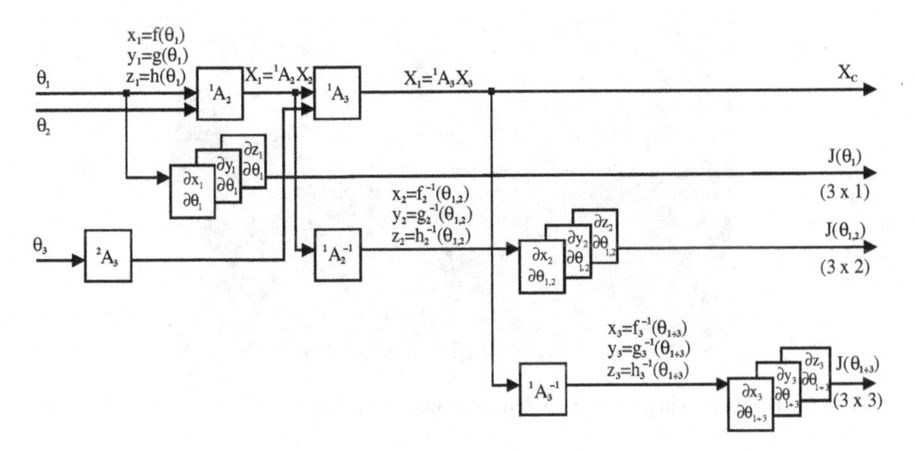


**Fig. 43.3** The data flux for OAH implementation with PLC multiprocessor systems

the role of MASTER communication by data flux through the ARCNET network (Fig. 43.3).

The active process (5) takes place within a sequential process with process (4), determining the value of the angular error $\delta\theta^k_i$ by inverted substitution.

## 43.3 Conclusions and Results

Mathematical modeling of the movement of the MERO walking robot has led to the development of two applications. For the transportation of nuclear material (Fig. 43.4) the imposed requirement is that the transported load maintain the same coordinate of the Z axis (vertical axis) throughout the movement.

**Fig. 43.4** MERO modular walking robot for Nuclear Transport

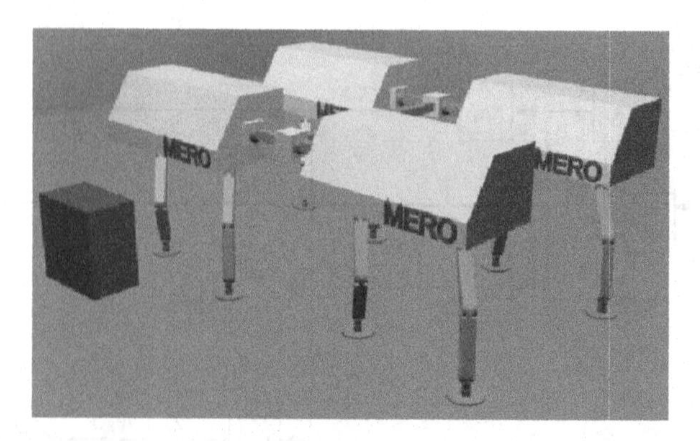


**Fig. 43.5** MERO walking robot with detour around obstacles

It is of note that after reaching an obstacle, the forward legs have a different movement from the rest so that the height of the transported load is constant. For a walking robot which moves in the presence of obstacles which need to be circumvented (Fig. 43.5), when reaching the object the robot changes direction by successive rotation of all legs. The results of this research have been materialized through the MERO modular walking robot (Fig. 43.6), which was built and tested at "Politehnica" University of Bucharest. Owing to the great computation speed of microprocessor systems and serial connection links for data transmission, the time necessary for establishing the inverted matrix is short enough to allow the robot control in real time, with no influence in performing the other programs.

The results show that the time necessary to perform the program for the MERO robot position control in Cartesian coordinates is 80% shorter. Moreover, the shorter execution time will ensure a faster feedback, allowing other programs to be performed in real time as well, such as the prehension force control, objects recognition, making it possible for the PC-OHA system to have a flexible and friendly human interface.

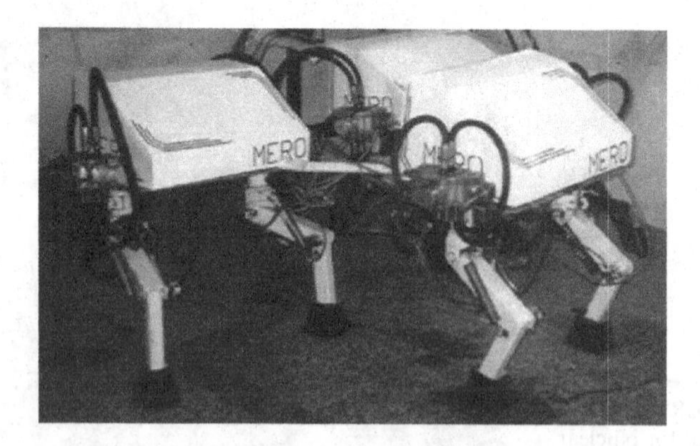


**Fig. 43.6** MERO modular walking robot

## References

1. Denavit J, Hartenberg RS (1955) A kinematic notation for lower-pair mechanism based on matrices. ASME J Appl Mech 22(2):215–221
2. Hiller M, Muller J, Roll U, et al (1999) Design and realization of the anthropomorphically legged and wheeled duisburg robot alduro. In: The Tenth World Congress on the Theory of Machines and Mechanisms, Oulu, Finland
3. Sidhu GS (1997) Scheduling algorithm for multiprocessor robot arm control. In: Proceedings of the 19th Southeastern Symp
4. Vladareanu L, Peterson T (2003) New concepts for the real time control of robots by open architecture systems. Mach Build 55(11), ISSN 0573-7419

# Chapter 44
# Mass Transfer Analysis in the Case of the EFG Method

**Liliana Braescu, Thomas F. George, and Stefan Balint**

**Abstract** The dependence of the $Nd^{3+}$ ion distribution on the pulling rate $v_0$ and on the capillary channel radius $R_{cap}$, in an $YVO_4$ cylindrical bar grown from the melt by the edge-defined film-fed growth (EFG) method with central capillary channel (CCC) shaper and melt replenishment is analyzed. The partial differential equations used for describing the growth process are those of the incompressible Navier-Stokes equations in the Boussinesq approximation and of the convection-conduction and conservative convection-diffusion equations. For a numerical solution, the finite-element numerical technique using COMSOL Multiphysics 3.3 software for a 2D axis-symmetric model is utilized. The Marangoni effect (i.e., surface tension driven flows due to the temperature gradient along the liquid free surface) is taken into account, and for its implementation the weak form of the boundary application mode is employed. The computations are carried out in the stationary case, for different pulling rates and capillary channel radii. The numerical results are compared with the experimental data [1].

## 44.1 Introduction

Single crystals of rare-earth vanadates possess various useful properties for laser and optical applications [4, 6, 7]. $Nd:YVO_4$ is one of the most efficient laser host crystals currently existing for diode-pumped solid state lasers.

$Nd:YVO_4$ single crystals were produced by many growth techniques (Czochralski, Verneuil, floating zone, laser-heated pedestal growth, top-seeded solution growth). Wide-range applications are still limited due to the serious difficulties in growing crystals with good quality; the homogeneity of the dopant concentration and free defects in the grown crystal are very important for optical applications. For solving the non-homogeneity problem, the

L. Braescu (✉)
Department of Computer Science, West University of Timisoara, Blv. V. Parvan 4, Timisoara 300223, Romania

N. Mastorakis et al. (eds.), *Proceedings of the European Computing Conference*, Lecture Notes in Electrical Engineering 28,
DOI 10.1007/978-0-387-85437-3_44, 

edge-defined film-fed growth (EFG) method has been considered as a prospective way to attain high-quality rare-earth vanadate single crystals.

In this paper, the $Nd^{3+}$ ion distribution is analyzed numerically in the framework of a stationary model, including incompressible fluid flow in the Boussinesq approximation, heat and mass transfer, and the Marangoni effect. The computations are made by using the 2D axis-symmetric hypothesis, for a cylindrical bar of radius $R = 1.5\times10^{-3}$ m grown with a pulling rate $v_0$ in the range $10^{-7}$–$10^{-5}$ m/s, using a meniscus height (liquid free surface) of $h = 0.5\times10^{-3}$ m. An EFG growth system with a central capillary channel shaper design [1, 2, 3] and a die radius $R_0 = 2\times10^{-3}$ m is considered for five values of the capillary channel radius $R_{cap} = 0.5, 0.75, 1, 1.25, 1.5 \times10^{-3}$ m; it is assumed that the melt level in the crucible is constant (continuous melt replenishment).

## 44.2 Problem Formulation

In order to find the stationary impurity distribution in the crucible, capillary channel and in the meniscus, the stationary incompressible Navier-Stokes in Boussinesq approximation (the temperature-dependent density appears only in the gravitational force term), stationary convection-conduction and stationary conservative convection-diffusion equations are considered:

$$\begin{cases} \rho_l(\bar{\mathrm{u}}\cdot\nabla)\bar{\mathrm{u}} = \nabla\cdot\left[-p\mathrm{I} + \eta\left(\nabla\bar{\mathrm{u}} + (\nabla\bar{\mathrm{u}})^T\right)\right] + \beta\rho_l\bar{\mathrm{g}}\cdot(T-\Delta T) \\ \nabla\cdot\bar{\mathrm{u}} = 0 \\ \nabla\cdot(-k\nabla T + \rho_l C_p T\bar{\mathrm{u}}) = 0 \\ \nabla\cdot(-D\nabla c + c\bar{\mathrm{u}}) = 0 \end{cases} \tag{44.1}$$

for which axisymmetric solutions are searched in the cylindrical-polar coordinate system $(rOz)$ (see Fig. 44.1). In the above system, the unknowns are: velocity vector $\bar{u} = (u,v)$, the temperature $T$, impurity concentration $c$, pressure $p$. The material parameters are: $\rho_l$ – reference melt density, $\beta$ – heat expansion coefficient, $\eta$ – dynamical viscosity, $k$ – thermal conductivity, $C_p$ – heat capacity, $D$ – impurity diffusion, and $\bar{g}$ – gravitational acceleration. The system (44.1) is considered in the two-dimensional domain whose boundary is from $\Omega_1$ to $\Omega_{10}$, as represented on Fig. 44.1(a).

The fluid density is assumed to vary with temperature as:

$$\rho(r,z) = \rho_l[1-\beta(T(r,z)-\Delta T)] \tag{44.2}$$

and the surface tension $\gamma$ in the meniscus is assumed to vary linearly with temperature as:

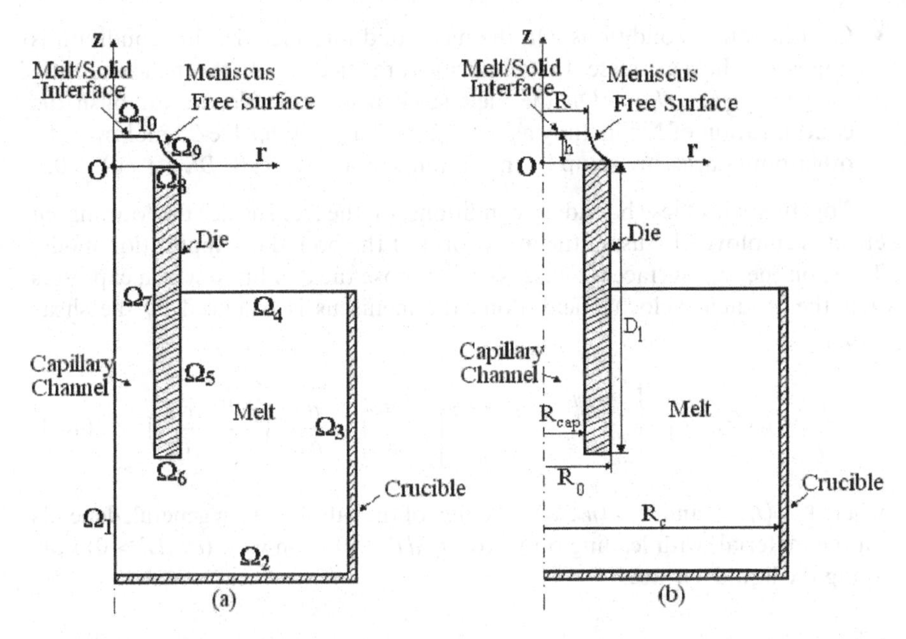


**Fig. 44.1** Schematic diagram of the EFG system and boundary regions used in the numerical model

$$\gamma(r,z) = \gamma_l + \frac{d\gamma}{dT}(T(r,z) - \Delta T) \tag{44.3}$$

where $\Delta T = (T_0 + T_m)/2$ is the reference temperature at the free surface, $\gamma_l$ is the surface tension at the temperature $\Delta T$ and $d\gamma/dT$ is the rate of change of surface tension with the temperature.

For solving the system (44.1), boundary conditions on $\Omega_1$ to $\Omega_{10}$ are imposed, with the $Oz$-axis being considered as a line of symmetry for all field variables:

- Flow conditions: On the melt/solid interface, the condition of outflow velocity is imposed, i.e., $\bar{u} = v_0\bar{k}$, where $\bar{k}$ represents the unit vector of the $Oz$-axis. On the melt level in the crucible, the inflow velocity condition is imposed (melt replenishment), $\overline{\mathrm{u}} = -\frac{\rho_s}{\rho_l} \cdot \frac{R^2}{R_c^2 - R_0^2} \cdot v_0\overline{\mathrm{k}}$. On the meniscus free surface, we set up the slip/symmetry condition, $\overline{\mathrm{t}} \cdot \left[-p\mathrm{I} + \eta\left(\nabla\overline{\mathrm{u}} + (\nabla\overline{\mathrm{u}})^T\right)\right]\overline{\mathrm{n}} = 0$, $\bar{u} \cdot \bar{n} = 0$, where $\bar{t}$ and $\bar{n}$ represent the tangent and normal vectors, respectively. The other boundaries are setup by the non-slip condition $\bar{u} = 0$.
- Thermal conditions: On the melt/solid interface, we set up the temperature $T = T_m$. On the free surface, we impose thermal insulation, i.e., $\overline{\mathrm{n}} \cdot (-k\nabla T + \rho_l C_p T\overline{\mathrm{u}}) = 0$. For the other boundaries, we have the temperature condition $T = T_0$.

- Concentration conditions: On the melt/solid interface the flux condition is imposed, which expresses that impurities are rejected into the melt according to $\frac{\partial c}{\partial \bar{n}} = -\frac{v_0}{D}(1 - K_0)c$. On the melt level in the crucible, we establish the concentration of Nd in alloy as $c = C_0 = 2$ at. % = 6.842e-22 mol/m$^3$. The other boundaries are setup by insulation/symmetry, $\bar{n} \cdot (-D\nabla c + c\bar{u}) = 0$.

Together with these boundary conditions, on the free surface the Marangoni effect is employed by using the weak form of the boundary application mode. Thus, on the free surface (meniscus), we impose the condition which expresses that the gradient velocity field along the meniscus is balanced by the shear stress:

$$\eta \cdot \begin{bmatrix} t_r & t_z \end{bmatrix} \cdot \begin{bmatrix} 2\frac{\partial u}{\partial r} & \frac{\partial u}{\partial z} + \frac{\partial v}{\partial r} \\ \frac{\partial u}{\partial z} + \frac{\partial v}{\partial r} & 2\frac{\partial v}{\partial r} \end{bmatrix} \cdot \begin{bmatrix} n_r \\ n_z \end{bmatrix} = \frac{d\gamma}{dT} \cdot \left( \frac{\partial T}{\partial r} + \frac{\partial T}{\partial z} \right) \tag{44.4}$$

where $\bar{t} = (t_r, t_z)$ and $\bar{n} = (n_r, n_z)$. The sign of the rate $d\gamma/dT$, in general, depends on the material, with leading outward ($d\gamma/dT < 0$) or inward ($d\gamma/dT > 0$) flow along the liquid surface.

## 44.3 Numerical Results

The computations are made in the axis-symmetric domain (Fig. 44.1) in the stationary case, for a Nd doped $YVO_4$ cylindrical bar having a radius of $R = 1.5 \times 10^{-3}$ m, grown with a pulling rate $v_0$ in the range $10^{-7}$–$10^{-5}$ m/s. The boundaries presented in Fig. 44.1 are determined from the particularities (characteristic elements) of the considered EFG growth system. Thus, a crucible with inner radius $R_c = 23 \times 10^{-3}$ m is considered, which is continuously fed with Nd-doped melted $YVO_4$ such that the melt height in the crucible is maintained constant at $45 \times 10^{-3}$ m. In the crucible, a die with the radius of $R_0 = 2 \times 10^{-3}$ m and length was introduced, such that 2/3 of the die $45 \times 10^{-3}$ m is immersed in the melt. In the die, a capillary channel is manufactured through which the melt climbs to the top of the die, where a low meniscus of height $h = 0.5 \times 10^{-3}$ m is formed. In the computations, five different radii of the capillary channel are considered: $R_{cap} = 0.5, 0.75, 1, 1.25, 1.5 \times 10^{-3}$ m.

The material parameters used in the numerical simulations and their significance for Nd:$YVO_4$ are given in Table 44.1 [4, 5, 6, 7].

The employed mesh is considered with maximum element size of 1e-3 and manually refined along the boundaries 8, 9 and 10 (maximum element size 1e-5). Thus, according to the considered geometries corresponding to $R_{cap} = 0.5, 0.75, 1, 1.25, 1.5 \times 10^{-3}$ m, we have 4823, 4589, 4231, 4153, and 3779 triangular elements. The stationary incompressible Navier-Stokes model for fluid flow, stationary heat transfer and mass transfer, and the weak form of the boundary condition imply for these meshes a total number of degrees of freedom equal to 44267, 41999, 38740, 37951, and 34592, respectively.

Table 44.1 Material parameters for Nd:$YVO_4$

| Nomenclature | | Value |
|---|---|---|
| $\beta$ | heat expansion coefficient (1/K) | 4.43e-6 |
| $c$ | impurity concentration (mol/m$^3$) | |
| $C_0$ | alloy concentration (at %) | 2 |
| $C_p$ | heat capacity (J/Kg×K) | 800 |
| $d\gamma/dT$ | rate change of the surface tension (N/ m×K) | −5e-5 |
| $D$ | impurity diffusion (m$^2$/s) | 2.4e-9 |
| $D_l$ | die length (m) | 45e-3 |
| $g$ | gravitational acceleration (m/s$^2$) | 9.81 |
| $h$ | meniscus height (m) | 0.5e-3 |
| $k$ | thermal conductivity (W/m×K) | 5.2 |
| $k_g$ | vertical temp. gradient [K/m] | 33000 |
| $K_0$ | partition coefficient | 0.5 |
| $\eta$ | dynamical viscosity (Kg/m×s) | 0.046 |
| $p$ | pressure (Pa) | |
| $R$ | crystal radius (m) | 1.5e-3 |
| $R_{cap}$ | capillary channel radius (m) | |
| $R_c$ | inner radius of the crucible (m) | 23e-3 |
| $R_0$ | die radius (m) | 2e-3 |
| $\rho_l$ | density of the melt (Kg/m$^3$) | 4200 |
| $\rho_s$ | density of the crystal (Kg/m$^3$) | 4200 |
| $\bar{u}$ | velocity vector | |
| $v_0$ | pulling rate (m/s) | |
| $T$ | temperature (K) | |
| $T_0$ | temperature at z = h (K) | |
| $T_m$ | melting temperature (K) | 2083 |
| $\Delta T$ | reference temperature (K) | |
| $z$ | coordinate in the pulling direction | |

The computed fluid flow in the meniscus, for three values of $R_{cap}$, is presented in Fig. 44.2 for the pulling rate $v_0$ = 1e-5 m/s.

This reveals that for a fixed pulling rate, the flow velocity decreases if $R_{cap}$ increases.

The maximum of the fluid flow is achieved on the meniscus free surface, and the arrows denoting the flow of the velocity field caused by the surface tension driven flow (outward flow because $d\gamma/dT < 0$) show that the Marangoni effect plays an important role in mass transfer, as it is reported by the crystal growers [1]. The impurities are moved from the lateral meniscus free surface, inside the meniscus near the melt/solid interface.

This combined effect of the buoyancy and Marangoni forces can be seen in Figs. 44.3 and 44.4 for two values of $R_{cap}$.

Computations show that for capillary radii less than the desired crystal radius, at the level $z = h$ (i.e., the melt/solid interface), the radial concentration increases, has a maximum, and after that decreases. The point where the radial concentration has a maximum depends on the pulling rate $v_0$, and in general, it

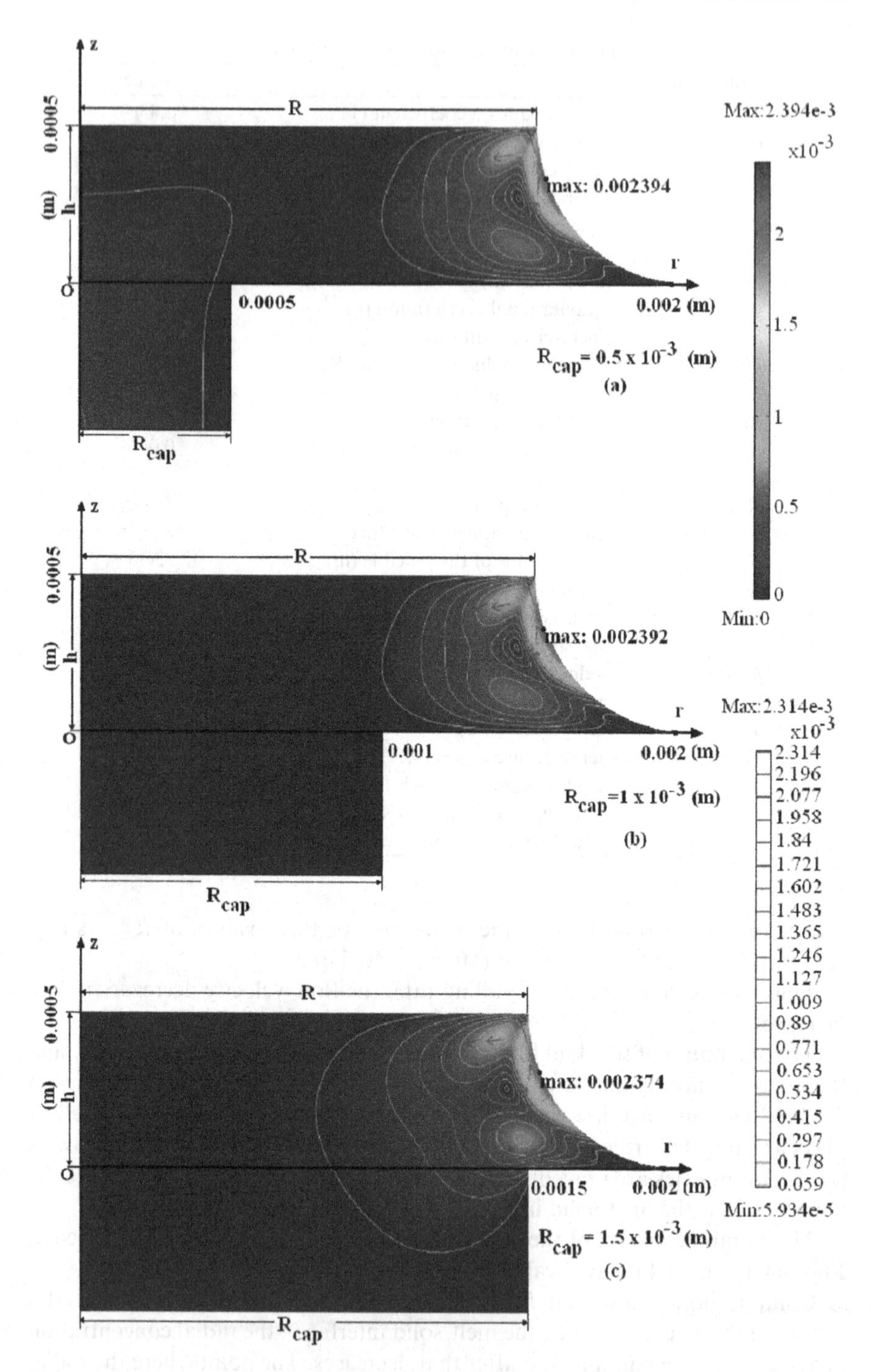


**Fig. 44.2** Fluid flow field in the meniscus for $v_0$ = 1e-5 m/s

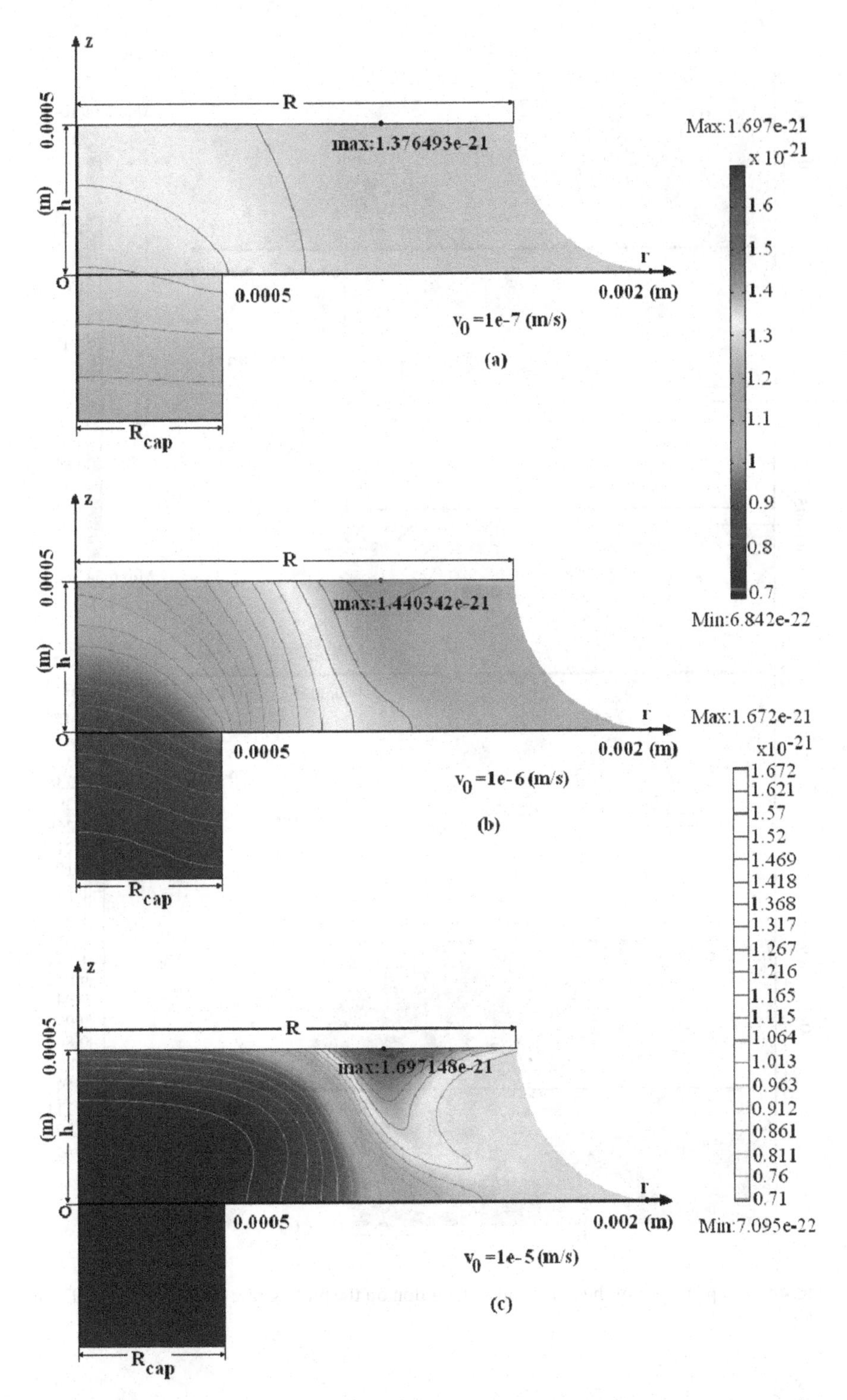


**Fig. 44.3** Dependence of the $Nd^{3+}$ ion distribution on the pulling rate for $R_{cap} = 0.5 \times 10^{-3}$ m

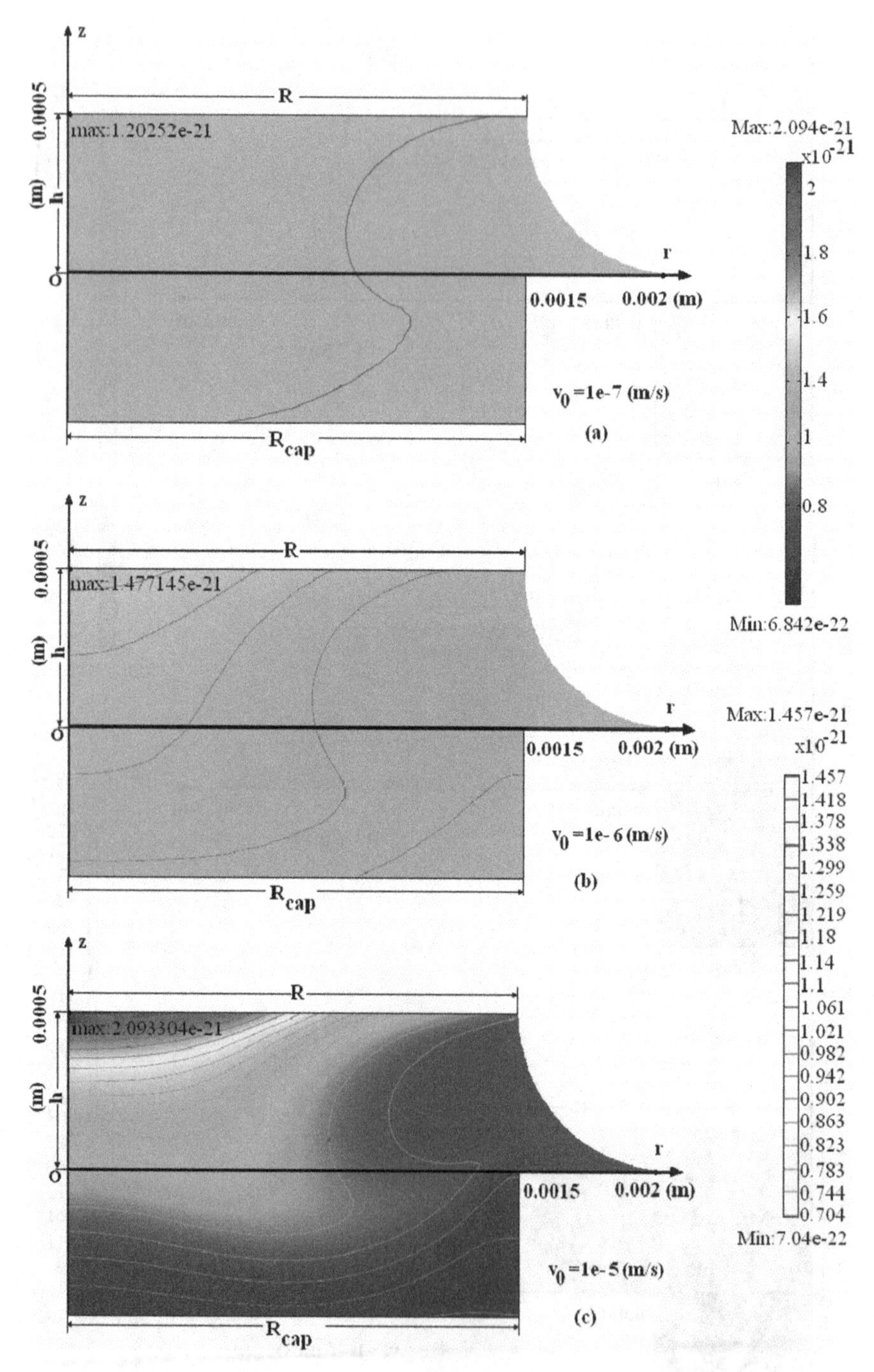


**Fig. 44.4** Dependence of the $Nd^{3+}$ ion distribution on the pulling rate for $R_{cap} = 1.5\times10^{-3}$ m

is located at a distance on the same order as the meniscus height from the external crystal surface. This behavior of the $Nd^{3+}$ concentration is in agreement with experimental data reported for another oxide (sapphire) cylindrical bar of radius $2.25\times10^{-3}$ m, grown in an EFG system with a CCC die of radius $R_0 = 2.35\times10^{-3}$ m and a capillary channel radius $R_{cap} = 0.5\times10^{-3}$ m.

Computations show that for capillary radii less than the desired crystal radius, at the level $z = h$ (i.e., the melt/solid interface), the radial concentration increases, has a maximum, and after that decreases. The point where the radial concentration has a maximum depends on the pulling rate $v_0$, and in general, it is located at a distance on the same order as the meniscus height from the external crystal surface. This behavior of the $Nd^{3+}$ concentration is in agreement with experimental data reported for another oxide (sapphire) cylindrical bar of radius $2.25\times10^{-3}$ m, grown in an EFG system with a CCC die of radius $R_0 = 2.35\times10^{-3}$ m and a capillary channel radius $R_{cap} = 0.5\times10^{-3}$ m.

Values of the impurity concentration at different points on the melt/solid interface and the maximum values $c_{max}(r,z) = c_{max}(r,h)$ are presented in Tables 44.2, 44.3, 44.4, 44.5, and 44.6.

**Table 44.2** $Nd^{3+}$ ion concentration on the melt/solid interface $z = h$ for $R_{cap} = 0.5\times10^{-3}$ m

| | $v_0$(m/s) | | |
|---|---|---|---|
| $(r;h)$ | 1e-7 | 1e-6 | 1e-5 |
| (0;0.5e-3) | 1.33e-21 | 1.09e-21 | 0.90e-21 |
| (0.5e-3;0.5e-3) | 1.35e-21 | 1.21e-21 | 0.96e-21 |
| (1e-3;0.5e-3) | 1.37e-21 | 1.43e-21 | 1.67e-21 |
| (1.5e-3;0.5e-3) | 1.37e-21 | 1.40e-21 | 1.34e-21 |

**Table 44.3** $Nd^{3+}$ ion concentration on the melt/solid interface $z = h$ for $R_{cap} = 0.75\times10^{-3}$ m

| | $v_0$(m/s) | | |
|---|---|---|---|
| $(r;h)$ | 1e-7 | 1e-6 | 1e-5 |
| (0;0.5e-3) | 1.35e-21 | 1.20e-21 | 0.99e-21 |
| (0.5e-3;0.5e-3) | 1.35e-21 | 1.26e-21 | 1.02e-21 |
| (1e-3;0.5e-3) | 1.37e-21 | 1.42e-21 | 1.65e-21 |
| (1.5e-3;0.5e-3) | 1.36e-21 | 1.38e-21 | 1.32e-21 |

**Table 44.4** $Nd^{3+}$ ion concentration on the melt/solid interface $z = h$ for $R_{cap} = 1.0\times10^{-3}$ m

| | $v_0$(m/s) | | |
|---|---|---|---|
| $(r;h)$ | 1e-7 | 1e-6 | 1e-5 |
| (0;0.5e-3) | 1.34e-21 | 1.26e-21 | 1.07e-21 |
| (0.5e-3;0.5e-3) | 1.34e-21 | 1.30e-21 | 1.11e-21 |
| (1e-3;0.5e-3) | 1.35e-21 | 1.40e-21 | 1.63e-21 |
| (1.5e-3;0.5e-3) | 1.35e-21 | 1.36e-21 | 1.27e-21 |

**Table 44.5** $Nd^{3+}$ ion concentration on the melt/solid interface $z = h$ for $R_{cap} = 1.25\times10^{-3}$ m

| $(r;h)$ | $v_0$(m/s) 1e-7 | 1e-6 | 1e-5 |
|---|---|---|---|
| (0;0.5e-3) | 1.27e-21 | 1.29e-21 | 1.11e-21 |
| (0.5e-3;0.5e-3) | 1.28e-21 | 1.32e-21 | 1.19e-21 |
| (1e-3;0.5e-3) | 1.29e-21 | 1.40e-21 | 1.61e-21 |
| (1.5e-3;0.5e-3) | 1.28e-21 | 1.36e-21 | 1.21e-21 |

**Table 44.6** Maximum $Nd^{3+}$ ion concentration $c_{max}(r,h)$ for different $v_0$ and $R_{cap}$

| $R_{cap}$ | $v_0$(m/s) 1e-7 | 1e-6 | 1e-5 |
|---|---|---|---|
| 0.5e-3 | $c_{max}$(0.00104,h) = 1.3764e-21 | $c_{max}$(0.00104,h) = 1.4403e-21 | $c_{max}$(0.00104,h) = 1.69714e-21 |
| 0.75e-3 | $c_{max}$(0.00102,h) = 1.3738e-21 | $c_{max}$(0.00103,h) = 1.4226 e-21 | $c_{max}$(0.00104,h) = 1.6740 e-21 |
| 1.0e-3 | $c_{max}$(0.00099,h) = 1.3545e-21 | $c_{max}$(0.00099,h) = 1.4084 e-21 | $c_{max}$(0.00100,h) = 1.6384 e-21 |
| 1.25e-3 | $c_{max}$(0.00097,h) = 1.2903 e-21 | $c_{max}$(0.00097,h) = 1.4041 e-21 | $c_{max}$(0.00097,h) = 1.6227 e-21 |

Computations show that the maximum $Nd^{3+}$ ion concentration $c_{max}(r,h)$ decreases if $R_{cap}$ increases or $v_0$ decreases.

For a capillary channel radius equal to the crystal radius (see Fig. 44.4), the buoyancy force will move this maximum to the axis of the crystal. In this case, the radial concentration decreases from the crystal axis to the lateral surface as can be seen in Table 44.7.

The $Nd^{3+}$ concentration on the axis $C_{ax}$ = c(0,0.5e-3), on the lateral surface of the crystal $C_{lat}$ = c(1.5e-3,0.5e-3) and the difference between them depend on the pulling rate and on the capillary radius. A measure of the non-homogeneity of the crystal is given by the absolute value of the difference ($C_{lat}$-$C_{ax}$), called radial segregation. This segregation is small when the pulling rate is small and the capillary channel radius is large.

**Table 44.7** $Nd^{3+}$ ion concentration on the melt/solid interface $z = h$ for $R_{cap} = 1.5\times10^{-3}$ m

| $(r;h)$ | $v_0$(m/s) 1e-7 | 1e-6 | 1e-5 |
|---|---|---|---|
| (0;0.5e-3) | 1.20e-21 | 1.47e-21 | 2.09e-21 |
| (0.5e-3;0.5e-3) | 1.19e-21 | 1.43e-21 | 1.89e-21 |
| (1e-3;0.5e-3) | 1.19e-21 | 1.35e-21 | 1.29e-21 |
| (1.5e-3;0.5e-3) | 1.18e-21 | 1.32e-21 | 1.01e-21 |

## 44.4 Conclusions

Mass transfer of $Nd^{3+}$ in a $YVO_4$ cylindrical bar is analyzed in the framework of a stationary model solved numerical using the finite element method. In this model, the surface tension driven flows due to the temperature gradient along the liquid free surface is included. The impurity distribution in the melt and in the crystal is computed as function of the pulling rate $v_0$ and of the capillary channel radius $R_{cap}$. The dependence of the impurity distribution on $v_0$ and $R_{cap}$ is established.

**Acknowledgment** We are grateful to the North Atlantic Treaty Organisation (Grant CBP.EAP.CLG 982530) for support of this project.

## References

1. Bunoiu O, Nicoara I, Santailler JL, Duffar T (2005) Fluid flow and solute segregation in EFG crystal growth process. J Cryst Growth 275:e799–e805
2. Braescu L, Balint S, Tanasie L (2006) Numerical studies concerning the dependence of the impurity distribution on the pulling rate and on the radius of the capillary channel in the case of a thin rod grown from the melt by edge-defined film-fed growth (EFG) method. J Cryst Growth 291:52–59
3. George TF, Balint S, Braescu L (2007) Mass and heat transport in Bridgman-Stockbarger and edge-defined film-fed growth systems. Springer Handbook of Crystal Growth, Defects and Characterization, Springer, Berlin Heidelberg New York
4. Hur MG, Yang WS, Suh SJ, Ivanov MA, Kochurikhin VV, Yoon DH (2002) Optical properties of EFG grown Nd:$YVO_4$ single crystals dependent on Nd concentration. J Cryst Growth 237–239:745–748
5. Kochurikhin VV (2007) Personal communications. Head of Growth Team, General Physics Institute, Moscow, Russia, http://webcenter.ru/~kochur/
6. Kochurikhin VV, Ivanov MA, Yang WS, Suh SJ, Yoon DH (2001) Development of edge-defined film-fed growth for the production of $YVO_4$ single crystals with various shapes. J Cryst Growth 229:179–183
7. Shur JW, Kochurikhin VV, Borisova AE, Ivanov MA, Yoon DH (2004) Photoluminescence properties of Nd:$YVO_4$ single crystals by multi- die EFG method. Opt Mater 26:347–350

# Chapter 45
# Information System Architecture for Data Warehousing

**Francisco Araque, Alberto Salguero, and Cecilia Delgado**

**Abstract** One of the most complex issues of the integration and transformation interface is the case where there are multiple sources for a single data element in the enterprise Data Warehouse (DW). There are many facets due to the number of variables that are needed in the integration phase. However we are interested in the temporal integration problem. This paper presents our DW architecture for temporal integration on the basis of the temporal properties of the data and temporal characteristics of the data sources. The proposal shows the steps for the transformation of the native schemes of the data sources into the DW scheme. In addition we present two new components for temporal integration: Temporal Integration Processor and Refreshment Metadata Generator, which will be used for integrating temporal properties of data and generating the necessary data for the later refreshment of the DW.

## 45.1 Introduction

The ability to integrate data from a wide range of data sources is an important field of research in data engineering. Data integration is a prominent theme in many areas and enables widely distributed, heterogeneous, dynamic collections of information sources to be accessed and handled.

Many information sources have their own information delivery schedules, whereby the data arrival time is either predetermined or predictable. If we use the data arrival properties of such underlying information sources, the Data Warehouse Administrator (DWA) can derive more appropriate rules and check the consistency of user requirements more accurately. The problem now facing the user is not the fact that the information being sought is unavailable, but rather that it is difficult to extract exactly what is needed from what is available.

F. Araque (✉)
Dpto. De Lenguajes y Sistemas Informáticos, University of Granada, 18071,
Granada (Andalucía), Spain
e-mail: faraque@ugr.es

N. Mastorakis et al. (eds.), *Proceedings of the European Computing Conference*, Lecture Notes in Electrical Engineering 28,
DOI 10.1007/978-0-387-85437-3_45, 

It would therefore be extremely useful to have an approach which determines whether it would be possible to integrate data from two data sources (with their respective data extraction methods associated). In order to make this decision, we use the temporal properties of the data, the temporal characteristics of the data sources and their extraction methods. Notice that we are not suggesting a methodology, but an architecture. Defining a methodology is absolutely out of the scope of this paper, and the architecture does not impose it.

It should be pointed out that we are not interested in how semantically equivalent data from different data sources will be integrated. Our interest lies in knowing whether the data from different sources (specified by the DWA) can be integrated on the basis of the temporal characteristics (not in how this integration is carried out).

The use of DW and Data Integration has been proposed previously in many fields. In [1] the Integrating Heterogeneous Tourism Information, the data sources problem is addressed using three-tier architecture. In [2] a Real-Time Decision Support System for space missions control is put forward using Data Warehousing technology. In [3] a multilevel security policies integration methodology to endow tightly coupled federated database systems with a multilevel security system is presented. In [4] a framework for quality-oriented DW management is exposed, where special attention is paid to the treatment of metadata. The problem of the little support for automatized tasks in DW is considered in [5], where the DW is used in combination with event/condition/action rules to get an active DW. In [6] an integrated decision support system from the perspective of a DW is exposed. Their authors state that the essence of the data warehousing concept is the integration of data from disparate sources into one coherent repository of information. Nevertheless, none of the previous works encompass the aspects of the integration of the temporal parameters of data.

In this article a solution to this problem is proposed. Its main contributions are: a DW architecture for temporal integration on the basis of the temporal properties of the data and temporal characteristics of the sources, a Temporal Integration Processor and a Refreshment Metadata Generator, that will be both used to integrate temporal properties of data and to generate the necessary data for the later DW refreshment.

The remaining part of this paper is organized as follows. In Section 45.2, the concept of DW and our previous related works are revised; in Section 45.3 our architecture is presented; finally, Section 45.4 summarizes the conclusions of this paper.

## 45.2 Federated Databases and Data Warehouses

Inmon [7] defined a Data Warehouse as "a subject-oriented, integrated, time-variant, non-volatile collection of data in support of management's decision-making process". A DW is a database that stores a copy of operational data with an optimized structure for query and analysis. The scope is one of the issues which defines the DW: it is the entire enterprise. In terms of a more limited scope, a new concept is defined: a Data Mart (DM) is a highly focused DW covering a single department or subject area. The DW and data marts are usually implemented using relational databases, [8] which define multidimensional structures. A *federated database system* (FDBS) is formed by different *component database systems*; it provides integrated access to them: they co-operate (inter-operate) with each other to produce consolidated answers to the queries defined over the FDBS. Generally, the FDBS has no data of its own, queries are answered by accessing the component database systems.

We have extended the Sheth and Larson five-level architecture [9], which is very general and encompasses most of the previously existing architectures. In this architecture three types of data models are used: first, each component database can have its own native model; second, a *canonical data model* (CDM) which is adopted in the FDBS; and third, external schema can be defined in different *user models*.

One of the fundamental characteristics of a DW is its temporal dimension, so the scheme of the warehouse has to be able to reflect the temporal properties of the data. The extracting mechanisms of this kind of data from the operational system will also be important. In order to carry out the integration process, it will be necessary to transfer the data of the data sources, probably specified in different data models, to a common data model, that will be the used as the model to design the scheme of the warehouse. In our case, we have decided to use an OO model as canonical data model, in particular, the object model proposed in the standard ODMG 3.0.

ODMG has been extended with temporal elements. We call this new ODMG extension ODMGT. This is also our proposal: to use for the definition of the data warehouse and data mart schema an object-oriented model as CDM, enhanced with temporal features to define loading of the data warehouse and data marts.

## 45.3 Architecture Extension with Temporal Elements

Taking papers [9, 10] as point of departure, we propose the following reference architecture (see Fig. 45.1):

**Native Schema.** Initially we have the different data source schemes expressed in its native schemes. Each data source will have, a scheme, the data inherent to the source and the metadata of its scheme. In the metadata we will have huge temporal information about the source: temporal data on the scheme, metadata on availability of the source, availability of the log file or delta if it had them, etc.

Some of the temporal parameters, explained in depth in [11], that we consider of interest for the integration process are:

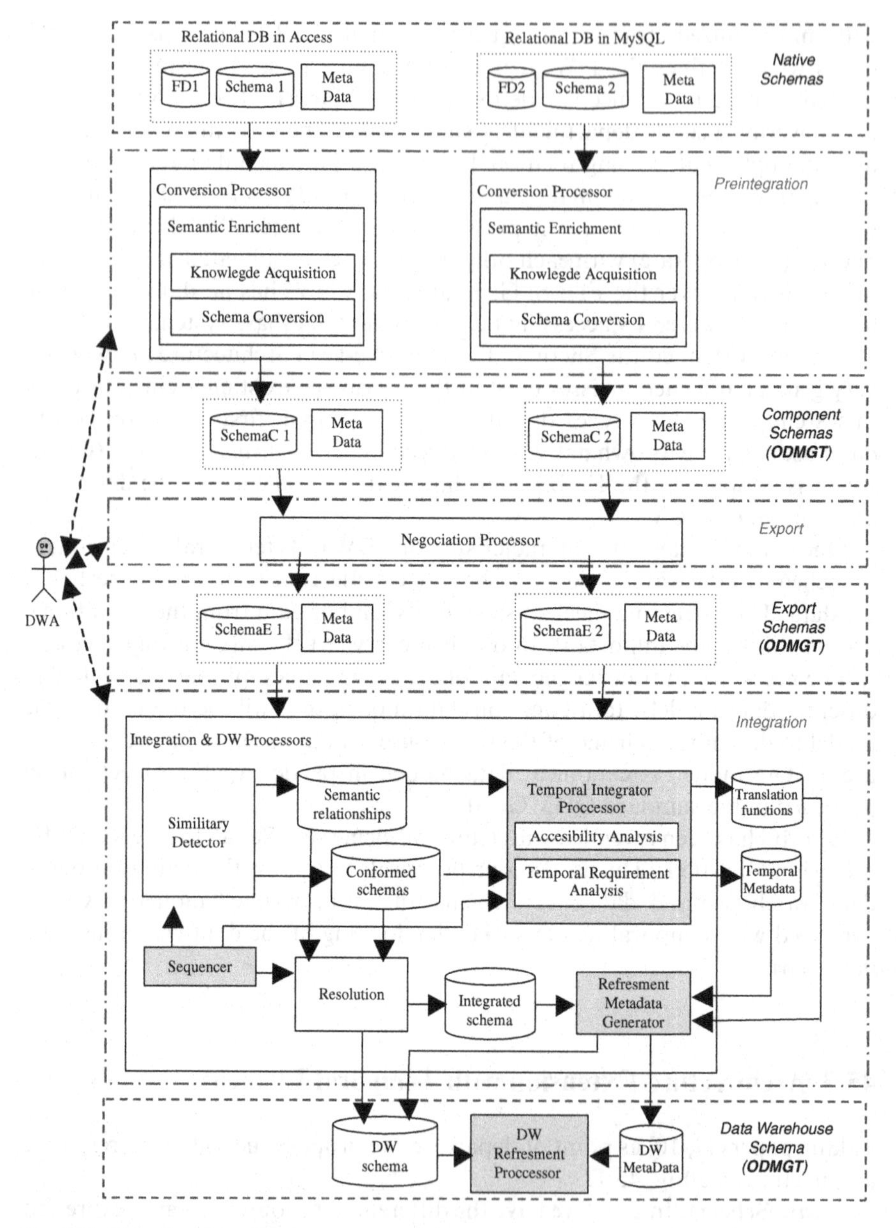


**Fig. 45.1** Functional architecture

- *Availability Window* (AW). Period of time in which the data source can be accessed by the monitoring programs responsible for data source extraction.
- *Extraction Time* (ET). Period of time taken by the monitoring program to extract significant data from the source.

- *Granularity* (Gr). It is the extent to which a system contains discrete components of ever-smaller size. In our case, because we are dealing with time, it is common to work with granules like minute, day, month and so on.
- *Transaction time* (TT). Time instant when the data element is recorded in the data source computer system. This would be the data source TT.
- *Storage time* (ST). Maximum time interval for the delta file, log file, or a source image to be stored.

**Preintegration.** In the *Preintegration* phase, the semantic enrichment of the data source native schemes is made by the conversion processor. In addition, the data source temporal metadata are used to enrich the data source scheme with temporal properties. We obtain the component scheme (CST) expressed in the CDM, in our case, ODMGT (ODMG enriched with temporal elements).

**Component and Export Schemas.** Apart from the five-scheme levels mentioned (Sheth and Larson 1990), three more different levels should be considered:

- *Component Scheme T(CST)*: the conversion of a Native Scheme to our CDM, enriched so that temporal concepts could be expressed.
- *Exportation Scheme T(EST)*: it represents the part of a component scheme which is available for the DW designer. It is expressed in the same CDM as the Component Scheme.
- *Data Warehouse Scheme*: it corresponds to the integration of multiple Exportation Schemes T according to the design needs expressed in an enriched CDM so that temporal concepts could be expressed.

From the CST expressed in ODMGT, the negotiation processor generates the export schemes (EST) expressed in ODMGT. These EST are the part of the CST that is considered necessary for its integration in the DW.

**Integration.** From many data sources EST schemas, the DW scheme is constructed (expressed in ODMGT). This process is made by the Integration Processor that suggests how to integrate the Export Schemes helping to solve semantic heterogeneities (out of the scope of this paper). In the definition of the DW scheme, the DW Processor participates in order to contemplate the characteristics of structuring and storage of the data in the DW.

Two modules have been added to the reference architecture in order to carry out the integration of the temporal properties of data, considering the extraction method used: the *Temporal Integration Processor* and the *Metadata Refreshment Generator*.

The *Temporal Integration Processor* uses the set of semantic relations and the conformed schemes obtained during the detection phase of similarities [12]. This phase is part of the integration methodology of data schemes. As a result, we obtain data in the form of rules about the integration possibilities existing between the originating data from the data sources (minimum granularity, if the period of refreshment must be annotated between some concrete values...). This information is kept in the Temporal Metadata Warehouse. In addition, as a result of the Temporal Integration process, a set of mapping functions is obtained. It identifies the attributes of the schemes of the data sources that are self-integrated to obtain an attribute of the DW scheme.

The *Metadata Refreshment Generato*r determines the most suitable parameters to carry out the refreshment of data in the DW scheme [13]. The DW scheme is generated in the resolution phase of the methodology of integration of schemes of data. It is in this second phase where, from the minimum requirements generated by the temporal integration and stored in the Temporal Metadata warehouse, the DW designer fixes the refreshment parameters. As result, the DW scheme is obtained along with the Refreshment Metadata necessary to update the former according to the data extraction method and other temporal properties of a concrete data source.

Obtaining of the DW scheme and the Export schemes is not a linear process. We need the Integration and Negotiation Processors collaborate in an iterative process where the participation of the local and DW administrators is necessary [12].

**Data Warehouse Refreshment.** After temporal integration and once the DW scheme is obtained, its maintenance and update will be necessary. This function is carried out by the DW Refreshment Processor. Taking both the minimum requirements that are due to fulfill the requirements to carry out integration between two data of different data sources (obtained by means of the Temporal Integration module) and the integrated scheme (obtained by the resolution module) the refreshment parameters of the data stored in the DW will be adjusted.

## 45.4 Conclusion

In this paper we have presented a DW architecture for the temporal integration on the basis of the temporal properties of the data and temporal characteristics of the data sources. We have described the new modules introduced to the architecture of reference: Temporal Integration Processor and Refreshment Metadata Generator. These modules are responsible for checking the temporal parameters of data sources and for determining the best refreshment parameters according to the DWA requirements, generating, thus, the necessary data for the later DW refreshment. We used ODMGT (an ODMG extension with temporal elements) as CDM. This work has been supported by the Andalucía Research Program under project GR2007/07-2 and by the Spanish Research Program under projects EA-2007-0228 and TIN2005-09098-C05-03.

## References

1. Haller M, Pröll B, Retschitzegger W, Tjoa AM, Wagner RR (2000) Integrating heterogeneous tourism information in TIScover – The MIRO-Web approach. In: Proceedings Information and Communication Technologies in Tourism, ENTER. Barcelona (España)
2. Moura Pires J, Pantoquilho M, Viana N (2004) Real-time decision support system for space missions control. In: International Conference on Information and Knowledge Engineering, Las Vegas, USA

3. Oliva M, Saltor F (2001) Integrating multilevel security policies in multilevel federated database systems. In: Thuraisingham B, van de Riet R, Dittrich KR, Tari Z (eds) Data and Applications Security: Developments and Directions, Kluwer Academic Publishers, Boston, pp 135–147
4. Vassiliadis P, Quix C, Vassiliou Y, Jarke M (2001) Data warehouse process management. Inf Syst 26:205–236
5. Thalhammer T, Schrefl M, Mohania M (2001) Active data warehouses: complementing OLAP with analysis rules. Data Knowl Eng 39(3):241–269
6. March ST, Hevner AR (2003) Integrated decision support: A data warehousing perspective. AIS SIGDSS Pre-ICIS Workshop Research Directions on Decision Support
7. Inmon WH (2002) Building the data warehouse. John Wiley, New York
8. Harinarayan V, Rajaraman A, Ullman J (1996) Implementing data cubes efficiently. In: Proc. of ACM SIGMOD Conference, Montreal
9. Sheth A, Larson J (1990) Federated database systems for managing distributed, heterogeneous and autonomous databases. ACM Comput Surv 22(3)
10. Samos J, Saltor F, Sistac J, Bardés A (1998) Database architecture for data ware-housing: An evolutionary approach. In: Quirchmayr G et al (eds) Proceedings of the International Conference on Database and Expert Systems Applications, Springer-Verlag, Vienna, pp 746–756
11. Araque F, Salguero AG, Delgado C, Garvi E, Samos J (2006) Algorithms for integrating temporal properties of data in Data Warehouse. In: 8th International Conference on Enterprise Information Systems (ICEIS), Paphos, Cyprus
12. Oliva M, Saltor F (1996) A negotiation process approach for building federated databases. In: Proceedings of 10th ERCIM Database Research Group Workshop on Heterogeneous Information Management, Prague, pp 43–49
13. Araque F, Samos J (2003) Data warehouse refreshment maintaining temporal consistency. In: 5th Intern. Conference on Enterprise Information Systems (ICEIS), Angers, France

# Chapter 46
# Dimensional Dynamics Identification of Reconfigurable Machine Tools

**Ciprian Cuzmin, Virgil Teodor, Nicolae Oancea, Vasile Marinescu, and Alexandru Epureanu**

**Abstract** The paper presents a new method of dimensional dynamics online identification to be used for the dimensional control of reconfigurable machining systems. This dimensional control is designed to be adaptive and predictive. The dimensional control of the reconfigurable machining system is a key action in achieving the desired quality of a final product. Since a decrease in deviation usually results in higher costs, the deviation compensation would be a better solution for improving the process accuracy. Deviation compensation occurring during the machining process requires a model that is able to describe the dimensional dynamics of a machining system, which stands for the relationship between the dimensional changing of part being machined and the process parameters. On the other hand, the machining system behavior is changing significantly in time even when a small number of parts are processed. This is the reason why dimensional dynamics must reveal the changing in time of the parts dimensional variation–process parameters relationship.

## 46.1 Introduction

The dimensional control of a machining process is a key action to obtain the desired quality of a final product. Two techniques of dimensional deviation control are used:

a. Diminishing the deviation by making the process parameters' values smaller or using more consecutive passing;
b. Deviation compensation by modifying the programmed tool-trajectory.

Since a decrease in deviation presents the major drawback of high costs, the deviation compensation seems to be a more suitable solution.

C. Cuzmin (✉)
Department of Manufacturing Engineering, "Dunărea de Jos" University of Galaţi, Galaţi, 800201, Romania
e-mail: ciprian.cuzmin@ugal.ro

N. Mastorakis et al. (eds.), *Proceedings of the European Computing Conference*, Lecture Notes in Electrical Engineering 28,
DOI 10.1007/978-0-387-85437-3_46, 

Deviation compensation requires deviation prediction based on a model able to describe the dimensional dynamics of the machining system [1, 2, 4]. This dimensional dynamics is in fact the relationship between the dimension deviation of the part being machined and the processing parameters. On the other hand, the behavior of the machining system changes significantly in time, even when small part batches are processed. This is the reason why dimensional dynamics must also take into account the changing in time of the relationship between the dimension part deviation and the processing parameters.

In literature a number of compensation techniques are described as follows:

a. by measuring each part immediately after having been machined, the difference between the target surface and the one achieved is determined; the constant component of the difference measured on to the last parts machined is compensated for at the next part;
b. modeling and compensation of some of the error components (e.g. thermal error) is performed [4, 5];
c. processing parameters are kept constant in order to obtain uniform deviation which can be further compensated for by the tool offset.

All the papers presented identify and compensate the error components, not the error as a whole [1, 3, 6, 8].

The purpose of the present paper is to develop a new method of online identification of the dimensional dynamics in order to be applied in the dimensional control of a Reconfigurable Machine Tool (RMT). This dimensional control is designed to be adaptive and predictive. The proposed method is experimentally tested in order to evaluate the full-scale performance. The paper has the following structure: Section 46.2 describes the identification algorithm of the dimensional dynamics; Section 46.3 shows the experimental research of the dimensional dynamics; finally, Section 46.4 covers relevant discussions and conclusions.

## 46.2 Algorithm for Online Identification of the Dimensional Dynamics

Let us consider that a RMT has been configured and the machining process is so changed as to be able to process another part of the some products family. As a consequence of these changes the RMT behaviour has been altered; therefore the model describing this behaviour is no longer valid [7]. For this reason, it is necessary for the RMT to be reidentified. When RMT is machining the part number **n** of the batch, it is first necessary to identify the relationship between dimensional deviation occured during part-machining and the RMT monitoring data found during machining of all the previous **n-1** parts.

In order to describe the stationary state of the system coresponding to a certain bin of the tool trajectory the following variables will be used:

- variables describing the mechanical field; these may represent components of the cutting force or the force the RMT is demanded in its different points; $\mathbf{F}^{(1)}$, $\mathbf{F}^{(2)}$, $\mathbf{F}^{(3)}$ (its average value in a bin width);
- variables describing the thermal field; these may be average values $\mathbf{T}^{(1)}$, $\mathbf{T}^{(2)}$, $\mathbf{T}^{(3)}$ of the temperature in different points of the RMT;
- variable **i** represents the count number of the current bin on the tool trajectory;
- variable **j** represents the count number of the parts group being processed whose RTM behavior and tool wear corresponding to a cartain bin remain unchaged. Numbering the parts forming the group has its origin in the current part, (e.g. if, during processing the parts **n** and **n-1** the RMT behavior and tool wear have not changed, then **n**, **n-1** parts represent the group numbered $\mathbf{k} = 1$, while parts **n-2**, **n-3** form a second group for $\mathbf{k} = 2$ and so on);
- deviation $\boldsymbol{\delta}$ coresponding to the bin of count number **i**.

The values of variables **F**,**T**and $\boldsymbol{\delta}$ will be scaled on their own variation in such way that all the variables are represented by integer values. For instance in Fig. 46.1 these variables have been scaled using six levels.

We define RMT state using an interger number set representing termal and mechanical fields, position and deviation.

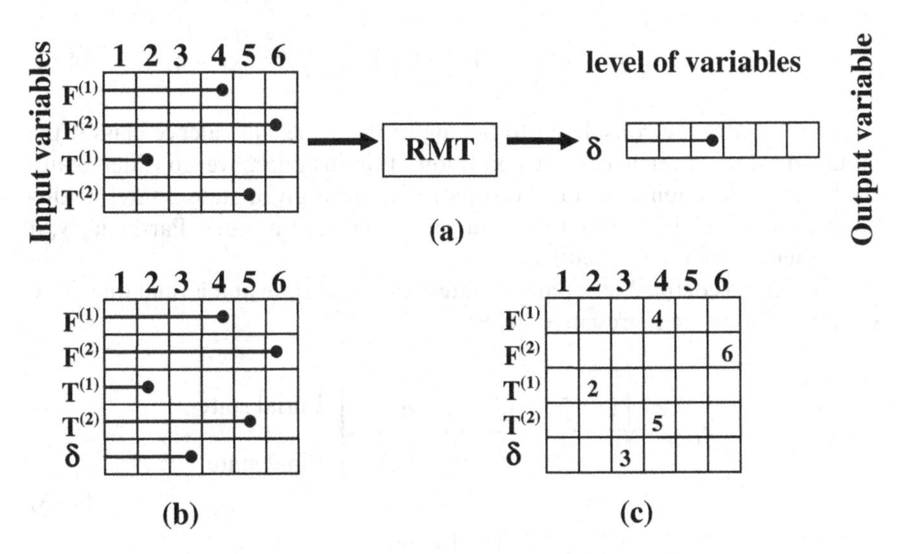


**Fig. 46.1** Three representations of the steady state of the system coresponding on the one tool path bin

The RMT state in a certain moment will be represented as follows:

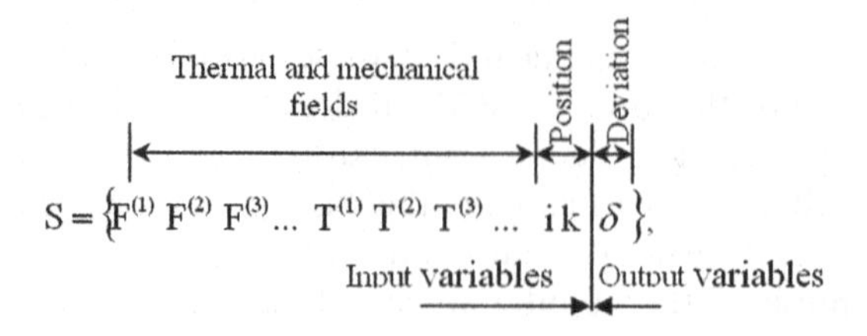


where $\mathbf{F^{(1)}}$, $\mathbf{F^{(2)}}$, $\mathbf{F^{(3)}}$ stand for the values reached by processing the signals received from the sensors in different points on the system; $\mathbf{T^{(1)}}$, $\mathbf{T^{(2)}}$, $\mathbf{T^{(3)}}$ represent the temperatures measured in different points on the system, **i** is the current bin count number, and **k** is the current number of the parts group processed with RMT behavior and tool wear considered unchanged.

For the sake of simplicity let us consider that the mechanical field is evaluated in one single point on the machine. In this case four states can be considered

$$S_1 = \{F_1 T_1 i_1 k_1 \delta_1\},\ S_2 = \{F_2 T_2 i_2 k_2 \delta_2\},$$
$$S_3 = \{F_3 T_3 i_3 k_3 \delta_3\}, S_4 = \{F_4 T_4 i_4 k_4 \delta_4\}.$$

In the states field the following metrixes can be defined: the difference between $S_1$ si $S_2$ states is

$$S_1 - S_2 = \{(F_1 - F_2), (T_1 - T_2), (i_1 - i_2), (k_1 - k_2), (\delta_1 - \delta_2)\} \quad (46.1)$$

and the distance between these states is:

$$d(S_1, S_2) = |F_1 - F_2| + |T_1 - T_2| + |i_1 - i_2| + |k_1 - k_2| \quad (46.2)$$

Several states are considered to belong to the same equivalence class if the input variables meet a certain condition. For instance, we consider states clasification depending on their distance from/to a given state. States at distance **a** from the given state form equivalence class of **a** order. Particularly, if **a** = 0, then the states are identical.

Transition is defined as a pair of states: an initial state and a final one. Two such transitions are represented below:

$$\tau_{(1,2)} = \begin{Bmatrix} F_1 & T_1 & i_1 & k_1 & \delta_1 \\ \hdashline F_2 & T_2 & i_2 & k_2 & \delta_2 \end{Bmatrix} \begin{matrix} \downarrow \text{Initial state} \\ \uparrow \text{Final state} \end{matrix}$$

$$\tau_{(3,4)} = \begin{Bmatrix} F_3 & T_3 & i_3 & k_3 & \delta_3 \\ F_4 & T_4 & i_4 & k_4 & \delta_4 \end{Bmatrix} \quad (46.3)$$

For given transitions we define the difference between two transitions:

$$\tau_{(1,2)} - \tau_{(3,4)} = \begin{Bmatrix} F_1 - F_3 & T_1 - T_3 & i_1 - i_3 & k_1 - k_3 & \delta_1 - \delta_3 \\ F_2 - F_4 & T_2 - T_4 & i_2 - i_4 & k_2 - k_4 & \delta_2 - \delta_4 \end{Bmatrix} \quad (46.4)$$

We consider that two transitions are equivalent if the difference is:

$$\tau_{(p,q)} - \tau_{(r,s)} = \begin{Bmatrix} 0 & 0 & a & b & \delta_p - \delta_r \\ 0 & 0 & a & b & \delta_q - \delta_s \end{Bmatrix} \quad (46.5)$$

where

$$\begin{aligned} a &= i_p - i_r = i_q - i_s \\ b &= k_p - k_r = k_q - k_s \end{aligned}.$$

Two equivalent transitions have the same input values variation (mechanical field (**F**), thermal field (**T**) and position (**i**, **k**)). As a result the output values variation will be the same as well:

$$(\delta_p - \delta_r) = (\delta_q - \delta_s). \quad (46.6)$$

Identification algorithm is illustrated in Fig. 46.2.

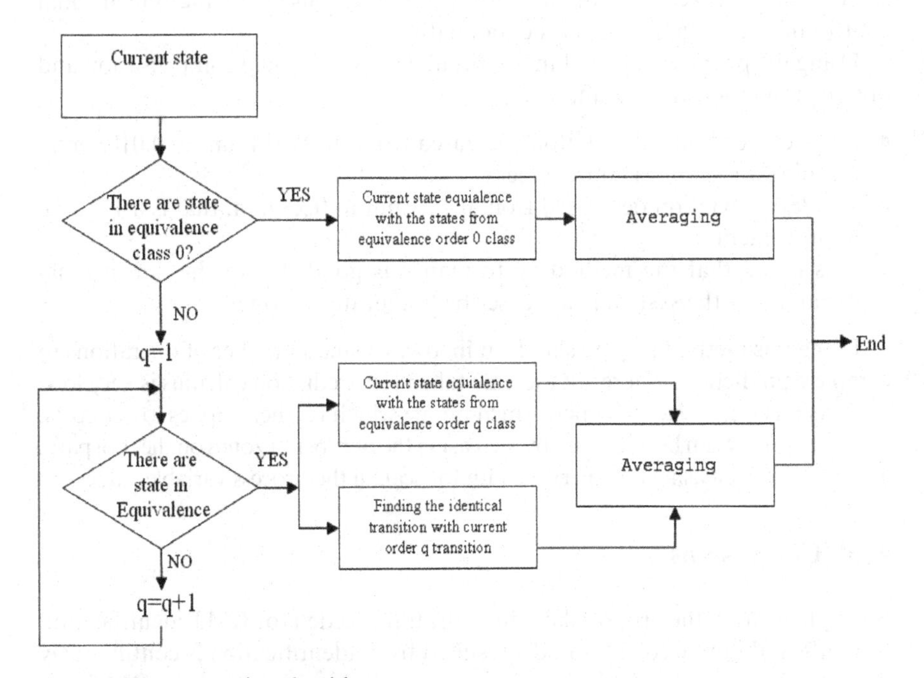


**Fig. 46.2** Data processing algorithm

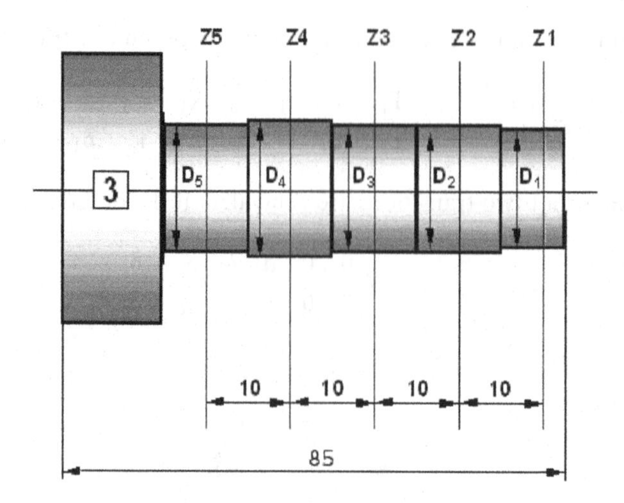


Fig. 46.3 Blank work piece

## 46.3 Experimental Research

A number of 70 parts batch has been processed (Fig. 46.3). The linear trajectory of the tool was divided into five zones. Dimensional control algorithm was applied separately in each zone. During processing the cutting force was measured. The average value of the cutting force was used for the dimensional control of one certain zone on the tool path.

Using the proposed algorithm, we simulated dimensional compensation and obtained the following results:

- the average value of deviation decreased from 0.837014 mm to 0.0103 mm, meaning an 81 times improvement;
- standard deviation decreased from 0.249 mm to 0.23 mm, that is a 1.1 time improvement.
- It is noted that the method performance is good due to the dimensional dynamics of the system being described accurately enough.

Using this method is quite simple, it involves a small number of operations to complete prediction; as a result the time taken for prediction calculus is very low. Dimension compensation is performed online which further implies lower costs.

The accuracy can be improved by increasing the number of zones on the tool path, as well as by decreasing the interval width for scaling the process variable values.

## 46.4 Conclusions

By implementing the proposed method, the time needed for RMT identification is significantly reduced. The model resulted from identification is continuously changing because it takes into account the evolution in time of the RMT and therefore it is able to describe more accurately the real behavior of the system at a certain moment.

We believe that the method advanced in this paper is able to improve the RMT behavior during parts machining process, and also to improve the machining process accuracy by error compensation.

The novelty of the proposed method is that, by considering the whole error as an entity, it successfully compensates for the entire deviation occurred during the process.

**Acknowledgments** This paper was developed under the research programs CEEX, grants 22 and 23, funded by the Romanian Ministry of Education and Research.

## References

1. Choudhury SK, Jain VK, Rama Krishna S (2001) On-line monitoring of tool wear and control of dimensional inaccuracy in turning. J Manuf Sci Eng 123(1):10–12
2. Dong C, Zhang C, Wang B, Zhang G (2002) Prediction and compensation of dynamic errors for coordinate measuring machines. J Manuf Sci Eng 124(3):509–514
3. Hong Yang, Jun Ni (2003) Dynamic modeling for machine tool thermal error compensation. J Manuf Sci Eng 125(2):245–254
4. Kim S, Landers RG, Ulsoy A (2003) Robust machining force control with process compensation. J Manuf Sci Eng 125(3):423–430
5. Lian RJ, Lin BF, Huang JH (2005) A grey prediction fuzzy controller for constant cutting force in turning. International J Mach Tools Manuf 45(9):1047–1056
6. Liang Steven Y, Hecker Rogelio L, Landers Robert G (2004) Machining process monitoring and control: The state-of-the-art. J Manuf Sci Eng 126(2):297–310
7. Mehrabi MG, Ulsoy AG, Koren Y, Heytler P (2000) Trends and perspectives in flexible and reconfigurable manufacturing systems. J Intell Manuf 13(2):135–146
8. Tseng PC, Ho JL (2002) A study of high-precision CNC lathe thermal errors and compensation. Int J Adv Manuf Technol, Springer London, 19(11):850–858

## References

# Chapter 47
# Compaction-Induced Deformation on Flexible Substrate

**B. Punantapong**

**Abstract** This paper describes the analytical method for evaluation of compaction-induced stresses and deformation on the thin layer of flexible substrate by using finite element analysis. The incremental placement and compaction of the thin layer of flexible substrate are based on a hysteretic model for residual stresses induced by multiple cycles of loading and unloading. The results showed that the large compaction load can be applied to a thin layer of flexible substrate and it achieves higher density effectively. The reinforcement of the layer also increases compaction efficiency, because it reduces the ratio between shear and vertical forces during the compaction process. The maximum vertical stress on the base of specimen usually decreased with higher compaction thickness. The reinforcement will acheive increased substrate stiffness under the compaction indenter and it initiates stress concentration. As a result, it maintains a higher vertical stress level on the base of the specimen that provides better compaction characteristics. Therefore, it can be concluded that the reinforcement is essential for achieving effective compaction on the thin layer of flexible substrate.

## 47.1 Introduction

The design of medical devices is a complex task and it's involved in the resolution of conflicts and compromise among the desired feature. One of the more inclusive is any substance or combination of substances synthetic or natural in origin, which can be used for any period of time, as a whole or part of a system which treats, augments or replaces tissue, an organ or function of the body. This definition must be extended because biomaterials are currently

B. Punantapong (✉)
Department of Industrial Physics and Medical Instrumentation, King Mongkut's Institute of Technology North Bangkok, Biomaterials and Nanotechnology Research Unit, 10800, Bangsue, Bangkok, Thailand

N. Mastorakis et al. (eds.), *Proceedings of the European Computing Conference*, Lecture Notes in Electrical Engineering 28,
DOI 10.1007/978-0-387-85437-3_47, 

being utilized as scaffolds for tissue-engineered devices (hybrids of synthetic or biologic scaffolds and living cells and tissue for vessels, heart valves, and myocardium).

For example, the cardiovascular biomaterials may contact blood (both arterial and venous), vascular endothelial cells, fibroblasts, as well as a number of other cells and a cellular matrix material that make up all biological tissue. At the same time, orthopedic biomaterials were the complex and important role of the cellular component of bone. Then the general properties of bone, ligament, tendon, and cartilage have been well characterized over the past century. So, orthopedic tissues are affected by both the stresses that they experience on a daily basis and as a result of trauma and disease processes [1]. Many of these injuries or pathologies require medical intervention that may be assisted through the use of engineered materials.

Thus biomaterials used in a clinical system are susceptible to a number of failure modes. Like all materials, mechanical failure is possible, particularly in implants (expected to exceed 10 years). However, advances in biomaterials have been focused on the design of dental and orthopedic implants, which are mainly structural applications. So, tremendous stiffness and strength improvement is not always the concern in the design of biomaterials and even less so for large weight savings. Other concerns, such as biocompatibility, precise property matching, mimicking natural structures, etc., can become more important, yet these too are areas where biomaterials offer much promise in device design.

Furthermore, it is very difficult to achieve good strength and stiffness for the flexible substrate of biomaterials. Hence, flexible substrate is not used for reinforced substrate structures, including solid-reinforced substrate. However, applying tensile reinforcement or compression with preloading and prestressing procedures only allows minimal deformation in the body of compacted substrate. Recently, de Boer et al. [2] proposed an engineering model that captures key grain scale phenomena in a manner that is consistent with both the experimentally measured bulk material behavior and grain contact mechanics. The model predictions agree well with experimental results for the explosion threshold of granular due to weak mechanical impact. Because the work was restricted to weak impact scenarios for which the average intergranular stress was much less than the bulk modulus of the pure phase solid, an incompressible solid was assumed. While this approach can provide meaningful statistical information about hot spot fluctuations, it requires extensive computational resources and time, and thus cannot be applied to engineering scale systems that contain in excess of a billion individual grains. This approach has proven to be successful in understanding the bulk granular material response induced by various impact conditions [3].

The present study investigates the effectiveness of a composite reinforcement on a thin layer of flexible substrate during compaction. Other simulation concepts like the finite element method mostly are used to investigate the phenomena of substrate layer interaction under quasi-static conditions like constant loading. So it is necessary to take into account the important influence of the layer on the dynamic behavior of substrate. Furthermore, the dynamical layer-load-changes enlarge, loading to an increased deformation of the surface. The following axle of the substrate will be excited by the back unevenness of the first axle, which again induces the excitation of the substrate.

## 47.2 Materials and Method

The finite element analysis methodology developed for analysis of stresses and deformations. The resulting from placement and compaction of layers of fill simulates the actual sequence of field operations in a number of sequential steps. These analytical procedures permit modeling of multiple cycles of compaction loading at any given fill stage with a single solution increment. In these analyses, compaction loading is realistically considered as a transient moving concentrated surficial load which may pass more times over some specified portion of the surface. Figure 47.1 shows that a spatially periodic loading (of period L) is applied to the surface of a horizontally layered profile of an elastic layer.

Therefore, a strip loading as a spatially periodic loading is applied to the surface of a horizontally layered profile of a flexible substrate. It is well known for such a periodic loading (or loading function). A Fourier series representation can be used Nemat-Nasser [4]. For instance, if the Cartesian coordinate in the horizontal direction is $x$ and the loading function is $p(x)$ then it is found that

$$p(x) = \sum_{n=0}^{\infty} \mathrm{P}^{(\mathrm{n})} \cos \alpha_n x \tag{47.1}$$

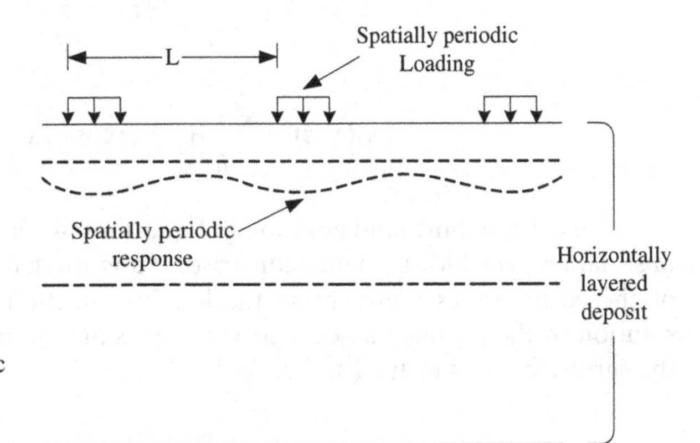


**Fig. 47.1** Spatially periodic loading on a horizontally layered deposit (adapted from [5])

where

$$P^n = \frac{2}{L}\int_0^L p(x)\cos\alpha_n x dx \quad (n>0)$$

$$P^n = \frac{1}{L}\int_0^L p(x) x dx \quad (n=0)$$

and $\alpha_n = 2n\pi/L$

The loading has been represented by the sum of periodic functions (in this case cosine functions because the loading function $p(x)$ was chosen to be an even function of $x$). It may be observed that, for such a spatially periodic loading, the displacements in the substrate below are also periodic. It is therefore possible to write the displacements as a series of periodic functions [5].

$$u_x = \sum_{n=0}^{\infty} U^{(n)} \sin\alpha_n x \tag{47.2a}$$

$$u_z = \sum_{n=0}^{\infty} W^{(n)} \cos\alpha_n x \tag{47.2b}$$

It will be observed that whereas a series of cosine functions is used to represent the displacement in the $z$ direction ($u_z$), a sine series is used for the lateral displacements in the lateral or $x$ direction ($u_x$). This follows from the observation that the lateral displacements are anti-symmetric about the axis of symmetry of each loaded area. The coefficients in the above series $U^{(n)}$, $W^{(n)}$ are of course different at different depths $z$ since both the displacement components vary throughout the depth of the layer.

$$u_x(x,z) = \sum_{n=0}^{\infty} U^{(n)}(z) \sin\alpha_n x \tag{47.3a}$$

$$u_z(x,z) = \sum_{n=0}^{\infty} W^{(n)}(z) \cos\alpha_n x \tag{47.3b}$$

However, the shorthand notation of Equations (47.2a and 47.2b) will be used in situations in which no confusion arises. Suppose that we now take one term of the cosine series representing the loading function $p(x)$, and obtain the solution to the problem associated with this single sinusoidal load applied to the surface of the substrate layer.

$$p(x) = P_n \cos\alpha_n x$$

Then

$$u_x = U_n(z) \sin \alpha_n x \tag{47.4a}$$

$$u_z = W_n(z) \cos \alpha_n x \tag{47.4b}$$

At the same time, we will be assumed that conditions of plane strain prevail so that there is no displacement in the $y$ direction and no variation of the field quantities with $y$. Then it is only necessary to consider stress and displacement fields having the form

$$u_x = U(\alpha, z) \sin \alpha x \tag{47.5a}$$

$$u_z = W(\alpha, z) \cos \alpha x \tag{47.5b}$$

$$\sigma_{xz} = T(\alpha, z) \sin \alpha x \tag{47.5c}$$

$$\sigma_{zz} = N(\alpha, z) \cos \alpha x \tag{47.5d}$$

$$\sigma_{xx} = H(\alpha, z) \cos \alpha x \tag{47.5e}$$

$$\sigma_{yy} = M(\alpha, z) \cos \alpha x \tag{47.5f}$$

since the solution for more complex load cases can be obtained by superposition of components given by Equations (47.5) and where it has been assumed that $\alpha$ stands for any particular value of $\alpha_n = 2n\pi/L$ and $U^{(n)} = U(\alpha_n, z)$, etc.

Let us now consider the determination of the field quantities in a particular layer, $l$ of the material [6]. If Equations (47.5) are substituted into the equilibrium equations it is found that

$$-\alpha H + \frac{\partial T}{\partial z} = 0 \tag{47.6a}$$

$$-\alpha T + \frac{\partial N}{\partial z} = 0 \tag{47.6b}$$

Similarly if Equations (47.5a, 47.5b, 47.5c, 47.5d, 47.5e, 47.5f) are substituted into Hooke's Law it is found that

$$\alpha U = \frac{1}{E_l}[H - \nu_l(M + N)] \tag{47.7a}$$

$$\frac{\partial W}{\partial z} = \frac{1}{E_l}[N - \nu_l(H + M)] \tag{47.7b}$$

$$\frac{\partial U}{\partial z} - \alpha W = \frac{2(1 + \nu_1)T}{E_1} \tag{47.7c}$$

$$M = \nu_l(H + N) \tag{47.7d}$$

where $E_l, \nu_l$ denote the values of Young's Modulus and Poisson's ratio for layer, $l$.

Equations (47.6) and (47.7) allow the non-zero stress and displacement to be expressed in terms of $N$ and thus

$$\begin{aligned} H &= -\frac{\partial^2 N}{\partial Z^2} \quad \text{and} \quad T = -\frac{\partial N}{\partial Z} \\ M &= \nu_l\left(N - \frac{\partial^2 N}{\partial Z^2}\right) \\ \alpha E^* U &= -\frac{\partial^2 N}{\partial Z^2} - \nu^* N \\ \alpha E^* W &= -\frac{\partial^3 N}{\partial Z^3} + (2 + \nu^*)N \end{aligned} \tag{47.8}$$

where

$$E^* = \frac{E_l}{1 - \nu_l^2}, \nu^* = \frac{\nu_l}{1 - \nu_l} \text{ and } Z = \alpha z$$

and

$$\frac{\partial^4 N}{\partial Z^4} - 2\frac{\partial^2 N}{\partial Z^2} + N = 0 \tag{47.9a}$$

If Equation (47.9a) is now solved, we obtain

$$N = X_1 C + X_2 Z + X_3 S + X_4 ZC \tag{47.9b}$$

The fundamental step in the finite layer technique is to determine the four constants $X_1, \ldots\ldots X_4$ in terms of boundary quantities. To be more precise, suppose that layer $l$ is bounded by the node planes $(z = z_l)$ and $(z = z_m)$ where $m = l + 1$ and that the subscripts $l$, $m$ indicate the value of a particular quantity on the indicated node plane. Then the solution can be used to determine $X_1, \ldots\ldots X_4$ in terms of $U_l$, $W_l$, $U_m$, $W_m$. Once $X_1, \ldots\ldots X_4$ are known they can be used to evaluate $T_l$, $N_l$, $T_m$, $N_m$ and so establish a relationship of the form

$$\begin{bmatrix} -T_l \\ -N_l \\ +T_m \\ +N_m \end{bmatrix} = \begin{bmatrix} k_{11} & k_{12} & k_{13} & k_{14} \\ k_{21} & k_{22} & k_{23} & k_{24} \\ k_{31} & k_{32} & k_{33} & k_{34} \\ k_{41} & k_{42} & k_{43} & k_{44} \end{bmatrix} \begin{bmatrix} U_l \\ W_l \\ U_m \\ W_m \end{bmatrix} \tag{47.10}$$

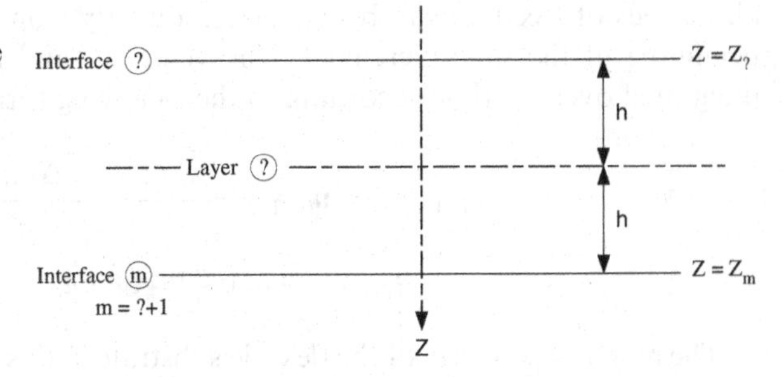


**Fig. 47.2** Single cell layer within the surface of flexible substrate

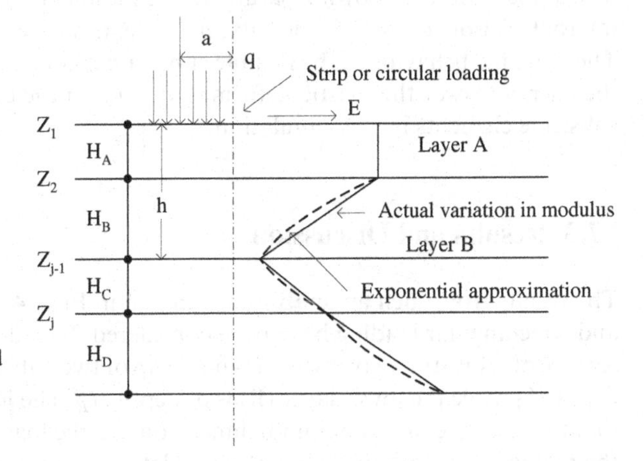


**Fig. 47.3** Non-homogeneous profile tested with exponential approximation (adapted from [6])

The matrix occurring in Equation (47.10) is called the layer stiffness matrix of the particular layer (layer, *l* in this case). Layer stiffness matrices can be used to construct solutions for layered deposits in exactly the same way as element stiffness matrices are used in conventional applications of the finite element method.

Consequently, we present a novel way to study the behavior of layer of flexible material when it received forces by compression. Thus FEM technique is necessary to use a dynamic solving algorithm where not only the displacements of the nodes, but as coming from the Newton's dynamic equilibrium the accelerations $\ddot{\mathbf{u}}$ for each node in the model are considered as Hiroma et al. [7]

$$\ddot{\mathbf{u}}_{(i)} = \mathbf{M}^{-1}(\mathbf{P}_{(i)} - \mathbf{I}_{(i)}) \tag{47.11}$$

where $\mathbf{M}$ is the mass matrix, $\mathbf{P}$ is the vector of the external forces (i.e. contact forces or nodal forces of neighboring elements) and $\mathbf{I}$ is the vector of internal forces on the basis of the stress inside the elements.

The mass matrix can be obtained by distributing the mass of each element on to its nodes. The created point mass belonging to a node is the sum of all partial masses of all the elements defined at this node. Using the explicit solving algorithm, available with ANSYS, the velocities $\dot{\mathbf{u}}$ and the displacement $\mathbf{u}$ for all degrees of freedom can be calculated directly from the information at the beginning of the time increment. The velocities and the displacements were integrated over the time according to the following formula.

$$\ddot{\mathbf{u}}_{(i-1/2)}+ = \dot{\mathbf{u}}_{(i-1/2)} + \frac{\Delta t_{(i+1)} + \Delta t_{(i)}}{2} .\ddot{\mathbf{u}}_{(i)} \quad (47.12)$$

$$\mathbf{u}_{(i+1)} = \mathbf{u}_{(i)} + \Delta t_{(i+1)}\dot{\mathbf{u}}_{(i+1/2)} \quad (47.13)$$

The material property of the flexible substrate in this simulation corresponds to granule with the following parameters: cohesion, $c = 0.029\ \mathrm{N/mm^2}$; angle of internal friction, $\varphi = 22.5°$; density, $\rho = 2.06\ \mathrm{t/mm^3}$; elasticity, $E = 20\ \mathrm{N/mm}$. The dynamic behavior of the flexible substrate is condensed on the damping by the energy loss of the plastic deformation and on the effect of mass-inertia of substrate elements in this simulation.

## 47.3 Results and Discussion

The results from such an analysis are shown in Figs. 47.4 and 47.5. A circular and a rectangular loading have been considered. Each loading is applied to the surface of a substrate layer which is made up of two sublayers; sublayer (A) is of depth $H_A$ while the lower layer (B) is of depth $H_B$. The loadings were chosen so that they had the same minimum dimension, i.e. the loading, $q$ was applied over the region $|x| < a$ (strip), $0 < r \leq a$ (circle), $|x| < a$, $|x| < b$ (rectangular). The material was assumed to be anisotropic. For layer A, $E_h / E_v = 1.5$, $G / E_v = 0.45$, $\nu_h = 0.25$, $\nu_{hv} = 0.3$, $\nu_{vh} = 0.2$ and for layer B, $E_h / E_v = 33$, $G / E_v = 0.5$, $\nu_h = 0.1$, $\nu_{hv} = 0.9$, $\nu_{vh} = 0.3$ (The subscripts $h$, $\nu$ indicate horizontal and vertical directions, respectively.).

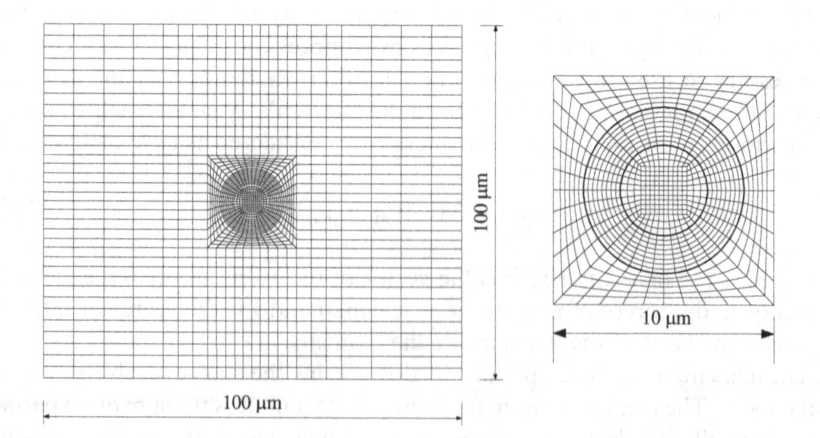


**Fig. 47.4** Surface pressure mobilization of single cell on the surface of flexible substrate

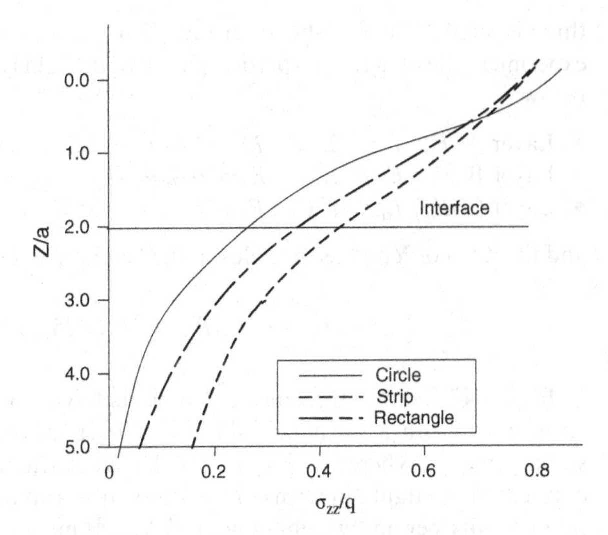


Fig. 47.5 Stresses in a substrate consisting of two anisotropic layers

The plots shown in Fig. 47.5 is for the vertical $\sigma_{zz}$ stresses along the axis of the loading ($x = y = r = 0$). There is a large difference in the vertical stress computed for each of the loading types. However, there is less difference in the horizontal stresses for this particular substrate profile.

All of the problems discussed thus far have been concerned with vertical loading, however there are many engineering problems which involve lateral or horizontal loadings applied to foundations. An example of a study is shown the result in Fig. 47.6, where a uniform shear, it is shown applied to the surface of a thin layered flexible substrate over a circular region of radius, $a$. The variation of vertical, $u_z$ and horizontal $u_x$ displacements with depth is calculated for the

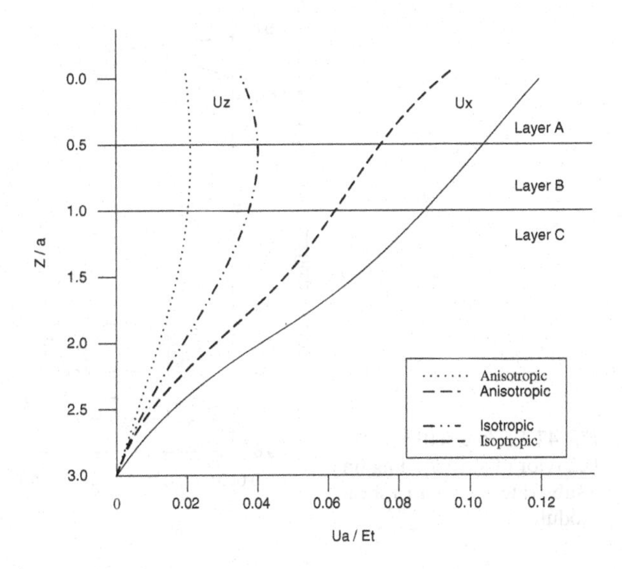


Fig. 47.6 Horizontal and vertical displacements in layered material

three layered system as shown in Fig. 47.6 ($x / a = 0.5$, $y / a = 0$). For this experiment the material properties of each of the sublayers A, B, C are described below:

- Layer A: $E_h / E_v = 2$, $G_v / E_v = 0.4$, $\nu_h = 0.3$, $\nu_{\nu h} = 0.2$
- Layer B: $E_h / E_v = 2$, $G_v / E_v = 0.4$, $\nu_h = 0.3$, $\nu_{\nu h} = 0.2$
- Layer C: $E_h / E_v = 1$, $G_v / E_v = 1/3$, $\nu_h = 0.5$, $\nu_{\nu h} = 0.5$

and the ratio of Young's modulus in the layers is assumed to be

$$(E_v)A : (E_v)B : (E_v)C = 25 : 5 : 1$$

Figure 47.7 shows the time-settlement behavior of strip loading on a substrate layer with a shear modulus which increases with depth (non-dimensional time, $\tau_0$ where $\tau_0 = c_0.t / l^2$). The coefficient of consideration, $c_0$ is defined on the figure in terms of the Poisson's ratio of the substrate, $\nu$ and its permeability per unit weight of water $k / \gamma_w$. It may be observed from the time settlement plots that as the rate of increase of shear modulus with depth $l_m / l$ becomes larger the final deflections become smaller and occur at a smaller time, $\tau_0$.

Problems involving the time-dependent deformation of materials under a constant applied loading may often be treated by assuming that the materials display viscoelastic behavior, i.e. that the material properties are themselves dependent on time. In this study, we used the assumption that the behavior of the material may be separated into a deviatoric and a volumetric behavior. That

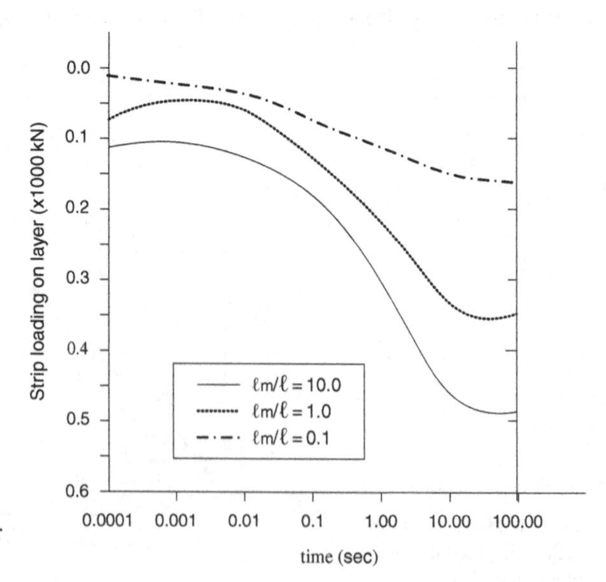


**Fig. 47.7** Time-settlement behavior of strip loading on a substrate layer with a shear modulus

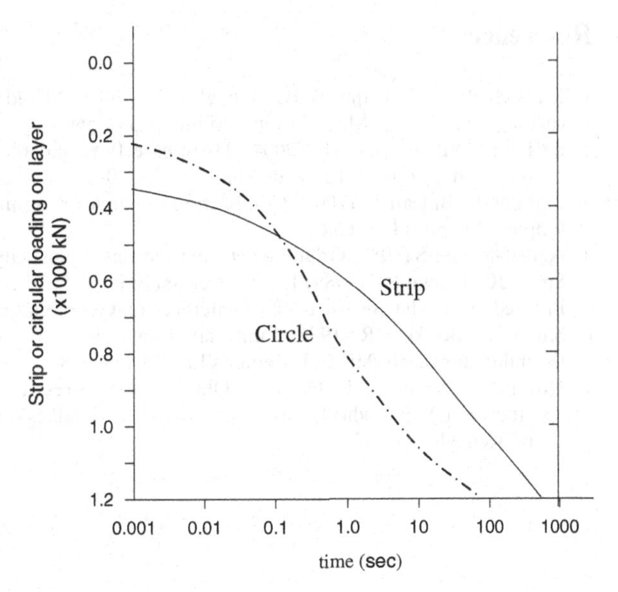


**Fig. 47.8** Deflection vs time for central point of a strip or circular loading resting on a layer material with a viscoelastic *upper* layer

is to say that there will be a different time-dependent response to a mean stress increase, then to a deviatoric or shear stress increase.

In Fig. 47.8 shows an example of the relation of deflection vs time for central point of a strip or circular loading resting on a layered material. It involves loading applied to a flexible substrate which consists of an upper viscoelastic layer and a lower elastic layer. Both materials were assumed to have a constant Poisson's ration, $\nu$, that of the upper layer being $\nu = 0.5$ (i.e. incompressible) and that of the lower layer $\nu = 0.3$. It was assumed that the deformation of the material was due to a time-dependent shear modulus.

## 47.4 Conclusions

In the paper, a two-dimensional FEM-model of the compaction-induced deformation on the thin layer of flexible substrate was presented. With this extension to the simulation concept, the new concept based on the dynamical FE-simulation was developed, that includes the possibility to investigate the dynamical load changes, caused by vertical oscillation of indenter and the according substrate deformation. It was pointed out that the indenter oscillation takes influence to the mobility of the substrate. Due to these oscillations, the dynamical compaction on the thin layer of substrate changes enlarges, leading to an increased deformation of the surface, which again induces the excitation of the substrate. Therefore, dynamic loads and substrate deformation influence mutually.

## References

1. Kaasschieter EE, Frijns A-JH, Huyghe JM (2003) Mixed finite element modeling of cartilaginous tissues. Math Comput Simul 61:549–560
2. de Boer R, Didwania AK (2004) Two-phase flow and the capillarity phenomenon in porous solids. Transport Porous Med 56:137–170
3. Philippe P, Bideau D (2002) Compaction dynamics of granular medium under vertical tapping. Erophys Lett 60:677
4. Nemat-Nasser S (1982) On finite deformation elasto-plasticity. J Solids Struct 18:857–872
5. Small JC, Brown PT (1988) Finite layer analysis of the effects of a sub-surface load. In: Proceedings of the 5th Aust-NZ. Conference on Geomechanics, Sydney, pp 123–127
6. Small JC, Booker JR (1986) Finite layer analysis of layered elastic materials using a flexibility approach Part 2. J Methods Eng 23:959–978
7. Hiroma T, Wanjii S, Kataoka T, Ota Y (1997) Stress analysis using FEM on stress distribution under a wheel considering friction with adhesion between a wheel and soil. J Terramech 34(4):35–42

# Chapter 48
# Robust and Accurate Visual Odometry by Stereo

**Aldo Cumani and Antonio Guiducci**

**Abstract** This work deals with the problem of estimating the trajectory of an autonomous rover by onboard passive stereo vision only (so-called *Visual Odometry*). The proposed method relies on bundle adjustment of tracked SURF feature points. Tests on real-world data show the effectiveness of this approach. In the case of cyclic trajectories, even greater accuracy is attained by periodically comparing the scene against a set of stored references.

## 48.1 Introduction

In the framework of research on Simultaneous Localisation and Mapping (SLAM), our group at INRIM has developed a visual odometry algorithm [1] which relies on the tracking of pointwise features extracted from the images acquired by an onboard binocular stereo head. At more or less regular intervals along the rover trajectory, its motion is estimated by robust bundle adjustment of the tracked features in the four images (two before and two after the motion).

Extensive tests have been performed using Shi-Tomasi-Kanade features [2], which however require small image displacements for a reliable tracking, and a stereo disparity cue (from a correlation disparity map) for stereo matching. We have therefore switched to SURF features [3], which appear to be more robust, allowing direct stereo matching and tracking with longer inter-frame steps. Moreover, the SURF features collected from an image appear to be a representation of the image contents sufficiently good for subsequent image-based matching.

Indeed, any localisation algorithm based on the integration of subsequent motion estimates is affected by error accumulation. In the case of cyclic

A. Cumani (✉)
Istituto Nazionale di Ricerca Metrologica, Str. delle Cacce, 91, I-10135
Torino, Italy
e-mail: cumani@inrim.it

N. Mastorakis et al. (eds.), *Proceedings of the European Computing Conference*, Lecture Notes in Electrical Engineering 28,
DOI 10.1007/978-0-387-85437-3_48, 

trajectories, i.e. with the rover returning to some previously visited place, it is important that the algorithm be able to recognize the latter situation and then perform some adjustment to its estimated pose.

This can be achieved by periodically storing the observed features and comparing the current features to the previously stored ones; when there is a sufficient number of matches, the bundle adjustment algorithm can then yield an estimate of the current rover pose, relative to the stored one, which is expected to be more accurate than the one obtained by the integration of the preceding steps along the path. This déjà vu feature has been incorporated in our algorithm, with rather encouraging results.

The following sections describe the algorithm in some detail, and presents some results from actual test runs.

## 48.2 The Visual Odometry Algorithm

Our algorithm relies on estimating the relative motion from corresponding features in the images acquired while the rover is moving. Such estimates are computed at *key frames*, whose spacing is a compromise between larger intervals, which are desirable for numerical accuracy, and the need for a sufficiently large number of features, which naturally tend to be lost either because of going out of view or because of matching errors. The accumulation of errors is overcome, in the case that the rover returns to some previously visited place, by recognizing that this situation occurs and correcting the estimated rover pose.

The algorithm can be summarized in the following steps:

- **Feature extraction.** Point features are extracted at each key frame and left-right matched.
- **Feature tracking.** Matched features are tracked in subsequent frames, where stereo matching is performed again.
- **Motion estimation.** The relative motion of the rover between the current frame and the last key frame is estimated by bundle adjustment of matched feature points.
- **Key frame declaration.** When the performance of the algorithm decays, due e.g. to the presence of too few matched points close to the rover, or to a failure of the bundle adjustment step either for numerical errors or because of an excessive number of outliers, a key frame is declared on the previous frame.
- **Cyclic correction.** Observed features and corresponding pose estimates are periodically saved to an archive. When the rover believes to be near some saved position (and with a similar heading), the detected features are compared to the stored ones. If enough matches are found, a bundle adjustment step between the current features and the stored ones yields the pose correction.

### 48.2.1 Feature Extraction and Tracking

The so-called *Speeded-Up Robust Features* [3] describe a distribution of Haar-wavelet responses within the neighbourhood of some interest points, the latter having been determined by a Hessian-based detector.

Being largely insensitive to scale and rotation, as well as to illumination, SURFs have proven quite robust and effective, both for what concerns stereo (left-right) matching, as well as for tracking along the sequence. While being in some way similar to the the SIFT features used by other researchers [4], SURFs have over the latter the distinct advantage of being unencumbered – at least for now – by patent issues.

Last but not least, in our experience the SURF features collected from an image appear to be a representation of the image contents sufficiently good for subsequent image matching, as explained in Section 48.2.3.

In order to get even stronger insensitivity to lighting conditions, in our implementation SURFs are not computed directly from the gray-level images, but from LOG filtered versions of the same. Both left-right and temporal matching relies on a feature similarity measure defined in terms of the Euclidean distance between feature descriptors. The matching algorithm can be improved by exploiting specific matching constraints (e.g. positive disparity for the stereo case), though in fact even the simplest approach already yields very good results. In our implementation, only stereo matching constraints are imposed, i.e. positive disparity and vertical distance under some threshold. No constraint is imposed instead on temporal matching.

Stereo matched point pairs are backprojected to a 3D point estimate, in the camera reference. This 3D estimate is used as a starting point for the bundle adjustment step. The features detected in the left image at a key frame are then tracked along the sequence up to the next key frame.

### 48.2.2 Motion Estimation

At every key frame, say $k+1$, except the first, we have a set of $N_k$ features $F_i$, left-right matched and tracked from the previous key frame $k$ to the next one $k+1$ (for the sake of simplicity, and without loss of generality, from now on we shall assume $k=1$ and drop the $k$ index). Let $\mathbf{X}_i = [x_i, y_i, z_i, t_i]^{\mathrm{T}}$ be the unknown 3D projective coordinates of feature $F_i$ in the reference of the left camera at frame 1. Note that the scale ambiguity on $\mathbf{X}_i$ can be fixed by taking $z_i = 1$ since all the points must be visible in the first left image, hence they have strictly positive depth $z$. Let $\mathbf{u}_{iq} = [u_{iq}, v_{iq}]^{\mathrm{T}}$ be the 2D image coordinates of the feature point in image $q \in 1L, 1R, 2L, 2R$, and $\mathrm{x}_{iq} = [su_{iq}, sv_{iq}, s]^{\mathrm{T}}$ the same in projective representation. Then

$$
\begin{aligned}
x_{i,1L} &= P_L \mathbf{X}_i \\
x_{i,1R} &= P_R M_S \mathbf{X}_i \\
x_{i,2L} &= P_L M_{12} \mathbf{X}_i \\
x_{i,2R} &= P_R M_S M_{12} \mathbf{X}_i
\end{aligned} \tag{48.1}
$$

where $P_L$ and $P_R$ are the 3×4 intrinsic camera matrices, and $M_S$, $M_{12}$ are 4×4 Euclidean transformation matrices representing the stereo (left-to-right) mapping and the motion (frame 1 to frame 2) mapping, respectively, and are of the form

$$
M = \begin{bmatrix} R & t \\ 0^T & 1 \end{bmatrix} = \begin{bmatrix} e^{[r]\times} & t \\ 0^T & 1 \end{bmatrix} \tag{48.2}
$$

where r is the vector representation of rotation R, i.e. $r/\|r\|$ = rotation axis, $\|r\|$ = rotation angle, and $[r]\times$ is the antisymmetric matrix representation of vector cross product by r.

Note that in the above both $P_L, P_R$ and the stereo transform $M_S$ are known from calibration data (in fact, in our implementation we pre-rectify the stereo pair so that $P_S \equiv I_{3\times3}$ and $t_S = [b, 0, 0]^T$ with $b$ the stereo baseline length). We can also take $P_{L,R} \equiv [I_{3\times3}, 0_3]$ in Equation (48.1) if the pixel-based feature coordinates are preliminarly transformed to *calibrated* coordinates using the actual intrinsic camera matrices. Now, for each pair of consecutive key frames we have measurements $u^*_{iq}$ of the quantities $u_{iq}$ corresponding to Equation (48.1), so we can define an error measure

$$
\begin{aligned}
J(p) &= \sum_i \sum_q f\left(\|e_{iq}\|^2\right) \\
&= \sum_i \sum_q f\left(\left\|u_{iq} - u^*_{iq}\right\|^2\right)
\end{aligned} \tag{48.3}
$$

parametrized as a function of the 6 + 3$N$ unknowns

$$
p = [r_{12}, t_{12}, x_1, y_1, t_1, \ldots x_N, y_N, t_N]^T \tag{48.4}
$$

with $r_{12}$, $t_{12}$ the generators of $M_{12}$. The visual odometry estimate is then given by the $r_{12}$, $t_{12}$ entries of the p value that minimizes $J$.

In the above, choosing $f(\cdot)$ = identity corresponds to standard least squares, which would be optimal for independent Gaussian distributed image errors. For reasons of robustness against outliers (mismatched points), it is however preferable to use a robust cost function such as the Lorentzian cost:

$$
f(e^2) = \log\left(1 + e^2/\sigma^2\right) \tag{48.5}
$$

with $\sigma$ chosen as a function of the expected image-plane error.

Since the first derivatives of $J$ with respect to p can be computed analytically, the minimization of $J$ can be achieved by a gradient-based method such as Levenberg-Marquardt. Moreover, for given $i$ the corresponding terms in Equation (48.3) only depend upon one of the triples $(x_i, y_i, t_i)$, which allows to use a sparse algorithm like the one described in [5, A4.3].

For better accuracy, the residuals $e_{iq}$ are compared against a threshold (proportional to the Lorentz $\sigma$), and those exceeding that threshold are marked as outliers. The bundle adjustment is then run again on the inliers only.

### *48.2.3 Cyclic Correction*

In order to be able to implement cyclic correction, the rover keeps an archive of scene descriptions. Each such description consists of the set of matched features in the left and right images, together with the estimated rover pose.

The description of the current scene is added to the top of the archive when:

- enough time is elapsed since the last entry in the archive (the top entry);
- the number of matched features between the frame to be added and the top entry is near zero.

This strategy is aimed to ensure that the scenes corresponding to the archive entries are all different from each other.

As soon as a new key frame is processed, the estimated rover pose is checked against those stored in the archive, starting from the bottom (i.e. oldest) entry. If an archive entry with similar pose is found, the features in the current frame and those stored in the archive entry are compared, and if enough matches are found a pose correction between the archived and the actual pose is computed by bundle adjustment. All the corrections computed while the rover remains near an archived position are saved in memory. When the rover leaves the neighborhood of that archived position, the saved correction corresponding to the frame with the highest number of matched features is picked and actually applied. The pose corrected this way is expected to be more accurate, as the error accumulated along the intervening path is exchanged with that from a single keyframe pair.

## 48.3 Experimental Results

### *48.3.1 Equipment and Setup*

The proposed method has been implemented in C and MATLAB, except for the SURF computation step, where we have used the authors' implementation available on the Web [3]. The algorithm has been tested on sequences of real images acquired by a small commercial rover, an ActivMedia Pioneer 3-DX, equipped with a suitable stereo head (Fig. 48.1).

Fig. 48.1 Detail of the Pioneer 3DX rover with stereo head

While previous tests had been done with a Videre Design STH-MDCS stereo head, a low-cost commercial product with a stereo baseline of 88 mm, for the tests reported here we have used a better quality, home-made stereo head. This head consists in a pair of Basler A312f digital cameras, equipped with 6 mm lenses and mounted on an aluminum slab, as shown in the figure. Each camera provides a stream of CCIR-size (720×576) 8-bit graylevel images on a IEEE1394 bus, and the two cameras can be accurately synchronized by an external trigger signal.

The cameras were mounted at a nominal center-to-center distance (stereo baseline) of 236 mm, and the head was calibrated using Bouguet's method [6], with a 0.6×0.6 m checkerboard pattern.

In order to have some ground-truth rover position data against which to compare the outcome of the visual odometry algorithm, the following setup was used. As shown in Fig. 48.2, three tennis balls were laid on the test field, approximately at the corners of an equilateral triangle (the fourth central

Fig. 48.2 The Pioneer 3DX rover on the test field. The tennis balls laid on the ground are used as reference points for the estimation of a ground-truth rover position independent from the visual odometry algorithm

ball visible in the figure is just for check). Their distances were manually measured to within ±1 mm, so defining a metric reference frame. As shown in the figures, another ball was rigidly attached to the stereo head, to be used as reference point for the rover position. Two compact digital cameras (one with 2048×1536 and one with 2592×1944 resolution), after intrinsic calibration again by Bouguet's algorithm, were placed on tripods and used to frame the scene at about 90° with respect to the center of the reference frame. At regular intervals, the robot was stopped and images of the scene were taken by the two cameras; the ball centers, manually extracted from each image pair, allowed then to estimate a ground-truth position of the rover in the chosen reference, to within an accuracy surely better than the ball diameter (about 0.06 m).

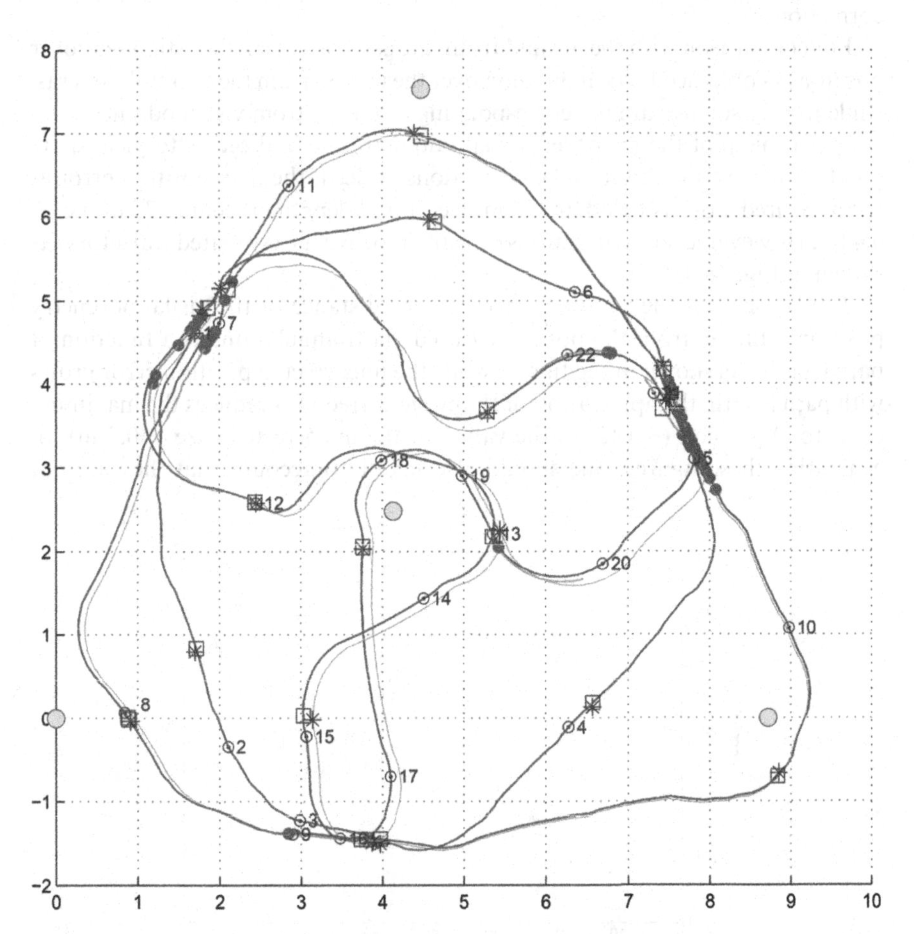


**Fig. 48.3** Estimated rover trajectory without (*light gray*) and with (*dark gray*) corrections. Coordinates in m. *Gray circles* indicate saved positions. *Dots* indicate points where the scene is checked against the archive, bigger ones are those with enough matches to allow computation of the correction

### 48.3.2 Test Results

Some results from a test tour are reported in the following. The rover was driven to perform multiple loops over the field framed by the two still cameras, returning several times, with the same heading, to three selected points (including the origin).

The total path length was about 78 m. In Fig. 48.3 the lighter curve is the rover path as estimated by the visual algorithm alone, i.e. without the déjà vu corrections, while the darker one takes into account the latter ones.

Numbered and circled points are the saved positions. The other dots on the trajectory are the positions where some match against the stored archive is attempted and the number of matches is deemed sufficient for computing the correction.

For comparison with the ground truth, the asterisks mark the estimated rover positions as obtained by triangulation over the images from the two still cameras, while nearby squares are the corresponding estimates from visual odometry.

From this plot the proposed visual odometry method seems to yield quite good results even without cyclic corrections; indeed, the max position error at the measured stops is less than 0.3 m over a path length exceeding 75 m (i.e. < 0.4%). However, cyclic corrections can still improve the estimated trajectory, as shown in Fig. 48.4.

The latter plots the position error, i.e. the distance of the visual odometry position estimate from the point measured via triangulation, as a function of path length. As shown in the figure, while the uncorrected position error grows with path length, the application of the cyclic corrections reduces the maximum error to about half (~0.15 m) the value for the uncorrected case (~0.3 m). In evaluating these figures, one should also take into account that they are not

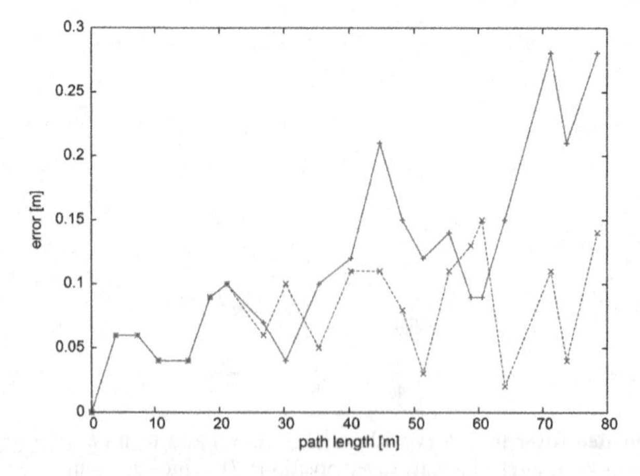


**Fig. 48.4** Position error as a function of path length, without corrections (*solid curve*) and with cyclic corrections (*dashed curve*)

much greater than the uncertainty of the ground truth values, which with our setup is of the order of several 0.01 m.

## 48.4 Concluding Remarks

The preceding discussion has shown that onboard passive stereo vision alone is able to yield an accurate estimate of the trajectory of the rover, even on an almost featureless ground like the one used in our tests. The error figures here reported are well under e.g. those reported by the Mars Rover team [7]. We have also presented a mechanism that, in case of cyclic trajectories, is able to automatically compensate the accumulated error.

It should be noted that our approach has some similarity with the one reported in [8]. Those authors, however, have a quite different approach (RANSAC) to the problem of outliers (wrong matches). Given that the feature detection/matching scheme adopted in our work typically yields rather few outliers, our approach, using a robust cost function, seems both effective and less computationally expensive.

There is however still room for improvements; for example, the current implementation finds corresponding image points only by comparing independently extracted interest points. Adding e.g. a step to refine feature position by local correlation (as with the Shi-Tomasi-Kanade features) could possibly yield even more accurate results. More extensive tests, especially on longer paths, are also needed to assess the practical usability of this method for unassisted navigation.

**Acknowledgement** The authors wish to thank their colleague Giorgio Quaglia for his invaluable help in setting up the stereo head and the related acquisition software.

## References

1. Cumani A, Guiducci A (2006) Stereo-based visual odometry for robust rover navigation. WSEAS Trans Circuits Syst 5(10):1556–1562
2. Shi J, Tomasi C (1994) Good features to track. In: Proceedings of the IEEE Conference on Computer Vision and Pattern Recognition, pp 593–600
3. Bay H, Tuytelaars T, Van Gool L (2006) SURF: Speeded up robust features. In: Proceedings 9th European Conference on Computer Vision, See also http://www.vision.ee.ethz.ch/~surf/
4. Se S, Lowe D, Little J (2001) Vision-based mobile robot localization and mapping using scale-invariant features. In: Proceedings of the IEEE International Conference on Robotics and Automation, pp 2051–2058
5. Hartley R, Zisserman A (2003) Multiple View Geometry in Computer Vision, 2nd edn. Cambridge University Press, Cambridge
6. Bouguet JY (2004) Complete camera calibration toolbox for MATLAB. Technical report, http://www.vision.caltech.edu/bouguetj/calibdoc/index.html

7. Cheng Y, Maimone M, Matthies L (2005) Visual odometry on the Mars exploration rovers. In: Proceedings Of the IEEE International Conference on Systems, Man and Cybernetics, 1:903–910
8. Sünderhauf N, Protzel P (2006) Towards using sparse bundle adjustment for robust stereo odometry in outdoor terrain. In: Proceeding of the International Conference on TAROS 2006, pp 206–213

# Chapter 49
# Financial Auditing in an E-Business Environment

L. D. Genete and Al. Tugui

**Abstract** This study has the purpose to specify the main changes determined by e-business in the field of financial audit. The importance that e-businesses have gained in the last years and also its potential have convinced many companies to embrace this method when completing their activity totally or partially. Under the circumstances the auditor's mission becomes more and more complex, being forced to handle equally both the specific verifications for financial activity and the information systems used, involving in this case the use of various and complex technologies. This study presents the audit's phases and their approach in an e-businesses environment.

## 49.1 Introduction

The technological development led to major changes in business process at the end of XX century and the beginning of the XXI, pointed mainly in the use at a large scale of the virtual business environment. This is proven to be the right direction for moving forward, to evolve, allowing companies to reduce their expenses, to intensify the activity's rhythm, to increase their products and services' quality gaining new markets and though to become global. These changes have determined, necessarily, modifications at the level of verifying the operations' legality, accuracy and precision, respectively their audit. Considering the statement above, on one half the audit of information systems became needful, to verify the information flows, the systems and applications used during the execution, and, on the other half, but highly related, the audit of the correctness of information in documents, independently of their format: paper support or electronic. It is obvious that technological progress stimulated not only the software developers and the end-users, but also the auditors, which aren't able to fulfil their daily mission without understanding the way a company uses information technology to achieve its

L.D. Genete (✉)
Business Information System Department, "Alexandru Ioan Cuza" University of Iasi, 22 Carol I Blvd., Ro-700505, Romania

N. Mastorakis et al. (eds.), *Proceedings of the European Computing Conference*, Lecture Notes in Electrical Engineering 28,
DOI 10.1007/978-0-387-85437-3_49, 

current activity [9]. The development of software applications like: Enterprise Resource Planning, Customer Resource Management, Supply Chain Management or the Internet, e-commerce and e-business determined a project for a new financial-accounting software system conceived and employed in a virtual environment, resumed by the concept of digital accounting. The mission of this new type of system is not only record and trace transaction information instantaneously but also cross-check internal and external documents automatically [14]. But at the same time, the auditors had to manage an important volume of information to be checked, resulting in the need for using specialized applications able to assist them during the audit verifications. All the changes mentioned above concurred to augment the specialists' interest to define new audit standards and to establish the specific activities in order to create the corresponding legislative environment [1, 2] providing the required security for online transactions and their adequate check, including domain specific devices [6, 16]. We have to mention that there are many cases when the companies' managers do not understand completely all the aspects related to e-business and the risks involved. The situation is more critical for the auditors having less time to outline an opinion according to the reality regarding the company's activity, involving the analysis of a massy information volume delivered from various departments, developed in different environments and saved in different formats. One of the most debated solutions of the last period of time is the continuous audit – allowing a continuous evaluation and supervision of checks, operations and risks [3, 7]. The overall impact of these complex and higher plane technologies is now visible on the modern day information system auditing processes and requires a relatively higher level of understanding by an auditor to succeed in the job [10]. It is necessary in these days for the auditor to have technical knowledge in order for him/her to accomplish his/her mission, confronted with new types of fraud, aspects related to the electronic signature, virtual transactions, without paper support leading to changes in the audit specific activities like evaluating risks, modifications at the control level and specific tests.

## 49.2 Impacts of e-Business on Audit Missions and Auditors

As mentioned above, the development of e-business applies for a reconsideration of the way to accomplish the audit, respectively a combination between the problems specific to the software systems' audit and the ones related to the financial audit. This requires from the auditors extensive knowledge regarding both the software systems involved and the information need to be mentioned in the financial-accounting documents, resumed in the financial statements at the end of the period. This study has the purpose of providing an answer for some elemental questions in this field like:

1. How is e-business affecting the financial audit of companies?
2. What aspects need to be followed by the auditors and what knowledge is requested in order to be solved?

3. What are the audit risks in an e-business environment?
4. What stages of an audit are experiencing changes when completing transactions in a virtual environment?

The answer to these questions will help the auditors to determine the problems that might appear during accomplishing audit missions and to improve the necessary skills suitable to the evolution of business environment in the XXI century.

### *49.2.1 Knowledge of the Client's Business and Planning Mission*

The online environment presents a mix of technological, human and physical elements and threats to business can arise from any of these elements [5]. In Fig. 49.1 it is presented the infrastructure for e-business and the connections between its components.

Knowledge of the client business and planning the mission are the first stages in auditing. It involves an analysis of the company's financial activities, also of the information system's components participating in its achievement (hardware, software, human resources etc.) and of the company's business environment. "Knowledge of the Business" is referring to both – the audited company's effective activity and the activity's environment. In the case of using practices specific for e-business, the auditor must consider the changes generated by their employment and the related risks proportionally with their effects on the company's financial situations. At this phase, the auditor must consider the following elements [8]:

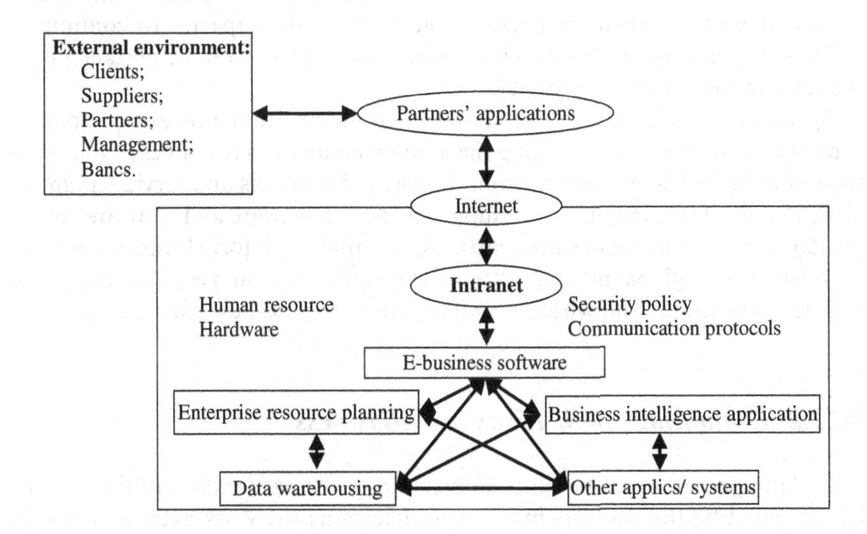


**Fig. 49.1** E-business infrastructure

- The activity branch and field for the audited company.
- E-business strategy.
- The extensions of e-business activities.
- External service suppliers of the company.

One of the essential aspects for online transactions having consequences for both – risks evaluation and verification procedures, is determined by immixture of other participants. Under the circumstances of using the Internet, the transactions are completed through a supplier of such services, or through a virtual, private network [15] which might determine the auditors to request information from the specific supplier in order to establish exactly how the information is circulating in the system, the viability and also the rapport of respecting the legal provisions, and the security policy established for each partner involved.

Considering the complexity and the plurality of used technologies, the different human resources employed in e-businesses, in planning the mission stage and knowledge of the client business, the auditor must complete verifications at least for hardware equipment, the communication network components, both internal and services suppliers, supports and environments used to save data, software components, and human resources employed.

To determine if a technology professional is needed, an audit planner is expected to consider these factors: complexity and usage of the firm's systems, significance of change to existing systems; extent of data sharing; extent of the entity's participation in e-business; and audit evidence available in digital form [12]. There are cases when he may consider he is not having enough information technology competence to complete verifications and to provide an opinion according to the reality. In this case, the auditor may call for one or more experts in the information technology field.

The business environment of the company's activity has an important role in "knowledge of the client business" stage because the capacity to continue a business depends not only on its members' skills, but also on the market where the services and products are sold.

Specific to e-business, for the external environment evaluation and company's capacity to integrate, is estimating the audited company's resources to adjust to the technological development, from the retailed products and services point of view, but also from the one of technologies used (hardware and software equipments) when contacting business partners, essential to perform business contacts in real-time on a global market, extremely dynamic, with no geographical limits, with less restrictions and where it is vital to trust the business partner.

### *49.2.2 Evaluating Audit Risks in E-Business*

Completing an e-business audit also demands a reconsideration of the risks to be evaluated by the auditors in order to determine the work extension and to establish the size of the verified samples. The approach might be considered

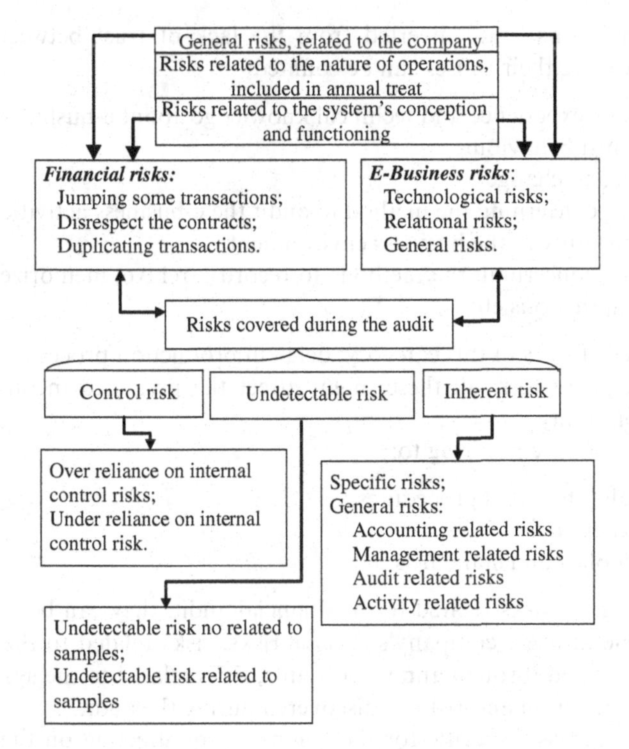


**Fig. 49.2** Audit risks in e-business

from two interdependent points of view: on one hand the risks related to the information system, and on the second one the risks related to the financial audit (Fig. 49.2).

If the risks related to the information system are to be considered, the state is that for the e-business the emission speed is much higher confronted to the traditional business, due to inexistent borders in cyberspace. There are more classifications of e-business risks as internal and external, human or machine generated, intended or not. Some authors have identified three main categories of risks: technology related risks, relational risks and general risks [13].

Technology related risks. These are associated with the access to the e-commerce infrastructure. Their presence might cause deviations from the performance targets expected in an electronic exchange "relationship". For instance, both the online seller and his/her partners may suffer hacker attacks. The security gap resulted from the hacker attack may lead to losing income or "image" for the seller. On the other hand, the information circulating between partners has vulnerable points: internal applications of partner companies, application interface, network connections, mail boxes etc. If these vulnerable points are operated it might reach to presenting no-updated information or misinformation of clients and, even if the situation interferes for a short period of time, they may sue the seller or ask for financial compensations.

Relational risks – are diverted from the lack of trust between business partners. Among their causes can be counted:

- The lack of experience and technical knowledge about e-business security.
- Opportunist behaviour.
- Resistance to changes.
- Ignorance concerning the method to audit the e-business activities.
- The uncertainty in the business environment.
- Ignorance concerning the methods to record (archive, memorize) the electronic transactions etc.

Examples of risks in this category: delay in production process, interrupted cash-flows, profit loss; all these might affect the predicted income and the business continuity.

General risks are referring to:

- Low quality business procedures.
- Environment risks.
- Lack of policy and standards.

Regarding the risks related to the financial audit, they can be classified as follows: general risks, company's specific risks, risks related to the nature of operations treated through annual accounts, risks related to the system's conception and function and risks undiscovered during the audit.

The general risks, specific for the company are affecting on the whole its operations. The auditor will analyze the company's activities and characteristics in order to get a better understanding of the potential risks able to take place and to identify the possible ones.

The risks related to the nature of operations treated through annual accounts are considering the manner of recording the three main data categories that the company is dealing: reiterative data resulting from daily activity, punctual data complementary to the reiterative ones, but distinguished at certain time periods, exceptional data, resulting from operations or decisions not specific to the daily activity: re-evaluations, mergers, reorganizations.

Risks related to the system's conception and functions are reported to conceiving the financial-accounting system so it allows detecting and correcting the errors, in serviceable time, especially through internal audits.

Risks undiscovered during the audit are affected by the auditor's choices regarding the procedures to use, the verifications' extension and dates.

There are risks mixing the two categories and crossed verifications are needed through the audited company-communication services suppliers – partners (customers, suppliers, banks, management etc). This range of risks determines the following measures to prevent/control the consequences:

- Verifying the customers' and suppliers' identity.
- Assure the transactions' integrity: validate the entries, prevent duplications or transactions' overlooks, respecting the terms of the contracts, prevent the

transactions' partial processing, the correct information transfer between involved systems, assure the back-up copies and secure the system.
- Establish compelling transaction terms regarding the deliveries, payments, cash, products and services reception, giving and receiving credits etc.
- Establish security policies in order to assure the information confidentiality and protection, protocols between the involved parties.

### *49.2.3 Specific Controls in Auditing E-Business*

Executing verifications during an e-business auditing follows some classical attributes, regulated through audit standards. There can be identified from this point of view the following two verification types: general controls and application controls (Fig. 49.3).

General controls regard the problems related to the fluent operating of the information system, including the systems' management, software programs' purchase and maintenance, physical and access security of equipments, programs and any saved data, planning the back-ups, conceiving and employing control devices integrated into the informatics equipments [4]. The general controls are pursuing two functions: on the one side the equipments and human resources management function, and on the other side the systems' proper functioning. In the e-business case, not only the internal function of the information system has the important role, but also the external connection and this involves the network audit – how the security standards are applied and the capacity to provide information from inside and outside the company.

Application controls track particular operations' verifications and must be executed for each account or operation category simulated by a software application. In the same category are also included the verifications for each program or module used to realize online transactions: sales, purchase, payments, cash, human resources management etc.

In the virtual business environment it is significant for the transaction generated from the company's website to be correctly processed inside the company's internal system [11], for instance the accounting one, the customers

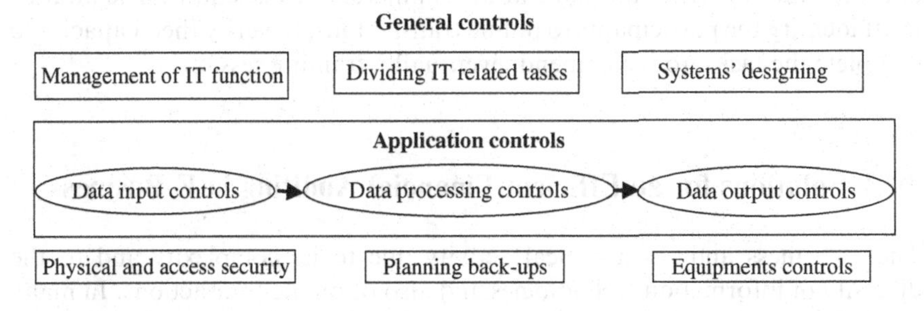


**Fig. 49.3** Controls categories in e-business

and supplier's relations management, the management of the state duties. There are cases when this one is not integrated onto the internal system and must carefully follow the transactions' capture and transfer in the accounting system since it's affecting [8]:

- The completeness and accuracy of transactions' processing and information storage.
- The timing of the recognition of sales revenues, purchases and other transactions.
- Identification and recording of disputed transactions.

Auditing e-business assumes more compelling controls of the hardware and software components, of the communication realized inside the company and with external partners and also of the human resources employed to manage these components and activities.

Hardware equipments are solutions used by dedicated systems during executing online transactions and sometimes are requiring specific technologies, concurring to the operations' accomplishment successfully, their analysis must be done for each one separately, but also from the point of view of interactions between them.

Network components are the resistance piece for completing online transactions considering that they are based on network communication through the Internet, Intranet and Extranet. At this level, the auditors must verify both the accuracy of the transactions and the execution speed, to identify possible intrusions, to verify the access rights and the degree of respecting them.

Saving supports where recorded clients, suppliers or other partner's data are, must guarantee to keep the information in proper conditions in order to be protected against natural dangers or human errors. With this purpose, companies must assure the execution and the storage in optimal conditions for original data and back-ups, especially when it comes to important information.

Software components are referring to both operation systems and specific applications (Enterprise Resource Planning, databases, e-business applications etc.). The audit must aim in this case each application, independently, but also the interaction between modules, the accuracy and the exhaustive of the data transfers.

Human resource is an essential component for any company, for both: the one charged with information technologies' development and maintenance and the other involved in financial or commercial transactions. The auditor must identify the participants to online transactions, to verify their capacity to complete the tasks, to recommend, eventually, training sessions.

## 49.3 Solutions for an Efficient Financial Auditing in E-Business

The e-business audit is a critical activity due to its complexity and to the diversity of information technologies and also of online transactions. In many cases the participants are coming from various environments, with different laws and standards. In order to increase the audit's efficiency in this field the following measures' categories are recommended:

- Unifying the legal settlements in the financial-accounting domain and the information technologies.
- Continuous training about information technologies for the auditors. In this case, a solution might be to ask for different auditors, information technologies' specialists, but it might also increase the time needed to complete an audit mission and the risks related to the skill to define an objective opinion according to the reality.
- Standardization of equipments and applications used to manage the company's activity and to establish contacts and contracts with the external environment, hard to accomplish due to the existent technologies diversity and fast development.

Building applications able to continuously supervise the activity, which need to be analyzed with attention by the external auditors.

In the present days it is common either to ask for auditors which are specialized in information technologies, either to specialized trainings for the existent auditors. Another solution, often applied, is to separate the information system auditing from the financial audit, which might lead to spreading the errors or to conflicts.

## 49.4 Conclusion

This study wanted to identify the main changes taking place during a financial auditing of e-business. Here are presented some conclusions which have been arrived at from the above:

- The extension of e-business to a global scale induces changes in all financial audits' phases.
- In order to reply to the market requests, the auditors' knowledge in the financial accounting domain has become insufficient, being necessary to add information technologies' capabilities related to e-business operations.
- The information technologies' function in the business process is essential and at the same time is conditioning the business' success in the virtual environment.
- The failure or the success of an audit mission is depending, in all of the situations, on the commitment of the company's management and employers for its accomplishment and also on the auditors' skills and knowledge.

There is need to mention that the present study shows only a few of the changes appearing during the audit missions, when passing from the classical approach for doing business to the virtual environment.

## References

1. AICPA, CICA (1999) Electronic Commerce Assurance Services Task Force. WebTrust Principles and Criteria for Business-to-Consumer Electronic Commerce, Version 1.1
2. AICPA, CICA (1999), Electronic Commerce Assurance Services Task Force. WebTrust-ISP Principles and Criteria for Internet Services Providers in Electronic Commerce, Version 1.0
3. Alles M, Kogan A, Vasarhelyi M (2002) Feasibility and economics of continuous assurance. Auditing: A J Pract Theory 21:125–138
4. Arens A, Loebbecke J, Elder J, Beasley S (2003) Auditing: An integrated approach. Arc Publishing House, Chisinau
5. Deshmukh A (2006) Digital accounting. The Effects of the Internet and ERP on Accounting, IRM Press, London
6. Helms G (2002) Traditional and emerging methods of the electronic assurance. CPA J 26–31
7. Hunton EJ, Wright MA, Wright S (2004) Continuous reporting and continuous assurance: Opportunities for behavioral accounting research. J Emerg Technol Account 1:91–102
8. IAPC (2002) IAPS 1013 – Electronic commerce: Effect on the audit of financial statements. http://www.paab.co.za/documents/doc_00301.pdf
9. Nearon HB (2001) Auditing e-business. CPA J
10. Pathak J (2003) Model for audit engagement planning of e-commerce. Int J Auditing 2:121–142.
11. Pathak J (2005) Information technology auditing. An evolving agenda. Springer-Verlag. Berlin
12. Pathak J, Lind R (2003) Audit risk, complex technology, & auditing processes. EDPACS 5:1–9
13. Ratnasingam P (2003) Inter-organizational trust for business-to-business e-commerce. IRM Press, London
14. Rezaee Z, Reinstein A (1998) The impact of emerging information technology on auditing. Managerial Account J 13:465–471
15. Yahaya Y, Little D, Onuh S (2002) Multiple case studies of total enterprise integration programs: lessons and benefits. Int J Manuf Technol Manage 3/4:283-302
16. Yu C, Yu H, Chou C (2000) The impacts of electronic commerce on auditing practices: An auditing process model for evidence collection and validation. Int J Intell Syst Account, Finance, Manage 3:195–216

# Chapter 50
# Reputed Authenticated Routing Ad-Hoc Networks Protocol

**Ahmed Sameh, Abdalla Mahmoud, and Sherif El-Kassas**

**Abstract** In this paper, we analyze one of the secure mobile ad hoc networks protocols, which is Authenticated Routing for Ad Hoc Networks (ARAN). Such protocol is classified as a secure reactive routing protocol, which is based on some type of query-reply dialog. That means ARAN does not attempt to continuously maintain the up-to-date topology of the network, but rather when there is a need, it invokes a function to find a route to the destination. Here, we detail the security attacks that the ARAN protocol defends against, criticize how an authenticated selfish node can disturb the network by dropping packets or by not cooperating in the routing functionality and propose a reputation-based scheme called Reputed-ARAN to detect and defend against selfish nodes.

## 50.1 Introduction

The Authenticated Routing for Ad Hoc Networks (ARAN) secure routing protocol proposed by Sanzgiri, Laflamme, Dahill, Levine, Shields and Belding-Royer [1] is seen as an on-demand routing protocol that detects and protects against malicious activities caused by other nodes and peers in the ad hoc environment. This protocol introduces authentication, message integrity and non-repudiation as part of a minimal security policy for the ad hoc environment. Our evaluation of ARAN shows that it has minimal performance costs for the increased security in terms of processing and networking overhead. However, by analyzing the protocol thoroughly, we figured out that an authenticated selfish node can disrupt the network by promising to cooperate in the routing functionality and then does not participate at all. This results in wasting the mobile ad hoc network bandwidth.

A. Sameh (✉)
Department of Computer Science, The American University in Cairo, Cairo, Egypt
e-mail: sameh@aucegypt.edu

N. Mastorakis et al. (eds.), *Proceedings of the European Computing Conference*, Lecture Notes in Electrical Engineering 28,
DOI 10.1007/978-0-387-85437-3_50, 

## 50.2 Security Analysis of ARAN

In this section, we will delve into two types of mobile ad hoc networks (MANETs) nodes: Malicious and selfish nodes. Then, we present a security analysis of ARAN protocol describing how it faces most of the malicious nodes attacks and fails to face selfish nodes.

Malicious nodes can disrupt the correct functioning of a routing protocol by modifying routing information, by fabricating false routing information and by impersonating other nodes. On the other side, selfish nodes can severely degrade network performances and eventually partition the network by simply not participating in the network operation [2]. Current ad hoc routing protocols do not address the selfishness problem and assumes that all nodes in the MANET will cooperate to provide the required network functionalities. In this section, we analyze the Authenticated Routing for Ad Hoc Networks by evaluating its robustness in the presence of the different attacks introduced in earlier subsection [1].

- Unauthorized participation: Since all ARAN packets must be signed, a node cannot participate in routing without authorization from the trusted certificate server. This access control therefore rests in the security of the trusted authority, the authorization mechanisms employed by the trusted authority, the strength of the issued certificates, and the revocation mechanism.
- Spoofed Route Signaling: Route discovery packets contain the certificate of the source node and are signed with the source's private key. Similarly, reply packets include the destination node's certificate and signature, ensuring that only the destination can respond to route discovery. This prevents impersonation attacks where either the source or destination node is spoofed.
- Fabricated Routing Messages: Since all routing messages must include the sending node's certificate and signature, ARAN ensures non-repudiation and prevents spoofing and unauthorized participation in routing.
- Alteration of Routing Messages: ARAN specifies that all fields of RDP and RREP packets remain unchanged between source and destination. Since both packet types are signed by the initiating node, any alterations in transit would be detected, and the altered packet would be subsequently discarded. Thus, modification attacks are prevented in ARAN.
- Denial-of-Service Attacks: Denial-of-service (DoS) attacks can be conducted by nodes with or without valid ARAN certificates. In the certificate-less case, all possible attacks are limited to the attacker's immediate neighbors because unsigned route requests are dropped. However, nodes with valid certificates can conduct effective DoS attacks by sending many unnecessary route requests and they will go undetected as the current existing ARAN protocol cannot differentiate between legitimate and malicious RREQs coming from authenticated nodes.

It is clear from the above mentioned security analysis of the ARAN protocol that ARAN is a secure MANET routing protocol providing authentication, message integrity, confidentiality and non-repudiation by using certificates infrastructure. As a consequence, ARAN is capable of defending itself against spoofing, fabrication, modification, DoS and disclosure attacks. However, erratic behavior can come from a malicious node, which will be defended against successfully by existing ARAN protocol, and can also come from an authenticated node. The currently existing ARAN secure routing protocol does not account for attacks that are conducted by authenticated selfish nodes as these nodes trust each other to cooperate in providing network functionalities. This results in that ARAN fails to detect and defend against an authenticated selfish node participating in the mobile ad hoc network. Thus, if an authenticated selfish node does not forward or intentionally drop control or data packets, the current specification of ARAN routing protocol cannot detect or defend against such authenticated selfish nodes. This weakness in ARAN specification will result in the disturbance of the ad hoc network and the waste of the network bandwidth. In the following section, we are proposing a solution to account for this type of attack.

## 50.3 The Proposed Reputation System Scheme: Reputed-ARAN

Performance of mobile ad hoc networks is well known to suffer from free-riding as there is a natural incentive for nodes to only consume, but not contribute to the services of the system [3]. So in the following subsections, we will introduce our newly designed reputation system scheme that we plan to integrate with the ARAN routing protocol, ending up having Reputed-ARAN.

Whereas most of the attacks performed by malicious nodes can be detected and defended against by the use of the secure routing ARAN protocol, as was explained earlier, there remain the attacks that an authenticated selfish node can perform.

There are two attacks that an authenticated selfish node can perform that the current ARAN protocol cannot defend against. To illustrate these two possible attacks that a selfish node can use to save its resources in a MANET communication, the attack-tree notation proposed by Bruce Schneier [4] that allows the categorization of attacks that lead an attacker to reach a specific goal is used. In Table 50.1, the attack tree that cannot be detected by current ARAN protocol is shown.

**Table 50.1** Attack Tree: Save own resources

| Attack tree: Save own resources |
|---|
| OR 1. Do not participate in routing |
|     1. Do not relay routing data |
|         OR 1. Do not relay route requests |
|             2. Do not relay route replies |
|     2. Do not relay data packets |
|         1. Drop data packets |

As shown in the above table, when nodes simply drop packets (case 1.1 and 2.1 in the attack tree), all the security features of ARAN fail to detect or defend against these attacks, as they focus only on the detection of malicious nodes' attacks and not the authenticated selfish nodes' attacks. ARAN protocol assumes that authenticated nodes are to cooperate and work together to provide the routing functionalities.

In this section, we are proposing a simple reputation based scheme to mitigate misbehavior and enforce cooperation in MANETs using ARAN protocol. We call this newly designed system: Reputed-ARAN. Different from global reputation based schemes like Confidant and Core, discussed in [5, 6], our solution uses local reputation only. Each node keeps only the reputation values of all its one-hop neighbors. Our work is partially following the same methodology that Prashant, Partha and Amiya used in their paper about reputation systems for AODV [7].

In the proposed reputation scheme, all the nodes in the mobile ad hoc network will be assigned an initial value of null (0) as in the Ocean reputation-based scheme [8]. Also, the functionality of the normal ARAN routing protocol in the authenticated route setup phase will be modified so that instead of the destination unicasts a RREP to the first received RDP packet of a specific sender only, the destination will unicast a RREP for each RDP packet it receives and forward this RREP on the reverse-path. The next-hop node will relay this RREP. This process continues until the RREP reaches the sender. After that, the source node sends the data packet to the node with the highest reputation. Then the intermediate node forwards the data packet to the next hop with the highest reputation and the process is repeated till the packet reaches its destination. The destination acknowledges the data packet (DACK) to the source that updates its reputation table by giving a recommendation of (+1) to the first hop of the reverse path. All the intermediate nodes in the route give a recommendation of (+1) to their respective next hop in the route and update their local reputation tables. If there is a selfish node in the route, the data packet does not reach its destination. As a result, the source does not receive any DACK for the data packet in appropriate time. So, the source gives a recommendation of (–2) to the first hop on the route. The intermediate nodes also give a recommendation (–2) to their next hop in the route up to the node that dropped the packet. As a consequence, all the nodes between the selfish node and the sender, including the selfish node, get a recommendation of (–2). The idea of giving (–2) to selfish nodes per each data packet dropping is due to the fact that negative behavior should be given greater weight than positive behavior. In addition, this way prevents a selfish node from dropping alternate packets in order to keep its reputation constant. This makes it more difficult for a selfish node to build up a good reputation to attack for a sustained period of time [9]. Moreover, the selfish node will be isolated if its reputation reached a threshold of (–40) as in the Ocean reputation-based scheme [8]. In the Table 50.2, the default Reputed-ARAN parameters are listed.

**Table 50.2** Reputed-ARAN Default parameters

| | |
|---|---|
| Initial Reputation | 0 |
| Positive Recommendation | +1 |
| Negative Recommendation | –2 |
| Selfish drop Threshold | –40 |
| Re-induction timeout | 5 minutes |

The proposed protocol will be structured into the following four main phases, which will be explained in the subsequent subsections: Route Lookup Phase, Data Transfer Phase, Reputation Phase, and Timeout Phase. This phase mainly incorporates the authenticated route discovery and route setup phases of the normal ARAN secure routing protocol. In this phase, if a source node S has packets for the destination node D, the source node broadcasts a route discovery packet (RDP) for a route from node S to node D. Each intermediate node interested in cooperating to route this control packet broadcasts it throughout the mobile ad hoc network; in addition, each intermediate node inserts a record of the source, nonce, destination and previous-hop of this packet in its routing records. This process continues until this RDP packet reaches the destination. Then the destination unicasts a route reply packet (RREP) for each RDP packet it receives back using the reverse-path. Each intermediate node receiving this RREP updates its routing table for the next-hop of the route reply packet and then unicasts this RREP in the reverse-path using the earlier-stored previous-hop node information. This process repeats until the RREP packet reaches the source node S. Finally, the source node S inserts a record for the destination node D in its routing table for each received RREP.

In Fig. 50.1–50.3, the route lookup phase is presented in details, illustrating the two phases of it, the authenticated route discovery phase and the authenticated route setup phase.

At this time, the source node S and the other intermediate nodes have many RREPs for the same RDP packet sent earlier. So, the source node S chooses the highly reputed next-hop node for its data transfer. If two next-hop nodes have the same reputation, S will choose one of them randomly, stores its information in the sent-table as the path for its data transfer. Also, the source node will start a timer before it should receive a data acknowledgement (DACK) from the destination for this data packet. Afterwards, the chosen next-hop node will again choose the highly reputed next-hop node from its routing table and will store its information in its sent-table as the path of this data transfer. Also, this chosen node will start a timer, before which it should receive the DACK from

| Next Hop | Reputation |
|---|---|
| A | 10 |
| B | 5 |

| Next Hop | Reputation |
|---|---|
| C | 20 |
| E | –5 |

**Fig. 50.1** A MANET Environment

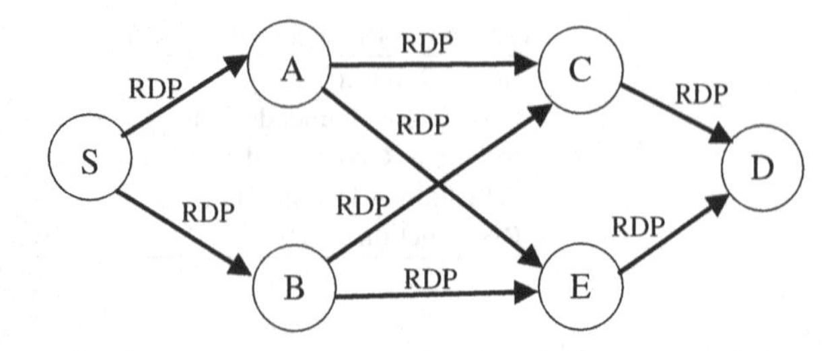


Fig. 50.2 Broadcasting RDP

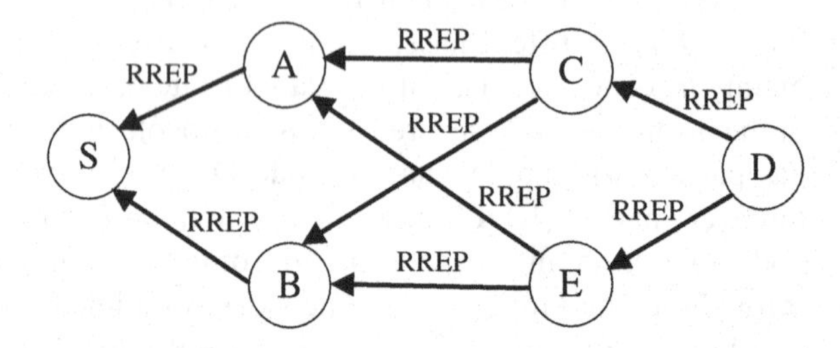


Fig. 50.3 Replying to each RDP

the destination for this data packet. This process continues till the data packet reaches the destination node D. And of course in this phase, if the data packet has originated from a low-reputed node, the packet is put back at the end of the queue of the current node. If the packet has originated from a high-reputed node, the current node sends the data packet to the next highly reputed hop in the route discovered in the previous phase as soon as possible. Once the packet reaches its destination, the destination node D sends a signed data acknowledgement packet to the source S. The DACK traverses the same route as the data packet, but in the reverse direction.

In Fig. 50.4, the data transfer phase is illustrated.

In this phase, when an intermediate node receives a data acknowledgement packet (DACK), it retrieves the record, inserted in the data transfer phase,

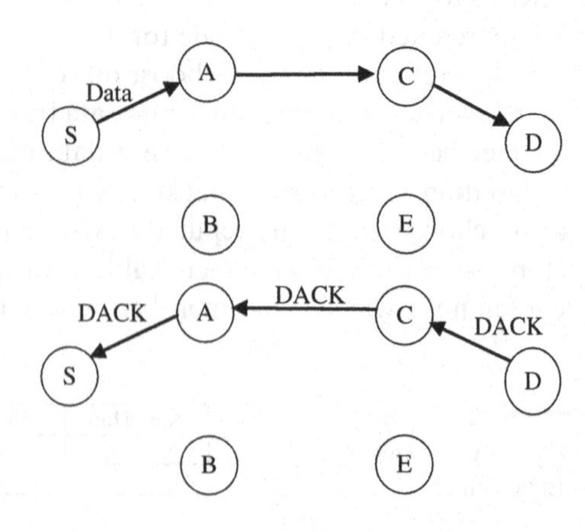


Fig. 50.4 Sending Data Acknowledgement for each received data packet

corresponding to this data packet then it increments the reputation of the next hop node. In addition, it deletes this data packet entry from its sent-table. Once the DACK packet reaches node S, it deletes this entry from its sent-table and gives a recommendation of (+1) to the node that delivered the acknowledgement.

In this phase, once the timer for a given data packet expires at a node, the node retrieves the entry corresponding to this data transfer operation returned by the timer from its sent-table. Then, the node gives a negative recommendation (–2) to the next-hop node and deletes the entry from the sent-table. Later on, when the intermediate nodes' timers up to the node that dropped the packet expire, they give a negative recommendation to their next hop node and delete the entry from their sent-table. As a consequence, all the nodes between the selfish node and the sender, including the selfish node, get a recommendation of (–2). Now, if the reputation of the next-hop node goes below the threshold (–40), the current node deactivates this node in its routing table and sends an error message RERR to the upstream nodes in the route. Then the original ARAN protocol handles it. Now, it is the responsibility of the sender to reinitiate the route discovery again. In addition, the node whose reputation value reached (–40) is now temporally weeded out of the MANET for five minutes and it later joins the network with a value of (0) so that to treat it as a newly joined node in the network.

## 50.4 Simulation

This section will be presenting the simulation environment, metrics and the results of the comparison between the proposed Reputed-ARAN and normal ARAN protocol.

All simulation experiments are developed and simulated on an Intel 1.4 GHz machine using Linux Red Hat 9.0 with 256 MB RAM and the network simulator Glomosim [10], a library-based sequential and parallel simulator for wireless mobile ad-hoc networks. Table 50.3 summarizes the different configuration values that are used in all the performed simulations.

**Table 50.3** Configuration values that are used in all the performed simulations

| Parameter | Value |
|---|---|
| MANET Area | 670*670 $m^2$ |
| Total number of nodes | 20 |
| % of Selfish Nodes Simulated | 0% up to 30% |
| Movement Pattern | Random Waypoint |
| Node Speed | 0 up to 10 m/s |
| Application | Five Constant Bit Rate (CBR) |
| Number of generated Packets | 1000 packets per CBR |
| Size of Packet | 512 bytes |
| Simulation Time | 15 min |

In order to compare the performance of Reputed-ARAN and normal ARAN, both protocols are run under identical mobility and traffic scenarios. First, an analysis of normal well-behaved ARAN network is done. Then, some uncooperative nodes are introduced to the normal ARAN network and analysis of the performance is done. Following that, the newly designed reputation-based scheme is added to normal ARAN, to become Reputed-ARAN, and performance is compared to normal ARAN. A comparison between both of these protocols using the metrics explained is presented in the following subsections.

This value represents the ratio of the total number of packets that reach their destination, to the total number of packets sent by the source. It is calculated according to this formula: Throughput = Packets Received/Packets Sent. The network throughput is directly influenced by packet loss, which may be caused by general network faults or uncooperative behavior.

This is the average delay between the sending of the data packet by the constant bit rate source and its receipt at the corresponding constant bit rate receiver.

In this experiment, the network throughput is being measured for the normal ARAN secure routing protocol and the Reputed-ARAN. The node speed and the percentage of selfish nodes participating in the mobile ad hoc network are varied to compare the results.

From the Fig. 50.5 and its corresponding table, it is clear that the lack of cooperation has a fatal effect on the efficient working of the MANET. This graph shows the dramatic fall in normal ARAN's network throughput with increasing percentage of selfish nodes. The different curves show a network of 20 nodes with different percentages of selfish nodes, from 0% up to 30%, and moving at different speeds.

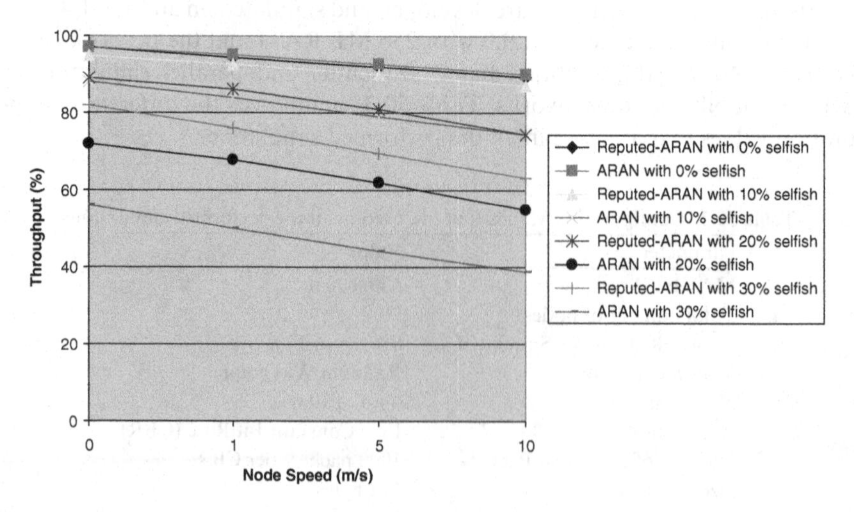


**Fig. 50.5** Effects of Selfish nodes on Network Throughput

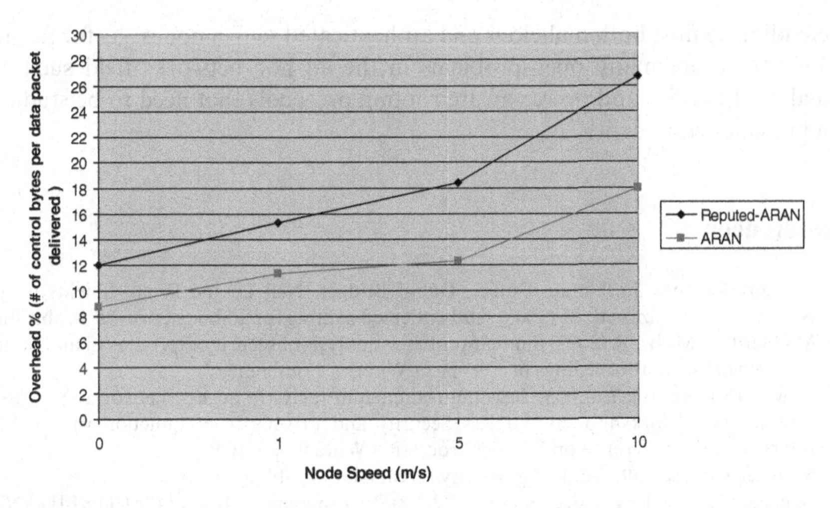


**Fig. 50.6** Overhead percentage

Fig. 50.6 shows the results of the overhead metric of both protocols: normal ARAN and Reputed-ARAN with different node speed. From the above graph and its corresponding table, it is clear that the newly proposed Reputed-ARAN protocol has a higher overhead than the normal ARAN secure routing protocol.

These extra RREPs are used later for the choice of the highly reputed next-hop node in the data transfer phase. Thus, when nodes are moving at speed of 10 m/s, the overhead percentage rises from 18%, in case of normal ARAN, to 26.7%, in the case of Reputed-ARAN. Though the overhead percentage added by the Reputed-ARAN is significant, this reputation-based scheme still improves considerably the network throughput, as shown in the previous section. In this experiment, the average end-to-end delay of data packets for the normal ARAN secure routing protocol and the Reputed-ARAN is measured. The percentage of selfish nodes participating in the mobile ad hoc network is varied to compare the results. The below figure shows the results of the average end-to-end delay of data packets metric of both protocols: normal ARAN and Reputed-ARAN with different percentage of selfish.

## 50.5 Conclusion and Future Directions

According to the many simulations that were performed, the newly proposed reputation-based scheme, built on top of normal ARAN secure routing protocol, achieves a higher throughput than the normal ARAN in the presence of selfish nodes. Thus, the proposed design, Reputed-ARAN, proves to be more efficient and more secure than normal ARAN secure routing protocol in defending against both malicious and authenticated selfish nodes. As for future work, there are many open-problems in the ad hoc networks field such as Quality of Service and power-aware routing protocols that need to be studied and researched in depth.

## References

1. Sanzgiri Kimaya, Laflamme Daniel, Dahill Bridget, Neil Levine Brian, Shields Clay, Belding-Royer Elizabeth M (2005) Authenticated routing for ad hoc networks. MobiCom
2. Michiardi P, Molva R (2002) Simulation-based analysis of security exposures in mobile ad hoc networks. In: Proceedings of European Wireless Conference
3. Dewan P, Dasgupta P (2003) Trusting routers and relays in ad hoc networks. In: First International Workshop on Wireless Security and Privacy in conjunction with IEEE International Conference on Parallel Processing Workshops (ICPP)
4. Schneier Bruce (1999) Modeling security threats. Dr Dobb's J
5. Buchegger Sonja, Le Boudec Jean-Yves (2002) Performance analysis of the CONFIDANT protocol: Cooperation of Nodes, Fairness In Dynamic Ad-hoc NeTworks. In: Proceedings of IEEE/ACM Symposium on Mobile Ad Hoc Networking and Computing, MobiHOC, Lausanne, CH

# Chapter 51
# Quantum Mechanical Analysis of Nanowire FETs

**Alireza Kargar and Rahim Ghayour**

**Abstract** In this paper, we present an analysis of a coaxial Nanowire FET by combining the analytical and numerical approaches. We achieve electrical characteristics of cylinder cal Nanowire FET and investigate the effects of changing metal/semiconductor Schottky barrier height and gate oxide thickness on electrical characteristics. First, we solve 1-D Laplace equation analytically to achieve the potential distribution within the wire. Then, we show the effect of changing the gate oxide thickness on the potential distribution. Second, by applying WKB approximation to tunneling problem and assuming ballistic current transport, we numerically achieve transfer characteristics of FET. Finally, we investigate the effect of varying Schottky barrier height (SB) on electrical characteristics of Nanowire FET.

## 51.1 Introduction

As the channel length of conventional planar metal oxide semiconductor field effect transistor shrinks into the nanometer scale, behavior of the device degrades mainly because of short channel effects due to a weakened gate control. In the Nanowire field effect transistor (NWFET), quantum effects such as the tunneling and confinement effects should be considered carefully to assess the device performance accurately. Also the ballistic transport can be regarded as making a dominant contribution to the total current. In particular, as the channel length of the NWFET shrinks, the source-to-drain direct tunneling current is expected to be increased significantly until it becomes the main source for the short channel effects in the NWFETs. On the other hand, semiconducting nanowire transistors are attracting attention due to their potential applications such as electronics

A. Kargar (✉)
Department of Electrical Engineering, School of Engineering, Shiraz University, Shiraz, Iran
e-mail: rghayour@shirazu.ac.ir

N. Mastorakis et al. (eds.), *Proceedings of the European Computing Conference*, Lecture Notes in Electrical Engineering 28,
DOI 10.1007/978-0-387-85437-3_51, 

[1] and bimolecular detection [2]. Also large ON-currents, large $I_{ON}/I_{OFF}$ ratios have been reported [1]. Such devices are typically fabricated in the plane of a substrate with either a top or bottom gate. It is well known that a fully surround gate structure, is optimal for electrostatic control over charge carries in the channel [3]. Further, most attempts on nanowire modeling assumed ballistic transport [4, 5].

In Section 51.2, first, by using an analytical approach, we present the electrostatics potential distribution of coaxial nanowire FET. Then in ballistic transport regime and application of WKB approximation, we numerically offer transfer characteristics. In Section 51.3, by looking at previous studies, we investigate the dependence of the carrier transport on the metal/semiconductor Schottky barrier height and gate oxide thickness.

## 51.2 Device Structure and Approach

Let us consider a coaxial gate NWFET as shown in Fig. 51.1. The channel between source and drain is considered a one-dimensional wire, located along the axis of the structure. The bias voltage of the gate is applied to the outer shell of the structure. The space between wire and outer shell is filled with insulator that is known as the gate oxide geometry.

The potential distribution of the structure is achieved by solving the Laplace's equation in cylindrical coordinates as:

$$\frac{1}{\rho}\frac{\partial}{\partial \rho}\left(\rho\frac{\partial U(\rho,z)}{\partial \rho}\right)+\frac{\partial^2 U(\rho,z)}{\partial^2 z}=0 \tag{51.1}$$

where $U(\rho,z)$ is the electrostatic potential distribution along the nanowire. Charges in the nanowire are neglected in the simulation. Thus, solving this equation analytically we achieve the potential distribution which is in exponential form.

Figure 51.2 demonstrates the potential distribution. The end contacts in Nanowire FET are considered the reservoirs of charge, from which the carriers are injected into the nanowire. The amount of injection depends on the transmission coefficient and the drain-source voltage of a Nanowire FET. The electrons which are coming from the source reservoir into the nanowire experience an

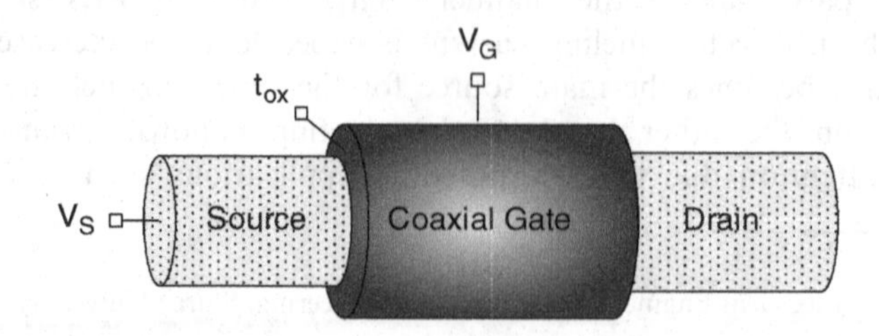


**Fig. 51.1** Schematic view of the CNWFET device structure. The gate forms the curved surface of the outer cylinder

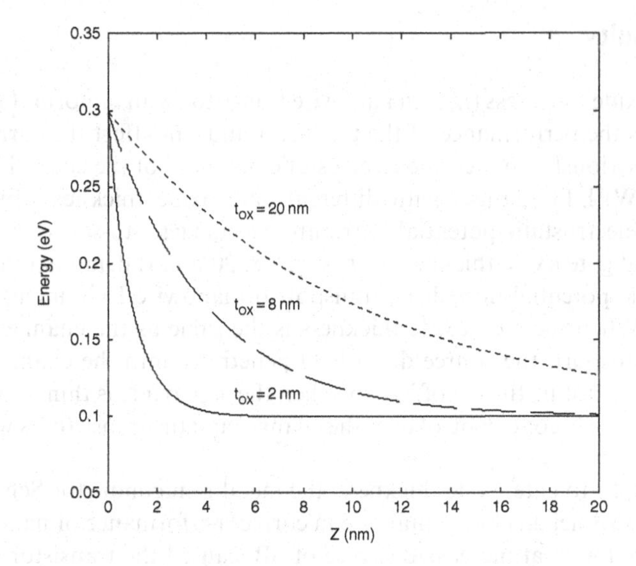


**Fig. 51.2** The potential distribution in the z direction for different gate oxide thickness and gate-source voltage is equal to 0.2 eV

exponential shape barrier which is the Schottky barrier between the wire and the source metal end, where it is demonstrated by Fig. 51.2. For a Schottky barrier, the Schrödinger equation cannot be precisely solved, so the Wentzel-Kramers-Brillouin (WKB) approximation is often the most widely used approach. Using this approach the transmission probability can be derived as [6]:

$$T(E) = \exp\left(-2\int_{z_i}^{z_f}\sqrt{\frac{2m_{eff}(U(z)-E)}{\hbar^2}}\right) \tag{51.2}$$

where $U(z)$ is the potential distribution, $E$ is its total energy, $\hbar$ is r-duced Planck's constant, $m_{eff}$ is the carrier effective mass, and $z_i$ and $z_f$ are the classical turning points. So, we analytically calculate transmission probability from Equation (51.2).

Assuming the ballistic transport, the channel current can be expressed using the Landauer-Büttiker formula as follows [7]:

$$I_{DS} = \frac{2q}{\pi\hbar}\int T(E)(f_S(E) - f_D(E))dE \tag{51.3}$$

where $f_S$ and $f_D$ are the Fermi-Dirac functions with Fermi level of the source and the drain, respectively. From a numerical point of view, we calculate the electrical characteristics by means of a simple numerical method from Equation (51.3).

## 51.3 Results

The gate oxide thickness ($t_{OX}$) in nanowire transistor is an important parameter that affects the performance of the transistor and can affect the current of the transistor seriously. In fact, the electrostatic potential of the coaxial Nanowire FET (CNWFET) changes with different gate oxide thickness. Figure 51.2 shows the electrostatic potential distribution at the gate-to-source voltage $V_{GS}$ = 0.2 V and gate oxide thickness of $t_{OX}$ = 2, 8, 20 nm. It demonstrates that the electrostatic potential in ballistic transport of nanowire FET intensively vary with $t_{OX}$. When the gate oxide thickness is the same as the channel length in ballistic transport, the source/drain field penetrates into the channel and the transistor cannot be turned off. Also, when the gate oxide is thin, however, the gate still has very good control over the channel and the transistor is well turned off [8].

In addition to gate oxide thickness, the metal/semiconductor Schottky barrier height (SB height) has a vital role in correct performance of nanowiretransistor so that a great increase/decrease of SB, caused the transistor cannot be well turned on/turned off. In Fig. 51.3 the transfer characteristics *($I_{DS}$-$V_{GS}$ plot)* are plotted for a drain-to-source voltage of $V_{DS}$=0.3 volt and different Schottky barrier heights (SB height).

Figure 51.4 shows the transfer characteristic, $I_{DS}$ versus $V_{DS}$, with different $V_{GS}$, SB = 0.3 eV and $t_{OX}$=2 nm so that it confirms, the fact which drain-source current increases with the larger amounts of gate-source voltage.

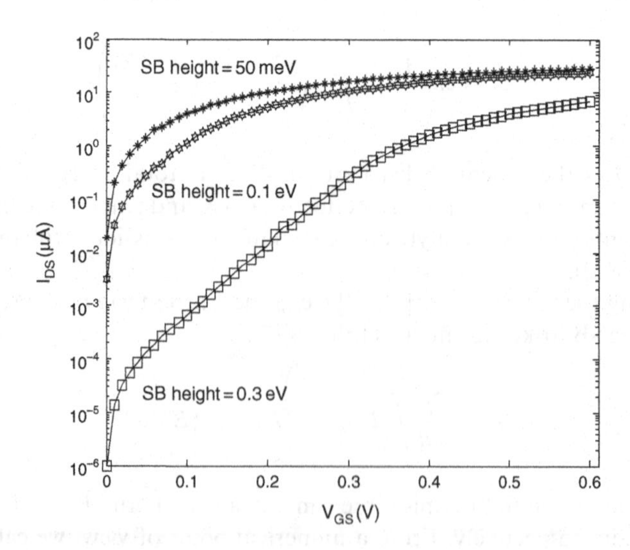


**Fig. 51.3** The transferred curves $I_{DS}$-$V_{GS}$ of the CNWFET with $V_{DS}$= 0.3 V, SB height of 50 meV, 0.1 eV, 0.3 eV and $t_{OX}$= 2 nm

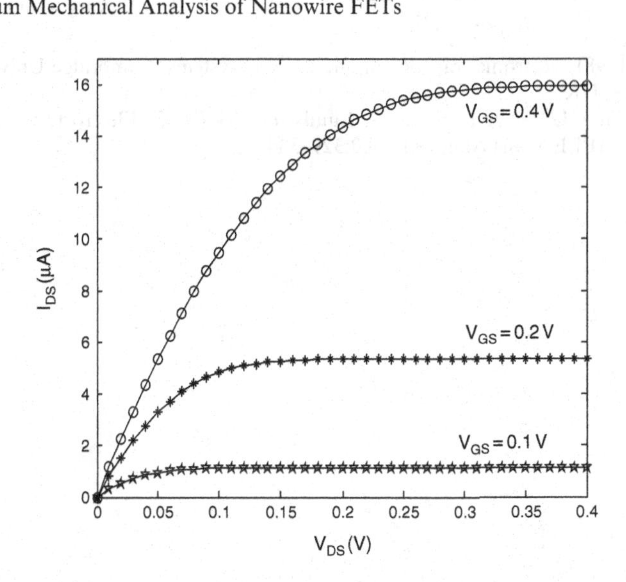


**Fig. 51.4** The computed $I_{DS}$-$V_{DS}$ curves of the CNWFET with $V_{GS}$= 0.1, 0.2, 0.4 V and $t_{OX}$= 2 nm

## 51.4 Conclusion

The coaxial nanowire FET (CNWFET) is presented and by applying an analytical-numerical approach the electrical characteristics are achieved. The ballistic transport is assumed for carrier transport such as [5] and for handling the tunneling problem the WKB approximation is suited. So, we demonstrate that in ballistic regime for CNWFET, two essential factors are the insulator thickness and Schottky barrier height so that, with optimization of these factors, the on-off current ratio and generally the electrical performance of this structure can be improved.

## References

1. Xiang J, Lu W, Hu Y, Wu Y, Yan H, Lieber CM (2006) Ge/Si nanowire heterostructure as high-performance field-effect transistors. Nature 441:489–493
2. Hahm J, Lieber CM (2004) Label-free detection of DNA hybridization using carbon nanotube network field-effect transistors. Nano Lett 103:921–926
3. Wang J, Polizzi E, Lundstrom M (2004) A three-dimensional quantum simulation of silicon nanowire transistors with the effective-mass approximation. J Appl 96:2192–2203
4. Chen J (2006) A circuit-compatible model for ballistic nanowire transistors. In: Proceeding Nanotechnology Conference, pp 850–853
5. Jimenes D, Saenz JJ, Iniguez B, Sune J, Marsal LF, Pallares J (2003) Unified compact model for the ballistic quantum wire and quantum well metal-oxide-semiconductor field-effect-transistor. J Appl Phys 94:1061–1068
6. Powell and Crasmann, Quantum Mechanics

7. Datta S (1998) Electronic transport in mesoscopic systems. Cambridge University Press, Cambridge, UK
8. Guo J, Wang J, Pollizi E, Datta S, Lundstrom M (2003) Electrostatics of nanowire transistors. IEEE Trans Nanotechnol 2:329–334

# Part VII
# Software Tools and Programming Languages

The purpose of this part is to provide a forum for software scientists to present and discuss recent researches and developments in the programming languages and software. The scope of the part is not restricted, it covers ongoing research in different countries related to software technology with an emphasis on programming languages and software tools e.g.: algorithms and data structures, programming languages and paradigms, software engineering, information systems, logic and declarative programming, parallel and distributed computing, embedded computing, etc. The papers include the impact of quality factors on the success of software development, language structure decompositions, embedded applications for driverless metro train, a post-mortem JavaSpaces debugger, stochastic numerical simulations using C#, visualization and analyses of multidimensional data sets using OLAP, design of a computer-based triangle chess machine, a taxonomy model for analyzing project-based learning using analytical hierarchy process , a framework of mixed model of a neural network and boolean logic for financial crisis prediction, Chinese document clustering using a self-organizing map based on botanical document warehouse, linear discriminant analysis in Ottoman alphabet character recognition, user requirements engineering and management in software development and tools for active filter synthesis.

# Chapter 52
# Impact of Quality Factors on the Success of Software Development

**M. Sepehri, A. Abdollahzadeh, M. Sepehri, and M. Goodarzi**

**Abstract** During the design of a system, software engineering is faced with many choices. Most of these choices involve the identification of a solution for a problem from a given set of alternatives. The alternative solutions generally only differ with respect to quality attributes. Quality factors have a very strong impact on the success of software development methodologies; therefore the alterative should be selected which satisfies the quality requirements best. This paper was undertaken to gathering popular methodologies such as SSADM (Structured System Analysis and Design) and Object-Oriented towards the impact of quality factors on methodology choice. In this paper, an empirical comparison was made between two mentioned methodologies used with a software project. The mentioned methodologies for project domain were compared using some quality factors such as maintainability, flexibility, reliability and usability. The obtained results indicated that investigation of quality factors such as maintainability, flexibility, reliability in Object-Oriented methodology was achieved in less time than SSADM, whereas on usability parameter users were clearly more satisfied with their experience with the SSADM.

## 52.1 Introduction

The engineering approach for constructing systems needs methodology and suitable tools to produce software. There is no general software development methodology powerful enough to handle all the variations of today's systems [11]. Methodology could be revised, improved and extended [6].

The main objective of this paper is to compare and evaluate two evaluation methodologies such as SSADM and Object-Oriented based on quality factors to discover the similarities and differences.

M. Sepehri (✉)
Department of Computer Engineering and IT, Islamic Azad University Fars Sciences and Research Branch, Fars, Iran

N. Mastorakis et al. (eds.), *Proceedings of the European Computing Conference*, Lecture Notes in Electrical Engineering 28,
DOI 10.1007/978-0-387-85437-3_52, 

Currently, the most widely used methodology for constructing a software system is SSADM. This methodology was introduced by Yourdon and Constantine [10], and has since been developed and modified by a number of authors [9, 13, 5, 4, 11]. SSADM methodology focuses on the flow of data through the software processes (functions) which each function is a basic design component.

Today, most information systems are built using Object-Oriented methodology [1, 7]. This methodology allows the programmer to build software from software parts called objects. An object contains both the data and functions that operate upon the data. An object can only be accessed via the functions it makes publicly available, so that the details of its implementation are hidden from other objects. By the adoption of Object-Oriented methodology, higher productivity, lower maintenance cost and better quality can be achieved.

Measurement is fundamental to any engineering discipline, and software engineering is, or should be, no exception. Software metrics deal with the measurement of the software product and process by which it is developed. The metrics are used to estimate/predict product costs and schedules and to measure productivity and product quality. So software metrics tell us something about the quality of software. In this paper, we discuss some quality metrics (Reliability, Maintainability, Usability and Flexibility) as quality parameters.

The next section briefly reviews some Distinguishing Characteristics of Object-Oriented methodology like encapsulation, inheritance and abstraction. Section three describes some quality factors that are used for evaluating the mentioned systems. Section four presents an Illustrative Case Study Problem. Section five shows the results of comparing SSADM and Object-Oriented used in an exciting software project and discusses the obtained results.

## 52.2 The Distinguishing Characteristics

The Object-Oriented methodology includes a set of mechanisms such as inheritance, encapsulation, and abstraction. Each of these characteristics is discussed briefly in the sections that follow:

### *52.2.1 Abstraction*

Abstraction is the best hope for dramatically the difficulty and cost of software development [7]. At a low level of abstraction, a more procedural orientation is taken. Finally, at the lowest level of abstraction, the solution is stated in a manner that can be directly implemented.

### *52.2.2 Encapsulation*

Encapsulation means that all of this information is packaged under one name and can be reused as one specification or program component. For Object-

Oriented systems, encapsulation encompasses the responsibilities of a class, including its attributes, and operation, and the states of the class as defined by specific attributes values [5].

### *52.2.3 Inheritance*

Inheritance is a mechanism that enables the responsibilities of one object to be propagated to other objects. Inheritance occurs throughout all level of a class hierarchy. Because Inheritance is a pivotal characteristic in many Object-Oriented systems, many Object-Oriented metrics focus on it [5].

## 52.3 Software Quality Factors

Software quality is a complex mix of factors that will vary across different applications and the customers who request them [5]. Software quality can be modeled in different ways and quality as a metric depends on the software we are measuring [2]. In this paper, we concentrate on some operational quality factors which are discussed briefly in the sections that follow:

### *52.3.1 Flexibility*

Flexibility is an effort required to modify an operational program. In this paper, the flexibility of two implemented systems was investigated with creating a new requirement.

### *52.3.2 Maintainability*

Maintainability is the ease with which a program can be corrected if an error is encountered, adapted if its environment changes, or enhanced if the customer desires a change in requirement. In this paper, maintainability of two implemented systems was investigated with creating code modification

### *52.3.3 Reliability*

Reliability metric as a quality factor attempts to measure and predict the probability of failures during a particular time interval, or the mean time to failure (MTTF). In this paper, the reliability of two implemented systems was investigated with creating faulty inputs.

### *52.3.4 Usability*

In early models of quality, McCall's and Boehms quality models [14], usability doesn't play a big part. But for end-users usability is perhaps the most important quality metric [2, 8]. For a large number of products, it is believed that usability becomes the final arbiter of quality. Usability metric as a quality factor attempts to measure the effort required to learn, operate, and prepare input, and interrupt the output of a program. In this paper, usability of two implemented systems was investigated with training users (staffs).

## 52.4 An Illustrative Case Study Problem

As a case study, in order to compare Structured System Analysis and Design Methodology (SSADM) and Object-Oriented methodologies, we decided to examine the software system produced by each of the methodologies for a single problem domain. For the problem domain, we take a project of Tehran exchange software system and investigate one part of it. The systems' overall purpose was the "management of monitoring factories and share investigation". First, the requirements analysis and system design was conducted using SSADM methodology with DFD diagrams by Easy Case software. Second, to complete the modeling, we performed a simulated Object-Oriented requirements and system design with UML diagrams by Rational Rose software. At last, we implemented the system with Delphi programming language, employing SQL Server database management system.

There is a fundamental difference between data flow and object models. In the data flow model, we concentrate on organizing system knowledge around a hierarchy of functions, with data organization playing a secondary role. In an object model, the primary organizing concept is that of an object and its embodied data (attributes) and functions (services). Because of this basic difference, it is natural to ask which the superior methodology for system modeling is.

In this paper, we are attempting the resolution of this question based on some quality factors such as flexibility, reliability, maintainability and usability.

## 52.5 Comparing the Two Methodologies

### *52.5.1 Preliminary Results*

This section presents quality factors (Reliability, Maintainability, Flexibility and Usability) in terms of the time needed to perform the task. In this section, we first review the results of comparing SSADM and Object-Oriented based on these factors and then discuss the obtained results.

#### 52.5.1.1 Reliability Results

In this paper, the reliability of two implemented systems was investigated with creating faulty inputs. Table 52.1 shows the number of failures detected during a particular time interval in two methodologies.

#### 52.5.1.2 Maintainability Results

In this paper, the maintainability of two implemented systems was investigated with creating code modification. Table 52.2 shows the time it takes to correct a change in two methodologies.

#### 52.5.1.3 Flexibility Results

In this paper, the flexibility of two implemented systems was investigated with creating a new requirement. Table 52.3 shows the time it takes this change in the analysis, design and coding.

**Table 52.1** Measurement of reliability in terms of time (Minutes)

| SSADM Methodology | Object-Oriented Methodology |
|---|---|
| 0.10 | 0.10 |
| 0.15 | 0.15 |
| 0.23 | 0.17 |
| 0.27 | 0.18 |

**Table 52.2** Measurement of maintainability in terms of time (Minutes)

| SSADM methodology | Object-oriented methodology |
|---|---|
| 37 | 20 |
| 45 | 30 |
| 40 | 40 |
| 85 | 60 |

**Table 52.3** Measurement of flexibility in terms of time

| SSADM methodology | Object-oriented methodology |
|---|---|
| 60 m | 40 m |
| 95 m | 55 m |
| 105 m | 70 m |
| 185 m | 85 m |

**Table 52.4** Measurement of training in terms of time

| Number of staffs | Methodology |
|---|---|
| 19 (59%) | SSADM |
| 9 (28.12%) | Object-Oriented |
| 4 (12.50%) | SSADM and Object-Oriented |

#### 52.5.1.4 Usability Results

In this paper, the usability of two implemented systems was investigated with training users (staffs). Learnability measured in time needed for learning perform tasks or required training. Thirty-two staffs participated in this study. There were 13 female and 19 male participants ranging in age from 22 to 38. Of the 32 participants, 19 learned SSADM in less time, nine learned Object-Oriented in less time and four learned approximately in equal time.

### *52.5.2 Discussions*

This section discusses the results that obtained in the previous section.

#### 52.5.2.1 Differences in Reliability

Results showed that the probability of failures during a particular time interval Object-Oriented methodology is fewer than SSADM. The obtained results of this research reveled that Object-Oriented is a suitable methodology based on reliability metric (Fig. 52.1).

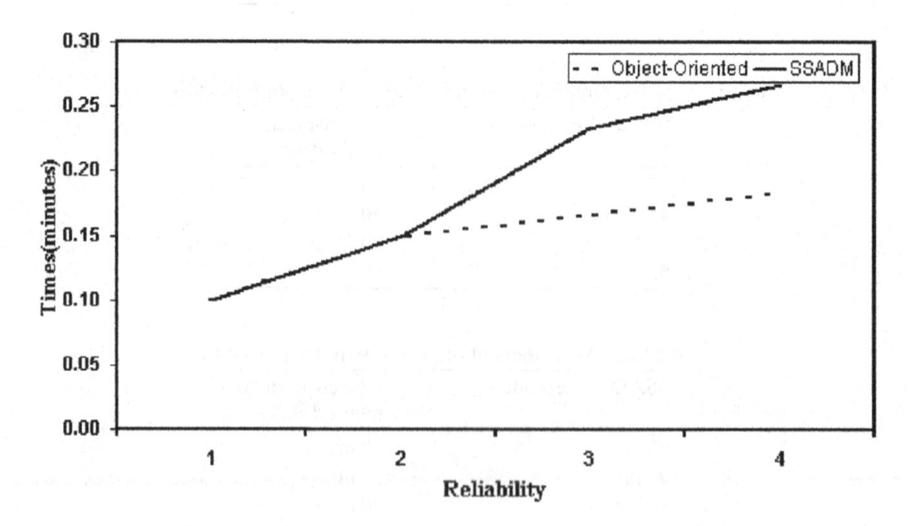


**Fig. 52.1** Comparing reliability of two methodologies

These results are because of some reasons as follows:

- Objects are reusable. Once they are designed, tested and built, objects can be reused in multiple systems and application. Therefore using reusable components can increase reliability.
- Coupling is already proven to be a useful criteria for reliability. Interfaces among encapsulated objects are simplified. An object that sends a message need not be concerned with the details of internal data structures. Hence, interfaces are simplified and the system coupling tends to be reduced so reliability increased.
- Encapsulation is the concept that objects should hide their internal contents from other system components and prevents the unauthorized access of an object's data.

#### 52.5.2.2 Differences in Maintainability

The obtained results of this research reveled that Object-Oriented is a suitable methodology based on maintainability metric (Fig. 52.2).

These results are because of some reasons as follows:

- Object-Oriented methodology consists of independent modules, so independent modules are easer to maintain and test because secondary effects caused by design or code modification are limited and error propagation is reduced.
- Encapsulation can cause the internal implementation details of data, and procedures are hidden from the outside world (information hiding). This reduces the propagation of side effects when changing occurs.

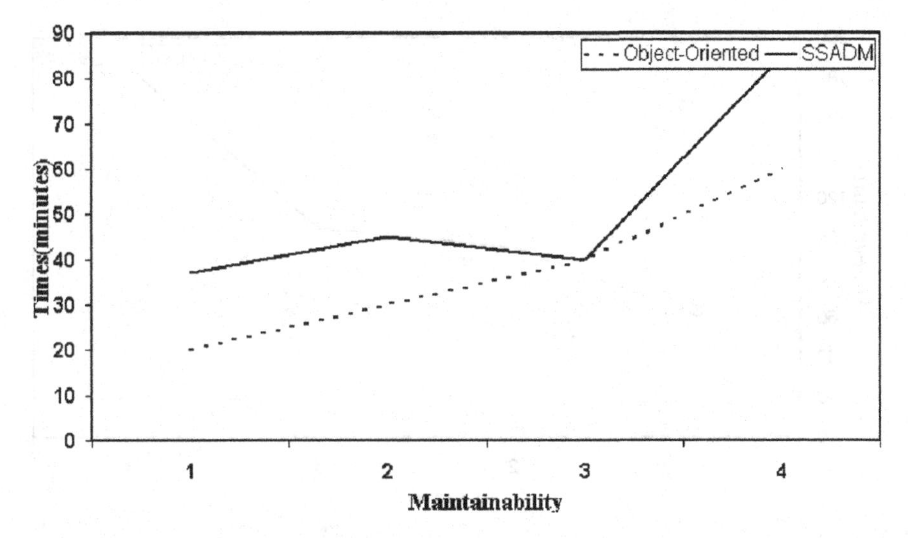


**Fig. 52.2** Comparing maintainability of two methodologies

- In Object-Oriented methodology, localization is based on objects, and the relationship between operations (functions) and classes is not necessarily one to one. So each object can be changed at any time without affecting the other objects.
- Objects are extensible. They can be changed easily without adversely impacting any previous application that used them.

#### 52.5.2.3 Differences in Flexibility

The obtained results showed that the Object-Oriented methodology handled creating a new requirement much faster than SSADM (Fig. 52.3).

These results are because of some reasons as follows:

- Reused of a subroutine or function decrease the need to create new code to perform the same task therefore reuse increases flexibility of software.
- In Object-Oriented methodology, better understanding established by the association of data with operations. By maintaining this tight association, maintainability of the software is facilitated by localizing the impact of change to specific parts of the system.
- Objects are extensible. They can be expanded easily without adversely impacting any previous applications that used them.

#### 52.5.2.4 Differences in Usability

The obtained results showed that time needed for learning perform tasks or required training in SSADM is much faster than Object-Oriented. The obtained

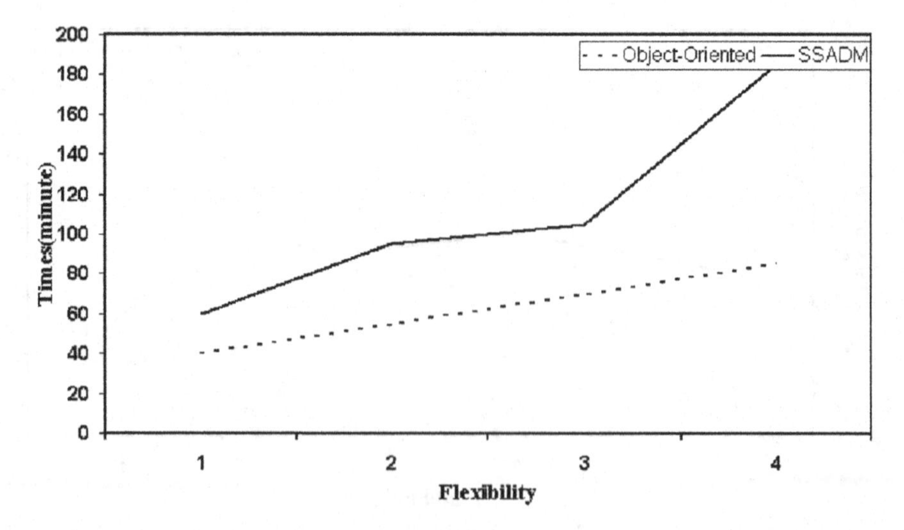


Fig. 52.3 Comparing flexibility of two methodologies

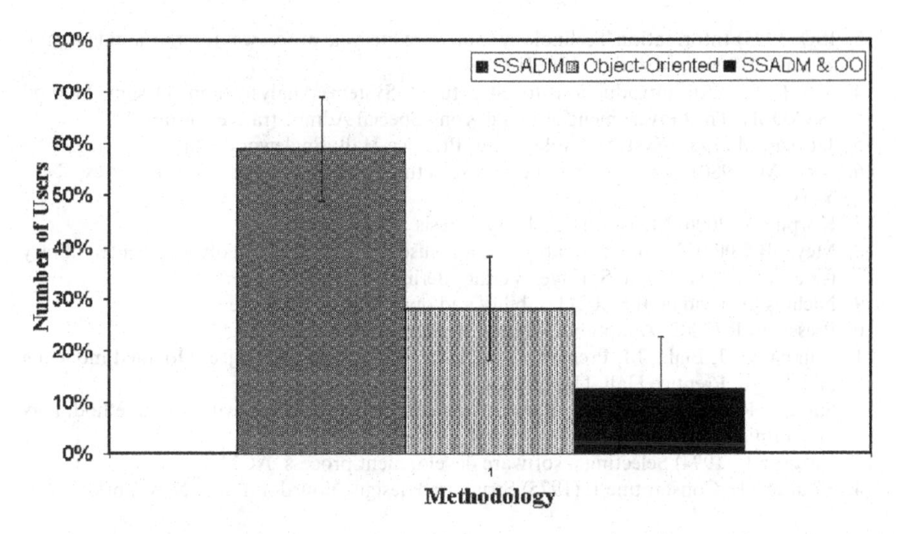


**Fig. 52.4** Comparing usability of two methodologies

results of this research reveled that SSADM is a suitable methodology based on Usability metric (Fig. 52.4).

These results are because of SSADM does not require very special skills and can easily be taught to the staff. Nowadays common modeling and diagramming tools are used; commercial case tools are also offered in order to be able to set up SSADM easily [3].

## 52.6 Conclusions

Selecting a suitable methodology for development of a system is difficult. However, some measurements are needed for software engineers in order to choose the suitable methodology. In this paper, we have defined some quality factors such as Reliability, Maintainability, Flexibility and Usability that are practical and easily computable. We introduced two types of methodologies like SSADM and Object-Oriented. In this paper, we compared the two mentioned methodologies based on quality factors to choose the suitable one. The obtained results of this research revealed that Object-Oriented is a suitable methodology based on Reliability, Flexibility and Maintainability, whereas SSADM is an appropriate methodology based on the usability factor.

## Refrences

1. Cox B (1987) Object Oriented Programming: An evolutionary approach. Addision-Welsey, Michigan
2. Fenton N, Lawrence S (1997) Software metrics – A rigorous et practical approach. PWS Publishing Company, Revised Printing, pp 337–361

3. ISO (1995) Information Technology- Software lifecycle processes. International Organization for Standardization, ISO/IEC
4. ITS (2003) An introduction to Structured System Analysis and Design Method (SSADM). The Government of Hong Kong Special Adminstrative Resion
5. Jackson M (1983) System development. Prentice-Hall, Englewood Cliffs
6. Jones M (1980) The practical guide to structured systems design. Yourdon Press, New York
7. Korpua A (2006) Measuring usability. Thesis
8. Meyer B (2003) The power of abstraction, reuse and simplicity: an object-oriented library for event-driven design. Springer-Verlag, Berlin
9. Nichols D, Twidale B (2002) Usability and open Source Software
10. Pressman R (2002) Adaptable Process Model
11. Rambaugh J, Blaha M, Premerlani W, Eddy F, Lorensen W Object Oriented modeling and design. Prentice Hall, Upper Saddle River
12. Sargent R, Behling P(2002) Back to systems delivery basics with five methodology imperatives.
13. Vazquez F (1994) Selecting a software development process. ACM
14. Yourdon E, Constantine L (1975) Structured design. Yourdon Press, New York

# Chapter 53
# Sisal 3.2 Language Structure Decomposition

**V.N. Kasyanov and A.P. Stasenko**

**Abstract** The functional programming system SFP under development at the Institute of Informatics Systems in Novosibirsk is aimed at supporting development of parallel computing applications that still offer high performance and portability. The paper describes equivalent transformations of the Sisal 3.2 programming language (based on Sisal 90) structures. These transformations are to decompose the complex language structures into more simple ones that can be directly expressed by the internal representation IR1, which is based on the intermediate form language IF1. Currently some description of similar transformations can be found in few works about Sisal 90 in the form of examples. These transformations are performed by the front-end compiler from Sisal 3.2 into IR1 and help to better understand its translation strategy. The paper also briefly describes IR1 languages.

## 53.1 Introduction

Using traditional methods, it is very difficult to develop high quality, portable software for parallel computers. In particular, parallel software cannot be developed on low cost, sequential computers and then moved to high performance parallel computers without extensive rewriting and debugging. Functional programming [1] is a programming paradigm, which is entirely different from the conventional model: a functional program can be recursively defined as a composition of functions where each function can itself be another composition of functions or a primitive operator (such as arithmetic operators, etc.). The programmer need not be concerned with explicit specification of parallel processes since independent functions are activated by the predecessor functions and the data dependencies of the program. This also means that control can be distributed. Further, no central memory system is inherent to the model

V.N. Kasyanov (✉)
A.P. Ershov Institute of Informatics Systems/Novosibirsk State University,
Novosibirsk, 630090, Russia

N. Mastorakis et al. (eds.), *Proceedings of the European Computing Conference*, Lecture Notes in Electrical Engineering 28,
DOI 10.1007/978-0-387-85437-3_53, 

since data is not "written" in by any instruction but is "passed from" one function to the next.

Functional language Sisal (Steams and Iterations in a Single Assignment Language) is considered as an alternative to Fortran language for supercomputers [2]. Compared with imperative languages (like Fortran), functional languages, such as Sisal, simplifies programmer's work. He has only to specify a result of calculations and it is a compiler that is responsible for mapping an algorithm to certain calculator architecture (including instructions scheduling, data transfer, execution synchronization, memory management and etc.). In contrast with other functional languages, Sisal supports data types and operators typical for scientific calculations.

At present, there are implementations of the Sisal 1.2 language for many supercomputers, e.g., SGI, Sequent, Encore Multimax, Cray X-MP, Cray 2, etc (see [3]). The Sisal 90 language definition [4] increases the language's utility for scientific programming. It includes language level support for complex values, array and vector operations, higher order functions, rectangular arrays, and an explicit interface to other languages like Fortran and C.

The Sisal 3.2 language that has been designed as the input language of the SFP system being under development at the Institute of Informatics Systems in Novosibirsk [6] is an extension of the Sisal 3.1 [5] which is based on the Sisal 90. Sisal 3.1 simplifies, improves, extends and more exactly specifies Sisal 90. Sisal 3.2 also incorporates ideas of enhanced module support, annotated and preprocessing of Sisal 3.0 [7]. Other extensions of Sisal 3.2 are function overloading and user-defined types, which allow user-defined operations.

The SFP system is aimed at supporting development of parallel computing applications that still offer high performance and portability. It works on a personal computer under Microsoft Windows 2000/XP and provides means to write and debug Sisal-programs regardless target architectures as well as to translate the Sisal-programs into optimized imperative programs, appropriate to the target execution platforms.

The SFP system uses intermediate language R1 [8], which is based on the intermediate form language IF1 [9]. IR1 is a language of hierarchical graphs [10] made up simple and compound computation nodes, edges and ports. Nodes correspond to computations. Simple nodes are vertices of underlying graph and denote operations such as add or divide. Compound nodes are fragments (or subgraphs) of underlying graph and represent compound constructions such as structured expressions and loops. Ports are vertices of an underlying graph that are used for input values and results of compound nodes. Edges show the transmission of data between simple nodes and ports; types are associated with the data transmitted on edges.

In the paper equivalent transformations of the Sisal 3.2 programming language structures are described. These transformations aim to decompose the complex language structures into more simple ones that can be directly expressed by the internal representation IR1. Currently some description of similar transformations can be found in few works about Sisal 90 in the form of examples.

These transformations are performed by our front-end compiler and help to better understand its translation strategy. The paper also briefly describes IR1 language.

The rest of the paper is structured as follows. Sections 53.2 and 53.3 briefly describe the most important *Select* and *Forall* compound nodes of the intermediate language IR1 that represent the most Sisal 3.2 language syntax constructions. Sections 53.4, 53.5, 53.6, 53.7, 53.8, 53.9, 53.10, 53.11, 53.12 present transformations of complex Sisal 3.2 language structures into more simple ones that can be directly represented by IR1.

## 53.2 The Compound Node Select

The compound node *Select* has an arbitrary number of input ports and a non-zero number of output ports. Let $N \geq 3$ be the number of its graphs. The input ports of all graphs are the same as the input ports of the containing compound node and directly receive values from them. All graphs except the first one have the same output ports as the output ports of the containing compound node. One of these $N$–1 graphs is chosen to supply values of its output ports to the output ports of the containing compound node.

The choice is based on the first graph, which has different semantics compared to its prototype from IF1. The first graph has $N$–2 Boolean output ports (edges that end in these ports have Boolean type), which are sequentially checked until the true value is found on the output port number $M$. In that case, graph number $M+1$ is chosen. If no *true* value is found, then the last graph is chosen. In the case of the erroneous Boolean value, the error values are placed on the output ports of the containing compound node.

It is clear that the *Select* compound node can directly represent the **if** expression of Sisal 3.2. If the **else** branch is omitted, then the last graph is generated to return error values of the corresponding types.

## 53.3 The Compound Node Forall

The compound node *Forall* uses the following groups of ports and has four graphs described in Table 53.1. Each port group consists of the same ports for each separate compound node *Forall.*

**Table 53.1** Port groups of the compound node *Forall* and its graphs

| Graph number | Graph name | Input port groups | Output port groups |
|---|---|---|---|
| | *Forall* | *C* | *R* |
| 1 | Initialization | *C* | *L* |
| 2 | Range generator | *C* | *D* |
| 3 | Loop body | *OL, D, C* | *L, L2* |
| 4 | Return statement | *OL, L, L2, D. C* | *R* |

- The group $C$ – the constants, imported to the compound node ports.
- The group $R$ – the result values, exported from the compound node ports.
- The group $L$ – the dependent loop values.
- The group $L2$ – the local loop values.
- The group $OL$ – the old dependent loop values.
- The group $D$ – the loop range generator values.

At the beginning, the output ports of the *initialization* graph are computed and their values are used as the values of the port group $OL$ on the first iteration (if any) of the *loop body* graph. The *loop body* graph computes its output ports, for each instance of the port group $D$, generated by the *range generator* graph. The *return statement* graph is computed after each loop iteration and after the last iteration, its output port values are used as the values of the compound node. Before the next iteration, the values from the port group $L$ are copied to the ports of the group $OL$.

The *return statement* graph contains the reduction nodes that can only (and only they can) supply values to the output ports of this graph. These reduction nodes directly correspond to one-dimensional reductions of Sisal 3.2 and have an additional memory associated with them between loop iterations.

The *range generator* graph also has the unique *Scatter* nodes that can only (and only they can) supply values to the output ports of this graph. The *Scatter* node has one input and two output ports. The input port has a type of the array or stream of the type $T$. The first output port has the type $T$ and the second output port has the integer type. The *Scatter* node sequentially emits a new array or stream element with its index for every loop iteration. If there are several *Scatter* nodes, then they emit new values simultaneously until at least one of them can do it, while the others that cannot, return the error values.

The compound node *Forall* can represent any one-dimensional **for** expression, which is controlled with a range.

## 53.4 Decomposition of the Case Expression

The conditional expression **case** is naturally decomposed into the conditional expression **if** with additional **elseif** branches. Since the **if** expression with arbitrary number of **elseif** branches can be already explicitly expressed by the IF1 language, its further simplification is not performed by the Sisal 3.2 front-end compiler and is not shown in this paper.

Every selection list of the **case** expression is transformed into one **if** or **elseif** condition using logical disjunction and conjunction operations over the comparison operation results: equality ( = ), "less than or equal to" (≤) and "greater than or equal to" ( ≥ ). For expressions "**case tag**" and "**case type**", the infix operation **tag** (**tag** function of Sisal 90) and the expression "**type**[...]" are used, respectively.

## 53.5 Decomposition of the multidimensional Loops

Let us consider the following $n$-ary $m$-dimensional loop, where each reduction corresponds to one loop dimension (for the simplicity of further notation):

**for** $D_1$ **cross** $D_2$ **repeat** $B$
 **returns** $RN_1$ **of** $RV_1$; . . .; $RN_n$ **of** $RV_n$
**end for**

The name $D_1$ denotes the loop range generator part without the operator **cross** and multidimensional indices of the **at** construction, the name $D_2$ denotes the remaining part of the range generator, the name $RN_i$ ($i \in 1,\ldots,n$) denotes the reduction name with possible initial values and the name $RV_i$ denotes reduction loop values. In that notation m-dimensional loop expression can be decomposed into the following two loop expressions of dimensions 1 and $m$–1:

**for** $D_1$ **repeat**
 $\overline{x_1}$ , . . . , $\overline{x_n}$:= **for** $D_2$ **repeat** $B$
  **returns** $RN_1$' **of** $RV_1$; . . .; $RN_n$' **of** $RV_n$
 **end for**
 **returns** $RN_1$" **of** $RV_1$; . . .; $RN_n$" **of** $RV_n$
**end for**

The name $\overline{x_i}$ here and any other name with overline without a special note later denote any unique name. The names $RN'_i$ and $RN''_i$ depend on the name $RN_i$ as shown in Table 53.2.

If the range generator contains multidimensional indices "$n$ **in** $S$ **at** $j_1$, . . ." before the operator **cross**, then the multidimensional loop can be represented in the following way:

**for** $D_3$ $n$ **in** $S$ **at** $j_1$, $D_4$ **repeat** $B$
 **returns** $RN_1$ **of** $RV_1$; . . .; $RN_n$ **of** $RV_n$
**end for**

**Table 53.2** Decomposition rules for multi-dimensional reductions, that show how to determine the names $RN'_i$ and $RN''_i$, which are used in this section before, from the name $RN_i$

| Value of the $RN_i$ name | $RN'_i$ | $RN''_i$ |
|---|---|---|
| Equals to value, product, least, greatest, catenate, "catenate(. . .)" or user-defined reduction. | $RN_i$ | value |
| Equals to "array[$k$]($i_1$,. . ., $i_k$)", where<br>– part "[$k$]" is optional and equals to "[m]" by default;<br>– last indices of the part "($i_1$,. . ., $i_k$)" are optional like this whole part and equal to 1 if omitted. | $k$>1<br>array[$k$–1]($i_2$, . . ., $i_k$)<br>$k$=1<br>array[1]($i_2$, . . ., $i_k$) | array($i_1$)<br>catenate($i_1$) |
| Equals to "stream[$k$]" where part "[$k$]" is optional and equals to "[$m$]" by default. | $k$>1 stream [$k$–1]<br>$k$=1 stream [1] | stream<br>catenate |

The name $D_3$ denotes the range generator part without the operator cross and multidimensional indices of the construction **at**, the name $S$ denotes the array or stream source of multidimensional indices, the name $D_4$ denotes the remaining part of the range generator. In that notation m-dimensional loop expression can also be decomposed into the following two loop expressions of dimensions 1 and $m$-1 (where the names $RN'_i$ and $RN''_i$ depend on the name $RN_i$ as shown in Table 53.2):

**for** $D_3$ $n$ **in** $S$ **at** $j_1$,**repeat**
  $\overline{x_1}$ ,. . ., $\overline{x_n}$:= **for** $n$ **in** $\overline{n_1}$ **at** $D_4$ **repeat** $B$
    **returns** $RN_1$' **of** $RV_1$; . . .; $RN_n$' **of** $RV_n$
   **end for**
  **returns** $RN_1$" **of** $\overline{x_1}$; . . .; $RN_n$" **of** $\overline{x_n}$
**end for**

## 53.6 Decomposition of the Array Element Selection

Let us represent the element selection expression from the array $A$ as "$A$ [*selection construction*]". If a selection construction does not have the cross (or comma) operator, then it can be represented as "$D_1$ **dot** $D_2$ **dot** . . . **dot** $D_m$", where $m \geq 1$ and all expressions $D_1$, $D_2$, . . ., $D_m$ are ranges (as required by the operator dot semantics). If $m = 1$ and the part $D_1$ is a singlet, then the array element selection operation can be represented directly in the IR1 and does not require further decomposition, otherwise the array element selection operation can be decomposed into the following one-dimensional loop:

**for** $x_1$ **in** $D_1$ **dot** $x_2$ **in** $D_2$ **dot** . . . $x_m$ **in** $D_m$
  $\overline{A_1}$:= $A$ [$x_1$, $x_2$ , . . ., $x_m$]
  **returns array of** $\overline{A_1}$
**end for**

The name $x_i$ (here and later) denotes any unique name, if the part $D_i$ does not have the form "*name* $N$ **in** $D_i$", and denotes the name $N$ otherwise. If the selection construction contains the operator **cross** (or comma), then it can be represented as "$S_1$, $S_2$, . . ., $S_m$ **cross** $C_1$" or "$D_1$ **dot** $D_2$ **dot** . . . **dot** $D_m$ **cross** $C_2$", where $S_1$, $S_2$, . . ., $S_m$ denote singlets and the names $C_1$ and $C_2$ denote the remaining parts of the selection construction and additionally the part $C_1$ does not begin with a singlet.

The array element selection operation beginning with a singlet can be decomposed into the following let expression (further decompositions need to be applied recursively):

**let** $\overline{A_1}$ := $A$ [$S_1$, $S_2$, . . ., $S_m$] **in** $\overline{A_1}$[ $C_1$] **end let**

The array element selection operation beginning with a range can be decomposed into the following one-dimensional loop (further decompositions need to be applied recursively):

**for** $x_1$ **in** $D_1$ **dot** $x_2$ **in** $D_2$ **dot** ... $x_m$ **in** $D_m$ **repeat**
    $\overline{A_1} := A\,[x_1, x_2, \ldots, x_m]$
    **returns array of** $\overline{A_1}[C_2]$
**end for**

The presented decomposition of the array element selection operation also explains the additional restriction, which is missed in Sisal 90 user manual, for the selection construction triplets with omitted parts: they must be placed as the first operand of the selection construction or just after the **cross** operator. In the range $D_1$ the first and second omitted triplet parts are explicitly represented via "liml($A$)" and "limh($A$)", correspondingly. In the ranges $D_2, \ldots, D_m$ the triplet parts cannot be omitted because there is no corresponding univocal array dimension available whose lower and upper bounds can be taken. In summary, any array element selection operation was decomposed into the array element selection with simple indices.

## 53.7 Array Element Replacement with a Singlet List Selection Construction

This section continues to use the notation of selection construction introduced before. The array element replacement expression in a general form looks like "*A*[*selection construction*:= *replacement construction*]".

As is shown further, any array element replacement expression can be decomposed into the array one-element replacement with simple index.

If the selection construction is a singlet list $S_1, S_2, \ldots, S_n$, then the replacement construction is allowed to be the expression $E_1, \ldots, E_t$ and the array element replacement operation is elementary represented as a composition of the following one-element replacement operations:

$$A[S_1, S_2, \ldots, S_n := E_1][S_1, S_2, \ldots, (S_n)+1 := E_2] \ldots [S_1, S_2, \ldots, (S_n)+(t\text{-}1) := E_t]$$

## 53.8 Array Element Replacement With a Scalar Replacement Construction

Let us consider the case then the selection construction is not a singlet list and the replacement construction is an expression of type of the n-th dimension of the array A, where n is the number of the selection construction ranges and singlets. In this case, the array element replacement operation can be decomposed into nested one-dimensional loops, obtained after the recursive application of the decompositions given below.

If the selection construction does not have the cross (or comma) operator, the array element replacement operation can be decomposed into the following one-dimensional loop:

**for** $x_1$ **in** $D_1$ **dot** $x_2$ **in** $D_2$ **dot** … $x_m$ **in** $D_m$
  $A$:= **old** $A$ [$x_1$, $x_2$ , …, $x_m$ := *replacement construction*]
  **returns value of** $A$
**end for**

The array element replacement operation beginning with a singlet can be decomposed into the following **let** expression (further decompositions need to be applied recursively):

**let** $\overline{A_1}$:= $A$ [$S_1$, $S_2$, …, $S_m$]
 **in** $\overline{A_1}$[$C_1$ := *replacement construction*]
**end let**

The array element replacement operation beginning with a range can be decomposed into the following one-dimensional loop (further decompositions need to be applied recursively):

**let** $\overline{A_1}$:= $A$
  **in for** $x_1$ **in** $D_1$ **dot** $x_2$ **in** $D_2$ **dot** … $x_m$ **in** $D_m$
   $\overline{A_2}$ := **old** $\overline{A_1}$[$x_1$, $x_2$ , …, $x_m$ ]
   $\overline{A_3}$ := $\overline{A_2}$[ $C_2$ := *replacement construction* ]
   $\overline{A_1}$ := **old** $\overline{A_1}$[$x_1$, $x_2$ , …, $x_m$ := $\overline{A_3}$]
   **returns value of** $\overline{A_1}$
  **end for**
**end let**

## 53.9 Array Element Replacement with Array Replacement Construction

Let us consider the case then the selection construction is not a singlet list and the replacement construction is an expression of type of a $k$-dimensional array of elements that have the type of the $n$-th dimension of the array A. In the considered case, k should be a sum of ranges in the selection construction minus the number of its dot operators. In this case, the array element replacement operation can be also decomposed into nested one-dimensional loops, obtained after the recursive application of decompositions given below.

If the selection construction does not have the cross (or comma) operator, the array element replacement operation can be decomposed into the following one-dimensional loop:

**let** $\bar{i} := 1$
  **in for** $x_1$ **in** $D_1$ **dot** $x_2$ **in** $D_2$ **dot** ... $x_m$ **in** $D_m$
   $A :=$ **old** $A$ [
     $x_1, x_2, \ldots, x_m :=$
     (*replacement construction*)$[\bar{i}]$
     ];
   $\bar{i} :=$ **old** $\bar{i} + 1$
   **returns value of** $A$
  **end for**
**end let**

The array element replacement operation beginning with a singlet can be decomposed into the following let expression (further decompositions need to be applied recursively):

**let** $\overline{A_1} := A[S_1, S_2, \ldots, S_m]$
  **in** $\overline{A_1}[C_1 :=$ *replacement construction*]
**end let**

The array element replacement operation beginning with a range can be decomposed into the following one-dimensional loop (further decompositions need to be applied recursively):

**let** $\overline{A_1} := A; \bar{i} := 1$
  **in for** $x_1$ **in** $D_1$ **dot** $x_2$ **in** $D_2$ **dot** ... $x_m$ **in** $D_m$
   $\overline{A_2} :=$ **old** $\overline{A_1}[x_1, x_2, \ldots, x_m]$
   $\overline{A_3} := \overline{A_2}[C_2 :=$ (*replacement construction*)$][\bar{i}]]$;
   $\overline{A_1} :=$ **old** $\overline{A_1}[x_1, x_2, \ldots, x_m := \overline{A_3}]$
   $\bar{i} :=$ **old** $\bar{i} + 1$
   **returns value of** $\overline{A_1}$
  **end for**
**end let**

## 53.10 Decomposition of the Where Expression

The Sisal 3.2 where expression is decomposed into nested one-dimensional loops in the following way, where $A$, $n$, $R$ and $I$ are some values

**for** $\overline{A_1}$ **in** $A$ **returns array of**
 **for** $\overline{A_2}$ **in** $\overline{A_1}$ **returns array of**
  ...
  **for** $I$ **in** $\overline{A_{n-1}}$ **returns array of**
    *expression* $R$
  **end for**
  ...
 **end for**
**end for**

## 53.11 Decomposition of the Vector Operations

All vector operations are decomposed into one-dimensional loops. An operation on multidimensional vectors is decomposed into a vector operation on vectors of lower dimensions.

Prefix and postfix operations on arrays *op*(*A*) decompose into:

**for** $\bar{i}$ **in** *A* **returns array** (liml(*A*)) **of** *op*($\bar{i}$) **end for**

Prefix and postfix operations on streams *op*(*S*) decompose into:

**for** $\bar{i}$ **in** *S* **returns stream of** *op*($\bar{i}$) **end for**

An infix operation op on two arrays $A_1$ and $A_2$ decomposes into:

**for** $\bar{i}_1$ **in** $\overline{A_1}$ **dot** $\bar{i}_2$ **in** $\overline{A_2}$
  **returns array of** $\bar{i}_1$ *op* $\bar{i}_2$
**end for**

An infix operation op on array *A* and stream *S* decomposes into:

**for** $\bar{i}_a$ **in** *A* **dot** $\bar{i}_s$ **in** *S*
  **returns stream of** $\bar{i}_a$ *op* $\bar{i}_s$
**end for**

An infix operation *op* on array *A* and scalar value *V* decomposes into:

**for** $\bar{i}$ **in** *A*
  **returns array** (*liml*(*A*)) **of** $\bar{i}$ *op* *V*
**end for**

An infix operation op on stream *S* and scalar value *V* decomposes into:

**for** $\bar{i}$ **in** *S*
  **returns stream of** $\bar{i}$ **op V**
**end for**

## 53.12 Conclusion

In the paper the intermediate language IR1 of the functional programming system SFP for supporting supercomputing has been briefly presented. During translation from Sisal 3.2 to the internal representation IR1 some complex Sisal 3.2 language structures need to be reduced to more unified objects of the IR1-language. These transformations have been shown in the terms of Sisal 3.2 by the decomposition of complex language structures into more simple ones that can be directly represented by IR1. The transformations are performed by the front-end compiler and help to better understand its translation strategy. They can be used also as a basis for formal description of semantics of the Sisal 3.2 language. For a general-purpose machine (without any special hardware support for the operations considered in this paper), the described transformations do not introduce unnecessary inefficiency and open additional optimization opportunities.

**Acknowledgments** The authors are thankful to all colleagues taking part in the SFP project. The work was partially supported by the Russian Foundation for Basic Research under grant N 07-07-12050.

## References

1. Backus J (1978) Can programming be liberated from the von Neumann style? Commun ACM 21:613–641
2. Cann D (1992) Retire Fortran? A debate rekindled. Commun ACM 35:81–89
3. Gaudiot J–L, DeBoni T, Feo J, Bohm W, Najjar W, Miller P (2001) The Sisal project: real world functional programming. Lect Notes Comput Sci 1808:45–72
4. Feo J, et al (1995) SISAL 90. In: Proc. High Performance Functional Computing, Denver, pp 35–47
5. Stasenko AP, Sinyakov AI (2006) Basic means of the Sisa 3.1 language (in Russian). Preprint N 132, A.P. Ershov Institute of Informatics Systems, Novosibirsk
6. Kasyanov VN, Stasenko AP, Gluhankov MP, Dortman PA, Pyjov KA, Sinyakov AI (2006) SFP – An interactive visual environment for supporting of functional programming and supercomputing. WSEAS Trans Comput 5:2063–2070
7. Kasyanov VN, Biryukova YuV, Evstigneev VA (2001) A functional language SISAL 3.0 (in Russian). In: Supercomputing support and Internet-oriented technologies, Novosibirsk, pp 54–67
8. Stasenko AP (2004) Internal representation of functional programming system SISAL 3.0 (in Russian). Preprint N 110, A.P. Ershov Institute of Informatics Systems, Novosibirsk
9. Skedzielewski SK, Glauert J (1985) IF1 – An intermediate form for applicative languages, version 1.0. Tech. Rep. M-170, Lawrence Livermore National Laboratory, Livermore, CA
10. Kasyanov VN, Lisitsyn IA (2000) Hierarchical graph models and visual processing. In: Proceedings of the International Conference on Software: Theory and Practice, 16th IFIP World Computer Congress, PHEI, Beijing, pp 179–182

# Chapter 54
# An Embedded Application for Driverless Metro Train

**Vivek Kumar Sehgal and Nitin**

**Abstract** We developed an embedded platform with 8-bit microcontroller 89C51 to implement driverless Metro Train. This sort of work is reported in India for the first time.

## 54.1 Introduction

Automated Software Controlled Train Simulation (ASCTS) is an example for implementation of a general-purpose rail vehicle application. The trains are equipped with a Central Processing Unit. A human administrator, if required can also access the controls. The train is programmed for a specific path. Each station is well-defined i.e. stoppage timing of the train and distance between stations is predefined. ASCTS is built over the Atmel processor 8051, an 8-bit microcontroller with 4 KB of on chip ROM, 128 Bytes of RAM, 2 timers and 32 I/0 pins. The motion of the train is controlled by a Stepper Motor and for display of messages intelligent Liquid Crystal Display is used.

Since we use software control, the benefit is that the hardware can be standardized i.e. hardware for a new model becomes a question of selecting units from a library of existing units [1].The design is under research and we propose to add more features to it also. The simulator would have more user-defined variables. Reports would be generated for multiple trains on a single track [2]. This sort of real-time implementation would pave a way to the success of the country by reducing time complexity as well as increasing reliability.

The rest of the paper is organized as follows: Section 54.2 describes the working of designed application and hardware used in it with a schematic diagram. Section 54.3 describes the hardware implementation for this embedded application. Section 54.4 describes the assembly language code and software for implementation and Section 54.5 describes challenges followed by conclusion and future scope.

V.K. Sehgal (✉)
Member IEEE and ACM, Department of ECE, Jaypee University of Information Technology, Waknaghat, Solan–173215, Himachal Pradesh, India
e-mail: vivekseh@gmail.com

N. Mastorakis et al. (eds.), *Proceedings of the European Computing Conference*, Lecture Notes in Electrical Engineering 28,
DOI 10.1007/978-0-387-85437-3_54, 

## 54.2 Experimental Set Up of Train Control System

The train system is functioned as according to the flow diagram shown in Fig. 54.1 and the embedded platform for this automated application is designed as according to the circuit shown in Fig. 54.2. Here we are taking examples of two stations, Shimla (2,800 m from sea level) and Kalka, two hill stations in

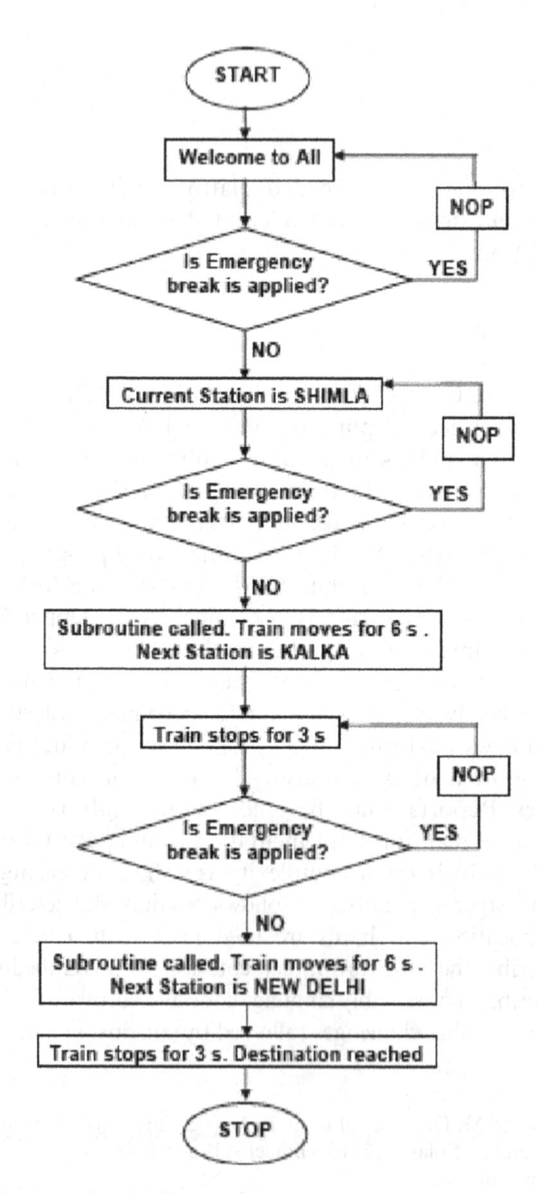


**Fig. 54.1** Flow diagram of embedded application

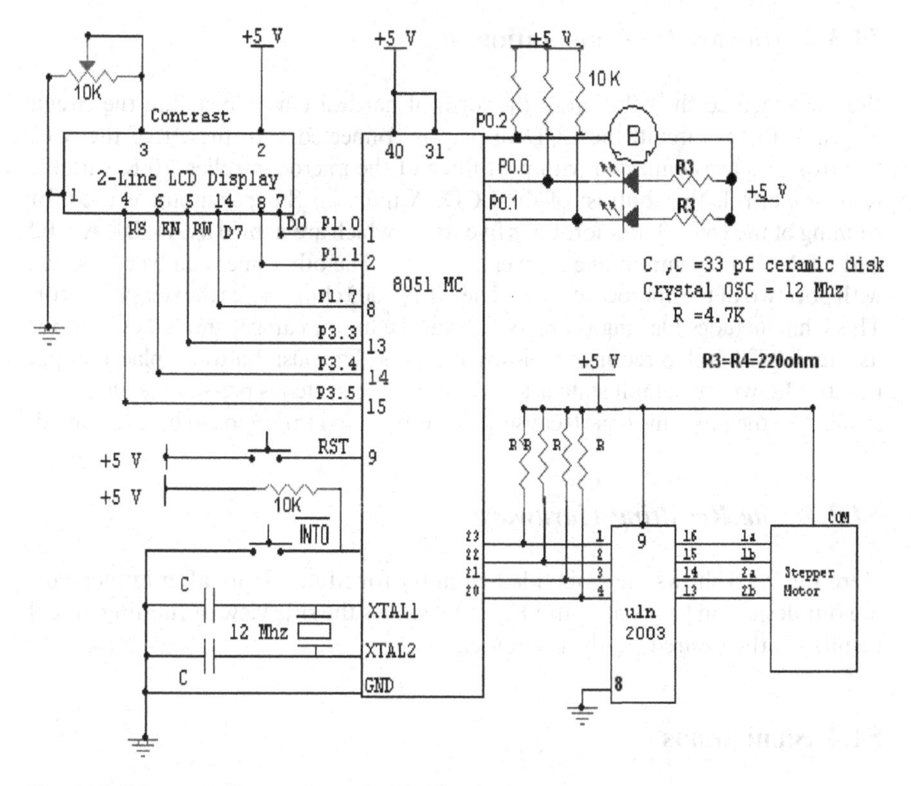


Fig. 54 2 Schematic diagram for embedded hardware

Himachal Pradesh in India. Initially the train starts at a station "Shimla", then blows a buzzer and starts moving for a predefined time of 6 s to next station "Kalka". Here the train stops for a predefined interval of 3 s and if emergency brakes are applied then it stays at the station till the brakes are released. Again, it starts moving to the last programmed station "New Delhi", halts for a predefined 3 s and now starts the upward journey or the round trip. An LED displays the direction of the movement. The cycle repeats until the train is stopped finally. Everything is depicted in the flowchart.

## 54.2.1 Hardware

The ASCTS hardware consists of

1. Atmel's AT89C51
2. 2×16 Liquid Crystal Display
3. ULN2003
4. A step down transformer: 220:9 for power supply
5. A Buzzer

## 54.3 Hardware Implementation

We implemented the solution in the form of hardware as depicted in the circuit diagram. In the circuit, the LCD display is connected with the P1 of the MC. Control lines are connected with port three of the microcontroller. 10 K variable resistor controls the contrast of the LCD. A unipolar Stepper motor is used for running of the train. This motor has five wires, which are named as A1, B1, A2, B2 and COM. The common line is given at +5 V. The other lines can be connected with port two of the microcontroller. The ULN 2003 chip derives the stepper motor. This Chip includes Darlington pairs, so that the motor can get enough current for its running. This chip requires pull-ups at inputs. The push button is placed at pin number 12, whose default state is logic 1 and when switch is pressed then logic 0 is applied on the pin. This logic 0 causes 8051 Embedded Processor to be interrupted.

### *54.3.1 The Resulting Hardware*

Here Fig. 54.3 shows the embedded circuitry for Metro Train after implementing our design on the board, and Fig. 54.4 shows the side view of running model results i.e. the desired hardware model

## 54.4 Simulations

The system programming is done using assembly language. The program handles things as:

1. Displaying interactive messages on the LCD.
2. Giving the right amount of rotations to the stepper motor at regular intervals.
3. Stopping the train for the required time interval.

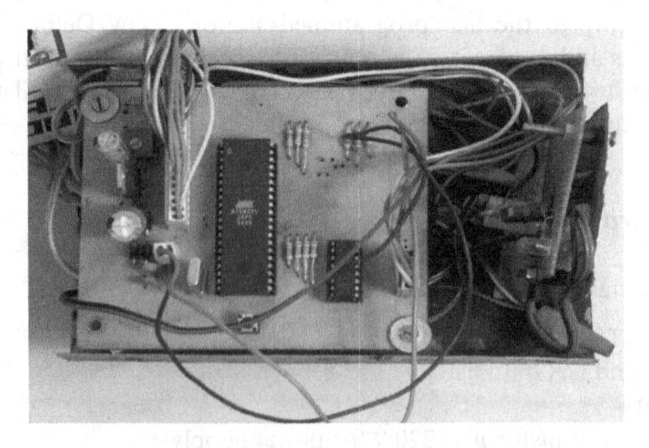

**Fig. 54.3** Embedded circuitry for MetroTrain

**Fig. 54.4** Side view of running model

4. Stopping the train as soon as the passenger activates the emergency braking system.
5. Controlling the forward as well as the backward motion of the train.
6. Calling the subroutines as is required.

### *54.4.1 Assembly Code (written on UMPS Version 2.1)*

The coding of the metro train prototype is given in the assembly language. The program main routines are the routines for running of the stepper motor in a forward direction as well as in a reverse direction. The routines for this purpose are *stepperf* and *stepperb*. The code also explains the various subroutines in detail that are called for correct display of interactive messages on the LCD as well as controlling the overall movement of the train on the desired path.

```
data equ  p1
busy equ p1.7
rs equ   p3.5
rw equ p3.4
en equ p3.3
bzr equ p0.2
ledf equ p0.0
ledb equ p0.1

org 400h
ajmp main
org 0003h
test: mov c,p3.2
jnc halt
setb bzr
reti
halt:
clr bzr ;v
on the  bzr
ajmp test

main:
mov ie,#00h
acall read
clr ledf      ;p1.0
acall delay

mov a,#01h
acall command; Now make
memory clear cursor home
mov dptr,#show1
acall read
setb.ex0   #####
mov a,#0c0h
acall command
mov dptr, #show3
acall read
acall.delay;;Stopage1
time 3 sec Shimla
loopa1:
    acall delay
    clr bzr
    acall delay

    mov a,#01h
    acall command
acall read
setb bzr
acall delay10
acall stepperf

mov a,#01h
acall command
mov dptr,#show1
acall read
mov a,#0c0h
acall command
mov dptr,#show4
acall read
acall delay ;Stopage2
time 3 sec Kalka
acall delay
clr bzr
acall delay

mov a,#01h
acall command
mov dptr,#show2
acall read
acall stepperf

mov a,#01h
acall command
mov dptr,#show1
acall read
mov a,#0c0h
acall command
mov dptr,#show5
acall read
acall delay
;Stopage3  time 3 sec
New Delhi
acall delay
clr bzr
acall delay

setb ledf   ; p1.0;off
led at  p1.0 for
forward journey
clr ledb  ;  p1.1 ; 0n
Led for back ward
journey
```

```
setb ea
setb ex0
here:
mov p2,#00h
acall ini
mov dptr,#show0
acall command
mov dptr,#show4
acall read
setb bzr
acall delay10
acall stepperb
mov a,#01h
acall command
mov dptr,#show1
acall read
mov a,#0c0h
acall command
mov dptr,#show4
acall read
acall delay ;Stopage2 time 3
sec Kalka
acall delay
clr bzr
acall delay
mov a,#01h
acall command
movdptr,#show2;display
ne Shimla
acall read
mov a,#0c0h
acall command
mov dptr,#show3
acall read
setb bzr
acall delay10
acall stepperb
mov a,#01h
acall command
mov dptr,#show1
acall read
mov a,#0c0h
acall command
mov dptr,#show3
acall read
setb ledb ;p1.1
ljmp here
;routine for stepper motor
;Delay Routine
delay:
push acc
push 00h
mov dptr,#show2
acall read
mov a,#0c0h
acall command
mov dptr,#show4
push 01h
push p0
push p1
mov r0,#0eh
loopr:
mov a,#0ffh
loopb:
mov b,#0ffh
loopa: djnz b,loopa
djnz 0e0h,loopb
djnz r0,loopr
pop p1
pop p0
pop 01h
pop 00h
pop acc
ret
;Delay stepper
delays:
push acc
push 00h
push 01h
push p0
push p1
mov a,#0ffh****
loopa1:
mov b,#0fh
loopb1:
djnz b,loopb1
djnz 0e0h,loopa1
pop p1
pop p0
pop 01h
pop 00h
pop acc
ret
delay10:
mov tmod,#01h
mov tcon,#00h
mov tl0,#0f0h
mov th0,#0f8h
setb tr0
no: jnb tf0,no
clr tr0
clr tf0
ret
mov a,#0c0h
acall command
mov dptr,#show5
acall read
setb bzr
acall delay10
Routine to read data from
prog mem
read:
nex: clr a
movc a,@a+dptr
cjne a,#'0',aga
sjmp down
aga: acall display
sjmp nex
down:
ret
stepper routine
stepperf:
push acc
push p1
mov a,#88h
mov r1,#04h
loop1:
mov r0,#0e0h
loop: mov p2,a
acall delays
rr a
djnz r0,loop
djnz r1,loop1
pop p1
pop acc
ret
stepperb:
push acc
push p1
mov a,#88h
mov r1,#04h
loop12:
mov r0,#0e0h
loop0: mov p2,a
acall delays
;LCD strobe subroutines
ini:
mov a,#38h
acall command
mov a,#38h
acall command
mov a,#38h
acall command
mov a,#38h
acall command
mov a,#01h
acall command
mov dptr,#show2
;display ne Kalka
acall read
mov a,#0c0h
mov a,#0eh
acall command
mov a,#06h
acall command
mov a,#01h
acall command
mov a,#80h
acall command
ret
command:
acall ready
mov data, a
clr rs
clr rw
setb en
clr en
ret
display:
acall ready
mov data,a
setb rs
clr rw
setb en
clr en
ret
ready:
clr en
mov data,#0ffh
clr rs
setb rw
wait:
clr en
setb en
jb busy,wait
clr en
ret
end
```

## 54.5 Challenges

The domain of railways represents a huge variety of possibilities for applications of software. The realm of automatic control interacts strongly with many facets of statistics gathering, planning, train scheduling, resource allocation etc. [3]. The field of computing science provides design tools, techniques, principles and methods that allow classical automatic control engineering designs to link up to more modern operations analysis, graph theoretic and combinatorial resource optimization [4].

## 54.6 Conclusion and Future Scope of this Model

We developed an embedded platform with an 8-bit microcontroller 89C51 to implement a driverless metro train. This sort of work is first of its kind in India. This sort of system could make the railways network more efficient and reliable.

Optimum throughput can be achieved through the mapping of tracks. Another feature of the system could be the implementation of vital remote controls on software. We need to define safe failures also in the control system. A safe failure occurs when the failure of a component in a processing system is detected and the effect of the failure emerges as a safe-side operation [4]. In the railway application, the safe-side failures are defined for the outputs of the controller. Failures in wayside control functions can be safely handled by shutting down the microprocessor-based controller. In addition, ad hoc methodologies to safety-critical design can be implemented in these systems to ensure safety and fault-tolerant operation.

**Acknowledgement** This model got the Third Prize at The Annual Techfest – TROIKA 2007, Sponsored by IEEE and IEE Student Chapters of Delhi College of Engineering at DCE, New Delhi in INDIA.

## References

1. Astrom P (1992) Control electronics on rail vehicles. In: Proceedings of the ASME/IEEE Spring Joint Railroad Conference, pp 107–116
2. Gordon SP, Leherer DG (1998) Coordinated train control and energy management control strategies. In: Proceedings of the 1998 ASME/IEEE Joint Railroad Conference, Philadelphia, PA, USA, pp 165–176
3. Raymond AJ (2000) Formal software techniques in Railway Systems. In: Proc. 9th IFAC Symposium on Control in Transportation Systems, Braunschweig, Germany, pp 1–12
4. Hachiga A (1993) The concepts and technologies of dependable and real time computer systems for Shinkansen train control. Dependable computing and fault tolerant systems, Springer Verlag, Berlin, pp 225–252
5. Mazidi MA, Mazidi JG (2006) 8051 microcontroller and embedded system. Prentice Hall

# Chapter 55
# A Post-mortem JavaSpaces Debugger

**Jason Howarth and Edward Stow**

**Abstract** We describe a post-mortem debugger called JavaSpaces Debugger (JSD) that is able to detect a class of properties that occur in a distributed JavaSpaces program. To detect a property using JSD, the user specifies a global predicate to be evaluated. This global predicate is divided into a series of local predicates that are evaluated by each system node. During execution, an event message is sent to the debugger whenever one of these local predicates is true. The debugger arranges event messages according to logical time, and reports whether the global predicate occurred. JSD can detect both weak and strong unstable predicates, using integer, boolean, or bytecode expressions.

## 55.1 Introduction

Distributed computing systems are characterized by the absence of a global time base, unpredictable message delays, and the dispersion of program state amongst nodes [10]. This makes it difficult to capture and analyse the state of such systems.

We describe JSD (JavaSpaces Debugger), a debugger able to overcome some of these limitations. JSD is used to detect a class of properties that occur in a distributed JavaSpaces program.

To detect a property using JSD, the user first specifies a global predicate to be evaluated. This global predicate is divided into a series of local predicates that are evaluated by each system node. During execution, an event message is sent to a global checker process whenever one of the local predicates is true. The checker process arranges these messages according to logical time, and reports whether the global predicate occurred.

J. Howarth (✉)
School of Computing and Mathematics, Charles Sturt University, Boorooma St.
Wagga Wagga, NSW, Australia, 2678

N. Mastorakis et al. (eds.), *Proceedings of the European Computing Conference*, Lecture Notes in Electrical Engineering 28,
DOI 10.1007/978-0-387-85437-3_55, 

JSD is used to detect global predicates classified as unstable. An unstable predicate is one that may alternate between true and false during execution. Unstable predicates may be either weak or strong [5, 6].

A strong unstable predicate, where true, defines a property that certainly occurred in an observed run of the monitored program. A weak unstable predicate, where true, specifies a property that may have occurred, but which cannot be proven. JSD can detect both strong and weak unstable predicates.

## 55.2 Background

The following sections contain a brief background on the technologies and algorithms used in our debugger.

### *55.2.1 JavaSpaces*

JavaSpaces is based on the shared-memory model of distributed computing [1]. Instead of cooperating directly, processes exchange Java objects through a network storage area, called a JavaSpace. This makes JavaS-paces different to the message-exchanging framework common to many distributed systems. JavaSpaces has been used for cluster computing [2], software agent systems [11] and distributed programming generally [3].

The main operations performed on a JavaSpace are *write, read, take*, and *notify*. The *write*method deposits an object in the JavaSpace. To *read* an entry, it is necessary to supply a template object to match against. If a match is found, a copy of the matched object is returned from the JavaS-pace. The *take* method acts the same as *read*, except the entry is deleted from the JavaSpace. The notify method allows a process to register interest in future entries that appear.

Although the JavaSpaces API (Application Programmer's Interface) is minimal, complex distributed algorithms can be constructed via the flow of objects in and out of JavaSpaces [4].

### *55.2.2 Jini*

JavaSpaces rests upon the Jini foundation, a technology used to create networked communities of software and hardware services. JavaSpaces is provided as a standard Jini service. Like all Jini services, JavaSpaces is used through its service proxy, which is stored at a Jini Lookup Service. The Lookup Service is a registry of all currently active services available within a Jini community.

To retrieve a service proxy, the lookup repository is searched for ser-vices matching certain criteria. For example, to retrieve the JavaSpaces proxy, a client might search the Lookup Service for a proxy with a name attribute of

'JavaSpaces'. Once the client has the proxy, it can make local method calls on this object to access the backend service.

JavaSpaces plays an important role in the overall Jini architecture by providing a network storage facility. This allows Jini clients to store objects centrally, as well as allowing clients to communicate via the flow of objects in and out of the JavaSpace.

### 55.2.3 Weak Unstable Predicates

We turn to an approach for detecting weak unstable predicates. Such predicates may be either conjunctive or disjunctive. A conjunctive predicate is of the form $s(i) \wedge s(j) \wedge s(k)$, where $s(i)$ represents the local state of process $i$, $s(j)$ the local state of process $j$, etc. This paper will focus only on conjunctive predicates. Disjunctive predicates are trivial to detect since only the local state must be monitored [5].

The algorithm [5] relies on each process checking its local portion of the predicate. For example, if the predicate is $x=1 \wedge y=2$ for two variables $x$ and $y$ held at processes $P_1$ and $P_2$ respectively, then process $P_1$ monitors $x$, and process $P_2$ monitors $y$. When a local predicate becomes true, the current vector clock timestamp is sent by the local process to a global checker process. Once computation has ended, the checker process evaluates the global predicate to see whether both local predicates could have occurred concurrently. The weak conjunctive predicate $a \wedge b$ is true if $a$ and $b$ occur concurrently in the program trace [5].

The implementation of this algorithm requires a special type of vector clock, called an lcmvector (last causal message vector) [5]. An lcmvector consists of a vector of $n$ integers, one for each process. Each process is responsible for maintaining its own vector. For a system with three processes, the lcmvector stored at process $Pi$ takes the form $[Ci, Cj, Ck]$, where $Ci$ is the clock that measures progress at $Pi$, and $Cj$ and $Ck$ are set to values advised by processes $Pj$ and $Pk$, respectively.

The operations permitted on an lcmvector are initialisation <init>, increment <incr>, and synchronization <synch>. Initially, the values in each vector are set to 0.

The <init> operation occurs at process startup, where each process $Pi$ sets its clock value to $i = 1$. Whenever a message is sent, the <incr> operation is performed. For a process $Pi$, this involves incrementing its clock by one ($Ci = Ci + 1$). When a message is received, the synchronize operation is invoked. This means that the receiving process performs the componentwise maximum operation, using as input its own clock and the clock of the process sending the message. To implement the algorithm, lcmvectors are passed between processes when a message is sent.

Operations on lcmvectors are depicted as occurring between states. Process events have the clock value that appears before the event. Hence in Fig. 55.1, events $a$ and $b$ both have the vector time stamp $\{1, 0\}$.

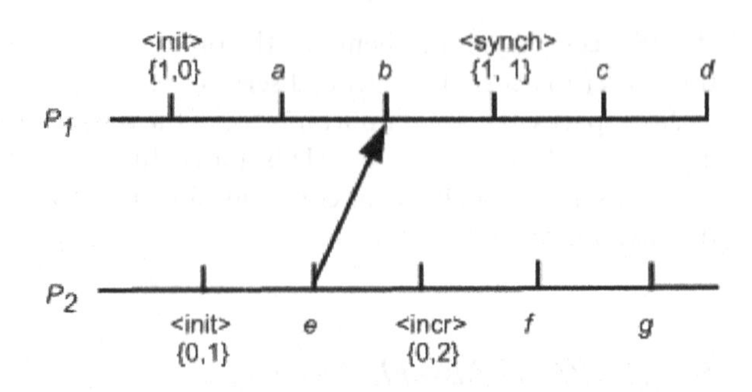


**Fig. 55.1** Events measured using lcmvectors

Vector clocks are compared using a type of vector clock algebra, based on the *happened before* [8] relation (⇒). Vector clocks are considered concurrent (⇔) if the happened before relation cannot be established.

Consider two clocks $V_1$ and $V_2$. Based on the values contained in these clocks, the following holds true:

$V_1 \Rightarrow V_2$ if $V_1 \mathrel{!=} V_2$ and if each value in $V_1$ is less than or equal to each value in $V_2$

$V_2 \Rightarrow V_1$ if $V_1 \mathrel{!=} V_2$ and if each value in $V_1$ is greater than or equal to each value in $V_2$

$V_1 \Leftrightarrow V_2$ if neither $V_1 \Rightarrow V_2$ nor $V_2 \Rightarrow V_1$

Applying these rules to the example in Fig. 55.1, we can assert that $e \Rightarrow c$, and that $c \Leftrightarrow f$.

We turn to the checker process to see how the global predicate is evaluated. The checker process enqueues clock values received as <debug> messages (Fig. 55.2) whenever a portion of the global predicate is detected locally. For example, if the global predicate is $x = 4 \wedge y = 3$, process $P_1$ sends a message to checker process $C$whenever its portion of the predicate is true (see Fig. 55.2); process $P_2$ does likewise.

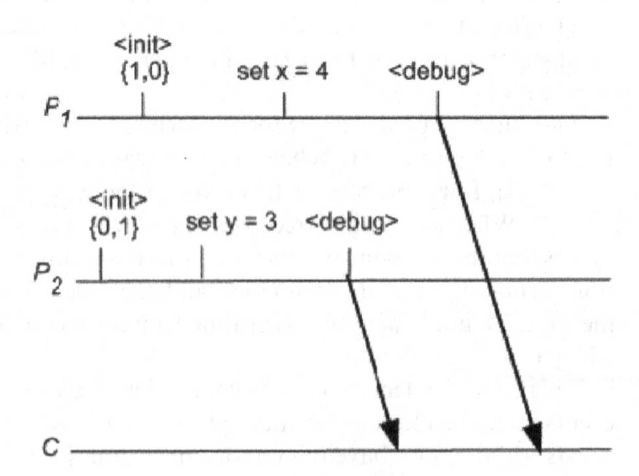


**Fig. 55.2** Time-stamped debug messages sent to the checker process

The messages sent to the checker process *C* contain the timestamps associated with detection of the local predicate. The checker process uses the vector clock comparison rules to decide if the predicate is true. In the case of a weak unstable predicate, the predicate $i \wedge j$ is true if $i \Leftrightarrow j$ [5].

### 55.2.4 Strong Unstable Predicates

Strong conjunctive predicates are of the form $s(i) \wedge s(j) \wedge s(k)$, where $s(i)$ represents the local state of process *i*, $s(j)$ the local state of process *j*, and so on. In detecting such predicates, we rely on the notion of *intervals* [6]. An interval is a pair of vector clocks that delineate a period during which the predicate was true. In the following discussion, *clock_lo* refers to the start, and *clock_hi* to the end of an interval.

Consider two processes with a boolean variable *ready_to_commit*. As-sume it must be determined if there is a point when these processes on Hosts *A* and *B* are simultaneously ready to commit. To achieve this, it must be determined if the intervals where the predicate is detected on each host overlap.

For example, assume host *A*is ready to commit during the interval {2, 1} to {4, 2}, and host *B* during the interval {0, 1} to {2, 2}. If these inter-vals overlap, then there must have been a time when the global predicate (i.e., when both hosts were simultaneously ready to commit) is also true.

To determine overlap, the rules for comparing vector clocks are applied to see if host *A.clock_lo* ⇒*host B.clock_hi AND host B.clock_lo* ⇒*host. clock_hi*. If so, these intervals have overlapped in (logical) time and the global predicate must have occurred [6]. In the example above, since {2, 1} ⇒{2, 2} and {0, 1} ⇒{4, 2}, these intervals overlap, and it is certain that the global predicate was true at some point in the computation.

## 55.3 JSD Architecture

A feature of JSD is the level of indirection added to the JavaSpaces architecture to capture debugging information. A typical JavaSpaces client interacts with the JavaSpace using the JavaSpaces proxy. In JSD, the debugged host is modified to load a different proxy, called the *debugging proxy*. The debugging proxy does not communicate directly with the JavaSpace; instead, it routes messages through another proxy-like object, *JSDebugSpace*.

*JSDebugSpace* is a custom-written Jini service used to filter messages intended for the real JavaSpace. Information is captured about these messages before they are passed to the JavaSpace for normal processing. Simi-larly, when a message is returned from the JavaSpace, it returns through *JSDebugSpace* before it is handed back to the debugged application. By this means, debugging control is interposed between the debugged pro-gram and the JavaSpace.

Both *JSDebugSpace*and the debugging proxy on each host maintain a lcmvector clock to order system events. These clocks are exchanged between the proxy and *JSDebugSpace*whenever a JavaSpaces call is made. By this means, clocks in the debugged system are kept current through the increment and synchronization methods described earlier.

To monitor the local predicate, the debugging proxy loads a DLL (dynamic link library) on the host being debugged. This DLL uses the JVMDI (Java Virtual Machine Debugger Interface) [7]. The DLL is passed details of the predicate to watch. Based on the contents of this predicate, it then registers for *watchpoint*events with the local JVM. When such an event occurs, the JVM notifies the DLL, which in turn notifies a checker process with the lcmvector reading at the local host. These readings are put in a queue maintained by the checker process. The contents of this queue are later used to reconstruct the event sequence that occurred during monitoring. Once the debugged program ends, the truth of the global predicate first entered by the user can be determined.

### 55.3.1 Predicate Detection Language

We have designed a simple language called JPL (JavaSpaces Predicate Language) to submit predicates to our debugger. JPL is used to express a predicate containing either integer or boolean values. A bytecode index can also be specified. An integer or boolean predicate takes the form *host-name: class name: variable operator result*[^]. The ^ symbol is the conjunctive operator, used to chain local predicates into a single global predicate.

To give an example, the following predicate seeks to determine if vari-ables x and y, at different hosts in the system, are ever equal to 10 at the same point in the computation: *Host A: Adder: x* = 10 ^ *Host B: Subtrac-tor: y* = 10. The local predicates in this example are *Host A: Adder: x* = 10 *and Host B: Subtractor: y* = 10. The variables *x* and *y* are declared in classes *Adder* and *Subtractor*, respectively.

JPL also allows predicates that contain more complex expressions. All of the basic mathematical operators are supported. In addition, a boolean predicate can use greater-than or less-than logic, such as *Adder:* $i > 5$ ^ *Subtractor:* $j < 3$.

The other type of predicate is the bytecode predicate, which has the structure *hostname: break: class name: method name: bytecode index* [^]. An example is Host *A: break: Adder: increment:* 44 ^ *Host B: break: Sub-tractor: decrement:* 23. This is asking: when class *Adder* hits bytecode in-dex 44 in the *increment* method, is it true that class *Subtractor* hits byte-code index 23 simultaneously in the *decrement* method? We believe bytecode predicates offer an advantage over integer and boolean predicates because a variable need not exist in the application being debugged to per-form predicate evaluation.

## 55.4 JSD Example

The example we describe is based on a simple chat program that consists of a Listener and two Talkers. To send a message to the Listener, a Talker must first hold an exclusive token *T*. This token is exchanged between each Talker via a JavaSpace. The Listener is notified through the JavaS-paces *notify* method whenever a message is written to space. Our concern here is with the token, which should not simultaneously be held by more than one Talker. The predicate for this is as follows: *Host A: Talker: has-Token = true* ∘ *Host B: Talker: hasToken = true*.

If this predicate is true, there is a problem with our algorithm because there is a chance that both hosts can talk at once. The setup is as follows. Hosts Sun and Moon each run a Talker process (Fig. 55.3). In this test, Sun.Talker was started first, followed by Moon.Talker. We assume that the token has already been written to the JavaSpace, and contains the name of the host who has first possession. This host takes the token, sends a message to the Listener (not shown), then releases the token again, addressed to the next recipient who performs the same set of actions.

After running the test, the following values appear in the local predicates screen of JSD:

*Sun*
Talker:hasToken = true {2,0,1}
*Moon*
Talker:hasToken = true {2,2,3}

This shows that each local predicate was true when the lcmvector readings were {2, 0, 1} and {2, 2, 3}, respectively. The three readings in each lcmvector indicate that three clocks were used in this debugging session, one for each host and one for the JavaSpace.

We are now in a position to determine whether the weak predicate was true during this run of the program. To do this, we must determine if it were possible for readings {2, 0, 1} and {2, 2, 3} to have occurred concurrently. When we compare these clocks, we find that {2, 0, 1} ⇒{ 2, 2, 3}, which means that the weak conjunctive predicate is false, at least under the scenario where host Sun is started first.

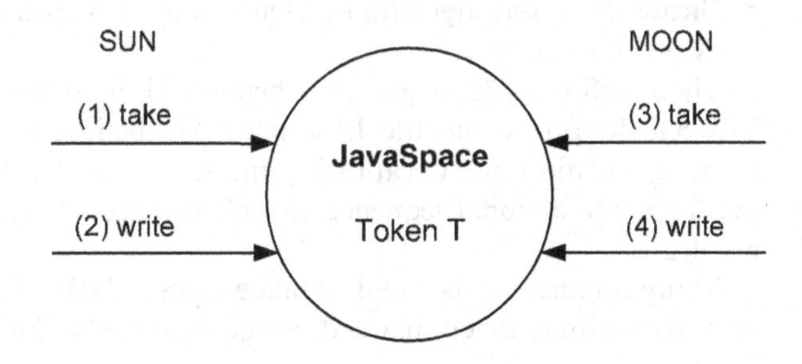


**Fig. 55.3** Hosts Sun and Moon take and write the token

In this example, the source code on both hosts contains the boolean variable *hasToken = true* for use in detecting the predicate. If this variable did not exist the predicate could not be assessed. Unfortunately, many distributed debuggers [3, 9] assume such a variable. JSD overcomes this limitation by allowing the user to specify a bytecode index instead of a variable when constructing the predicate.

We next apply a strong conjunctive predicate to the same scenario – this time to see if it was definitely the case that, at some point in the computation, Sun had the token and Moon did not. The predicate to express this is *Sun: Talker: hasToken = true ^ Moon: Talker: hasToken = false*. Only one round of token passing is used, and Moon is started before Sun. Once the test is run, JSD reports the following intervals:

*Sun*
Talker: hasToken = true [{2,1,1} {3,1,2}]
*Moon*
Talker:hasToken = false [{0,1,0} {2,2,3}]

JSD reports that the strong conjunctive predicate is true because intervals {[2, 1, 1] [3, 1, 2,]} and {[0, 1, 0] [2, 2, 3]} overlap. In general, how-ever, due to the loosely-coupled nature of JavaSpaces, a strong conjunctive predicate will never be true unless the processes named in the predicate are both active in the same JavaSpace at the same (physical) time. A time overlap only occurs when both processes are engaged in a common task – that is, when both simultaneously interact with the JavaSpace. This makes strong conjunctive predicates much less useful than weak conjunctive predicates in a JavaSpaces environment.

## 55.5 Conclusion

We have described JSD, a debugger able to detect both weak and strong unstable predicates in a JavaSpaces program, using the algorithms of [5, 6]. *MaX*[9], a distributed debugger framework for a Java/CORBA environment, also uses these algorithms. There are several important differences between *MaX*and our debugger, including the ability of JSD to detect bytecode predicates. Our design also offers fewer overheads than *MaX*because only one process per machine is needed to run both the original program and the predicate detection algorithms. Hence, our debugger is in process, whereas *MaX* is out-of-process.

The JavaSpaces debugger described in [3] is used to capture message activity between the proxy and the JavaSpace. We believe our approach is superior because, in addition to capturing message activity, JSD can arrange these messages into a global sequence, as well as detect strong and weak conjunctive predicates.

Many enhancements could be made to JSD. Debugging data could be saved in XML format to compare debugging sessions. This would offer a more complete picture than is currently provided. Any predicate judged against this additional data would have more validity than one judged against a single program run. The architecture might also be adapted to an RMI or CORBA environment so that debugging on each host is directed by stub objects, rather than by a Jini proxy.

**Acknowledgments** The authors gratefully acknowledge the valuable comments provided on an early draft of this paper by Dr. Barney Dalgarno of Charles Sturt University, Wagga Wagga.

## References

1. Attiya H (1998) Distributed computing theory. In: Zomaya A (ed) Parallel and distributed computing handbook. McGraw-Hill, New York
2. Batheja J, Parashar M (2001) Adaptive cluster computing using JavaSpaces. Research Report, Dept. of Electrical and Computer Engineering, Rutgers University, New Jersey
3. Bishop P, Warren N (2003) Javaspaces in practice. Addison-Wesley, London
4. Freeman E, Hupfer S, Arnold K (1999) JavaSpaces: Principles, patterns and practice. Addison-Wesley, Reading, Massachusetts
5. Garg V, Waldecker B (1994) Detection of weak unstable predicates in distributed programs. IEEE Trans Parallel Distrib Systems 5(3):299–307
6. Garg V, Waldecker B (1996) Detection of strong unstable predicates in distributed programs. IEEE Transactions on Parallel and Distributed Syst 7(12):1323–1333
7. Java Platform Debugger Architecture, Sun Microsystems Inc (2002) last viewed May 10, 2007 <http://java.sun.com/j2se/1.4.2/docs/guide/jpda/>
8. Lamport L (1978) Time, clocks and the ordering of events in a distributed sys-tem. Commun ACM 21(7):558–65
9. Otta M (2001) A distributed debugger framework applicable in a Java/CORBA environment. European Research Seminar on Advances in Distributed Systems, last viewed May 1, 2007 http://www.cs.unibo.it/ersads/program.html
10. Stoller S (1997) Detecting global predicates in distributed systems with clocks. In: Proceedings of the 11th International Workshop on Distributed Algorithms (WDAG97), Springer-Verlag, London, pp 185–199
11. Wang A (2002) Using JavaSpaces to implement a mobile multi-agent system. Research Report, Norwegian University of Science and Technology, Trondheim

# Chapter 56
# Stochastic Numerical Simulations Using C#

**Edita Kolářová and Csaba Török**

**Abstract** This paper considers numerical solutions of stochastic differential equations. The stochastic Euler scheme and the stochastic Milstein scheme are examined. We present two different approaches to the implementation of these schemes in the MS .NET Framework. The first one is based on matrices and the second on delegates. The schemes and the codes are illustrated with an example.

## 56.1 Introduction

The effects of intrinsic noise within physical phenomena are ignored when using deterministic differential equations for their modelling. Stochastic differential equations (SDEs) include a random term which describes the randomness of the system as well. A general scalar SDE has the form

$$dX_t = F(t, X_t)dt + G(t, X_t)dW_t,$$

where $F: \langle 0, T\rangle \times R \to R$ is the drift coefficient and $G: \langle 0, T\rangle \times R \to R$ is the diffusion coefficient. $W_t$ is the so called Wiener process (see [2]), a stochastic process representing the noise. We can represent the SDE in the integral form

$$X_t = X_0 + \int_{t_0}^{t} F(s, X_s)ds + \int_{t_0}^{t} G(s, X_s)\, dW_s,$$

where the first integral is an ordinary Riemann integral. Since the sample paths of a Wiener process do not have bounded variation on any time interval, the second integral cannot be a Riemann-Stieljtes integral. K. Itô proposed a way to overcome this difficulty with the definition of a new type of integral, a stochastic integral which is now called the Itô integral (see [2]). Later

E. Kolářová (✉)
Brno University of Technology, Technická 8, 616 00 Brno, Czech Republic

N. Mastorakis et al. (eds.), *Proceedings of the European Computing Conference*, Lecture Notes in Electrical Engineering 28,
DOI 10.1007/978-0-387-85437-3_56, 

R. L. Stratonovich proposed another kind of stochastic integral, called the Stratonovich integral.

## 56.2 Numerical Solutions of SDEs

There are two main approaches to numerical solutions. The first one is based on numerical methods for ordinary differential equations (see [5]). The second one uses more information about the Wiener process (see [1]). We give here the formulae for the simplest examples of these two approaches, the Euler and the Milstein methods.

A general N-dimensional Itô SDE with an M-dimensional Wiener process $\mathbf{W}_t = (W_t^1, \ldots, W_t^M)$ can be written componentwise as

$$dX_t^i = F^i(t, \mathbf{X}_t)dt + \sum_{j=1}^{M} G^{i,j}(t, \mathbf{X}_t)dW_t^j,$$

$i = 1, \ldots, N$, where $\mathbf{X}_t = (X_t^1, \ldots, X_t^N)$. Let the stochastic process $\mathbf{X}_t$, $t_0 \leq t \leq T$ be the solution of the multidimensional SDE on $t_0 \leq t \leq T$ with the initial value $\mathbf{X}_{t_0} = \mathbf{X}_0$. Let us consider an equidistant discretization of the time interval $t_n = t_0 + nh$, where $h = \frac{T-t_0}{n} = t_{n+1} - t_n$ and the corresponding discretization of Wiener processes $\Delta W_n^j = W_{t_{n+1}}^j - W_{t_n}^j = \int_{t_n}^{t_{n+1}} dW_s^j, j = 1, \ldots, M$. To be able to apply any stochastic numerical scheme, first we have to generate the random increments of the Wiener processes as independent Gauss random variables with $E[\Delta W_n^j] = 0$ and $E[(\Delta W_n^j)^2] = h$.

**The Euler scheme** for the i-th component of the multidimensional SDE has the form

$$X_{n+1}^i = X_n^i + F^i(t_n, \mathbf{X}_n)h + \sum_{j=1}^{M} G^{i,j}(t_n, \mathbf{X}_n)\Delta W_n^j.$$

**The Milstein scheme** for the i-th component of the multidimensional SDE has the form

$$X_{n+1}^i = X_n^i + F^i h + \sum_{j=1}^{M} G^{i,j}\Delta W_n^j + \sum_{j_1,j_2=1}^{M} L^{j_1} G^{i,j_2} \int_{t_n}^{t_{n+1}} \int_{t_n}^{t} dW_s^{j_1} dW_t^{j_2}$$

for $i = 1, \ldots, N$, where the operators $L^j$ are defined by $L^j = \sum_{i=1}^{N} G^{i,j} \dfrac{\partial}{\partial x_i}$.

The Euler scheme converges with strong order $\gamma = \frac{1}{2}$ and the stochastic Milstein scheme converges with strong order $\gamma = 1$ (see [1]).

## 56.3 Numerical Experiments Using C#

In this section we present two different approaches to implementation of the Euler and Milstein methods for numerical solution of SDEs in the MS .NET Framework. The first one is based on matrices and the second one on delegates. To make matrix manipulation and visualization simpler we used the component library LinAlg described in [3]. LinAlg is a set of classes that enables vectorial programming and incorporates a wide range of numerical, statistical and graphical methods.

**Matrix arguments.** The following codelines present the implementation in MS Visual C# of the Euler and Milstein methods described in Section 56.2 for the Itô SDE of form $d\mathbf{X}_t = \mathbf{F}\mathbf{X}_t\,dt + \mathbf{G}\mathbf{X}_t\,dW_t$, where $X_t$ is an $r$ dimensional vector, $F$ and $G$ are $r \times r$ matrices and $W_t$ is a one dimensional Wiener process:

```
static public MatrixD Euler (VectorD X0, MatrixD F,
MatrixD G, double h, int n, VectorD w)
{   MatrixD xM = new MatrixD (X0.Length, n);
    VectorD xMj = new VectorD (X0);
    xM.ColumnsSet (0, xMj); double dw;
    for (int j=1; j  <n; j++)
    {       dw = w[j] - w[j-1];
            xMj += F* xMj* h + G* xMj* dw;
            xM.ColumnsSet (j, xMj);
    }   return xM;
}
static public MatrixD Milstein (VectorD X0, MatrixD F,
MatrixD G, MatrixD DG, double h, int n, VectorD w)
{   MatrixD xM = new MatrixD (X0.Length, n);
    VectorD xMj = new VectorD (X0);
    xM.ColumnsSet (0, xMj);
    double dw;
    for (int j=1; j<n; j++)
    {    dw = w [j] - w[j-1];
    xMj += F* xMj* h + G* xMj* dw + 0.5* DG* G* xMj* (dw* dw-h);
        xM.ColumnsSet(j, xMj);
    }   return xM;
}
```

where the approximant **xM** of $\mathbf{X}_t$ is a matrix and its $k$-th column contains the approximation of $\mathbf{X}_t$ in $t_k$.

**Delegate arguments.** The C# methods, Euler(...) and Milstein(...), presented above, use matrices as arguments to solve the Itô SDE $d\mathbf{X}_t = \mathbf{F}\mathbf{X}_t\,dt + \mathbf{G}\mathbf{X}_t\,dW_t$. Fortunately the MS .NET Framework enables, due to the concept of delegate, the implementation of methods with function arguments that solve the more general equation $d\mathbf{X}_t = \mathbf{F}(t, \mathbf{X}_t)dt + \mathbf{G}(t, \mathbf{X}_t)dW_t$. Delegates are type-safe function pointers that can be used to program events, multithreads, asynchron operations or remoting (see [4]). Now we describe shortly in four steps the implementation of the Euler scheme with function arguments (the steps for the Milstein scheme are similar).

**Step 1:** declare a public delegate FuncVec with a given signature (double and VectorD arguments) and VectorD return type:

```
public delegate VectorD FuncVec (double t, VectorD X);
```

**Step 2:** implement a method EulerDelegate with arguments of FuncVec delegate type:

```
static public MatrixD EulerDelegate(VectorD X0, FuncVec
F, FuncVec G,
double h, int n, VectorD w)
  {  MatrixD xM = new MatrixD (X0.Length, n);
     VectorD xMj = new VectorD (X0);
     xM.ColumnsSet (0, xMj);
     double dw;
     for (int j=1; j<n; j++)
     {    dw = w[j] - w[j-1];
          xMj += F(0, xMj)*h + G(0, xMj)*dw; // Dele-
gate-function
argument use
          xM.ColumnsSet(j, xMj);
     }return xM;
  }
```

**Step 3:** declare new VectorD functions **F**(...), **G**(...) with the return type and signature of the FuncVec delegate:

```
static VectorD F(double t, VectorD X)
{         return -0.5*(new VectorD(X[0], X[1]));
          // F corresponds to the example
 }
static VectorD G(double t, VectorD X)
{         return new VectorD(X[1], -X[0]);
          // G corresponds to the example
}
```

**Step 4:** call the method EulerDelegate(...) with the corresponding wrapped functions F and G as arguments:

```
xy = EulerDelegate(x0, new FuncVec(F),
new FuncVec(G), h, n, w);
```

**Example.** We illustrate the use of these methods on the Itô SDE:

$$dx_t = -0.5x_t\,dt + y_t\,dw_t, dy_t = -0.5y_t\,dt - x_t\,dw_t.$$

This SDE allows to compare the numerical solution with the analytical solution $x_t = cosw_t$ and $y_t = sinw_t$.

The **Euler approximation** of $x$ and $y$ over $\langle 0, 1\rangle$ is computed by the codelines:

```
int n = 100, kernel = 1900317;
double T = 1, h = T/n;
VectorD x0 = new VectorD(0, 1);
MatrixD f, g, xy;
f = MatrixD.CreateFromVector (2,2, -0.5, 0, 0,-0.5);
g = MatrixD.CreateFromVector (2,2, 0, 1,-1, 0);
VectorD wd = Wiener.Create (h, n, kernel);
xy = Euler(x0, f, g, h, n, wd);
```

Figure 56.1 shows the generated approximation $w$ of the Wiener process $w_t$ and the approximants $x$ and $y$ of $x_t$ and $y_t$.

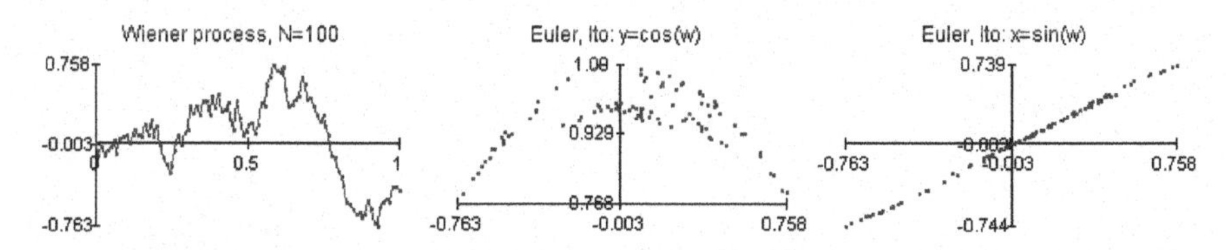


**Fig. 56.1** The approximation of $sin(w_t)$ and $cos(w_t)$ by the Euler method

Figure 56.2 shows the approximants $x$ and $y$ based on the Milstein schneme.

```
MatrixD dg= MatrixD.CreateFromVector (2,2, 0, 1,-1,
0); xy=Milstein(x0, f, g, dg, h, n, wd);
```

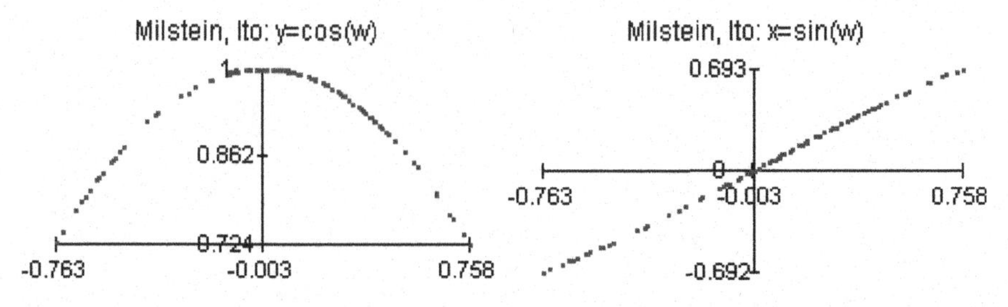


**Fig. 56.2** The approximation of $sin(w_t)$ and $cos(w_t)$ by the Milstein method

**Acknowledgments** This research has been supported by the Czech Ministry of Education in the frame of MSM 0021630529 Research Intention Inteligent Systems in Automation.

## References

1. Kloeden P, Platen E, Schulz H (1997) Numerical solution of SDE through computer experiments. Springer-Verlag, New York
2. Oksendal B (1995) Stochastic differential equations – An introduction with applications. Springer-Verlag, New York
3. Török Cs (2004) Visualization and data analysis in the MS .NET Framework. Communication of JINR, Dubna, E10-2004-136
4. Török Cs et al (2002) Professional Windows GUI Programming: Using C#. Wrox Press Inc, Chicago, ISBN 1-861007-66-3
5. Török Cs (1994) Numerical solution of linear stochastic differential equations. Comput Math Appl 27(4):1–10

# Chapter 57
# Visualization and Analyses of Multi-Dimensional Data Sets Using OLAP- A Case Study of a Student Administration System

**Che-Chern Lin, Fun-Chi Hsu, Chun-Hung Wu, Chia-Jui Hsu, Chia-Hsun Wu, Chung-Ping Lee, and Hung-Jen Yang**

**Abstract** In this paper, we introduce the visualization and analyses of multidimensional data sets using on-line analytic processing (OLAP). We discuss three fact table schemas for building data cubes for displaying different views and the levels of details of date sets Five operations of OLAP are used for manipulating data cubes including slice, dice, roll-up, drill-own, and rotation operations. By the five operations one can perform extraction, aggregation, sub-dividing, changing perspective of information stored in OLAP data cubes. We present a case study to explain the above operations. In the proposed case study, a student administration system was developed. To build data cubes, we established a constellation schema consisting of three fact tables and four dimension tables. We also demonstrate the five operations on the data cubes and discuss how these operations are applied to analyze data.

## 57.1 Introduction

Recently with the development of information technologies, huge data can be handled by computers. Based on databases, many data analysis techniques are developed and applied in different areas. Data warehouse is one of the popular techniques where multidimensional data structures are established by extracting a different perspective from related normalized data tables in an operational database. To build a data warehouse, people use data cubes to represent the multidimensional data structure. Fact tables are utilized to describe the data structures of the data cubes. With the capabilities of handling and displaying the multidimensional data sets, data warehouse applications are successfully utilized in data analyses and decision support systems and therefore widely

C.-C. Lin (✉)
National Kaohsiung Normal University, 116 Heping First Road, Kaohsiung City, Taiwan (R.O.C.)
e-mail: cclin@njknucc.nknu.edu.tw

N. Mastorakis et al. (eds.), *Proceedings of the European Computing Conference*, Lecture Notes in Electrical Engineering 28,
DOI 10.1007/978-0-387-85437-3_57, 

applied in many fields such as business, engineering, education, medicine, etc. Data warehouses are normally utilized to analyze off-line data for making decisions. On-line analytic processing (OLAP) is a similar technique but focuses on analyzing real time data. It retrieves data from an on-line database. Data mining is another information technique to cluster data by creating association rules. It is also widely applied in different areas such as business, engineering, education, and medicine.

A cubic-wise balance method for privacy preservation has been proposed to furthermore make closely estimation of data without actual data access [1]. MSMiner, an OLAP development software has been generally discussed including design requirements, system architecture, and an application example [2]. Based on unified modeling language (UML), Prat et al. proposed a data warehouse design methodology with three phases: conceptual, logic, and physical phases [3]. A data warehouse application on site determination was implemented by Ahmad et al. where an OLAP-based decision support system was built to select the most appropriate residential area in Miramar, Florida, U.S.A.. Moon et al. discussed the effectiveness on cost estimation of OLAP-based management systems [4]. Efficient view selection methods for data warehouse were proposed to improve the query time and data storage space [5].

In this paper, we apply the data warehouse technique in education and provide a case study on a student administration system where OLAP data cubes were generated for manipulating data.

## 57.2 Design of OLAP Cubes

Relational databases uses normalized tables where relationships among these tables are used to aggregate data from different data and hence get integrated information for views and reports. A table in a relational database is two-dimensional, consisting of fields (columns) to define data attributes of the table and records (rows) to describe the field values for a particular data instance. Normalizations are utilized to generate the normalized tables in a relational database from an original unnormalized data set. During the normalization procedure, tables are sequentially divided into smaller tables without losing information. After normalizations, tables are said to be in normal forms. Five normalizations are used to generate five normal forms from first normal form (1NF) to fifth normal form (5NF). To implement the relationships among these tables in a database, people use join operations to integrate different data from different tables using "keys". Structured query language (SQL) is used to practically realize the join operations using a "select" command. Two keys are generally used in tables: primary key (PK) and foreign key (FK). A PK is used to identify a particular record and to get the rest of the field values associated with this particular record. A FK is utilized to link the PK of another table to get more information form the linked table.

A table in a relational database only shows the information related to the table. Using SQL commands gets some partially integrated data. For decision making, people need total information to analyze data and make decisions. Data warehouses are applied to integrate data from a database in support of subject-oriented data analyses for decision making. Figure 57.1 shows a data warehouse process model [6]. In the figure, a data warehouse obtains its data by using an ETL (extract, transform, and load) routine from different ways including external data, operational database and independent marts. To build a data warehouse, people use fact tables to aggregate different data from dimension tables which provide necessary information for establishing the data warehouse. Similar to a PK, a dimension key is used to link a fact table to a dimension table associated with the dimension key. Three types of fact table schemas are often used to build data cubes including: star, snowflake, and constellation schemas. The star schema is the simplest structure of fact table where a single fact table is located at the center of the structure, linking each dimension table with a corresponding dimension key. To decrease data redundancy, a snowflake schema is used to further divide some dimension tables. A constellation schema is another structure where more than one fact table is located at the center of the structure. In a constellation schema, dimension tables can be multiply linked to fact tables.

Based on the techniques of building data warehouses, OLAP is a query-based approach to handle real time data sets for data analyses. Five operations are often applied to perform OLAP including: slice, dice, dill-down, roll-up, and rotation operations. A slice is an operation to perform data retrieval on a single dimension of an OLAP cube. A dice operation is similar to a slice operation but

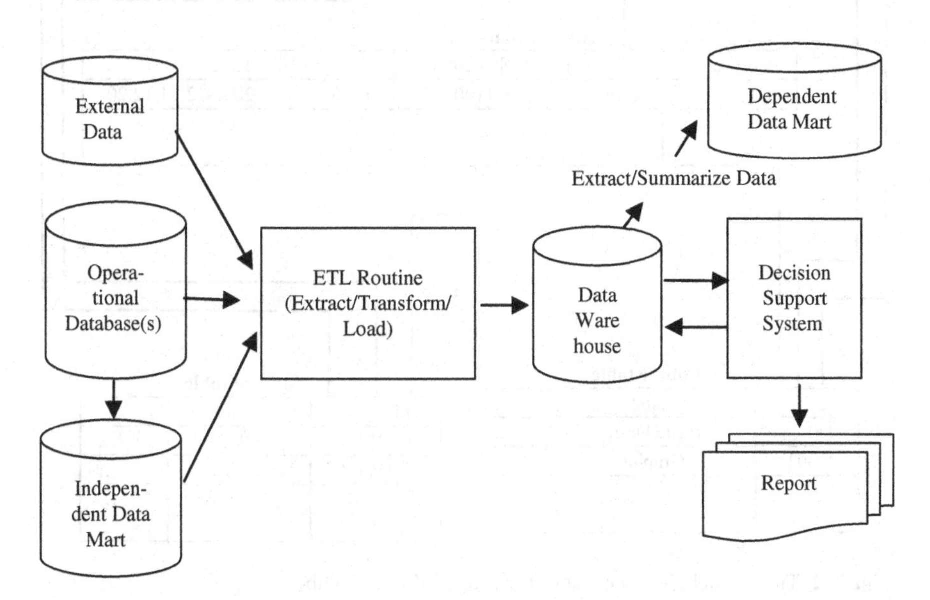


**Fig. 57.1** A data warehouse process model (taken from [6])

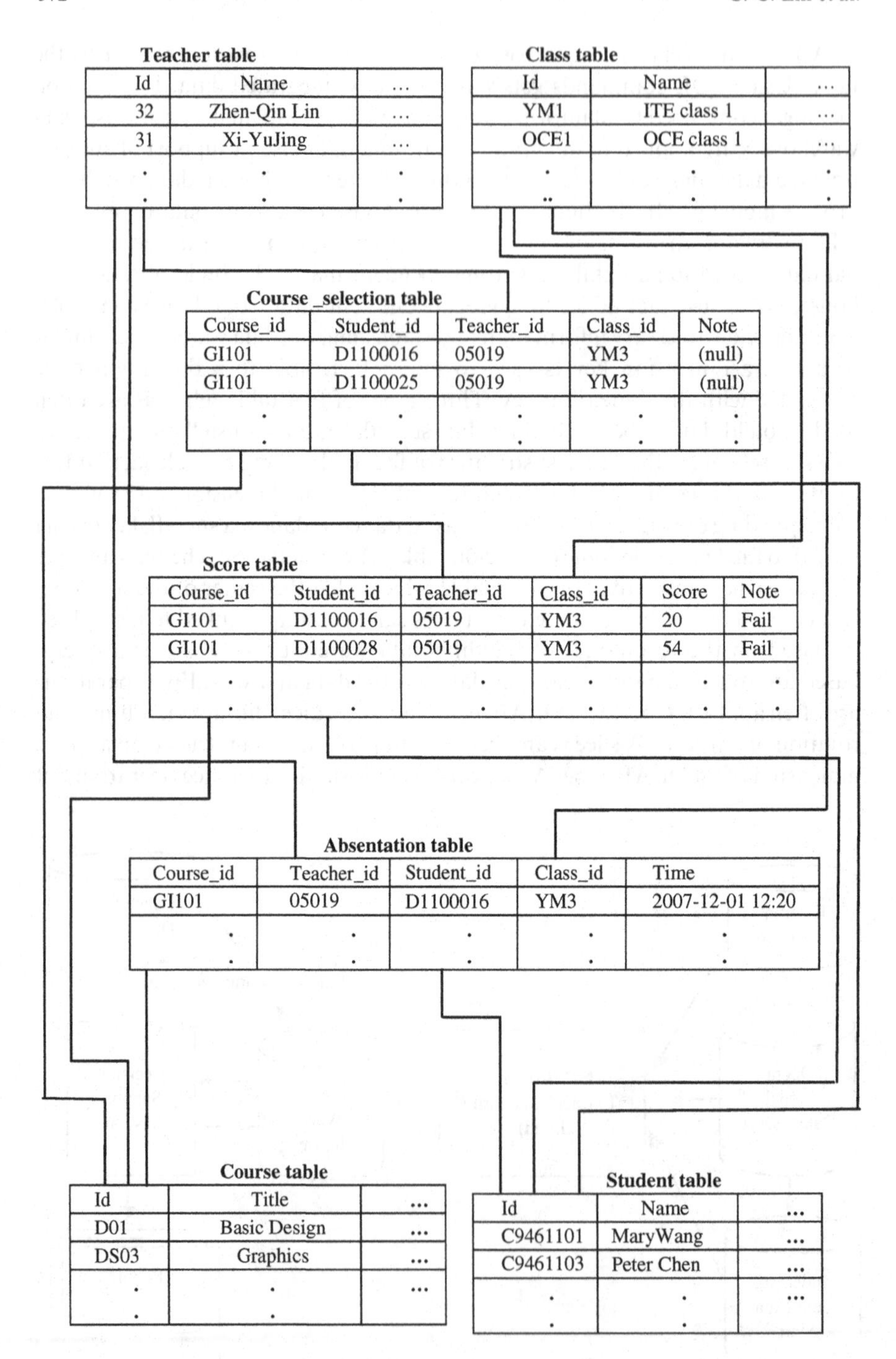


**Fig. 57.2** The constellation schema of the proposed OLAP cubes

focuses on selecting data on two or more dimensions. A roll-up operation performs date generalization by aggregating data on one particular dimension on an OLAP cube. A drill-down operation, a reverse process of roll-up, is used to obtain more detailed data for a particular dimension on an OLAP cube. A rotation simply selects another perspective of an OLAP cube by rotating the displayed dimension without aggregation, sub-dividing, and extraction of the data.

## 57.3 Case Study

To demonstrate visualization of multidimensional data sets, we present an example of a student administration system. Figure 57.2 shows the schema of the proposed example where a constellation schema is used to construct OLAP cubes .Three fact tables are used to form the center parts of the constellation schema including course selection table, score table, and absentation table. In addition to the fact tables, four dimension tables are also utilized to form the OLAP cubes including teacher, class, course and student tables. The detailed fields of the fact tables and dimension tables are also displayed in Fig. 57.2.

## 57.4 Conclusion

We introduced the visualization and analyses of multidimensional data sets using OLAP. We also discussed three fact table schemas for building data cubes. Five operations are discussed for manipulating data cubes including slice, dice, roll-up, drill-own, and rotation operations. We presented an example of a student administration system to explain the above operations. To demonstrate the example, we established OLAP cubes using a constellation schema.

As for further studies, we suggest extending the system with more tables and building more complex schema to get more information.

## References


1. Liu Y, Sung SY, Xiong H (2000) A cube-wise balance approach for privacy preservation in data cubes. Inf Sci 176:1215–1240
2. Shi Z, Huang Y, He Q, Xu L, Liu S, Qin L, Jia Z, Li J, Huang H, Zhao L (2007) MSMiner-a developing platform for OLAP. Decis Support Syst 42:2016–2028
3. Prat N, Akoka J, Comyn-Wattiau I (2006) A UML-based data warehouse design method. Decis Support Syst 42:1449–1473
4. Ahmad I, Azhar S, Lukauskis P (2004) Development of a decision support system using data warehousing to assist builders/developers in site selection. Autom Constr 13:525–542
5. Moon SW, Kim JS, Kwon KN (2007) Effectiveness of OLAP-based cost data management in construction cost estimate. Autom Constr 16:336–344
6. Roiger RJ, aGeatz MW (2003) Data mining: A tutorial-based primer. Addsion Welsley, Pearson Education, Inc., Chapter 6

# Chapter 58
# Design of a Computer-Based Triangle Chess Machine

**Che-Chern Lin, Chia-Jung Hsu, Ming-Jun Chan, Chih-Hung Chung, Chong-Cheng Hong, and Hung-Jen Yang**

**Abstract** Computer-based chess machines have been built based on tree searches and parallel computing. Several papers have discussed the computer implementations of different types of chess games such as chess (western), Chinese chess (Xiangqi), Japanese chess (Shogi), and Hex. This paper focuses on the implementation of a triangle chess machine. We first make a brief explanation on the popular chess games and their related studies. We then describe the rules for playing a triangle chess. Basically, the triangle chess machine proposed in this paper considers all possible configurations of chess patterns and then makes moves based on these patterns. We present the software modules of the machine and discuss the computation issues related to this study. Conclusions and the directions for future studies are also given at the end of this paper.

## 58.1 Introduction

Recently with the promotion of computational speed and data storage, computers have stronger capability to handle more complicated computing. Computer programs for different types of chess games have been built successfully in the past decades including the developments of chess (western), Chinese chess (Xiangqi), Japanese chess (Shogi), Hex, and Go. Xiangqi, a Chinese chess, is very popular in the worldwide Chinese community. It uses an 8×8 grid-like chessboard played by two players, "*Black*" and "*Red*" [1]. Each of them owns a territory in a space roughly half the size of the chessboard and has fifteen pieces. These pieces are categorized into seven different types of roles with different moving regulations, respectively. The seven roles are King, Assistant, Elephant, Horse, Chariot,

C.-C. Lin (✉)
National Kaohsiung Normal University, 116 Heping First Road, Kaohsiung, Taiwan (R.O.C)
e-mail: cclin@nknucc.nknu.edu.tw

N. Mastorakis et al. (eds.), *Proceedings of the European Computing Conference*, Lecture Notes in Electrical Engineering 28,
DOI 10.1007/978-0-387-85437-3_58, 

Gunner, and Pawn [1]. The player who eliminates all of the opponent's pieces wins the game. Similar to Xiangqi, Shogi is a Japanese chess played by two players, called "*Black*" and "*White*" [2]. It uses a 9x9 grid-like chessboard. Each of the players has 20 pieces. Hex is also a two-player game played on a hexagonally grid-like chessboard with regularly an 11x11 rhombus [3]. The two players, named "*Red*" and "*Blue*", have their own sides, respectively. During a Hex game, each player uses his pieces to make a link from his side to the opponent's side. The first player to make a successful implementation of the link wins the game. Go is another popular chess in East Asia. It also has two players, "*Black*" and "*White*", played on an 18x18 grid-like chessboard [4]. During a Go game, each player uses his pieces to form his own territory by removing the opponent's pieces tightly surrounded by his pieces [4]. After an endgame, the player who has the maximum area of territory wins the game.

Basically a chess machine uses searching and parallel computing techniques. However, it requires computers with faster CPU speed and more memory capacity. Using rule-based machines needs to transfer the chess experts' strategies into rules by which to determine the next steps. It is sometimes difficult to generate and implement these rule-based chess machines.

Originally derived from Deep Thought, Deep Blue, developed at IBM Watson Research Center, is a successful computer-based chess machine which defeated a World Chess Champion in 1997 [5]. It basically used parallel computing techniques and searching trees to deal with massive data processing. The computational algorithms developed for Deep Blue calculate all possible moves and might result in massive data manipulations. After calculating the possible moves, Deep Blue then uses evaluation functions to measure the possible scores for these moves and makes a decision for the next steps. Deep Blue not only used the computer software techniques but also hardware techniques. It is so far the most successful chess machine implemented by computer technologies. Iida et al. compared the similarity and the differences between Chess and Shogi, and described the techniques used to build a computer-based shogi machine, including opening and endgame plays, searching algorithm, tactical strategies, and position evaluation. [2]. Wu and Beal used endgame database construction method to implement a computer-based Chinese chess machine and demonstrated its performance in reducing the size of the database [1]. Anshelevich presented a hierarchical method for building a Hex chess machine where he used deduction rules to recursively compute the Hex positions from the beginning of a game [6].

In this paper, we present a case study of implementation of a computer-based triangle chess machine. We also discuss the rules for playing a triangle chess game and the design methodology of this chess machine. Finally we give conclusions and suggest the directions for future studies.

## 58.2 Implementation

The rules for playing a triangle chess are very simple and straightforward. Figure 58.1 shows a triangle chess machine where fifteen circles are arranged in a triangle chessboard. During a triangle chess game, one should select one or more circles (up to five circles) in each step. The selected circles in each step must form a line. This is called the "*co-linear*" rule. The player who selects the last circle is the loser.

We used Visual C++ to implement the triangle chess machine. The machine provides a user interface for users to play a triangle chess game and a replay function to repeat the whole process of the game. To implement the machine, we need some software modules (subroutines) to perform the functionalities. There are mainly four modules for doing this. The first module is to validate if the circles selected by a user (the opponent) fit the co-linear rule. We then call this module the "co-linear checking" module. The second module is to determine the pattern of the configuration of the triangle chess after the opponent selects his/her circles. The second module also makes a responding selection of circles (automatically done by the machine) to react to the opponent selection. This module is called pattern determination/circles selection module. The third module is called replay module serving to record all steps in a triangle game and to perform replay functionality if necessary. The fourth module is called "user interface" module providing a graphic interfaces for playing a triangle chess game. Figure 58.2 shows the four functional modules and their relationships. The four modules are explained in detail below.

1. Co-linear checking module: The co-linear checking module uses tree structures to determine if circles selected by the opponent are co-linear. Each circle in a triangle chessboard has its own searching tree by which to examine if selected circles satisfy the co-linear rule. Each searching tree describes all possible co-linear paths for its corresponding circle. If the opponent selects

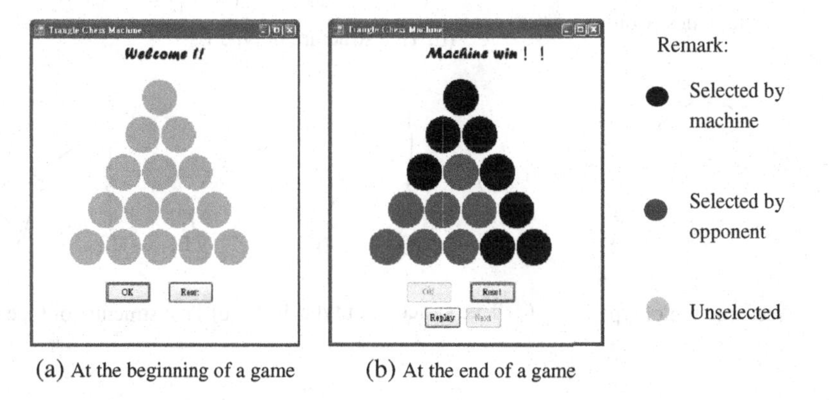


(a) At the beginning of a game (b) At the end of a game

**Fig. 58.1** The triangle machine

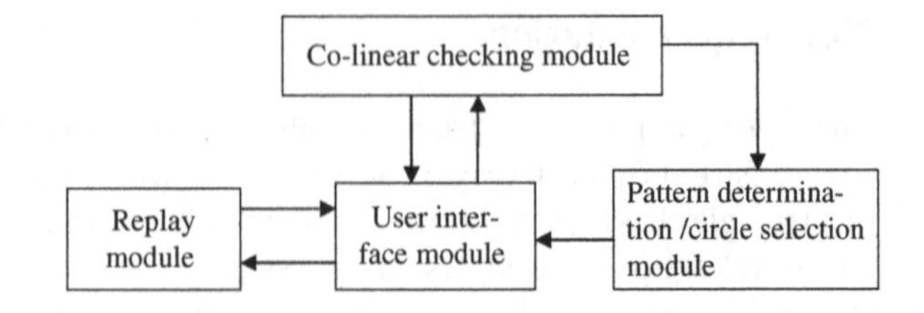


**Fig. 58.2** The functional modules in the chess machine

his/her circles, the module examines if the selection of circles is on any of the co-linear paths. Figure 58.3 shows a triangle chess configuration and the associated tree structures. For convenience, we number the 15 circles in a triangle chessboard from 1 to 15, as indicated in Fig. 58.3(a). By the property of symmetry, the 15 circles are categorized into four types based on the geometrical property. They are shown as follows:

Type 1: circles 1, 11, and 15.
Type 2: circles 2, 3, 7, 10, 12, and 14.
Type 3: circles 4, 6 and 13
Type 4: circles 5, 8 and 9.

The sampled examples for the four types of trees are also shown in Fig. 58.3. It is important to note that the selected circles must be sorted before using a tree since the user (opponent player) may make their selections of circles in any order.

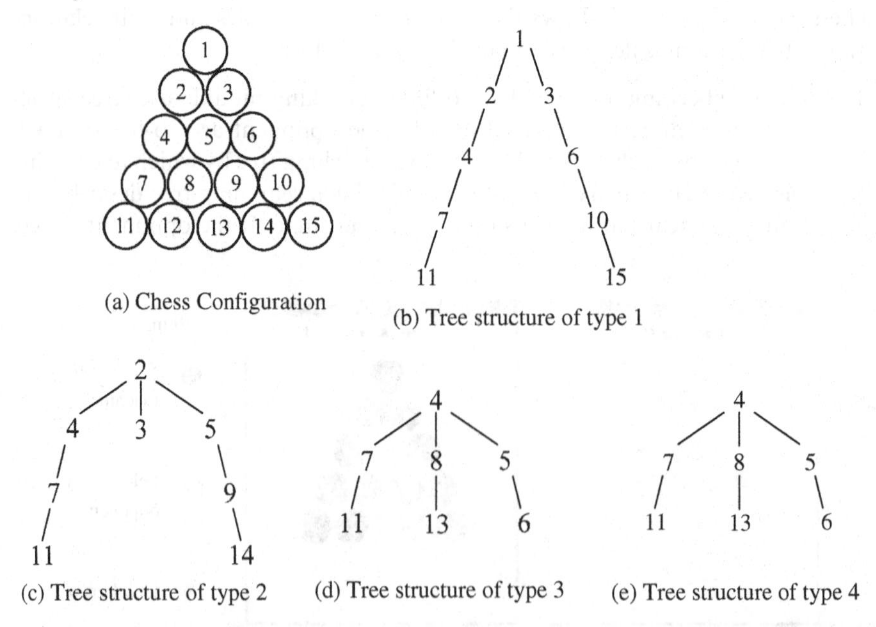


**Fig. 58.3** Trees to examine the co-linear rule

2. Pattern determination/circles selection module: This module stores all possible configurations of patterns for a triangle chess game. There are 15 circles in a triangle chessboard and each of the circles has two stages: selected (= 1) and unselected (= 0). We then have $2^{15}$ (=32768) combinations of chess configurations. Of the 32,768 combinations, only 7,795 patterns satisfy the co-linear rule. We therefore use the 7,795 patterns as pattern-matching data source. Each of the patterns is associated with a response configuration which determines the next move (done by the machine). Since the values of the 15 circles are binary (0 and 1), we then encode each of the patterns in a two-byte short integer. The response patterns are also encoded in two-byte short integers. We use a two-dimensional array to store these short integers.
3. Replay module: The replay module records all steps played in a game and replays them if a user (opponent) needs to know the history of the game.
4. User Interface module: The user interface module provides a playing platform for an opponent player.

## 58.3 Conclusions

We made a brief explanation on several popular chess games and their computer implementations. We also presented a case study of implementation of a computer-based triangle chess machine. Basically, the chess machine considers all possible configurations of chess patterns and then makes a selection of circles based on these patterns. We demonstrated the software modules of the machine and discussed the issues related to this study.

Using decision rules to implement a chess machine might be an interesting issue for future studies since it can significantly reduce searching spaces.

## References

1. Wu R, Beal DF (2001) Solving Chinese chess endgames by database construction. Inf Sci 135:207–228
2. Iida H, Sakuta M, Rollason J (2002) Computer Shogi. Artif Intell 134:121–144
3. http://en.wikipedia.org/wiki/He_(board_game)
4. http://en.wikipedia.org/wiki/Go_(board_game)
5. Champbell M, Hoane Jr AJ, Hsu F (2002) Deep Blue. Artif Intell 134:57–83
6. Anshelevich VV (2002) A hierarchical approach to computer hex. Artif Intell 134:101–120
7. Müller M (2001) Global and local tree search. Inf Sci 135:187–206
8. Ewerhart C (2002) Backward induction and the game-theoretic analysis of chess. Games Econ Behav 39:206–214

# Chapter 59
# A Taxonomy Model for Analyzing Project-Based Learning Using Analytical Hierarchy Process (AHP): A Case Study of Design of a Beer Company Website in an Elementary School

**Chung-Ping Lee, Che-Chern Lin, Rong-Jyue Fang, Howard Lo, and Ming-Chih Wu**

**Abstract** This paper presents a taxonomy model which applies analytical hierarchy process (AHP) to project-based learning. To implement the proposed model, a website has been built for the sixth-graders of an elementary school for their project-based learning. The learning target is knowledge about beer. The learning materials were provided by a local beer company. Six factors are considered as the inputs of the AHP including company introduction, characteristics of beer of the company, company visiting, beer promotion, beer manufacturing, and learning activity sharing. To investigate the importances among the six factors, we interviewed five experts who have at least seven years of teaching experience in an elementary school in Taiwan. The AHP is utilized to generate the weights (importances) for the input variables for the taxonomy model.

## 59.1 Introduction

Analytical hierarchy process (AHP) is developed for decision-making [1–4]. Basically, an AHP model is a hierarchically layered structure. Figure 59.1 shows a three-layer AHP [3]. Nodes are used in each layer to represent factors or dimensions. One utilizes weights to link two nodes in the adjacent layers. Mappings are performed from inputs to outputs sequentially. The mapping function is given by the follow formula [3]

$$y_j = \sum_{i=1}^{n} w_{ij} x_i \tag{59.1}$$

C.-P. Lee (✉)
Department of Industrial Technology Education, National Kaohsiung Normal University, Taiwan (R.O.C.)
e-mail: cl87369@yahoo.com.tw

N. Mastorakis et al. (eds.), *Proceedings of the European Computing Conference*, Lecture Notes in Electrical Engineering 28,
DOI 10.1007/978-0-387-85437-3_59, 

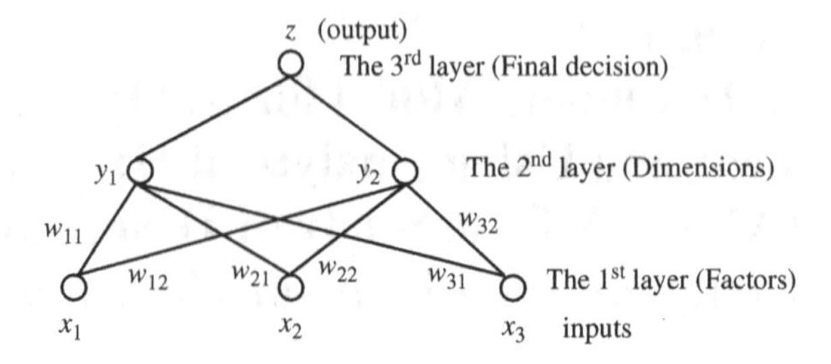


**Fig. 59.1** A simple three-layer AHP (taken from [3])

where

$y_j$ = the output for node $j$ in a particular layer,
$x_i$ = the output of node $i$ in the previous layer of this particular layer,
$w_{ij}$ = the weight linking node $j$ in the particular layer and the node $i$ in the previous layer,
$n$ = the number of node in the previous layer.

Project-based learning (PBL) is a model that organizes learning materials from different projects. The projects are complex tasks that involve learners in design, problem-solving, decision making, investigative activities, etc. [5]. Instructors provide learners some activities to analyze information on learning topics, compare other cases with their problems, and develop an optimal method for the task [6]. The task-oriented approach is used to find directions of problem solving by providing different practical problems. Finally, learners must gather and organize relevant information in order to implement their learning projects.

In this paper, we propose a taxonomy model which applies a two-layer AHP to PBL. We introduce the AHP procedure and explain how AHP is utilized to PBL on beer. A website whose contents were provided by a local beer company has been built for the sixth-graders of an elementary school for their learning on beer. Six factors were determined using a brainstorming procedure including company introduction, characteristics of beer of the company, company visiting, beer promotion, beer manufacturing, and learning activity sharing. Five experts were selected to answer specially designed questionnaires. Finally we present the evaluation model for the taxonomy.

## 59.2 The AHP Procedure

The AHP procedure of the propose model is depicted as follows [1–4, 7].

*Step 1: Define the problem and determine the goal.* The goal of this study is to select the useful messages and knowledge from the forum of the website for PBL on beer. This is done by the collection of information through the Internet, group discussion, and special homework related to this study.

**Table 59.1** The definition and explanation of AHP 9-point scale (taken from [3, 4])

| Intensity of Relative Importance | Definition |
|---|---|
| 1 | Equal Importance |
| 3 | Moderate important of one over another |
| 5 | Essential or strong important |
| 7 | Demonstrated importance |
| 9 | Absolute importance |
| 2, 4, 6, 8 | Intermediate values between the two neighboring scales |

*Step 2: Select factors for the model.* We used brainstorming to determine the relative importance for establishing evaluation criteria of the elements in the hierarchy. Nine sixth-graders in an elementary school were selected to carry out the brainstorming and generate the structure of factors using a free software package, Xebece. Xebece produces structures of brainstorming by the combination of catalogues, attachment of documents, and XML documents [8]. After the brainstorming, six factors were considered to build the proposed model including company introduction, characteristics of beer of the company, company visiting, beer promotion, beer manufacturing, and learning activity sharing.

*Step 3: Design the questionnaire.* The questionnaire was designed to make all possible pair-wise comparisons among the factors. Table 59.1 shows a typical nine-point scale for an AHP questionnaire [3, 4]. The questionnaire was designed to measure all possible importance ratios among the factors. Table 59.2 shows a simple example of the questionnaire where three factors are selected: factors A, B, and C. In the table, factor A is twice as important as factor B, the ratio of factors A to B is 2:1. Row 1 is corresponding to the ratio of factors A to B. In row 1, we then mark "✓" in the cell associated with a value of 2 (closed to the factor of A). Similarly, the importance ratio of factor A to factor C is 3:1. The importance ratio of factor B to factor C is 1:5. Both of them are shown in Table 59.2 (rows two and three).

*Step 4: Use the questionnaire to collect the experts' opinions.* After collecting questionnaires from experts, one uses a "***matrix of importance ratios***" to describe the results of pair-wise comparisons. Equation 59.2 shows the matrix

**Table 59.2** A simple example of questionnaire (taken from [3])

| Factor | 9 | 8 | 7 | 6 | 5 | 4 | 3 | 2 | 1 | 2 | 3 | 4 | 5 | 6 | 7 | 8 | 9 | Factor |
|---|---|---|---|---|---|---|---|---|---|---|---|---|---|---|---|---|---|---|
| A (row 1) | | | | | | | | ✓ | | | | | | | | | | B |
| A (row 2) | | | | | | | ✓ | | | | | | | | | | | C |
| B (row 3) | | | | | | | | | | | | | ✓ | | | | | C |

**Table 59.3** Random Index [1]

| n | 1 | 2 | 3 | 4 | 5 | 6 | 7 | 8 | 9 | 10 |
|---|---|---|---|---|---|---|---|---|---|---|
| RI | 0.00 | 0.00 | 0.58 | 0.90 | 1.12 | 1.24 | 1.32 | 1.41 | 1.45 | 1.49 |

Remark: *n* is the number of fact

of importance ratios associated with Table 59.2 [3]. The matrix is a symmetrical and reciprocal matrix due to pair-wise comparisons.

$$\begin{array}{c} \\ A \\ B \\ C \end{array}\begin{array}{c} \begin{array}{ccc} A & B & C \end{array} \\ \begin{bmatrix} 1 & 2 & 3 \\ 1/2 & 1 & 1/5 \\ 1/3 & 5 & 1 \end{bmatrix} \end{array} \tag{59.2}$$

*Step 5: Test the consistency.* The Consistency Index (CI) is utilized to express the degree of consistency. Saaty [4] defined the consistency index (CI) as follows:

$$CI = \frac{\lambda_{\max} - n}{n - 1} \tag{59.3}$$

where $\lambda_{\max}$ is the maximum eigenvalue of the matrix of the importance ratios and $n$ is the number of factors.

Accordingly, Saaty [4] defined the Constituency Ratio (CR) as follows:

$$CR = \frac{CI}{RI} \tag{59.4}$$

where Random Index (RI) is given by Table 59.3. If the value of the consistency ratio CR is less than or equal to 0.1 the questionnaire is considered to be acceptable. If CR is greater than 0.1, the questionnaire fails.

**Table 59.4** Eigenvalue and consistency indices of experiment

| Experts | $\lambda_{\max}$ | CI | CR |
|---|---|---|---|
| Experts 1 | 7.15 | 0.23 | 0.19 |
| Experts 2 | 6.00 | 0 | 0.00* |
| Experts 3 | 7.30 | 0.26 | 0.21 |
| Experts 4 | 6.33 | 0.07 | 0.05* |
| Experts 5 | 7.18 | 0.24 | 0.19 |

Remark: * indicates CR $\leq$0.1 (passing the consistence tests)

## 59.3 Experimental Setup and Results

In this study, five experts from an elementary school in Taiwan were selected. All of them have at least seven years of teaching experience. The results of five experts' questionnaires and consistency tests are shown in Table 59.4. Only two questionnaires (Expert twp and Expert four) passed the consistency tests.

We took geometrical averages of the weights from the questionnaires which passed the consistency tests.

The evaluation formula for the taxonomy model is obtained by taking the normalized eigenvector associated with the maximum eigenvalue of the matrix of importances ($\lambda_{\max}$). Eq. 8.5 demonstrates the evaluation formula

$$y = 0.071x_1 + 0.106x_2 + 0.232x_3 + 0.068x_4 + 0.228x_5 + 0.228x_6 \quad (59.5)$$

where

$x_1$ = company introduction,
$x_2$ = characteristics of beer of the company,
$x_3$ = company visiting,
$x_4$ = beer promotion,
$x_5$ = beer manufacturing,
$x_6$ = learning activity sharing.

In Equation (59.5), the higher value of $y$ means the stronger recommendation for taxonomy.

## 59.4 Conclusions

We proposed a taxonomy model which applied AHP to PBL. The learning target of the PBL is the knowledge of beer. The learners were the sixth-graders of an elementary school. To implement the model, we built a website cooperating with a local beer company. Six factors were determined using a brainstorming procedure including company introduction, characteristics of beer of the company, company visiting, beer promotion, beer manufacturing, and learning activity sharing. Five experts were selected to answer the specially designed questionnaires. Finally we used the AHP procedure to generate the evaluation formula for the taxonomy model.

Concerning the directions of further studies, the Delphi technique may be considered to collect more experts' opinions on the project. Besides, using fuzzy logic to describe the factors might be an interesting topic for future study. Finally, we suggest using more layers in the AHP structure to collect more detailed items for further analyses.

## References

1. Saaty TL (1990) How to mark a decision: the analytic hierarchy process. Eur J Oper Res 48(1):9–26
2. Saaty TL (2003) Decision-making with the AHP: Why is the principal eigenvector necessary. Eur J Oper Res 145:85–91
3. Huang C, Lin Y, Lin C (2007) An evaluation model for determining insurance policy using AHP and fuzzy logic: case studies of life and annuity insurances. In: The 8th WSEAS international conference on fuzzy systems, Vancouver, Canada, pp 126–131
4. Saaty TL (1980) The Analytic Hierarchy Process. Revised and extended 1988, McGraw-Hill, New York
5. John TW (2000) A review of research on project-based learning. T. A. Foundation, California
6. Brunettti AJ, Petrell RJ, Sawada B (2003) Team project-based learning enhances awareness of sustainability at the University of British Columbia, Canada. Int J Sustainability High Educ 4(1):210–217
7. Lee SK, Yoon YJ, Kim JW (2007) A study on making a long-term improvement in the national energy efficiency and GHG control plans by the AHP approach. Energy Policy 35:2862–2868
8. Kennke R (2007) Xebece: Visualize and organize information easily. Retrieved 6.1, from http://xebece.sourceforge.net/

# Chapter 60
# A Framework of Mixed Model of a Neural Network and Boolean Logic for Financial Crisis Prediction

**Che-Chern Lin, Ghuan-Hsien Mao, Howard Lo, and Hung-Jen Yang**

**Abstract** This paper proposes a framework for pattern classifications. This framework includes two neural networks and a logic circuit. The first neural network serves to classify regular instances, and the second one handles the overlapped instances caused by the first neural network. The proposed framework integrates the two neural networks by the logic circuit and then makes the final classifications. We provide two methodologies using the proposed framework including financial crisis prediction and principle components analysis. In the financial crisis prediction case, we use the financial ratios of individual companies as the inputs of the first neural network and the macroeconomic indicators as the inputs of the second neural network. In the principle components analysis case, we use the attributes in a major factor group as the inputs of the first neural network and those in a minor factor group as the inputs of the second neural network.

## 60.1 Introduction

Traditionally, people use statistical methods to analyze data, build models, and make decisions (prediction, estimation, and classification). Instead of statistical methods, recently artificial intelligent algorithms are widely utilized in different areas. The most popular artificial intelligence algorithms include data mining algorithms, fuzzy systems, genetic algorithms, and neural networks. Data mining algorithms use computer programs to find relationships among different attributions by establishing association rules. Based on an iterative procedure to find the optimal combinations of items, the apriori algorithm is the most popular one to general association rules for categorical items. Basket analyses are successful applications for businesses to analyze the

C.-C. Lin (✉)
National Kaohsiung Normal University, 116 Heping First Road, Kaohsiung City, Taiwan (R.O.C.)
e-mail: cclin@nknucc.nknu.edu.tw

N. Mastorakis et al. (eds.), *Proceedings of the European Computing Conference*, Lecture Notes in Electrical Engineering 28,
DOI 10.1007/978-0-387-85437-3_60, 

behaviors of customers. Fuzzy expert systems mimic the thinking processes of human beings where verbal expresses are used to generate rules and make decisions. Unlike traditional crisp values, fuzzy variables use membership functions to describe the marching degree of how inputs fit a particular fuzzy set. Two types of membership functions are often adopted in fuzzy systems: triangular and trapezoidal functions. In general, a fuzzy expert system consists of three parts: a fuzzy base, an inference engine, and user interfaces. The fuzzy base consists of a set of fuzzy rules usually obtained by domain experts. The inference engine is the mechanism to make conclusions by firing the fuzzy rules. The user interfaces allow users to define membership functions, compose fuzzy rules and display outcomes. Genetic algorithms use an evolutionary procedure to find optimal solutions. Like biological evolutionary process, genetic algorithms use fitness function to evaluate the performance for particular generations and use genes (chromosomes) as input strings. In genetic algorithms, selections are used to get better offspring and mutations are utilized to mimic the random changing of genes. Neural networks use computer programs to simulate the structure of human brains and the cognition processes of human beings. A neural network is usually a layered structure with nodes in each layer. Weights are used to connect two nodes between two adjacent layers and updated by training algorithms. The back-propagation algorithm is one of the most popular updating algorithms where the gradient descent method is utilized to minimize the errors. The training algorithm uses a set of well-known examples, called training set, to backwardly adjust weighs from outputs to inputs. This procedure is called training procedure. After the training procedure, neural networks are ready to classify unknown instances based on the well-trained weights. Neural networks are widely applied to financial applications. A large number of studies related to financial crisis prediction can be found in [1–6].

## 60.2 The Framework

Figure 60.1 shows the diagram for the proposed framework. There are two neural networks and a logic circuit in Fig. 60.1. The first neural network functions serve to classify regular instances and the second one handles the overlapped instances caused by the first neural network. For a two-class classification problem (Class I and Class II), the classification results are determined in Table 60.1.

To handle the overlapped instances, we use the second neural network to furthermore classify these overlapped instances. We select these overlapped instances as the training data for the second neural network. The logic circuit is used to integrate the two neural networks and get the final classification results. Table 60.2 is the truth table for the logic circuit of the proposed framework.

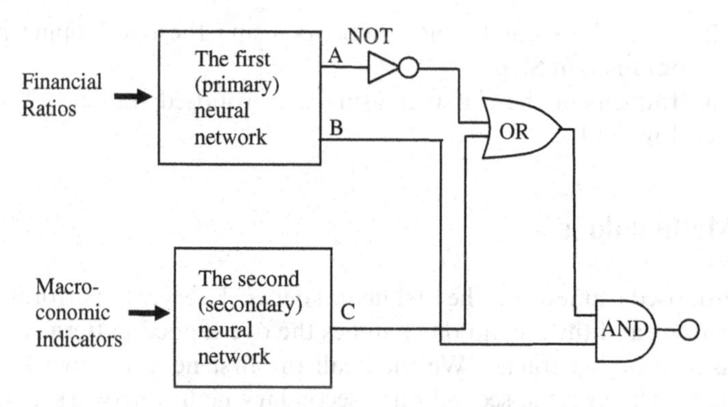


**Fig. 60.1** The proposed framework

**Table 60.1** The classification results of a two-class classification problem

| A | B | Class assigned |
|---|---|---|
| 0 | 0 | Undetermined |
| 0 | 1 | Class I |
| 1 | 0 | Class II |
| 1 | 1 | Overlapped |

**Table 60.2** The truth table for the logic circuit

| Input | | | Output |
|---|---|---|---|
| A | B | C | Y |
| 0 | 0 | 0 | 0 |
| 0 | 0 | 1 | 0 |
| 0 | 1 | 0 | 1 |
| 0 | 1 | 1 | 1 |
| 1 | 0 | 0 | 0 |
| 1 | 0 | 1 | 0 |
| 1 | 1 | 0 | 0 |
| 1 | 1 | 1 | 1 |

Remark: Y=1 represents Class I, and Y= 0 represents Class II.

The simplified Boolean expression for Table 60.2 is given by

$$Y = B(\overline{A} + C) \tag{60.1}$$

The logic circuit for Equation 60.1 is also shown in Fig. 60.1.
In brief, the procedure of the proposed framework is described as follows:
Step 1: Train the first neural network. The overlapping instances are the patterns with A = 1 and B = 1.

Step 2: Train the second neural network using the overlapping instances obtained in Step 1.
Step 3: Implement the classifier using the proposed framework shown in Fig. 60.1.

## 60.3 Methodologies

In the proposed framework, the first neural network serves to perform the main classifications and the second one handles the overlapped patterns to improve the classification accuracies. We then call the first neural network "primary neural network" and the second one "secondary neural network". Below, we present two methodologies applying the proposed framework.

Methodology one: Establish the primary and secondary neural networks using well-defined dimensions of attributes. If the dimensionality of the variables was well-defined, one can use the variables in major dimension as the inputs of the primary neural network, and the minor dimension as the input of the secondary neural network. Financial crisis prediction is one of the applications of using methodology one [6]. Previous studies have indicated that the financial crisis can be predicted using neural networks based on two groups of variables: financial ratios for individual companies and macroeconomic indicators. The financial ratios affect the financial crisis more strongly than the macroeconomic indicators. We then use the financial ratios as the inputs for the primary neural network and the macroeconomic indicators as the inputs for the secondary neural network. Mao used this framework to predict the financial crises for public electronic companies in Taiwan and got better results than regular neural networks [6]. The financial ratios and macroeconomic indicators used in Mao's study are shown in Table 60.3 [6].

**Table 60.3** The financial ratios and macroeconomic indicators used for financial crisis prediction [6]

| Financial ratio | Macroeconomic indicators |
|---|---|
| Return on total assets | Gross Domestic Product |
| Growth rate in sales | Export |
| Growth rate in operating profits | Import |
| Growth rate in total assets | Annual increasing rate of wages |
| Growth rate in equity | Employment rate |
| Growth rate in fixed assets | Unemployment rate |
| Current ratio | Economic growth rate |
| Quick ratio | Annual increasing rate of loans and investments |
| Debt ratio | Rediscount rate |
| Times interest earned ratio | Interest rate in money market |
| Accounts receivable turnover | Exchange rate |
| Inventory turnover | |
| Total asset turnover | |
| Debt- to-Equity Ratio | |

Methodology two: The second methodology is to use statistical techniques to separate variables into two groups according their importance: the major affecting and minor affecting groups before using the proposed framework. Consider a set of instances with n attributes. Based on the importance of the affecting degree of the n attributes, one can use statistical methods to divide the n attributes into two groups: the major factor group with higher affecting degrees and minor factor group with lower affecting degrees. Principle components analysis is one of the popular methods to do this. After dividing the input attributes, one can use the attributes in major factor group as the input of the primary neural network and those in minor factors as inputs of the secondary neural network.

## 60.4 Conclusions

We introduced the artificial intelligence algorithms and their applications. We proposed a framework for pattern classifications. This framework includes two neural networks and a logic circuit. The first neural network serves to classify regular instances, and the second one handles the overlapped instances caused by the first neural network. The proposed framework integrates the two neural networks by the logic circuit and then makes the final classifications. We provided two methodologies using the proposed framework including financial crisis prediction and principle components analysis. In the financial crisis prediction case, we used the financial ratios of individual companies as the inputs for the primary neural network and the macroeconomic indicators as the inputs for the secondary neural network. In the principle components analysis case, we used the attributes in the major factor group as the input of the primary neural network and those in minor factors as the inputs of the secondary neural network.

As for further studies, people might add fuzzy logics into the proposed framework to furthermore classify overlapped patterns. Using more neural networks for the framework might be interesting research topics for further works.

## References

1. Altman EI (1968) Financial ratios, discriminant analysis and the prediction of Corporate bankruptcy. J Finance 589–609
2. Altman EI, Marco GV, Varetto F (1994) Corporate distress diagnosis: Comparisons using linear discriminant analysis and neural networks (The Italian Experience). J Bank Finance 18:505–529
3. Beaver WH (1966) Financial ratios as predictors of failure. J Accoun Res 71–111
4. Blum M (1974) Failing company discriminant analysis. J Account Res 12:1–25
5. Coats PK, Fant LF (1993) Recognizing financial distress patterns using a neural network tool. Financ Manage 12:142–155
6. Mao C (2002) A study on financial crisis forecasting using a mixed model of neural network and digital logic manipulation — A case study of public electronic companies. Master Thesis, Shu-Te University, Kaohsiung County, Taiwan

# Chapter 61
# Chinese Document Clustering Using Self-Organizing Map-Based on Botanical Document Warehouse

**Howard Lo, Che-Chern Lin, Rong-Jyue Fang, Chungping Lee, and Yu-Chen Weng**

**Abstract** The exponential growth of information has made an overflow situation in the sea of information. It had created difficulties in the search for information. An efficient method to organize the query of information and assist users' navigation is therefore particularly important. In this paper, we applied Self-Organizing Map (SOM) algorithm to cluster Chinese botanical documents onto a two-dimensional map. Each botanical document has been regarded as bags of words, and transferred into plain text respectively. We applied term frequency and inverse term frequency to extract key terms from documents as the input of SOM. 892 Chinese botanical documents have been projected onto a 2D map to assist users' navigation. In our experimental results, the lowest recall was 0.71 for Polygonaceae documents and the highest recall rate was 0.94 for Amaranthaceae documents. The lowest precision rate was 0.81 for Umbelliferae documents, and the highest precision rate was one hundred percent for Convolvulaceae and Cruciferae documents.

## 61.1 Introduction

The Self-Organizing Map (SOM) algorithm, which is developed by Kononen [1], is largely used in images and documents clustering. The SOM algorithm works best with browsing tasks, especially in a very large document storage, such as a document warehouse [2].

The aim of this study is to apply SOM algorithm to cluster those botanical data in a document warehouse into a two-dimensional map, in order to fulfill the naïve users' information needs. We exploited the discrimination values to cull important terms from the document warehouse, and transferred the

H. Lo (✉)
Department of Industrial Technology Education, National Kaohsiung Normal University, Kaohsiung City 806, Taiwan
e-mail: howardlab@gmail.com

N. Mastorakis et al. (eds.), *Proceedings of the European Computing Conference*, Lecture Notes in Electrical Engineering 28,
DOI 10.1007/978-0-387-85437-3_61, 

unstructured documents into vectorized tuples as inputs of the SOM algorithm. The results will be projected onto a two-dimensional SOM map.

When considering retrieval performance evaluation, we adopted the recall rate and precision rate as our measurements.

The rest of the paper is organized as follows: Section 61.2 describes related works and the SOM algorithm. Section 61.3 describes the research steps and model of our research. The evaluation and experiment results are discussed in Sections 61.4 and 61.5 reaches a conclusion and discusses possible future research.

## 61.2 Related Works

According to the survey of Sullivan [3], eighty percent of the necessary information for the decision maker came from unstructured data, such as documents and only twenty of them came from structured data, such as a database.There is a need to deal with documents.

### *61.2.1 Dealing With Unstructured Documents*

Since the 1950 s, lots of researchers have been trying to deal with unstructured documents with computer technologies. In 1961, Doyle used an association map to describe documents. In 1967, Borko proposed an idea to index and classify documents using computer technologies [4]. In 1980, Porter [5] proposed an algorithm to remove suffixes automatically from words in English and restore the words to their original form. In 1983, Salton recommended a vector space model (VSM) to help users compare the similarities between documents [6–9]. Salton's VSM has made progress in information retrieval from documents' indexing and classifying, to comparison of similarities. Their studies have proved one thing: it's possible to deal with documents using computer technologies.

Artificial neural networks are currently attracting particular interest as methodologies to deal with unstructured data. Chen [10] applied Hopfield network to classify terms in the documents. Nigam proposed the use of maximum entropy techniques for text classification in his paper [11]. Mladenic [12] applied a näive Bayesian classifier to select features from documents. Chen et al. [13] exploited a fair feature-subset selection algorithm and an adaptive fuzzy learning network to classify on-line documents.

Since Kohonen introduced the SOM in the early 1980 s [14], the SOM algorithm has attracted a great deal of interest among researchers in a wide variety of domains. In 1996, Lagus et al. [15] applied SOM to develop a tool called WEBSOM to cluster documents. Kohonen et al. mapped 684,0568 patent abstracts onto a SOM with 1002240-node [16]. In 2002, Vicente et al. [17]

exploited vector space model to extract key terms from documents as an input of SOM. In 2003, Smith et al. developed a LOGSOM system using Kohonen's self-organizing map to organize web pages onto a two-dimensional map [18]. In 2004, Azcarraga et al. [19] analyzed the relative weight distribution of the SOM weight vectors and took advantage of some characteristics of the random projection method to select keywords in order to extract meaningful labels for each cluster on the map.

### *61.2.2 Self Organizing Map Algorithm*

The SOM is an artificial neural network model based on unsupervised learning, introduced by Kohonen in the early 1980 s. The SOM is a tool for automatically arranging high dimensional statistical data onto a two-dimensional map. Therefore, it is able to transfer sophisticated, non-linear statistical relationships among high dimensional data into a simple, low dimensional map.

In our study, we adopted Smith's revised SOM algorithm [18]. The algorithm is summarized as follows.

- **Initialize input nodes, out nodes, and connection weights**. Initialize a small random value for $W_{ij}$ as seeds. Assign $N_k$ as neighborhood size. Assign a random value between zero and one for parameter function $\alpha(t)$and$\sigma^2(t)$. In the study, we assigned $\alpha(0) = 0.1$and $\sigma^2(0) = 1$.
- **Calculate the distance with Euclidean distance**. Present an input vector $x_i$ to represent document $x_i$ and calculate the similarity $d$ (distance) of the input vector to the weights $w$ of each node $j$ using Equation 61.1.

$$d_j = \|x - w_j\| = \sqrt{\sum_i^n (x_i - w_{ij})^2} \quad (61.1)$$

- Select the node with minimum Euclidean distance as the winner node $k$.
- Update the weight. Applying Equation 61.2 to adjust the weight related to the input layer and winning node, and its neighboring nodes.

$$w_{ij}(t+1) = w_{ij}(t) + c(x_i - w_{ij}(t)) \quad (61.2)$$

where $c = \alpha(t)\exp(-\|r_i - r_k\|/\sigma^2(t))$, and $r_i - r_k$ is the number of nodes, to represent the physical distance between node j and winning node $k$.
- **Repeat the step above, until the weights have stabilized**. In each epoch, increase t by 1, and decrease the neighborhood size, $\alpha(t)$and$\sigma^2(t)$by Equation 61.3.

$$\alpha(t) = \alpha(0)N_k(t)/N_k(0) \quad (61.3)$$

$N_k$: size of neighborhood.

- **The final stage.** The iterative step will be terminated after all the vectors of the documents have been presented to the training network, and mapped onto a two-dimensional map.

### *61.2.3 Users' Information Need*

According to the researches of Baeza-Yates [20], there are three seeking techniques for most users to fulfill their information needs.

The first technique is browsing. Marchionini defined browsing behavior as "*an exploratory, information seeking strategy that depends upon serendipity and especially appropriate for ill-defined problems and for exploring new task domains*" [21].

The second technique is searching. According to the definition of Sim, searching is "*an analytical process, is a planned activity with a specific goal, such as to find a particular fact*" [22].

The third technique is visualization. Visualization interfaces, such as Windows, menu, icon, and dialog boxes, make the computer system more friendly and easy to use. The goal of this study is to apply SOM algorithm to cluster Chinese documents onto a two-dimensional SOM. With this map, it can assist users' browsing and visualization needs in information retrieval.

## 61.3 Methodology

In our research, we regarded documents as a bag of words. We ignored the order of the words as well as any punctuations or structural information, but retained the frequencies of every word. Our research steps are as follows.

- **Acquiring documents**. We attained 892 botanical documents from Herbarium, Research Center for Biodiversity, Academia Sinica, in Taiwan (HAST) [23] and Taiwan Agricultural Chemicals and Toxic Substances Research Institute, Council of Agriculture (TACTRI/COA) [24] as our training data. The documents are all in HyperText Markup Language (html) format, and needed to be transferred to plain text.
- **Pre-processing documents**. All tags in the botanical documents needed to be removed and transferred to plain text as training data.
- **Segmenting Chinese terms**. CKIP was a Chinese term segmentation system, developed by Institute Linguistics, Academia Sinica, Taiwan [25]. We applied the CKIP system to extract Chinese terms from sentences.
- **Extracting Chinese key terms**. We removed the stopwords from the documents, and revised the errors which segmented by CKIP with our revising system developed by Java language.

- **Calculating term weight**. Calculate the weight of every key term. We applied Equation 61.4 to calculate the term weight.

$$w = (tf \times idf) \tag{61.4}$$

where $idf = \log\left(\frac{N}{df}\right)$, $W$: the term weight, $tf$: the term frequency, $df$: the term frequency shows in the document collection, $idf$: the inverse $df$.
- **Vectorizing documents**. After the term extracting and weight calculating, the 892 botanical documents will transfer into vectorized statistical data shown as Equation 61.5.

$$D_i = (W_1, W_2, W_3, \ldots, W_n) \tag{61.5}$$

where $D_i$ represents the 892 documents, and $W$ represents the term weight.
- **Clustering documents with SOM**. The Vectorized documents will be used as input data of SOM algorithm. After the training processes, the input data will be mapped to the Kohonen layer.
- **Projecting documents onto a SOM map**. The vectorized data $D_1 \ldots D_i$ will be projected onto a two-dimensional grid map. This map can represent the relationships among documents. The features of adjacent nodes will be similar to the centroid node.

## 61.4 Results

In this study, we applied Vector Space Model, developed by Salton to vectorize the 892 botanical documents into statistical data as the input value of SOM algorithm. We assigned 20×20 nodes in our research. The learning rate would be 01, 0.5, and 0.9; learning epochs would be 100, 200, 400, and 500.

### *61.4.1 Discussion*

Figure 61.1 represents an output map of SOM. Each number on the cell points to the amount of botanical documents clustered into this cell. There are two documents in cell (10, 4). Those features of these two documents were "葉對生 (Phyllotaxy: Opposite) 長橢圓形 (Leaf shape: Oblong) 葉尖銳 (Apex: Acuminate) 全緣 (Leaf margin: Entire) 花白色 (flower: White) 球形 (Inflorescence: sphere) 腋生 (Inflorescence: axillary)". When we look up the botanical illustration, we found they are the same plant, but using different Chinese names. Both of them are "Alternanthera philoxeroides (Mart.) Griseb" in the scientific name of "長梗滿天星" and the nickname of "空心蓮子草" in Chinese. They have been clustered into the same area correctly.

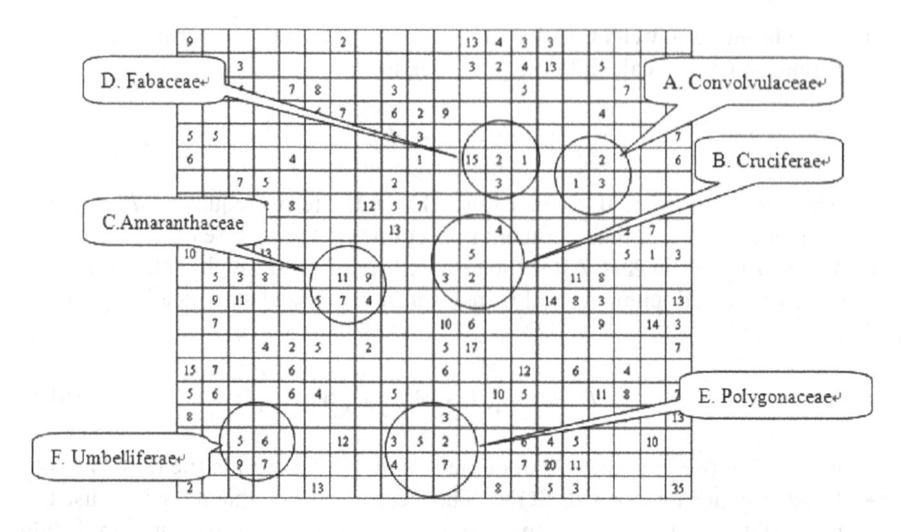


**Fig. 61.1** Analysis of botanical document map

## *61.4.2 Evaluation*

To evaluate the efficiency of an information retrieval system, Baeza-Yates [23] proposed recall rate and precision rate as the measurements. Equation 61.6 is the formula for calculating recall rate, and Equation 61.7 is the formula for calculating precision rate.

$$\text{Recall} = \frac{|\text{number of relevant Docs retrieved}|}{|\text{number of relevant Docs in collection}|} \tag{61.6}$$

$$\text{Precision} = \frac{|\text{number of relevant Docs retrieved}|}{|\text{number of Docs retrieved}|} \tag{61.7}$$

Figure 61.1 depicts the analysis of a botanical document map, and Table 61.1 shows the result of our study.

**Table 61.1** Analysis of recall and precision rate

| Area | Family | Related docs been retrieved | Related docs in collection | Retrieve d docs | Recall rate | Precision rate |
|---|---|---|---|---|---|---|
| A | Convolvulaceae | 6 | 8 | 6 | 0.75 | 1 |
| B | Cruciferae | 14 | 15 | 14 | 0.93 | 1 |
| C | Amaranthaceae | 32 | 34 | 36 | 0.94 | 0.89 |
| D | Fabaceae | 19 | 25 | 21 | 0.76 | 0.90 |
| E | Polygonaceae | 22 | 31 | 24 | 0.71 | 0.92 |
| F | Umbelliferae | 22 | 30 | 27 | 0.73 | 0.81 |

In our experimental results, the lowest recall was 0.71 for Polygonaceae documents and the highest recall rate was 0.94 for Amaranthaceae documents. The lowest precision rate was 0.81 for Umbelliferae documents, and the highest precision rate was one hundred percent for Convolvulaceae and Cruciferae documents.

## 61.5 Conclusion

In this work we applied SOM algorithm to classify the Chinese botanical documents onto a two-dimensional map. The method performed a completely automatic and unsupervised clustering of the botanical documents.

The result shown in Fig. 61.1 depicted a lowest recall rate 0.71 for Polygonaceae and lowest precision rate 0.81 for Umbelliferae. The experimental results depicted a good performance in recall and precision rate. This botanical document map can assist users to browse information with visualization SOM map.

In the future research we hope to adopt other artificial neural networks to cluster or classify unstructured data, and improve the efficiency of information retrieval.

## References

1. Kohonen T, Kaski S, Lagus K, Salojärvi J, Honkela J, Paatero V, Saarela A (2000) Self-organization of a massive document collection. IEEE Trans Neural Netw 11:574–585
2. Fang R-J, Lo H, Weng Y-C, Tsai HL (2007) Mobile learning system using multi-dimension data warehouse concept-based on botanical data. In: Proceeding of the 6th WSEAS International Conference on Applied Computer Science, Hangzhou, China
3. Sullivan D (2001) Document warehousing and text mining. John Wiley & Sons, Inc .
4. Trybula WJ (1999) Text mining and knowledge discernment: An exploratory investigation. Ph.D., The University of Texas at Austin
5. Porter M (1980) Readings in information retrieval. Morgan Kaufmann, San Francisco
6. Salton G, Wong A, Yang CS (1975) A vector space model for automatic indexing. Commun ACM 18:613–620
7. Salton G, McGill M (1983) Introduction to modern information retrieval. McGraw Hill, New York
8. Salton G, Buckley C (1988) Term-weighting approaches in automatic text retrieval. Inf Process Manag 24:513–532
9. Salton G, Singhal A (1995) Automatic text browsing using vector space model. Department of Computer Science, Cornell University
10. Chen H, Ng T (1995) An algorithmic approach to concept exploration in a large knowledge network (Automatic Thesaurus Consultation): Symbolic Branch-and-Bound Search vs. Connectionist Hopield Net Activation. J Am Soc Inf Sci 45:348–369
11. Nigam K, Lafferty J, McCallum A (1999) Using maximum entropy for text classification. In: IJCAI-99 Workshop on Machine Learning for Information Filtering, Stockholm, Sweden, pp 61–67
12. Mladenic D, Grobelnik Marko (2003) Feature selection on hierarchy of web documents. Decision Support Sys 35:45–87

13. Chen C-M, Lee H-M, Tan C-C (2006) An intelligent web-page classifier with fair feature-subset selection. Eng Appl Artif Intell 19:967–978
14. Kohonen T (1995) Self-organizing maps. Springer, Berlin, Heidelberg
15. Lagus K, Honkela T, Kaski S, Kohonen Teuvo (1996) Self-organizing maps of document collections: A new approach to interactive exploration. In: Proceedings of the Second International Conference on Knowledge Discovery & Data Mining
16. Kohonen T, Kaski S, Lagus K, Salojärvi J, Honkela J (2000) Self organization of a massive document collection. IEEE Trans on Neural Netw 11:574–585
17. Guerrero VP, Moya-Anegón FL, and Herrero-Solana V (2002) Automatic extraction of relationships between terms by means of Kohonen's algorithm. Libr Inf Sci Res 24:235–250
18. Smith KA, Ng A (2003) Web page clustering using a self-organizing map of user navigation patterns. Decision Support Sys 35:245–256
19. Azcarraga AP, Y Jr TN, Tan J, Chua TS (2004) Evaluating keyword selection methods for WEBSOM text archives. IEEE Trans Knowl Data Eng 16:380–283
20. Baeza-Yates R, Ribeiro-Neto B (1999) Modern information retrieval.
21. Marchionini G, Shneiderman B (1988) Finding facts vs. browsing knowledge in hypter-text systems. IEEE Comput 70–79
22. Sim SE, Clarke CLA, Holt RC, Cox AM (1999) Browsing and searching software architectures. In: Proceedings of the International Conference on Software Maintenance, Oxford, England
23. T. Academia Sinica (2007) Herbarium, research center for biodiversity. Available at http://hast.sinica.edu.tw
24. Hsu LM, Chiang MY (2007) Lawn weeds in Taiwan. Available at http://www.tactri.gov.tw/htdocs/plant/lawnweed/index.asp
25. Institute Linguistis (2007) CKIP Chinese term segmentation system. Academia Sinica

# Chapter 62
# Linear Discriminant Analysis in Ottoman Alphabet Character Recognition

**Zeyneb Kurt, H. Irem Turkmen, and M. Elif Karsligil**

**Abstract** This paper proposes a novel Linear Discriminant Analysis (LDA) based Ottoman Character Recognition system. Linear Discriminant Analysis reduces dimensionality of the data while retaining as much as possible of the variation present in the original dataset. In the proposed system, the training set consisted of 33 classes for each character of Ottoman language alphabet. First the training set images were normalized to reduce the variations in illumination and size. Then characteristic features were extracted by LDA. To apply LDA, the number of samples in train set must be larger than the features of each sample. To achieve this, Principal Component Analysis (PCA) were applied as an intermediate step. The described processes were also applied to the unknown test images. K-nearest neighborhood approach was used for classification.

## 62.1 Introduction

Within the last years, rapidly growing interest in document archiving and retrieval systems increased attention to character recognition systems. Numerous approaches have been proposed for optical or handwritten character recognition mostly in Latin-based alphabets. On the contrary, some alphabets as Arabic, Chinese, and Japanese have picture-like characters which require more complex recognition algorithms to identify. This paper presents an Ottoman alphabet recognition system. Ottoman characters are the characters of an alphabet that was formed by adding some new letters from Farce and Turkish alphabet to the Arabian alphabet. Besides, Ottoman alphabet has been improved with its original features and has become an art due to its various scriptural formats.

There are few implementations of automated Ottoman alphabet recognition. Atici and Yarman-Vural applied chain coding to characters to propose a

Z. Kurt (✉)
Department of Computer Engineering, Yildiz Technical University,
34349 Besiktas/Istanbul, Turkey
e-mail: zeyneb@ce.yildiz.edu.tr

N. Mastorakis et al. (eds.), *Proceedings of the European Computing Conference*, Lecture Notes in Electrical Engineering 28,
DOI 10.1007/978-0-387-85437-3_62, 

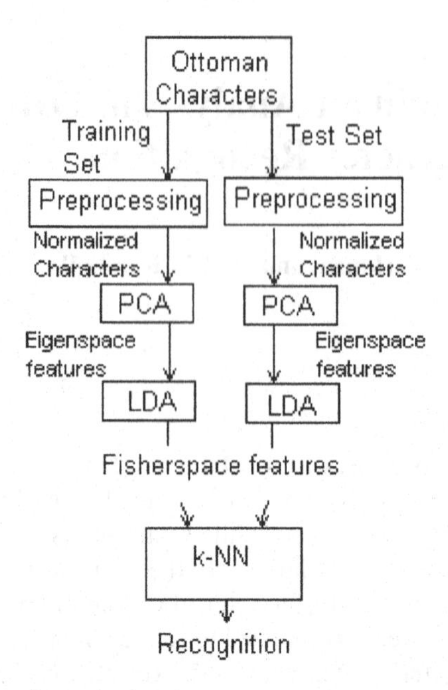


**Fig. 62.1** Block diagram of proposed system

Hidden Markov Model based Ottoman alphabet recognition system [1]. Ozturk, A.Gunes proposed a method for recognition of isolated Ottoman characters by neural networks [2].

We propose a new global-feature-based approach for Ottoman alphabet recognition by using Linear Discriminant Analysis (LDA). LDA projects the data onto a lower dimensional space while preserving as much of the class discriminatory information as possible.

Ottoman script is fundamentally cursive. Since this paper focuses on the character recognition, the segmentation of the connected characters has not been studied. In Fig. 62.1, the block diagram of the proposed system is illustrated. The outline of the paper is as follows: Section 62.2 presents preprocessing steps of the system. Section 62.3 describes extraction of the characteristics features using LDA. In Section 62.4, decision steps of the system are described. Finally in section 62.5, experimental results of our implementation are given. Moreover, the performance of the proposed system is discussed.

## 62.2 The Preprocessing

Ottoman characters have many common points with each other in their main shapes, whereas they differ in detail. In the proposed system, each Ottoman character is considered as a class. Our main goal is to discuss the influence of the components that points out the within class similarities and between class differences on success rate.

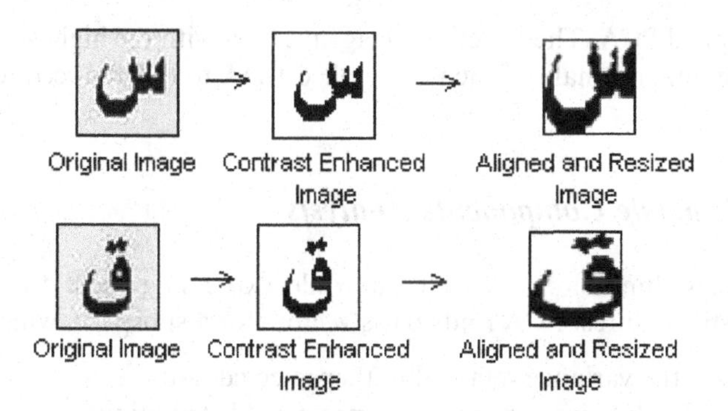


**Fig. 62.2** Examples of the preprocessing steps

Since our purpose is to measure the performance on recognition of isolated characters, the classification success of separate characters is scrutinized.

The objective of LDA is to find out the most efficient combinations to split up multiple classes and to maximize between class differences. Thus, it is widely used in several applications such as face recognition, speech recognition, image summarization and data classification [5].

As a precondition of LDA implementation, the size of each sample must be less than the number of samples in the training set. Since the training set is inconvenient with that precondition, dimension reduction should be applied. For this purpose PCA is applied to training set as an intermediate step.

The main purpose of preprocessing is to reduce the variation in size and illumination. The preprocessing steps are applied both to the training and testing images.

In this study, 256 gray level images with a light colored background were used. By the contrast enhancement, not only were the characters slimmed down and their identification improved, but also the potential noises in the images were eliminated. Moreover, the images were aligned and resized.

Figure 62.2, shows the preprocessing steps applied on images. For each image, the inside of the frame which was drawn by using the tangents to the maximum and minimum points of characters on x and y coordinates is taken into consideration.

After the alignment process, images may be in several sizes. In this study, each character was normalized to $32 \times 32$ pixels.

## 62.3 The Feature Extraction

Each Ottoman character has some features that are similar to others and different from others. In this study LDA was used to extract the characteristic features of Ottoman characters. To implement the LDA, the number of samples in train set must be larger than the features of each example. To achieve this, first we applied PCA. The selection of eigenvectors with the highest eigenvalues reduces the dimensionality. Then we applied LDA to this reduced feature set.

### 62.3.1 Principle Components Analysis

PCA reduces dimensionality of data while expressing most the important characteristic features. PCA finds basis vectors for a subspace, which:

- maximizes the variance retained in theprojected data
- or (equivalently) gives uncorrelated projecteddistributions
- or (equivalently) minimizes the least square reconstruction error[3].

Each of the p number of images whose size is n*m is assumed to be [n*m]*1 sized column vectors that correspond to the original data [4].

$$\Omega = X^{*}X^{T} \tag{62.1}$$

The covariance matrix ($\Omega$) is defined by the Equation 62.1 where X is the mean matrix subtracted vector whose size is [n*m]*p, where there are p image in the training set. Because of the multidimensionality, ($\Omega$) is calculated using (62.2), instead of (62.1).

$$\Omega = X^{T*}X \tag{62.2}$$

The eigenvalues and associated eigenvectors are obtained as in (62.3) where $\Omega$ is the covariance matrix, $\lambda$ is the set of eigenvalues and V is the associated eigenvectors.

$$\Omega^{*}V = \lambda^{*}V \tag{62.3}$$

We do not have to use the whole acquired eigenvectors. Therefore the eigenvectors are sorted by descending order and only the selected number of highest eigenvalues are used.

PCA was applied to both training data and test data. The acquired projections in the eigenspace were given to LDA as input.

### 62.3.2 Linear Discriminant Analysis

LDA aims to determine the linear combinations of the features which can separate the objects and events into more than one class. These combinations, which were obtained after applying LDA, can be used as a linear classifier or it can be used to decrease the dimension for the classification step.

LDA maximizes between-class scatter while minimizing the within-class scatter [6].

The projections in the eigenspace, which were obtained by PCA and belonged to the alphabet's each character, were vectorized. The mean vector (M) of the alphabet's character set was calculated. The mean vectors of each character class (Mi) were calculated and for each class the mean vector of the class was subtracted from each character in a class. Let m be the number of the samples in each class, n be the number of classes and A be the mean-centered eigenspace projections.

The scatter matrices for each class were acquired by:

$$S_1 = A_{11}{}^* A_{11^T} + \ldots + A_{1m}{}^* A_{1m^T} \quad (62.4)$$

$$S_n = A_{n1} * A_{n1^T} + \ldots + A_{nm} * A_{nm^T} \quad (62.5)$$

The within class scatter matrix $S_w$ was obtained by (62.6) where $M_i$ is the mean vector of $i^{th}$ class and M is the mean vector of whole characters in the train set.

$$S_w = S_1 + S_2 + \ldots + S_n \quad (62.6)$$

The between class scatter matrix $S_B$ was built by:

$$S_B = 2^* \Sigma(M_i - M) \quad (62.7)$$

The eigenvectors (V) and eigenvalues ($\lambda$) were calculated as:

$$S_B{}^* V = \lambda^* S_w{}^* V \quad (62.8)$$

The eigenvectors were sorted by their associated eigenvalues in descending order and the first n-1 eigenvectors were kept. Fisher space projections of each class in the train set were obtained by using these eigenvectors.

## 62.4 Classification

Each test image was first projected into eigenspace and then into Fisher space. The k-Nearest Neighborhood was used for classification of unknown test images. The system performance was evaluated by applying 10-fold cross-validations [7].

## 62.5 Experimental Results

We conducted experiments using a database of 10 sample images of 33 Ottoman alphabet characters. The samples, gathered from electronic resources were both handwritten and in press printed format. In k-nearest neighborhood, k was selected as 3 by 10-fold cross-validation.

Figure 62.3(a) illustrates true classification of three test examples.

Figure 62.4 illustrates false classification of three test examples. As it can clearly be seen, the test examples are very similar to the characters of the incorrect classes.

As given in Table 62.1 success ratio of our Ottoman Character Recognition system is 88%.

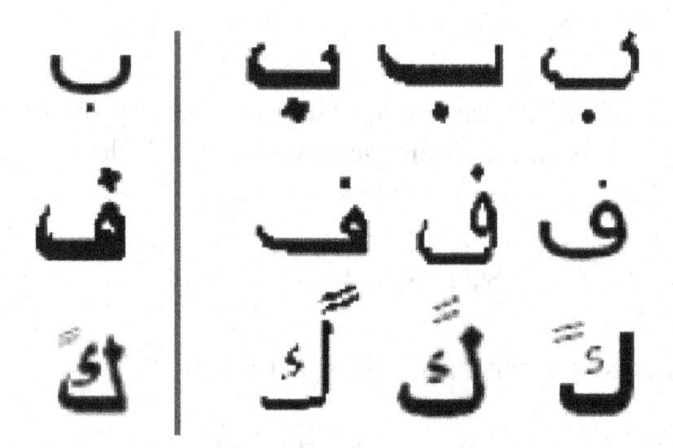

**Fig. 62.3** **(a)** Three samples for the test data which were recognized correctly. **(b)** The train set samples of these characters respectively

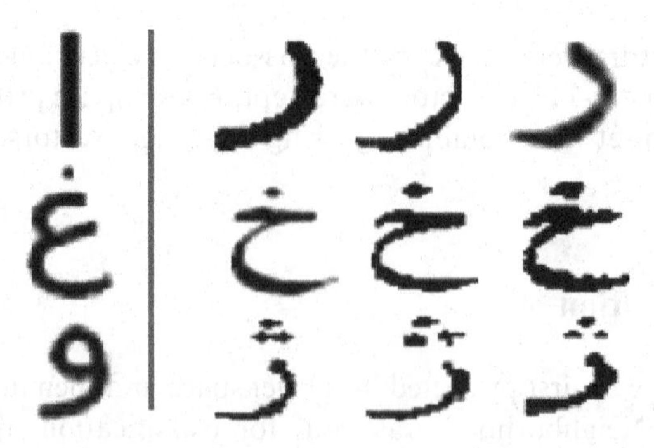

**Fig. 62.4** **(a)** Three samples for the test data which were recognized incorrectly. **(b)**Train set samples for the decision classes of the system for incorrectly recognized letters

Table 62.1 The success ratio of the processed system

| | |
|---|---|
| True Classification | 290 |
| False Classification | 40 |
| Total Number of Characters | 330 |
| Recognition Ratio | 88% |

## 62.6 Conclusion

This study presents a Linear Discriminant Analysis based automatic Ottoman Alphabet Character Recognition System. This approach retains class separability while reducing the dimensionality. As the performance of our proposed system is very promising for the recognition of individual characters, in the future work we will study a system for the recognition of cursive characters in scripts. This will yield to processing and recognition of a large number of historical documents and we will be able to archive these documents.

## References

1. Atici AA, Yarman-Vural FT (1997) A heuristic algorithm for optical character recognition of Arabic script. Signal Process 62(1):87–99
2. Ozturk A, Gunes S, Ozbay Y (2000) Multifont Ottoman character recognition. In: Proceedings of the 7th IEEE Int. Conf. on Electronics Circuits and Systems (ICECS), Jounieh, Lebenon, pp 945–949
3. Turk M, Pentland A (1991) Eigenfaces for recognition. J Cog Neurosci 3(1)
4. Turk M, Pentland A (1991) Face recognition using eigenfaces. In: Proceedings of the IEEE Conference on Computer Vision and Pattern Recognition, Maui, Hawaii, USA, pp 586–591
5. Belhumeur PN, Hespanha JP, Kriegman DJ (1996) Eigenfaces vs. Fisherfaces: Recognition using class specific linear projection. In: Proceedings of the 4th European Conference on Computer Vision, ECCV'96, Cambridge, UK, pp 45–58
6. Etemad K, Chellappa R (1997) Discriminant analysis for recognition of human face images. J Opt Soc Am A 1724–1733
7. Kung SY, Mak MW, Lin SH (2004) Biometric authentication: A machine learning approach. Prentice Hall

# Chapter 63
# User Requirements Engineering and Management in Software Development

Zeljko Panian

**Abstract** Requirements engineering is an often underutilized discipline in software development. Yet the economic benefits – lowered development costs, faster development and better products that delight customers – have plainly been demonstrated in study after study. In this paper, the case is made for requirements engineering as an integral part of any product lifecycle. Metrics are provided to show the positive effect of requirements engineering and requirements management (RM), from individual developers to the business to the customer. A software product is examined as an example of how to bring effective requirements management into product development.

## 63.1 Introduction

Many software problems arise from shortcomings in the ways that people acquire, document, agree on, and modify the product's requirements. Typical problem areas include informal information gathering, implied functionality, inadequately defined requirements, and a casual change process.

Most people wouldn't ask a construction contractor to build a custom multi-hundred-thousand-dollar house without extensively discussing their needs and desires and refining the details progressively. Homebuyers also understand that making changes carries a price tag.

However, people blithely gloss over the corresponding issues when it comes to software development. One of the widely quoted CHAOS reports from The Standish Group relates the consequences of casual approaches to requirements engineering [12]. Year after year, lack of user input, incomplete and changing requirements are the major reasons why so many information technology projects fail to deliver all of their planned functionality on schedule and within budget.

Z. Panian (✉)
Faculty of Economics and Business, University of Zagreb, Croatia

N. Mastorakis et al. (eds.), *Proceedings of the European Computing Conference*, Lecture Notes in Electrical Engineering 28,
DOI 10.1007/978-0-387-85437-3_63, 

Nor do requirements issues only plague the early stages of a software project. One study at a large defence contractor found that 54 percent of all the errors were discovered *after* unit testing was complete, and that 45 percent of these were requirements or design errors [4]. Other studies also indicate that 40–60 percent of all defects found in a software project can be traced back to errors made during the requirements stage.

Nonetheless, many organizations still practice ineffective methods for this essential project activity. The typical outcome is an expectation gap, a difference between what developers think they are supposed to build and what customers really need.

## 63.2 Stakeholders' Interests

Nowhere more than in the requirements process do the interests of all the stakeholders in a software or system project intersect. These stakeholders include end-users, customers acquiring the product, requirements analysts, project managers, developers, testers, regulators and auditors, manufacturing staff, sales and marketing, and field support or helpdesk staff.

Handled well, this intersection can lead to exciting products, delighted customers, and fulfilled developers. Handled poorly, it is the source of misunderstanding, frustration, and friction that undermine the product's quality and business value.

Requirements are the foundation for all the subsequent software development and project management activities. Therefore, all stakeholders must be committed to following an effective requirements process. This process includes activities for:

- Requirements development (eliciting, analyzing, specifying, and validating requirements).
- Requirements engineering (dealing with the requirements once you have them in hand).

Developing and engineering requirements is hard! There are no simple shortcuts or magic solutions. But a toolkit of techniques, coupled with tools to store and manage requirements, provides a powerful defence against the expectation gap.

## 63.3 Requirements are Interrelated

If John Donne had been a business analyst rather than a seventeenth century poet, he might have said, "No requirement is an island." Requirements have a natural parent–child relationship because high-level user requirements can be decomposed into lower level functional requirements that, in turn, generate specific test requirements.

For example, the requirement “Generate e-mail confirmation to customer on-line order” would produce a number of distinct functional requirements (“Verify e-mail,” “Transmit message,” “Receive returned e-mail message,” etc.), each of which needs to be implemented in a code module and requires testing.

Understanding these relationships is essential to planning and executing adequate test coverage. And when analyzing a proposed change to a requirement – or test procedure – it should be known what other requirements may be affected by the change.

Traceability between requirements has to be maintained since this provides a useful tool for the analyst or software designer when trying to asses the software product quality. For example:

- Users can define multiple types of requirements and maintain relationships between them.
- Requirement types can be customized for each project.
- Trace-to and trace-from views can display the parent –child relationship between requirements.
- Requirements can be organized into folders for easier navigation.
- Stakeholders need to be automatically informed via e-mail or requirements traceability alerts when a requirement changes.
- Audit trail can track which values are changed and when.

But, traceability between requirements only solves part of the problem. It is also essential to follow individual requirements through the application lifecycle to test defects and release. That means testers and quality managers should be able see what and how much testing is planned for each requirement. They can than analyze the status of the testing to see how much has been completed and how many defects have been found. And when defects are found, it lets them see which requirements are affected.

## 63.4 The Need for Requirements Management

The major consequence of requirements problems is rework – doing something again that you thought was already done. Rework can consume 30–50 percent of total development cost [3] and requirements errors account for 70–85 percent of the rework cost [10]. As illustrated in Fig. 63.1 it costs far more to correct a defect that is found late in the project than to fix it shortly after it was created [7].

Preventing requirements errors and catching them early, therefore, has a huge leveraging effect on reducing rework.

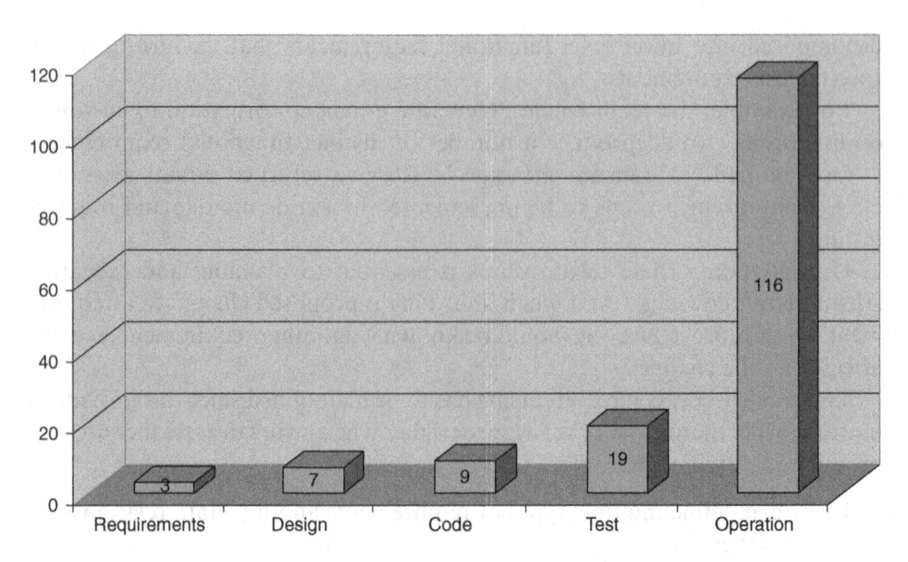


**Fig. 63.1** Relative cost to correct a requirement defect depending on when it is discovered

Shortcomings in requirements cause no end of headaches on software projects. For example, creeping requirements contribute to schedule and budget overruns. Now, change is going to happen. Requirements change because initial elicitation activities are imperfect, because business needs evolve, and because customer expectations change once they see the product beginning to take shape.

Project plans and commitments therefore need to anticipate change, rather than expecting that an initial stage of gathering requirements is all that's needed. Requirements creep, though, is the uncontrolled growth of requirements well beyond the original scope boundary.

The way in which requirements changes are incorporated into the product can degrade quality. Changes need to be introduced at the highest level of abstraction they affect and follow the links in the traceability chain to see the impact of each proposed change. Traceability data stored in an appropriate requirements management tool greatly facilitates this impact analysis.

Many projects suffer from gold-plating. Developers might add unnecessary features they think the users will like, or users request excessively complex systems. Tracing each functional requirement back to its origin – such as a user task, a business rule, or a business objective – helps solve this problem.

The result of all of these kinds of requirements problems is to slow the pace of software development. It doesn't do you any good to work very efficiently on the wrong requirements. If you don't get the requirements right, it doesn't matter how well your team executes the rest of the project.

## 63.5 Accelerating Development

A sign in a high-school chemistry lab reads, “If you don’t have time to do it right, when will you have time to do it over?” [1].

To convince itself of the benefits, a firm should think about how much effort on their projects is misspent because of miscommunications, misunderstandings about requirements, missing information, or misstated requirements. The case for emphasizing solid requirements practices is an *economic argument*, not a philosophical or technical argument.

A solid foundation of agreed-upon business requirements and project objectives establishes a shared vision, goals, and expectations. Frequent and intimate user involvement makes software development a collaborative partnership between IT and the customer community. It also reduces the chance of system rejection upon rollout.

It is frustrating for everyone to release a product only to find that essential functionality is missing and other capabilities are included that no one will ever use. Good requirements processes ensure that the functionality built enables users to perform essential business tasks. Well-specified requirements also define achievable quality expectations, letting the team implement both the capabilities and the characteristics that will make the users happy.

While typical projects spend perhaps 10 percent of their effort on requirements, investing more has a big payoff. A study of 15 projects in the telecommunications and banking industries revealed that the most successful projects spent 28 percent of their resources on requirements engineering [8].

The average project devoted 15.7 percent of its effort and 38.6 percent of its schedule to requirements engineering. National Aeronautics and Space Administration (NASA) projects that invested more than 10 percent of their total resources on requirements development had substantially lower cost and schedule overruns than projects that devoted less effort to requirements [9].

In a European study, teams that developed products more quickly devoted more of their schedule and effort to requirements than did slower teams, as shown in Table 63.1.

Not all of the requirements development effort should be allocated to the beginning of the project. Don’t think of it as a “requirements phase,” but rather a set of requirements-development activities that are threaded

**Table 63.1** Investing in Requirements [2]

| | Effort devoted to requirements (%) | Schedule devoted to requirements (%) |
|---|---|---|
| Faster Projects | 14 | 17 |
| Slower Projects | 7 | 9 |

throughout the project's lifecycle. Projects that follow an incremental lifecycle will work on requirements in every iteration through the development process.

Agile development projects take an incremental approach, rapidly building small portions of the product so users can refine their needs and developers can keep up with changing business demands. Agile projects will have frequent but small requirements development efforts [11]. Regardless of how large an increment your team tackles, they need to understand the requirements for that increment.

## 63.6 Requirements Development and Management

The goal of requirements *development* is to deduce, capture, and agree upon a set of functional requirements and product characteristics that will achieve the stated business objectives. Requirements *management,* then, involves everything that happens with the requirements in the product lifecycle once the requirements are developed, as shown in Fig. 63.2.

The demarcation between requirements development and requirements management is the defining of a requirements baseline. A baseline is a snapshot in time that represents the agreed-upon set of requirements for a specific product release.

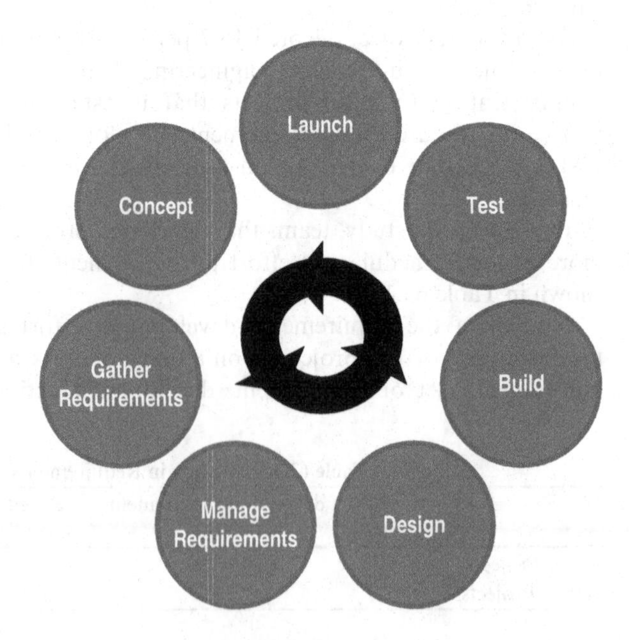


**Fig. 63.2** Product life cycle

## 63.7 Strategies and Best Practices for Better User Requirements Management

A variety of strategies and practices can help software development teams bridge communication gaps and do a better job of understanding, documenting and communicating customer needs. Figure 63.3 synthesizes several best practices we evidenced in our research, and yet in the categories of elicitation, analysis, specification, validation, and management.

- *Requirements Elicitation* – Defining the product vision and project scope; identifying stakeholders, customers and users; choosing elicitation techniques; exploring user scenarios.
- *Requirements Analysis* – Creating visual analysis models; building and evaluating prototypes; prioritizing requirements using previously selected attributes.
- *Requirements Specification* – Looking for ambiguities in text-based requirements; storing requirements in a central, usually shared database for better communication; tracing requirements into design, code and tests.

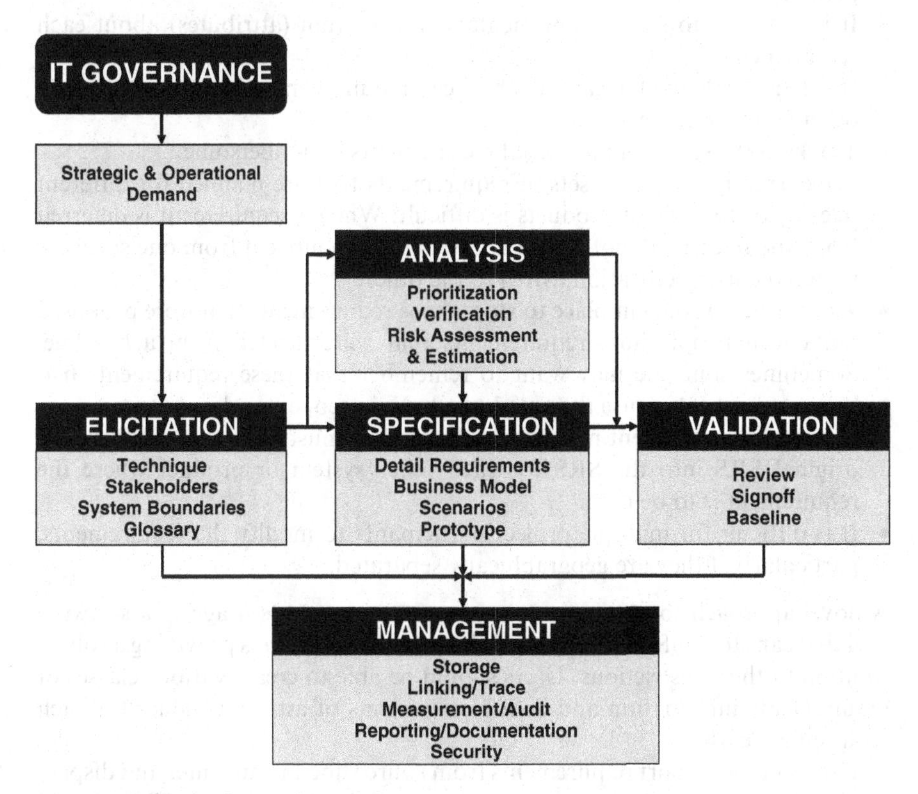


**Fig. 63.3** Strategies for requirements management

- *Requirements Validation* – Reviewing the requirements; creating test cases from requirements.
- *Requirements Management* – Managing requirements versions; adopting a change control process; performing requirements change impact analysis; storing requirements attributes; tracking the status of each requirement.

## 63.8 Traditional vs. Novel Approach to RM

The traditional requirements engineering and management approach is to create documents that contain business, user, functional, and non-functional requirements written in natural language, preferably using the organization's standard templates.

But, it appears that the traditional document-based approach to storing requirements has numerous limitations [5]:

- It is difficult to keep the documents current and synchronized.
- Communicating changes to all affected team members is usually a manual process.
- It is not easy to store supplementary information (attributes) about each requirement.
- It is hard to define logical links between requirements of various types and other system elements.
- Tracking the status of individual requirements is cumbersome.
- Concurrently managing sets of requirements that are planned for different releases or for related products is difficult. When a requirement is deferred from one release to another, an analyst needs to move it from one software requirements specification (SRS) to the other.
- There is no convenient place to store those requirements that were proposed but rejected, nor those requirements that were deleted from a baseline. Sometimes someone may want to remember that these requirements had been proposed, because they might be back in scope one day.
- Reusing a requirement means that the analyst must copy the text from the original SRS into the SRS for each other system or product where the requirement is to be used.
- It is difficult for multiple project participants to modify the requirements, particularly if they are geographically separated.

A novel approach to requirements management suggests usage of a software tool that can store information in a multi-user database, thus providing a robust solution to these restrictions. Users should be able to create various classes of requirements information and define unique sets of attribute values for each requirement class.

Users should import requirements from source documents, filter and display the database contents, and export requirements in various formats. They should also define traceability links between pairs of objects stored in the database, as well as connecting requirements to items stored in other software development tools.

The following are some of the tasks such a tool should help users to perform:

- *Tracking requirements status* – Collecting the requirements into a database lets the user know how many discrete requirements he/she has specified for the product. Tracking the status of each requirement during development supports the overall status tracking of the project. A project manager has good insight into project status if he or she knows that, e.g., 55 percent of the requirements committed to the next release have been verified, 28 percent have been implemented but not verified, and 17 percent are not yet fully implemented.
- *Communicate with stakeholders* – The tool should permit team members to discuss requirements issues electronically through threaded and interactive conversations. Additionally, e-mail messages can automatically be triggered to notify affected individuals when a specific requirement is modified or a change is proposed. Making the requirements accessible on-line can save travel costs and reduce document proliferation.
- *Store requirements attributes* – Possible attributes that provide a richer understanding of each requirement include priority, status, verification method, origin, person responsible, planned release, and so on. Everyone working on the project must be able to view the attributes and selected individuals be permitted to update their values [6]. The tool should generate system-defined attributes such as the date a requirement was created and its current version number, and it lets you define additional attributes of various data types. One can sort, filter, or query the database to display subsets of the requirements that have specific attribute values. For example, a user might ask to see a list of all the requirements planned for release 2.3 that were assigned to Julie and have a current status of Implemented.
- *Manage versions and changes* – The project should define a requirements baseline. Software tool should provide flexible baselining functions and maintains a history of the changes made to every requirement. You can record the rationale behind each change decision and revert to a previous version of a requirement if necessary. The tool should also contain a built-in change-proposal system that links change requests directly to the affected requirements.
- *Facilitate impact analysis* – The appropriate software tool should enable requirements tracing by letting the user define links between different kinds of requirements, between requirements in different subsystems, and between individual requirements and related system components – for example, designs, code, tests, and user documentation. These links then help when analyzing the impact that a proposed change will have on a

specific requirement by identifying other system elements the change might affect. It is also a good idea to trace each functional requirement back to its origin or parent so that you know where every requirement came from.

- *Control access* – An organization can define access permissions for individuals or groups of users. The software tool should enable information sharing with a geographically dispersed team through a Web interface to the database. The database uses requirement-level locking to permit multiple users to update the database contents concurrently.
- *Reuse requirements* – Storing requirements in the tool's database facilitates reusing them in multiple projects or subprojects. Requirements that logically fit into multiple parts of the product description can be stored once and referenced when necessary to avoid duplicating requirements.

In short, the software tool should be a powerful complement to a robust requirements engineering process. The tool alone will not guarantee that high-quality requirements are written, that the system designer has talked with the right user representatives or that he/she has set the right priorities. However, the tool can keep a project team aligned on the specified requirements for each release, so they all agree when they are done and there will not be so many surprises when they get there.

## 63.9 Assessing Return on Investment

Our research has convinced us that estimating the payback that better requirements can bring to the organization requires considering the following questions:

- What fraction of an organization's own development effort is expended on rework (due to miscommunication)?
- How much does a typical customer-reported defect cost the organization?
- How much costs them a defect found in system testing?
- What fraction of user-reported defects and what fraction of defects discovered during system testing originate in requirements errors?
- How much of the organization's costs – such as defect correction and unplanned enhancements – can be attributed to missing requirements or other requirements inadequacies?

Aside from obvious benefits, improving requirements approaches leads to other valuable – though less tangible – outcomes. Fewer miscommunications on a software project reduce the overall level of disorder. Less disorder lowers unpaid overtime, increases team morale, boosts employee retention and improves the team's chances of delivering on time.

Best of all, these benefits ultimately lead to higher customer satisfaction.

## 63.10 Conclusion

Software development involves roughly 50 percent computing and 50 percent communication. Unfortunately, most IT professionals and teams are better at computing – user requirements, however, are almost entirely about communication.

Because there are many links in the requirements communication chain, a breakdown in any of these links leads to significant problems. For example, if an analyst misunderstands stakeholder input about requirements, if important requirements information does not surface, or if an analyst and developer do not share the same understanding about requirements, the resulting product fails to meet customers' needs.

The inevitable outcome of requirements errors is time-consuming and costly rework. Experts report that rework can consume 30–40 percent of the total effort expended on a software project. Our study has indicated that nearly 50 percent of the defects identified on software projects observed can be traced back to errors in the requirements.

Our analysis of potential return on investment from requirements suggests that requirements errors can consume between 70 and 85 percent of all project rework costs. It can cost up to 110 times more to correct a requirements defect found in operation that it would if that same defect has been discovered during requirements assessment and definition.

To convince itself of the benefits, a firm should think about how much effort on their projects is misspent because of miscommunications, misunderstandings about requirements, missing information, or misstated requirements.

Requirement engineering and management is a developing discipline that tries to solve those problems. A novel approach to requirements engineering and management suggests usage of a software toolkit that can store information in a multi-user database, thus providing a robust solution to these restrictions.

## References

1. Bates J (2005) Business in real time – Realizing the vision. DM Review, 05/2003 (http://www.dmreview.com/)
2. Blackburn JD, Scudder GD, Van Wassenhove LN (1996) Improving speed and productivity of software development: A global survey of software developers. IEEE Trans Software Eng 22(12):875–885
3. Boehm BW, Papaccio PN (1998) Understanding and controlling software costs. IEEE Tran Software Eng 14(10):1462–1476
4. Davis AM (1993) Software requirements: Objects, functions, and states. Prentice Hall PTR, Upper Saddle River (NJ)
5. Ellis K (2006) The executive briefing on the issues surrounding getting business requirements right. http://www.digital-mosaic.com, May 2
6. Frankl A, King T (2005) How to detect requirements errors – A guide to slashing software costs and shortening development time. http://www. n8systems.com, July 22

7. Grady RB (1999) An economic release decision model: insights into software project management. In: Proceedings of the Applications of Software Measurement Conference, Orange Park FL, pp 227–239
8. Hofmann HF, Lehner F (2001) Requirements engineering as a success factor in software projects. IEEE Software 18(4):58–66
9. Hooks IF, Farry KA (2001) Customer-centered products: creating successful products through smart requirements management. In: Proceedings of the AMACOM Conference, Miami FL, pp 207–215
10. Leffingwell D (1997) Calculating the return on investment from more effective requirements management. Am Programmer 10(4):13–16
11. Light M (2005) Agile requirements definition and management will benefit application development. http://www.gartner.com, Apr. 18
12. The Standish Group Staff (2005) The CHAOS report on software design. The Standish Group, Palo Alto (CAL)

# Chapter 64
# SSAF – A Tool for Active Filter Synthesis

**Elena Doicaru, Lucian Bărbulescu, and Claudius Dan**

**Abstract** In this paper a software tool used to determine the topological elements of active analog continuous-time filters is presented. The filter is described by an arbitrary transfer functions expressed as $t(s) = p(s)/e(s)$, where $e(s)$ is the natural mode (pole) polynomial and p(s) is the transmission zero polynomial. The goal is to obtain filters having small sensitivities, low noise and high dynamic ranges. The synthesis method implemented in this program is based on the intermediate transfer functions (IFs) method.

## 64.1 Introduction

In the VLSI era, as mixed analog/digital circuits of increasing complexity are implemented into a single IC, the need for low-sensitivity, low-noise, high performance filters steadily increases. A major application for continuous-time filters is direct signal processing, especially for medium dynamic range applications, when high speed and/or low power operation are needed. For this reason we developed a software tool for automate synthesis of analog continuous-time filters – SSAF (Automatic *S*tructural *S*ynthesis of *A*ctive Analog Continuous-Time High-order *F*ilters). The software tool was developed in C++ using Visual C++ 2005 programming environment and can be used to determine the topological elements of active analog continuous-time filters, described by an arbitrary transfer function in the form $t(s) = p(s)/e(s)$, where $e(s)$ is the natural mode (pole) polynomial (of order $n$) and $p(s)$ is the transmission zero polynomial (of order $m$). The goal is to obtain small sensitivities, very low noise and a good dynamic range.

The synthesis method selected is the intermediate transfer functions (IFs) method [4]. The method consists in obtaining an $n$-th transfer function, $t(s)$, by

E. Doicaru (✉)
Faculty of Automation, Computer Science and Electronics, University of Craiova str. Decebal, Nr.5, 200646 Craiova Romania
e-mail: dmilena@electronics.ucv.ro

N. Mastorakis et al. (eds.), *Proceedings of the European Computing Conference*, Lecture Notes in Electrical Engineering 28,
DOI 10.1007/978-0-387-85437-3_64, 

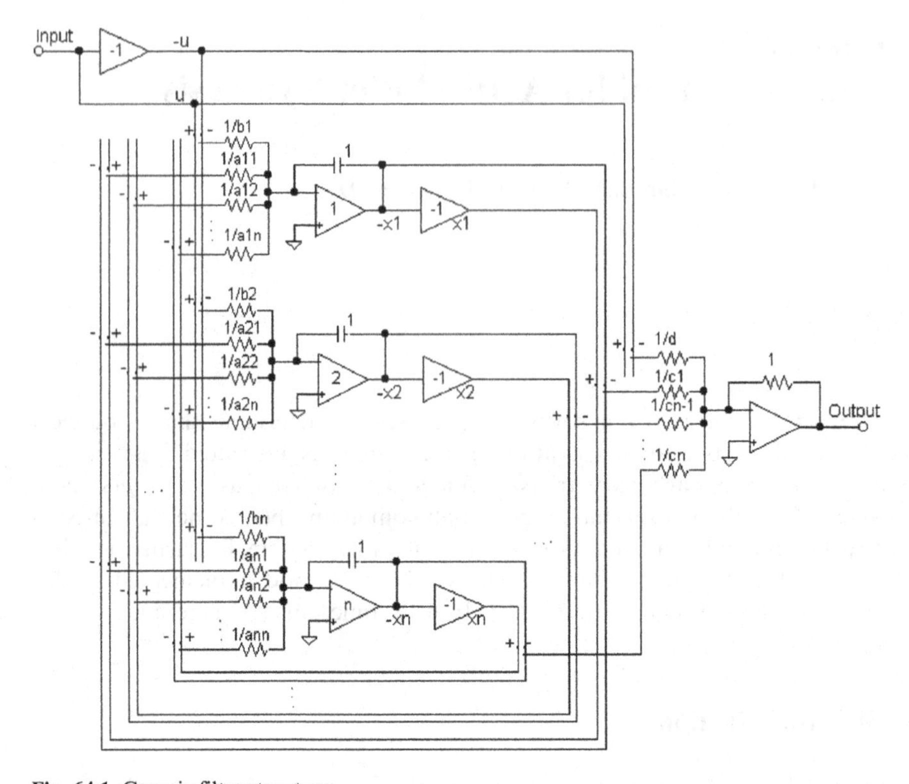


**Fig. 64.1** Generic filter structure

means of a structure of $n$ resistively interconnected integrators using the linearly independent IFs selected from the given $t$(s) following the general structure presented in Fig. 64.1. The IFs have identical denominator polynomials, $e$(s), and arbitrary numerator polynomials of degree less than $n$. The program selects the IFs that are able to provide optimum performance structures. The significant advantage of this method consists in the possibility to optimize the filter performance at the abstract level of Ifs; using other methods only topological level optimization is possible.

## 64.2 Program's Structure and Algorithm

The filter structure presented in Fig. 64.1 can be described by a state-variable formulation:

$$\mathrm{s} \cdot \mathbf{x}(\mathrm{s}) = \mathbf{A} \cdot \mathbf{x}(\mathrm{s}) + \mathbf{b} \cdot u(\mathrm{s}) + \varepsilon(\mathrm{s}), y(\mathrm{s}) = \mathbf{c}^{\mathrm{T}} \cdot \mathbf{x}(\mathrm{s}) + d \cdot u(\mathrm{s}) \quad (64.1)$$

In Equation (64.1) vector **x**(s) represents the circuit state (integrators output voltages), matrix **A** describes the interconnections between the *n* integrators, vector **b** contains the coefficients that multiply the input signal *u*(s), vector **c** contains the coefficients required to form the output, scalar *d* is the coefficient of the feed-through component and $\epsilon$(s) is the vector containing the noise components at the integrator inputs.

The dual sets of IFs, {*fi*(s)} and {$g_i$(s)}, are given by:

$$\begin{aligned} f_i(\mathrm{s}) &= x_i(\mathrm{s})/u(\mathrm{s}); \mathbf{f}(\mathrm{s}) = (\mathrm{s}\cdot\mathbf{I}-\mathbf{A})^{-1}\cdot\mathbf{b} \\ g_i(\mathrm{s}) &= y(\mathrm{s})/\varepsilon_i(\mathrm{s}); \mathbf{g}^{\mathrm{T}}(\mathrm{s}) = \mathbf{c}^{\mathrm{T}}\cdot(\mathrm{s}\cdot\mathbf{I}-\mathbf{A})^{-1} \end{aligned} \quad (64.2)$$

The {$f_i$(s)} set contains the transfer function from the filter input to the integrator outputs and the {$g_i$(s)} set can be physically interpreted as the integrators noise gains.

From these sets we can obtain the {**A**, **b**, **c**, *d*} topological parameters using the following relationships:

$$\begin{aligned} &\mathbf{A}=\mathbf{F}\cdot\mathbf{E}\cdot\mathbf{F}^{-1},\mathbf{b}=\mathbf{F}\cdot\mathbf{1},\mathbf{c}^{\mathrm{T}}=\mathbf{t}^{\mathrm{T}}\cdot\mathbf{F}^{-1},d=t_{n+1},t(\mathrm{s})=\mathbf{t}^{\mathrm{T}}\cdot\mathbf{v}(\mathrm{s})+t_{n+1} \\ &\mathbf{f}(\mathrm{s})=\mathbf{F}\cdot\mathbf{v}(\mathrm{s}),\mathbf{g}(\mathrm{s})=\mathbf{G}\cdot\mathbf{v}(\mathrm{s}),\ v_i(\mathrm{s})=1/(\mathrm{s}-e_i),i=\overline{1,\ n},\mathbf{G}^{\mathrm{T}}=\mathbf{H}\cdot\mathbf{F}^{-1} \end{aligned} \quad (64.3)$$

where **t** is a vector containing the *n* residues of *t*(s) at the poles, $t_{n+1}$is the residue at s = $\infty$, **F** is a matrix containing the residues of the *f* functions at the poles, **G** is a matrix of the residues of the *g* functions, *ei* are the roots of *e*(s), **E** is the diagonal matrix having the roots *ei* as its elements and **H** is a diagonal matrix formed from the residues of *t*(s). The sensitivities of the **A**, **b**, **c**, *d* parameters, the integrator gain $\gamma i$ and the operational amplifier gain $\mu i$(s) hardly depend by the IFs set [4]. In the terms of {**A**, **b**, **c**, d} elements the gain of the integrator *i* is proportional to the *i* row sum of system (64.1.a). From classical sensitivities results [3] can be demonstrated that the minimum sensitivity realization has:

$$f_i(\mathrm{s})\cdot g_i(\mathrm{s}) = -(1/n)\cdot[\mathrm{d}\,t(\mathrm{s})/\mathrm{ds}], \forall i=\overline{1,\ n} \quad (64.4)$$

If we assume that the noise signals, modeled by $\epsilon$(s), have white spectra with equal densities $N_i^2$, the output noise will have a power spectrum and a rms level given by:

$$P_{n0}(\omega) = N_i^2\cdot\sum_i |g_i(\mathrm{j}\cdot\omega)|^2,\ \|P_{n0}(\omega)\| = N_i^2\cdot\sqrt{\sum_i \|g_i(\mathrm{j}\cdot\omega)\|_2^2} \quad (64.5)$$

The mathematical analysis emphasizes the fact that the best realizations are those obtained by orthogonal, orthonormal and derivative IFs sets. The SSAF program generates all these types of IFs sets. The general structure of the program is presented in Fig. 64.2.

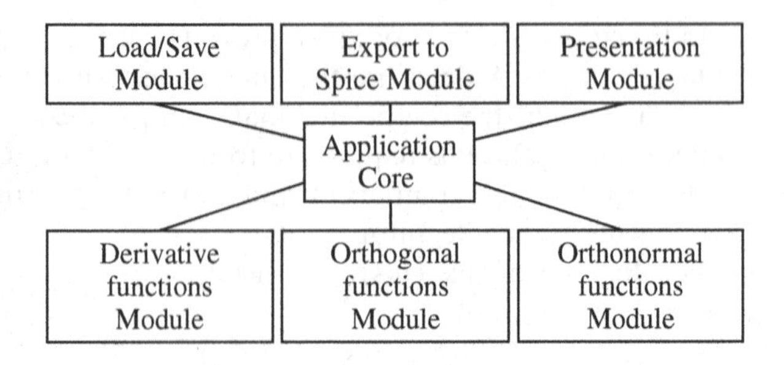


**Fig. 64.2** Application structure

An IFs set is orthogonal if the all functions of the set are linearly independent and their inner product is zero. An IFs set is orthonormal if all *fi*(s) are orthogonal and $\|f_i\|_2 = 1, \forall i = \overline{1,n}$. Such sets can be obtained from any set of independent {*ui*(s)} by using the Gram-Schmidt ortho-normalization procedure [5]. The inner products, $u_i \bullet v_j$, and the norm of functions, $\|v_j\|_2 = \sqrt{v_i \bullet v_i}$, required by the Gram-Schmidt procedure, are determined using the residue theorem. This is possible because of rational form of functions. So the program uses the following relations:

$$u_i \bullet v_j = \int_{-\infty}^{\infty} R_{ij}(\omega) \cdot \mathrm{d}\omega = 2 \cdot \pi \cdot \mathrm{j} \cdot \sum_{\substack{m \\ \mathrm{Im}(pol_m)>0}} rez(R_{ij}, pol_m) \tag{64.6}$$

The function residues are determined in conformity with classical definition. The orthonormal functions are $f_i = v_i / \|v_i\|_2$.

Then the topological elements are determined in conformity with relation (64.3). Their sensitivities and the filter noise are evaluated and the following sensitivity and noise criteria are listed:

$$\begin{aligned} SA &= \max_{i,j,\omega \in R_a} \mathrm{Real}(\mathrm{S}_{A_{ij}}^{t(\mathrm{s})}), Sb = \max_{i,\omega \in R_a} \mathrm{Real}(\mathrm{S}_{b_i}^{t(\mathrm{s})}), Sc = \max_{i,\omega \in R_a} \mathrm{Real}(\mathrm{S}_{c_i}^{t(\mathrm{s})}), \\ Sd &= \max_{\omega \in R_a} \mathrm{Real}(\mathrm{S}_{d}^{t(\mathrm{s})}), Sd = \max_{\omega \in R_a} \mathrm{Real}(\mathrm{S}_{d}^{t(\mathrm{s})}), S\gamma = \max_{i,\omega \in R_a} \mathrm{Real}(\mathrm{S}_{\gamma_i}^{t(\mathrm{s})}) \\ P_{zg} &= \left| \|P_{n0}(\omega)\|_2 \right|_{N_i=1} = \sqrt{\sum_i \|g_i(\mathrm{j} \cdot \omega)\|_2^2} \end{aligned} \tag{64.7}$$

Since sensitivities are proportional to {*fi*(s)} it results that various sensitivities (magnitudes and phases) go to zero in the pass-band. In a ladder only magnitude sensitivity is zero.

From Equation (64.4) one can see that if the minimum sensitivity realization exists, then {*fi*(s)} must divide *t'*(s). In the SSAF program, {*fi*} are chosen so that the factors of *t'*(s) missing from *fi*(s) (that could be the factors of *gi*(s) in the case of minimum-sensitivity realizations) can appear in *fn-i* [1].

Unlike orthogonal and orthonormal IF's that provide a single set of *n* functions *fi*, in the DIF's case we can obtain a lot of *n* functions sets. The program generates all DIF's sets and selects one or more DIF's sets, following the user options. The first step in the DIF's sets determination is to obtain the derived transfer function $t'(s)$ and its roots. The DIF's can't be generated in two cases: (1) if the transmission zero polynomial order of $t'(s)$ is less than $n$; (2) if in all sets of IF's we have at least a function with the transmission zero polynomial order greater than $n$. The next step is to separate the real roots from the complex roots. The conjugated complex roots are grouped in binomials. Then, the monomials and binomials are binary coded. This binary code facilitates the simultaneous generation of *fi* and *fn-1* DIF's of a set. The number of monomials and binomials possible combinations is reduced by generating only the combinations that yield polynomials with order less than $n$. These combination variants, representing all possible physical variants, are combined once again in $n/2$ groups.

## 64.3 SSAF Program – Implementation and Operation

As presented previously, the application contains several modules, each with its own role, as follows:

- *Load/Save Module:* this module is used to save or load a filter to of from a file. The files must have the extension ".mil" and their format is proprietary and text-based.
- *Export to Spice Module:* this module is used to export the filter into a spice-recognized file, thus offering the possibility to obtain a simulation.
- *Presentation Module:* this module is used to obtain several information about the filter such as sensitivity, residuum etc. Also a schematic view of the filter can be obtained using this module.
- *Derivative function module:* is used to compute the derivative functions for the filter.
- *Orthogonal function module:* is used to compute the orthogonal functions for the filter.
- *Orthonormal function module:* is used to compute the orthonormal functions for the filter.
- *Application Core*: this module is the bridge between all previous modules. It contains the application GUI and links to the functions implemented by all the previous modules.

Each of the previous modules is implemented as a set of C+ + classes. All modules exchange data with the application core in both ways. One piece of data that is very important for the entire application is the transfer function represented as two polynomials. Each polynomial is saved within the application as a template class. The template for this class was needed to represent the polynomial coefficients. Usually those are real numbers, but the class allows the usage of complex values too.

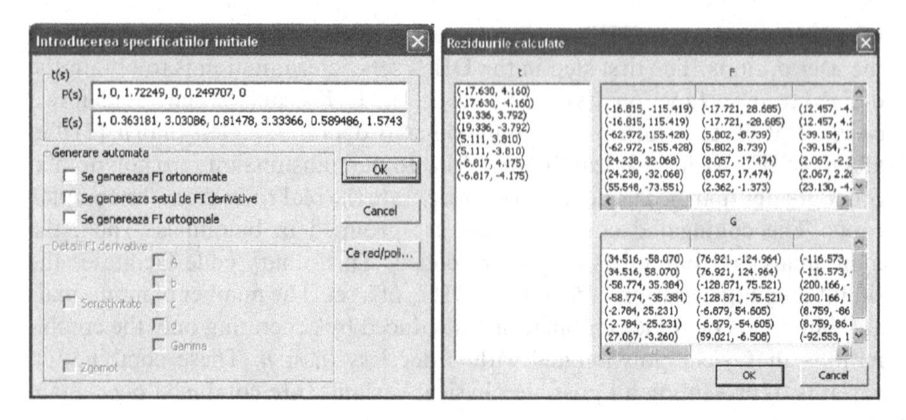


**Fig. 64.3** Coefficients' specification and computed residuum

Working with the application is a very easy task. After starting it the user can either choose to create a new filter or to open an existing one. The creation of a filter can be performed in two ways: either by entering the coefficients for the P(s) and E(s) polynomials (see Fig. 64.3), or by specifying their degrees and complex zeroes. After the required data is entered, the user may decide which type of functions he/she wants to generate. The application will generate the functions and the result will be displayed in a list form. The user can view information about IFs sets or about the transfer function t(s), such as the Residuum (see Fig. 64.3), the user-friendly representation of the transfer function and its derivative. The elements of the topological set {**A**, **b**, **c**, d} and their computed sensitivity are also given. Another feature is the application's ability to generate a Spice-compatible file, based on which a simulation of the filter can be performed. In this way the user can further test the filter before deciding to implement in using real hardware.

## 64.4 Conclusion

The paper presented a C + + code program for the synthesis of continuous-time analog filters. The synthesis method is based on the selection of an appropriate set of intermediate transfer functions. This way the filter sensitivity, noise and dynamic range properties are easily determined from these intermediate functions. Therefore, it is possible to optimize the sensitivity, noise and dynamic range by the appropriate choice of intermediate functions. The synthesis program is easy to use and the user can select the optimum realization in rapport with one of the following performances: **A**, **b**, **c** and d sensitivities, noise and dynamic range. The solution helps the user to define better filters without using any expensive hardware components. He/she can generate several implementations for a filter and then, after examining the results using a simulation environment, can choose the best solution for his/her needs. The user-friendly interface is very intuitive and the help integration offers further information about the usage of SSAF.

**Acknowledgement** This work was supported by the CNCSIS Research Project no. 29C/ 08.05.2007, theme no. 5, CNCSIS code no.602.

## References

1. Doicaru E (1998) The optimum automatic synthesis at active analogue continuous-time high-order filters. In: 6th International Symposium on Automatic Control and Computer Science, Volume II Computer Science, Iasi, Romania, pp 182–189
2. Press WH, Teukolski SA, Vetterling WT, Flannerry BP (1992) Numerical recipes in C: The art of scientific computing. Cambridge University Press, Cambridge
3. Snelgrove WM, Sedra AS (1980) A novel synthesis method for state-space active networks. In: Proceedings of the Mid-west Symposium. Circuits and Systems, Toledo
4. Snelgrove WM, Sedra AS (1986) Synthesis and analysis of state-space active filters using intermediate transfer functions. IEEE Trans Circuits Syst CAS 33:287–300
5. Sabac IG (1981) Special Mathematics (in roumanian). Editura Didactica si Pedagogica, Bucuresti

# Part VIII
# Theoretical Approaches in Information Science

Information science has, for a long time, been drawing on the knowledge produced in psychology and related fields. This is reasonable, for the central issue in information science concerns individual users navigating information spaces such as libraries, databases, and the Internet. Because information science is about the definition and location of information, information seeking is the fundamental problem in information science, while other problems, such as document representation, are subordinate. In this part we present some theoretical approaches in the information science. The papers cover the complexity of the auxetic systems, new verification of reactive requirement for Lyee method, a tree derivation procedure for multivalent and paraconsistent inference, some results about feedback stabilization of matrix, Maxwell geometric dynamics, fractional right ideals, concurrent password recovery of archived files, solutions for an integral-differential equation, development of genetic algorithms and C-method for optimizing a scattering by rough surface.

# Chapter 65
# On the Complexity of the Auxetic Systems

**Ligia Munteanu, Dan Dumitriu, Ştefania Donescu, and Veturia Chiroiu**

**Abstract** Two major levels of complexity are discussed in a way of understanding the structure and processes that define an auxetic system. The auxeticity and structural complexity is interpreted in the light of Cosserat elasticity which admits degrees of freedom not present in classical elasticity, i.e. the rotation of points in the material, and a couple per unit area or the couple stress. The Young modulus evaluation for a laminated periodic system made up of alternating aluminum and an auxetic material is an example of computing complexity.

## 65.1 Introduction

Materials with a negative Poisson ratio are auxetic materials. The term auxetic comes from the Greek word *auxetos*, meaning *that which may be increased.* Instead of getting thinner like an elongated elastic band, the auxetic material grows fatter, expanding laterally when stretched. All the major classes of materials (polymers, composites, metals, ceramics honeycomb structures, reticulated metal foams, re-entrant structures, the skin covering a cow's teats, certain rocks and minerals, living bone tissue) can exist in the auxetic form. Love [1] presents an example of cubic single crystal pyrite as having a Poisson's ratio of −0.14, and he suggests the effect may result from a twinned crystal. An *auxetic system* is composed from different materials with different properties with a new mechanical architecture based on tailored properties [2–5]. The idea is to transform a non-auxetic material into auxetic forms as foams or cellular materials, or to employ new architecture techniques for auxetic materials. The classical mechanics fail in describing the behavior of the auxetic material, because the chiral effects imply a fifth rank modulus tensor, which changes under an inversion. The chirality effects imply the first major level of

L. Munteanu (✉)
Continuum Mechanics Department, Institute of Solid Mechanics, Ctin Mille 15, Bucharest, 010141, Romania

N. Mastorakis et al. (eds.), *Proceedings of the European Computing Conference*, Lecture Notes in Electrical Engineering 28,
DOI 10.1007/978-0-387-85437-3_65, 

complexity, i.e. *the structural complexity*. This structural complexity is described by Cosserat elasticity [6]. Phenomena associated with micropolar elasticity are likely to be of larger magnitude, and therefore of greater interest in materials such as auxetic systems with larger scale structural features [7–9]. The second level of complexity refers to *the computing complexity*. The estimation of the macroscopic Young modulus for a laminated periodic structure made up of alternating aluminum and an auxetic material layers, by using the homogenization technique, is an example of the computing complexity.

## 65.2 Structural Complexity

Consider a chiral Cosserat medium, in a Cartesian coordinates system $(x,y,z)$. The equations of motion for the case without body forces and body couples are [10–12]

$$\sigma_{kl,k} - \rho \ddot{u}_l = 0, m_{rk,r} + \varepsilon_{klr}\sigma_{lr} - \rho j \ddot{\varphi}_k = 0. \tag{65.1}$$

Here $\sigma_{kl}$ is the stress tensor, $m_{kl}$ is the couple stress tensor, $u$ is the displacement vector, $\varphi_k$ is the microrotation vector which in Cosserat elasticity is kinematically distinct from the macrorotation vector $r_k = 1/2\varepsilon_{klm}u_{m,l}$, and $\varepsilon_{klm}$ is the permutation symbol. Note that $\varphi_k$ refers to the rotation of points themselves, while $r_k$ refers to the rotation associated with the movement of nearby points. In (65.1) $\rho$ is the mass density and $j$ the microinertia. The constitutive equations are

$$\begin{aligned} \sigma_{kl} &= \lambda e_{rr}\delta_{kl} + (2\mu + \kappa)e_{kl} + \kappa\varepsilon_{klm}(r_m - \varphi_m) \\ &+ C_1\varphi_{r,r}\delta_{kl} + C_2\varphi_{k,l} + C_3\varphi_{l,k}, \end{aligned} \tag{65.2}$$

$$\begin{aligned} m_{kl} &= \alpha\varphi_{r,r}\delta_{kl} + \beta\varphi_{k,l} + \gamma\varphi_{l,k} + C_1 e_{rr}\delta_{kl} + \\ &(C_2 + C_3)e_{kl} + (C_3 - C_2)\varepsilon_{klm}(r_m - \varphi_m), \end{aligned} \tag{65.3}$$

where $e_{kl} = 1/2(u_{k,l} + u_{l,k})$ is the macrostrain vector, $\lambda$ and $\mu$ are Lamé elastic constants, $\kappa$ is the Cosserat rotation modulus, $\alpha, \beta, \gamma$, the Cosserat rotation gradient moduli, and $C_i, i = 1, 2, 3$, are the chiral elastic constants associated with noncentrosymmetry. For $C_i = 0$ the equations of isotropic micropolar elasticity are recovered. For $\alpha = \beta = \gamma = \kappa = 0$, (65.2) is reduced to the constitutive equations of classical isotropic linear elasticity theory [13–15]. Consider the case of the laminated plates made up of a periodic layering of sheets normal to the direction $x$ of wave propagation, each of elastic material with constant properties. For simplicity, the particular 2D case in which all quantities depend only on $x$ and $z$ is considered. Let $\mathbb{F} = \{\sigma_{kl}, m_{kl}, u_k, \varphi_k, k, l = 1, 2, 3\}$ be a set composed of the asymmetric tensors $\sigma_{kl}, m_{kl}, k, l = 1, 2, 3$, and the vectors $u_k, \varphi_k$. We call $\mathbb{F}$ an *elastodynamic state* on the bounded medium, if it satisfies (65.1),

(65.2), and (65.3) and appropriate initial conditions. There is one-by-one transformation, which transforms the elastodynamic state $\mathbb{F}$ into another elastodynamic state $\hat{\mathbb{F}}= \{\hat{\sigma}_{kl}, \hat{m}_{kl}, \hat{u}_k, \hat{\varphi}_k\}, k,l = 1,2,3,$ composed by the symmetric tensors $\hat{\sigma}_{kl}, \hat{m}_{kl}, k,l = 1,2,3,$ and by the vectors $\hat{u}_k, \hat{\varphi}_k$, satisfying (65.1), (65.2), and (65.3). The state $\mathbb{F}$ can be decomposed in $\hat{\mathbb{F}} = \hat{\mathbb{F}}_1 + \hat{\mathbb{F}}_2$ , where

$$\hat{\mathbb{F}}_1 = \{\hat{\sigma}_{11}, \hat{\sigma}_{13}, \hat{\sigma}_{33}, \hat{m}_{22}, \hat{u}_1, \hat{u}_3, \hat{\varphi}_2\}, \hat{\mathbb{F}}_2 = \{\hat{\sigma}_{22}, \hat{m}_{11}, \hat{m}_{13}, \hat{m}_{33}, \hat{u}_2, \hat{\varphi}_1, \hat{\varphi}_3\}.$$

After a proper combination, the following equations in $\hat{u} = (\hat{u}_1, \hat{u}_2, \hat{u}_3)$ and $\hat{\varphi} = (\hat{\varphi}_1, \hat{\varphi}_2, \hat{\varphi}_3)$ are found

$$(\lambda + 2\mu + \kappa)\nabla\nabla\hat{u} - (\mu + \kappa)K_0^2\nabla \times \nabla \times \hat{u} + \kappa(1 - K_0^2)\nabla \times \hat{\varphi} = \rho\ddot{\hat{u}}, \quad (65.4)$$

$$(\alpha + \beta + \gamma)\nabla\nabla\hat{\varphi} - \gamma K_0^2\nabla \times \nabla \times \hat{\varphi} + \kappa(1 - K_0^2)\nabla \times \hat{u} - 2\kappa(1 - K_0^2)\hat{\varphi} = \rho j\ddot{\hat{\varphi}}, \quad (65.5)$$

with the coupling coefficient $K_0$ defined as

$$K_0^2 = 1 + \frac{(C_1 + C_2 + C_3)^2}{(\lambda + 2\mu + \kappa)(\alpha + \beta + \gamma)}. \quad (65.6)$$

One can see that (65.4) and (65.5) are decoupled into two sets of equations in $\hat{\mathbb{F}}_1$ and respectively $\hat{\mathbb{F}}_2$. We will concentrate only on the set of equations corresponding to $\hat{\mathbb{F}}_1$, the other set being solved in a similar way.

## 65.3 Computing Complexity

To solve the Equations (65.4), (65.5), and (1.6) with proper initial conditions, let us apply the Laplace and Fourier transforms, as follows

$$\tilde{v}_1'' = \frac{b_s}{a^2}\tilde{v}_1 + \frac{\mathrm{i}\xi(1 - a^2)}{a^2}\tilde{v}_3' + \frac{s_4^*}{a^2}\tilde{\phi}_2', \tilde{v}_3'' = c_s\tilde{v}_3 + \mathrm{i}\xi s_4^*\tilde{\phi}_2 + \mathrm{i}\xi(1 - a^2)\tilde{v}_1',$$

$$b_s = \xi^2 + \frac{p^2}{s_1 + s_2}, c_s = a^2\xi^2 + \frac{p^2}{s_1 + s_2}, d_s = \xi^2 + \frac{2c_1^2\kappa(1 - K_0^2)\mu}{\omega^2\gamma K_0^2} + \frac{\rho^2}{s_4}.$$

The eigenvalue problem is obtained by taking the solutions of the form

$$W(\xi, z, p) = X(\xi, p)\exp(qz), \quad (65.7)$$

where $W(\xi, z, p) = \{\tilde{v}_1, \tilde{v}_3, \tilde{\phi}_2\}$. The characteristic equation becomes

$$q^3 - \lambda_1 q^2 + \lambda_2 q + \lambda_3 = 0. \quad (65.8)$$

The roots of (65.8) are $q_i, i = 1,2,3$, with real parts positive. The eigenvector $X(\xi,p)$ is

$$X_{i1}(\xi,p) = \begin{vmatrix} a_i q_i \\ b_i \\ -\xi \end{vmatrix}, X_{i2}(\xi,p) = \begin{vmatrix} a_i q_i^2 \\ b_i q_i \\ -\xi q_i \end{vmatrix}, i = 1,2,3.$$

Consequently, the solutions (65.7) can be written as

$$W(\xi,z,p) = \sum_{i=1}^{3} B_i X_i(\xi,p) \exp(q_i(\xi,p)z),$$

where $B_i, i = 1,2,3$, are arbitrary constants.

## 65.4 Case Study and Conclusions

Let us consider a laminated 2D composite plate which occupies the region $x \in [0,L], z \in [-c,c]$, and made up by alternating the $N$ aluminum and auxetic material layers, normal to the direction $x$ of wave propagation (Fig. 65.1). The layers are parallel, planar and periodic, the displacements being continuous across them. The length of each layer is $l$. The interfaces between layers are located at $nl, n = 1,2,\ldots,N$, each joint having two faces identified by + and −. The coordinates are chosen so that the waves lie in the ($x$,$z$) plane. The plate is assumed to support in plane strains and waves running in the $x$-direction. The Bécus homogenization technique is applied via multiple scale expansion [16].

We are interested in knowing the influence of the Cosserat rotation modulus $\kappa$, the Cosserat rotation gradient moduli $\alpha, \beta, \gamma$, and the chiral elastic constants $C_i, i = 1,2,3$, on the effective Young' modulus value of the laminated plate. The

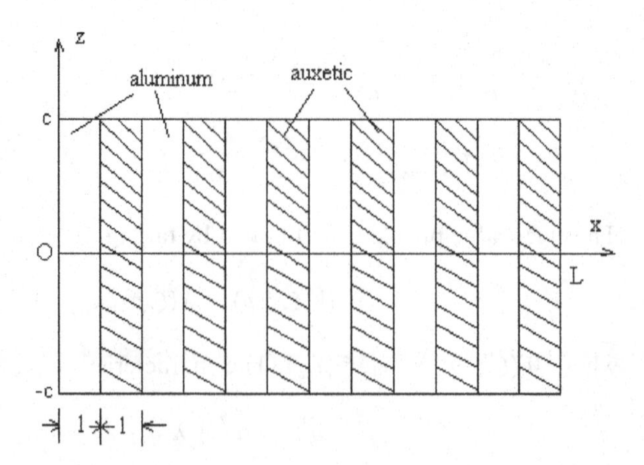


**Fig. 65.1** The composite plate

material constants for this laminated composite are periodic functions of $x$, i.e. $\tilde{C}(x+P)=\tilde{C}(x)$, where the period $P$ equals to the length of the basic cell for the composite $P=2l$, with $2\,l$ the period represented by the length of the basic cell for the composite.

The Bécus homogenization via multiple scale expansion yields

$$E=F(\tilde{C}')+E', E'=\gamma+\frac{1}{2}\tilde{p}^2, \tilde{p}^2=\frac{2\kappa_{aux}}{(K'2_0-1)}, K'2_0^{=}1+\zeta,$$

$$\gamma=\frac{(2\mu'+\kappa')(3\lambda'+2\mu'+\kappa_{aux})}{(2\lambda'+2\mu'+\kappa_{aux})},$$

$$\zeta=\frac{(C_{1aux}+C_{2aux}+C_{3aux})^2}{(\lambda'+2\mu'+\kappa_{aux})(\alpha_{aux}+\beta_{aux}+\gamma_{aux})}.$$

Here $\tilde{C}_{al}$ are the aluminum constants and $\tilde{C}_{aux}$ the auxetic constants. The function $F(\tilde{C}')$ is numerically determined only. In the simulation, the Young's moduli of aluminum and respectively of the auxetic material are 109 GPa, and 1.55 GPa. We observe that the Young's modulus is increasing with respect to the volume fraction $\theta$ from 2 GPa to about 140 GPa, having a maximum value for $\theta=0.75$. For $\theta$ above this value, the Young's modulus is decreasing with respect to $\theta$ from 140 GPa to about 90 GPa.

**Acknowledgement** The authors acknowledge the financial support of the National University Research Council (NURC-CNCSIS) Romania, Grant nr. 55/2007 and Postdoctoral CEEX Grant nr. 1531/2006.

## References

1. Love AEH (1926) A treatise on the mathematical theory of elasticity. 4th ed., Dover, New York
2. Lakes RS (1991) Experimental micro mechanics methods for conventional and negative Poisson's ratio cellular solids as Cosserat continua. J Eng Mater Technol 113:148–155
3. Lakes RS (1987) Foam structures with a negative Poisson's ratio. Science 235:1038–1040
4. Lakes RS (1986) Experimental microelasticity of two porous solids. Int J Solids Struct 22:55–63
5. Chiroiu V (2004) Identification and inverse problems related to material properties and behaviour. In: Chiroiu V, Sireteanu T (eds) Topics in Applied Mechanics. Ed. Academiei Bucharest, 1:83–126
6. Cosserat E and F (1909) Theorie des Corps Deformables. Hermann et Fils, Paris
7. Hlavacek M (1975) A continuum theory for fibre reinforced composites. Int J Solids Struct 11:199–211
8. Hlavacek M (1975) On the effective moduli of elastic composite materials. Int J Solids Struct 12:655–670
9. Berglund K (1982) Structural models of micropolar media. Mechanics of Micropolar Media, World Scientific, Singapore

10. Eringen AC (1968) Theory of micropolar elasticity. In: Liebowitz R (ed) Fracture. Academic Press, 2:621–729
11. Mindlin RD (1964) Microstructure in linear elasticity. Arch Rat Mech Anal 16:51–78
12. Mindlin RD (1965) Stress functions for a Cosserat continuum, Int J Solids Struct 1:265–271
13. Gauthier RD (1982) Experimental investigations on micropolar media. Mechanics of Micropolar Media, World scientific, pp 395–463
14. Teodorescu PP, Munteanu L, Chiroiu V (2005) On the wave propagation in chiral media. New Trends in Continuum Mechanics, Ed. Thetha Foundation, Bucharest, pp 303–310
15. Teodorescu PP, Badea T, Munteanu L, Onisoru J (2005) On the wave propagation in composite materials with a negative stiffness phase. New Trends in Continuum Mechanics, Ed. Thetha Foundation, Bucharest, pp 295–302
16. Bécus GA (1979) Homogenization and random evolutions: Applications to the mechanics of composite materials. Q Appl Math XXXVII(3):209–217c

# Chapter 66
# New Verification of Reactive Requirement for Lyee Method

**Osamu Arai and Hamido Fujita**

**Abstract** Software development in general lacks simplicity and richness in expressing appropriately the requirement, as well related supportive tools. Despite the advances in this field, the solutions have still not overcome and the proposed way of thinking is far from resolving the problems on software development and maintenance. Recently, a new promising methodology called Lyee has been proposed. It aims to generate programs automatically from user requirement. Program structure in terms of Lyee was formalized. However, when requirements are of the reactive type, the validation mechanism has not proposed yet to show correctness or appropriateness of the running requirement. For this purpose, the properties that requirement and operation should have are described as a proposition in terms of temporal logic and are verified by such logic. New verification of reactive requirement for Lyee Method puts together as a design rule.

## 66.1 Introduction

If requirement is well-defined and appropriately expressed, software methodologies can realize them correctly for system built up. In order that the program runs correctly and ends, we should be able to clarify that requirement and the operation should have a certain property for evaluation purposes.

The Lyee program generation principle [1] is applicable to Lyee requirement specification. This means that the program is either providing the user with the correct computation results, or an indication to the user that the specification is incorrect. In Lyee methodology, requirement specification is described using Word and Process Root Diagram which is automatically expanded to Routing Vector (a kind of Word). The correctness of Lyee in the fixed-point setting [2], extension of this for parallel and distributed computing [3] is already discussed.

O. Arai (✉)
Faculty of Software and Information Science, Iwate Prefectural University, Morioka, Iwate, Japan
e-mail: arai@fujita.soft.iwate-pu.ac.jp

N. Mastorakis et al. (eds.), *Proceedings of the European Computing Conference*, Lecture Notes in Electrical Engineering 28,
DOI 10.1007/978-0-387-85437-3_66, 

Lyee requirement and program structure are formalized [4–6]. However, we think that almost all of Lyee style program behaves as reactive system program. In the case with multi-screen and database management system or external equipments, the behavior of the system is dependent on the external employment such as human or external equipment reaction. Therefore, it is necessary that we apply another style of verification for such reactive requirement. As an example, Lyee requirements representing “The Light Control Case Study” [7] are discussed. In the case of Process Control System such as the energy saving system of buildings, the users’ intention (requirement) is to save energy. Such intention (requirement) cannot verify without using the process model.

In this paper, we consider the whole system including not only the structure of its program but also its external requirement and the operation of humans to behave as a *reactive system*. We formalize the whole system by process model, which consist of *processes*, *states* and *shared variables* as a bulletin board type communication tool. For this purpose, a language called L is defined by shared variables declaration and process definition embedded in its syntax. The semantics is represented by the *labeled transition system* extracted from the program. By constructing the *Temporal Frame* (TF) that is the semantics in terms of temporal logic extracted from the calculation sequence of that system, the linkage of semantics of temporal logic and that of L, is presented. Based on this provision, the property, (i.e., users’ intention) which the requirement specification should have, is described as a proposition in temporal logic. Such a proposition is verified by the theorem of temporal logic, and its derivation rule.

The remainder of this paper is organized as follows. In Section 66.2, we introduce Lyee Methodology. In Section 66.3, we define the program of reactive system in terms of language L. We discuss the semantics of a reactive system in terms of language L and the semantics of temporal logic for validation purposes. In Section 66.4, we verify those properties in terms of a Reactive program. In Section 66.5, we discuss classification of condition and fairness, and transition of state and transition of SF in PRD in Lyee method. As a summary of the discussion, the design rule is shown. Finally, Section 66.6 provides concluding remarks on this work, and related future work.

## 66.2 Lyee Methodology and Requirement

In the Lyee methodology, the requirement specification is described using Word and Process Route Diagram which is routing every SF. SF (scenario function) is a unit in which inputs, process computations and outputs are reached in fixed-point representation. If SF is an object for screen, then the output is displayed waiting for input. With pushing the enter button after a value of variable input, it is restarted and processing is performed. Using transition of screen or the database management system (DBMS), two or more SF are required. One SF is assigned to every single button of screen and to every single table of database management system. Accordingly, Process Route Diagram (PRD) is constructed. PRD expresses transition of two or more screens, the correlation of a screen and DBMS, etc. as transition between SFs. A routing vector is defined for this purpose. Transition of two or more SFs is defined as PRD. For more details on Lyee, refer to [1].

In the case of screen and keyboard communication, the Lyee requirement almost have reactive property in itself or whole system including environment such as human operation have such property.

## 66.3 Reactive Systems

In recent years, not only the Lyee requirement, but the computing system itself has developed into an on-line system style computation. Such systems are called a reactive system. Functional modeling cannot capture a reactive system feature. Therefore, it is necessary to investigate all states during calculation. Since the sequence of these states continues indeterministically, a program does not have deterministic output value at the time of program end, so the formulization by input/output relation is not simple. A program continues reacting over input from the outside.

To confirm whether the reactive system runs correctly as intended, even if such a system itself does not have concurrent property, the whole system, which consists of system and its exteriors have concurrent property.

### *66.3.1 Composition of Reactive System*

A reactive system consists of the following.

- Process: The subject that calculates by running a program is called a process. The reactive system consists of two or more calculation subjects called process, and each process runs in parallel, communicating with other processes.
- State: The global running status of a process that is named as *state* is used to grasp the internal state of the system. The state includes not only allocation of value to variable but also working situation of process.
- Inter-process communication: Communication is done by shared variables or channels [4] that is representing message exchange among processes. We use a bulletin board type communication that uses the shared variable as communication means.

### *66.3.2 Reactive System on Lyee Specification*

We define a program L for reactive system and give its syntax and semantics for formal representation. The program has declaration of shared variables and definition of processes. Process definition has two attributes such as state and action. In declaration of a shared variable, this program declares to use the variable shared between each process. The other part of this program defines the action of a process. A process changes its state, by updating shared variables. An action of a certain process is defined as in the following: when a process is in state $\alpha$, if the value of a shared variable fulfills conditions B, performing substitution A of an assigned value to a shared variable then such process transfers itself to state $\beta$. The below formula shows such representation:

$$\alpha \underset{B \Rightarrow A}{\overset{t}{\rightarrow}} \beta$$

where t is the name of action. Here after, the program expressed using this representation we call process model. The conditional expression B about the value of a shared variable is called the execution conditions of t. The substitution of the value to a shared variable is called the command A of t. In Lyee Method structure, Scenario Function SF which has pallets W04-W02-W03 [1] in Fig. 66.1 can be expressed by process model in Fig. 66.2. Pallet W02 has two processes; one is computer (C), the other is external event such as human operation (H). If processes have more than two outputs, to describe which output should be selected, Process Route Diagram is needed. Generally, output is selected by

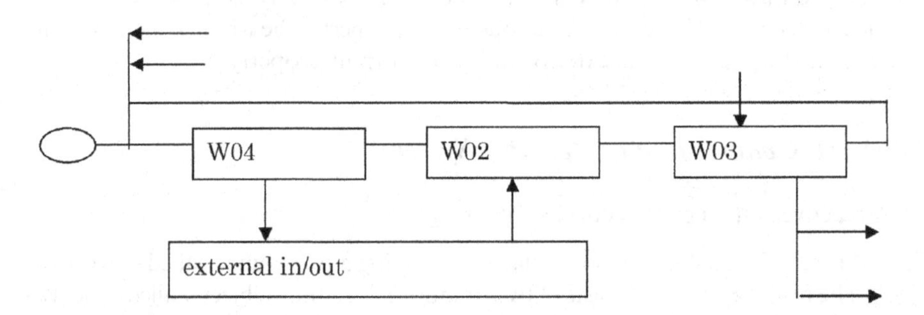


Fig. 66.1 Lyee program

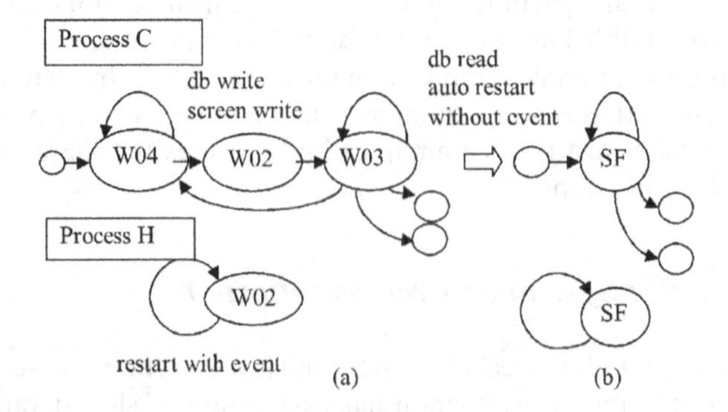


Fig. 66.2 Inner of SF

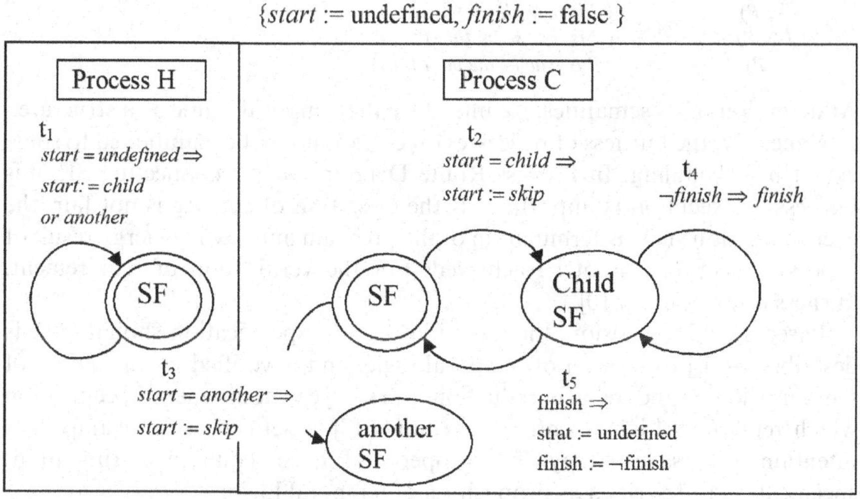

Fig. 66.3 Reactive System Program

human operation or executed result of W03. The left part (a) of Fig. 66.2 can be summarized to right part (b). This is because the variable domain for routing is limited inside of SF; W04-W02-W03. The requirement of a reactive type system could be described as a transition of SF which is shown in (b).

A sample program of reactive system type is shown as Fig. 66.3 in full description, where two variables *start* and *finish* (includes initial value) are defined by {*start*: = undefined, *finish*: = false}. Two processes named H (Human) and C (Computer) react with each other using shared variables *start* and *finish*. At state *SF*, when shared variable *start* is false, process H transfer its state from *a* to *itself* changing *start* variable.

Then Process C can transfer its state from *SF* to *Child SF* without changing any variables. At state *Child SF*, Process C can transfer its state from *Child SF* to *itself* changing variable *finish* from false to true when after memory reading is complete. In the case to transfer another SF, variable *start* have another value.

## 66.4 Verification Lyee Requirement by Temporal Logic

To treat this representation more formally, we introduce a formal language L and temporal logic [8–10]. In the definition of program C in language L, its syntax and semantics is defined using Labeled Transition System (LTS). We introduce temporal logic to verify requirement described as a proposition of temporal logic. Time modality connector $\Box, \Diamond, \bigcirc$ means

| | |
|---|---|
| $P: = p$ | *(The notation showing a proposition)* |
| $\mid(\neg P)\mid(P \to Q)$ | *(not P)(if P then Q)* |
| $\mid(P\ \ Q)\mid(P \vee Q)$ | *(P and Q)(P or Q)* |

|(□ $P$) (*P is true allways*)
|(◇ $P$) (*P is true some time*)
|(○ $P$) (*P is true at the next time*)

At definition of its semantics, we introduce the temporal frame as a structure.

Generally, the fairness of processes execution should be maintained to some extent by scheduling. In Process Route Diagram, at the connecting SF, this fairness of execution is important. If the execution of routing is not fair, the user's intention such as termination of the program and saving energy result of process execution cannot be achieved. For the verification of requirement, fairness is introduced [10].

Based on this provision, the intention that the specification should have is described as a proposition of temporal logic, and is verified by the axiom of temporal logic, and related reduction rules. Lyee requirement specification which represents Light Control Case Study [7] is defined as an example; its intention is to save energy. The property of Lyee requirement (intention) specification is described as proposition of temporal logic.

### 66.4.1 Verification by Temporal Logic

Verification that a C program in terms of L-program satisfies a property P that is described by expression in temporal logic, means that P is universally valid in C. In this case, we argue about the property "*to be universally valid in the temporal frame made from the calculation sequence of LTS extracted from C*"; although it is not universally valid in any other temporal frame. For this reason, the following theorems are provided.

**Theorem 1** *Suppose that*$P_i, ..., P_n| =_{IM}$ *Pis true in a set IM of temporal frame and logical expressions*$P_1, \cdots, P_n$ *and P. Then,*$Q_1, \cdots, Q_m| =_{IM} Q$ *is true if Q can be extracted from*$Q_1, \cdots, Q_m$ *by adding* $P_1, \cdots, P_n| - P$ *as reduction rule.*

*IM* is the set of a temporal frame. Moreover, the following general theorems are used for our use.

**Theorem 2** *Let P, Q be logical formula.*$P \rightarrow Q$ *is universally valid formula on C, when the statement [If P is realized before the execution of* $\tau$ *and then* $\tau$ *is performed, and Q will be true after execution of* $\tau$ *.] is realized to arbitrary transition* $\tau$ *in terms ofLTS that is extracted from Program C.*

Generally, liveness property is to prove the statement [a property Q will be satisfied at a certain time]; therefore the rules below are used.

If P is satisfied then P or Q will be satisfied at a certain time. If P continues to be satisfied then $\tau$ can be done at a certain time. If P is satisfied before execution of $\tau$ then Q is satisfied after execution. Therefore, it will be warranted that if P is satisfied then Q will be satisfied at a certain time. What formulized above as reduction rule of liveness shown below. This used to verify the property of specification that is written as a proposition in temporal logic.

**Theorem 3 (rule of liveness)** The following are satisfied when the set $\{\tau\}$ which consists of transition $\tau$ of the LTS extracted from Program C are a weak fairness set or a strong fairness set on C.

$$\models_{\{M_{C,\sigma} | \sigma \in [[C]]_{J,F}\}} \frac{\begin{array}{l} P \rightarrow \circ(P \vee Q), \\ P \rightarrow (Q \vee enabled\_\tau), \\ P \rightarrow \circ(executed\_\tau \rightarrow Q) \end{array}}{P \rightarrow \diamond Q}$$

$\models_{\{M_{C,\sigma} | \sigma \in [[C]]_{J,F}\}}$ means that to the temporal frame $M_{C,\sigma}$ $P \rightarrow^{\diamond} Q$ is universally valid in terms of $C_{J,F}$ under strong and weak fairness where $\sigma$ is a calculation sequence belonging to $[[C]]_{J,F}$.

**Theorem 4 (rule of chain)** *The following is satisfied on arbitrary natural number i.*

$$\begin{array}{l} P\_0 \rightarrow \diamond Q \\ P\_1 \rightarrow \diamond(P\_0 \vee Q) \\ \vdots \\ P\_i \rightarrow \diamond(P\_0 \vee P\_1 \vee \cdots \vee P\_i - 1 \vee Q) \\ \models (P\_0 \vee P\_1 \vee \cdots \vee P\_i) \rightarrow \diamond Q \end{array}$$

These two Theorems 3 and 4 are proven by deduction. According to the Theorem 1, we can use the formula, which used $\vdash$ instead of $\models$.

## 66.4.2 Example

Our Example is a part of program specification which represents the Light Control Case Study [7]. This case study is saving energy and security control of a building. Here we use energy saving part of this case study. That is « If nobody is in the room, light is turned off in fixed time». The Program is shown in Fig. 66.4. Related processes and its states are shown in Table 66.1.

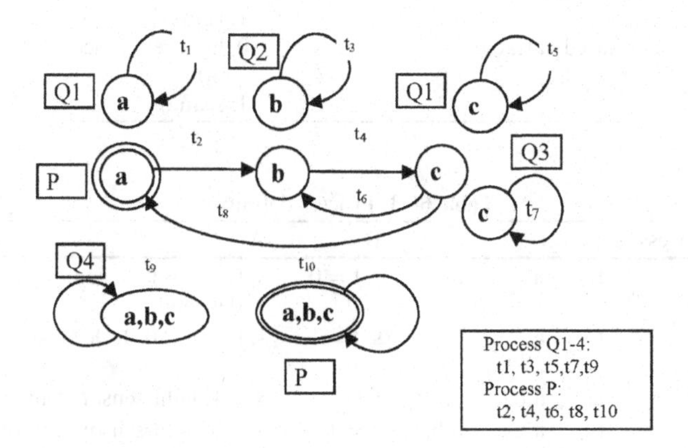


**Fig. 66.4** Example of program light control case

**Table 66.1** Process and state of example

| name | description |
|---|---|
| Process Q1 | human detector on |
| Process Q2 | human detector off |
| Process Q3 | deemed-timer time up |
| Process Q4 | light sensor scanning timer time up (repeat) |
| Process P | computer whose role is governor |
| State a | nobody in one room |
| State b | somebody in the room |
| State c | deemed-timer running |
| State a,b,c | deemed somebody in the room |

Shared variables shared by processes which control state transition and common variables that are used plural states commonly are defined. Declaration of shared variables and common variables are shown in Table 66.2. The state which means somebody deemed in the room made by events of different equipments is introduced. Process Definition is shown in Table 66.3. Since the variable which controls transition is expressed as a shared variable, then this variable is specified in command A. The shared variable is cleared by transition before turning into its initial state. s, t, z, x = 0, 0, 0, 0. States (SFs) c have two outgoing transitions (routes). In c, transition t6 and t8 are selectable. In this case, fare selection is needed. In order to save energy, program selects transition from c to a in a certain time.

**Table 66.2** Declaration of shared and common variables

| Declaration of shared and common variables | Description |
|---|---|
| shared variable<br>s = {0,1} | 0: nobody in<br>1: somebody in |
| shared variable<br>t = {0,1} | 0: deemed-timer stopped<br>1: running |
| shared variable<br>x = {0,1} | 0: light sensor scan off<br>1: scan on |

**Table 66.3** Process definition

| Process | $\tau$ | $\alpha$ | $\beta$ | B | A |
|---|---|---|---|---|---|
| Q1 | $t_1$ | a | a | s, t = 0, 0 | s, t, z = 1, 0, 0: some one detected |
| P | $t_2$ | a | b | s, t = 1, 0 | s, t, z = 1, 0, 1: deemed |
| Q4 | $t_9$ | a,b,c | a,b,c | x = 0 | x = 1: light-sensor-scan on |
| P | $t_{10}$ | a,b,c | a,b,c | x = 1 | x = 0: light-sensor-scan off<br>control room light on z |

### 66.4.3 Proof of Example

The property which specification should have is described, as a proposition in terms of temporal logic, and is verified by the axiom of temporal logic and its derivation rule. The requirement that represents Light control case study is defined as an example. It is to verify that, "*its L-program surely save energy*".

In this verification, we clarify the conditions of fairness, which requirement and operation should have. Liveness property of "*save energy*" is influenced by fairness property. Verification of Liveness is the proof of logical formula deploying fairness sets (defined in Section 66.2). If people go out of the room, a timer will surely be finished after a while and the light will be put out. That is, fairness is required between the frequency of incomings and outgoings of people, and interval setting of a timer.

As the proposition of temporal logic, [*save energy*] is expressed as: $at\ P.c \rightarrow^{\diamond} at\ P.a$.

At this time, Processes P and Q are located c, and uses shared variables $t$ (initial value of which is $\{t := 0\}$). For this reason, here after, [$at\ P.c \rightarrow^{\diamond} at\ P.a$] is verified.

Using the state $c$ of process P and the value of variable $t$, $at\ P.c$ is *classified* as in the following two cases:

$$at\ P.\ c = ((at\ P.\ c \wedge t = 0) \tag{66.1}$$

$$\wedge\ (at\ P.c \wedge t = 1) \tag{66.2}$$

Since $t = \{0, 1\}$, this classification is comprehensively inclusive and exclusively not overlapped internally in this case. Since the value of $t$ is 0 in the case of (66.1), Process P can approach into $a$ immediately. On the other hand, (66.2) is far from reaching the end In order to return to $a$, at timer is over, input 0 for $t$ is required in Process Q.

(Apply Chain rule)

The name $P_0, P_1$ is given to each term according to the order near the goal ($at\ P.a$) for (66.1) to (66.2).

$$P_0 \equiv atP.c \wedge \mathrm{t} = 0$$

$$P_1 \equiv atP.c \wedge \mathrm{t} = 1$$

The following reduction will be obtained when the chain rule of the Theorem 4 is applied to such logical formula.

$$\begin{aligned} &P_0 \rightarrow^{\diamond} at\ P.a, \\ &P_1 \rightarrow^{\diamond} (P_0 \vee at\ P.a), \\ |-\quad &(P_0 \vee P_1) \rightarrow^{\diamond} at\ P.a \end{aligned}$$

Liveness property is verifiable when the premise of rule from (66.1) to (66.2) is derived because $(P_0 \vee P_1)$ is *at P.c*.

(Apply Live rule)

Logical expressions $P_0$, $P_1$named above are derived using the temporal logic shown below.

(1) Derivation of (66.1)

Since $P_0$ is *at* $P.c \wedge t = 0$, when this is true, $\langle P, t_8 \rangle$ is performed at a certain time and the state of P should be set to *a*. That is, fair execution is required for $t_8$ as mentioned above. Therefore, in applying the live rule of Theorem 3 to $P_0 \rightarrow^{\Diamond}$ *at P.a* and $\langle P,t_8 \rangle$, the following derivation can be obtained:

$$P_0 \rightarrow \mathrm{o}(P_0 \vee atP.a), \tag{66.3a}$$

$$P_0 \rightarrow (atP.a \vee enabled_{\langle P,t_8 \rangle}), \tag{66.3b}$$

$$P_0 \rightarrow o(executed_{\langle P,t_8 \rangle} \rightarrow atP.a)| =_{\{M_{C,\sigma} | \sigma \in [[C]]_{T,F}\}} P_0 \rightarrow atP.a \tag{66.3c}$$

The logical expressions 66.3a, 66.3b, 66.3c are the premise of this derivation and verified as below.

(Verification of (66.3a))

In $P_0$, since states of process P and Q are both $c = 0$ and $t = 0$, possible transition is restricted to $\langle P,t_8 \rangle$; (case 1) and the idling transition $\tau_I$; (case 2).

Case 1:When $\langle P,t_8 \langle$ is performed, since variable *at P.c* is true, the formula (a) is verified using Theorem 2.

Case 2:When $\tau_I$ is performed, since $P_0$ is true, the formula (66.3a) is true by Theorem 2.

Therefore, formula (66.3a) is verified. (66.3b), (66.3c) are verified same way.

Therefore, it is verified that formula (66.1), (i.e., $P_0 \rightarrow^{\Diamond}$ *at P.a*) turns into a universally valid formula of L-program. (66.1) is verified same way.

Since it is verified that (66.1) and (66.2) are universally valid formulas, the liveness property *at* $P.c \rightarrow^{\Diamond}$ *at P.a* which is our objective, is verified.

□

Consequently, when satisfying the following requirements, the property that energy saving is performed correctly is verified under the fair execution.

1. Using the combination of shared variables, a special classification is necessary to carry out such execution. In this classification, the condition must be comprehensively inclusive and exclusively defined. This means that the conditions of transition of SF is defined only once all over the domain of a vector.
2. It is demanded that fair execution (i.e., Timer time up should surely be occurred some time) need to be done.

## 66.5 Discussions

From the state transition diagram (STR), extract SF pattern of STR according to event. Unify asynchronously simultaneous events (SF pattern of STR) to one SF. This is the SF division rule. One Parent SF is a group of events that may happen there. A screen (SF) has some buttons, i.e. events, in human interface with screen and keyboard, machine (sensor, actuator) interface event. If these events happen asynchronously, it needs to be received from one parent SF. That is, a group of events in which one of these events happens and the other event does not happen, compose one parent SF. Event transfers process from one state to another state (event driven). Therefore, more than one state exists in one parent SF.

### *66.5.1 Shared Variable and Common Variable*

In Process Control Application, there are many events such that communication that include Internet, timer, human interface with screen and keyboard, machine (sensor, actuator) interface event. If these events happen asynchronously, it needs to be received from one parent SF. That is, a group of events in which one of these events happens and the other event does not happen, compose one parent SF. Event transfers process from one state to another state (i.e., are event driven). Therefore, more than one state exists in one SF. The variable in one SF that is used commonly by plural states is called common variable.

### *66.5.2 Classification of Condition and Fairness*

In classification of transition condition, using the combination of shared variables needs fair classify. All conditions must be comprehensively inclusive and exclusively defined. This means that the conditions of transition of SF is defined only once all over domain of variable. Fair execution should be done. In the example, timer should surely time-up some time.

### *66.5.3 Expression as a Design Rule*

The program in terms of Lyee structure is correctly performed only after a definition andthe operation of requirement of Lyee are executed correctly. Calculation conditions of words and the conditions of transition of SF are the reference of requirement correctness.

PRD of Lyee specification is expressed in an L-program by considering SF as a state using the process model. Since a state is an atomic action, SF cannot be partitioned into more than one atomic action. If SFs are unified and enlarged, an excessive calculation will be carried out and its execution speed may become slow. It is important to extract the boundary of SF appropriately. For this reason, it is effective to apply the process model proposed here for safety checking.

In order to apply the above discussion result from now on, it puts together as a design rule.

1. Describe a state transition diagram based on L-language.
2. Verify Property which design intention (requirement) wants.
3. Repeat (1) and (2) until the state transition is fixed.
4. Divide system to program (PRD).
5. Divide program (PRD) to SF based on the SF division rule above.
6. Draw PRD (transition between SF) based on state transition diagram.
7. Realize state as a set of common variables in SF by Lyee method such as a memory table, action vector, and word.

Terms from 1 to 3 are about process model, terms from 4 to 7 is about Lyee Methodology. According to this design rule, we have developed full of the Light Control Case Study correctly and successfully by Lyee Method.

## 66.6 Conclusion

To verify the reactive property of requirement that multi-screen, database management and process control system have, we should apply to a reactive style of verification.

As an example, the requirement that represents Othello game is defined to verify that this program calculates correctly and surely stops anyway. Consequently, we clarify the conditions of fairness in which the execution conditions of requirement and operation should have.

In classification presented at the proof of the example, conditions must be comprehensively inclusive and exclusively defined. Moreover, operation should be executed fairly. These requirements are common in the PRD of Lyee methodology.

Especially, in Process control example, the Light Control Case Study, the trace of logic is difficult, and so maintainability is poor, the concept of state of process model is effective as an approach on visualizing and catching the program structure. When using a structured program such as Lyee or PROLOG, etc. for reactive application with many state transitions, this model is effective.

In Lyee Methodology, requirement specification is not verified. To use this methodology for actual system development, system design intention is desired to be verified beforehand. Using the process model, we can verify other reactive property of intention (requirement) such as safety, partial correctness, avoidance of dead lock, mutual exclusion of critical resource, and total correctness in the future.

## References


1. Negoro F (2000) Principle of Lyee software. In: Proceedings of 2000 International Conference on Information Society in the 21st Century (IS2000), Aizu, Japan, pp 441–446
2. Gorlatch S (2003) The Lyee programming model: Analysis correctness in a fixed point setting. In: Fujita H, Johannesson P (eds) New Trends in Software Methodologies, Tools and Techniques. IOS Press, pp 214–224
3. Gorlatch S (2004) Declarative programming with Lyee for distributed systems. In: Fujita H, Gruhn V (eds) New Trends in Software Methodologies, Tools and Techniques. IOS Press, pp 129–137
4. Fujita H, Mejri M, Ktari B (2004) A process algebra to formalize the Lyee methodology. Knowledge-Based Sys 17(7–8):263-282, ISSN 0950-7051
5. Arai Osamu, Fujita Hamid (2003) Mathematical structure model for Word-Based Program. Knowledge-Based Sys 16(7–8):399–411
6. Arai O, Fujita H (2006) Verification of the Lyee requirement. In: Fujita Hamid, Mejri Mohamed (eds) New Trends in Software Methodologies, Tools and Techniques. IOS Press, (Proceedings of the 5th SoMeT_06, Quebec), pp 340–361
7. Queins S et al. (2000) Requirement engineering: The light control case study. J Universal Comput Sci 6(7)
8. Manna Z, Pnueli A (1991) The temporal logic of reactive and concurrent systems. Sringer-Verlag
9. Manna Z, Pnueli A (1995) The temporal verification of reactive system; Safety. Sringer-Verlag
10. Manna Z, Pnueli A (1996) The temporal verification of reactive system; Progress. Draft

# Chapter 67
# A Tree Derivation Procedure for Multivalent and Paraconsistent Inference

**David Anderson**

**Abstract** The use of trees as a derivation procedure, which is a refinement of Beth's semantic tableaux method and Gentzen's sequent calculus, is an elegant and well established technique. Despite widespread acceptance and application, little effort has been made to extend the use of tree derivation procedures to multivalent alternatives to classical logic and none at all, so far as I am aware, to paraconsistent systems. The purpose of this paper is to outline briefly a single tree derivation procedure which was designed to work with a particular multi-valued paraconsistent system: $\text{Epsilon}_{442}$ [1], but will, with slight modification, work generally.

## 67.1 Valid Arguments

In classical systems, a valid argument may be distinguished from an invalid argument by virtue of the fact that the former has no counter examples. Hence, if an exhaustive search fails to produce a counter example we may be certain that an argument is valid and otherwise invalid. The first method for carrying out such a search which most of us encounter is the 'truth table' but a much more elegant approach is the construction of logical trees.

In classical propositional logic the familiar rules will suffice Fig. 67.1.

## 67.2 Tree Construction

A basic process for constructing a tree derivation is as follows:

Step 1: Arrange the premises in a column one above the other.

Step 2: To the column formed in Step 1, add the NEGATION of the conclusion.

D. Anderson (✉)
School of Computing, University of Portsmouth, Portsmouth PO1 2EG, UK
e-mail: david.anderson@port.ac.uk

N. Mastorakis et al. (eds.), *Proceedings of the European Computing Conference*, Lecture Notes in Electrical Engineering 28,
DOI 10.1007/978-0-387-85437-3_67, 

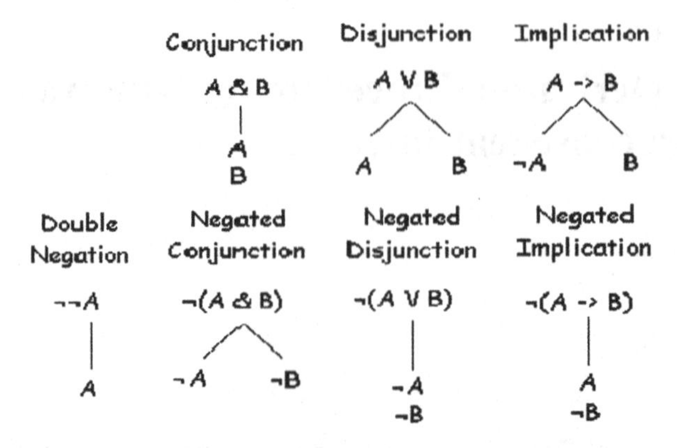


**Fig. 67.1** Tree derivation rules for classical logic

Step 3: Using the decomposition rules presented above simplify each of the expressions (or nodes) as far as possible.

Step 4: When a node has been simplified, it is no longer of interest and may be crossed out, ticked or otherwise identified as no longer forming an active part of the structure.

A trivial example follows. Given A → B, A |- B, the application of steps 1–4 yield the Fig. 67.2.

Each path from top to bottom through the tree we have created represents a possible valuation demonstrating the satisfiability of the set of statements under consideration.

The test for argument validity is carried out in:

Step 5: We put a cross at the bottom of each branch which contains any given statement AND its negation. A branch so constituted is called closed and any tree where all of its branches are closed is itself said to be closed.

A valid inference will always give rise to a closed tree and invalid inferences produce trees with one or more open branches. This is because the decomposition rules are simply ways of displaying the conditions under which a given complex expression may be true and, since we have included the negation of the conclusion as part of the structure, each branch represents a possible counter-

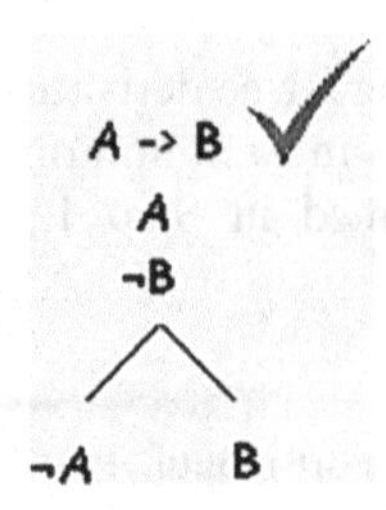


**Fig. 67.2** An example of inference using tree derivation with classical logic

example to the original inference. A closed tree is one where all the possible counter-examples have been eliminated and hence demonstrates validity.

## 67.3 Multivalent Tree Construction

If this general approach is to be extended to cover multi-valued systems there are a number of considerations which need to be addressed. First, we need to exercise care over the use of the negation sign, '$\neg$'. In the system just described $\neg$ stands as a marker for falsity (its absence indicates truth) as well as a logical operator. In a multi-valued system matters are apt to be more complicated and it will prove more effective to avoid any possible ambiguity and use a new piece of notation to indicate the valuation being asserted at any given point on the tree. To this end, I will employ subscripted parentheses containing the valuation. Thus to indicate that A is true we would employ the following: $A_{(\text{true})}$, and for A is false, we would use $A_{(\text{false})}$. In what follows I will be outlining a system designed to handle a four-valued system (more details of which are given in the Appendix) whose possible valuations are 1,2,3 and 4 hence $A_{(1)}$, $A_{(2)}$, $A_{(3)}$ and $A_{(4)}$ will be used to make clear which of the available valuations A takes at any given point.

Another consideration to bear in mind is that multi-valued systems give rise to greater complications concerning how to interpret valuations. In particular, decisions about which valuations are to be designated, anti-designated or non-designated mean that what counts as a counter-example is not as clear-cut as in the bivalent system. One possible approach to this would be to construct a new tree for each of the sorts of counter example but this leads to a more cumbersome system and the preferred approach is to rely on a single tree at the expense of a slightly more convoluted tree derivation process. The rules given in Fig. 67.3 were developed for Epsilon 442, in which {1,3} are designated values, {2} is anti-designated and {4} is non-designated. The full matrices for Epsilon 442 are given in the Appendix.

How do we arrive at exactly these rules and how would we develop rules for a different system?

Rather than painstakingly going through each structure in turn I will take a single example in order to demonstrate the principle. Take$^{\mathbf{(A\&B)}}{}_{(4)}$. To devise the appropriate decomposition structure we need to consult the Epsilon 442 matrix for conjunction (Fig. 67.4):

We can readily see that (A & B) is assigned the value four in just five cases:

i. the first operand has a value of $_{(1)}$ and the second has a value of $_{(3)}$.
ii. the first operand has a value of $_{(1)}$ and the second has a value of $_{(4)}$.
iii. the first operand has a value of $_{(3)}$ and the second has a value of $_{(1)}$.
iv. the first operand has a value of $_{(4)}$ and the second has a value of $_{(1)}$.
v. the first operand has a value of $_{(4)}$ and the second has a value of $_{(4)}$.

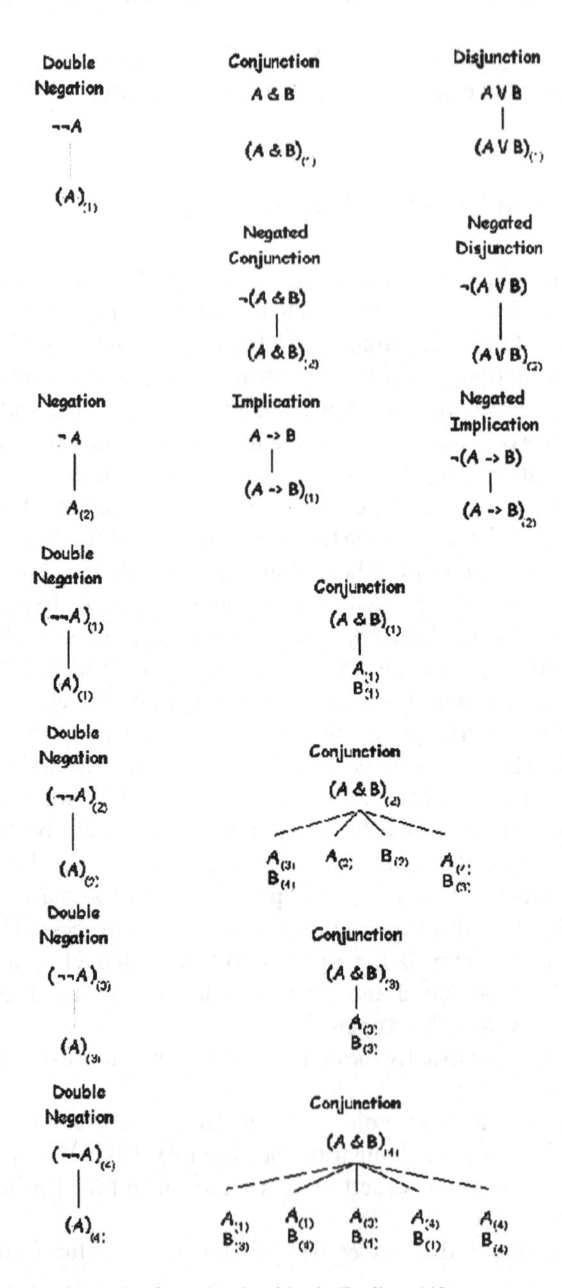


**Fig. 67.3** Tree derivation rules for a 4-valued logic Epsilon 442

And these five cases are captured precisely by the decomposition rule where each of the alternatives is represented by its own branch. Each of the other structures is arrived at in a similar fashion. Of course, different matrices would give rise to a different decomposition rule but the process remains straightforward.

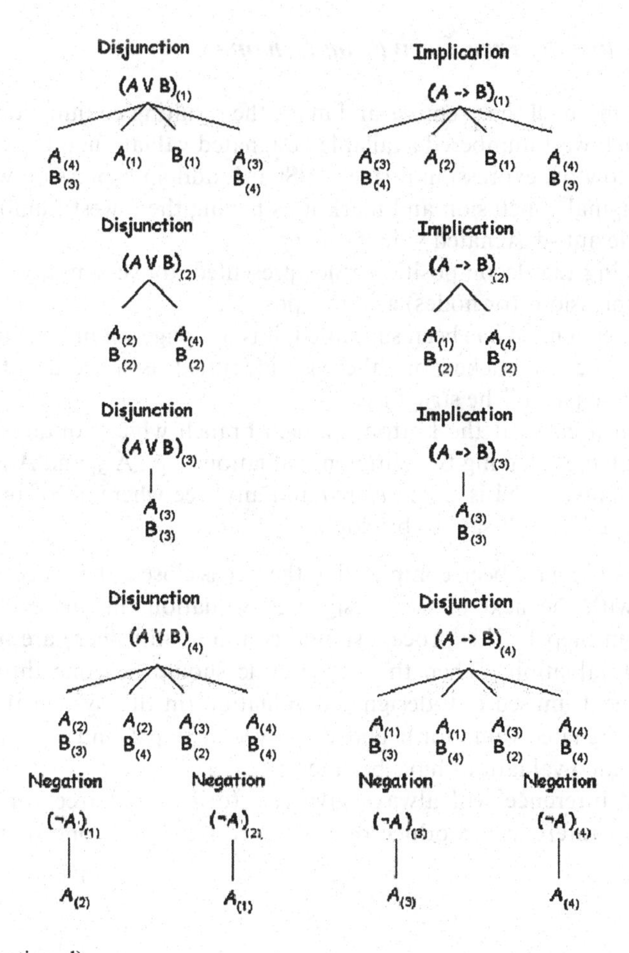


**Fig. 67.3** (continued)

| AND | 1 | 2 | 3 | 4 |
|---|---|---|---|---|
| 1 | 1 | 2 | 4 | 4 |
| 2 | 2 | 2 | 2 | 2 |
| 3 | 4 | 2 | 3 | 2 |
| 4 | 4 | 2 | 2 | 4 |

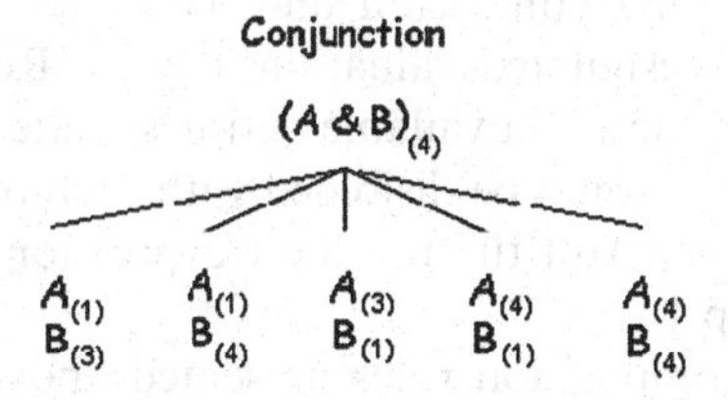


**Fig. 67.4** Devising an appropriate tree structure from an Epsilon 442 matrix

### 67.3.1 Multiple Tree Derivation Scheme

Step 1: Conjoin all the premises and mark the resulting conjunction as having the lowest numbered available designated valuation, e.g. $_{(1)}$.

Step 2: Below the expression formed in Step 1, add an expression which is the original conclusion and mark it as having the lowest numbered available anti-designated valuation, e.g. $_{(2)}$.

Step 3: Using the decomposition rules presented above simplify each of the expressions (or nodes) as far as possible.

Step 4: When a node has been simplified, it is no longer of interest and may be crossed out, ticked or otherwise identified as no longer forming an active part of the structure.

Step 5: Put a cross at the bottom of each branch which contains any given statement having two different valuations e.g., $A_{(3)}$ and $A_{(1)}$ A branch so constituted is called *closed* and any tree where all of branches are closed is itself said to be closed

When steps 1–5 have been completed, if the tree is closed, the process should be repeated with the next unused designated valuation (in this case $_{(3)}$) being employed in Step 1. This process should continue until there are no untested designated valuations. Then the whole cycle should be gone through again using the next unused anti-designated valuation (in this system there is only one anti-designated valuation) being employed in Step 2, and so on until all the anti-designated valuations have been tested.

A valid inference will always give rise to a closed tree for every test and invalid inferences produce one or more trees with one or more open branches.

### 67.3.2 Single Tree Derivation Scheme

Step 1: Conjoin all the premises and mark the resulting conjunction as having the lowest numbered available designated valuation, e.g. $_{(1)}$. Repeat the process with each remaining unused available designated valuation, the resulting conjunctions should be disjoined with each other to form one complex disjunction.

Step 2: Mark the original conclusion and as having the lowest numbered available anti-designated valuation, e.g. $_{(2)}$. Repeat the process with each remaining unused available anti-designated valuation, the resulting expressions should be disjoined with each other to form one complex disjunction. Add the resultant expression below the expression formed in Step 1.

Step 3: Using the decomposition rules presented above simplify each of the expressions (or nodes) as far as possible.

Step 4: When a node has been simplified, it is no longer of interest and may be crossed out, ticked or otherwise identified as no longer forming an active part of the structure.

Step 5: Put a cross at the bottom of each branch which contains any given statement having two different valuations e.g., $A_{(3)}$ and $A_{(1)}$ A branch so constituted is called *closed* and any tree where all of the branches are closed is itself said to be closed.

A valid inference will always give rise to a closed tree and invalid inferences produce one or more open branches.

It is worth noting that the tree derivation scheme suggested here handles a multivalent system which is also paraconsistent and in this respect it is, to the best of my knowledge, unique.

## Appendix: Matrices for Epsilon 442

NB. In Epsilon 442, {1,3} are designated values, {2} is anti-designated and {4} is non-designated.

| NOT | |
|---|---|
| 1 | 2 |
| 2 | 1 |
| 3 | 3 |
| 4 | 4 |
| **¬(¬P & ¬Q)** | |

| AND | 1 | 2 | 3 | 4 |
|---|---|---|---|---|
| 1 | 1 | 2 | 4 | 4 |
| 2 | 2 | 2 | 2 | 2 |
| 3 | 4 | 2 | 3 | 2 |
| 4 | 4 | 2 | 2 | 4 |
| **¬(P & ¬Q)** | | | | |

| OR | 1 | 2 | 3 | 4 |
|---|---|---|---|---|
| 1 | 1 | 1 | 1 | 1 |
| 2 | 1 | 2 | 4 | 4 |
| 3 | 1 | 4 | 3 | 1 |
| 4 | 1 | 2 | 1 | 4 |

| COND | 1 | 2 | 3 | 4 |
|---|---|---|---|---|
| 1 | 1 | 2 | 4 | 4 |
| 2 | 1 | 1 | 1 | 1 |
| 3 | 1 | 4 | 3 | 1 |
| 4 | 1 | 2 | 1 | 4 |

## References

1. Anderson CDP (2002) Developing a framework for investigating inconsistency handling in automated reasoning. In: 6th World Multi-Conference on Systemics, Cybernetics and Informatics (SCI 2002), Orlando, Florida

# Chapter 68
# On the Feedback Stabilization of Matrix

JianDa Han, Zhe Jiang, YiWen Zhao, and YiYong Nie

**Abstract** In this paper, several strategies for feedback stabilization of matrix are presented via the numerical stabilization of polynomial and matrix.

## 68.1 Introduction

It is important to guarantee the stability of a dynamic system. In order to design a stable control system, the numerical stabilization of the system has become quite an interesting problem.

Every state equation of the linear time-invariant system can be described as follows,

$$\dot{x}(t) = Ax(t) + Bu(t) \tag{68.1}$$

For a system with $p$ inputs, and $n$ state variables, $A$ and $B$ are, respectively, $n \times n$ and $n \times p$ constant matrices. In state feedback, the input vector $u$ is given by $u = r - Kx$, where $K$ is a $p \times n$ gain matrix with real elements $k_{ij}$. This is the constant gain negative state feedback. We may suppose $r = 0$ since only stability of the state feedback system is investigated. Substituting $u = -Kx$ into the state equations of (68.1) yields

$$\dot{x} = (A - BK)x = \begin{pmatrix} a_{11} - v_{11} & a_{12} - v_{12} & \cdots & a_{1n} - v_{1n} \\ a_{21} - v_{21} & a_{22} - v_{22} & \cdots & a_{2n} - v_{2n} \\ & & \cdots & \\ a_{n1} - v_{n1} & a_{n2} - v_{n2} & \cdots & a_{nn} - v_{nn} \end{pmatrix} x, \tag{68.2}$$

J. Han (✉)
State Key Laboratory of Robotics, Shenyang Institute of Automation, Academia Sinica. 110016 Shenyang, China

N. Mastorakis et al. (eds.), *Proceedings of the European Computing Conference*, Lecture Notes in Electrical Engineering 28,
DOI 10.1007/978-0-387-85437-3_68, 

where

$$v_{ij} = \sum_{s=1}^{p} b_{is}k_{sj}, i,j = 1, 2, \cdots, n. \tag{68.3}$$

The matrix $A - BK$ is referred to as stable if its each eigenvalue has negative real-part. The system (68.1) is asymptotically stable if and only if the matrix $A - BK$ is stable; see, e.g., [3, 5, 6]. The matrix $A - BK$ is referred to as critical if the largest real-part of eigenvalues is equal to zero. If the system (68.1) is critical, then the second approximation may be constructed such that (68.1) is stable or unstable, see, e.g., [5]. So that the critical matrix is rather considered unstable. A useful matrix for control system must have sufficient stability, namely, the largest real-part of eigenvalues must be less than or equal to an appropriate negative number.

Regarding each $k_{sj}$ in (68.3) as a parameter, we can appropriately select $k_{sj}(s = 1, \cdots, p, j = 1, \cdots, n)$ such that the feedback system (68.2) is stable. This is referred to as *feedback stabilization*. Evidently, simultaneously selecting $k_{sj}$ is too complicated for general $A$ and $B$. Fortunately, in order to stabilize the matrix $A$ by the state feedback matrix $K$, we need only to investigate some simple cases of $B$ and $K$.

The stability of a given matrix is equivalent to the stability of its eigenpolynomial. In fact, using numerically stabilized elementary transformations we can first reduce a matrix to upper Hessenberg form, and then reduce upper Hessenberg form to Frobenius-like form, and finally we can obtain the eigenpolynomial. In the computation process, the reduction from initial matrix to Frobenius-like form may be numerically stable for initial perturbation and rounding errors; see [2, 6, 7].

In the same reason, the largest real-part of roots of the eigenpolynomial must be less than or equal to an appropriate negative number.

In this paper, so-called *numerical stabilization* of polynomial and matrix means to regulate the changeable coefficients of polynomial or the changeable elements of matrix such that the largest real-part of roots or eigenvalues, $\xi$, is less than or equal to an appropriate negative number, the *stabilized-threshold* $\theta$,

$$\xi \leq \theta, \tag{68.4}$$

while original polynomial or original matrix is possibly unstable ($\xi > \theta$) or over stable ($\xi \ll \theta$) in itself, see [1].

The stabilized-threshold $\theta$ is a key to numerical stabilization. If (68.4) holds for polynomial or matrix after stabilization, then the polynomial or matrix is properly stable, otherwise unstable or over stable. Clearly $\theta < -2^{-t}$, where $t$ is the mantissa digits of computer, because $2^{-t} \approx 0$ in computer. Moreover, since all the stability criteria are based on the calculations of polynomial coefficients or matrix elements, $\theta$ must be dependent on the accumulation of relative rounding-errors in the computational process. Unfortunately, the accumulation of rounding-errors is a statistical process. We can only estimate a error bounds too high, even if the algorithm is numerically stable.

The error analyses of the numerically stable algorithms, including the reduction from initial matrix to Frobenius-like and the quasi-Routh array, show that the error bounds $O(n2^{-t})$ relates to elements of $A - BK$. This is inexpedient. In order to avoid including elements of $A - BK$, we rather regard an enlarged error bounds as the stabilized-threshold $\theta$ in (68.4). That is,

**Hypothesis one** *For the numerically stable algorithms with $O(n^3)$ flops, including the reduction from initial matrix to Frobenius-like and the quasi-Routh array, the stabilized-threshold in* (68.4)

$$\theta = -n^3 2^{-t} \tag{68.5}$$

where $n$ is dimension of the system (68.1), and $t$ is the mantissa digits of computer.

If $t = 52$ and $n = 50, 100, 150$, then $\theta \approx -2.7756 \times 10^{-11}$, $-2.2204 \times 10^{-10}$, $-7.4940 \times 10^{-10}$, respectively. If $t = 32$ and $n = 30, 40, 50$, then $\theta \approx -6.2864 \times 10^{-6}$, $-1.4901 \times 10^{-5}$, $-2.9104 \times 10^{-5}$ respectively. Practical computation shows that the stabilized-threshold given by (68.5) is appropriate for numerical stabilization of polynomial and matrix.

It is easy to see a simple strategy for feedback stabilization of (68.2).

**Proposition two** *The proper stabilization of the matrix (68.2) may be always realized by the origin-shifted transformations. Namely the matrix $A$ in (68.1) may be always stabilized by the diagonal feedback matrix $K = g_d I$, where $I$ denotes unit matrix and $g_d$ is a real number (diagonal gain).*

*Proof.* Let $P(\lambda) = \det(A - \lambda I)$ be the eigenpolynomial of the matrix $A$ in (68.1), where $\det(\cdot)$ denotes determinant of matrix, and let $l$ be the bounds of the eigenvalues of $A$. Using the origin-shifted transformation, we know that $P(\lambda + l) = \det(A - (\lambda + l)I) = \det((A - lI) - \lambda I)$ is stable and $P(\lambda - l) = \det((A + lI) - \lambda I)$ unstable.

Let $s_0 = 0$. If $P(\lambda + s_0)$ is unstable or critical, then let $s_1 = s_0 + l/2$, otherwise let $s_1 = s_0 - l/2$, then investigate the stability of $P(\lambda + s_1)$. If $P(\lambda + s_1)$ is unstable or critical, then let $s_2 = s_1 + l/2^2$, otherwise let$s_2 = s_1 - l/2^2$, then investigate the stability of $P(\lambda + s_2)$. $\cdots$.. If $P(\lambda + s_{r-1})$ is unstable or critical, then let $s_r = s_{r-1} + l/2^r$, otherwise let $s_r = s_{r-1} - l/2^r$. We may always make $P(\lambda + s_r)$ stable or critical, $P(\lambda + s_r - l/2^r)$ unstable, and $P(\lambda + s_r + l/2^r)$ stable. When $r$ is big enough, $l/2^r \le -\theta$, and the origin-shifted transformation may be terminated. Thus, the largest real-part of matrix eigenvalues, $\xi \approx s_r$, may be found by a serial origin-shifted transformations.

Let $B = I$, $K = g_d I$ and $g_d = s_r - \theta$ in (68.2). Clearly $P(\lambda + s_r - \theta) = \det((A - BK) - \lambda I)$ is stable, and $\xi \approx g_d + \theta$. The matrix $A - BK$ is properly stable.

With numerical determination of matrix stability, Proposition two gives a basic approach to feedback stabilization. And the length of numerically experimental interval decreases with experimental times in exponent. However, the diagonal feedback matrix $K = g_d I$ requires $n$ inputs with the same gain parameter $g_d$. This is possibly inconvenient. We must look for more strategies for feedback stabilization.

## 68.2 Numerical Stabilization of Polynomial and Matrix

In this paragraph, we list all the useful results of numerical stabilization of polynomial and matrix; see [1].

Let

$$F(\lambda) = -det(A - \lambda I) = \lambda^n + a_1\lambda^{n-1} + a_2\lambda^{n-2} + \cdots + a_{n-1}\lambda + a_n \qquad (68.6)$$

be the monic polynomial with $n$ degree. The determining coefficients $\alpha_i = \alpha_{i-1}\alpha_{i-2}/(\alpha_i\alpha_{i+1})$, $i = 1, \cdots, n-2, a_0 = 1$. Using the determining coefficients, we can present several strategies for stabilization of the point-wise polynomial (68.6).

**Proposition three** *If some invariable coefficients of (68.6) are given such that a determining coefficient $\alpha_i \leq 0$ or $\alpha_i \geq 1, 1 \leq i \leq n-2$, at least, then the polynomial (68.6) is impossible to be stabilized.*

Generally, if there are many invariable coefficients in (68.6), then, in Proposition 3, it easily happens that (68.6) is impossible to be stabilized.

Suppose now (68.6) may be stabilized, then $0 < \alpha_i < 1$, $(i = 1, 2, \cdots, n-2)$. In order to look for more efficient strategies for numerical stabilization of (68.6), we utilize the sufficient conditions which are nearing critical condition, the optimum criteria of polynomial stability.

**Hypothesis four** *The polynomial (68.6) is considered a properly stabilized polynomial if one of the following two equalities holds, where and $\alpha_i = \alpha_{i-1}\alpha_{i-2}/(\alpha_i\alpha_{i+1})$ and $0 < \alpha_i < 1$*

*(i)* $\max\{\alpha_1, \alpha_2, \cdots, \alpha_{n-2}\} = 0.46557$;

*(ii)* $\max\{\alpha_1 + \alpha_2, \alpha_2 + \alpha_3, \cdots, \alpha_{n-2} + \alpha_{n-1}\} = 0.88988$.

Via Kharitonov's theorem, see [4], the stability of monic interval polynomial

$$F(\lambda) = \lambda^n + a_1\lambda^{n-1} + a_2\lambda^{n-2} + \cdots + a_{n-1}\lambda + a_n, \underline{a}_i \leq a_i \leq \bar{a}_i, i = 1, 2, \cdots, n, \quad (68.7)$$

is equivalent to the stability of the following four point-wise polynomials:

$$\begin{aligned} F_1(\lambda) &= \lambda^n + \underline{a}_1\lambda^{n-1} + \bar{a}_2\lambda^{n-2} + \bar{a}_3\lambda^{n-3} + \underline{a}_4\lambda^{n-4} + \underline{a}_5\lambda^{n-5} + \cdots, \\ F_2(\lambda) &= \lambda^n + \bar{a}_1\lambda^{n-1} + \bar{a}_2\lambda^{n-2} + \underline{a}_3\lambda^{n-3} + \underline{a}_4\lambda^{n-4} + \bar{a}_5\lambda^{n-5} + \cdots, \\ F_3(\lambda) &= \lambda^n + \underline{a}_1\lambda^{n-1} + \underline{a}_2\lambda^{n-2} + \bar{a}_3\lambda^{n-3} + \bar{a}_4\lambda^{n-4} + \underline{a}_5\lambda^{n-5} + \cdots, \\ F_1(\lambda) &= \lambda^n + \bar{a}_1\lambda^{n-1} + \underline{a}_2\lambda^{n-2} + \underline{a}_3\lambda^{n-3} + \bar{a}_4\lambda^{n-4} + \bar{a}_5\lambda^{n-5} + \cdots. \end{aligned} \quad (68.8)$$

Each polynomial of (68.8) has $n-2$ determining coefficients $\alpha_i$ $(i = 1, \cdots, n-2)$, while $\underline{\alpha}_i = \underline{a}_{i-1}\underline{a}_{i+2}/(\bar{a}_i\bar{a}_{i+1})$, and $\bar{\alpha}_i = \bar{a}_{i-1}\bar{a}_{i+2}/(\underline{a}_i\underline{a}_{i+1})$ are the least one and the largest one respectively.

**Proposition five** *If some invariable intervals of (68.7) are given such that a determining coefficient of (68.8),* $\underline{\alpha}_i \leq 0$ *or* $\bar{\alpha}_i \geq 1, 1 \leq i \leq n-2$*, at least, then the interval polynomial (68.7) is impossible to be stabilized.*

Suppose (68.7) may be stabilized, then $0 < \underline{\alpha}_i \leq \bar{\alpha}_i < 1 (i = 1, \cdots, n-2)$.

**Proposition six** *If each point-wise polynomial of (68.8) is stabilized by Hypothesis four, then the interval polynomial (68.7) is stabilized numerically.*

If the changeable elements of the matrix $K$ in (68.2) is arbitrarily distributed, then one of the most reliable strategies for numerical stabilization of $A - BK$ is to reduce from $A - BK$ to the Frobenius-like form and to apply stabilized strategies of the eigenpolynomial with a lot of numerical experiments. Here, the numerical stability of each numerical experiment is important, otherwise the coefficients of eigenpolynomial are possibly distorted by accumulation of rounding-errors. Fortunately, the numerically stable algorithm of reduction from $A - BK$ to Frobenius-like form has been found with stabilized elementary similarity transformations, see [6, 7].

Since the stability of $A$ is only dependent on $n$ invariants, it is reasonable to assume that there are only $n$ changeable elements in $n^2$ elements of $BK$ at most. Generally, with elementary similarity transformations, these $n$ elements may be put on the first row of matrix. Therefore we assume further that the $n$ changeable elements are distributed on the first row of$BK$. The numerical stabilization of $A - BK$ becomes more efficient on these assumptions.

**Proposition seven** *If the changeable elements of $A - BK$ in (68.2) are on the first row, then the numerical stabilization of the matrix may be realized by the stabilized strategies of polynomial, Proposition three and Hypothesis four.*

## 68.3 Feedback Stabilization of Interval Controllable Canonical Form

Suppose $A = A_c$, $p = 1$, $B = B_c = (0, \cdots, 0, 1)^T$and $K = K_c = (k_1, \cdots, k_n)$ in (68.1) and (68.2), where the superscript $T$ denotes transposition and $A_c$ is an interval controllable canonical form given by

$$A_c = \begin{pmatrix} 0 & 1 & 0 & \cdots & 0 & 0 \\ 0 & 0 & 1 & \cdots & 0 & 0 \\ & & & \cdots & & \\ 0 & 0 & 0 & \cdots & 0 & 1 \\ a_1 & a_2 & a_3 & \cdots & a_{n-1} & a_n \end{pmatrix} \alpha_i \in [\underline{a}_i, \bar{a}_i], i = 1, 2, \cdots, n. \quad (68.9)$$

The $n$-dimensional state equation of controllable canonical form with single-input is as follows,

$$\dot{x}(t) = A_c x(t) + B_c u(t). \quad (68.10)$$

After feedback $u = -K_c x$ of single-input, the state equations (68.10) becomes

$$\dot{x} = (A_c - B_c K_c)x = \begin{pmatrix} 0 & 1 & \cdots & 0 \\ 0 & 0 & \cdots & 0 \\ & & \cdots & \\ 0 & 0 & \cdots & 1 \\ a_1 - k_1 & a_2 - k_2 & \cdots & a_n - k_n \end{pmatrix} x, \quad (68.11)$$

where $k_i, i = 1, 2, \cdots, n$, are regarded as undetermined parameters for stabilization of $A_c - B_c K_c$. The matrix $A_c - B_c K_c$ is still a canonical form.

Clearly $(A_c - B_c K_c)^T$ is a Frobenius-like form. Its monic eigenpolynomial is

$$P_c(\lambda) = \lambda^n + (k_n - a_n)\lambda^{n-1} + \cdots + (k_2 - a_2)\lambda + (k_1 - a_1). \quad (68.12)$$

Therefore all the stabilized strategies of polynomial may be applied to the interval controllable canonical form $A_c - B_c K_c$ in (68.11).

**Proposition eight** *For the interval controllable canonical form (68.9), we may set* $k_{n-1} = k_{n-1}^c$ *and* $k_n = k_n^c$ *and then select* $k_1, \cdots, k_{n-2}$ *such that the interval controllable canonical form* $A_c - B_c K_c$ *in (68.11) is stable.*

*Proof.* Setting $k_n > a_n$ and $k_{n-1} > a_{n-1}$ in any point-wise polynomial of (68.12) and using Proposition three and Hypothesis four(i), we can obtain the following selected intervals of other undetermined parameters, $n - 2 \geq n + 2 - 2i$, $n + 1 - 2i \geq 2$,

$$0 < k_{n+2-2i} - a_{n+2-2i} \leq (k_n - a_n)(\alpha^{i-1}(k_{n-1} - a_{n-1}))^{i-1}, \quad (68.13)$$

$$0 < k_{n+1-2i} - a_{n+1-2i} \leq (\alpha^{i-1}(k_{n-1} - a_{n-1}))^i, \quad (68.14)$$

where $\alpha = \max\{\alpha_1, \alpha_2, \cdots, \alpha_{n-2}\} \leq 0.46557$. Evidently, from (68.13) and (68.14), $k_j \geq \bar{a}_j - \theta (j = 1, 2, \cdots, n)$, where $\theta$ is the stabilized-threshold. Let

$k_j = \bar{a}_j + c_j(c_j \geq -\theta, j = 1, 2, \cdots, n)$. Then $k_j - a_j = \bar{a}_j - a_j + c_j$ and $k_j - a_j \in [c_j, d_j + c_j]$, where $d_j = \bar{a}_j - \underline{a}_j$ denotes the length of $j$-th interval. The coefficient intervals of (68.12) are $[c_j, d_j + c_j], j = 1, 2, \cdots, n$. Let $d = \max\{d_1, d_2, \cdots, d_n\}$.

It is easy to see that, from (68.14), the maximal value of function $f(i) = (\alpha^{i-1}(k_{n-1} - a_{n-1}))^i$ occurs close to some $i^*, 4 \leq 2i^* \leq n$. This requires $\alpha^{3-2i^*} < k_{n-1} - a_{n-1} < \alpha^{-2i^*-1}$, namely $\alpha^{3-2i^*} < c_{n-1} < \alpha^{-2i^*-1} - d_{n-1}$. We hope also that, from (68.13), the maximal value of function $g(i) = (k_n - a_n)(\alpha^{i-1}(k_{n-1} - a_{n-1}))^{i-1}$ occurs close to $i^*$. This requires $\alpha^{4-2i^*} < c_{n-1} < \alpha^{-2i^*} - d$. Therefore we get

$$\alpha^{3-2i^*} < c_{n-1} < \alpha^{-2i^*} - d_{n-1}. \tag{68.15}$$

Suppose $\alpha^{-2i^*} + d_{n-1} > \alpha^{3-2i^*}$. Then $i^*$ may be equal to $2, 3, \cdots$ for $d_{n-1} < 20.160198$. Let $m = \lfloor (n+1)/2 \rfloor$ be the biggest integer less than or equal to $(n+1)/2$, and

$$c_{n-1}^* = (d + c^o)^{1/i^*} / \alpha^{i^*-1}, i^* \leq m, \tag{68.16}$$

$$c_n^* = (d + c^e)/(\alpha^{i^*-1} c_{n-1}^*)^{i^*-1} = (d + c^e)(d + c^o)^{1-1/i^*}, \tag{68.17}$$

where $c^o = \max\{c_{n-1}, c_{n-3}, \cdots\}$ and $c^e = \max\{c_n, c_{n-2}, \cdots\}$ are the prearranged parameters which must satisfy, with (68.15),

$$\begin{gathered} \alpha^{-(i^*-2)} < (d + c^o)^{1/i^*} < \min\{\alpha^{i^*-1} c^o, \alpha^{-(i^*+1)} - \alpha^{i^*-1} d_{n-1}\}, \\ (2 \leq i^* \leq m), \end{gathered} \tag{68.18}$$

and

$$d \leq ((d + c^o)^{1-1/i^*} - 1)c^e, (m \geq 3). \tag{68.19}$$

The inequalities (68.18) and (68.19) yield, with $c^o + d = \alpha^{-i^*-u}$ and $\alpha^{-u} - \alpha^{-u/i^*} \geq \alpha^{i^*} d$,

$$\begin{gathered} \max\{\alpha^{-i^*(i^*-2)} - d, \alpha^{i^*-u/i^*} + \theta\} < c^o < \\ \alpha^{-i^*(i^*+1)}(1 - \alpha^{2i^*} d_{n-1})^{i^*} - d < \alpha^{-i^*(i^*+1)} - d, \end{gathered} \tag{68.20}$$

and

$$c^e \geq d/((d + c^o)^{1-1/i^*} - 1). \tag{68.21}$$

Besides $c^o$ and $c^e$ satisfy (68.20) and (68.21) respectively, they must satisfy, with (68.16) and (68.17),

$$\alpha^{(m-1)(m-i^*)}(d+c^o)^{(m-i^*)/i^*}(d+c^e)>d, \tag{68.22}$$

and

$$\alpha^{m(m-i^*)}(d+c^o)^{m/i^*}>d \tag{68.23}$$

for $n=2m$, or

$$\alpha^{(m-1)(m-1-i^*)}(d+c^o)^{(m-1)/i^*}>d \tag{68.24}$$

for $n=2m-1$. It is easy to see $i^*>(m-1)/2$ from (68.20), (68.23) and (68.24). Let $i^*=\lceil m/2 \rceil$ be the least integer bigger than or equal to $m/2$. Then $\alpha^{-i^*(m-i^*)}<\alpha^{-i^*(i^*+1)}$.

Let

$$c^o=\alpha^{-i^*(m-i^*)}\max\{1,d^{i^*/(m-1)}\}-d-\theta. \tag{68.25}$$

Then (68.20), (68.23) and (68.24) hold for $m\geq 4$ and $d\leq 4.6$. Let

$$c^e=\max\{\alpha^{-(i^*-1)(m-i^*)},\alpha^{-(i^*-1)(m-i^*)}d^{(i^*-1)/(m-1)},2d\}-d-\theta. \tag{68.26}$$

Then (68.21) and (68.22) hold for $m\geq 3$. Substituting (68.25) and (68.26) into (68.16) and (68.17) respectively, we can obtain the representations of $c^*_{n-1}$ and $c^*_n$, which are only dependent on$\alpha$,$d$and $m$. For instance, with $m\geq 5$,

$$c^*_{n-1}=\alpha^{-(m-1)}d^{1/(m-1)}-\theta, c^*_n=1, \tag{68.27}$$

where $1\leq d\leq 4.6$. When $d<1$, the equality (68.27) is still right if $d$ is regarded as 1.

Setting $k_{n-1}=k^c_{n-1}=\bar{a}_{n-1}+c^*_{n-1}$ and using the Kharitonov's theorem, we may select $c_{n-3},c_{n-5},\cdots,c_{n-2m+3},c_{n-2m+1}$, namely $k_{n-3},k_{n-5},\cdots,k_{n-2m+3},k_{n-2m+1}$, satisfying (68.14) because of (68.15). For instance, via (68.8), $c_{n-3}\leq(\alpha(d_{n-1}+c^*_{n-1}))^2$and $d_{n-3}+c_{n-3}\leq\alpha(c^*_{n-1})^2$, so that

$$c_{n-3}\approx\alpha(c^*_{n-1})^2-d_{n-3}.$$

Also via (68.8), $d_{n-5}+c_{n-5}\leq(\alpha^2(d_{n-1}+c^*_{n-1}))^3$ and$c_{n-5}\leq(\alpha^2c^*_{n-1})^3$, so that

$$c_{n-5}\approx\min\{(\alpha^2c^*_{n-1})^3,(\alpha^2(d_{n-1}+c^*_{n-1}))^3-d_{n-5}\}.$$

Similarly, setting $k_{n-1}=k^c_n=\bar{a}_n+c^*_n$ and using the Kharitonov's theorem, we may select $c_{n-2},c_{n-4},\cdots,c_{n-2m+4},c_{n-2m+2}$, namely $k_{n-2},k_{n-4},\cdots,k_{n-2m+4},k_{n-2m+2}$, satisfying (68.13) because of (68.16). For instance, via (68.8), $c_{n-2}\leq(c^*_n+d_n)\alpha c^*_{n-1}$ and $d_{n-2}+c_{n-2}\leq c^*_n\alpha c^*_{n-1}$, so that

$$c_{n-2} \approx c_n^* \alpha c_{n-1}^* - d_{n-2}.$$

Also via (68.8), $d_{n-4} + c_{n-4} \leq (c_n^* + d_n)(\alpha^2 c_{n-1}^*)^2$ and$c_{n-4} \leq c_n^*(\alpha^2 c_{n-1}^*)^2$, so that

$$c_{n-4} \approx \min\{c_n^*(\alpha^2 c_{n-1}^*)^2, (c_n^* + d_n)(\alpha^2 c_{n-1}^*)^2 - d_{n-4}\}.$$

After $c_{n-3}, \cdots, c_1$ are represented by (68.8), (68.14) and (68.13), we should test and verify the stability of (68.12) with Kharitonov's theorem and Hypothesis 4(i). If the stability cannot be guaranteed, then $\alpha$ should be reduced gradually. If $d_i = d, i = 1, \cdots, n$, then we need only to test and verify

$$\begin{aligned}
\max\{\alpha_1\} &= (c_{n-2} + d)/(c_{n-1}^* c_n^*) \leq 0.46557, \\
\max\{\alpha_2\} &= (c_n^* + d)(c_{n-3} + d)/(c_{n-1}^* c_{n-2}) \leq 0.46557, \\
\max\{\alpha_{n-3}\} &= (c_2 + d)(c_5 + d)/(c_3 c_4) \leq 0.46557, \\
\max\{\alpha_{n-2}\} &= (c_1 + d)(c_4 + d)/(c_2 c_3) \leq 0.46557
\end{aligned}$$

because of (68.14) and (68.13).

*Example 1* Suppose $n = 10$ and

$$\begin{aligned}
&a_1 \in -9 \pm 0.8, a_2 \in -11 \pm 0.8, a_3 \in -20 \pm 0.8, a_4 \in -4 \pm 0.8, a_5 \in -6 \pm 0.8, \\
&a_6 \in -10 \pm 0.8, a_7 \in -5 \pm 0.8, a_8 \in -7 \pm 0.8, a_9 \in -1 \pm 0.8, a_{10} \in -2 \pm 0.8
\end{aligned}$$

in (68.9). Clearly $m = 5$, $d = 1.6$, $i^* = 3$ and $d_1 = \cdots = d_{10} = d$.

In this example, via (68.27), $c_9^* = 1.6^{1/4}\alpha^{-4}$ and $c_{10}^* = 1$.

According to Proposition 8 we obtain

$$\begin{aligned}
&c_8 = \alpha c_9^* - 1.6 = 1.6^{1/4}(\alpha^{-3} - 1.6^{3/4}), \\
&c_7 = (\alpha c_9^*)^2 - 1.6 = 1.6^{1/2}(\alpha^{-6} - 1.6^{1/2}), c_5 = 1.6^{3/4}\alpha^{-6}, \\
&c_4 = 1.6^{3/4}\alpha^{-3} - 1.6 \ c_3 = (\alpha^3 * c_9^*)^4 - 1.6 = 1.6(\alpha^{-4} - 1), \\
&c_2 = (\alpha^4 * c_9^*)^4 = 1.6, \\
&c_1 = (\alpha^4 * (c_9^* + 1.6))^5 - 1.6 = (1.6^{1/4} + 1.6\alpha^4)^5 - 1.6.
\end{aligned}$$

The $\alpha$ should also satisfy

$$\begin{aligned}
&1.6^{1/4}\alpha^{-3}/(1.6^{1/4}\alpha^{-4}) = \alpha \leq 0.46557, \\
&2.6 * 1.6^{1/4}\alpha^{-2}/(1.6^{1/4}\alpha^{-3} - 1.6) \leq 0.46557, \\
&(1.6^{1/4} + 1.6\alpha^4)^5 * 1.6^{3/4}\alpha^{-3}/1.6/(1.6\alpha^{-4} - 1.6) \leq 0.46557, \\
&3.2 * (1.6^{3/4}\alpha^{-6} + 1.6)/(1.6\alpha^{-4} - 1.6)/(1.6^{3/4}\alpha^{-3} - 1.6) \leq 0.46557,
\end{aligned}$$

Namely

$$\alpha \leq 0.1790653846(1 - 1.6^{3/4}\alpha^3),$$
$$\alpha(1 + 1.6^{3/4}\alpha^4)^5 \leq 0.46557(1 - \alpha^4),$$
$$\alpha(1 + 1.6^{1/4}\alpha^6) \leq 0.232785(1 - \alpha^3 - 1.6^{1/4}\alpha^4 + 1.6^{1/4}\alpha^7).$$

Evidently $\alpha = 0.46$ cannot satisfy the above inequalities. Since $\alpha < 0.46$, $1.6^{3/4}\alpha^4 \ll 1$, we get approximately

$$\alpha + 0.2547426292\alpha^3 \leq 0.1790653846, \alpha \leq 0.1776,$$
$$\alpha + 0.46557\alpha^4 + 5 * 1.6^{3/4}\alpha^5 < 0.46557,$$
$$\alpha + 3.737604604\alpha^4 < 0.47557, \alpha \leq 0.38,$$
$$\alpha + 0.232785\alpha^3 + 0.26180925\alpha^4 \leq 0.232785, \alpha \leq 0.2292.$$

Select $\alpha = 0.1776$. According to Proposition 6 we have thus $c_9^* = 1130.467678664$ and $c_1 \approx 0.2122628$, $c_2 \approx 1.6$,

$$c_3 \approx 1606.6299, c_4 \approx 252.3575, c_5 \approx 45334.8536769,$$
$$c_6 \approx 1271.417385, c_7 \approx 40307.4184, c_8 \approx 199.1710597.$$

Therefore

$$\begin{aligned}
&k_{10} - a_{10} \in [1, 2.6],\\
&k_9 - a_9 \in [1130.467678664, 1132.067678664],\\
&k_8 - a_8 \in [199.1710597, 200.7710597],\\
&k_7 - a_7 \in [40307.4184, 40309.0184],\\
&k_6 - a_6 \in [1271.417385, 1273.017385],\\
&k_5 - a_5 \in [45334.8536769, 45336.4536769],\\
&k_4 - a_4 \in [252.3575, 253.9575],\\
&k_3 - a_3 \in [1606.6299, 1608.2299],\\
&k_2 - a_2 \in [1.6, 3.2],\\
&k_1 - a_1 \in [0.2122628, 1.8122628]
\end{aligned} \tag{68.28}$$

in the interval polynomial (68.12). It is easy to test and verify that the interval polynomial (68.12) with coefficients (68.28) is proper stable since

$$\max\{\alpha_1, \cdots, \alpha_8\} \leq \max\{\alpha_2\} \approx 0.46557$$

The parameters of feedback stabilization in this example are as follows,

$$k_1 = -7.9877372, k_2 = -8.6, k_3 = 1587.4299, k_4 = 249.1575,$$
$$k_5 = 45329.6536769, k_6 = 1262.217385, k_7 = 40303.2184,$$
$$k_8 = 192.9710597, k_9 = 1130.267678664, k_{10} = -0.2.$$

## 68.4 Single Feedback Stabilization

If $B = (b_1, b_2, \cdots, b_n)^T$ and $K = (k_1, \cdots, k_n)$ in $n$-dimensional state equation with single-input (68.10), then the state equation of single feedback becomes

$$\dot{x} = (A - BK)x = \begin{pmatrix} a_{11} - b_1k_1 & a_{12} - b_1k_2 & \cdots & a_{1n} - b_1k_n \\ a_{21} - b_2k_1 & a_{22} - b_2k_2 & \cdots & a_{2n} - b_2k_n \\ & & \cdots & \\ a_{n1} - b_nk_1 & a_{n2} - b_nk_2 & \cdots & a_{nn} - b_nk_n \end{pmatrix} x, \quad (68.29)$$

where $k_i (i = 1, \cdots, n)$ are undetermined parameters. We can appropriately select $k_i (i = 1, \cdots, n)$ such that the system (68.29) is stable. However, simultaneously selecting $k_i$ is evidently too complicated for numerical experiments. For general $A$ and $B$, we can select single $k_i$ successively.

Let $K = (0, \cdots, 0, k_s, 0, \cdots, 0)$ in (68.29). Then (68.29) is simplified into

$$\dot{x} = (A - BK)x = \begin{pmatrix} a_{11} & \cdots & a_{1s} - b_1k_s & \cdots & a_{1n} \\ a_{21} & \cdots & a_{2s} - b_2k_s & \cdots & a_{2n} \\ & & \cdots & & \\ a_{n1} & \cdots & a_{ns} - b_nk_s & \cdots & a_{nn} \end{pmatrix} x. \quad (68.30)$$

We want to appropriately select $k_s$ such that the stability of system (68.30) is improved with all possible. Let $E_{1s}$ be the matrix commuting $1 - s$ row/column. Then $\bar{A}_{ks} = (E_{1s}A_{ks}E_{1s})^T$ has the same stability as $A_{ks}$. The changeable elements of $\bar{A}_{ks}$ have been put on the first row of $\bar{A}_{ks}$. With the algorithm given by [1, 6], $\bar{A}_{ks}$is reduced to the Frobenius-like form.

$$F_l = \begin{pmatrix} 0 & 0 & 0 & \cdots & 0 & p_0 \\ 2^{\beta_2} & 0 & 0 & \cdots & 0 & p_1 \\ & 2^{\beta_3} & 0 & \cdots & 0 & p_2 \\ & & & \cdots & & \\ & & & & 2^{\beta_n} & p_{n-1} \end{pmatrix}, \quad (68.31)$$

where $p_{n-1} = p_{n-1}^0 + \gamma_{n-i}k_s, i = 1, \cdots, n.$ Let $r_{n-i} = \sqrt{(p_{n-i}^0)^2 + \gamma_{n-i}^2}$. Then $p_{n-i} = r_{n-i}(p_{n-i}^1 + \gamma_{n-i}^1 k_s), i = 1, \cdots, n.$

Each determining coefficient of (68.31) has become the following ratio

$$\alpha_i = \frac{r_{n-i+1}r_{n-i-2}}{r_{n-i}r_{n-i-1}} \frac{\eta_{i0} + \eta_{i1}k_s + \eta_{i2}k_s^2}{\xi_{i0} + \xi_{i1}k_s + \xi_{i2}k_s^2}, (r_n = 1, 1 \le i \le n-2).$$

Using Proposition 3 and Hypothesis 4(i), we obtain that

$$f_i(k_s) = p_{n-i}/r_{n-i} = p_{n-i}^1 + \gamma_{n-i}^1 k_s < 0, i = 1, \cdots, n, \tag{68.32}$$

where every $f_i(k_s) 1 \le i \le n$ is a normalized linear function, and that

$$g_i(k_s) = q_{i0} + q_{i1}k_s + q_{i2}k_s^2 \le 0, i = 1, \cdots, n-2, \tag{68.33}$$

where $q_{ij} = (r_{n-i+1}r_{n-i-2})/(r_{n-i}r_{n-i-1})\eta_{ij} - 0.46557\xi_{ij}, j = 0, 1, 2$. The necessary condition (68.32) is easily satisfied if all $\gamma_{n-i}^1 \ne 0$. The sufficient condition (68.33) usually is $n-2$ conics which give the conclusion whether $k_s$ can be selected from some real interval$[-l, l]$.

If there is an appropriate real number $k_s^*$ such that both (68.32) and (68.33) are satisfied, and

$$y(k_s^*) = \max\{y(k_s) = \max\{g_i(k_s) \le 0 | 1 \le i \le n-2, k_s \in [-l, l]\}\}, \tag{68.34}$$

then the system (68.30) may be properly stabilized by single feedback $k_s^*$. If there is no $k_s \in [-l, l]$ such that (68.32) and (68.33) are satisfied, then we can select the best $k_s^*$ such that

$$\begin{aligned} z(k_s^*) = \min\{z(k_s) = \min\{f_j(k_s), g_i(k_s)| \\ 1 \le j \le n, 1 \le i \le n-2, k_s \in [-l, l]\}\}, \end{aligned} \tag{68.35}$$

therefore the stability of (68.30) is improved with all possible. The $k_s^*$ is referred to as *gain of single feedback stabilization.*

For general matrices $A$ and $B$, we first design the gain of single feedback stabilization $k_1^*$ for $A_1 = A$. If $A_{k_1^*}^1$ is still unstable, then let$A^2 = A_{k_1^*}^1$. For $A^2$we design the gain of single feedback stabilization $k_2^*$ by (68.35) or (68.34), and so on. If $A_{k_{s-1}^*}^s$ is still unstable, then let $A^s = A_{k_{s-1}^*}^s$. For $A^s$ we design the gain of single feedback stabilization $k_s^*$ by (68.35) or (68.34). (If $s = n$ and $k_s^*$ is given by (68.35), then go on the next cycle of design, $s + 1 = 1$.) We may expect that $A^{s+1} = A_{k_s^*}^s$ will be properly stabilized, or that the stability of $A^{s+1}$ cannot be improved any more by single feedback$BK$. ($k_1^* = k_2^* = \cdots = k_n^* = 0$.) Usually, if $k_1^* \approx k_2^* \approx \cdots \approx k_n^* \approx 0$ in one cycle of design and $A^{s+1}$ is still unstable, then $A$ cannot be stabilized by single feedback $BK$.

It is possible that the effect of the above single feedback stabilization is different with a different order of single feedback stabilization. Generally speaking, the single column feedback (68.30) is hard to stabilize a matrix.

*Example 2* Suppose a general matrix is given by

$$A_8 = \begin{pmatrix} 1 & 2 & 1 & 4 & 5 & 3 & 2 & 1 \\ 1 & 3 & 2 & 5 & 3 & 2 & 3 & 2 \\ 2 & 4 & 1 & 7 & 2 & 1 & 3 & 1 \\ 3 & 5 & 6 & 1 & 3 & 4 & 1 & 2 \\ 2 & 7 & 3 & 2 & 1 & 2 & 5 & 1 \\ 4 & 5 & 1 & 3 & 7 & 3 & 2 & 6 \\ 3 & 3 & 5 & 7 & 1 & 2 & 4 & 5 \\ 2 & 2 & 4 & 1 & 3 & 2 & 2 & 1 \end{pmatrix}. \tag{68.36}$$

and

$$B_8 = (2, 4, 5, 7, 8, 1, 3, 5)^T. \tag{68.37}$$

For $A_8$ and$B_8$, gains of single feedback stabilization

$$k_1^* = 3.98, k_2^* = 2.05, k_3^* = 2.0, k_4^* = -1.54,$$
$$k_5^* = -0.34, k_6^* = -0.13, k_7^* = -0.07, k_8^* = 0.29$$

are selected by (68.35) in the first cycle of single feedback stabilization, while $A_8^{s+1}$ is still unstable. We consider that the stability of $A_8^{s+1}$ cannot be improved by single feedback $B_8K^*$.

On the other hand, we know, via the Gerschgorin's disk theorem, that $G_8 = A_8 - 31I$ is possibly over stable, where $I$ denotes unit matrix. The matrix $G_8$ can be made a properly stable matrix by single feedback stabilization.

For instance, with (68.32), (68.33), (68.34) we may design the gain of single feedback stabilization $g_1^*$ for the over stable matrix $G^1 = G_8$, and obtain $g_1^* \geq -2.18945802807$. It is easy to test and verify that $G^1_{g_1^*}$ is properly stable if $g_1^* = -2.189458$. In fact, the maximal real part of eigenvalues (a real eigenvalue) $\approx -0.000000088903$ for $G^1_{g_1^*}$. According to Hypothesis 1, $G^1_{g_1^*}$ is a properly stable matrix.

## 68.5 Feedback Stabilization of Hessenberg Form

Suppose $A = H, p = 1, B = B_h = (b, 0, \cdots, 0)^T, K = K_h = (k_1, \cdots, k_n)$ in (68.1) and (68.2), where $H$ is a Hessenberg form which can be reduced to from general matrix using stabilized elementary similarity transformations, see [7], and is given by

$$H = \begin{pmatrix} h_{1,1} & h_{1,2} & h_{1,3} & \cdots & h_{1,n-1} & h_{1,n} \\ h_{2,1} & h_{2,2} & h_{2,3} & \cdots & h_{2,n-1} & h_{2,n} \\ & h_{3,2} & h_{3,3} & \cdots & h_{3,n-1} & h_{3,n} \\ & & & \cdots & & \\ & & & & h_{n,n-1} & h_{n,n} \end{pmatrix}. \quad (68.38)$$

Without loss of any generality, we may assume that $h_{i,i-1} \neq 0 (i = 2, \cdots, n)$ in (68.38). Notice that there is no need to commute row/column for reduction to Frobenius-like form from Hessenberg form.

After feedback $u = -K_h x$ of single-input, the state equations $\dot{x}(t) = Hx(t) + B_h u(t)$ become

$$\dot{x} = (H - B_h K_h)x = \begin{pmatrix} h_{1,1} - bk_1 & h_{1,2} - bk_2 & \cdots & h_{1,n} - bk_n \\ h_{2,1} & h_{2,2} & \cdots & h_{2,n} \\ & h_{3,2} & \cdots & h_{3,n} \\ & & & \cdots \\ & & & h_{n-1,n} \\ & & & h_{n,n} \end{pmatrix} x, \quad (68.39)$$

where $k_i, i = 1, 2, \cdots, n$ are regarded as undetermined parameters for stabilization of $H - B_h K_h$. The matrix $H - B_h K_h$ is still a Hessenberg form and its changeable elements are located on the first row.

Proposition seven shows that the matrix in (68.39) may be stabilized by solving $n$ linear inequalities and solving $n - 2$ inequalities of products of two linear combinations. For the Hessenberg form $H - B_h K_h$, the solution can be found if we investigate firstly $n$ linear inequalities.

Via the stabilized elementary similarity transformations and the scaling strategy, the single row feedback matrix $H - B_h K_h$ in (68.39) is reduced to the Frobenius-like form (68.31). In the computational process, the variables $h_{1,1} - bk_1, h_{1,2} - bk_2, \cdots, h_{1,n} - bk_n$ are linearly propagated to the last column of (68.31). We have thus

$$\begin{aligned} &2^{\theta_1} p_{n-1} = 2^{\theta_1}(p_{n-1}^0 - bk_1) < 0, \\ &2^{\theta_2} p_{n-2} = 2^{\theta_2}(p_{n-2}^0 + \gamma_{2,1}k_1 + \gamma_{2,2}k_2) < 0, \cdots, \\ &2^{\theta_s} p_{n-s} = 2^{\theta_s}(p_{n-s}^0 + \textstyle\sum_{j=1}^{s} \gamma_{s,j}k_j) < 0, \\ &2^{\theta_n} p_0 = 2^{\theta_n}(p_0^0 + \textstyle\sum_{j=1}^{n} \gamma_{n,j}k_j) < 0, \\ &(\theta_1 = 0, \theta_2 = \beta_n, \cdots, \theta_s = \beta_n + \cdots + \beta_{n-s+2}, \cdots, \theta_n = \beta_n + \cdots + \beta_2), \end{aligned} \quad (68.40)$$

where $p^0_{n-s}(1 \le s \le n)$ and $\gamma_{s,j}(2 \le s \le n, 1 \le j \le n)$ are given by invariable elements of $H$. If $\gamma_{s,s} \ne 0(s = 2, \cdots, n)$, then the linear inequalities (68.40) have infinitely many solutions.

**Proposition nine** *The Hessenberg matrix $H - B_hK_h$ can be stabilized by the strategies of polynomial, Proposition 3 and Hypothesis 4(i).*

*Proof.* We first prove that if $\gamma_{s,s} \ne 0(s = 2, \cdots, n)$ in (68.40), then the Hessenberg matrix $H - B_hK_h$ can be stabilized by the strategies of polynomial, Proposition 3 and Hypothesis 4(i).

In fact, under the assumption, it is easy to see that each $k_s(1 \le s \le n)$ of (68.40) can be solved successively for any negative real array $(p_{n-1}, \cdots, p_0)^T$. Therefore the $n-2$ quadratic inequalities (68.33) can be satisfied if the left-hand side of (68.40)

$$2^{\theta_1}p_{n-1}, 2^{\theta_2}p_{n-2}, \cdots, 2^{\theta_{n-1}}p_1, 2^{\theta_n}p_0 \tag{68.41}$$

are, respectively, equal to the coefficients of a stable polynomial or properly stable polynomial. This polynomial is referred to as *polynomial of pole-zero assignment*.

For instance, $-(\lambda + 1)^n$ is such a polynomial of pole-zero assignment, the coefficients (68.41) may be, respectively, equal to the corresponding coefficients of $-(\lambda + 1)^n = -(\lambda^n + n\lambda^{n-1} + \cdots + n\lambda + 1)$, namely $p_{n-1} = -n/2^{\theta_1}$, $p_{n-2} = -0.5n(n-1)/2^{\theta_2}, \ldots, p_1 = -n/2^{\theta_{n-1}}$, $p_0 = -1/2^{\theta_n}$. We have thus

$$\begin{aligned} &p^0_{n-1} - bk_1 = p_{n-1} < 0, p^0_{n-2} + \gamma_{2,1}k_1 + \gamma_{2,2}k_2 = p_{n-2} < 0, \cdots, \\ &p^0_{n-s} + \textstyle\sum_{j=1}^{s}\gamma_{s,j}k_j = p_{n-s} < 0, \cdots, p^0_0 + \sum_{j=1}^{n}\gamma_{n,j}k_j = p_0 < 0, \end{aligned} \tag{68.42}$$

The linear equations (68.42) have unique solution $K_h = (k_1, \cdots, k_n)$ for given coefficients (68.41).

Clearly $b \ne 0$ in the linear equations (68.42). Via the algorithm of reduction from Hessenberg matrix to Frobenius-like form, see [1], we see that

$$\gamma_{s,s} = 2^{\beta_2 + \cdots + \beta_{n-s+1} - \beta_{s+1} - \cdots - \beta_n}\bar{b}_s \ne 0, 2 \le s \le n-1, \gamma_{n,n} = \bar{b}_n \ne 0,$$

since, with (68.38), $b_s = -b2^{-\beta_2 - \cdots - \beta_s}h_{2,1} \cdots h_{s,s-1} \ne 0, 2 \le s \le n$. This complete the proof of the proposition. □

Notice that if the coefficients (68.41) are given by properly stable polynomial of pole-zero assignment, then the stability of stabilized matrix $H - B_hK_h$ is highly sensitive to any perturbation of$K_h$. Moreover, the solving process is possibly numerically unstable because of $\gamma_{s,s}$. Therefore we should test and verify the stability of $H - B_hK_h$ after $K_h$ is found up. It would be best if the coefficients (68.41) are given by stable polynomial of pole-zero assignment.

*Example 3* Suppose that $H_8$ is the Hessenberg form reduced to from $A_8 - B_8 K^1$, where $A_8$ and$B_8$ are (68.36) and (68.37) respectively, and

$$K^1 = (3.98, 2.05, 2.0, -1.54, -0.34, -0.13, -0.07, 0.29)$$

is the feedback gain selected by (68.35) in the first cycle of single feedback stabilization of Example 2. $H_8$ is still unstable.

We can stabilize $H_8$ by Proposition 9. Let $B_h = (4, 0, 0, 0, 0, 0, 0, 0)^T$ and $-(\lambda + 1)^8$ be the polynomial of pole-zero assignment. Solving (68.42) we get

$$\mathrm{K_h} \approx (2.30750, 1.07202, 0.44005, 0.51673, 0.38066, 0.09244, -0.04457, 0.85180).$$

It is easy to test and verify the stability of $H_8 - B_h K_h$. Since $K_h$ is an approximate solution of (68.42), the stability of matrix $H_8 - B_h K_h$ has been different from that of the companion matrix of $-(\lambda + 1)^8$.

The matrix $H_8 - B_h K_h$ is stable but not properly stable. We want to make it properly stable matrix by the single feedback stabilization $B_h K_h^1$, where $K_h^1 = (k_1^1, 0, \cdots, 0)$. Using (68.32) and (68.33), we find that the selected interval of gain $k_1^1$ is very small:$-0.000008555 \le k_1^1 \le 0.000016050$. It is easy to test and verify that both $H_8 - B_h K_h - B_h K_1^*$ and $H_8 - B_h K_h - B_h K_1^{**}$ are properly stable, where $K_1^* = (-0.000008555, 0, \cdots, 0)$ and $K_1^{**} = (-0.000016050, 0, \cdots, 0)$ respectively. For $H_8 - B_h K_h - B_h K_1^*$, the maximal real part of eigenvalues (a real eigenvalue) $\approx -0.0000048051$. For$H_8 - B_h K_h - B_h K_1^{**}$, the maximal determining coefficient $\approx 0.465551$. Therefore the properly stable conclusion may be come to from Hypotheses 1 and 4(i) respectively.

We now make $H_8 - B_h K_h$ properly stable matrix by the single feedback stabilization $B_h K_h^2$, where $K_h^2 = (0, k_2^2, 0, \cdots, 0)$. Using (68.32) and (68.33), we also find that the selected interval of gain $k_2^2$ is very small: $-0.0000217707 \le k_2^2 \le 0.0000113129$. It is easy to test and verify that both $H_8 - B_h K_h - B_h K_2^*$ and $H_8 - B_h K_h - B_h K_2^{**}$ are properly stable, where $K_2^* = (0, -0.0000217707, 0, \cdots, 0)$ and $K_2^{**} = (0, 0.0000113129, \quad 0, \cdots, 0)$ respectively. For $H_8 - B_h K_h - B_h K_2^*$, the maximal determining coefficient $\approx 0.46556914$. For $H_8 - B_h K_h - B_h K_2^{**}$, the maximal real part of eigenvalues (a real eigenvalue) $\approx -0.00000009157$. Therefore the properly stable conclusion may be come to from Hypotheses 4(i) and 1 respectively.

Successively we make $H_8 - B_h K_h$ properly stable matrix by the single feedback stabilizations $B_h K_h^3, \cdots, B_h K_h^8$. Similarly we find that all the selected intervals of gains are very small.

These numerical experiments show us that $H_8 - B_h K_h$ is almost a properly stable Hessenberg matrix.

Since a general matrix may be reduced to Hessenberg form, the single row feedback described in this paragraph is easy to stabilize a matrix. That is,

**Proposition ten** *Assume that, in the state equation (68.1), A may be reduced to n -order Hessenberg form for any similarity transformation commuting row/ column, and there is a column* $(b_{1l}, b_{2l}, \cdots, b_{n,l})^T$ *in the control matrix B such*

*that* $b_{1l} = 0, \cdots, b_{s-1,l} = 0, b_{s,l} = b \neq 0, b_{s+1,l} = 0, \quad \cdots, b_{nl} = 0, (1 \leq s \leq n)$. *Then we can design the single row feedback matrix*

$$K = \begin{pmatrix} 0 & 0 & \cdots & 0 \\ & & \cdots & \\ 0 & 0 & \cdots & 0 \\ k_{l1} & k_{l2} & \cdots & k_{\text{ln}} \\ 0 & 0 & \cdots & 0 \\ & & \cdots & \\ 0 & 0 & \cdots & 0 \end{pmatrix} \tag{68.43}$$

such that $A - BK$ is stable or properly stable.

*Proof.* It is easy to see that, via (68.1), $u = -Kx$ and (68.43),

$$A_s^{k_{lj}} = A - BK = \begin{pmatrix} a_{11} & a_{12} & \cdots & a_{1n} \\ & & \cdots & \\ a_{s-1,1} & a_{s-1,2} & \cdots & a_{s-1,n} \\ a_{s,1} - bk_{l1} & a_{s,2} - bk_{l2} & \cdots & a_{s,n} - bk_{\text{ln}} \\ a_{s+1,1} & a_{s+1,2} & \cdots & a_{s+1n} \\ & & \cdots & \\ a_{n1} & a_{n2} & \cdots & a_{nn} \end{pmatrix}.$$

The gain parameters $k_{lj} = (j = 1, \cdots, n)$ appear only in the $s$-th row of $A - BK$. Let $E_{1s}$ be the matrix commuting $1 - s$ row/column. Clearly $\bar{A}_1^{k_{lj}} = E_{1s} A_s^{k_{lj}} E_{1s}$ has the same stability as $A_s^{k_{lj}}$. The changeable elements of $A_s^{k_{lj}}$ have been put on the first row of $\bar{A}_1^{k_{lj}}$.

According to the assumption of proposition, $\bar{A}_1^{k_{lj}}$ may be reduced to $n$-order Hessenberg form by the stabilized elementary similarity transformations. After this reduction is completed, elements of the first row have become

$$w_{11} = a_{ss} - bk_{ls}, w_{12} = h_{12} - b(k_{lj^2} + \sum_{i=3}^{n} \alpha_{2i} k_{lj^i}),$$

$$\cdots, w_{1t} = h_{1t} - b(k_{lj^t} + \sum_{i=t+1}^{n} \alpha_{ti} k_{lj^i}), \cdots,$$

$$w_{1,n-1} = h_{1,n-1} - b(k_{lj^{n-1}} + \alpha_{n-1,n} k_{lj^n}), w_{1n} = h_{1n} - bk_{lj^n},$$

where $h_{1t}$ and $\alpha_{ti}(|\alpha_{ti}| \leq 1)$ are given by the elements $a_{ij}$ of $A$, and $j^2, j^3, \cdots j^n$ is a permutation of $1, \cdots, s-1, s+1, \cdots, n$.

We now regard $k_{ls}, (k_{lj^2} + \sum_{i=3}^{n} \alpha_{2i} k_{lj^i}), \cdots, (k_{lj^{n-1}} + \alpha_{n-1,n} k_{lj^n}), k_{lj^n}$ as $k_1, k_2, \cdots k_{n-1}$ in (68.39) respectively, and apply Proposition 7 to the Hessenberg form of $\bar{A}_1^{k_{lj}}$. The following equations

$$k_{ls} = k_1^*, k_{lj^2} + \sum_{i=3}^{n} \alpha_{2i} k_{lj^i} = k_2^*, \cdots, k_{lj^t} + \sum_{i=t+1}^{n} \alpha_{ti} k_{lj^i} = k_t^*, \cdots,$$
$$k_{lj^{n-1}} + \alpha_{n-1,n} k_{lj^n} = k_{n-1}^*, k_{lj^n} = k_n^* \tag{68.44}$$

have thus been obtained, where $k_1^*, k_2^*, \cdots, k_n^*$ is the solution of (68.42).

The linear equations (68.44) have clearly unique solution. □

## 68.6 Conclusion

Several efficient strategies for feedback stabilization of matrix have been presented via the numerical stabilization of polynomial and matrix.

The basic strategy is to give successively the gains of negative state feedback stabilization with numerical experiments on stability of the Frobenius-like matrix.

Proposition two shows that any matrix may always be stabilized properly by diagonal feedback matrix with a suitable gain. If some diagonal elements of matrix aren't allowed to change too big, then the strategies of single feedback stabilization may be considered.

If a matrix is the point-wise or interval controllable canonical form, then Proposition three, Hypothesis four, and Proposition eight are the most efficient stabilized strategies.

Propositions nine and ten give efficient stabilized strategies for Hessenberg form and general matrix respectively.

If a matrix is over stable, then it can be made into properly stable matrix by single column feedback stabilization.

## References

1. Han JD, Jiang Z, Nie YY (2007) Numerical stabilization of polynomial and matrix. IMA J Math Control Inform 24:473–482
2. Higham Nicholas J (2002) Accuracy and stability of numerical algorithms, 2nd ed.. SIAM, Philadelphia
3. Huang L (2003) Foudations of the theory of stability and robustness. Chinese Academy of Sciences, Acta Press, Beijing (in Chinese)
4. Kharitonov VL (1978) Asymptotic stability of an equilibrium position of a family of systems of linear differential equation. Differential'nye Uravneniya 14:1483–1485 (in Russian)
5. Malkin IG (1958) Theory of motion stability. Chinese Academy of Sciences, Acta Press, Beijing (translation in Chinese)
6. Nie YY, Nie MH, Yang ZJ (2006) Numerical determination for stability of Hessenberg matrix, dynamics of continuous, discrete and impulsive systems (DCDIS). Series A Math Anal 13(2):792–803
7. Wilkinson JH (1965) The algebraic eigenvalue problem. Oxford University Press, U.K.

# Chapter 69
# Maxwell Geometric Dynamics

**Constantin Udrişte**

**Abstract** This paper describes the least squares approximations of the solutions of Maxwell PDEs via their Euler-Lagrange prolongations. Section 5.1 recalls the variational problem in electrodynamics. Section 69.2 describes the Ibragimov-Maxwell Lagrangian. Section 69.3 gives the Ibragimov-Udrişte-Maxwell Lagrangian and finds its Euler-Lagrange PDEs system. Section 5.4 finds the Euler-Lagrange PDEs system associated to Udrişte-Maxwell Lagrangian, and shows that the waves are particular solutions. Section 69.5 studies the discrete Maxwell geometric dynamics. Section 69.6 performs Von Neumann analysis of the associated difference scheme. Section 69.7 addresses an open problem regarding our theory in the context of differential forms on a Riemannian manifold. Section 69.8 underlines the importance of the least squares Lagrangian.

## 69.1 Variational Problem in Electrodynamics

The classical variational problem in electrodynamics provides a Lagrangian for the wave PDE

$$\Delta A - \frac{1}{c^2}\frac{\partial^2 A}{\partial t^2} = 0$$

regarding the vector potential $A$ of the electromagnetic field, but not for the Maxwell PDEs. This Lagrangian contains the following ingredients:

– the covariant four-vector potential

C. Udrişte (✉)
University Politehnica of Bucharest, Department of Mathematics, 060042, Splaiul Independentei 313, Bucharest, Romania
e-mail: udriste@mathem.pub.ro

N. Mastorakis et al. (eds.), *Proceedings of the European Computing Conference*, Lecture Notes in Electrical Engineering 28,
DOI 10.1007/978-0-387-85437-3_69, 

$$\phi_i = (-cA, \phi);$$

- the electromagnetic tensor field

$$F_{ij} = \frac{\partial \phi_j}{\partial x^i} - \frac{\partial \phi_i}{\partial x^j}$$

- or in matrix form

$$(F_{ij}) = \begin{pmatrix} 0 & -cB_z & cB_y & -E_x \\ cB_z & 0 & -cB_x & -E_y \\ -cB_y & cB_x & 0 & -E_z \\ E_x & E_y & E_z & 0 \end{pmatrix};$$

- the Lorentz metric with the diagonal components (1,1,1, − 1);
- the corresponding contravariant tensor $F^{ij}$ obtained by changing the signs of the elements $F_{ij}$ in the fourth row and fourth column (rising the indices with the Lorentz metric)

All together determine the "least squares" electromagnetic Lagrangian

$$L = -\frac{1}{4} F_{ij} F^{ij}.$$

Udrişte [4–15] built a least squares Lagrangian whose Euler-Lagrange PDEs are prolongations of the Maxwell PDEs. Ibragimov [2] built a new kind of Lagrangian just for the Maxwell PDEs.

## 69.2 Ibragimov-Maxwell Lagrangian

In vacuum, modelled as $R^3 \times R = \{(x, y, z, t)\}$, the Maxwell linear PDEs

$$\frac{1}{c}\frac{\partial E}{\partial t} = curlH, divE = 0, \frac{1}{c}\frac{\partial H}{\partial t} = -curlE, divH = 0$$

are over-determined. Indeed, they contains six dependent variables, namely, three components of the electric field $E = (E^1, E^2, E^3)$ and three components of the magnetic field $H = (H^1, H^2, H^3)$ and eight equations. On the other hand, it is obvious that the PDEs $div\ E = 0$ and $div\ H = 0$ hold at any time provided they are satisfied at the initial time $t = 0$. Hence, they can be used as initial conditions.

Let us set $t' = ct$ and take $t'$ as a new $t$. Also we consider the determined linear PDEs system

$$curlE + \frac{\partial H}{\partial t} = 0, curlH - \frac{\partial E}{\partial t} = 0. \quad (69.1)$$

According Ibragimov [2], let us introduce six new dependent variables (Lagrange multipliers) by the vector fields $V = (V^1, V^2, V^3)$ and $W = (W^1, W^2, W^3)$ and the Lagrangian

$$L = V \cdot (curlE + \frac{\partial H}{\partial t}) + W \cdot (curlH - \frac{\partial E}{\partial t}). \quad (69.2)$$

Of course, the Euclidean scalar product on $R^3$ can be replaced with a suitable Riemannian metric. The associated Euler-Lagrange PDEs are

$$\frac{\delta L}{\delta V} = curlE + \frac{\partial H}{\partial t} = 0, \quad \frac{\delta L}{\delta W} = curlH - \frac{\partial E}{\partial t} = 0$$
$$\frac{\delta L}{\delta E} = curlV + \frac{\partial W}{\partial t} = 0, \quad \frac{\delta L}{\delta H} = curlW - \frac{\partial V}{\partial t} = 0.$$

Since these PDEs coincide for $V = E, W = H$, the differential operator in PDEs (69.1) is self-adjoint. Consequently, setting $2\,V = E, 2\,W = H$, one obtains the *Ibragimov-Maxwell Lagrangian* [2]

$$L_1 = \frac{1}{2}[E \cdot (curlE + \frac{\partial H}{\partial t}) + H \cdot (curlH - \frac{\partial E}{\partial t})]$$

or, in coordinates,

$$L_1 = \frac{1}{2}[E^1(E^3_y - E^2_z + H^1_t) + E^2(E^1_z - E^3_x + H^2_t) + E^3(E^2_x - E^1_y + H^3_t)$$
$$+ H^1(H^3_y - H^2_z - E^1_t) + H^2(H^1_z - H^3_x - E^2_t) + H^3(H^2_x - H^1_y - E^3_t)]$$

The Euler-Lagrange PDEs produced by the *Ibragimov-Maxwell Lagrangian* $L_1$ are just the Maxwell PDEs (69.1).

## 69.3 Ibragimov-Udrişte-Maxwell Lagrangian

The Maxwell PDEs system (69.1) and the Euclidean scalar product determine a *least squares Lagrangian* [4–15]. This can be obtained also from (69.2) setting $2\,V = curlE + \frac{\partial H}{\partial t}$ and $2\,W = curlE + \frac{\partial H}{\partial t}$, i.e.,

$$L_2 = \frac{1}{2}\left[\left(curlE + \frac{\partial H}{\partial t}\right) \cdot \left(curlE + \frac{\partial H}{\partial t}\right) + \left(curlH - \frac{\partial E}{\partial t}\right) \cdot \left(curlH - \frac{\partial E}{\partial t}\right)\right]$$

or

$$L_2 = \frac{1}{2}\left[||curl\ E + \frac{\partial H}{\partial t}||^2 + ||curl\ H - \frac{\partial E}{\partial t}||^2\right].$$

In coordinates we can write

$$L_2 = \frac{1}{2}[(E_y^3 - E_z^2 + H_t^1)^2 + (E_z^1 - E_x^3 + H_t^2)^2 + (E_x^2 - E_y^1 + H_t^3)^2$$
$$+ (H_y^3 - H_z^2 - E_t^1)^2 + (H_z^1 - H_x^3 - E_t^2)^2 + (H_x^2 - H_y^1 - E_t^3)^2].$$

The Ibragimov-Udrişte-Maxwell Lagrangian $L_2$ must be accompanied by the PDEs $divE = 0$ and $divH = 0$ as initial conditions.

**Theorem.** *If* $divE = 0$ *and* $divH = 0$ *, then the Euler-Lagrange PDEs associated to* $L_2$ *(Euler-Lagrange prolongation of the PDEs system* (69.1)) *are*

$$\Delta E + \frac{\partial^2 E}{\partial t^2} = 2\frac{\partial}{\partial t}(curlH), \quad \Delta H + \frac{\partial^2 H}{\partial t^2} = -2\frac{\partial}{\partial t}(curlE).$$

## 69.4 Udrişte-Maxwell Lagrangian

Now let us consider the over-determined Maxwell PDEs system

$$curlE + \frac{\partial H}{\partial t} = 0, curlH - \frac{\partial E}{\partial t} = 0, divE = 0, divH = 0. \tag{69.3}$$

This PDEs system and the Euclidean scalar product determine the *least squares Udriş te-Maxwell Lagrangian* (see [4–15])

$$L_3 = \frac{1}{2}\left[||curlE + \frac{\partial H}{\partial t}||^2 + ||curlH - \frac{\partial E}{\partial t}||^2 + (divE)^2 + (divH)^2\right].$$

The expression of $L_3$ in coordinates is

$$L_3 = \frac{1}{2}\Big[(E_y^3 - E_z^2 + H_t^1)^2 + (E_z^1 - E_x^3 + H_t^2)^2 + (E_x^2 - E_y^1 + H_t^3)^2$$
$$+(H_y^3 - H_z^2 - E_t^1)^2 + (H_z^1 - H_x^3 - E_t^2)^2 + (H_x^2 - H_y^1 - E_t^3)^2$$
$$+(E_x^1 + E_y^2 + E_z^3)^2 + (H_x^1 + H_y^2 + H_z^3)^2\Big].$$

**Theorem.** *The Euler-Lagrange PDEs associated to* $L_3$ *(Euler-Lagrange prolongation of the Maxwell PDEs system* (69.3)) *are*

$$\Delta E + \frac{\partial^2 E}{\partial t^2} = 2\frac{\partial}{\partial t}(curlH), \Delta H + \frac{\partial^2 H}{\partial t^2} = -2\frac{\partial}{\partial t}(curlE).$$

Consequently the Euler-Lagrange PDEs of $L_2$ subject to $divE = 0$ and $divH = 0$ are the same with the Euler-Lagrange PDEs of $L_3$.

A solution to the electromagnetic PDEs system in the last theorem takes the form of *waves* $E(r, t) = E_0 e^{i(\omega t - k \cdot r)}$ and $H(r, t) = H_0 e^{i(\omega t - k \cdot r)}$, where $t$ is *time* (in seconds), $\omega$ is the *angular frequency* (in radians per second), $k = (k_x, k_y, k_z)$ is the *wave vector* (in radians per meter) and $r = (x, y, z)$. Of course, we accept that the electric field, the magnetic field, and direction of wave propagation are all orthogonal, and the wave propagates in the same direction as $E_0 \times H_0$. The wave vector $k$ is related to the *angular frequency* by $||k|| = \frac{\omega}{c} = \frac{2\pi}{\lambda}$, where $||k||$ is the *wave number* and $\lambda$ is the *wave length*.

## 69.5 Discrete Maxwell geometric dynamics

The theory of integrators for Lagrangian dynamics shows that instead of discretization of Euler-Lagrange PDEs we must use a discrete Lagrangian and then discrete Euler-Lagrange equations. Moreover, the discrete Euler-Lagrange equations associated to a least squares Lagrangian can be solved successfully by the Newton method which proves to be convergent for a convenient step (see [10–12]).

The discretization of the Lagrangian $LL_3$ can be performed by using the *centroid rule* which consists in the substitution of the point $(x, y, z, t)$ with the step $h = (h_1, h_2, h_3, h_4)$, of the point $p = (E^1, E^2, E^3, H^1, H^2, H^3)$ with

$$\frac{p_{klmn} + p_{k+1lmn} + p_{kl+1mn} + p_{klm+1n} + p_{klmn+1}}{5}$$

and of the *partial velocities* $p_\alpha, \alpha = 1, 2, 3, 4$ by

$$\frac{p_{k+1lmn} - p_{klmn}}{h_1}, \quad \frac{p_{kl+1mn} - p_{klmn}}{h_2},$$
$$\frac{p_{klm+1n} - p_{klmn}}{h_3}, \quad \frac{p_{klmn+1} - p_{klmn}}{h_4}.$$

One obtains a discrete Lagrangian

$$L_d : R^4 \times (R^6)^5 \to R,$$

$$L_d(u_1, u_2, u_3, u_4, u_5) = L(h, \frac{u_1 + u_2 + u_3 + u_4 + u_5}{5}, \frac{u_2 - u_1}{h_1}, \frac{u_3 - u_1}{h_2}, \frac{u_4 - u_1}{h_3}, \frac{u_5 - u_1}{h_4}).$$

This determines the 4-dimensional discrete action

$$S : R^4 \times (R^6)^{(N_1+1)(N_2+1)(N_3+1)(N_4+1)} \to R,$$

$$S(h, A) = \sum_{k=0}^{N_1-1} \sum_{l=0}^{N_2-1} \sum_{m=0}^{N_3-1} \sum_{n=0}^{N_4-1} L(h, p_{klmn}, p_{k+1lmn}, p_{kl+1mn}, p_{klm+1n}, p_{klmn+1}),$$

where

$$A = (p_{klmn}), k = 0, \ldots, N_1, l = 0, \ldots, N_2,$$
$$m = 0, \ldots, N_3, n = 0, \ldots, N_4.$$

We fix the step $h = (h_1, h_2, h_3, h_4)$. The *discrete variational principle* consists in the characterization of the table $A$ for which the action $S$ is stationary, for any family

$$p_{klmn}(\varepsilon) \in R^6$$

with

$$k = 0, \ldots, N_1 - 1, \quad l = 0, \ldots, N_2 - 1,$$
$$m = 0, \ldots, N_3 - 1, n = 0, \ldots, N_4 - 1,$$
$$\varepsilon \in I \subset R, \quad 0 \in I, \quad p_{klmn}(0) = p_{klmn}$$

and fixed elements

$$p_{0lmn}, p_{N_1lmn}, p_{k0mn}, p_{kN_2mn}, p_{kl0n}, p_{klN_3n}, p_{klm0}, p_{klmN_4}.$$

The discrete variational principle is obtained using the first order variation of $S$. In other words the point $A = (p_{klmn})$ is stationary for the action $S$ if and only if

$$\sum_{\xi} \frac{\partial L}{\partial p_{klmn}^{i}}(\xi) = 0,$$

where $\xi$ runs over five points,

$$(p_{klmn}, p_{k+1lmn}, p_{kl+1mn}, p_{klm+1n}, p_{klmn+1})$$
$$(p_{k-1lmn}, p_{klmn}, p_{k-1\,l+1mn}, p_{k-1\,lm+1n}, p_{k-1lmn+1})$$
$$(p_{kl-1mn}, p_{k+1\,l-1mn}, p_{klmn}, p_{kl-1m+1n}, p_{kl-1mn+1})$$
$$(p_{klm-1n}, p_{k+1\,lm-1n}, p_{kl+1m-1n}, p_{klmn}, p_{klm-1n+1})$$
$$(p_{klmn-1}, p_{k+1lmn-1}, p_{kl+1mn-1}, p_{klm+1n-1}, p_{klmn})$$

and

$$i = 1, \ldots, 6; \quad k = 0, \ldots, N_1 - 1, \quad l = 0, \ldots, N_2 - 1,$$
$$m = 0, \ldots, N_3 - 1, \quad n = 0, \ldots, N_4 - 1.$$

Giving the elements

$$p_{0lmn}, p_{N_1lmn}, p_{k0mn}, p_{kN_2mn}, p_{kl0n}, p_{klN_3n}, p_{klm0}, p_{klmN_4},$$

the previous algebraic system is solved by the *Newton method* which is contractive for a small step $h$ (see [10–12]).

## 69.6 Von Neumann Analysis

To verify the stability of the previous finite difference scheme, we pass to the *frequency domain*, through what is called *Von Neumann analysis* [1]. For that

1. we accept a uniform grid spacing in all the spatial coordinates, i.e., $h = h_1 = h_2 = h_3$, and an unbounded spatial domain;
2. we denote by $\tau = h_4$ the time step and we define the *space step/time step ratio* to be $\nu_0 = \frac{h}{\tau}$.

Also, to simplify, we consider the real-valued grid functions of the type

$$E^1 = E^1(x,z,t), \quad E^2 = E^3 = 0$$
$$H^1 = 0, \quad H^2 = H^2(y,z,t), \quad H^3 = 0.$$

The general case $(3+1)D$ is similar and it will be analyzed in a furthercoming paper. Also, for the partial derivatives we can use several other forward difference relations or backward difference relations or central difference relations of various orders. Each case could also include error analysis.

Now the least squares Lagrangian $L_3$ is given by

$$2L_3 = (E_z^1 + H_t^2)^2 + (H_z^2 + E_t^1)^2 + (E_x^1)^2 + (H_y^2)^2.$$

The associated discrete Lagrangian

$$\begin{aligned} 2h^2 L_d =& (p^1_{klm+1n} - p^1_{klmn} + (p^5_{klmn+1} - p^5_{klmn})\nu_0)^2 \\ &+ (p^5_{klm+1n} - p^5_{klmn} + (p^1_{klmn+1} - p^1_{klmn})\nu_0)^2 \\ &+ (p^1_{k+1lmn} - p^1_{klmn})^2 + (p^5_{kl+1mn} - p^5_{klmn})^2, \end{aligned}$$

where $p^1 = E^1, p^5 = H^2$, produces the discrete Euler-Lagrange equations

$$\begin{aligned} & 4p^1_{klmn} - p^1_{klm+1n} - p^1_{k+1lmn} - p^1_{k-1lmn} - p^1_{klm-1n} \\ & +(-p^5_{klm+1n} + p^5_{klmn} - (p^1_{klmn+1} - p^1_{klmn})\nu_0)\nu_0 \\ & +(-p^5_{klmn+1} + p^5_{klmn} + p^5_{klm-1n+1} - p^5_{klm-1n})\nu_0 \\ +(p^5_{klm+1n-1} - & p^5_{klmn-1} + (p^1_{klmn} - p^1_{klmn-1})\nu_0)\nu_0 = 0 \end{aligned}$$

$$\begin{aligned} & 4p^5_{klmn} - p^5_{klm+1n} - p^5_{kl+1mn} - p^5_{kl-1mn} - p^5_{klm-1n} \\ & +(-p^1_{klm+1n} + p^1_{klmn} - (p^5_{klmn+1} - p^5_{klmn})\nu_0)\nu_0 \\ & +(-p^1_{klmn+1} + p^1_{klmn} + p^1_{klm-1n+1} - p^1_{klm-1n})\nu_0 \\ +(p^1_{klm+1n-1} - & p^1_{klmn-1} + (p^5_{klmn} - p^5_{klmn-1})\nu_0)\nu_0 = 0. \end{aligned}$$

We need a 3D *discrete spatial Fourier transform* which can be obtained via the substitutions

$$p^1_{klmn} \to P^1_n(\alpha,\beta,\gamma)e^{j(\alpha+\beta+\gamma)h}, p^5_{klmn} \to P^5_n(\alpha,\beta,\gamma)e^{j(\alpha+\beta+\gamma)h},$$

where $(\alpha,\beta,\gamma)$ denotes the *radian spatial wave vector*. We find a system of second order difference equations (*digital filters*)

$$\begin{aligned}
-P^1_n e^{j\gamma h} + 4P^1_n - (P^5_{n+1} - P^5_n)v_0 \\
+(-P^5_n e^{j\gamma h} + P^5_n - (P^1_{n+1} - P^1_n)v_0)v_0 \\
-P^1_n e^{-j\alpha h} - P^1_n e^{-j\gamma h} + (P^5_{n+1} - P^5_n)v_0 e^{-j\gamma h} \\
-(-P^5_{n-1}e^{j\gamma h} + P^5_{n-1} - (P^1_n - P^1_{n-1})v_0)v_0 = 0,
\end{aligned}$$

$$\begin{aligned}
-P^5_n e^{j\gamma h} + 4P^5_n - (P^1_{n+1} - P^1_n)v_0 \\
+(-P^1_n e^{j\gamma h} + P^1_n - (P^5_{n+1} - P^5_n)v_0)v_0 \\
-P^5_n e^{-j\alpha h} - P^5_n e^{-j\gamma h} + (P^1_{n+1} - P^1_n)v_0 e^{-j\gamma h} \\
-(-P^1_{n-1}e^{j\gamma h} + P^1_{n-1} - (P^5_n - P^5_{n-1})v_0)v_0 = 0
\end{aligned}$$

that need its stability checked. For this purpose we introduce the *z-transforms* $F^1(z,\alpha,\beta,\gamma), F^5(z,\alpha,\beta,\gamma)$ and we must impose that the *poles* of the recursion do not lie outside the unit circle in the z-plane. To simplify, we accept the initial conditions $P^1_0 = 0, P^5_0 = 0$. One obtains the homogeneous linear system

$$a_{11}F^1 + a_{12}F^2 = 0, \quad a_{21}F^1 + a_{22}F^2 = 0,$$

where

$$\begin{aligned}
a_{11} &= 4z - v_0^2 - v_0^2z^2 + 2v_0^2z - 2zcos(\gamma h) - 2zcos(\alpha h) \\
a_{22} &= 4z - v_0^2 - v_0^2z^2 + 2v_0^2z - 2zcos(\gamma h) - 2zcos(\beta h) \\
a_{12} = a_{21} &= -v_0(z^2 - 2z + 1) \\
&+v_0(z^2 - 2z + 1)cos(\gamma h) + jv_0sin(\gamma h) - jv_0z^2sin(\gamma h).
\end{aligned}$$

The poles are the roots of the characteristic equation

$$a_{11}a_{22} - a_{12}^2 = 0,$$

with the unknown $z$. Explicitly, we have

$$a_4 z^4 + a_3 z^3 + a_2 z^2 + a_1 z + a_0 = 0,$$

Where

$$\begin{gathered}\bar{a}_4 = a_0 = v_0^2(v_0^2 - (e^{j\gamma h} - 1)^2)\\ \bar{a}_3 = a_1 = 2v_0^2(cos(\alpha h) + cos(\beta h) + (e^{j\gamma h} - 1)^2 - 2 - 2v_0^2)\\ a_2 = \bar{a}_2 = -2(2(v_0^2 - 1)cos^2(\gamma h)\\ +2v_0^2(cos(\alpha h) + cos(\beta h) - cos(\gamma h)) - 4v_0^2 - 3v_0^4 - 8\\ +4(cos(\alpha h) + cos(\beta h) + 2cos(\gamma h)) - 2\sum_{cyclic} cos(\alpha h)cos(\beta h))\end{gathered}.$$

Since in our case $|a_4| = |a_0|$, the Viète relation $z_1 z_2 z_3 z_4 = \frac{a_0}{a_4}$ shows that the condition $|z_i| \leq 1$ is equivalent to $|z_i| = 1,\ i = 1, 2, 3, 4$. Some general theorems on equations with roots of absolute value unity can be found in [3].

**Case one**. For $Im(a_4) \neq 0$, our equation is a reciprocal complex equation. Consequently, if $z_1$ is a root of the equation, then $\frac{1}{\bar{z}_1}$ is also a root. Therefore, a particular quartic complex reciprocal equation having roots only on unit circle is of the form

$$ae^{-j(\theta_1+\theta_2)}(z - e^{j\theta_1})^2(z - e^{j\theta_2})^2 = 0, \quad a \in R\backslash\{0\}, \quad \theta_1, \theta_2 \in [0, 2\pi).$$

Identifying, we obtain the conditions

$$\begin{gathered}\bar{a}_4 = a_0 = ae^{j(\theta_1+\theta_2)}, \bar{a}_3 = a_1 = -2a(e^{j\theta_1} + e^{j\theta_2})\\ a_2 = a(e^{j(\theta_1-\theta_2)} + e^{j(\theta_2-\theta_1)} + 4).\end{gathered}$$

If $a = |a_0|$, then the existence of $\theta_1, \theta_2 \in R$ is equivalent to the system

$$2|a_0| = |a_1|, -8|a_0|a_0 + 4a_2 a_0 - a_1^2 = 0.$$

If $a = -|a_0|$, then the existence of $\theta_1, \theta_2 \in R$ is equivalent to the system

$$2|a_0| = |a_1|, 8|a_0|a_0 + 4a_2 a_0 - a_1^2 = 0.$$

These constraints on the *time step* $\tau$, the *grid spacing h*, and *spatial wave vector* $(\alpha, \beta, \gamma)$ describe the *stability regions* where our scheme is *marginally stable*.

**Case two**. For $Im(a_4) = 0$ we find $h = \frac{\pi}{\gamma}$ or $h = \frac{2\pi}{\gamma}$ and all the coefficients of the equation are real numbers. To avoid for a moment the complications, here we discuss only the case when $a_3 = 0, a_2 = 0$, i.e., the case of the biquadratic equation $z^4 + 1 = 0$. The equations $a_3 = 0, a_2 = 0$ lead to

$$cos(\alpha h), cos(\beta h) \in \left\{ v_0^2 - 1 \pm \frac{\sqrt{2v_0^2 - 24v_0 + 64}}{2} \right\}.$$

The condition $cos(\alpha h) = cos(\beta h) = 1$ gives the ratio $v_0^2 = \sqrt{68} - 6$; the condition $cos(\alpha h) = cos(\beta h) = -1$ produces the ratio $v_0^2 = \sqrt{28} - 2$. All these conditions ensure that our scheme is marginally stable, and over the stability region, the previous constraints give us a time step $\tau$, in terms of the grid spacing $h$, and implicitly in terms of the wave number $\gamma$.

The Von Neumann analysis of the general case will be presented in another paper.

## 69.7 Open Problems

Let us formulate the Udrişte-Maxwell theory in terms of differential forms [10]. In this sense, it is well-known that $E, H$ are *differential 1-forms*, $J, D, B$ are *differential 2-forms*, $\rho$ is a *differential 3-form*, and the *star operator* from $D = \varepsilon * E$, $B = \mu * H$ is the *Hodge operator*. If $d$ is the *exterior derivative operator*, and $\partial_t$ is the *time derivative operator*, then the Maxwell's equations for static media are coupled PDEs of first order,

$$dE = -\partial_t B, dH = J + \partial_t D, dD = \rho, dB = 0$$

defined on $R^3 \times R$. Furthermore, some real problems require to replace the *Euclidean manifold* $(R^3, \delta_{ij})$ with a *Riemannian manifold* $(R^3, g_{ij})$. In this context the least squares Lagrangian can be written

$$2L_4 = ||dE + \partial_t B||^2 + ||dH - J - \partial_t D||^2 + ||dD - \rho||^2 + ||dB||^2.$$

*Find interpretations for the extremals of $L_4$ which are not solutions of Maxwell equations. Analyze the influence of the Riemannian metric on the associated discrete Maxwell geometric dynamics. Can we ameliorate the convergence or the stability conditions?*

## 69.8 Conclusions

This paper studies the extremals of the least squares Lagrangian associated to Maxwell PDEs. Between these extremals we have the classical waves, but also other solutions whose interpretation is an open problem. The discrete Maxwell geometric dynamics can produce relevant results since the Newton method for discrete Euler-Lagrange PDEs is contractive in the case of least squares Lagrangian and the Von Neumann analysis proves the stability of the finite difference scheme in reasonable conditions.

**Acknowledgements** Partially supported by Grant CNCSIS 86/ 2007 and by 15th Italian-Romanian Executive Programme of S&T Co-operation for 2006-2008, University Politehnica of Bucharest.

We are very appreciative to Prof. Dr. Nikos Mastorakis, Prof. Dr. Valeriu Prepeliţă and Lecturer Romeo Bercia for their suggestions regarding the improvement of numerical aspects for the least squares approximations via Von Neumann analysis.

## References

1. Bilbao S (2004) Wave and scattering methods for numerical simulation. John Wiley and Sons, Chicester, UK, p 368
2. Ibragimov NH (2006) Integrating factors, adjoint equations and Lagrangians. J Math Anal Appl 318(2):742–757
3. Kempner AJ (1935) On the complex roots of algebraic equations. Bull Amer Math Soc 41(12):809–843
4. Udrişte C (1999) Nonclassical Lagrangian dynamics and potential maps. In: Conference in Mathematics in Honour of Professor Radu Roşca on the occasion of his Ninetieth Birthday, Katholieke University Brussel, Katholieke University Leuwen, Belgium, http://xxx.lanl.gov/math.DS/0007060 (2000)
5. Udrişte C (2000) Solutions of ODEs and PDEs as potential maps using first order Lagrangians. Centenial Vrânceanu, Romanian Academy, University of Bucharest, http://xxx.lanl.gov/math.DS/0007061 (2000); Balkan J Geometry Appl (2001) 6(1):93–108
6. Udrişte C (2003) Tools of geometric dynamics. Buletinul Institutului de Geodinamică, Academia Română, 14(4):1–26; In: Proceedings of the XVIII Workshop on Hadronic Mechanics, honoring the 70-th birthday of Prof. R. M. Santilli, the originator of hadronic mechanics, University of Karlstad, Sweden, June 20–22, 2005; Valer Dvoeglazov, Tepper L. Gill, Peter Rowland, Erick Trell, Horst E. Wilhelm (Eds) Hadronic Press, International Academic Publishers, December 2006, ISBN 1-57485-059-28, pp 1001–1041
7. Udrişte C (2003-2004) From integral manifolds and metrics to potential maps. Atti dell'Academia Peloritana dei Pericolanti, Classe I di Scienze Fis. Mat et Nat 81–82:1–16, C1A0401008
8. Udrişte C (2005) Geodesic motion in a gyroscopic field of forces. Tensor, N. S. 66(3):215–228
9. Udrişte C (2006) Multi-time maximum principle. Short communication at International Congress of Mathematicians, Madrid
10. Udrişte C, Ferrara M, Opriş D (2004) Economic geometric dynamics. Monographs and Textbooks 6, Geometry Balkan Press
11. Udrişte C, Postolache M (2000) Least squares problem and geometric dynamics. Italian J Pure Appl Math 15:77–88
12. Udrişte C, Postolache M, Ţevy I (2001) Integrator for Lagrangian dynamics. Balkan J Geometry Appl 6(2):109–115
13. Udrişte C, Ţevy I (2007) Multi-time Euler-Lagrange-Hamilton theory. WSEAS Trans Math 6(6):701–709
14. Udrişte C (2007) Multi-time controllability, observability and bang-bang principle. In: 6th Congress of Romanian Mathematicians, Bucharest, Romania
15. Udrişte C, Ţevy I (2007) Multi-Time Euler-Lagrange Dynamics. In: Proceedings of the 7th WSEAS International Conference on Systems Theory and Scientific Computation (ISTASC'07), Athens, Greece, pp 66–71

# Chapter 70
# Fractional Right Ideals

**Ulrich Albrecht**

**Abstract** Fractional ideals play a central role in the investigation of integral domains. This paper extends this important concept to right non-singular rings of finite right Goldie-dimension. We show that a satisfactory extension is only possible if $R$ is a semi-prime right and left Goldie-ring.

There are several ways to extend the concept of torsion-freeness from modules over integral domains to arbitrary non-commutative rings. The most straightforward approach towards such a generalization is to call a right module $M$ over a ring $R$ *torsion-free in the classical sense* if, for all non-zero $x \in M$ and all regular $c \in R$, one has $xc \neq 0$. Here, $c \in R$ is *regular* if it is neither a right nor a left zero-divisor. To overcome the inherent limitations of this generalization, Goodearl introduced the notion of non-singularity, which is perhaps the most successful extension of torsion-freeness to non-commutative rings (see [6] and [8] for details). The *singular submodule* of a right $R$-module $M$ is $Z(M) = \{x \in M | xI = 0$ for some essential right ideal $I$ of $R\}$. The module $M$ is *non-singular* if $Z(M) = 0$ and *singular* if $Z(M) = M$, while the ring $R$ itself is *right non-singular* if $R_R$ is a non-singular module. The right non-singular rings are precisely those rings $R$ which have a regular, right self-injective maximal right ring of quotients which is denoted by $Q^r$. Furthermore, a submodule $U$ of a right

$R$-module $M$ is $S$-closed if $M/U$ is non-singular. A right non-singular ring is *a right Utumi-ring* if every $S$-closed right ideal of $R$ is a right annihilator. The right and left Utumi-rings are precisely the right and left non-singular rings for which $Q^r = Q^\ell$ [6, Theorem 2.35].

For submodules $U$ and $V$ of $Q^r_R$, set $(U : V)_\ell = \{q \in Q^r | qV \subseteq U\}$ and $(U : V)_r = \{q \in Q^r | Vq \subseteq U\}$. Moreover, $ann_r(S) = \{r \in R | Sr = 0\}$ is the right annihilator of a subset $S$ of $Q^r$. If $M$ is a right (left) $R$-module, then $M^*$ denotes the

U. Albrecht (✉)
Department of Mathematics, Auburn University, Auburn, AL 36849, USA
e-mail: albreuf@mail.auburn.edu

N. Mastorakis et al. (eds.), *Proceedings of the European Computing Conference*, Lecture Notes in Electrical Engineering 28,
DOI 10.1007/978-0-387-85437-3_70, 

left (right) $R$-module $\mathrm{Hom}_R(M, R)$. A $R$-module $M$ is *torsion-less* if the natural map $\Psi_M : M \to M^{**}$ is a monomorphism. It is *reflexive* if $\psi_M$ is an isomorphism.

An essential submodule $U$ of $Q^r_R$ is *a fractional right ideal* if $IU \subseteq R$ for some essential left ideal $I$ of $R$. While it is easy to see that the collection of fractional right ideals is closed with respect to finite sums and finite intersections, other of the properties of fractional ideals over an integral domain, which were discussed in [5], may fail to carry over to the more general setting of non-singular rings. For instance, finitely generated submodules of the field of quotients of an integral domain are fractional ideals. However, this need not be the case if $R$ is a right non-singular ring as the next result shows.

**Theorem one.** The following are equivalent for a right and left non-singular ring R:

a) ${}_RR$ is essential in ${}_RQ^r$.
b) Every essential, finitely generated submodule of $Q^r_R$ is a fractional right ideal.
c) $(R : U)_\ell$ contains an essential left ideal for all finitely generated submodules $U$ of $Q^r$.
d) If $U$ is a fractional right ideal and $V$ is a submodule of $Q^r$ containing $U$ such that $V/U$ is finitely generated, then $V$ is a fractional right ideal.
e) If $U$ is a fractional right ideal, then so is $U + qU$ for all $q \in Q^r$.

If, in addition $R$ has finite right Goldie dimension, then it satisfies any (and hence all) of the above if and only it is a right and left Utumi-ring. Moreover, every fractional right ideal $U$ can be embedded into a free module and has the property that $U^*$ is reflexive and a fractional left ideal.

*Proof. a)* $\Rightarrow$ *d)*: Suppose that $U$ is a fractional right ideal, and select $q_1, \ldots, q_n \in Q^r$ such that $V = U + q_1R + \ldots + q_nR$. Since ${}_RR$ is essential in ${}_RQ^r$ by a), there is an essential left ideal $I$ of $R$ with $Iq_i \subseteq R$ for all $i = 1, \ldots, n$. Moreover, $JU \subseteq R$ for some essential left ideal $J$ of $R$. Then, $K = I \cap J$ is an essential left ideal $R$ with $KV = KU + Kq_1R + \ldots + Kq_nR \subseteq R$.

To see *d)* $\Rightarrow$ *c)*, let $U$ be a finitely generated submodule of $Q^r_R$. Since $R$ is a fractional right ideal, $U + R$ is fractional by d). Select an essential left ideal $I$ with $I(U + R) \subseteq R$.

Since *c)* $\Rightarrow$ *b)* is clear, we turn to the proof of *b)* or *e)* $\Rightarrow$ *a)*: In either case, $qR + R$ is a fractional right ideal for all $q \in Q^r$. Consequently, there is an essential left ideal $K$ of $R$ with $Kq \subseteq R$; and $Q^r/R$ is a singular left $R$-module. Hence, ${}_RR$ is essential in ${}_RQ^r$ provided that $Q^r$ is non-singular as a left $R$-module. To see this, assume that $Iq = 0$ for some essential left ideal $I$ of $R$. Select an essential right ideal $J$ of $R$ such that $qJ \subseteq R$. For every $r \in J$, we have $0 = Iqr$. Since $R$ is left non-singular, $qr = 0$. But then, $qJ = 0$ yields $q = 0$ because $R$ is right non-singular.

*a)* $\Rightarrow$ *e)*: Select an essential left ideal $I$ of $R$ such that $IU \subseteq R$. Since $R$ is an essential in ${}_RQ^r$, $I$ is essential in ${}_RQ^r$, and there is an essential left ideal $J$ of $R$ with $Jq \subseteq I$. Then, $I \cap J$ is an essential left ideal of $R$ with $(I \cap J)(U + qU) \subseteq IU + JqU \subseteq IU \subseteq R$.

Furthermore, if $R$ has finite right Goldie-dimension, then $Q^r$ is semi-simple Artinian by [8, Proposition XIII.3.3]. Then, [6, Proposition 2.27 and Theorem 2.30] guarantee that $Q^r$ is a maximal left ring of quotients of $R$ since ${}_RR$ is essential in ${}_RQ^r$. Therefore, $R$ is a right and left Utumi-ring. Another application of [8, Proposition XIII.3.3] yields that $R$ also has finite left Goldie-dimension.

To show that every fractional right ideal $U$ has the described properties, observe that $M^* \cong (R : M)_\ell/(0 : M)_\ell$ for all submodules $M$ of $Q^r_R$ since $Q^r_R$ is right self-injective. Because $(0 : M)_\ell = 0$ if $M$ is essential in $Q^r_R$, we can identify $M^*$ and $(R : M)_\ell$. Since $R$ is a right and left Utumi-ring $R$, we may also assume $N^* = (R : N)_r$ for all essential submodules $N$ of ${}_RQ$. In particular, $U^* = (R : U)_\ell$. Since $U^*$ therefore contains an essential left ideal, the essential right ideal $U \cap R$ satisfies $U \cap R \subseteq U^{**} = (R : U^*)_r$. Hence, $U^*$ is a fractional left ideal. By symmetry, $U^{**}$ is a fractional right ideal of $R$, and $U^{***} = (R : U^{**})_\ell \subseteq Q^r$. By the Third Dual Theorem in [7], $U^*$ is a direct summand of $U^{***}$. On the other hand, since $U^*$ is a fractional left ideal, it is essential in $U^{***}$. This is only possible if $U^* = U^{***}$, i.e. $U^*$ is reflexive.

To see that $U$ is isomorphic to a submodule of a free module, it suffices to show that a torsion-less module $N$ of finite Goldie-dimension can be embedded into a finitely generated free module since $U \subseteq U^{**} \subseteq Q^r$. Suppose that $N$ cannot be embedded into a finitely generated free module. Choose an index-set $I$ such that $N \subseteq R^I$, and let $\pi_J : R^I \to R^J$ be the canonical projection with kernel $R^{I \setminus J}$ whenever $J$ is a subset of $I$. For $0 \neq x_0 \in N$, select a finite subset $J_0$ of $I$ with $\pi_{J_0}(x_0) \neq 0$, and consider $U_0 = N \cap R^{I \setminus J_0}$. Since $x_0 \notin U_0$, we have $N/U_0 \neq 0$. Furthermore, $N/U_0$ is isomorphic to a submodule of $R^{J_0}$. Consequently, $U_0$ is an $S$-closed submodule of $M$. Moreover, $U_0 \neq 0$ since otherwise $N \subseteq R^{J_0}$, a contradiction. Choose $0 \neq x_1 \in U_0$ and a finite subset $J_1 \supseteq J_0$ of $I$ with $\pi_{J_1}(x_1) \neq 0$. Then, $U_1 = N \cap R^{I \setminus J_1}$ is a non-zero, $S$-closed submodule of $N$ which is contained in $U_0$. Inductively, we obtain an infinite descending chain $N \supseteq U_0 \supseteq \ldots \supseteq U_n \supseteq \ldots$ of $S$-closed submodules of $N$ with $U_n \neq U_{n+1}$ for all $n$. Thus, $N$ has infinite Goldie-dimension, a contradiction.

While a submodule of the field of quotients of an integral domain is fractional if and only if it is isomorphic to an ideal, this does not have to hold in the non-commutative setting, even if $R$ is a finite dimensional right and left Utumi-ring as the next result shows. We want to remind the reader that two submodules $U$ and $V$ of $Q^r$ are *quasi-isomorphic* if there are essential monomorphisms $a : U \to V$ and $\beta : V \to U$.

**Theorem two.** The following are equivalent for a right non-singular ring $R$ which has finite right Goldie-dimension:

a) R is a semi-prime right and left Goldie-ring.
b)

   i) All finitely generated essential submodules of $Q^r_R$ are fractional right ideals.
   ii) A submodule $Q^r$ is a fractional right ideal if and only if it is isomorphic to an essential right ideal.

c) Whenever $U$ is a fractional right ideal, then a submodule $V$ of $Q^r_R$ containing $U$ is quasi-isomorphic to $U$ if and only if there is a submodule $W$ of $Q^r_R$ containing $V$ such that $W/U$ is finitely generated.
d) Finitely generated, essential submodules of $Q^r_R$ quasi-isomorphic.

*Proof. a)* $\Rightarrow$ *b)*: The ring $R$ has a semi-simple Artinian classical left and right ring of quotients, $Q$, which is also its maximal right and left ring of quotients [6, Theorem 3.35]. Hence, all finitely generated essential submodules of $Q_R$ are fractional right ideals by Theorem one. Moreover, if $U$ is a fractional right ideal, then there exists an essential left ideal $I$ of $R$ with $IU \subseteq R$. By a), there is a regular element $c \in R$ with $c \in I$. Then, $cU$ is a right ideal of $R$, and $cU \cong U$.

Conversely, assume that $U$ is a submodule of $Q_R$ which is isomorphic to an essential right ideal of $R$. There is $q \in Q$ such that left multiplication by $q$ is an essential monomorphism from $U$ into $R$. Consequently, $dim\ U = dim\ qU = dim\ R = dim\ Q < \infty$ denotes the Goldie-dimension of an $R$-module $M$. Hence, $U$ is an essential submodule of $Q_R$. Then, $U \cap ann_r(q) = 0$ yields $ann_r(q) = 0$. Thus, there are regular elements $c, d \in R$ with $q = d^{-1}c$. Consequently, $cU \subseteq dR \subseteq R$ yields that $(R : U)_\ell$ contains the essential left ideal $Rc$, proving b).

*b)* $\Rightarrow$ *c)*: By [6, Lemma 3.33], all regular elements of $R$ are invertible in $Q^r$ since $R$ has finite right Goldie-dimension. To see that $Q^r$ is the classical left ring of quotients of $R$, select $q \in Q^r$. By i), $U = R + qR$ is a fractional right ideal, for which one can find an essential monomorphism $a : U \to R$ because of ii). Since $Q^r_R$ is injective, $a$ is left multiplication by some $c \in Q^r$. Then, $cU \subseteq R$ yields that $c$ is an element of $R$. It is right regular because $0 = cs = a(s)$ for some $s \in R$ implies $s = 0$ since $a$ is one-to-one. By [6, Lemma 3.33], $c$ is regular and invertible in $Q^r$. If $r = cq \in R$, then $q = c^{-1}r$ by what has been shown. Thus, $Q^r$ is the classical left ring of quotients of $R$. Since $Q^r$ is semi-simple Artinian, $R$ is a semi-prime left Goldie-ring. However, a semi-prime right non-singular ring which has finite right Goldie-dimension is right Goldie [6, Corollary 3.32].

Observe that there is a regular element $d$ of $R$ such that $dU$ is an essential right ideal of $R$ since $U$ is a fractional right ideal. If $V$ is quasi-isomorphic to $U$, then there is a monomorphism $a : V \to U$. We can find $r, c \in R$ with $c$ regular such that $a$ is left multiplication by $q = c^{-1}r$. Since $a$ is one-to-one, $ann_r(r) \cap V = 0$. However, $V$ is essential in $Q^r$ since it contains $U$. Thus, $r$ is regular. Hence, $V \subseteq r^{-1}dU \subseteq r^{-1}cd^{-1}R$. Let $W = U + r^{-1}cd^{-1}R$. Conversely, assume that $U$ and $V$ are contained in a submodule $W$ of $Q^r$ such that $W/U$ is finitely generated. By Theorem 1, $W$ is a fractional right ideal.

*c)* $\Rightarrow$ *d)*: Let $M$ and $N$ be essential, finitely generated submodules of $Q^r_R$. Then, $M \cap N \cap R$ is a fractional right ideal. Since $W = M + N$ is a finitely generated submodule of $Q^r$ which contains $M$ and $N$, we obtain that $M$ and $N$ are quasi-isomorphic to $M \cap N \cap R$ by c).

*d)* $\Rightarrow$ *a)*: Since $R$ has finite right Goldie-dimension, every essential right ideal $I$ of $R$ contains an essentially finitely generated right ideal $J$. By d), there is a monomorphism $a : R \to J$. Clearly, $c = a(1)$ is a right regular element of $R$

which is regular since $R$ has finite right Goldie-dimension. By [6, Theorem 3.35], $R$ is a semi-prime right Goldie ring whose classical right ring of quotients is $Q^r$.

To show that $R$ is a left Goldie-ring, it remains to verify that it satisfies the left Ore-condition. For $r \in R$ and a regular element $c$ of $R$, consider $M = R + rc^{-1}R$. By b), there is a monomorphism $\lambda : M \to R$. We can find $d \in Q^r$ such that $\lambda(x) = dx$ for all $x \in M$. Then, $dM \subseteq R$ yields $dR \subseteq R$, and $d \in R$. Moreover, $0 = dr\lambda(r)$ yields $r = 0$, and $d$ is right regular. However, right regular elements of semi-prime right Goldie-ring are regular. Finally, there is $s \in R$ with $drc^{-1} = s$, i.e. $dr = sc$. Thus, $Q^r$ is also the classical left ring of quotient of $R$, and $R$ is left Goldie.

However, not much can be said without the finiteness condition on *dim* $R$:

*Example one.* Let $F$ be a field and $I$ be an infinite set. Consider the ring $R = \oplus_I F + (\ldots, 1, \ldots)F$. Then, $R$ is a commutative non-singular ring whose maximal ring of quotients is $Q = \Pi_I F$ [6, Theorem 4.13]. Observe that $J = \oplus_I F$ is an essential ideal of $Q$ which is contained in $R$. Therefore, all submodules $U$ of $Q$ are fractional ideals of $R$ since $JU \subseteq R$. However, $Q$ obviously is not isomorphic to a right ideal of $R$. Thus, $R$ satisfies condition i), but not condition ii) of part b) of the last theorem. On the other hand, the ring $Q$ satisfies both conditions.

*Example two.* Let $R$ be a ring without zero divisors which has infinite left Goldie dimension, but right Goldie dimension 1. The classical right ring of quotients $Q_{cl}$ of $R$ is a division algebra, which also is the maximal right ring of quotients of $R$ [6, Chapter 3.D]. Since $Q_{cl}$ is not the maximal left ring of quotients of $R$, ${}_RR$ cannot be essential in ${}_RQ$ [6]. Consequently, $R$ does not satisfy the first condition. Nevertheless, it satisfies the second condition of b): If $U$ is a fractional right ideal of $R$, then there is a non-zero element $q$ of $Q_{cl}$ with $cU \subseteq R$. Left multiplication by $q$ is a monomorphism from $U$ to $R$ since $Q^r$ is a division algebra.

A ring $R$ is *right fully bounded* if every essential right ideal contains a two-sided ideal which is essential as a right ideal. A submodule of $Q^r_R$ is a *two-sided submodule* of $Q^r$ if it is a submodule of ${}_RQ^r$ too.

**Corollary one.** *Let $R$ be a right and left fully bounded, semi-prime right and left Goldie-ring. The following are equivalent for a two-sided submodule $A$ of $Q_R$:*

a) *$A_R$ is a fractional right ideal of $R$.*
b) *${}_RA$ is a fractional left ideal of $R$.*

*Proof.* By symmetry, it suffices to show that (a) implies (b). Since $A_R$ is a fractional right ideal of $R$, there is a regular element $c$ of $R$ such that $cA \subseteq R$. Because $R$ is right fully bounded, the essential right ideal $cR$ contains a two-sided ideal $I$ which is essential as a right ideal. Hence, $I$ contains a regular element $d$. Then, $Rd \subseteq I \subseteq cR$ yields $A \subseteq c^{-1}R \subseteq Rd^{-1}$. Consequently, $Ad \subseteq R$. Moreover, $A$ is an essential submodule of ${}_RQ$ by [2].

Finally, we show that $R$ cannot always be embedded into a fractional right ideal.

*Example three.* There exists a right and left Utumi-ring $R$ of finite Goldie-dimension which contains a two-sided ideal $I$ of $R$ which is essential as a right ideal such that $R$ is not isomorphic to a submodule of $I$.

*Proof.* Consider the ring $R = \left\{ \begin{pmatrix} n & x \\ 0 & y \end{pmatrix} | n \in Z, x, y \in Q \right\}$. Since the right annihilator of every element of $R$ is generated by one of the idempotents

$$e_1 = \begin{pmatrix} 1 & 0 \\ 0 & 0 \end{pmatrix}, e_2 = \begin{pmatrix} 0 & 0 \\ 0 & 1 \end{pmatrix} \text{ and } e_{(x)} = \begin{pmatrix} 0 & x \\ 0 & 1 \end{pmatrix}$$

of $R$ where $x \in Q$, $R$ is right non-singular [4]. Because the additive group of $R$ is a torsion-free abelian group of rank 3, $R$ has finite right and left Goldie-dimension. However, this guarantees that $R$ is left non-singular too [4, Lemma 8.4]. Furthermore, $R$ is a subring of $Q = Mat_2(Q)$ in such a way that it is essential as a right and as a left submodule. By [6], $Q$ is the maximal right and left ring of quotients of $R$. Therefore, $R$ is a right and left Utumi-ring.

Finally let $I = \begin{pmatrix} 0 & Q \\ 0 & Q \end{pmatrix}$ which is an essential right ideal of $R$, and a left ideal of $Q$. Obviously, $R$ cannot be embedded into $I$ since $R$ has rank 3 as an abelian group while $I$ has rank 2.

## References

1. Albrecht U, Dauns J, Fuchs L (2005) Torsion-Freeness and Non-Singularity over Right p.p-Rings. J Algebra 285:98–119
2. Albrecht U (2007) Two-sided submodules of $Q^r$. Houst J Math 33(1):103–123
3. Anderson F, Fuller K (1992) Rings and Categories of Modules. Graduate Texts in Mathematics 13, Springer Verlag
4. Chatters AW, Hajarnavis CR (1980) Rings with Chain Conditions. Pitman Advanced Publishing 44, Boston, London, Melbourne
5. Fuchs L, Salce L (2001) Modules over Non-Noetherian Domains. AMS 84
6. Goodearl K (1976) Ring Theory. Marcel Dekker, New York, Basel
7. Jans J (1963) Rings and Homology. Holt, Rinehart, and Winston
8. Stenström B (1975) Rings of Quotients. Lecture Notes in Math. 217, Springer Verlag, Berlin, Heidelberg, New York

# Chapter 71
# A Distributed Approach for Concurrent Password Recovery of Archived Files

**Esra Celik, Z. Cihan Taysi, and A. Gokhan Yavuz**

**Abstract** Compressing files is a common practice for several reasons, such as keeping related files in an archive for easy handling or saving valuable storage space. Furthermore, encryption of compressed files is highly encouraged for privacy and security reasons, but loss of the encryption password will result in a total loss of the original data. Also, for cryptoanalysis purposes, one may be required to work with encrypted-compressed data. In this paper, we describe a distributed implementation for concurrent password recovery of archived files in Zip format, which is one of the most popular types of encrypted-compressed archived files. The results obtained from the sample implementation, namely Distributed Password Recovery Platform (DPRP), show that we have achieved a data independent, scalable implementation with optimum domain decomposition and communication overhead.

## 71.1 Introduction

An archive file is a bundle of files packaged together. Separate files are archived because it is easier to handle a single bundled file while moving in local disk or sending via e-mail. Most archived files are also compressed (shrunk in size) during the archiving process. File compression can be made with a special software utility to reduce the file size along with archiving. The same software must be used to decompress the file before use. Compressed file types can be distinguished by the file extension. The most common types are .ZIP, .TAR, .PKG and .ARJ for PCs; (.SIT for Macintosh). Furthermore, encryption of compressed files is highly encouraged for privacy and security reasons, but the loss of the encryption password will result in total loss of original data [1].

The Zip file format is a popular data compression and archival format. A Zip file contains one or more files that have been compressed, to reduce

E. Celik (✉)
Department of Computer Engineering, Yildiz Technical University, Besiktas/Istanbul, Turkey

N. Mastorakis et al. (eds.), *Proceedings of the European Computing Conference*, Lecture Notes in Electrical Engineering 28,
DOI 10.1007/978-0-387-85437-3_71, 

their file size or stored as-is and can be encrypted with Zip encryption methods. The most popular Zip compression utility WinZip [2] currently uses three types of encryption methods; WinZip 2.0 compatible encryption, 128-bit and 256-bit AES encryption [2]. The newer AES encryption methods are stronger than older WinZip 2.0 compatible encryption methods as they have longer encryption keys. Although extending the encryption key length makes the Zip file safer, the number of different password combinations has still the same value, therefore the performance of a password attack is not directly effected from the key length.

The improvement of today's processors makes it easier to recover passwords in less time than older ones, maybe not years but months. However this time can still be shortened by the way described in this paper.

There are several commercial Zip password recovery tools like Advanced Zip Password Recovery Tool [3], Visual Zip Password Recovery Processor [4], Zip Password Recovery [5] and so on but none of these tools have concurrent operation support.

In this paper we focused on partitioning the overall problem into separate tasks, allocating tasks to processors and synchronizing the tasks to get the desired result on time benefits.

## 71.2 Encrypted-Compressed Zip Archive Files

A standard encrypted Zip file has 12 bytes header prepended to the compressed data [6]. A checksum approach is used to create this header with three 32-bit keys. Although Adler-32 and CRC-32 (Cyclic Redundancy Check) are both 32-bit checks, the CRC-32 can take on any 32-bit value (232 possibilities), whereas Adler-32 [7] is limited to 655212 possibilities [8]. During encryption the key values are initialized using the supplied encryption password. After each byte is encrypted, the keys are then updated using pseudo-random number generation techniques in combination with the same checksum algorithm [6].

With all printable characters of the ASCII alphabet [9] one can produce 957 different passwords. This possibility is significantly smaller than checksum key combination possibilities. Thus we will focus on solving the problem over a password recovery rather than recovery of checksum keys.

### *71.2.1 Brute Force Attack*

A brute force attack consists of trying every possible code, combination, or password until the right one is found [10]. Brute force attack is a systematical numeration of all possibilities of a character set. The difficulty of a brute force attack depends on several factors, such as: the length of the password, number of possible values for each digits of the password, and the processing unit' power. Nevertheless a brute force attack will always succeed, eventually. However, brute force attacks against systems with sufficiently long password sizes may require billions of years to complete [10].

### *71.2.2 Dictionary Attack*

A dictionary attack consists of trying "every word in a given dictionary" as a possible password for an encrypted message [11]. The dictionary contains possibilities which are most likely to succeed. This method is very useful for recovering passwords that are almost remembered.

There are two methods of improving the success of a dictionary attack. The first method is to use a larger dictionary or more dictionaries. Technical dictionaries and foreign language dictionaries will increase the overall chance of discovering the correct password.

The second method is to perform string manipulation on the dictionary. For example, if the dictionary contains the word "Dpassword", common string manipulation techniques will try the word backwards (drowssap), with common number-letter replacements (p4ssw0rd), or with different capitalization (Password) [11].

## 71.3 Distributed Password Recovery

One of the first steps in parallelizing a program is to break it into discrete chunks of work that can be distributed to multiple tasks, as known as decomposition or partitioning. Also the amount of communication is another important issue when parallelizing a problem. Communications frequently impose some type of synchronization between processes, which causes processes to spend more time on waiting than doing work. Data dependency should also be given attention as it is one of the primary inhibitors to parallelism. The idle time of processes must be minimized, thus they must be kept as busy as possible doing their assigned work which infers a load balanced solution to a given problem [12].

Our distributed approach for concurrent password recovery of archived files which is called Distributed Password Recovery Platform (DPRP) surely considered the design issues which are mentioned above. The sample implementation of DPRP is done on a Beowulf Cluster [13].

DPRP partitions the password recovery problem by scattering the total range of possible passwords to the nodes where each node runs an exact copy of password recovery process which is shown in Fig. 71.1.

All DPRP processes running across the nodes generate and test the correctness of the passwords independently from other processes. In DPRP, communication overhead is significantly lower than the processing overhead as DPRP processes only need to communicate to notify each other if the correct password is found.

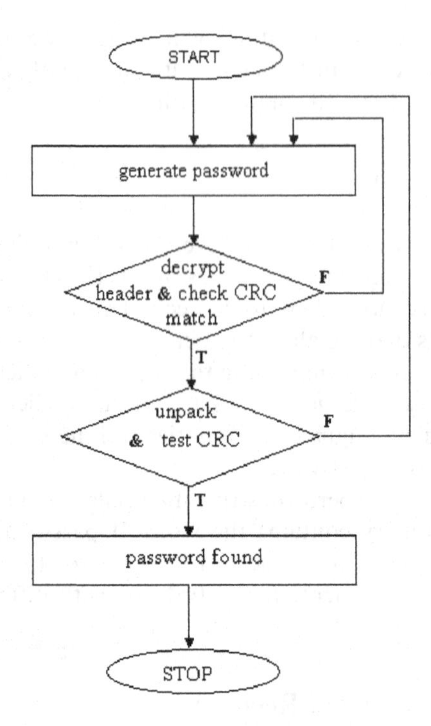


**Fig. 71.1** Main outline of the password recovery process e-mail: esra@ce.yildiz.edu.tr

Once a DPRP process starts iterating over its assigned range of possible passwords, it does not require any data from other processes, thus a maximum degree of data dependency is achieved.

Data independency and subtle communication of DPRP processes make DPRP an excellent candidate for load balancing and scalibity. Increased number of DPRP nodes will yield decreased recovery time.

As it is seen above, Zip password recovery is clearly convenient for parallelization. Now we will focus on how we partitioned the data and implemented the brute force and dictionary attack methods.

At first, all processes determine their own start and end point of brute force or dictionary attacks. Each of them generates and tries candidate passwords which are in their own range. While processes try candidate passwords sequentially, they test if any other process has found the correct password. It is known that messaging decreases the performance and should not be done when it is not very necessary or when it can be avoided. In DPRP in order to minimize messaging overhead, processes does not call the test message function in every cycle. This check is done only after a dynamically determined number of cycles. This number will be calculated over the current length of the generated password.

### *71.3.1 Achieving Data Independency for Brute Force Attack*

When DPRP processes run in brute force attack mode, they themselves partition the given set of characters into equal chunks and each process generates candidate passwords from the global set of characters starting with the characters contained in their own chunk.

Figure 71.2 outlines the decomposition of a given character set into chunks and the generation of candidate passwords. Pseudo code for this operation is given in below.

1. Calculate number of characters in chunks
2. Calculate the start point by using process id and chunk size
3. Calculate the end point by using start point and chunk size
4. Generate the first candidate password

Splitting the data with this method also increases the possibility to reach the correct password in less time (Fig. 71.3).

### *71.3.2 Achieving Data Independency for Dictionary Attack*

In dictionary attack mode, each DPRP process uses candidate passwords within its calculated part of a common dictionary file which resides on a shared storage. Pseudo code for processes to calculate the password range and get the first password from a dictionary is given below.

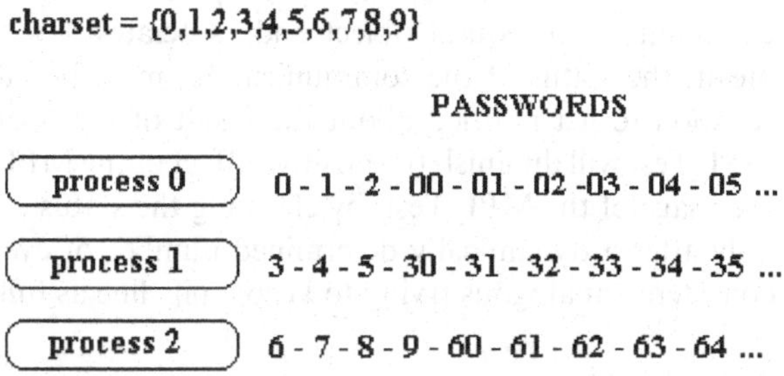

**Fig. 71.2** Password generationFigure

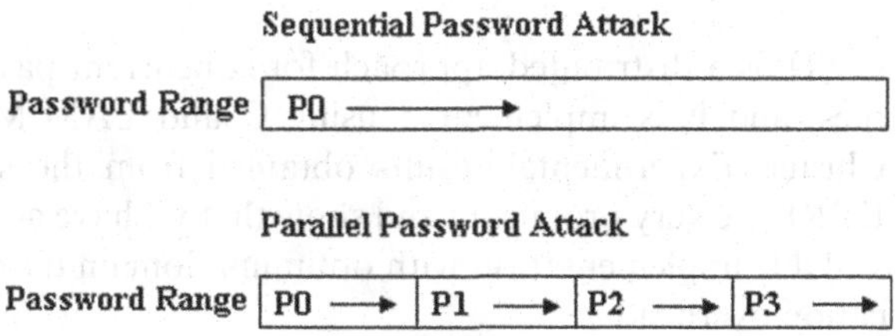

**charset = {0,1,2,3,4,5,6,7,8,9}**

| | PASSWORDS |
|---|---|
| process 0 | 0 - 1 - 2 - 00 - 01 - 02 -03 - 04 - 05 ... |
| process 1 | 3 - 4 - 5 - 30 - 31 - 32 - 33 - 34 - 35 ... |
| process 2 | 6 - 7 - 8 - 9 - 60 - 61 - 62 - 63 - 64 ... |

**Fig. 71.3** Partitioning the candidate passwords into chunks

**Sequential Password Attack**

**Password Range** | P0 ⟶ |

**Parallel Password Attack**

**Password Range** | P0 ⟶ | P1 ⟶ | P2 ⟶ | P3 ⟶ |

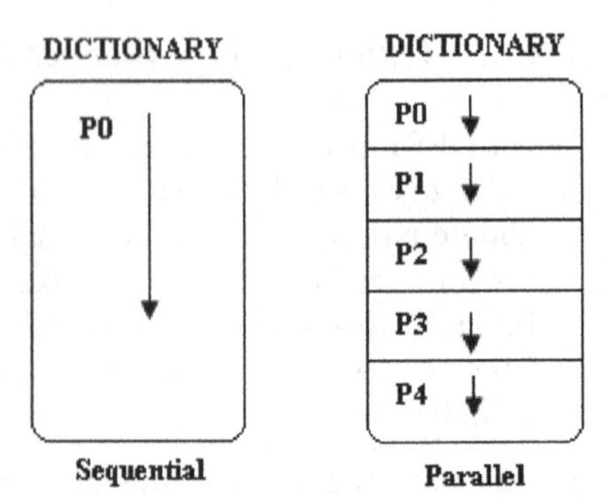


**Fig. 71.4** Partitioning the dictionary for processes

1. Get number of passwords in dictionary
2. Calculate the starting line for process by using process id
3. Calculate the number of candidate password for each process
4. Get passwords line by line

Each process opens the dictionary file, finds their first and total number of lines, gets and tries the passwords in their own range (Fig. 71.4).

### *71.3.3 Coding and Performance Issues*

C programming language and LAM/MPI [14] library were used to implement the DPRP. In order to increase the concurrency of DPRP processes, each DPRP process makes use of non-blocking MPI_Irecv [15] call to check whether another DPRP process has found the correct password. MPI_Irecv function allocates a communication request object and associates it with a request handle, which means the status of the communication, must be queried using MPI_Test [15] function to get notified about the result of the operation. As excessive use of MPI_Test will diminish the positive affect of the MPI_Irecv usage, we minimized the usage of the MPI_Test, by checking the status of other DPRP process once only after a dynamically determined number of cycles. This behaviour can be considered analogous trying to keep a pipeline as full as possible.

## 71.4 Conclusion

DPRP is a distributed approach for concurrent password recovery of archive files, and it is implemented using C and LAM/MPI library on a Beowulf Cluster. Experimental results obtained from the sample implementation of DPRP are very promising and show that we have achieved a data independent, scalable implementation with optimum domain decomposition and communication overhead.

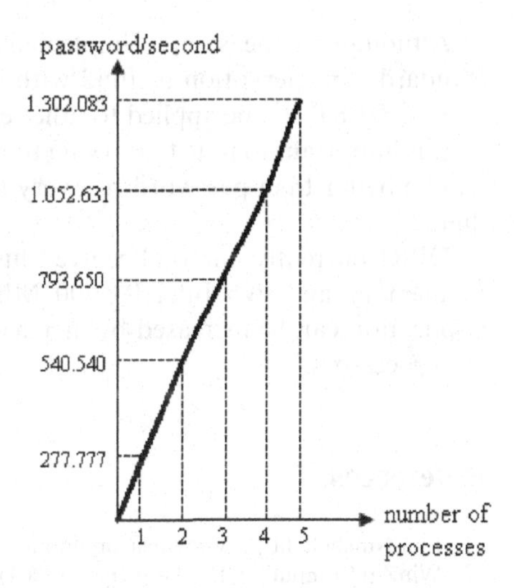


Fig. 71.5 Number of passwords generated and test per second

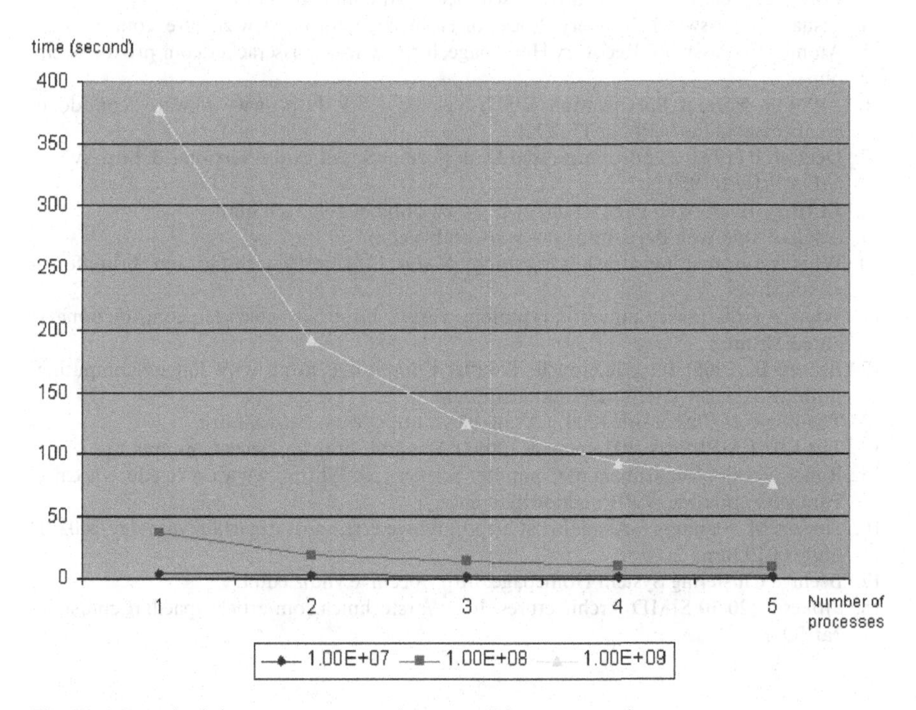


Fig. 71.6 Required time to generate and test candidate passwords

Figures 71.5 and 71.6 clearly shows how well DPRP is scaled, as the number of passwords that can be generated per second increases promotional to the increased number of cooperating DPRP processes.

Although in the sample implementation of DPRP we have focused on standard Zip encryption method with CRC-32 and Adler-32 checksum algorithms, DPRP can be applied to other encrypted-compressed data formats.

It is important to note that trying to recover the encryption keys [16] directly will be better than password recovery if the password length is more than 14 digits.

DPRP currently runs on Beehive Cluster [17] which has five nodes with 2 GB of memory and two Intel P3 700 Mhz processors.The performance of the application can be increased by increasing the number of nodes or improving the processors.

## References


1. In a Nutshell. http://www.telacommunications.com/nutshell/achive.htm g
2. WinZip Computing, Inc. Homepage (2007) http://www.winzip.com/3. Advance Zip Password Recovery Homepage, http://www.elcomsoft.com/azpr. html
4. Visual Zip Password Recovery Processor Homepage, http://www.zipcure. com/
5. Atomic Zip Password Recovery Homepage, http://www.apasscracker.com/pro ducts/zip. php
6. ZIP File Format Specification (2007) Version 6.3.1, http://www.pkware. com/documents/casestudies/APPNOTE.TXT
7. Deutsch P (1996) ZLIB Compressed Data Format Specification version 3.3. http://www. ietf.org/rfc/rfc1950.txt
8. ZLIB Software Web Page (2006) http://www.zlib.net/zlib_tech.html
9. ASCII Table Web Page, http://www.asciitable.com/
10. What is a brute force attack? Cryptology FAQs, http://www.tech-faq.com/ brute-force-attack.shtml
11. What is a dictionary attack? Cryptology FAQs, http://www.tech-faq.com/ dictionary-attack.shtml
12. Barney B (2006) Introduction to Parallel Computing. http://www.llnlgov/computing/ tutorials/parallel_comp/#DesignPartitioning.
13. The Beowulf Cluster Site (2007) Available at http://www.beowulf.org/
14. The LAM/MPI Project Homepage (2007) Available at http://www.lam-mpi.org
15. Rance Necaise Department of Computer Science (2007) http://www.cswlu.edu/~necaise/ ParallelComputing/MPIFunctionReference.
16. Stevens M, Flanders E Available at http://archive.cert.uni-stuttgart.de/vuln-dev/2003/02 /msg00019.html
17. Beehive Clustering System Homepage. http://beehive.yildiz.edu.tr
18. Stokes J (2000) SIMD Architectures. http://arstechnica.com/articles/paedia/ cpu/simd. *ars[Q7]

# Chapter 72
# Solutions for an Integral-Differential Equation

**Nikos E. Mastorakis and Olga Martin**

**Abstract** In this paper we find a continuous, non-constant and periodical solution for an integral-differential equation with the variable coefficients. This is compared with the solution obtained with the help of the finite differences method.

## 72.1 Introduction

The main problem in the nuclear reactor theory is to find the neutrons distribution in the reactor, hence their density. This is the solution of an integral-differential equation named the neutron transport equation. Many authors paid attention to this problem and its applications [1, 3, 9, 11], but an exact solution for this equation there is only for any particular cases.

Using the theory of the integral depending on parameter and Cauchy principal value of the improper integral involving unbounded functions [4, 12], we provide a periodical analytical solution for the one-dimensional stationary problem (72.1), where $\varphi$ is the neutron density and $g$ is the source function. The neutron moves to a direction, which makes the $\nu$ angle with $Ox$ axis and $y = \cos\nu$. We prove that, if the function $g$ is the sum of a uniformly convergent series, then the solution $\varphi$ is the sum of a uniformly convergent series, too.

With the method of the finite differences and the trapezoidal approximation [2, 5, 6, 10], we obtain a numerical solution for our problem and this is compared with the analytical solution.

N.E. Mastorakis (✉)
Military Institutes of University Education, Hellenic Naval Academy, Greece
e-mail: http://www.wseas.org/mastorakis

N. Mastorakis et al. (eds.), *Proceedings of the European Computing Conference*, Lecture Notes in Electrical Engineering 28,
DOI 10.1007/978-0-387-85437-3_72, 

## 72.2 Problem Formulation

Let us now consider the integral-differential equation of the neutron transport theory:

$$y\frac{\partial\varphi(x,y)}{\partial x} = \int_{-1}^{1} \varphi(x,\xi)d\xi + g(x,y) \qquad (72.1)$$

where $u$ is a continuous and non-constant function defined on the domain $D = [0,H][-1,1]$. We find the form of function $g(x, y)$ so that, the solution $u$ to be periodical function with respect to the variable $x$. If the period is $H \in \mathbf{Z}$, then we have

$$\varphi(x,y) = \varphi(x+H,y), \forall y \in [-1,1] \qquad (72.2)$$

The equation (72.1) may be written in the form

$$y\frac{\partial\varphi(x,y)}{\partial x} = v(x) + g(x,y) \qquad (72.3)$$

Where

$$v(x) = \int_{-1}^{1} \varphi(x,\xi)d\xi \qquad (72.4)$$

Integrating (72.3) with respect to the variable $x$ we obtain for $\forall x\in[0, H]$:

$$u(x,y) = \frac{1}{y}\int_{0}^{x} v(\xi)d\xi + \frac{1}{y}\int_{0}^{x} g(\xi,y)d\xi + c(y) \qquad (72.5)$$

Now we get

$$v(x) = \int_{-1}^{1} c(y)dy + \int_{0}^{x} v(\xi)d\xi \int_{-1}^{1} \frac{1}{y}dy + \int_{-1}^{1} \frac{1}{y}\left(\int_{0}^{x} g(\xi,y)d\xi\right)dy \qquad (72.6)$$

The improper integral, $\int_{-1}^{1} \frac{dy}{y}$, diverges, but there exists its Cauchy principal value given by the formula

$$v.p.\int_{-1}^{1} \frac{dy}{y} = \lim_{\varepsilon\to 0}\left(\int_{-1}^{-\varepsilon} \frac{dy}{y} + \int_{\varepsilon}^{1} \frac{dy}{y}\right) = 0$$

and

$$v(x) = \int_{-1}^{1} c(y)dy + \int_{-1}^{1} \frac{1}{y}\left(\int_{0}^{x} g(\xi, y)d\xi\right)dy$$

Replacing $v(x)$ in (72.5) we have

$$\begin{aligned}\varphi(x, y) =& c(y) + \frac{x}{y}\int_{-1}^{1} c(y)dy + \frac{1}{y}\int_{0}^{x} g(\xi, y)d\xi \\ &+ \frac{1}{y}\int_{0}^{x}\left(\int_{-1}^{1}\frac{1}{y}\left(\int_{0}^{\xi} g(t, y)dt\right)dy\right)d\xi\end{aligned} \tag{72.7}$$

and because $u(x,y)$ is a continuous function on $D$, we choose $c(y)$ so that

$$\int_{-1}^{1} c(y)dy = 0 \tag{72.8}$$

This means that $c(y)$ may be of the form

$$c(y) = D_1 y^{2n-1}, n = 1, 2, \ldots \tag{72.9}$$

or

$$c(y) = D_2 \sin(2\pi n y), n \in \mathbf{Z} \tag{72.10}$$

or

$$c(y) = D_1 \cos(\pi n y), n \in \mathbf{Z} \tag{72.11}$$

where $D_i$ are constants. Let us now consider

$$c(y) = y \tag{72.12}$$

and let

$$g(x, y) = \sum_{k=1}^{\infty}(A_k(y)\cos(2k\pi x) + B_k(y)\sin(2k\pi x)) = \sum_{k=1}^{\infty} g_k(x, y) \tag{72.13}$$

Then

$$\varphi(x, y) = \sum_{k=1}^{\infty}\varphi_k(x, y) + c(y)$$

For a fixed $y$, the series (72.13) is uniformly convergent on any interval $(0, x)$, therefore the series can be integrated term wise from zero to $x$ and the next integral from (72.7) becomes:

$$\frac{1}{y}\int_0^{\xi} g_k(t,y)dt = \frac{1}{y}(A_k(y)\frac{\sin(2k\pi t)}{2k\pi}\Big|_{t=0}^{t=\xi} - \frac{B_k(y)}{2k\pi}\cos(2k\pi t)\Big|_{t=0}^{t=\xi}) =$$

$$= \frac{A_k(y)\sin(2k\pi\xi)}{2k\pi y} + \frac{B_k(y)\sin^2(k\pi\xi)}{k\pi y}$$

The last term from (72.7) will be of the form

$$\frac{1}{y}\int_0^x [\int_{-1}^1 \frac{1}{s}(\int_0^{\xi}(A_k(s)\cos 2k\pi t + B_k(s)\sin 2k\pi t)dt)ds]d\xi =$$
$$= \frac{\sin^2 k\pi x}{2k^2\pi^2 y}\int_{-1}^1 \frac{A_k(s)}{s}ds + \frac{1}{y}(\frac{x}{2k\pi} - \frac{\sin 2k\pi x}{4k^2\pi^2})\int_{-1}^1 \frac{B_k(s)}{s}ds \tag{72.14}$$

and the continuity condition of $\varphi$ on $D$ leads to

$$\int_{-1}^1 \frac{B_k(s)}{s}ds = 0. \tag{72.15}$$

Then, we obtain for each index $k$:

$$\varphi_k(x,y) = \frac{1}{y}\int_0^x (A_k(y)\cos 2k\pi\xi + B_k(y)\sin 2k\pi\xi)d\xi +$$
$$+ \frac{\sin^2 k\pi x}{2k^2\pi^2 y}\int_{-1}^1 \frac{A_k(s)}{s}ds = \frac{A_k(y)\sin 2k\pi x}{2k\pi y} + \tag{72.16}$$
$$+ \left(\frac{B_k(y)}{k\pi y} + \frac{1}{2k^2\pi^2 y}\int_{-1}^1 \frac{A_k(s)}{s}ds\right)\cdot \sin^2 k\pi x$$

Also, in view of the continuity of the $\varphi_k(x,y)$ in any point $y$Î$[-1; 1]$, it is necessary that

$$\frac{B_k(y)}{k\pi y} + \frac{1}{2k^2\pi^2 y}\int_{-1}^1 \frac{A_k(s)}{s}ds = 0.$$

Let us now consider the functions $A_k$ of the form

$$A_k(y) = y^k, k \geq 1 \tag{72.17}$$

and then $B_k(y)$ becomes

$$B_k(y) = -\frac{1}{2k\pi}\int_{-1}^{1}\frac{A_k(s)}{s}ds \quad \Rightarrow B_k(y) = \begin{cases} 0, & \text{if } k \text{ even} \\ -\frac{1}{k^2\pi}, & \text{if } k \text{ odd} \end{cases} \tag{72.18}$$

Finally, from (72.7), (72.12), (72.16), (72.17) and (72.18) we obtain

$$\varphi(x,y) = y + \sum_{k=1}^{\infty} y^{2k-2} \cdot \frac{\sin 2(2k-1)\pi x}{2(2k-1)\pi} \tag{72.19}$$

for

$$g(x,y) = \sum_{k=1}^{\infty}\left(y^{2k-1}\cos 2(2k-1)\pi x - \frac{\sin 2(2k-1)\pi x}{(2k-1)^2\pi}\right). \tag{72.20}$$

We now prove that the functional series (19) is uniformly convergent on (0; 1). Let us consider $H = 1$ and $y$ fixed in [−1; 1]. We remind the Dirichlet's test:

"*If series* $\sum_{n=1}^{\infty} f_n(x)$, $f_n$: *DÌ* $\mathbf{R}\to\mathbf{R}$ *is of the form* $\sum_{n=1}^{\infty} \alpha_n(x)v_n(x)$, *the sequence* $\{S_n(x)\}$ *of the partial sums for the series* $\sum_{n=1}^{\infty} v_n(x)$ *is equal bounded on BÌ D and* $\{\alpha_n(x)\}$ *is a sequence which is uniformly convergent to zero on B, then its is uniformly convergent on B.*"

Let

$$S_n(x) = \sum_{k=1}^{n} \sin 2(2k-1)\pi x \quad \Rightarrow$$

$$|\sin 2\pi x + \sin 6\pi x + \sin 10\pi x + \ldots. + \sin 2(2n-1)\pi x| =$$

$$= \frac{\sin^2(2n\pi x)}{|\sin 2\pi x|} \leq \frac{1}{|\sin 2\pi x|},$$

$\forall x \in \left[\frac{\varepsilon}{2\pi}, 1-\frac{\varepsilon}{2\pi}\right] = B \subset [0;1]$, where $\varepsilon > 0$ is very small in comparison with the unity and $n$ a natural number. Then, exists $M = 1/\sin\varepsilon > 0$, such that $|S_n(x)| < M, \forall n \in \mathbf{N}, \forall x \in \left[\frac{\varepsilon}{2\pi}, 1-\frac{\varepsilon}{2\pi}\right] \subset [0;1]$.

- If $y$ is fixed, $y \in [-1; 1]$, $\left\{\frac{y^{2(k-1)}}{2k-1}\right\}_k$, $k \in \mathbf{N}^*$, is a sequence which converges to zero and $\sum_{k=1}^{n} \sin 2(2k-1)\pi x$ is equal bounded series. It results from Dirichlet's test, that series (72.19) is uniformly convergent on $B$.

- Let us now consider in (72.19), $x$ fixed. The numbers series $\sum_{k=1}^{\infty} \sin 2(2k-1)\pi x$ has the sequence of the partial sums bounded by $\frac{1}{|\sin 2\pi x|}, \forall x \in \left[\frac{\varepsilon}{2\pi}, 1-\frac{\varepsilon}{2\pi}\right] \subset [0;1]$ and the sequences of functions, $\left\{\frac{y^{2(k-1)}}{2k-1}\right\}_k, k \in \mathbf{N}^*$ are uniformly convergent to zero on $[-1; 1]$.

Thus, the uniform convergence of the functional series (72.19) is proved, too.

Using (72.20), the source function, $g(x, y)$ must to be periodically with respect to the variable $x$ and the following result can be derived.

**Theorem** Let $\varphi$ a non-constant and continuous solution on $D$ for one-dimensional transport equation (72.1). If the source function, $g(x, y)$ is a periodic function with respect to $x$, with $H = 1$, of the form

$$g(x,y) = \sum_{k=1}^{\infty} \left( y^{2k-1} \cos 2(2k-1)\pi x - \frac{\sin 2(2k-1)\pi x}{(2k-1)^2 \pi} \right) \qquad (72.21)$$

then the following function

$$\varphi(x,y) = y + \sum_{k=1}^{\infty} y^{2k-2} \cdot \frac{\sin 2(2k-1)\pi x}{2(2k-1)\pi} \qquad (72.22)$$

is a periodic solution with respect to $x$ of the equation (72.1).

■ Using (72.19), the solution $\varphi$ is expressed as a uniform convergence series. Since all terms of this series are continuous with continuous derivatives and the series of derivatives converges uniformly on $D$ (see the prove for $\varphi$), it can be integrated and differentiated term-wise. Taking $g$ and $\varphi$ of the form (72.19) and (72.20), the equation is immediately verified. ■

## 72.3 Numerical Method

Further, to find the solution of the problem (72.1) – (72.2) a numerical algorithm is presented. Let a domain $D = [0, H] \times [-1, 1]$ be, $H = 1$ and a grid:

$$\Delta_D = \left\{ (x_i, y_j) \middle| x_{i+1} = x_i + h, y_{j+1} = y_j + \tau \right\}$$

where $i = 0,1,.., N, j = 0,1,..., M$-1, $x_0 = 0$, $x_{N+1} = 1$, $y_0 = -1$, $y_M = 1$. In the network we have: $N = 2$, $M = 5$, $h = 1/3$ and $\tau = 2/5$ and the coordinates of the nodes $P(x_i, y_j) =$ **ij** are presented in the Fig. 72.1.

The function $g$ is calculated by (72.20) (summation index $k = 1$):

$$g(x_i, y_j) = y_j \cos 2\pi x_i - \frac{\sin 2\pi x_i}{\pi}$$

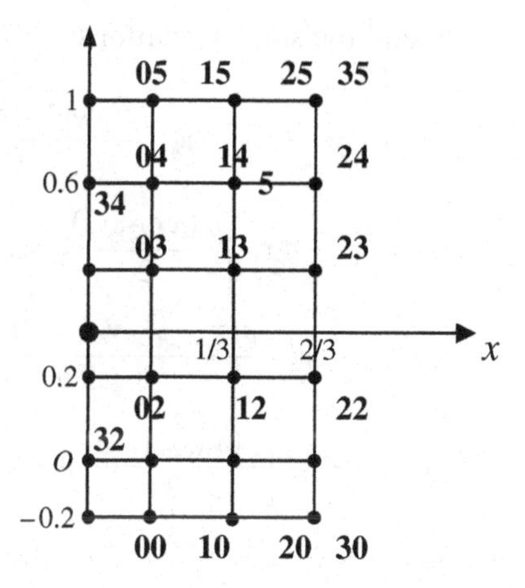


**Fig. 72.1** The coordinates of the nodes $P(x_i, y_j) = \mathbf{ij}$

and the surface $z = g(x, y)$ is presented in the Fig. 72.2.

Using the method of finite difference and the trapezoidal rule for the integral calculus we obtain a system of linear equations, which depends on $N$ and $M$.

1. If $N$ is an even natural number we have
   - for $1 \leq i \leq N/2$:

$$y_k \cdot \frac{\varphi(x_{i-1}, y_k) - \varphi(x_{i+1}, y_k)}{2h} - \tau \cdot \Big(\frac{\varphi(x_i, y_0)}{2} + \\ + \varphi(x_i, y_1) + \ldots + \varphi(x_i, y_{M-1}) + \frac{\varphi(x_i, y_M)}{2}\Big) = g(x_i, y_k)$$

   - for $N \leq i \leq N/2 + 2$

$$y_k \cdot \frac{\varphi(x_{i+1}, y_k) - \varphi(x_{i-1}, y_k)}{2h} - \tau \cdot \left[\frac{\varphi(x_i, y_0)}{2} + \ldots + \frac{\varphi(x_i, y_M)}{2}\right] \\ = g(x_i, y_k)$$

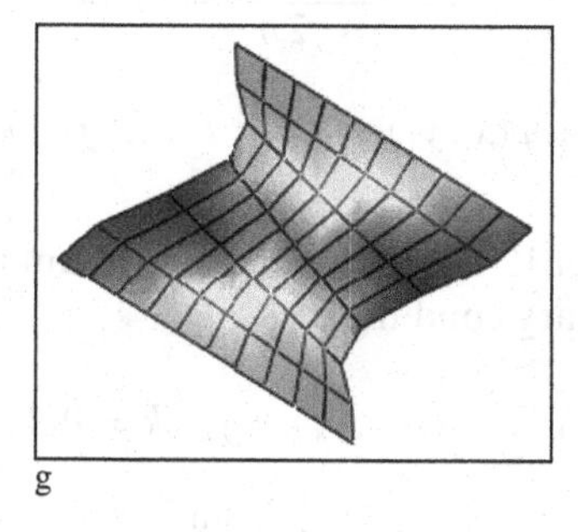


**Fig. 72.2** The surface $z = g(x, y)$

and the similar relations:

$$y_k \cdot \frac{\varphi(x_{N/2-1}, y_k) - \varphi(x_{N/2+1}, y_k)}{2h} -$$

$$- \tau \cdot \left[\frac{\varphi(x_{N/2}, y_0)}{2} + \ldots + \frac{\varphi(x_{N/2}, y_M)}{2}\right] = g(x_{N/2}, y_k)$$

$$y_k \cdot \frac{\varphi(x_{N/2+2}, y_k) - \varphi(x_{N/2}, y_k)}{2h} -$$

$$- \tau \cdot \left[\frac{\varphi(x_{N/2+1}, y_0)}{2} + \ldots + \frac{\varphi(x_{N/2+1}, y_M)}{2}\right] = g(x_{N/2+1}, y_k)$$

$$k \in 0, 1, .., M.$$

2. If $N$ is an odd natural number we have
   - for $1 \leq i < (N + 1)/2$:

$$y_k \cdot \frac{\varphi(x_{i-1}, y_k) - \varphi(x_{i+1}, y_k)}{2h} - \tau \cdot (\frac{\varphi(x_i, y_0)}{2} + \varphi(x_i, y_1) +$$

$$+ \ldots + \varphi(x_i, y_{M-1}) + \frac{\varphi(x_i, y_M)}{2}) = g(x_i, y_k)$$

   - for $(N + 1)/2 < i \leq N$:

$$y_k \cdot \frac{\varphi(x_{i+1}, y_k) - \varphi(x_{i-1}, y_k)}{2h} -$$

$$- \tau \cdot \left[\frac{\varphi(x_i, y_0)}{2} + \ldots + \frac{\varphi(x_i, y_M)}{2}\right] = g(x_i, y_k)$$

   - and for $i = (N + 1)/2$:

$$y_k \cdot \frac{\varphi(x_{i+1}, y_k) - \varphi(x_{i-1}, y_k)}{2h} -$$

$$- \tau \cdot \left[\frac{\varphi(x_i, y_0)}{2} + \varphi(x_i, y_1) + \ldots\right] = g(x_i, y_k), k = 0, 1, \ldots, M.$$

We add at the systems of 12 linear equations in 12 unknowns that corresponding to **1** or **2**, the boundary conditions:

$$\varphi(x_0, y_k) = \varphi(x_{N+1}, y_k) = y_k., k = 0, 1, \ldots, M,$$

$$\varphi(x_i, 1) = 1 + \frac{\sin 2\pi x_i}{2\pi};$$

**Table 72.1** Comparison between the approximate values and the values of the analytical solution

| | $\varphi$ analytical | $\varphi$ numerical |
|---|---|---|
| $\varphi_{10}$ | −0.8 | −0.6 |
| $\varphi_{20}$ | −1.1 | −0.8 |
| $\varphi_{11}$ | −0.46 | −0.32 |
| $\varphi_{21}$ | −0.7 | −0.5 |
| $\varphi_{12}$ | −0.06 | −0.1 |
| $\varphi_{22}$ | −0.3 | −0.2 |
| $\varphi_{13}$ | 0.3 | 0.2 |
| $\varphi_{23}$ | 0.06 | 0.1 |
| $\varphi_{14}$ | 0.7 | 0.5 |
| $\varphi_{24}$ | 0.46 | 0.32 |
| $\varphi_{15}$ | 1.1 | 0.8 |
| $\varphi_{25}$ | 0.8 | 0.6 |

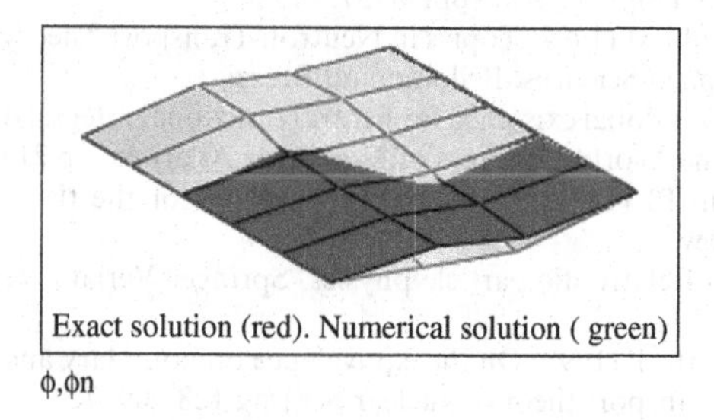


**Fig. 72.3** The surfaces that correspond to the exact solution

$$\varphi(x_i, -1) = -1 + \frac{\sin 2\pi x_i}{2\pi}, \quad i = 1, 2, \ldots, N.$$

The approximate values of the solution are compared with the values of the analytical solution and the results there are in the Table 72.1:

In the Fig. 72.3 are presented the surfaces that correspond to the exact solution, $\varphi_{\ analytical}$ (red color) and the numerical solution, $\varphi_{\ numerica}$ (green color).

## 72.4 Conclusions

The comparative study of the analytical and numerical periodical solution of (72.1) shows that the errors of the approximate solution decrease if in the series (72.20) of $g$, we consider $k \leq 2$. Also, we conclude from the numerical integration method that the values of the absolute errors in the presented algorithm are of the $h^2$ and $\tau^2$ orders.

## References


1. Ackroyd RT, Gashut AM, OA Abuzid O (1996) Discontinuous variational solutions for the neutron diffusion equation. Ann Nucl Energy 23:1289–1294
2. Case KM, Zweifel PF (1967) Linear transport theory. Addison-Wesley, Massachusetts
3. Davis WR (1970) Classical fields, particles and the theory of relativity. Gordon and Breach, New York
4. Glasstone S, Kilton C (1982) The elements of nuclear reactors theory. Van Nostrand, Toronto, New York, London
5. Marchouk G (1980) Méthodes de Calcul Numérique, Édition MIR de Moscou
6. Marchouk G, Shaydourov V (1983) Raffinement des Solutions des Schémas aux Différences. Édition MIR de Moscou
7. Marciuk G, Lebedev V (1971) Cislennie Metodî v Teorii Perenosa Neitronov. Atomizdat, Moscova
8. Martin O (1992) Une Méthode de Résolution de l'Équation du Transfert des Neutrons. Rev Roum Sci Tech Méc Appl 37:623–646
9. Mastorakis N, Martin O (2005) About the numerical solution of stationary transport equation. WSEAS Trans Inf Sci Appl 9:1373–1380
10. Mokhtar-Kharroubi M (1997) Topics in Neutron Transport Theory. Series on Advances in Math. For Applied Sciences, Bellomo and Brezzi
11. Ntouyas SK (1997) Global existence for neutral functional integral differential equations. In: Proceedings 2nd World Congress of Nonlinear Analysts, pp 2133–2142
12. Parton VZ, Perlin PI (1984) Mathematical methods of the theory of elasticity. MIR Publishers Moscow
13. Pilkuhn H (1980) Relativistic particle physics. Springer Verlag, New York, Heidelberg, Berlin
14. Rahnema F, Ravetto P (1998) On the equivalence on boundary and boundary condition perturbations in transport theory. Nuclear Sci Eng 128:209–222

# Chapter 73
# Development of Genetic Algorithms and C-Method for Optimizing a Scattering by Rough Surface

**N. Lassouaoui, H. Hafdallah Ouslimani, and A. Priou**

**Abstract** We present in this paper the design of a frequency selective surface (FSS) with developed curvilinear coordinate based method (C-method) and Genetic Algorithms (GAs). The C-method analyzes the rough surface and transforms the boundary-value problem to scalar Eigen equation that is solved in the spectral domain. In addition, the study of the reflectivity of a stack constituted by this rough surface and dielectric layers necessitates the development of the S-matrix algorithm and shooting method. Because the GAs are powerful in the exploration of the research spaces, in the hope to minimize the reflectivity over the studied frequency band, the synthesis with GAs is used and allows for the determination of the appropriate geometry characteristics and material properties. We present numerical results in the X band frequency.

## 73.1 Introduction

Analysis of wave scattering from periodic structures is important for various applications in the fields of electromagnetic, optic and acoustic. Rigorous methods have been developed and used [1, 3, 5, 7]. The curvilinear coordinate method (C-Method) is now well established for analyzing the periodic surfaces; it is known to provide both simplicity and versatility [1, 5, 7]. It is used to transform the complicated boundary-value problems to the simpler, with the hope that the basic equation of the problem will remain solvable in the new space. The TE and TM fields are derived from the same scalar Eigen equation in which the geometry appears with derivative of the profile function.

Genetic Algorithms have become very popular for the solutions of electromagnetic design problems, here in; we propose to use them for finding the geometry characteristics and material properties of the frequency selective surface with ensuring the minimization of the reflectivity over the X-frequency band.

N. Lassouaoui (✉)
University Paris X, Nanterre, Pôle Scientifique et Technique de Ville d'Avray, Groupe Electromagnétisme Appliqué, 50 rue de Sèvre 92410, Ville d'Avray, France

N. Mastorakis et al. (eds.), *Proceedings of the European Computing Conference*, Lecture Notes in Electrical Engineering 28,
DOI 10.1007/978-0-387-85437-3_73, 

This paper will proceed as follows: In Sect. 73.2, we present the analysis of the frequency selective surface and the compute of the reflection. By Sect. 73.3, we give the synthesis with GAs. Then, we present the numerical results of the study. Finally, we give the conclusions and perspectives.

## 73.2 Compute the Reflectivity

We consider a structure with a media invariant along the *z*axis in the Cartesian coordinate system ($Oxyz$), periodic along $x$ axis with period $d_x$. The surface ($S$) is described by an equation: $y = a(x)$. We suppose that this structure is illuminated by a plane wave $u_i$ propagated in the vacuum with an incidence angle $\theta_i$ from the normal and $k_i$ is wave number.

The considered diffraction problem is to find the scattered field which satisfies [3, 5, 6, 7]:

- Helmholtz equation in each media;
- Quasi-periodicity of the field $U$, where $U$ is $E_z$ in TE mode or $H_z$ in the TM mode;
- The boundary condition ($N$ is the normal vector on the surface):

$$U(x, a(x)) = 0 \text{ in TE mode and } \left.\frac{dU(x,y)}{dN}\right|_{y=a(x)} = 0 \text{ in TM mode} \quad (73.1)$$

- Radiation condition is that outside certain vicinity of contour $L$, the scattered field must be presented with Expansion of Rayleigh [5, 7].

It is necessary to simplify Equation (73.1) without disturbing other conditions. For that, we seek a coordinate system which coincides with the surface $S$.The C-method is simplest one and translates the coordinate system by:

$$x = u, y = v + a(x), z = w \quad (73.2)$$

The boundary conditions become simple, since$v = 0$ is the profile surface where the tangential components of the fields are continuous.

The Maxwell' equations in the new coordinate system have been derived [1, 5], and the propagation equation in both regions *v>0* and *v<0* is:

$$\begin{pmatrix} k_j^2 + \frac{\partial^2}{\partial u^2} & 0 \\ 0 & 1 \end{pmatrix} \begin{pmatrix} U_j \\ R_j \end{pmatrix} = \begin{pmatrix} i(\frac{\partial}{\partial u} a' + a' \frac{\partial}{\partial u}) & 1 + a'2 \\ 1 & 0 \end{pmatrix} \times \frac{1}{i} \frac{\partial}{\partial v} \begin{pmatrix} U_j \\ R_j \end{pmatrix} \quad (73.3)$$

where $j = 1, 2$ ($j = 1$ corresponds to periodic media and $j = 2$ to vacuum).

The solution $\begin{pmatrix} U_j \\ R_j \end{pmatrix}$ can be searched with an $\exp(i\rho\nu)$ dependence, where $\rho$ will be a diagonal matrix whose elements are the Eigen values of the matrix $M$, which is deduced by Equation (73.3) with $d([U_j/R_j])/_{dv} = M[U_j/R_j]$, $\lfloor U_j \rfloor$, and $\lfloor R_j \rfloor$ contain the pseudo-Fourier coefficients of $U_j$ and $R_j$.

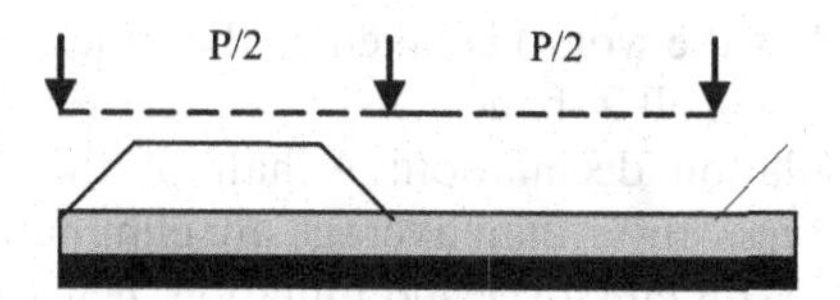


**Fig. 73.1** Stack with metallic periodic media and homogeneous dielectric layer backed by a perfect electrical conductor

By Equation (73.3), we note that Eigen value problem is the same for both TE and TM polarizations, which represents one of the main interests of the method; only the matching of the solution at the interface $v = 0$ will differ. However, initial values at $v = 0$ are unknown. So, we used the shooting method [5] which is used to solve the boundary-value differential problem and allows deducing the transmission matrix $T_g$ of the grating [5].

Here in, we are interesting by the stack composed by a periodic media (upper media) and a dielectric media (on the bottom) backed by a perfect electrical conductor (Fig. 73.1). The transmission matrix of a homogeneous dielectric layer $T_h$ is well known [6]. With *S*-matrix propagation algorithm [5] and the various matrices $T_g$ and $T_h$, we deduce the reflectivity of stack

Knowing how we compute the reflectivity, in what follows, we present how we do the synthesis of this structure with GAs

## 73.3 Synthesis with Genetic Algorithms

GA is started with a set of solutions (population). They work by iteratively applying genetic operators to the population to form a new one. This is motivated by a hope that the new population will be better than the old one. The solutions are selected according to the fitness; the more suitable have the greater chance to reproduce. The genetic process is repeated until some condition is satisfied. Our genetic procedures are:

1. Initial population is generated randomly. To delimit the research space, we define the thresholds for each variable. Also, the theory of Maxwell equations imposes energy constraints, indeed, imaginary parts of the dielectric and magnetic constants must be positive [6, 7].
2. Coding step contains information about the problem. For the periodic media, we use a copper of a thickness of 100 μm, the period *P* is unknown. We take the same width $P_x$ for the two triangles and rectangle in metallic part, then, $P = 6P_x$ (Fig. 73.1). The dielectric layer is defined by its thickness *T*, the real and imaginary parts of the dielectric and magnetic constants ($\varepsilon_j = \varepsilon_j' + i\varepsilon_j''$ and $\mu_j = \mu_j' + i\mu_j''$). So, we represent the candidates by real arrays of size 6, since there are 6 variables.
3. Evaluation associates to each chromosome its reflection in dB which defines its adaptation to the problem and guides to the optimum. To optimize over X-band, we compute the reflectivity $R_i$ for various frequencies, and the

maximal *Ri* (which is the worst) is taken as the fitness, the goal is to ensure the minimization along all X-band.

4. Selection by population decimation: A half of the population with the candidates with fitness lower than average are eliminating.
5. Genetic operators: With crossover and mutation, new candidates are created from the survival to replace those eliminated. There are no criteria for choosing between the crossover operators, then, from generation to generation, with equal probability, we select one operator between (mitosis, mitosis with chain reversal, N-point crossover and channeled crossing over) [2, 4]. The mutation is used to maintain the genetic diversity. If a gene is selected for mutation, then its numerical value is randomly changed by adding (or multiplying with) a random value $\chi$. To decrease mutation effect according to the generations [2, 4], we propose a rate *Pm* which decreases exponentially according to the generations.
6. Stopping criteria: We chose either a maximum number of iterations; or the best fitness value being less than a threshold; or no improvement in the best fitness value for a number of generations.

## 73.4 Application

We present the results of optimization with GAs of the stack of Fig. 73.1 for incidence normal. The research space is defined by: $(\varepsilon'_j, \mu'_j) \in ]-4, 4[$, $(\varepsilon''_j, \mu''_j) \in [0, 4[$, $T \in [0.6, 10]$ *mm* and $P \in [25, 35]$ *mm*.

Figure 73.2(a) shows the decrease of fitness according to the generations. Figure 73.2(b) shows the reflectivity in X-band of the final solution. After 50 generations, we reach a reflectivity of –39 dB

For the homogeneous layer, the obtained characteristics are: $\varepsilon = -2.900 + i2.3151$, $\mu = 3.400 + i\,0.8805$ with the thickness of 5.3408 mm, it can be realized by a metamaterial structure. The obtained period of the rough surface is 27.5433 mm

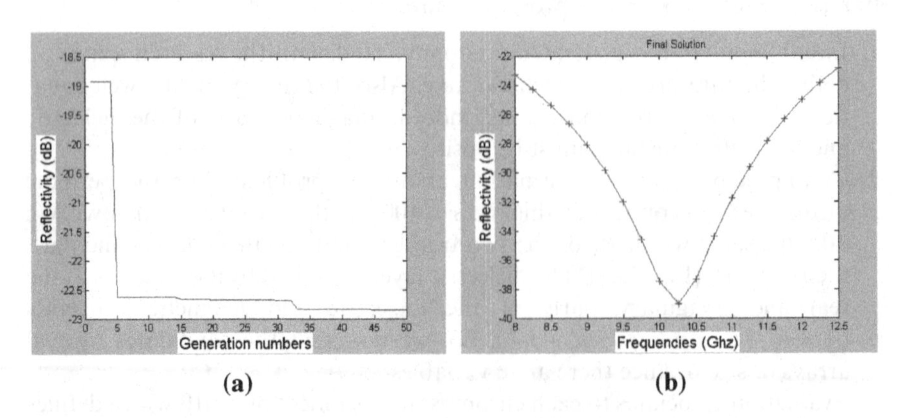


**Fig. 73.2** (**a**) Evolution of the fitness according to the generations, (**b**) obtained solution at generation 50

## 73.5 Conclusions

In this paper, a curvilinear coordinate based method (C-method) and genetic algorithms (GAs) are developed for designing a frequency selective surface with the hope of the minimization of the reflectivity along the X-frequency band.

The reflectivity is used as fitness function to direct the research for deducing the geometry characteristics and material properties which minimize the reflectivity over X-frequency band. We chose the GAs because they are powerful for problems with a large number of dimensions and in which the solution space is unknown. Good results are obtained

In future works, the obtained FSS structure will be simulated with 3D electromagnetic Ansoft simulator and realized using metamaterial layer for the homogeneous layer

## References

1. Granet G, Edee K, Felbacq D (2002) Scattering of a plane wave by rough surfaces: a new curvilinear coordinate system based approach. Prog Electromagnetics Res PIER 37:235–250
2. Goldberg DE (1989) Genetic Algorithms in search, optimization and machine learning. Addison-Wesley
3. Kong JA (2000) Electromagnetic wave theory. EMW Publishing, Boston
4. Mitchell M (1996) Introduction to Genetic Algorithms. MIT press, Cambridge, MA
5. Neviere M, Popov E (2003) Light propagation in Periodic Media, Differential theory and design. Marcel Dekker, New York
6. Petit R (1992) Electromagnetic waves in radio electricity and optics. Masson
7. Poyedinchuk AY, Tuchkin YA, Yashina NP (2006) C-Method: Several aspects of spectral theory of gratings. Prog Electromagnetics Res PIER 59:113–149

# Author Index

# Subject Index

## Lecture Notes in Electrical Engineering

*(Continued from Page ii)*

Fundamental Numerical Methods for Electrical Engineering
Rosloniec, Stanislaw
ISBN: 978-3-540-79518-6 , Vol. 18

RFID Security and Privacy: Concepts, Protocols, and Architectures
Henrici, Dirk
ISBN: 978-3-540-79075-4, Vol. 17

Advances in Mobile and Wireless Communications: Views of the 16th IST Mobile and Wireless Communication Summit
Frigyes, István; Bito, Janos; Bakki, Péter (Eds.)
ISBN: 978-3-540-79040-2 , Vol. 16

Informatics in Control Automation and Robotics: Selected Papers from the International Conference on Informatics in Control Automation and Robotics 2006
Andrade Cetto, J.; Ferrier, J.-L.; Pereira, J.M.C.D.; Filipe, J. (Eds.)
ISBN: 978-3-540-79141-6 , Vol. 15